Numerical Dating in Stratigraphy

PART I

Numerical Dating in Stratigraphy

PART I

Edited by
Gilles S. Odin
Chargé de Recherches—CNRS
Université Pierre et Marie Curie, Paris

A Wiley–Interscience Publication

JOHN WILEY & SONS
Chichester · New York · Brisbane · Toronto · Singapore

Library of Congress Cataloging in Publication Data
Main entry under title:

Numerical dating in stratigraphy.

'A Wiley–Interscience publication.'
1. Geological time—Addresses, essays, lectures.
I. Odin, Gilles S.
QE508.N84 551.7′01 81-14792
ISBN 0 471 10085 4 (Part 1) AACR2
ISBN 0 471 10086 2 (Part 2)
ISBN 0 471 90022 2 (Set of both parts)

British Library Cataloguing in Publication Data

Numerical dating in stratigraphy

1. Geology, Stratigraphic
I. Odin, Gilles S.
551.7 QE651

ISBN 0 471 10085 4 (Part 1)
ISBN 0 471 10086 2 (Part 2)
ISBN 0 471 90022 2 (Set of both parts)

Printed in the United Kingdom at The Universities Press (Belfast) Ltd.

Contents

* Contributions to the International Geological Correlation Programme (IGCP), Project 133.

Section IV: Utilization of high-temperature rocks as chronometers

CALIBRATING THE TIME SCALE

Section V: The Cambrian to Triassic times

Section VI: The Jurassic to Palaeogene times

PART II

ABSTRACTS FOR A REVISION OF THE PHANEROZOIC TIME SCALE

Contributors

C. J. ADAMS — *Institute of Nuclear Sciences, DSIR, Lower Hutt, New Zealand*
Chapter: 7

J. A. S. ADAMS — *Department of Geology, Rice University, Houston, Texas 77001, USA*
NDS: 163

M. AFTALION — *Scottish Universities' Research and Reactor Centre, National Engineering Laboratory, East Kilbride, Glasgow G75 0QU, Scotland*
NDS: 134, 135

F. ALBARÈDE — *École Nationale Supérieure de Géologie and C.R.P.G., B P 20, 54501 Vandoeuvre cédex, France*
Chapter: 10

R. L. ARMSTRONG — *Department of Geological Sciences, University of British Columbia, 6339 Stores Road, Vancouver, BC, Canada V6T 2B4*
Chapters: 7, 28; NDS: 170–184

J. AUBOUIN — *Département de Géologie Structurale, Université Pierre et Marie Curie, 4, place Jussieu, 75230 Paris cédex 05, France*

B. AUVRAY — *Institut de Géologie, Université de Rennes, Campus de Beaulieu, 35042, Rennes cédex, France*
NDS: 249

H. BAADSGAARD — *Department of Geology, University of Alberta, Edmonton, Canada T6G 2E3*
Chapter: 22; NDS: 126, 127, 129

G. P. BAGDASARYAN — *Institut de Géologie, rue Barekamountian 24a, 375019 Yerevan, Armenia, USSR*
Chapter: 7

A. K. BAKSI — *National Geophysical Research Institute, Uppal Road, Hyderabad 500007, India*
Chapter: 7

K. BALOGH — *Institute of Nuclear Research, 18/C Bem Tér, H-4001 Debrecen PF 51, Hungary*
Chapter: 7

I. L. BARNES — *Isotope Analysis S°, National Bureau of Standards, A25 Physics Bag, Washington DC 20234, USA*
Chapter: 7

B. BATTAIL — *Muséum National d'Histoire Naturelle, Institut de Paléontologie, 8, rue Buffon, 75005 Paris, France*
NDS: 186, 187

J. C. BAUBRON — *Bureau de Recherches Géologiques et Minières, SGN, B P 6009, 45060 Orléans cédex, France*
NDS: 200, 213–216

R. D. BECKINSALE — *Institute of Geological Sciences, 64 Grays Inn Road, London WC1X 8NG, UK*
NDS: 122, 241, 243

L. BENDA — *Niedersächsisches Landesamt Bodenforschung, Postfach 510153, 3000 Hannover 51, Federal Republic of Germany*
NDS: 245

M. BIELSKI — *Department of Geology, The Hebrew University, Jerusalem, Israel*
NDS: 251

G. BIGNOT — *Laboratoire de Micropaléontologie, Université Pierre et Marie Curie, 4, place Jussieu, 75230 Paris cédex 05, France*
NDS: 37

B. J. Bluck — *Department of Geology, University of Glasgow, Glasgow G12 8QQ, Scotland*
NDS: 134, 135

N. A. I. M. Boelrijk — *Laboratorium voor Isotopen-Geologie, De Boelelaan 1085, Amsterdam 11, The Netherlands*
Chapter: 7

E. H. Bon — *Billiton International Metals, Ocean Minerals, Inc, 465 N Bernardo Avenue, Mountain View, San Diego, California 94043, USA*
Chapter: 27

F. P. Bonadonna — *CNR, Laboratorio geocronologia, Via Cardinale Maffi 36, 56100 Pisa, Italy*
Chapter: 7

M. G. Bonhomme — *Institut de Géologie, 1, rue Blessig, 67084 Strasbourg cédex, France*
Chapters: 7, 17

C. Cassignol — *Centre des faibles radioactivités, Laboratoire mixte CEA/CNRS, B P 2, 91190 Gif-sur-Yvette cédex, France*
Chapters: 7, 9

C. Cavelier — *Bureau de Recherches Géologiques et Minières, SGN, B P 6009, 45060 Orléans cédex, France*
NDS: 215

P. Cepek — *Bundesanstalt für Geowissenschaften und Rohstoffe, Postfach 510153, 3000 Hannover 51, Federal Republic of Germany*
NDS: 226

L. Chanin † — *Laboratory of Geochronology, Institute of Geology, Ac Staromonetny pereulok 35 IGEM, Moscow 109017, USSR*
Chapter: 7

J. E. T. Channell — *Lamont Doherty Geological Observatory, Palisades, NY 10964, USA*
Chapter: 5

† Deceased

R. CHARLOT † — *Institut de Géologie, Campus de Beaulieu, B P 25 A, 35042 Rennes cédex, France*
NDS: 130, 133, 229, 250

J. CHARVET — *UER des Sciences Fondamentales et Appliquées, Domaine Universitaire, 45056 Orléans, France*
NDS: 158

C. A. CINGOLANI — *Divisão Geologia, Museo de la Plata, Paseo del Bosque, 1900 La Plata, Argentina*
NDS: 186, 187

N. CLAUER — *Institut de Géologie, 1, rue Blessig, 67084 Strasbourg cédex, France*
Chapter: 13

M. CONARD — *Faculté des Sciences, Route de Laval, 72017 Le Mans cédex, France*
Chapter: 16; NDS: 73, 74

D. CURRY — *Geology Department, University College, Gower St., London, WC1E 6BT, UK*
Chapters: 34, 37; NDS: 1–7, 9–14, 16–21, 23–25, 27–35, 37–39, 69, 71, 72, 90, 91, 159, 160, 212

G. H. CURTIS — *Department of Geology and Geophysics, University of California, Berkeley, California 94720, USA*
NDS: 120

P. E. DAMON — *Room 232, Geology Building, Department of Geosciences, University of Arizona, Tucson, Arizona 85721, USA*
NDS: 232

H. A. F. DE SOUZA — *Département de Minéralogie, 13, rue des Maraîchers, 1211 Genève 4, Switzerland*
Chapter: 24; NDS: 165–169, 230–232

M. H. DODSON — *Department of Earth Sciences, The University, Leeds LS2 9JT, UK*
Chapter: 14

F. DORÉ — *Département de Géologie, Esplanade de la Paix, 14032 Caen cédex, France*
Chapter: 35/2; NDS: 121

† Deceased

R. E. DRAKE

Department of Geology and Geophysics, University of California, Berkeley, California 94720, USA
NDS: 120

E. ELEWAUT

Vrije Universiteit Brussel, Department of Geochronology, Pleinlann 2, B-1050 Brussels, Belgium; present address: Zwijmaardse steenweg 488, B-9000 Gent, Belgium
NDS: 77–84

G. FAURE

Department of Geology and Mineralogy, The Ohio State University, 107 Mendenhall Laboratory, 125 South Oval Mall, Columbus, Ohio 43210, USA
Chapter: 4

M. FLISCH

Abteil Isotopen-geologie, Universität Bern, Erlachstrasse 9A, CH-3012 Bern, Switzerland
Chapter: 8

P. D. FULLAGAR

Department of Geology, University of North Carolina, Chapel Hill, North Carolina 27514, USA
NDS: 219–225, 236

N. H. GALE

Department of Geology and Mineralogy, The University, Oxford OX1 3PR, UK
Chapters: 6, 23, 25, 26, 37; NDS: 237–239, 241–243

K.-H. GEORGI

Deutsche Texaco AG, Hauptlaboratorium, D-3101 Wietze, Federal Republic of Germany
NDS: 144, 211

P. Y. GILLOT

Centre des faibles radioactivités, Laboratoire mixte CEA/CNRS, B P 2, 91190 Gif-sur-Yvette, France
Chapters: 7, 9

A. GLEDHILL

Department of Earth Sciences, The Open University, Walton Hall, Milton Keynes MK7 6AA, UK
Chapter: 7

D. GOUJET

Muséum National d'Histoire Naturelle, 8, rue Buffon, 75005 Paris, France
NDS: 125

K. Govindaraju — *Centre de Recherches Pétrographiques et Géochimiques, CO no. 1, 54500 Vandoeuvre-lès-Nancy, France*
Chapter: 7

F. Gramann — *Niedersächsisches Landesamt für Bodenforschung, Postfach 510153, 3000 Hannover 51, Federal Republic of Germany*
NDS: 40, 41, 89, 124, 228

J.-P. Groetzner — *Institut für Geologie und Paläontologie, Universität Hannover, Callinstrasse 15A, 3000 Hannover 1, Federal Republic of Germany*
NDS: 144

A. N. Halliday — *Scottish Universities' Research and Reactor Centre, Isotope Geology Unit, East Kilbride, Glasgow G75 0QU, Scotland*
NDS: 134, 244

B. U. Haq — *Woods Hole Oceanographic Institution, Woods Hole, Massachusetts 02543, USA*
Chapter: 2

R. Harakal — *Department of Geological Sciences, University of British Columbia, Vancouver, BC, Canada V6T 2B4*
Chapter: 7

W. Harre — *Bundesanstalt für Geowissenschaften und Rohstoffe, Postfach 510153, 3000 Hannover 51, Federal Republic of Germany*
Chapter: 7

W. B. Harris — *University of North Carolina, PO Box 3725, Wilmington, North Carolina 28406, USA*
Chapter: 33

B. Hartung — *Institut für Geologie und Paläontologie, Technische Universität, Pockelstrasse 14, 3300 Braunschweig, Federal Republic of Germany*
NDS: 218

E. H. HEBEDA — *Laboratorium voor Isotopen-Geologie, De Boelelaan 1085, Amsterdam 11, The Netherlands*
Chapter: 7

K. S. HEIER — *Norges Geologiske Undersokelse, Pb 3006, 7001 Trondheim, Norway*
NDS: 240

K. N. HELLMANN — *Nelkenstrasse 16, 6800 Mannheim 31, Federal Republic of Germany*
NDS: 196

J. C. HUNZIKER — *Mineralogisch-petrographisches Institut, Erlachstrasse 9A, Universität Bern, 3012 Bern, Switzerland*
Chapters: 7, 31

C. O. INGAMELLS — *Amax Co, 5950 McIntyre Street, Golden, Colorado 80401, USA*
Chapter: 7

K. KAWASHITA — *Instituto de Geosciencias, Cidade Universitária, Cx Postal 20899, São Paulo, Brazil*
Chapter: 7

E. KEMPER — *Bundesanstalt für Geowissenschaften und Rohstoffe, Postfach 510153, 3000 Hannover 51, Federal Republic of Germany*
NDS: 143

W. J. KENNEDY — *University Museum, Oxford OX1 3PW, UK*
Chapters: 32, 37; NDS: 60, 62–68, 70, 76, 85, 86, 96, 98, 99, 115–117, 119, 139, 140, 164

E. KEPPENS — *Vrije Universiteit Brussel, Département de Géochronologie, Pleinlaan 2, B 1050, Brussels, Belgium*
Chapter: 12; NDS: 77–84, 159, 160, 212

E. KISS — *Research School of Earth Sciences, The Australian National University, PO Box 4, Canberra, ACT 2600, Australia*
Chapter: 7

H. KREUZER — *Bundesanstalt für Geowissenschaften und Rohstoffe, Postfach 510153, 3000 Hannover 51, Federal Republic of Germany*
Chapters: 7, 16; NDS: 40, 41, 89, 93–98, 100, 101, 123, 124, 143–149, 162, 211, 217, 218, 226–228, 245–247

J. R. LANCELOT — *Laboratoire de Géochimie Isotopique, USTL, 34060 Montpellier cédex, France*
NDS: 136

M. A. LANPHERE — *Geological Survey, Isotope Geology, MS18, 345 Middlefield Road, Menlo Park, California 94025, USA*
NDS: 118, 128

J. F. LERBEKMO — *Department of Geology, University of Alberta, Edmonton, Canada T6G 2E3*
Chapter: 22; NDS: 126, 127, 129

R. LÉTOLLE — *Département de Géologie Dynamique, Université Pierre et Marie Curie, 4, place Jussieu, 75230 Paris cédex 05, France*
Chapter: 30

E. LIPPARINI — *CNR, Laboratorio di Geologia Marina, Bologna, Italy*
NDS: 49–52

L. E. LONG — *Department of Geology, University of Texas, Austin, Texas 78712, USA*
Chapter: 7; NDS: 150–153

C. D. LONGMAN — *Department of Geology, University of Glasgow, Glasgow G12 8QQ, Scotland*
NDS: 135

R. M. MACINTYRE — *Scottish Universities' Research and Reactor Centre, East Kilbride, Glasgow G75 0QU, Scotland*
NDS: 134

L. G. MARSHALL — *Field Museum of Natural History, Roosevelt Road, Lake Shore Drive, Chicago, Illinois 60605, USA*
NDS: 120, 138

A. McAlpine
Shell Exploration and Development, Shell-Mex House, London WC2, UK
NDS: 244

I. McDougall
Research School of Earth Sciences, The Australian National University, PO Box 4, Canberra, ACT 2600, Australia
Chapter: 7

F. W. McDowell
Department of Geological Sciences, University of Texas, Austin, Texas 78712, USA
Chapter: 7; NDS: 141, 142

H. Mehnert
US Geological Survey, Federal Center, Box 25046, Denver, Colorado 80225, USA
Chapter: 7

A. Mifdal
Institut de Géologie, Université de Rennes, Campus de Beaulieu, 35042 Rennes cédex, France
Chapter: 35/5

J. G. Mitchell
School of Physics, The University, Newcastle-upon-Tyne NE1 7RU, UK
NDS: 244

R. Montigny
Institut de Physique du Globe, Université Louis Pasteur, 5, rue Descartes, 67084 Strasbourg cédex, France
Chapter: 7; NDS: 158, 188

J. P. Morton
Department of Geology, University of Texas, Austin, Texas 78712, USA
NDS: 150–153

J. Mutterlose
Institut für Geologie und Paläontologie, Universität Hannover, Callinstrasse 15A, 3000 Hannover 1, Federal Republic of Germany
NDS: 162

J. D. Obradovich
US Geological Survey, Federal Center, Box 25046, Denver, Colorado 80225, USA
NDS: 92, 102–111, 157, 161

G. S. Odin — *Département de Géologie Dynamique, Université Pierre et Marie Curie, 4, place Jussieu, 75230 Paris cédex 05, France*
Chapters: 1, 3, 7, 14–20, 26, 30–32, 34, 35/1, 37; NDS: 1–48, 53–76, 85–93, 96–117, 119, 120, 125, 129, 131, 132, 138–142, 154–156, 163, 164, 186, 187, 200, 210, 213, 214, 228, 236, 240

E. Paproth — *Geologisches Landesamt Nordrhein-Westphalen, De Greiff Strasse 195, Postfach 1080, 4150 Krefeld, Federal Republic of Germany*
Chapter: 35/5

P. Pasteels — *Laboratoire de Géochronologie, Vrije Universiteit Brussel, Pleinlaan 2, 1050 Brussels, Belgium*
Chapters: 7, 12; NDS: 77–84, 121

K. Perch-Nielsen — *Geologisches Institut, ETH, Sonnegstrasse 5, Zürich CH 8006, Switzerland*
Chapter: 35/7

A. Person — *Département de Géologie Dynamique, Université Pierre et Marie Curie, 75230 Paris cédex 05, France*
Chapter: 21

J. J. Peucat — *Université de Rennes I, Avenue du Général Leclerc, Campus de Beaulieu, 35042 Rennes cédex, France*
NDS: 229, 250

B. Pomerol — *Laboratoire de Géologie, Université de Paris Val de Marne, Avenue du Général de Gaulle, 94010 Créteil cédex, France*
NDS: 38, 59, 75

K. Požaryska — *Polska Akademia Nauk, Zaklad Paleobiologii, Al. Swiski i Wiguri 93, PL-02-089 Warsaw, Poland*
NDS: 247

H. N. A. Priem — *ZWO Laboratorium voor Isotopen Geologie, De Boelelaan 1085, 1081 HV Amsterdam, The Netherlands*
Chapter: 27; NDS: 137

A. Rabitz — *Geologisches Landesamt Nordrhein-Westphalen, De Greiff Strasse 195, 1450 Krefeld, Federal Republic of Germany*
NDS: 94, 95

F. Radicati — *CNR, Laboratorio geocronologia, Via Cardinale Maffi 36, 56100 Pisa, Italy*
Chapter: 7

P. Rat — *Institut des Sciences de la Terre, 6 Bd Gabriel, Université de Dijon, 21100 Dijon, France*
NDS: 70

M. Renard — *Laboratoire de Géologie des Bassins Sédimentaires, Université Pierre et Marie Curie, 4, place Jussieu, 75230 Paris cédex 05, France*
Chapter: 3

D. C. Rex — *Department of Earth Sciences, The University, Leeds LS2 9JT, UK*
Chapters: 7, 19

J. R. Richards — *Research School of Earth Sciences, The Australian National University, PO Box 4, Canberra, ACT 2600, Australia*
NDS: 210, 233–235

S. Ritzkowski — *Geologisch-Paläontologisches Institut, Goldschmidtstrasse 3, 3400 Göttingen, Federal Republic of Germany*
NDS: 217

M. Robardet — *Institut de Géologie, Université de Rennes I, Campus de Beaulieu, BP 25A, 35042 Rennes cédex, France*
Chapters: 35/3, 35/4

F. Robaszynski — *Faculté Polytechnique de Mons, Laboratoire de Géologie, rue de Houdain 9, B7000 Mons, Belgium*
NDS: 77–83

J. P. Rolley — *Entima, 6, avenue de Clavière, 30107 Alès cédex, France*
NDS: 188

C. C. RUNDLE — *Institute of Geological Sciences*, 64–78 *Grays Inn Road, London WC1X 8NG, UK*
Chapter: 7; NDS: 122, 189–192

C. SAVELLI — *CNR Laboratorio di Geologia Marina, Bologna, Italy*
Chapter: 7; NDS: 49–52

A. SCHREINER — *Geologisches Landesamt Baden-Württemberg, Albertstrasse 5, 7800 Freiburg, Federal Republic of Germany*
NDS: 100

E. SEIBERTZ — *Niedersächsisches Landesamt für Bodenforschung, Postfach 510153, 3000 Hannover 51, Federal Republic of Germany*
NDS: 95, 211, 226, 227

T. J. SHEPHERD — *Isotope Geology Unit, Institute of Geological Sciences, 64 Grays Inn Road, London WC1X 8NG, UK*
NDS: 122

K. SHIBATA — *Geological Survey, 1-1-3 Higashi, Yatabe, Ibaraki 305, Japan*
NDS: 185

R. SIEGENTHALER — *Mineralogisch-petrographisches Institut, Erlachstrasse 9A, Universität Bern, 3012 Bern, Switzerland*
Chapter: 36

J. SONET — *Centre de Recherches Pétrographiques et Géochimiques, CO no. 1, 54500 Vandoeuvre-lès-Nancy, France*
Chapter: 7

N. SPJELDNAES — *Department of Palaeoecology, Aarhus University, Aarhus DK 8000, Denmark*
NDS: 131, 132

F. F. STEININGER — *Institute of Palaeontology, University of Vienna, Universitätstrasse 7–11, A1010 Vienna, Austria*
Chapter: 35/8

D. STORZER — *Laboratoire de Minéralogie, Muséum National d'Histoire Naturelle, 61, rue Buffon, Paris 75005, France*
Chapter: 11

B. SUNDVOLL — *Mineralogisk Museum, Sars' gate 1, N Oslo 5, Norway*
NDS: 240

I. L. TAILLEUR — *Office of National Petroleum Reserve in Alaska, Geological Survey, 345 Middlefield Road, Menlo Park, California 94025, USA*
NDS: 248

A. THIERMANN — *Geologisches Landesamt Nordrhein-Westphalen, De Greiff Strasse 195, 4150 Krefeld, Federal Republic of Germany*
NDS: 145, 146

R. THUIZAT — *Institut de Physique du Globe, Université Louis Pasteur, 5, rue Descartes, 67084 Strasbourg cédex, France*
NDS: 158

O. VAN BREEMEN — *Scottish Universities' Research and Reactor Centre, National Engineering Laboratory, East Kilbride, Glasgow G75 0QU, Scotland*
NDS: 135

C. VERGNAUD GRAZZINI — *Département de Géologie Dynamique, Université Pierre et Marie Curie, 4, place Jussieu, 75230 Paris cédex 05, France*
Chapter: 3

P. VIDAL — *Institut de Géologie, Université de Rennes I, Campus de Beaulieu, 35042 Rennes cédex, France*
NDS: 130, 133

C. H. VON DANIELS — *Niedersächsisches Landesant für Bodenforschung, Postfach 510153, 3000 Hannover 51, Federal Republic of Germany*
NDS: 89

A. J. WADGE — *Institute of Geological Sciences, Ring Road, Halton, Leeds, UK*
NDS: 241, 243

G. A. WAGNER — *Max-Planck-Institut für Kernphysik, Saupfercheckweg, 6900 Heidelberg 1, Federal Republic of Germany*
Chapter: 11

J. A. WEBB — *Department of Geology, University of Melbourne, Parkville 3052, Australia*
Chapter: 29; NDS: 193–195, 197–209

W. L. WEISS — *Bundesanstalt für Geowissenschaften und Rohstoffe, Postfach 510153, 3000 Hannover 51, Federal Republic of Germany*
NDS: 226

E. WELIN — *Naturhistoriska Riksmuseet, Isotop-geologi, S 10405, Stockholm 50, Sweden*
Chapter: 7

M. WESTPHAL — *Institut de Physique du Globe, Université Louis Pasteur, 5, rue Descartes, 67084 Strasbourg cédex, France*
NDS: 188

T. R. WORSLEY — *Department of Geology, Ohio University, Athens, Ohio 45701, USA*
Chapter: 2; NDS: 8, 15, 22, 36, 55, 56, 112–114

W. J. ZACHARIASSE — *Institut voor Aardwetenschappen, Rijksuniversiteit, Budapestlaan 4, Pb 80021, NL 350BTA, Utrecht, The Netherlands*
NDS: 218

H. ZAPFE — *Institut für Paläontologie der Universität, Universitätstrasse 7, A 1010 Vienna, Austria*
Chapter: 35/6

J. L. ZIMMERMANN — *Centre de Recherches Pétrographiques et Géochimiques, CO no. 1, 54500 Vandoeuvre-lès-Nancy, France*
Chapters: 7, 18

Foreword

The present work under the editorship of Gilles S. Odin is most welcome.

Absolute (numerical) geochronology has an ever-increasing importance in the Earth Sciences, although its relationship to traditional stratigraphical chronology is not always clearly understood. As a result, there have been both uncertainty and some distrust regarding its use.

It has therefore become essential to review the calibration of the time scale in relation to the stratotypes of the accepted stratigraphical scale. In view of the fact that this scale is essentially based on Europe, it was important that, following a detailed analysis of fundamental principles and problems (particularly in the use of sedimentary minerals as geochronometers), the book should present an exhaustive analysis of the Phanerozoic by comparing geochronological data with the stratigraphical data from the fossil-bearing sequences on which classical stratigraphical chronology is founded.

There is always something paradoxical about defining, for example, the numerical age of the Eocene epoch on North American volcanic rocks whose age cannot be definitely confirmed as Eocene by palaeontological studies, or about establishing the numerical age of the subdivisions of the Triassic on the basis of sequences from the Gondwana continent, when such formations are so far removed from the classic type sections of the Alps.

Apart from the obvious value of being able to give numerical ages to geological units, it is no less important to know how accurate these ages are likely to be. In addition to providing a necessary and detailed re-evaluation of the numerical ages in question, this work therefore analyses the likely margins of error involved in these estimates of age. The new approach does much to explain the differences which exist between the various previous attempts at a geological chronology.

This work therefore fills an important gap and, in doing so, should reconcile the often conflicting views of specialists in stratigraphy and geochronology, as well as giving new impetus to these two approaches to the measurement of geological time.

JEAN AUBOUIN
Membre de l'Institut

Préface

Le présent ouvrage édité par Gilles S. Odin est le bienvenu.

La géochronologie absolue occupe dans les Sciences de la Terre une place de plus en plus grande sans que chacun ne mesure toujours exactement quels sont ses rapports avec la chronologie stratigraphique traditionnelle. Il en résulte une certaine incertitude quant à l'usage qui en est fait, voire un certain malaise.

Il était donc essential de revoir la calibration de l'échelle des temps en fonction des stratotypes de l'échelle stratigraphique communément admise. Celle-ci étant essentiellement européenne, il était important que, après une première partie consacrée au rappel à la discussion et à l'établissement des fondements, en particulier pour ce qui regarde l'utilisation des minéraux des sédiments en tant que geochronomètres, l'ouvrage édité par M. Odin comporte une analyse exhaustive des temps phanérozoïques par une confrontation des données de la Géochronologie avec celles de la Stratigraphie des gisements fossilifères sur lesquelles est fondée la Chronologie stratigraphique classique.

Car il y a toujours quelque chose de paradoxal à définir, par exemple, l'âge numérique de l'Eocène sur des gisements dont on ne soit pas sûr qu'ils sont bien éocènes lorsqu'on s'adresse à des roches volcaniques d'Amérique du Nord ou celui des subdivisions du Trias d'après des gisements du continent du Gondwana si éloigné des formations types des Alpes occidentales.

S'il était utile de connaître l'âge des formations géologiques il ne l'était pas moins de savoir dans quelle mesure cette connaissance était approchée. Tout en réévaluant clairement ces âges, ce qui constitue un nécessaire progrès ce travail analyse et précise les limites de confiance de ces estimations ce qui est plus nouveau et explique bien des imprécisions, parfois des conflits entre les différentes chronologies géologiques.

Voilà donc une lacune comblée par un ouvrage qui devrait réconcilier les stratigraphes et les géochronologistes et redonner à ces deux approches de la mesure du temps un élan nouveau.

JEAN AUBOUIN
Membre de l'Institut

Preface

During the period 1968 to 1980, a group of French sedimentologists have accumulated a large amount of data on the geology of formations yielding glaucony, a chronometer which is widespread in the well studied sequences of Western Europe. As a member of that group, the editor of this book has applied these data to construct a numerical time-scale for the Mesozoic and Cenozoic eras. Between 1975 and 1980 this research, which is still ongoing, has been undertaken as part of a project in the International Geological Correlation Programme, which has studied both methodological and practical problems. About 20 laboratories all over the world have been involved in the research, either as part of IGCP Project 133 or independently. Bentonites, glauconies, lavas and plutons have been analysed using potassium–argon, rubidium–strontium, uranium–lead and fission-track methods.

In this book, the data now available from the research for Cambrian to Palaeogene times are presented and compared with earlier data, with many of which they disagree. As far as possible, those geochronologists actually working with the dating techniques were invited to contribute their results. Most were able to accept and have provided contributions based on their published data, in many cases supplemented by more recent information and interpretations. Gaps in existing knowledge were identified and research was initiated, giving the editor an opportunity to undertake isotopic studies on specific problems in various host laboratories including Paris, Berne, Strasbourg, Orsay, Leeds and Hanover over the past eight years. This book provides a synthesis of this research into many geological horizons and includes much new information which is placed into context with classic geochronological data.

The book is divided into two Parts or volumes for ease of reference and handling. Part I comprises 34 chapters on methodology and calibrating the time scale. Under 'Methodology' (Sections I to IV) are gathered the methodological results which form the basis for the principles used in selecting age data discussed in the chapters on calibrating the time scale. It gives an up-to-date review of current opinion relating to the evaluation of the uncertainties involved when calibrating the time scale. There is an examination of stratigraphical correlation, analytical errors and geochemical problems, with emphasis on the last, as this is probably the most complex question of the three and the one least understood by geologists. A particular attempt has been made to explain that an *apparent age* or a series of apparent ages obtained by means of a chronometer (a whole-rock, a mineral or a mixture) can only be used as a calibration point on the time scale after a thorough analysis and proper interpretation of a sufficient amount of stratigraphical, geochemical and analytical information. At the end of each chapter on methodology there is a short résumé in French of the most important points, for the benefit of French-speaking readers.

Under the heading 'Calibrating the Time Scale' in Part I (Sections V and VI) are gathered chapters discussing the time scale itself on the basis of the criteria set out in the chapters on methodology. Reference is also made to data used for the time scale proposed by Harland *et al.* (1964) and revised by Harland and Francis (1971). These data are referred to using the abbreviation PTS. Many of Harland's data have now been superseded due to the development of more reliable measuring techniques for the same samples or to the use of samples whose biostratigraphical relations are better established. The more recent and more accurate age data are presented in Part II of the book and referred to by the abbreviation NDS. All relevant data are considered in the discussions proposing a new time scale. Each stratigraphical boundary is in fact defined by the time interval in which it may be located with some degree of accuracy, so long as its geological definition is not modified. The results are presented in the form of tables and summary diagrams, which it is hoped will make the information clear even to non-English-speaking readers.

Part II, comprising mainly the collection of 251 Abstracts, was put into a separate volume in order to make it easier to consult the Abstracts as they are mentioned throughout the first Part of the book. Also included in Part II are a table for the calculation of potassium–argon ages, using the new decay constants adopted in Sydney and here abbreviated as ICC (for International Geological Congress of Sydney Conventional Constants); a scale summarizing the time periods studied; and a complete Bibliography of all the works mentioned in the book. Each of the 251 Abstracts was written after a discussion between stratigrapher and geochronologist and attempts to summarize all available data in a common format. The data are then used to support (or else discredit) the formation in question as a reference point for calibrating the time scale. Very often the conclusions arrived at in the Abstracts are quite different from those appearing in earlier publications, which they therefore should supersede. It should be pointed out that the techniques for isotope analysis were not really developed until after 1960 and that considerable progress has been made since 1971.

Finally, it is hoped that the time scale proposed in this book, being more precise and better founded than the preceding ones, will serve as a good *instrument* to geologists for evaluating the duration measured in years of geological phenomena such as biological, geochemical or sedimentological changes as well as the date, duration and speed of tectonic or magmatic events.

GILLES S. ODIN

Methodology

Numerical Dating in Stratigraphy
Edited by G. S. Odin

1
Introduction: Uncertainties in evaluating the numerical time scale

GILLES S. ODIN

1 INTRODUCTION

Studying rocks and minerals by radiometric dating for calibration of the stratigraphic column raises specific questions. These may be reviewed by evaluating the actual uncertainties that must be assigned to the analytically obtained apparent age. An *apparent* age is first an analytical ratio, i.e. radiogenic product *versus* radioactive producer, and has the dimension of a duration. The probable numerical age of a deposit in the stratigraphic column is the *interpretation* of this analytical ratio. The more indefinite the non-analytical information, the more uncertain the interpretation of the apparent age.

Four kinds of uncertainties must be taken into account for a correct interpretation of the analytical data. The more obvious ones are correlation between the chronometer used and the stratigraphic column and the definition of this stratigraphic column; this defines the *stratigraphical uncertainties.* The amount of information available about the isotopic geology of the chronometer at the moment of its formation and deposition enables one to recognize specific uncertainties known as the *genetic uncertainties.* The subsequent history of the chronometer between its deposition and sampling will also affect the radiogenic *versus* radioactive ratio; these are *historical uncertainties* and must also be accounted for. Finally, the *analytical uncertainties* are the uncertainties involved in representativity or otherwise of the analytical data in regard to the chronometer used and the sampled formation.

The first section of the volume covers methodology and examines these different *possibilities for error* (uncertainties), emphasizing those which are specific to the present subject. As far as possible, an attempt is made to quantify such uncertainties, in the hope of correcting the usual 'plus or minus' given with the numerical ages and which generally takes into account only the analytical uncertainties, or even the analytical reproducibility alone.

Henceforth it must be kept in mind that unfortunately these uncertainties are accumulative, each of them leading to a successively inaccurate age compared to the apparent age.

2 THE STRATIGRAPHICAL UNCERTAINTIES

In a recent review of the dating of the Silurian stratigraphic sequence, Spjeldnaes (1978) defines three types of relationship between a physically dated sample and a palaeontologically dated sequence: (1) pluton—sedimentary deposits; (2) volcanics—sedimentary lenses; (3) undisturbed fossiliferous sequences with datable materials, e.g. bentonites, glauconitic levels. Figure 1 demonstrates a new version of this definition having three complementary proposals: there are periods of discontinuity of varying duration in the relationship between the physically dated sample and the palaeontologically dated series; there are more or less correlatable palaeontologically dated series; and the zero time of the chronometer may vary in the accuracy of its location compared with the deposited sequence. Four main possibilities are presented as examples and various others exist. Usually magmatic chronometers are bracketed by two significant breaks and then important stratigraphical uncertainties occur in dating these rocks for calibration of the time scale; useful data are exceptional. As an example, the reader may wish to examine the data obtained in British Columbia on Late Triassic and Early Jurassic plutonic rocks (see Armstrong, this volume, and abstracts NDS 171–184). In spite of numerous geochronological studies of these Canadian batholiths, which have provided extensive documentation of diverse plutons, it is almost impossible from these data to accurately define even one stratigraphical boundary. However, as few better data are available at the moment it is necessary to compile and use what exist.

Volcanic continental chronometers are sometimes characterized by an underlying discontinuity but the most significant difficulty lies in research on accurately correlatable faunas. Frequently there is no clear relationship with a guide fauna and this results in the necessity to eliminate these rocks from the calibration of the time scale. On the other hand, interpretation of age frequently depends on the current opinion about the significance of the continental fauna. For example, the very interesting data obtained from Cenozoic volcanics in North America (Evernden *et al.*, 1964), although palaeontologically quite well documented with mammals, are of little use for calibration of the time scale due to the fact that it is not possible to achieve an accurate correlation with European mammals and classical stages (Evernden and Evernden, 1970). In fact, correlations can always be attempted, but the stratigraphical uncertainties quickly become more important than the

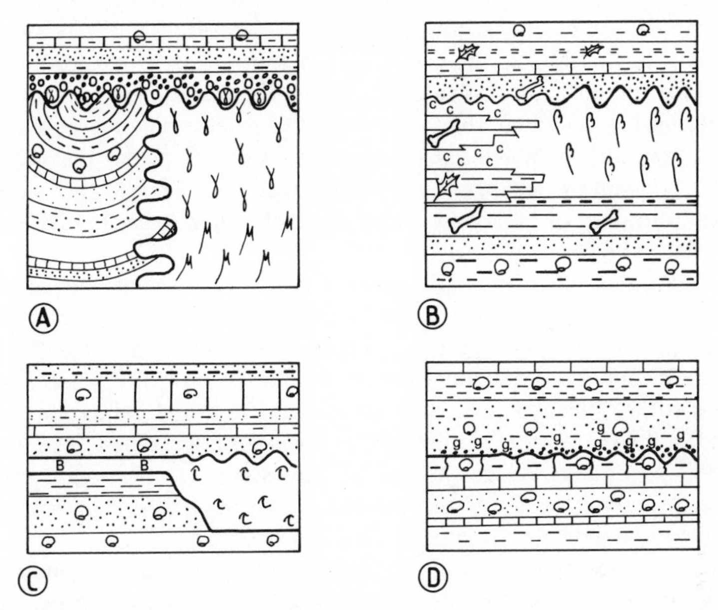

Figure 1. Relationships between radiometrically and palaeontologically dated rocks. A. A plutonic chronometer is correlated with the palaeontological sequence with an uncertainty of two important discontinuities. The closure of the minerals used (the chronometers) is not instantaneous. A plutonic rock (γ) will be more correlatable than a metamorphic rock (μ). B. A continental volcanic chronometer is generally correlated to the biostratigraphic sequence with uncertainty due to frequent discontinuous deposits. The lower and upper faunal (floral) sequences are continental; the correlation with the marine planktonic sequence requires one or two steps of admitted equivalences. Cinerites (c) can be correlated better than lava flows (β). C. A marine volcanic chronometer (bentonite, B, or tuff, τ) is generally well correlated with the biostratigraphic sequence. D. A marine sedimentary chronometer (glaucony, g) is often itself palaeontologically dated. The closure of an evolved chronometer is generally contemporaneous with the beginning of the upper deposited level (see Figure 2).

analytical ones and the result is too inaccurate for comparison with data obtained from better correlated chronometers, even when the latter are *a priori* regarded as less reliable.

A marine volcanic chronometer, and especially the various types of ash deposits, is generally well defined inside the local marine faunal sequence. Stratigraphical uncertainties may be absent.

An authigenic marine sedimentary chronometer is frequently accurately included in the local faunal sequence. However, in many examples a discontinuity of greater or lesser importance occurs below or at the dated level due to the non-instantaneous genesis of authigenic marine minerals: glauconitic minerals (see Odin and Hunziker, this volume).

The knowledge of the stratigraphic column itself and the *itinerary of correlation* (Sigal, 1977) to the normalized planktonic biozonations results in specific uncertainties. In particular, the quite dogmatic view of 'normalized international biozonation' which has persisted in past years is now moderated by new fundamental research in palaeontology: species may be asynchronous; endemism changes the appearance of species; the speed of evolution (duration of biozones) varies along the time scale; faunal assemblages replace the fossil guides; . . . .

It is clear that the presently available biozonations must be used very cautiously for correlations over large distances, and in some cases there is no guideline defining an internationally recognized boundary. The Eocene–Oligocene boundary is a good example of the lack of any palaeontologic reference with which all researchers agree (see Curry and Odin, this volume, and NDS 40–41) and the difficulty is even greater when it comes to estimating its numerical age. Moreover, the study of stratotypic areas or outcrops frequently produces incomplete or incompatible definitions which lead one to abandon the original definitions of stages.

According to Sigal (1977), an accepted correlation must be interpreted carefully due to the numerous *steps of approximation* involved in proceeding from a local well-established faunal sequence related to the dated chronometer to the pelagic 'normalized' sequence established in a second palaeogeographic environment, and further to the stratotypic lithostratigraphic sequence which is often neritic or epicontinental.

From step to step, the placing of the initially well-correlated sample becomes increasingly inaccurate. For his part, breaking with tradition, Sigal prefers to establish firm regional biozonations. A possible way to reduce these uncertainties is first to radiometrically date the typical sections and to place only limited confidence in the so-called internationally accepted faunal sequences. This stratigraphic question is discussed further in the chapter by B. U. Haq and T. R. Worsley.

Because the problem of the correlation of lithostratigraphic levels using fossils is always an open question, it appeared useful to review other currently available possible means of correlation in this volume. Recent developments in the analysis of stable isotopes in rocks are summarized in the chapter by Odin, Renard and Vergnaud Grazzini, where it is shown that isotopes may be useful markers in particular levels of the Phanerozoic series. In addition, recent progress in magnetostratigraphy (J. E. T. Channell, this volume) may greatly help in correlating poorly documented series. However, one must emphasize that the above-quoted methods apply to *relative stratigraphy* and cannot be used as a substitute for radiometric dating. Tentative suggestions of this kind were made by several pioneers (Heirtzler *et al.*, 1968; Sclater *et al.*, 1974) assuming a constant rate of spread of the sea-floor. These very interesting tentative correlations have often been misused because researchers have forgotten that the initial proposals were based on the numerical ages available at the moment that these 'magnetic scales' were established. The accumulation of more recent and obviously better radiometric dates must *ipso facto* change the initially proposed numerical ages in the magnetic scales, even if, as a result, the initially assumed rates of spread must be modified. Most frequently, the recognition of a single or several changes in polarity is of little help due to the nature of the registered events, which give only a binary signal. One magnetozone differs from another in duration only. However, if this duration is exceptional, the registration of this signal change, locally in the series, becomes remarkably useful. There are two famous examples: the 'Kiaman long reversal interval', which covers the whole Permian and part of the Carboniferous (Irving and Parry, 1963), more than 50 Ma; and the 'long normal interval' within the Cretaceous system, which probably covers nearly 30 Ma. In the light of the presently available information the palaeomagnetic reversal scales may be regarded as useful, firstly for stratigraphic correlations, and secondly as a possible means of *approximation for extrapolating* between numerically dated levels. These extrapolations must remain a secondary means of evaluating numerical ages which is useful where no radiometric dating is available.

3 THE GENETIC UNCERTAINTIES

Depending on the nature and origin of the chronometer used, the uncertainties that arise concerning the genetic time call for completely different information. The most frequently used chronometers are: minerals separated from magnetic rocks, minerals separated from bentonites or lava flows, total volcanic and magmatic rocks and authigenic sedimentary minerals.

Two genetic problems occur when minerals separated from magmatic rocks are used. The blocking temperature (the time zero) is established a very long time after the intrusion and varies with the cooling rates. This problem can be solved by specific study of the dated pluton. The time differs depending on the isotopes measured and mineral used. The second problem is the possibility of inheritance of radiogenic product, for example excess argon in pneumatolytic rocks which can be studied with specific new analytical data. These questions are discussed by N. H. Gale and may also arise where volcanic rocks are used.

The minerals separated from bentonites are generally dated by the K–Ar method and recently provided some interesting information on the time scale (see for example NDS 103–111). The specific conditions of genesis of these bentonites are discussed in the chapter by A. Person.

As bentonites are frequently ash layers deposited and altered in the marine environment, two main genetic uncertainties arise. Firstly, minerals extracted may have been inherited from an older volcanic formation remobilized by a new volcanic event (zircons). Secondly, alteration of the ash and of the included mineral particles may occur in the sea-water. The change which converts an ash into a bentonite may lead to apparent ages which are different from the time of deposition, especially in biotite, a mineral in which cationic exchange is easier than in any other mica. This alteration of biotite may produce ages which can be both much younger and much older than the age of deposition (see for example NDS 54, 57). A good criterion for this alteration is the potassium content; K-poor biotite therefore must automatically be suspected of poor reliability. The most important feature to remember here is that the minerals used in bentonites have necessarily crystallized *before their deposition in the sedimentary sequence.* These specific uncertainties are examined by H. Baadsgaard and J. F. Lerbekmo. The volcanic rocks known as tuffs have characteristics similar to those outlined above. In the absence of minerals, glassy shards have been tested, but the results are less reliable due to recrystallization processes (Evernden and Evernden, 1970).

Lava flows are also frequently tested for radiometric dating as compared with the stratigraphic column. There are various possible uses, depending on the fraction considered as the chronometer. Whole-rocks are used when the sample quantity is small, but many uncertainties may be eliminated by using several fractions of separated minerals (Cassignol and Gillot, this volume). Considering the genetic uncertainties alone, excess argon has rarely been observed, especially in whole-rocks, while a few authors think an initial deficit in ^{40}Ar relative to ^{36}Ar is possible. These uncertainties may be estimated using isochron plots or step-by-step extraction of both radiogenic and radioactive isotopes, as shown by F. Albarède.

The genetic uncertainties linked with the use of sedimentary chronometers, clays or glaucony, may be quite important. They have been investigated during the last few years especially in European laboratories, by the K–Ar method. The first uncertainty, that of the possibility of inheritance, is itself due to two phenomena. (1) Reworking of older deposits occurs in epicontinental basins and may be recognized by sedimentological regional studies; it is examined in the chapter by G. S. Odin and D. C. Rex. (2) The possibility of inheritance is related to the process of the genesis of glaucony in an initial substrate, itself probably rich in radiogenic isotopes (Figure 2). In this latter case the inheritance will diminish progressively throughout the evolutionary process, and its advancement is shown by X-ray diffraction studies. This problem is examined using the K–Ar method by G. S. Odin and M. H. Dodson.

The second kind of uncertainty linked with the genetic time is the definition of the zero time, i.e. the closure of the chronometer with regard to its environment in the sediment. One must always remember that the

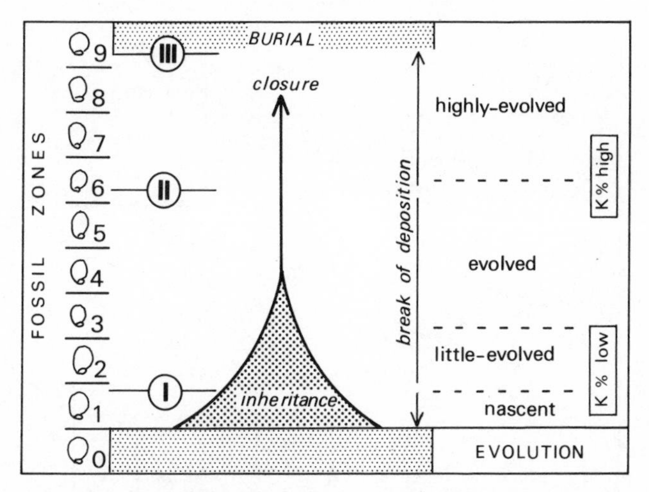

Figure 2. The genetic uncertainties related to glauconies: burial may arise at any moment from the beginning of the process of 'verdissement'. In case I, early burial, there is a short break in deposition; inheritance of radiogenic isotopes from the substrate may arise. In case II, inheritance is absent; the closure of the chronometer arises *after* burial, during the deposition of biozone 7. In case III, burial occurs after the closure of the glauconitic minerals during a long break in deposition. The fossils of zones 1–8 may have been dissolved and *the apparent age will predate* the fossils of zone 9.

closure of a glaucony necessarily always occurs *after the deposition of* the glauconitized substrates and generally during the burial process; the measured apparent age is generally, but not always, rather characteristic of the fauna deposited *above* the glauconitic formation. The question is developed in the chapter quoted above and exemplified in the chapter by J. C. Hunziker and G. S. Odin.

The use of the Rb–Sr method on various clay minerals still remains empirical, as shown by N. Clauer, although the first general laws may be proposed in accordance with local examples (see chapter by E. Keppens and P. Pasteels).

A comparison of the qualities of the various chronometers used in dating the stratigraphic column is proposed schematically in Figure 3. Logarithmic scales are used to show the genetic uncertainties as a function of the stratigraphical uncertainties which may arise. In relation to these two problems the best chronometers are near the zero.

Volcanic rocks may in fact cover the whole diagram but bentonites may be 'perfect' in favourable cases; glauconies may be acceptable if the genetic uncertainty can be considered as low with regard to the time since deposition. According to our findings the use of this chronometer is not recommended for rocks of post-oligocene age due to inaccuracy in setting its zero

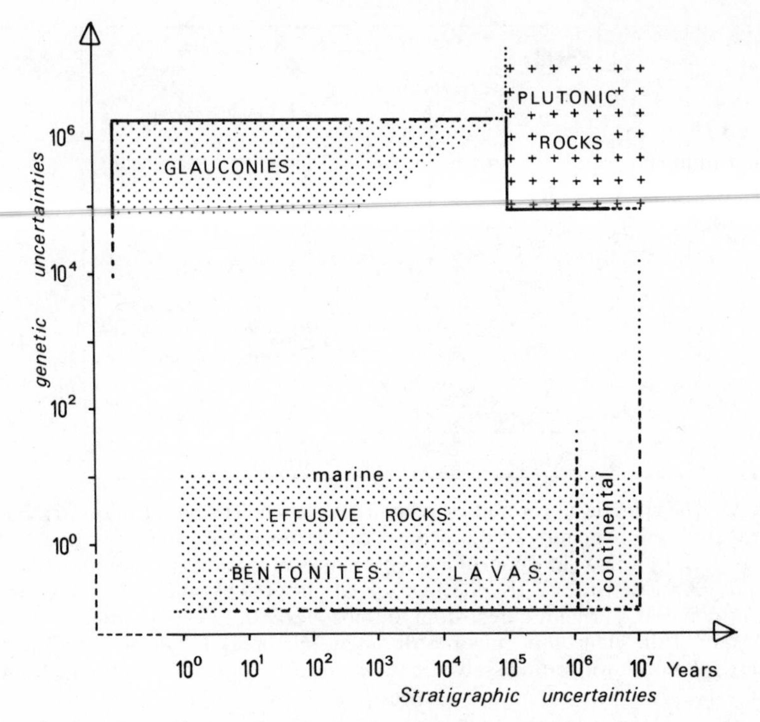

Figure 3. Stratigraphical and genetic uncertainties related to the use of various chronometers: with regard to these two kinds of uncertainty, the best chronometers are the nearest to the x–y intercept.

time although analytically there is no major problem (see Appendix by Keppens).

Plutonic rocks can generally be regarded as providing a means of calibrating the time scale where a $\pm$ 1–5 Ma does not significantly change the analytical uncertainties; for post-Triassic rocks, and even for older series, the results obtained remain tentative.

In conclusion, one can emphasize the fact that for each part of the time scale the type of useful chronometer will be different. Volcanic rocks are essentially useful for Neogene dating when easy correlation with marine fauna is possible, glauconies may be used for the Late Palaeogene to Early Mesozoic, plutonic rocks are utilizable for the Palaeozoic series, and bentonitic minerals are used essentially for the Carboniferous to Oligocene series but in general there is no reason to limit their use before or after this period.

4 THE HISTORICAL UNCERTAINTIES

The historical uncertainties may be divided into two groups, thermodynamic events and superficial changes, each one acting specifically with various chronometers. Schematically, high-temperature rocks and Rb–Sr equilibrium will be more affected by superficial conditions such as weathering, while sedimentary chronometers and the K–Ar equilibrium will be more affected by minor thermodynamic events.

Each material used as a chronometer and each isotope analysed is studied separately with regard to the conditions encountered during its history. In this volume most of the authors quoted above study the genetic and historical uncertainties both together for each kind of chronometer, and in this case they are termed *geochemical uncertainties.*

To summarize, we may recall that the isotopic equilibria in a crystallized plutonic rock will be little modified during its history if the sampled rock is fresh. A bentonite may be weathered at the surface although the presence of clay often protects the minerals from the meteoritic water. The action of possible weathering may be shown up by comparison between the ages obtained for two minerals such as sanidine and biotite. The results obtained, if identical, will be a good indication of low historical as well as genetic uncertainty. The use of whole-rock basalts or glassy shards will lead to greater historical uncertainties due to the frequently observed processes of recrystallization of the vitric phase. This tends to rejuvenate the K–Ar apparent ages, and a careful examination of thin sections becomes an integral part of the interpretation of apparent age for these particular chronometers. Chronometers measured with fission track dating are the most susceptible to apparent age changes; D. Storzer and G. A. Wagner demonstrate that a process of 'correction' must systematically be applied for certain chronometers. The absence of any examination of the process of correction by the analyst must lead to suspicion with regard to the accuracy of the proposed results.

The two groups of historical uncertainties (thermodynamic and superficial) may affect the glaucony K–Ar apparent ages. Tectonic events may lower the ages, as shown in the chapter by M. Conard *et al.* Examples of natural or experimental leaching are also examined in the chapter by G. S. Odin and C. Rex for the K–Ar glaucony ages. As a consequence of this, the first step in a dating process is to select an undisturbed outcrop for which the *geological history* is well known. The second step is the choice of a chronometer least susceptible to apparent age disturbance.

5 THE ANALYTICAL UNCERTAINTIES

New methods and techniques are currently being developed for measuring numerical ages as well as for correlating stratigraphic levels, as is shown above. The possible applications of these new methods are often presented with much enthusiasm. Various contributors to this volume demonstrate that a minimum of care must be taken when using these data: D. Storzer and G. A. Wagner discuss the nature of the uncertainties arising during the use of the promising fission track method; F. Albarède examines the use of the 39–40 technique; further, G. Faure presents an up-to-date view of the variation of the $^{87}Sr/^{86}Sr$ ratio in the seas during Phanerozoic times.

The fundamental question of the decay constants is solved today by the use of a set of International Conventional Constants (ICC) which were adopted in 1976. These constants are presented by N. H. Gale and are exclusively used in this volume.

However, a convention has not yet been established to define the fission rate of ^{238}U. Although the most recent studies indicate a rate of decay of nearly $8.43 \times 10^{-17} a^{-1}$, recently published papers always use as low a value as $6.85 \times 10^{-17} a^{-1}$. The question is only partly solved by using mineral standards.

With reference to high-temperature rocks, the usual K–Ar technique is now well known. Considerable progress has been made from the technical point of view especially during the last decade: *precision* of measurement, vacuum technique, speed of measurement. In consequence, it is now *much more pertinent to try to obtain several new analytical results rather than to discuss outdated individual analyses*, which often are biased, for analytical precision.

However, dating of the stratigraphic column also necessitates *accuracy*, which is another problem that may be verified by interlaboratory calibration. It appeared to me, therefore, that the review of interlaboratory standards which I present in this volume would constitute a useful contribution. Very valuable data are proposed by W. Harre and H. Kreuzer; M. Flisch also reassembles useful data. With regard to the calibration of the time scale, this question is significant if we consider that the reproducibility of apparatus is commonly better than 1%, while the absolute calibration may change by more than 2% from one laboratory to another.

Unfortunately, no well-known interlaboratory standard is currently available for the 40–39 dating technique nor for fission track dating.

As far as the use of sedimentary chronometers is concerned, and especially the K–Ar/glaucony measurements, many speculations have developed in the literature since the period 1959–1963. New experiments have been developed in the last few years which have been designed to examine the conditions under which argon is lost *in vacuo* or under atmospheric pressure, as demonstrated in this volume by G. S. Odin and M. G. Bonhomme, and J. L. Zimmermann and G. S. Odin. These new results show that moderate conditions of preliminary heating of the samples may assist good measurement without altering the apparent age. It should be noted that high preliminary heating temperature (more than 180°C) may change the apparent age of particular whole-rock basalts, a fact that is often ignored.

Rb–Sr/glaucony measurements are greatly aided by preliminary acid leaching, which removes adsorbed common strontium from the glauconitic pellets. The precision of the measurements is then much better: see E. Keppens and P. Pasteels.

For these analytical uncertainties related to glaucony dating, the lack of agreement on the question has led to the proposal of recommendations applying to all stages from sampling to measurement: see the chapter on the use of glaucony by G. S. Odin.

For dating of a glaucony as well as for any other chronometer, the use of a 2σ analytical uncertainty (95% confidence level) was recommended to all authors as being the most realistic. The calculation of apparent ages and above all of analytical uncertainty is a rather subjective matter: the variety of calculation systems is large. As an example, the apparent age of Albian glauconies presented by Odin and Hunziker has been calculated on the basis of various possibilities (see Table 1) using the $^{40}Ar/^{36}Ar$ and $^{40}K/^{36}Ar$ measured and calculated ratios. The cubic regressions according to York (1969) or Brooks *et al.* (1972) and the arithmetic mean do not give similar figures, although they are near. Moreover, as it is known that the initial 40/36 ratio was normal in the environment of genesis of the samples and that it was also normal in the various processes contaminating the analysed samples, the influence of fixing this ratio has been tested. Brooks *et al*'s cubic regression is more sensitive to this addition of the 'initial' 40/36 ratio. York's regression is not as affected by the newly added data. The ± given for the arithmetic mean is calculated using

$$2\left\{\frac{\sum_1^n (x_n - \bar{x})^2}{n-1}\right\}^{1/2}$$

Table 1. K–Ar age of Albian glauconies according to various methods of calculation (original data in Odin and Hunziker, this volume, Table 2).

Method of integration	Data used	Initial ratio calculated $^{40}Ar/^{36}Ar$	Apparent age (Ma, ICC)
Cubic regression (Brooks *et al*; 1972)	11 glauconies	304±34	98.6±2.1
Cubic regression (Brooks *et al.*, 1972)	11 glauconies +1×initial (295.5)	295.6±2.7	99.0±1.3
Cubic regression (York, 1969)	11 glauconies	319±16	97.8±0.7
Cubic regression (York, 1969)	11 glauconies +1×initial (295.5)	308±12	98.2±0.5
Cubic regression (York, 1969)	11 glauconies +3×initial (295.5)	302±8 ↓	98.5±0.4
Arithmetic mean	11 glauconies (7 samples)	295.5 assumed	99.0±1.1

where x_n is an individual value and $\bar{x}$ the arithmetic mean value. This corresponds to two standard deviations.

In fact, the best possible calculation cannot replace an experimental test of the actual uncertainty arising in dating a stratigraphic level (and thus adding the genetic and historical uncertainties to the strictly analytical process). This actual reproducibility at the level of a sampled formation is tested by the use of different fractions of the same sample or of different samples in the formation. This experimental test is useful for all types of chronometer and will be more conclusive when done in several laboratories.

For glaucony dating purposes it is generally preferable to choose the best possible green grains according to outcropping, palaeogeographic conditions and their crystallographic properties, eliminating *a priori* all less favourable samples: for example where too many 'open' layers (swelling layers) are present. Then, one may make a comparison between the apparent ages obtained on two different outcrops of equivalent stratigraphic age in order to arrive at a conclusive age for the dated stratigraphic bed.

The method proposed by H. Kreuzer in Gramann *et al.* (1975), and independently used in this volume by W. B. Harris, applies to different leaching of cleaning treatments of a glaucony, a correct representative age being assumed when differently treated and untreated samples or different size fractions give equivalent results (see the abstracts by H. Kreuzer). This method may be compared with the step-by-step 39/40 Ar technique. This *a posteriori* criterion of choice (after isotopic analysis), which does not take into account the supposed correct age in judging the value of the apparent age, may also be considered as a correct method but needs many more analyses and often leads to the gathering and publishing of geologically meaningless results. This is a situation that stratigraphers sometimes find difficult to understand.

6 CONCLUSION

For an accurate interpretation of analytical data, it is often necessary to increase the number of measurements. Thanks to the ongoing accumulation of results, the genetic, historical and analytical uncertainties will decrease.

Further, it appears that a convenient study of the environment of the chronometer used in both space and time is an *integral part* of any correct analysis. This environmental study may only be achieved by the use of methods and reasoning which are specific to the type of chronometer used. It remains of the first importance to be aware of the uncertainties that may arise *before* the analytical dating process, to try to eliminate most of those data whose discrepant ages are interpreted *a posteriori* as being incorrect due to the use of the initially unreliable chronometers. Within a given sequence all apparent ages are correct if the measurements are carefully done but the *interpretation* of these ages may be difficult to a varying degree. *An informed preliminary choice of chronometer will make the future use of the analytical data easier.*

During the 20 years that the numerical time scale has been intensively studied, there has been much progress in both geochronological and stratigraphical research. The aim of this book is essentially to assess recent progress in geochronology. On this basis there are at least four reasons for current revision of the numerical time scale.

(1) More data are available on more samples.

(2) Decay constants have been revised and the same ones are accepted in all laboratories (except for fission tracks).

(3) Analytical processes are far more reliable and accurate; the new data must then supersede the old.

(4) We now have a better idea than before of how to use the isotopic data, especially on sedimentary chronometers, a fact which is extensively discussed in this volume.

Acknowledgements

The author is greatly indebted to the publishers for improvement of the English version of this chapter. The accumulation of much of the data used in this volume was possible thanks to the help and permission of the International Geological Correlation Programme (IGCP). Project 133 of this Programme, 'Geochronology of Mesozoic and Cenozoic sediments', was accepted in 1975. The first leader of the project, Professor Paul Pasteels, agreed one year later to transfer his responsibilities to me. This volume is one of the results of the researches done under the aegis of IGCP Project 133 and has been supplemented with data resulting from the personal interest of the editor in all subjects relating to the calibration of the time scale.

Résumé

La datation radiométrique des roches destinées à l'élaboration de l'échelle des temps pose des questions spécifiques. L'objet daté doit être *bien repéré stratigraphiquement*; l'âge mesuré doit être aussi proche que possible du moment du dépôt de la formation, ceci comprend à la fois deux conditions: l'âge apparent initial doit être nul et est lié aux *conditions de genèse*; les teneurs en isotope mesurées doivent être strictement liées à la loi de décroissance radioactive naturelle tout au long de *l'histoire du matériel* choisi comme chronomètre après son inclusion dans les sédiments. Enfin les *conditions analytiques* doivent être bien connues autant en ce qui concerne la représentativité des aliquotes soumises à l'analyse qu'en ce qui concerne la précision et la justesse des mesures (étalonnage de l'appareil de mesure).

J'ai défini ainsi quatre types d'incertitudes: (1) *stratigraphique*; (2) *génétique*; (3) *historique*; (4) *analytique.* Les incertitudes génétiques et historiques peuvent être groupées en une incertitude géochimique qui dépend de chaque type de chronomètre (sédimentaire, volcanique, plutonique). C'est la *combinaison de toutes ces incertitudes* qui donne l'incertitude réelle sur un âge repère.

C'est l'examen de ces quatre types d'incertitude qui a été conseillée et tentée dans chacune des contributions de ce volume qui est, en quelque sorte, un essai sur les incertitudes mises en jeu lors de l'établissement de l'échelle des temps.

(Manuscript received 12–1–1981)

Section I

Methods of correlation

Numerical Dating in Stratigraphy
Edited by G. S. Odin

2
Biochronology—Biological events in time resolution, their potential and limitations

BILAL U. HAQ and THOMAS R. WORSLEY

1 INTRODUCTION

Biochronology utilizes the irreversible process of organic evolution to order the sequence of 'events' that help in the subdivision of geological time and the documentation of earth history. This aspect of geochronology exploits the sequence of biostratigraphic events (usually the first and last appearances of taxa in continuous sedimentary sequences) and their inferred (or interpolated) ages to arrive at a utilitarian chronology. Biochronology thus goes one step further than biostratigraphy, which only aims at recognizing fossil features that uniquely characterize a known stratigraphic level in a sedimentary section, without inherent chronological significance.

The purpose of this chapter is a general discussion of the basic problems inherent in the practice of biochronology and to briefly review the state of the art of planktonic biochronology, and its potential and limitations for wide geographical correlations.

For supplemental reading the reader is referred to the excellent discussion of biochronology by Berggren and van Couvering (1978), illustrated by reference to the isochronous appearances of the planktonic foraminifer *Globigerina nepenthes* and the three-toed horse *Hipparion* in marine and continental sequences, respectively.

The major tools for dating sediments are radiochronology (isotopic decay) and biochronology, which are used both directly and in calibrating the sequence of reversals of the earth's magnetic field (magnetostratigraphy—see Channell, this volume). Recently, stable-isotopic stratigraphy (oxygen and carbon isotopic fluctuations) has also achieved limited usefulness (see Odin, Renard and Vergnaud Grazzini, following chapter). The resolutions of all chronostratigraphies other than radiochronology are interdependent. There are numerous logistic and methodological problems inherent in each approach (discussed elsewhere in this volume) which limit their usefulness to varying extents. Fossils, however, are the most widespread and abundant datable features in Phanerozoic sediments and thus, by default rather than choice, biochronology achieves the status of the most readily available tool for long-distance correlations.

2 CHARACTERISTICS OF FOSSILS USEFUL TO BIOCHRONOLOGY

The *species* is the basic taxonomic unit of all biochronology; its definition remains one of the most troubling problems in palaeontology, and therefore in biochronology, as only morphology is available to the biostratigrapher, and changes in morphology form the basis of all biochronological subdivisions. These changes include (1) stratigraphic appearance or disappearance of a morphotype, (2) changes in abundance of a morphotype, (3) changes in the morphology of a morphotype (either shape or size). The degree of biochronological resolution is therefore an intimate function of the degree to which palaeontologists can effectively *communicate the changes they observe* to other palaeontologists. The first appearance of a particular morphotype due to immigration from another area, especially in the case of planktonic species, is considered, geologically speaking, sufficiently rapid for no distinction of this event to be usually made from its appearance due to evolution from an ancestral morphotype. The first appearance datum (FAD) and the last appearance datum (LAD) are by far the most utilitarian and most easily communicated type of morphological change information. For this reason, FADs and LADs have come to dominate global biochronological subdivision. Furthermore, this method of subdivision works well even for broadly defined morphotypes and is useful over great distances within the range of the morphotypes.

The establishment of a utilitarian range end-point biochronology for any fossil group is begun first by identifying the FADs and the LADs of key taxa that are relatively widely distributed in contemporaneous sequences. Ages may be assigned to these datum levels through direct or indirect calibration, but this is an added feature that is not essential to basic biostratigraphic operations. Invalid calibrations do not (should not) invalidate correlations. From the position of the datum levels in a stratigraphic sequence with two or more radiometric dates, or magnetostratigraphy if locally well calibrated, and the rates of sedimentation between levels of known ages, an age can be estimated for each datum enclosed within the sequence. When direct calibration is not available for age estimation, which is quite often the case, then ages of datum levels can be estimated through their stratigraphic relationships with calibrated datum levels of other fossil groups occurring in the sequence that have been dated in sequences elsewhere.

Repetitive changes in abundance of a morphotype through time are easily quantified and can be used to generate 'wiggly lines' or abundance profiles. Corresponding wiggles between and among different sites form the basis for a type of correlation analogous to that seen in stable-isotopic stratigraphy or magnetostratigraphy. Unfortunately, their utility in long-range correlation is minimal because the relative abundance of a morphotype in a sample depends not only on changes in its own abundance but *on changes in the abundance of other taxa* in the sample. Therefore one can never be sure of the exact cause of the change in relative abundance of the morphotype in question, and if the change has a multitude of causes that are a function of environmental parameters that vary differently in differing directions, the correlation might well be sharply diachronous, especially over long distances.

Continuous change in morphotypic character through time forms the basis for describing evolutionary lineages and seemingly should be one of the most useful parameters for long-range correlations. Unfortunately, morphometric measurement is one of the most tedious tasks in palaeontology and most practising biochronologists are loath to do it. Futhermore, palaeoecologic parameters exert a strong overprint on evolutionary changes so that ecophenotypic variation often disguises evolutionary change. Therefore the lineage approach, while remaining one of the basic tools of the trade, has never been able to consistently offer high-resolution, long-range correlation because of the confusion between palaeoecologic and evolutionary influences on morphology. Furthermore, ecophenotypic variation across climatic belts has led to pronounced taxonomic confusion when it is demonstrated that two separately named species are in reality ecophenotypic variants of a single interbreeding population, so that the supposed evolution from one to the other is in fact reversible and therefore diachronous. Changes in size of morphotypes are analogous to changes in shape and offer the same problems.

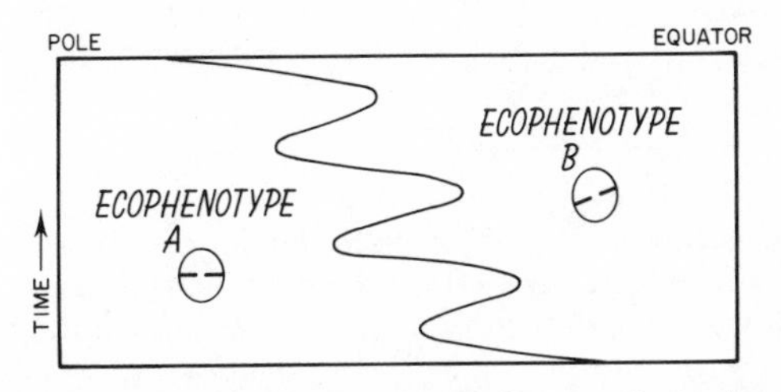

Figure 1. The diachronous distribution of ecophenotypic variants of a species in different geographical areas that can be mistaken for an evolutionary lineage in oscillating climates.

In Figure 1, the ecophenotypic replacement of variant A by variant B of the same species throughout a long-term climatic amelioration resembles evolution, but a correlation based on the change from A to B would be diachronous. Such replacements have been demonstrated for living species and are surely present in the stratigraphic record; the degree of their recognizability is strictly dependent upon whether other correlation criteria have sufficient resolution to demonstrate the diachroneity.

3 CHARACTERISTICS OF THE PRESERVED FOSSIL RECORDS

The above discussion assumes that the stratigraphic record faithfully mirrors earth history; in theory, a perfectly preserved record could be perfectly sampled so that this history would be unambiguous. We now consider what happens when stratigraphic discontinuities, imperfect preservation (due to dissolution and diagenesis), or imperfect or incomplete sampling of the record enter the picture. Figure 2 represents the effects of an unconformity or other stratigraphic discontinuity (i.e. change in lithology, sedimentation rate, etc.) on the three basic forms of biochronological criteria: range end-point, 'wiggly line' and morphological variation. Figure

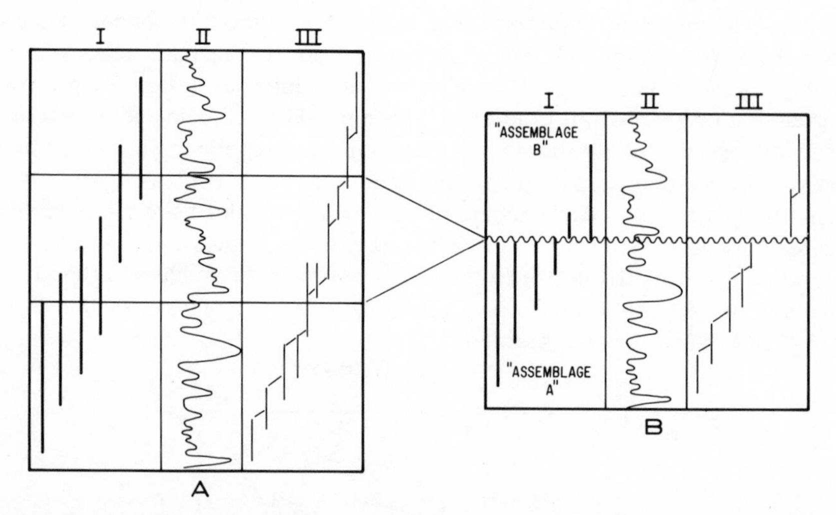

Figure 2. The effects of a stratigraphic discontinuity on observed ranges of species (I), abundance curves (wiggly line) (II) and evolutionary lineage (III). A, complete section; B, section with a mid-portion missing. Note the stratigraphic juxtaposition ('pile-up') of the range end-points in the discontinuous section (BI), giving the assemblages below and above the discontinuity totally different character. The discontinuity has a similar effect of removing mutually common or 'linking' taxa from the lineage. The wiggly line fails to detect the unconformity on its own.

2A represents a continuous section and 2B is a nearby section of similar character but which contains an unconformity. The range end-point approach to biochronological correlation clearly shows a continuous taxonomic turnover in the continuous section but exhibits a 'pile-up' of FADs and LADs at the unconformity. Such pile-ups are in fact one of the easiest ways to recognize unconformities in the absence of other criteria of lithological character. In many deep-sea sections where no lithological changes are observable, pile-ups represent the only way to recognize unconformities.

In the past, many of the changes at unconformity pile-ups have been used as the limits of 'assemblage zones' as well as other stratigraphic boundaries such as the lower and upper limits of the stages, not to mention the classic system boundaries. Obviously this led to much confusion when long-range correlations were attempted, because the stratigraphic magnitude and locations of the unconformities bounding the zones or stages change with locality, thereby leading to 'mixed assemblages' and diachronous correlations based on arbitrary criteria about which no consensus could be achieved. Fortunately, as a continuous data base is being slowly constructed and synthesized, both for land-based and deep-sea sections, the bases and tops of the unconformity-bounded stages are now being fitted into a continuous synthesized sequence.

As can be seen from Figure 2, the 'wiggly line' approach fails completely to detect the unconformity. Many miscorrelations have resulted from improperly matched peaks and troughs of such wiggly lines when the presence of an unconformity went undetected. The lineage approach, on the other hand, is influenced by condensed or 'speeded-up' biostratigraphy at the unconformity, perceived as 'evolutionary bursts', which are in practice sometimes used to define the boundaries of lineage zones. However, as in the case of the assemblage zone, the sharpness of the transition is artificial and blurs as one correlates laterally to where the interval is continuous, and the question of where to place the now imaginary transitional 'burst' leads to endless debate and controversy.

4 SAMPLING OF STRATIGRAPHIC SECTIONS

Sampling stratigraphic sections presents another series of problems to the biochronologist. Ironically, end-point stratigraphy (using FADs and LADs) is most dependent on the recognition of species exactly in the places where they become most difficult to detect (i.e. at the ends of their ranges). Figure 3A represents the ranges of three species (A, B and C) for which a continuous stratigraphic section has been sampled at uniform intervals using uniform search criteria for the species in each (e.g. counting 300 specimens or searching for 20 minutes). As we can see, the true sequence of events is FAD A, FAD B, FAD C, LAD A, LAD B, LAD C. However, the sequence

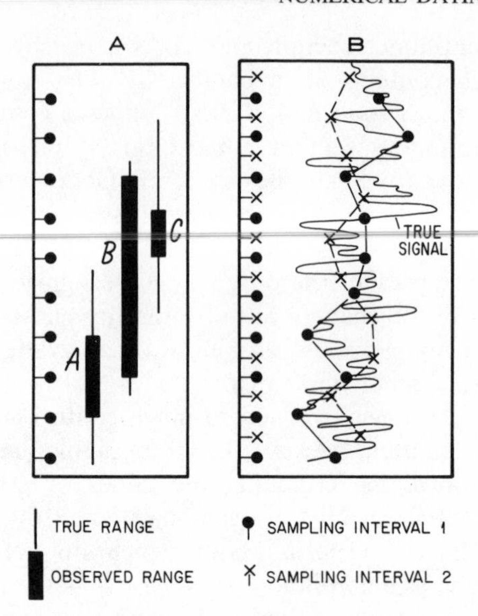

Figure 3. The effects of sampling and search criteria upon the interpretation of stratigraphic record.

as determined by the sampling and search strategy used is very different: namely, FAD A, FAD B, LAD A, FAD C, LAD C, LAD B. Predictably, the palaeontologic literature is replete with such discrepancies, and the problem is compounded when different investigators work on different sections using different sampling and search criteria, and have slightly different taxonomic concepts of the species in question.

Figure 3B represents a continuous section with an extremely well preserved 'wiggly line' signal that has been sampled using a uniform sampling frequency (represented by closed circles) much larger than the frequency of the 'wiggly line' signal. In this case the true signal has actually been aliased (sampled in such a way that a trend bearing no relation to the true signal emerges), and the sample data are virtually meaningless with respect to the information content of the true signal. This is clearly shown when a second uniform sampling interval (represented by Xs) is chosen at points halfway between the original points, which generates a trend bearing no relation to that of the first attempt. Obviously, if a series of such 'wiggly lines' is generated for several widely spaced sections, the mating of peaks and troughs in the 'wiggly lines' becomes an artform that in many cases bears little, if any, relation to reality, and the correlations upon which it is based are again diachronous.

5 STRATIGRAPHICALLY DISPLACED FOSSILS

Much paranoia exists in the micropalaeontological community (and especially among nannofossil palaeontologists because of the extreme small size and propensity to displacement of nannofossils) about the serious deleterious effects of displaced fossils upon biochronology. This fear is for the most part quite unfounded, as so elegantly expressed by Shaw (1963) in his now classic book *Time in Stratigraphy*. Shaw notes that if a fossil is sufficiently stratigraphically (temporarily) displaced for current biochronological resolution to recognize it as being displaced, no harm is done. If, on the other hand, it is displaced so slightly that current biochronological resolution cannot recognize its displacement, its secondary position will fall within the 'noise level' of resolution and again do no real harm.

However, the effects of slightly displaced fossils (i.e. *via* bioturbation, penecontemporaneous slumping, selected winnowing, etc.) seriously perturb the fine-scale fidelity of the palaeoclimatic and palaeoenvironmental signal contained in the stratigraphic record. For example, deep-sea sediments accumulate at an average rate of about 1 cm/10^3 yr and deep-sea benthonic organisms burrow to an average depth of 30 cm. These factors have the effect of defocusing the stratigraphic record in average deep-sea sections to a resolution of about 30,000 years, thereby destroying short-term palaeoenvironmental signals such as delgaciation events or short-term climatic signals such as the 'little Ice Age'. The only two ways to circumvent such defocusing and attain very high resolution is to work on sections that either are deposited under anoxic bottom conditions so that they contain no significant bottom fauna to perturb the record (i.e. fiords, the Black Sea, etc.) or to work on sections having relatively high sedimentation rates (i.e. greater than 1 m/10^3 y), so that bioturbation defocuses the record to a much lesser extent. Unfortunately, sections deposited under anoxic bottom conditions are relatively uncommon in the geological record and sections having very high sedimentation rates usually have their fossil content exceedingly diluted with terrigenous detritus, making biostratigraphic work extremely difficult. In either case, palaeoenvironmental–palaeoclimatic reconstructions become cost-inefficient and have only been attempted with a few time intervals dealing with a very small fraction of the record (see e.g. CLIMAP, 1976).

6 BIOCHRONOLOGY AND CROSS-LATITUDINAL CORRELATIONS

As Berggren and van Couvering (1978) point out and the above discussion clearly shows, the biostratigraphic, lithological, geochemical or magnetostratigraphic criteria used for local correlations of strata are often physically discontinuous and subject to iterative confusion. Therefore, most long-distance correlations are essentially geochronological. And because biochrons are more readily available and more common than other geochrons, biochronology remains the staple tool of geochronology. However, there are limitations to the usefulness of biochronological datum events when attempting cross-latitudinal correlations. The applicability of the datum events is limited by the geographical distribution of taxa; for instance, many low-latitude taxa are ecologically excluded from higher latitudes. The high-stress environments of the high latitudes inhibit development of diversified assemblages, which are populated with a few cosmopolitan, robust and well-adapted eurythermal taxa with long stratigraphic ranges that are marginally, if at all, useful in biochronology. Climate has been shown to be one of the major factors controlling biotic diversities—during times of climatic amelioration diversities increase and during climatic deterioration they decline (Haq, 1973). It is only during climatic ameliorations, when elements of the more diverse, low-latitude assemblages expand into higher latitudes, that biochronologies established for temperate latitudes can become applicable in high latitudes.

By the same token, the lowered diversities during colder intervals would reduce the biochronological resolution capability due to dearth of datum events, and the resolution would increase during warmer periods because of greater availability of biochrons due to higher evolutionary turnover (Haq, 1973). The changes in the climate, therefore, control the biostratigraphic/biochronological resolution and explain the differences in duration of biostratigraphic zones at different times in the stratigraphic record.

If the frequency and geographical extent of datum events are climate-dependent, a certain amount of diachronism is to be expected between the evolutionary first appearance of a taxon in one area and its migration to another region. From a geochronological perspective, the prochoresis (spreading out) of species is almost instantaneous (Berggren and van Couvering, 1978) within the climatic/ecological niche of certain taxa (especially planktonic microbiota). Datum events can then be considered isochronous only within the normal geographical (tolerance) range of taxa, and thus should be used with caution beyond these limits, where they may be time-transgressive. For fossil species, without modern analogues, the determination of 'normal geographical range' can be difficult, if not impossible. Quantitative biogeographical studies can help define the 'normal' distributional limits of taxa and perhaps provide a solution to the high and low latitude correlation problems.

The concept of biogeographical acme events and their usefulness in low to high latitude correlations has been discussed by Haq and Lohmann (1976) and more recently by Haq (1980a). It involves the use of quantitative palaeobiogeographical data to identify synchronous distributional events between stratigraphic sections. Within this time framework, the synchroneity

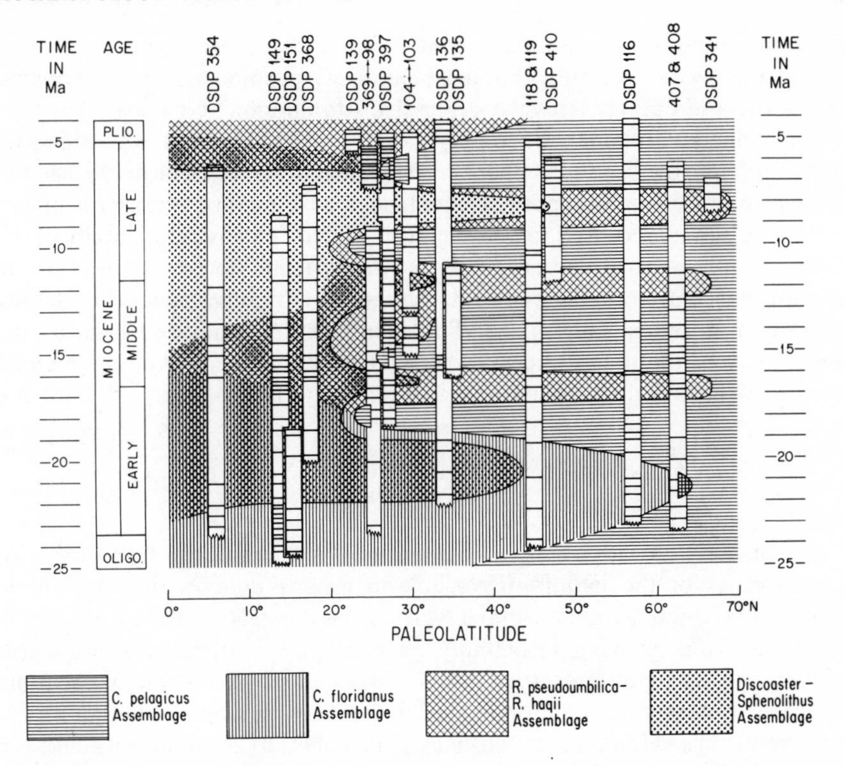

Figure 4. Cycles of latitudinal shifts of calcareous nannoplankton assemblages in the North Atlantic Ocean during the Miocene interpreted as a response to major fluctuations in climate. The major shifts of relatively warmer, mid-latitude assemblages into higher latitudes can be used for the refinement of biochronological scale in the higher latitudes from which marker, low-latitude taxa are normally excluded. (After Haq, 1980a)

of biochronological levels may be evaluated and high- and low-latitude sequences may be more precisely correlated.

Essentially, the procedure involves four steps:

(1) Delineation of cycles of latitudinal shifts of assemblages through quantitative biogeographical data in stratigraphic sequences (see Figure 4);

(2) Approximate correlation of these cycles in various sequences by the usual (i.e. less refined in higher latitude) biostratigraphic methods;

(3) Refinement of this approximate time framework by using the maximum changes in migration cycles as time-correlation points; and

(4) Evaluation, with this time framework, of the synchroneity of datum levels.

The fundamental assumption in this concept is that the times of maximum environmental change, recorded by the major floral and faunal shifts through latitudes, are essentially contemporaneous. This scheme is analogous to the method of local time correlation by position within bathymetric cycles recorded in transgressive–regressive stratigraphic sequences. The time levels defined by the maxima in migrationary cycles (Figure 4) are similar to, but more narrowly defined than, acme or peak zones, which are characterized only by the 'exceptional' abundance of a taxon rather than by the maximum abundance due to geographical shift of an environmentally sensitive assemblage. The precision with which acme horizons approximate synchronous surfaces depends on the ability to unambiguously identify maximum latitudinal migrations. This identification is affected both by the complexity of the migrationary cycles and by the resolution of the method used to delineate them (see Haq and Lohmann, 1976; Haq, 1980a,b).

7 CENOZOIC BIOCHRONOLOGY—STATE OF THE ART

In this section we briefly describe the 1980 status of Cenozoic biochronologies of the major microplankton groups used in time resolution of marine sediments. Due to the advent of the Deep Sea Drilling Project in 1968, and the growing availability of relatively continuous stratigraphic sequences with well-preserved microfossils, there has been a quantum improvement in the biochronological resolution capability of major microfossil groups. However, much work still needs to be done to achieve a uniform refinement in biochronologies of the different groups, and especially in the documentation of synchroneity/diachroneity of datum events in different latitudes. The need for better chronology in the study of cyclic phenomena or detailed time series is self-evident in the rapid growing field of palaeooceanography.

The Cenozoic planktonic foraminiferal biochronology is the best documented of all the major microfossil groups. Furthermore, the direct correlations of planktonic foraminiferal datum events with magnetostratigraphy, and in some cases with radiometric dates from selected horizons, has established a refined and utilitarian time scale for this group in the low latitudes for approximately the last 6 Ma. Prior to 6 Ma, the zones and datum events are assigned ages by indirect correlations *via* siliceous microplankton events that have been palaeomagnetically dated, or through the few radiometric dates (particularly in the Palaeogene) in stratotype sections of European stages that form the basis of the time-stratigraphic hierarchy. The Late Neogene biochronology and time scale of planktonic foraminifera and other groups (including nannofossils, radiolaria and diatoms) was summarized by Ryan *et al.* (1974) and Berggren and van Couvering (1974). The known biochronological relationships of Palaeogene datum levels form the basis of the revised Palaeogene time scale presented by Hardenbol and Berggren (1978); this scale was discussed by other authors with respect to numerical ages (Odin 1978b; see also Curry and Odin, this volume). Numerous other studies have appeared recently in which foram datum levels have been either directly or indirectly correlated with magnetostratigraphy (see e.g. Keller, 1980; 1981).

As far as calcareous nannofossils are concerned, an accurate and fairly refined time scale by first-order correlation with magnetostratigraphy is now available for the last 8 Ma for low- and mid-latitude sequences. A summary of the approximate ages of the nannofossil datum levels by second-order correlations (or extrapolation) as far back as 22 Ma was included in Ryan *et al.* (1974). Gartner (1973) presented the ages of numerous important nannofossil datum events of the last 6 Ma and a refined Pleistocene biochronology (Gartner, 1978) by direct correlations with magnetostratigraphy (Figure 5).

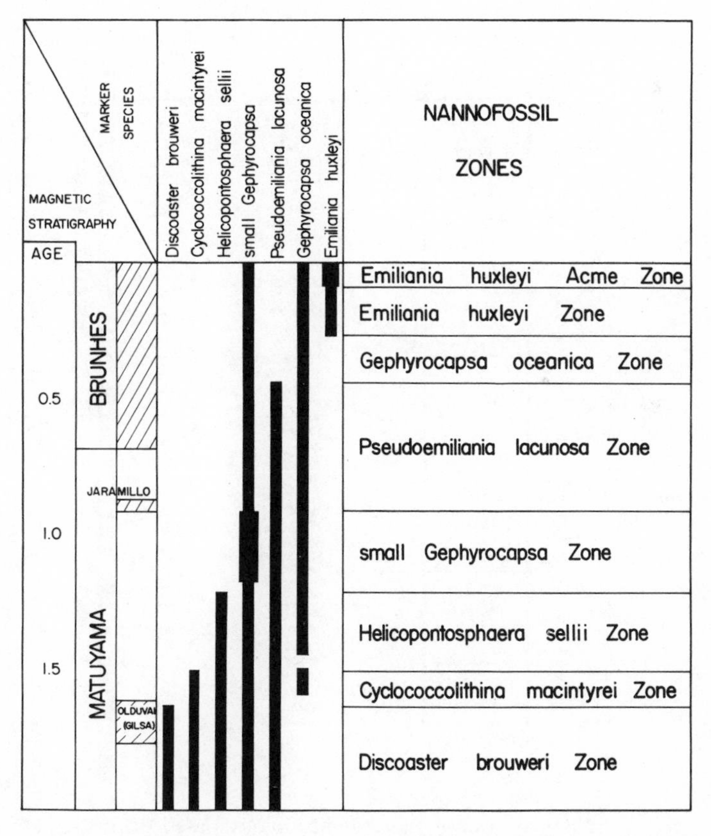

Figure 5. A refined Pleistocene biostratigraphic zonation of calcareous nannoplankton acheived by direct, first-order correlation of datum events with magnetostratigraphy. (From Gartner, 1973, figure courtesy of S. Gartner, reproduced by permission of *Marine Micropaleontology*.)

The biochronological framework has recently been extended by first-order correlation with magnetostratigraphy to 8 Ma (Haq *et al.*, 1980). Prior to 8 Ma, ages of datum levels have been assigned by second- or third-order (through their relationship to planktonic foraminiferal datum levels) correlations.

Siliceous microplanktonic datum events have been more extensively dated by first-order correlations with magnetostratigraphy. In the case of radiolaria, a refined biochronology has been extended down to about 25 Ma (latest Oligocene, see Figure 6) in the equatorial Pacific sequences (Theyer *et al.*, 1978). Mid-latitude regions are generally silica-free in the Cenozoic and there has been little work on radiolarian biochronology of these latitudes. In the higher latitudes some Quaternary radiolarian datum events have been

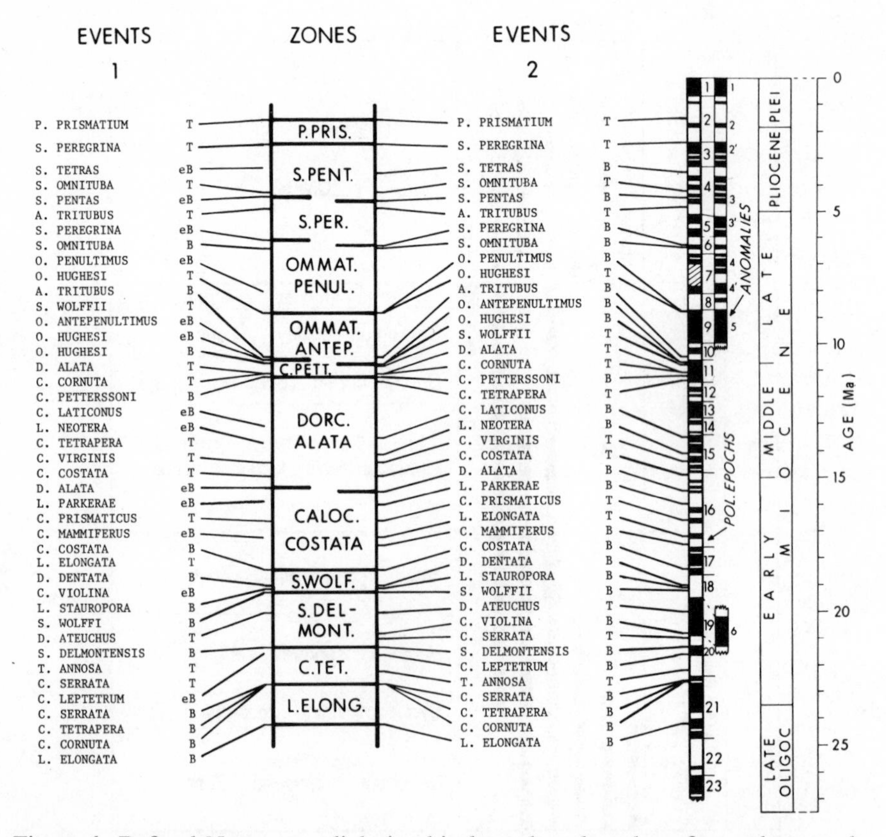

Figure 6. Refined Neogene radiolarian biochronology based on first-order correlation of datum events with magnetostratigraphy of cores from the equatorial Pacific Ocean. T, top of the range of species; B, bottom of the range of species. (After Theyer *et al.*, 1978, figure courtesy of F. Theyer.)

assigned accurate ages through magnetostratigraphy (Hays and Opdyke, 1967), and one radiolarian datum has been shown to be globally synchronous, validated by oxygen-isotopic stratigraphy (Hays and Shackleton, 1976).

Diatoms are most abundant in high-nutrient areas, such as the equatorial upwelling regions and along continental margins where water masses from high latitudes converge. Because of this strong dependence on water mass properties, most diatom species are provincial, which limits their usefulness in high-to low-latitude correlations. A fairly refined mid-Miocene to Pleistocene diatom biochronology, by first-order correlation to magnetostratigraphy, is now available for equatorial latitudes (Figure 7). For the higher latitudes a lower-resolution, but utilitarian biochronology is also available for the Late Miocene to Pleistocene by means of second-order correlations (Burckle, 1978).

Within the Pleistocene and part of the Neogene the stratigraphy based on major fluctuations in the stable-isotopic ratios of oxygen and carbon has been used for accurate time resolution to varying degrees of success, and it holds promise for better geochronological resolution as far back as the Palaeogene (see Odin, Renard and Vergnaud Grazzini, following chapter). For example, within the Pleistocene, Thierstein *et al.* (1977) used the oxygen-isotopic stratigraphy (assisted by magnetostratigraphy) established by Shackleton and Opdyke (1973) to assign accurate ages to three of the Pleistocene nannofossil datums and to document the relative synchroneity of two events and the diachroneity of the third (Figure 8). Similarly, Haq *et al.* (1980) used a prominent and permanent decrease in $\delta^{13}C$ of benthonic foraminifera in the Late Miocene sequences of the Pacific, Indian and Atlantic Oceans to document the relative synchroneity of a number of Late Miocene phytoplanktonic datums (Figure 9). Numerous other marked and widely distributed oxygen-isotopic events have been recognized in the Palaeogene, which may prove to be useful tools for further geochronological resolution.

These examples clearly show the need for greater effort towards refinement of various biochronological frameworks of different microfossil groups. The plans by the Deep Sea Drilling Project for hydraulic piston coring, which recovers relatively long, undisturbed cores containing detailed biostratigraphic information and which, in addition, are suitable for palaeomagnetic measurements, promise to go a long way in resolving the problem of direct, first-order correlation of potential biochrons with magnetostratigraphy and perhaps with isotopic stratigraphy. The 1980s will probably see a major breakthrough in the development of refined and accurate time scales based on marine microbiochronologies, pushed closer to the theoretical limits of resolution discussed earlier in this essay. Hydraulic piston coring, however, becomes more difficult in older, more consolidated sediments, and may thus help resolve only the Late Palaeogene and younger

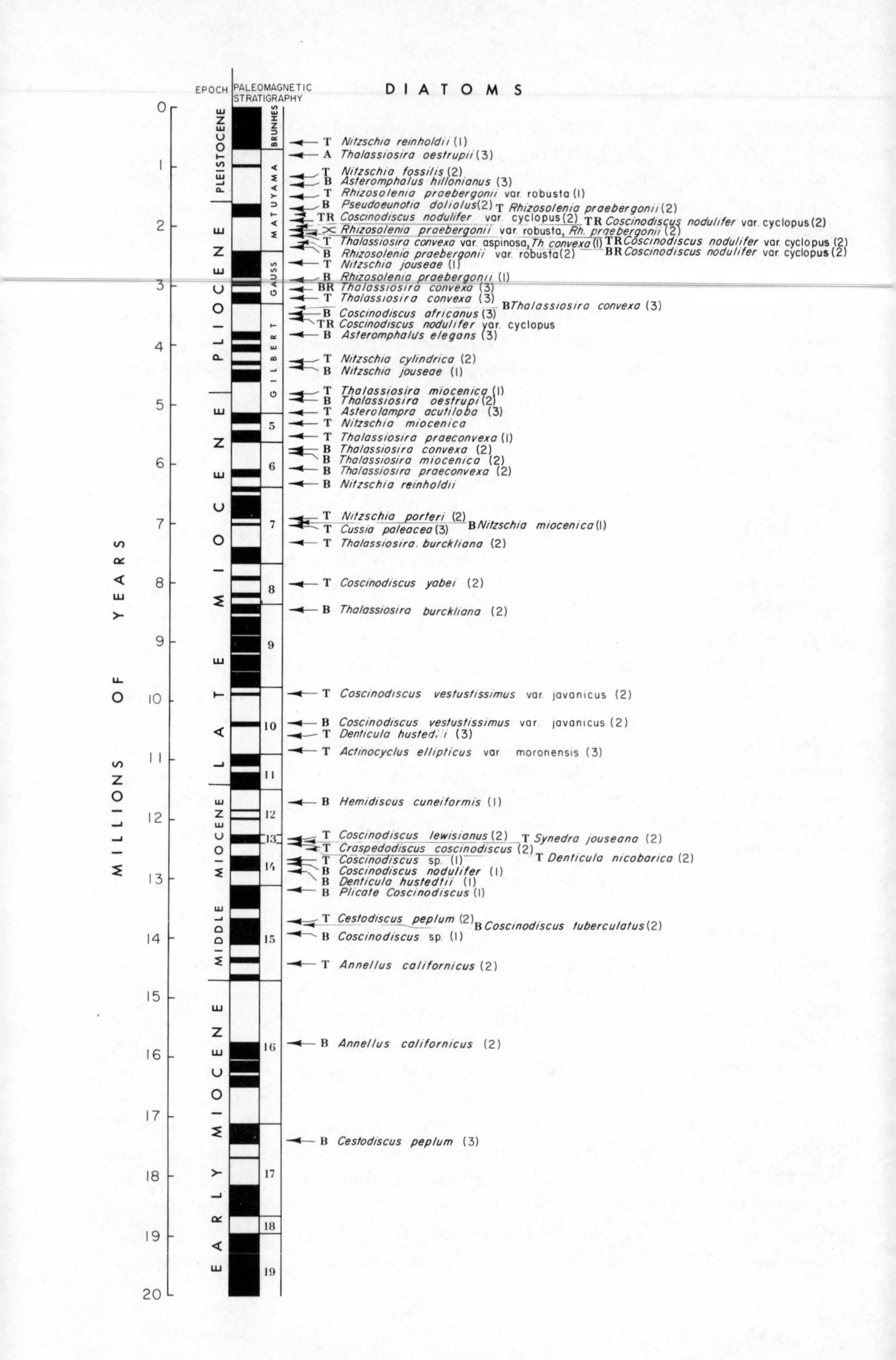

EPOCH
PALEOMAGNETIC STRATIGRAPHY
DIATOMS
MILLIONS OF YEARS
0
1
2
3
4
5
6
7
8
9
10
11
12
13
14
15
16
17
18
19
20
PLEISTOCENE
PLIOCENE
LATE MIOCENE
MIDDLE MIOCENE
EARLY MIOCENE
BRUNHES
MATUYAMA
GAUSS
GILBERT
5
6
7
8
9
10
11
12
13
14
15
16
17
18
19
T Nitzschia reinholdii (1)
A Thalassiosira oestrupii (3)
T Nitzschia fossilis (2)
B Asteromphalus hillonianus (3)
T Rhizosolenia praebergonii var. robusta (1)
B Pseudoeunotia doliolus (2)
T Rhizosolenia praebergonii (2)
TR Coscinodiscus nodulifer var. cyclopus (2)
TR Coscinodiscus nodulifer var. cyclopus (2)
Rhizosolenia praebergonii var. robusta, Rh. praebergonii (2)
T Thalassiosira convexa var. aspinosa, Th. convexa (1)
TR Coscinodiscus nodulifer var. cyclopus (2)
B Rhizosolenia praebergonii var. robusta (2)
BR Coscinodiscus nodulifer var. cyclopus (2)
T Nitzschia jouseae (1)
B Rhizosolenia praebergonii (1)
BR Thalassiosira convexa (3)
T Thalassiosira convexa (3)
B Thalassiosira convexa (3)
B Coscinodiscus africanus (3)
TR Coscinodiscus nodulifer var. cyclopus
B Asteromphalus elegans (3)
T Nitzschia cylindrica (2)
B Nitzschia jouseae (1)
T Thalassiosira miocenica (1)
B Thalassiosira oestrupi (2)
T Asterolampra acutiloba (3)
T Nitzschia miocenica
T Thalassiosira praeconvexa (1)
B Thalassiosira convexa (2)
B Thalassiosira miocenica (2)
B Thalassiosira praeconvexa (2)
B Nitzschia reinholdii
T Nitzschia porteri (2)
B Nitzschia miocenica (1)
T Cussia paleacea (3)
T Thalassiosira. burckliana (2)
T Coscinodiscus yabei (2)
B Thalassiosira burckliana (2)
T Coscinodiscus vestustissimus var. javanicus (2)
B Coscinodiscus vestustissimus var. javanicus (2)
T Denticula hustedtii (3)
T Actinocyclus ellipticus var. moronensis (3)
B Hemidiscus cuneiformis (1)
T Coscinodiscus lewisianus (2)
T Synedra jouseana (2)
T Craspedodiscus coscinodiscus (2)
T Denticula nicobarica (2)
T Coscinodiscus sp. (1)
B Coscinodiscus nodulifer (1)
B Denticula hustedtii (1)
B Plicate Coscinodiscus (1)
T Cestodiscus peplum (2)
B Coscinodiscus tuberculatus (2)
B Coscinodiscus sp. (1)
T Annellus californicus (2)
B Annellus californicus (2)
B Cestodiscus peplum (3)

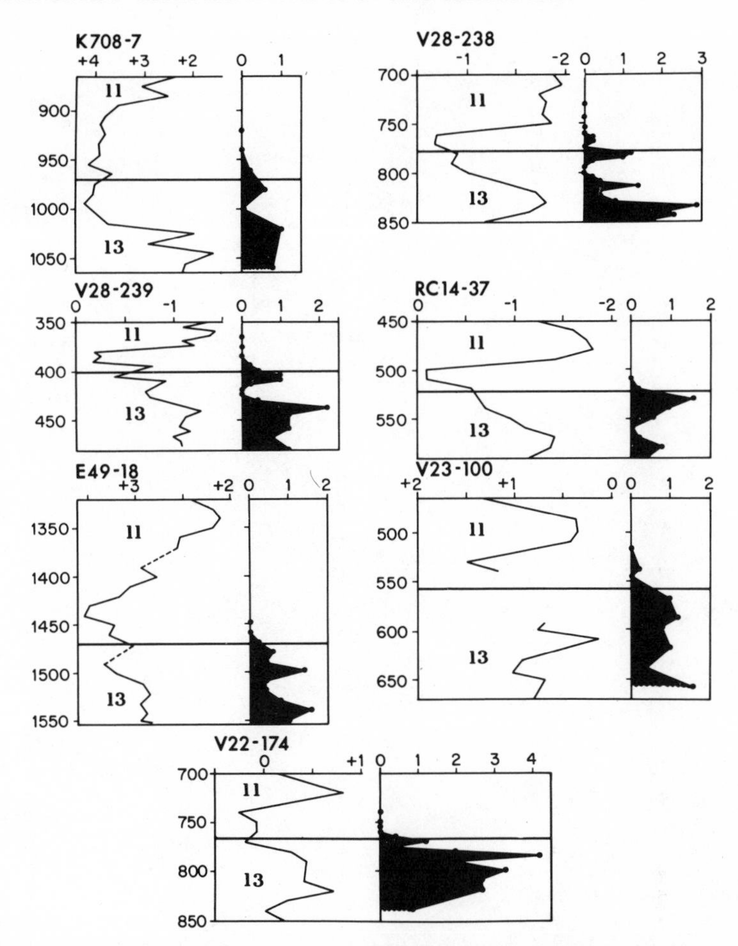

Figure 8. An example of the use of stable-isotopic stratigraphy in resolving biochronological problems, documenting the global synchroneity of the extinction datum of nannoplankton *Pseudoemiliania lacunosa*. This datum, (represented by horizontal lines in each section) representing the first sharp decrease in the abundance of the species (right columns, shaded black), was found to consistently occur in oxygen-isotopic stage 12 (of Shackleton and Opdyke, 1973) in cores from high to low latitudes and in different oceans. (From Thierstein *et al.*, 1977, figure courtesy of H. Thierstein with permission of the Geological Society of America.)

Figure 7. Neogene equatorial Pacific diatom biochronology based on both first-order and second-order (through radiolarian datum events) correlations of datum events with magnetostratigraphy. T, top of the range of species; B, bottom of the range of species; TR & BR, top and bottom of recurring taxa; A, upward increase in abundance; X, reversal in dominance of two species. Numbers (1–3) after the species names refer to reliability index of the datum, 1 being most reliable for long-distance correlations and 3 least reliable. (Data and figure courtesy of L. Burckle.)

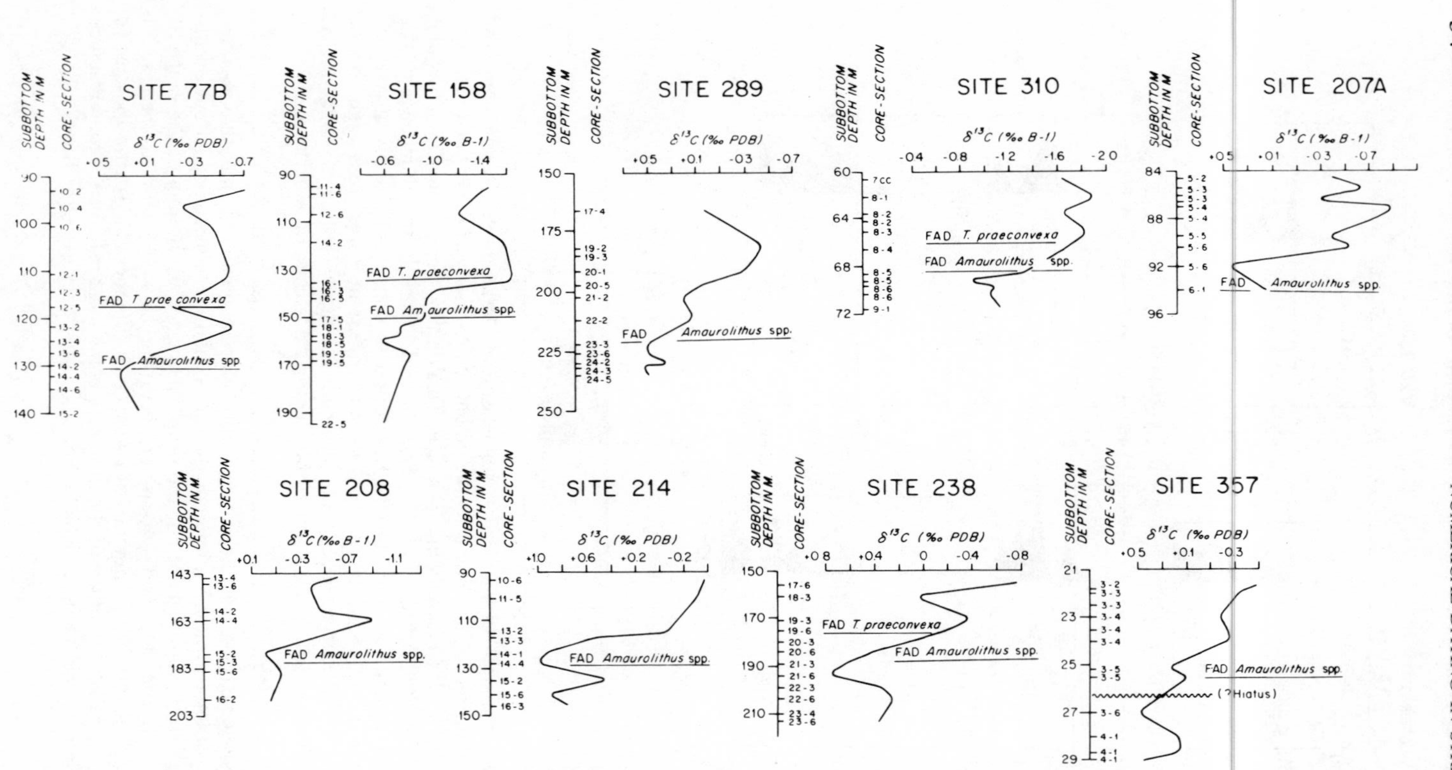

Figure 9. The use of a prominent, global decrease in carbon-isotopic values of benthic foraminifera in the Late Miocene to ascertain the synchroneity of the first appearance datum of nannoplankton species of genus *Amaurolithus* and the diatom *Thalassiosira praeconvexa*. (From Haq *et al.*, 1980, reproduced by permission of the Geological Society of America.)

biochronological problems, with limited usefulness in the earlier Cenozoic and Mesozoic intervals. In addition to achieving the desired accuracy and resolution in biochronology, these undisturbed cores obtained in key areas of the oceans will also lend themselves to refined palaeoenvironmental studies, which are the ultimate aim of all stratigraphic exercises.

Acknowledgements

This chapter was reviewed by John van Couvering, David Johnson and Brian Tucholke. The authors are indebted to Lloyd Burckle, Stefan Gartner, Fritz Theyer and Hans Thierstein for providing some of the illustrative material presented here. Our research is supported by grants from the National Science Foundation, Division of Submarine Geology and Geophysics, under CENOP grant numbers OCE78-18286 and OCE78-82514. This is Woods Hole Oceanographic Institution number 4742.

Résumé du rédacteur

Cette contribution est une brève revue des potentialités et des limites de la biostratigraphie à l'aide des fossiles planctoniques. Les auteurs constatent que la résolution de la méthode est fonction de la possibilité qu'a le paléontologiste de communiquer les changements de morphotype qu'il observe. L'apparition (FAD), la dernière observation (LAD) sont les critères les plus aisés à établir, l'abondance est utile; ces critères sont cependant fonction de conditions locales. De plus, les lacunes de dépôt créent des coupures artificielles. En fonction de la densité de l'échantillonnage, donc du temps de recherche effectivement consacré à l'étude, les coupures obtenues sont plus ou moins représentatives. Une nouvelle difficulté liée aux très petits fossiles est la grande facilité qu'ils ont d'être déplacés dans la séquence sédimentaire; ceci amoindrit la finesse des coupures biostratigraphiques réalisable. L'utilisation des évènements repères est limitée à la répartition géographique possible des espèces mises en jeu. Spécialement le facteur latitude limite clairement les possibilités de corrélation car la survie des espèces est souvent liée à un domaine de température limité. Ceci peut entraîner un diachronisme dans l'apparition d'une espèce en fonction de l'évolution temporelle du climat.

Ces données fondamentales sont enfin examinées dans le cas concret du Cénozoïque.

(Manuscript received 10-7-1980)

Numerical Dating in Stratigraphy
Edited by G. S. Odin

3
Geochemical events as a means of correlation

GILLES S. ODIN, MAURICE RENARD and COLETTE VERGNAUD GRAZZINI

1 INTRODUCTION

1.1 Means of correlation

An evaluation of the Phanerozoic time scale needs geologically well-established events to allow a wide range of stratigraphic correlations. Classically, two means are available: one based on sedimentology, which schematically leads to *lithostratigraphy*, and the other one based on palaeontology, which leads to *biostratigraphy*. The possibilities and limitations of biostratigraphic events as a mean of correlation are discussed in the preceding chapter by Haq and Worsley.

We will briefly record the possibilities of lithostratigraphy using sedimentological characters of the lithologic units as markers for geological events. But the main purpose of this chapter is to explore the possibilities of more hidden properties of the rocks. We will see later (chapter by J. Channell) that there exists another possibility of correlation of the stratigraphic series: palaeomagnetism. Just as palaeomagnetism led to the construction of a *magnestostratigraphy*, we think that the use of trace elements and isotopic ratios may lead to a *geochemical stratigraphy*.

1.2 Causes of the geological events

As a first approach, the presence of a marker in the stratigraphic column can reflect two kinds of phenomenon relative to the earth's crust. Outside the earth's crust, climatic changes indirectly lead to variations readable in the lithologic succession (changes in sea-level, change in geochemical reactions in the sea and in the earth . . .); the falling of extra-terrestrial material, if sufficiently abundant and dispersed, may leave a useful direct imprint in the lithologic succession.

Inside the earth's crust, tectonic activity generates a more or less direct imprint (volcanic rocks, deposition of orogenic detrital sediments, but also sudden arrivals of specific elements sometimes of unusual isotopic composition). In the following sections we will try to specify the origin of each geological event used as a marker.

1.3 Qualities of these events

In each case, a good marker (i.e. geological event) will be useful if its qualities of *synchronism* and *duration* are favourable. In the question of synchronism, this depends on what possibilities of checking it by different systems of marker are available. The question in considering duration is: should the event used as a marker be considered as instantaneous or not? This very much depends on the age of the series under study. We can say that a 'good' geological event is an event whose duration is less than 1% of its age: for example, 100 years for a 10,000 years old rock, 100,000 years for a Miocene rock (10 Ma from now), or 1 Ma for an Albian rock (100 Ma from now).

Practical examples of these geological events can be subdivided into three types: sedimentological events, trace element events and stable-isotope events.

2 SEDIMENTOLOGICAL EVENTS

2.1 Lithologic facies as time markers

Geologists have recognized specific facies which may help for purposes of correlation. The correlations obtained are frequently imperfect although the observed facies were related to general climatic or tectonic changes.

As an example, the *Red Sandstones* formations, a facies composed of reddy pink sandstones with feldspar and sometimes gypsum indicating a desert climate, have been identified twice in the Palaeozoic sequence of NW Europe: the Old Red Sandstones from the Devonian age; the New Red Sandstones from the Permian age. They can only be encountered in a limited area from England to Finland, but there their original characteristics are very useful to correlate the very small outcrops which subsist after the erosion of a formation which probably covered the whole Caledonian area. In both cases (Old and New Red Sandstones) the facies is characteristic of a complete system: 40–45 Ma. In other cases facies are reduced to small horizons of which the specific extent, synchronism and duration are probably equivalent to many regional biozones.

However, the use of such horizons may lead to questionable interpretation, as shown by the example of the '*glauconie grossière*' (g.g. on Figure 1),

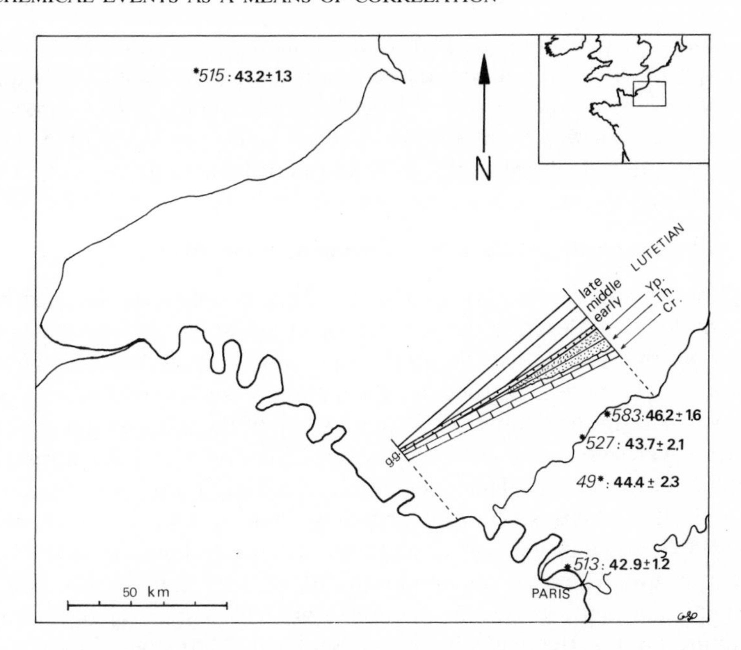

Figure 1. K–Ar apparent ages of Early Lutetian glauconies in the Paris Basin. Cr, Cretaceous; g. g., *glauconie grossière* facies; Th, Thanetian; Yp, Ypresian.

a glauconitic horizon (several decimetres thick) lying at the 'base' of the Lutetian of the Paris Basin. This well-known facies underlies various Lutetian calcareous facies which have been independently identified as being from different levels within the stage. In turn, the '*glauconie grossière*' overlies either Ypresian sands (Yp) or Cretaceous chalk (Cr). Taking into account the fact that fossil remains are frequently reworked there, it cannot be determined whether the horizon is synchronous or not. Five samples were radiometrically dated (see NDS 29–33). The results (Figure 1) seem to show a trend to the younger apparent age when the facies is overlain by younger rocks (samples 583, 527, 49, 513). However, the sample 515 dredged from the English Channel was thought to be Early Lutetian in age according to its palaeogeographic (uncertain) situation and its numerical age is on the young side.

As a general rule one must be very careful when using a facies like glaucony or phosphate as marker beds. These marine authigeneses are the result of *local* sedimentological conditions rather than the reflection of a given *general* environment. They are related with a common marine environment and their patchy distribution both horizontally and vertically in the Recent sea-bottom sediments cannot be regarded as a good criterion of synchronism (Odin and Létolle, 1980; Odin and Matter, 1981).

In summary, the lithologic facies are generally not accurate for correlation purposes: they cannot be regarded as instantaneous; they are frequently not very extensive and are not reliable because analogous facies may be of different ages. However, in the absence of guide fossils or of a detailed alternative study they are useful as a first approximation.

2.2 Volcanogenetic horizons as a means of correlation

Tephra, bentonites, ash layers, tonsteins, are the main words used by the sedimentologist to describe the occurrence of volcanogenetic products interbedded in the sediments. Through their mode of deposition they may practically always be regarded as the result of instantaneous events. As explained later in the chapter by Person, some of them led to the deposition of several centimetres of ash over several hundreds of kilometres (as an example, the Mount St Helen eruption, USA in 1980), sometimes over thousands of kilometres (Quizapu eruption, Chile, in 1931), or even all over the world (Perbuatan volcano at Krakatoa Island, Indonesia, in 1883). The Quizapu eruption poured out something like 8×10^9 tons of ash, while the most important known volcanic phenomenon, with respect to the volume of rock poured out, is the explosion of the Santorin, 3500 years ago, as a result of which it has been calculated that 160×10^9 tons of pumice were deposited in the Mediterranean area.

These volcanogenetic horizons are commonly employed as useful reference points inside basins where different facies were deposited. This is the case in Ordovician and Carboniferous sediments of Europe. In the example of the continental Carboniferous sediments of northern France, Bouroz (1967, 1972) has shown that the preceding palaeontological correlations were (slightly) erroneous. For example, the *Nevroleptis ovata* species which defined the Westphalian D appeared slightly earlier in the western side than it did 350 km further to the east.

Radiometric dating can also be applied to these volcanogenetic layers. With the necessary precautions (see chapter by Baadsgaard and Lerbekmo, below), the obtained ages may be used either for constructing the numerical time scale or, on the other hand, for accurately (±1–4%) locating these levels in the stratigraphic column.

Finally, the problematic correlation between terrestrial and marine sediments (and faunas) may be established through the study of volcanic material. Several authors have tried to correlate radiometrically well-dated continental *lavas* with *ash layers* observed in the neighbour basin (i.e. lavas from the Palaeocene from England and Greenland with ash from the North Sea Basin). To our knowledge, the identity and synchronism of the two volcanic materials has not yet been fully demonstrated. This is certainly related to the fundamental difference in the chemical nature of the volcanic reservoir of a volcano which flows (lava) and that of a volcano which explodes (ash).

In conclusion, volcanic events can be accurate because of their instantaneity. However, they generally leave a fairly small imprint in the sedimentary series both in thickness and extent. They become a reliable means of correlation when an adequate lateral and vertical study has led to a definite identification of an individual volcanic event.

2.3 Clay mineral distribution as a means of correlation

We have observed the disappearance of a small proportion of kaolinite at the boundary between the Ypresian (*sensu stricto* A. Dumont, 1851) and the regional Paniselian stage in the Belgian Basin (Figure 2). This limit is difficult to identify in the field: both are glauconitic clayey sands. Several authors were doubtful as to the existence of a distinct Paniselian unit, but in all the samples we analysed we observed this difference in kaolinite content. This criterion may tentatively be taken to indicate a regional sedimentological cut. This kind of change can rarely be used to any great extent. The phenomenon may be related to an erosional change, a change in hydrography or a change in sea-level. For some authors, then (Smoot, 1960), the kaolinite is deposited first, near the continent, and may disappear further.

But clay mineral distribution is also related to climatic conditions on the continent which feeds the oceanic sedimentation. As a result, the well-known distribution of the clay minerals on the present sea bottom is partly a

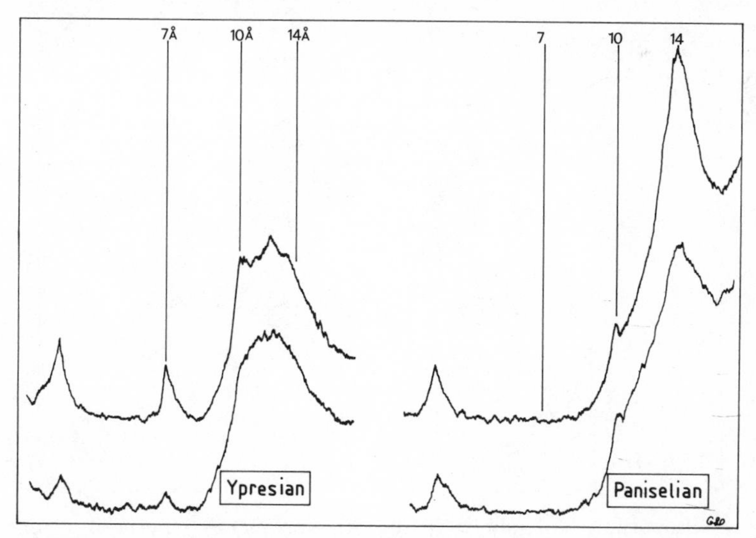

Figure 2. Diffractograms of clay fractions in the Belgian Basin. The absence of kaolinite may be considered as a marker of the Paniselian local stage.

latitudinal distribution. This was documented by several authors: Heezen *et al.* (1960), Biscaye (1965), Griffin *et al.* (1968) and Rateev *et al.* (1969), after whom we propose (Figure 3) the proportion of kaolinite in the clay fraction of the Recent sediments of the Atlantic Ocean.

Kaolinite (with gibbsite of equivalent origin in the soils) appears as a clay characteristic of the tropics. Its concentration clearly diminishes at higher latitudes. One may also note that the highest concentrations are in the immediate vicinity of the continents, as noted above. Chlorite is inversely distributed. Smectites, on the other hand, are not latitudinally distributed due to the superposition of smectites inherited from the continent and of authigenic smectites developed in the sea through the alteration of *in situ* volcanogenetic material.

A palaeoclimatic correlation system was successfully applied by H. Chamley (1971) in the Quaternary sediments of the Mediterranean Basin. This author uses the colour of the mud, the index of crystallinity of illite and chlorite, . . . to draw climatic curves which have been easily improved by

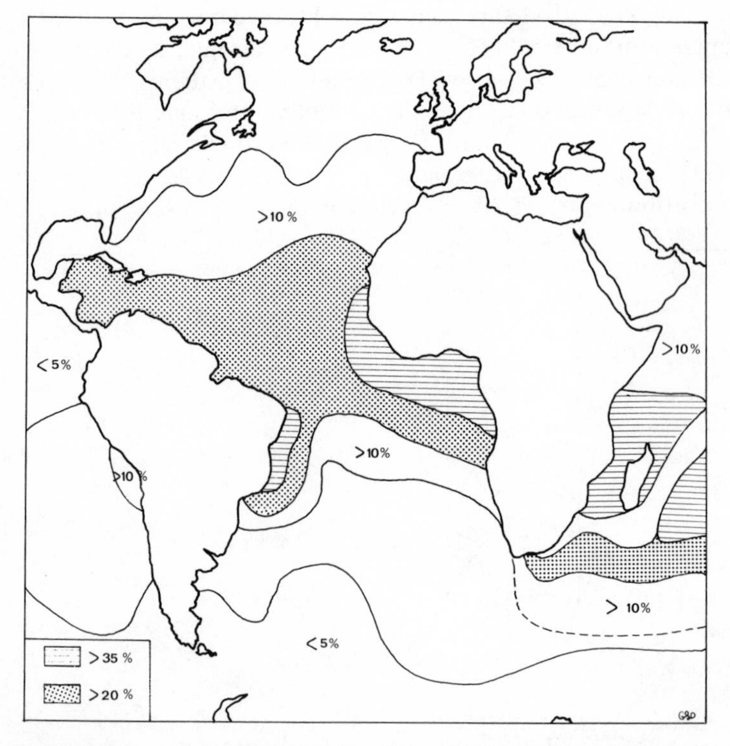

Figure 3. Proportion of kaolinite in the Atlantic Ocean superficial sediments, a scheme according to dates from Biscaye (1965), Griffin *et al.* (1968) and Rateev *et al.* (1969).

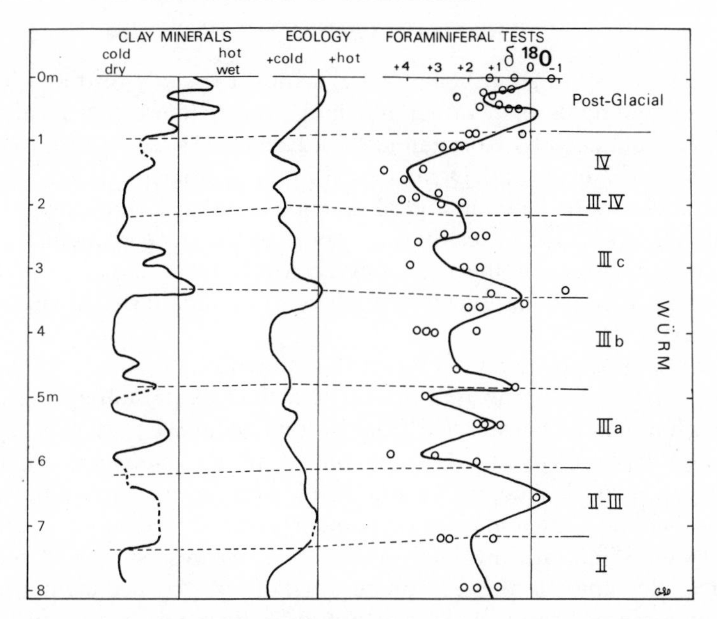

Figure 4. Curves showing the climatic interpretation of clays compared with other criteria (from Chamley, 1971, redrawn and reproduced by permission of the author and *Sciences géologiques Bulletin, Strasbourg*).

comparison with similar curves obtained from foraminifera on the one hand and pteropodes or oxygen isotopic ratio on carbonate on the other (Figure 4). The agreement is fairly good.

But the quality and usefulness of such curves are directly related to the speed and frequency of climatic changes. They will be useful only for palaeogeographically equivalent latitudes. In practice it will not be easy to correlate separate basins due to changes in local conditions (see for example the very different proportions of kaolinite on the two sides of central America, (Figure 3).

2.4 Conclusion

As a conclusion to this study on the use of sedimentological characteristics as a means of correlation, it may be noted that, most frequently, the data obtained will only allow accurate correlation *inside* a basin. However, in certain circumstances the variations which may be related with climatic events may extend further. As far as volcanogenetic horizons are concerned, these are of particular interest because they may be radiometrically dated; thus they will have more interesting applications and must be particularly sought out.

3 TRACE ELEMENT EVENTS

After giving rise to great hopes in the sixties, the study of trace elements in sedimentary rocks then rather fell into disuse. Interesting results were nevertheless obtained by Kinsman (1969), Renard (1972; 1975b), Land and Hoops (1973), Veizer and Demovic (1974) and Pomerol (1976) working with pure or purified mineralogical fractions. At first, the emphasis was naturally placed on the facies-marker aspect of the trace elements (salinity, continentality, origin of the sedimentary constituents, climatic indicators, etc., . . .), but it is now no longer impossible to think in terms of chronological markers.

However, the steps taken so far in this direction are still at the hesitant stage and there can be no question at this time of presenting correlation tables such as can be constructed from isotopic data. We shall consider here only those trace elements that are linked to the *carbonate fraction in sedimentary rocks*, dealing, on the one hand, with those elements that reveal variations in the sedimentary environment and, on the other hand, with those elements that are involved in diagenetic processes. This example of the trace elements in the carbonate fraction of the sedimentary rocks represents only one facet of the research in sedimentary geochemistry. We have chosen this example here because the *chronological aspect* of the data has been especially investigated in recent studies.

3.1 Chronology based on environment markers

The analysis involves the use of chemical elements that reveal changes in the sedimentary environment (salinity, water chemistry, nature and origin of supply, etc. . . .) and/or climatic changes (in temperature, rainfall, etc., . . .). This, in fact, constitutes a refined form of lithostratigraphy. The use of these markers will thus only be of value within a given basin and can only be extended with difficulty into neighbouring areas. Moreover, as with lithological variations, it is by no means evident that such variations are contemporaneous.

3.1.1 An example from the neritic and coastal-margin zone: the Tertiary sequence in the Paris Basin (Renard, 1975b)

The core drilled at Mont Pagnotte (Oise, France) cut through some thirty metres of fairly monotonous marly limestone (ranging in age from the Bartonian—Marinesian substage—to the Early Stampian) that corresponds to the lateral facies of the Paris Gypsum. Geochemical study of the carbonates (Sr, Na, K, Mg) enables four major breaks to be made (Figure 5). The variations in these elements can, with some reservations, be interpreted in

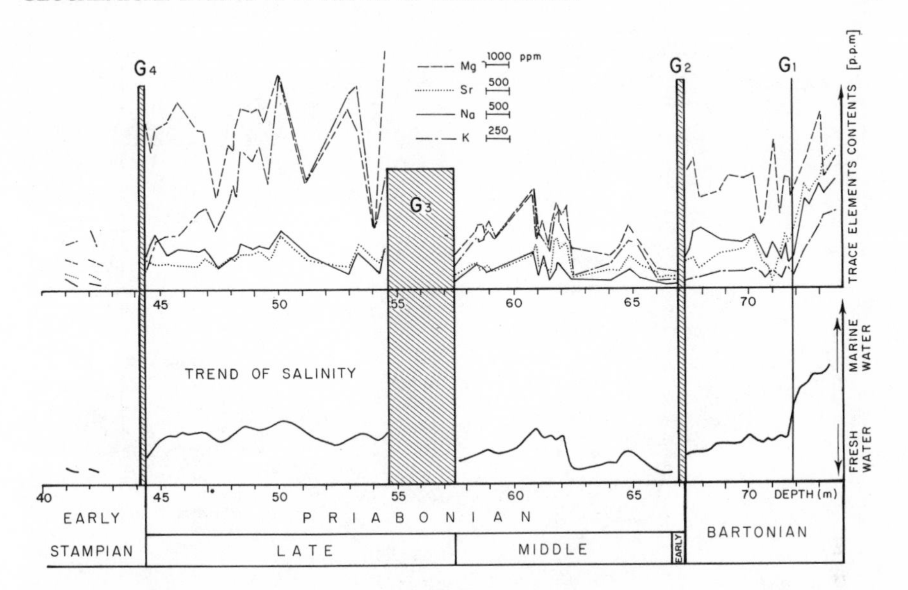

Figure 5. Evolution of the salinity curve (resulting from strontium, sodium, potassium and magnesium concentration curves) in the Eocene–Oligocene section of the Paris Basin (Mt Pagnotte, Oise, France). Interpretation of the geochemical breaks (G_1, G_2, G_3, G_4) is given in text.

terms of salinity variations within the depositional environment. The first break G_1 is very clearly marked and occurs within the Marinesian. It separates the basal part, where marine influences are strong, from the summit, where desalination is most marked. The transition between the two zones is relatively sudden. The second break G_2 corresponds to the Bartonian–Priabonian boundary and marks a more pronounced desalination that is maintained throughout the Middle Priabonian. The third break G_3, on the boundary between the Middle and the Late Priabonian, indicates a slight return of marine influences. Lastly, the fourth break G_4 marks the advent of a clear continental episode throughout the basin. At any given site, therefore, the construction of a geochemical stratigraphic succession is quite possible. But what is the importance of these breaks? Can they be extended to the rest of the basin?

The first break G_1 may be taken as an example. This marks a break in salinity within the Marinesian and its existence is also confirmed by the appearance of Charophyta (Renard and Riveline, 1973). The break can be found in different places (Figure 6) in the basin (Marines, Auvers, Mont-Pagnotte), whilst the Verzenay (Marne) section probably only represents rocks situated above this boundary. It seems, however, that the break is only

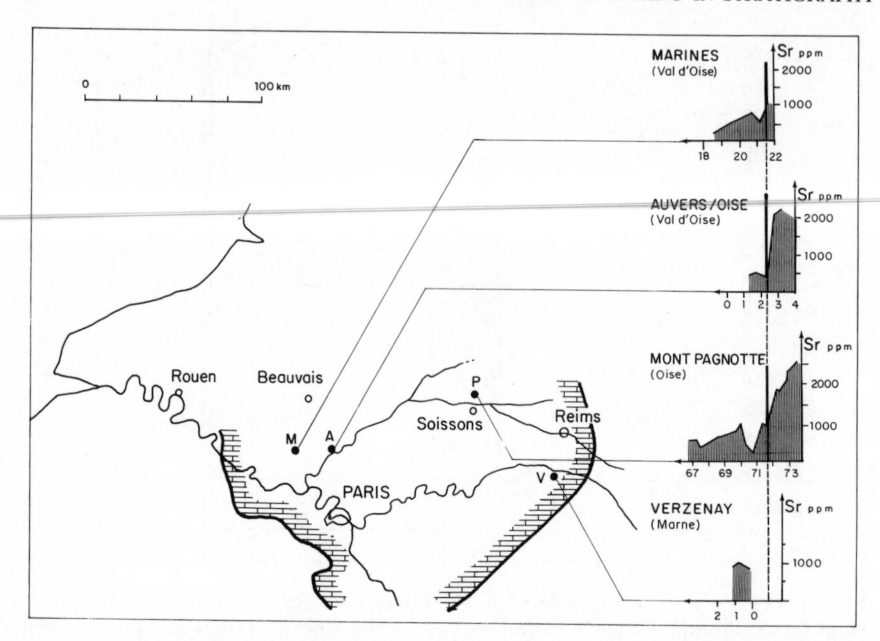

Figure 6. Trial of stratigraphic correlations based on the variation of salinity (shown by strontium content decrease) in the Marinesian sediments of the Paris Basin. The black line indicates the extension limit of the Marinesian Gulf.

clearly marked in the centre of the Marinesian Gulf (Auvers, Mont-Pagnotte). The facies show greater uniformity on the edges of the gulf (Marines) and the basal zone of marine influences is even absent altogether in places. It must, however, be clearly recognized that the fact that this break can be found in different places in the basin does not imply that the event occurred at the same time. *The same ambiguity exists, of course, in the evolution of the faunas* in these environments with variable salinity.

3.1.2 An example from the pelagic environment: the North Atlantic

The role of hydrothermalism and submarine vulcanicity in the distribution of iron and manganese in marine sediments is now widely recognized. However, the influence of these features seems to act on different scales. On a large scale, the correlation is acknowledged between the presence of metalliferous sediments and the activity of the Mid-Ocean Ridges. On a smaller scale, it seems that this Mid-Ocean Ridge activity is registered both subtly and a great distance away by the carbonates (whose iron and manganese contents are largely controlled by this activity). Thus, in the Bay of Biscay (site DSDP 398D, W Portugal and 400A, N Biscay), a very marked increase has been demonstrated in the iron and manganese content of the sediments during the Albian (Renard, *et al.*, 1979a,b). This indicates intense hydrothermal activity at this time linked to a major phase of extension, which is in agreement with the geophysical data (Williams, 1975). There was another anomaly during the Eocene, consisting, however, only of an increase in the manganese content (the iron content was scarcely, if at all, affected). The site was at that time so far distant from the hydrothermal centre that it received only the most soluble element. The detailed distribution of manganese (Figure 7) between the Maastrichtian and the Quaternary is very similar in the two sites. Within the limits set by the stratigraphic locations of the samples that are not always clear between the sites, the curves seem to be synchronous and a geochemical chronology may thus be established.

In the Maastrichtian, a decrease in manganese content is first observed during the *Lithraphidites quadratus* nannofossil zone, the minimum being centred on the passage from *L. quadratus* to *Tetralithus muras*. Then there is a slight rise during the *T. muras* zone, a stabilization during NP 1, another drop in content at the end of NP 2 (which probably continues during the beginning of NP 3), a rise during NP 3, and a stabilization in the content

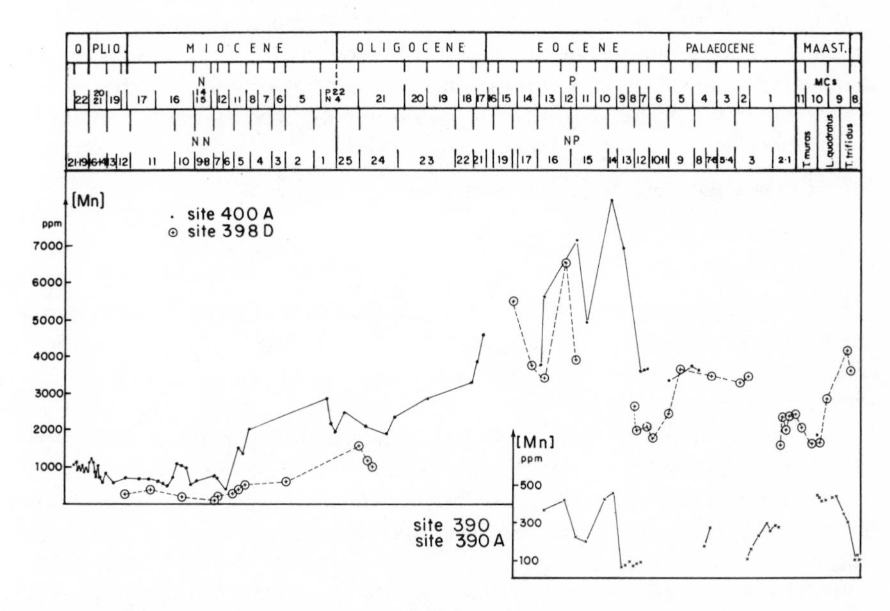

Figure 7. Manganese content in CH_3COOH-soluble fraction of pelagic sediments at site 400A (N Biscay), site 398D (W Portugal) and site 390-390A (Blake plateau). Note the similarity of the curves: high levels and low levels occur in the same planktonic foraminiferal zones at the three sites.

from the close of NP 3 until NP 9. The curve then shows a decrease in zones NP 10, 11, 12 and a sudden rise, beginning at the NP 12–NP 13 boundary, up to a very marked peak in NP 14. There is then a sudden fall in the middle of NP 15, and a quick rise at the close of this zone up to a second major peak. This is followed by a decrease in content up to the end of NP 16 and another rise to a third peak (probably lower than the first two) with a summit probably in NP 19 or NP 20. There is then a fairly rapid drop in NP 21 which becomes more gradual in NP 22, 23 and the beginning of NP 24. A slight rise during NP 25 is interrupted by a decrease on the NP 25–NN 1 boundary. It is followed by a slow, regular decline from NN 2 to NN 4 and a sudden drop during NN 5 reaching a minimum centred on NN 6–NN 7. The content then remains low with slight undulations. This result cannot at present be extended because data are still lacking. However, the similarity between the curves suggests that this is a fairly widespread feature that would be valid for the North Atlantic.

The study of the Blake Basin in site 390A (Renard *et al.*, 1978) seems to reveal an identical evolution during the Eocene (Figure 7), although the manganese content is 10 to 15 times lower than in sites 398D and 400A. these lower absolute values may be explained by the greater distance from the Mid-Ocean Ridge and perhaps also by a local current pattern counteracting the influences from the ridge.

It seems therefore that a quite subtle and detailed geochemical stratigraphic sequence may indeed be constructed based on variations in the manganese content, with a certain number of positive and negative occurrences which would be synchronous within a whole basin.

3.2 Chronology based on diagenetic markers

3.2.1 Principle

Here the evolution of a geochemical marker is studied in relation to the diagenetic development. There is therefore in this case a more or less direct relationship with the time factor. Strontium is an element that is very sensitive to carbonate diagenesis. The idea of trying to use this '*strontium loss*' as a chronological marker is already fairly old. It seems to be a fairly useful method on a large scale. Figure 8 shows the results of different studies of strontium losses with time. For detailed study, however, the idea should be modified. It is not, in fact, obvious that the curve $Sr = f(t)$ has any real validity, for two main reasons:

(a) The curve seems closely related on the one hand to the presence of metastable forms (aragonite, high magnesian calcite) that are rich in strontium in the Late Tertiary and Quaternary samples and, on the other hand, to the occurrence of greater quantities of dolomite (a carbonate with a low

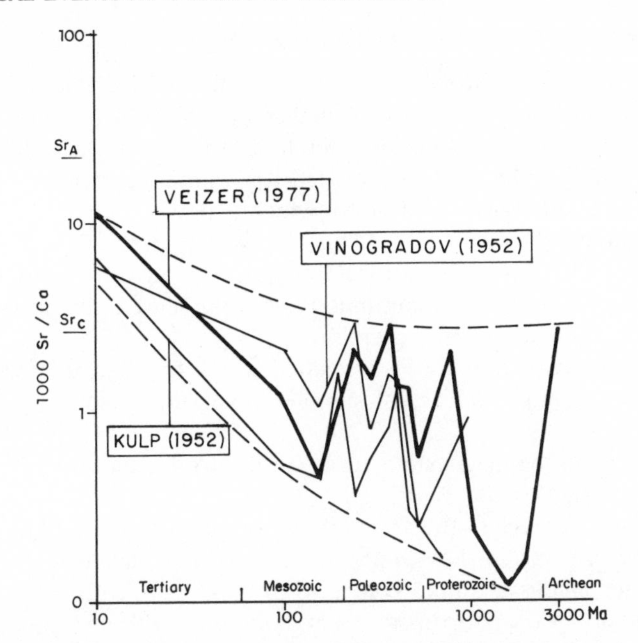

Figure 8. Variation of 1000 Sr/Ca ratio of carbonate rocks with age. (From Veizer, 1977. Reproduced by permission of the Society of Economic Paleontologists and Mineralogists.) Sr_A, aragonite in equilibrium with sea-water; Sr_C, calcite in equilibrium with sea-water.

strontium content) in the older samples. Recent work by Veizer (1977) pursuing these studies, while excluding the previously described factors, fails to bring out an obvious relationship between strontium content and time.

(b) In these studies samples of very different origins have been mixed together. But the strontium content of a carbonate also depends on the sedimentary environment and is closely linked to the facies. Thus the original content can be very variable before the slightest diagenesis takes place. It must, however, be noted that all these different studies agree in finding a fairly linear strontium/time relationship from the Recent until the Early Mesozoic. Thereafter the curve becomes much more complex for reasons which are hard to explain.

Should all these limitations induce us to abandon all hope of constructing a chronology based on strontium loss? It seems, in fact, that, if comparable things are compared—especially as concerns the depositional environment and diagenesis (controlled by marine or continental waters)—then a number of interesting results may be obtained. With this aim in view, a major distinction must be made between the neritic zone on the one hand and the pelagic zone on the other.

3.2.2 Recent (post-Neogene) neritic sediments

In this case, variations in trace elements will be used that are linked to mineralogical changes (inversion of high magnesian calcite and aragonite). Several examples have been described, chiefly from Barbados and the Mediterranean, but the best documented study in terms of chronology is still that by Gavish and Friedman (1969), which is devoted to the sediments on the Israelian coast (Figure 9). It shows similar losses of strontium and magnesium with time. The strontium loss is relatively regular but the magnesium loss is more discontinuous. A number of observations must, however, be made about this example:

—The regularity of the curves is related to the fact that they were traced from average points. The actual samples reveal quite considerable variations.

—The break in the magnesium curve springs from the early disappearance of high magnesian calcite.

—The disappearance of the aragonite is not marked, as might be expected, by a change of slope in the strontium curve. This is linked to the fact that the aragonite in this example comes from Mollusca (a group producing a carbonate that is very poor in strontium compared with inorganic precipitation). In other case (Pingitorre, 1978; Friedman and Brenner, 1977), where the aragonite is derived from corals, a break in the slope of the strontium curve is observed.

The difficulties in extending the method to the comparison of sequences from different places can thus be clearly appreciated. There is uncertainty as to the original mineralogical nature of the sediments, the types of producers concerned and, lastly, the chemical composition of the interstitial water, that varies in both space and time. This complexity in the neritic environment both in terms of sediment genesis and in terms of diagenesis makes it hard to extend these results profitably to older sequences.

3.2.3 Pelagic sequences

A great majority of these difficulties seem to be overcome in the case of the pelagic sequences.

—The carbonate producers vary but little: chiefly planktonic foraminifera and Coccolithophorideae. These two groups, moreover, only produce one single type of carbonate (low magnesian calcite).

—Diagenesis will continue as long as the sediments remain below water, influenced by interstitial waters whose composition is fairly close to that of sea-water (Manheim and Sayles, 1974).

Recent studies (Lorens *et al.*, 1977; Renard, 1979) have revealed a fractionation favouring strontium but not magnesium for calcites produced

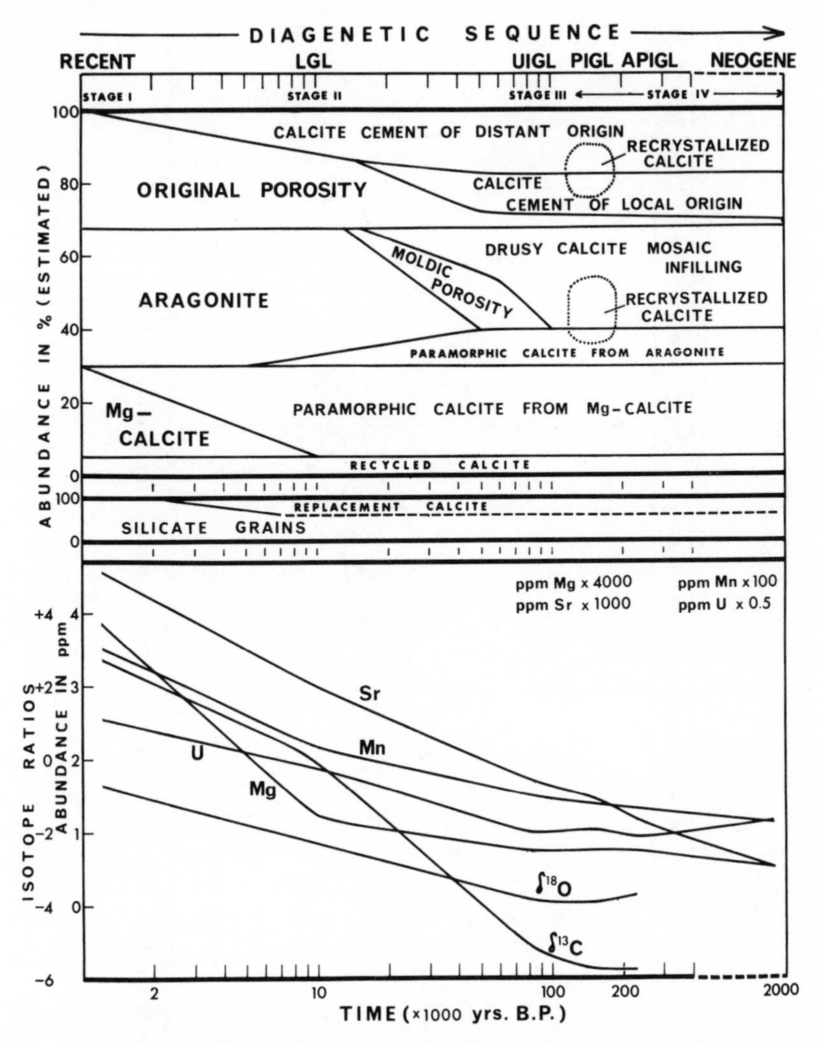

Figure 9. Summary of diagenetic textural, mineralogical and geochemical changes with progressive age in the carbonate sediments of the Mediterranean coast of Israel. (From Gavish and Friedman, 1969. Reproduced by permission of the Society of Economic Paleontologists and Mineralogists.)

by planktonic foraminifera and Coccolithophorideae compared with inorganic precipitation. Thus, when the mineralogical rearrangements and recrystallizations accompanying the ooze → chalk → limestone transformation in the pelagic sequence (Schlanger *et al.*, 1973), take place (excrescent crystals on nannofossils and cements), the diagenetic calcite will be richer in Mg and poorer in Sr compared with that emanating from foraminifera and

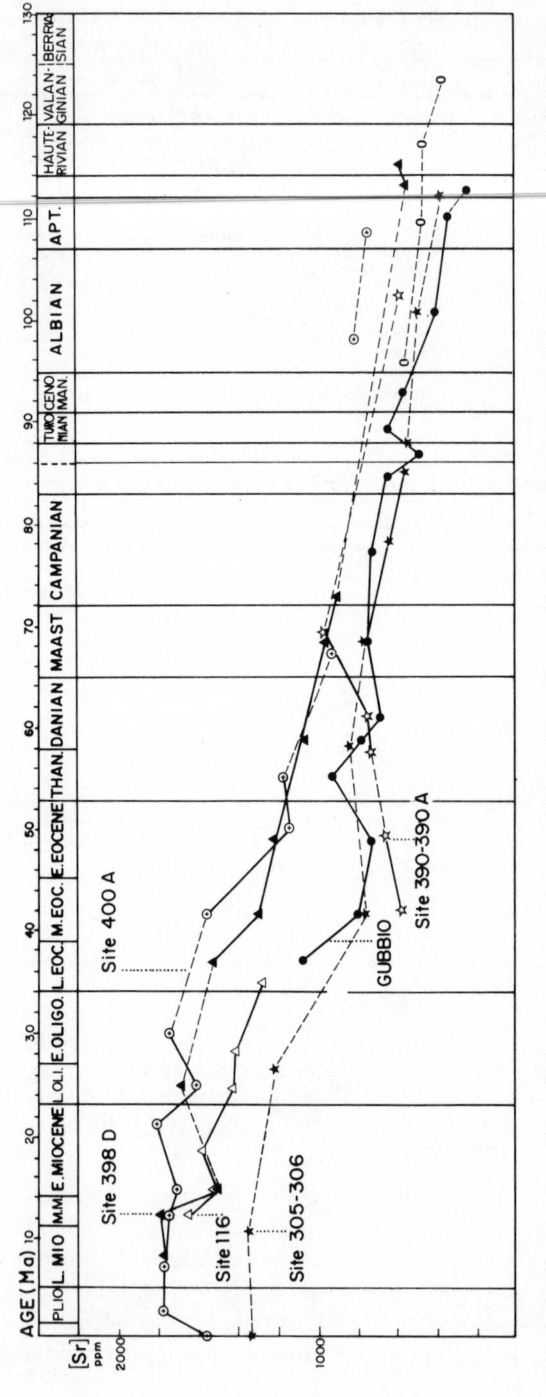

Figure 10. Strontium loss with progressive age in the HCl-soluble fraction of pelagic sediments of North Atlantic (sites 400A, 398D, 390-392A, 116), Pacific (sites 305-306) and Gubbio sections (Umbria, Italy).

Coccolithophorideae (Renard, 1979). The diagenetic sequence will thus be marked by an increase in magnesium and a loss of strontium.

Thus, in Figure 10 showing the curves of strontium in relation to time for sites 390 A, 398 D and 400 A (Renard *et al.* 1978, 1979a,b), site 116 (Rabussier, 1980) and the Gubbio section (Renard, 1979), it is clear that the fluctuations in content during the Miocene vary on either side of a mean horizontal line representing the variations in the oceanic environment. These increase in the Quaternary and thus represent the deterioration in the climate. In contrast, further down the stratigraphic column, a new phenomenon may be observed from the Late Oligocene downwards. The oscillations in content occur on either side of an mean line that slopes from a content of some 1500 ppm in the Eocene down to a content of 800 ppm in the Campanian. Diagenesis therefore only seems to become the dominant feature in relation to strontium content from samples dating at least from the Oligocene. In such diagenetic phenomena, as is shown by the results from several different sites (Table 1), it seems that the age (i.e. the time spent in a given diagenetic environment) is more important than the depth in the core (i.e. burial/sediment weight).

It must be noted that the example of the Gubbio section, which is above sea-level at present, seems to indicate that late atmospheric diagenesis does not play a major part in the strontium content in the pelagic sequences.

The conclusions, however, must not be taken too far. Although it is possible to trace a curve $[Sr]=f(t)$ applicable to different pelagic sequences (Figure 10), a given strontium content still cannot be directly linked to a given age:

(a) On the one hand, diagenesis does not entirely obliterate the variations in strontium content that are caused *either* by fluctuations in the sedimentary environment *or* by the latitudinal position of the sites analysed (e.g. site 116 in relation to site 400 A). In the case of fluctuations in the sedimentary environment, it seems, for example, that in the Tethys zone, represented by

Table 1. Strontium content in carbonate samples of same age from diverse locations and with different burial depths.

Site	Age	Depth in core (m)	Sr content (ppm)
392 A	Campanian	50	700–900
390 A	Campanian	100	800–900
400 A	Maastrichtian	750	900–1000
398 D	Maastrichtian	800	800–1000
Gubbio	Maastrichtian	Surface	700–900

Gubbio and site 390, there was an important variation at the Cretaceous–Tertiary boundary. For reasons that are still little understood, this produced very low values in the Palaeocene and Eocene and affected the curve of diagenesis in these sites.

(b) On the other hand, early submarine diagenesis (nodular limestones, hard-grounds, etc., . . .) ends up with the same results (strontium loss, manganese increase) as progressive late diagenesis. This has the effect of 'ageing' the sample. It is therefore necessary to be able to distinguish between the two types of diagenesis before any strontium contents can be compared from a chronological point of view.

By the end of this section perhaps more problems have been raised than have been resolved. It must be stressed nevertheless that a chronological approach to problems of sedimentary geochemistry is likely to bear fruit in the long run.

4 STABLE ISOTOPE EVENTS

4.1 Introduction

Oceans constitute an homogeneous reservoir, thanks to the relatively short mixing time of oceanic waters. The isotopic composition of ocean elements has varied through geological times and undergone large excursions; these variations are registered by sediments, through chemical or biological precipitations of parts of the dissolved elements. This is particularly evident for sulphur, carbon and oxygen elements, the isotopic composition of which ($\delta^{13}C$, $\delta^{18}O$ and $\delta^{34}S$) is recorded by sulphates and carbonates. And since the time constant of oceanic mixing is much less than a thousand years (Broecker, 1979) (for instance the waters of the western basin of the Atlantic Ocean are replaced on the time scale of only a hundred years), we expect isotopic changes of oceanic water and its dissolved elements to be spatially synchronous.

The establishment of sulphur, carbon and oxygen isotopic composition curves allows such changes to be used as stratigraphic markers, provided they obey some minimal conditions:

of being recognizable on a worldwide scale, i.e. in all oceans;

of being short in *duration*, i.e. visualized as a shift on isotopic curves: as stated previously, this relative duration will depend on the age of the studied series;

of being immediately localizable on a chronological scale, biostratigraphic, palaeomagnetic, radiometric, or other.

Chemical, physicochemical and biological effects influence the reactions of isotopic exchanges in equilibrium conditions. For all effects the trend is to

Table 2 Abundances and natural isotopic ranges of carbon, oxygen and sulphur stable isotopes. Asterisks refer to heavy isotopes used for stratigraphy.

Element	Z	A	Geochemical abundances (ppm)	Relative isotopic abundances (%)	Range of natural variations (‰)
C	6	12	230	98.99	
		13*		1.108	90
		14		10^{-12}	
O	8	16	470,000	99.759	
		17		0.037	
		18*		0.203	110
S	16	32	470	95.0	
		33		0.76	
		34*		4.22	140
		36		0.014	

decrease with increasing temperatures. The present sea-water isotopic compositions for sulphur, carbon and oxygen elements are shown in Table 2.

4.2 Natural isotopic effects of geological relevance

4.2.1 Sulphur isotopic variations

From a geological point of view, the most important isotopic effect results from *bacterial sulphate reduction.* Since bacteria preferentially metabolize the light isotope species, the H_2S produced is depleted in ^{34}S and *the remaining sulphate becomes enriched in* ^{34}S relative to the original sulphate by mass balance effect. This reaction induces an average isotopic fractionation of 25‰ (Harrison and Thode, 1958; Thode, 1964; Kaplan and Rittenberg, 1964; Kemp and Thode, 1968; Nielsen, 1979). The bacterial fractionation is responsible, for instance, for the very high δ values of sulphates in caprocks of the salt domes in Louisiana and Texas; the reduction operates on an anhydrite with a $\delta^{34}S$ close to +15‰ and results in a sulphate with a $\delta^{34}S$ ranging from +15 to +70‰.

Sedimentary sulphides display large variations around a relatively low average value. In large open systems, the isotopic composition of sulphur may be relatively uniform.

The oxidation of the bacterial H_2S occurs with a very low sulphur isotopic fractionation thus producing a sulphate depleted in ^{34}S relative to the original sulphate (Nakai and Jensen, 1964). *Precipitation of evaporites* from dissolved oceanic sulphate introduces an isotopic fractionation of +1.65‰

(Thode and Monster, 1965). The δ values of marine sulphates are thus lower than those of evaporites. In some cases, local influences may be superimposed on those effects, such as sulphide or sulphate addition from continental origin or deep sulphur addition through volcanic emanations.

4.2.2 *Carbon isotopic variations*

The $\delta^{13}C$ of deep secreted calcites primarily reflects the $^{13}C/^{12}C$ ratio of the CO_2 dissolved in deep waters. $\delta^{13}C$ values will be modified if the rate of cycling through the exchangeable CO_2 reservoir changes. The two main causes of such a change are: (1) a modification of the volume of one part of the exchangeable CO_2 reservoir, for instance an increase or a decrease in the continental biomass. Such a modification, being extra-oceanic, must be felt equally in all oceans and the $\delta^{13}C$ must change everywhere in the same way; (2) a change in the residence time of the deep-water masses: a longer residence time favours the conversion through oxidation of ^{13}C-depleted carbon from marine organic matter to dissolved bicarbonate. The longer the residence time, the lower the $\delta^{13}C$.

In surface waters, the exchange between dissolved CO_2 and the air varies according to latitudes. The ratio $^{13}C/^{12}C$ of the dissolved CO_2 decreases from equatorial latitudes to high latitudes. In the vicinity of continents, water runoff brings diluted waters enriched in organic matter with a low $^{13}C/^{12}C$, which can influence the $^{13}C/^{12}C$ ratio of superficial sea-waters.

4.2.3 *Oxygen isotopic variations in marine carbonates*

The $^{18}O/^{16}O$ variations of marine carbonates reflect the temperature and the $^{18}O/^{16}O$ ratio of the water in which they precipitate. Both the temperature and sea-water $^{18}O/^{16}O$ ratio vary with the climate. Atmospheric water vapour is depleted in heavy isotopes relative to the sea-water from which it evaporates (Epstein and Mayeda, 1953; Dansgaard, 1964). Conversely, condensation from a limited amount of vapour results in an enriched condensate and a residual vapour depleted in the heavy isotopes relative to the initial vapour. As a result, there is a correlation between the $^{18}O/^{16}O$ ratio of precipitations and the temperature of the air: the cooler the air, the lighter the rain. In Greenland and in the Antarctic area, precipitations are very depleted in heavy isotopes, relative to sea water: $-34.5‰$ in the crest of the Greenland ice sheet and $-58‰$ at the pole of relative inaccessibility in Antarctica (Lorius, 1974, Dansgaard *et al.*, 1975).

Therefore, with any ice-cap building or any increase in the amount of ice, water extracted from the ocean will be more and more depleted in heavy isotopes while *the remaining ocean water becomes isotopically heavier.* This

^{18}O enrichment is wholly reflected in marine carbonates. When the temperature of secretion of carbonates decreases, their $\delta^{18}O$ increases. Thus, heavy oxygen isotope ratios indicate cold climate. Conversely, temperature increases and ice-cap melting will be reflected by light oxygen isotope ratios.

Biogenic calcites—mainly those of foraminiferal tests, molluscan shells or nannoliths—are the most commonly used for palaeo-environmental reconstructions. Bulk carbonates can also be successfully analysed when calcitic organisms are not available, especially in the case of pelagic muds. Depending on what calcitic isotopic recorder is used, the interpretation of the analytical results will differ:

—*Benthonic foraminifera.* Before reflecting any ice accumulation on the continents or at the pole, $^{18}O/^{16}O$ ratios of benthonic foraminifera should reflect the variations of deep-water temperatures. A drop in these temperatures might correlate with the first formation of cold bottom waters. In fact, ocean bottom water is formed when cold dense water sinks at temperatures close to freezing and fills the deeper parts of all oceans. Cold dense water may be formed by the action of the atmosphere at the sea surface (e.g. today in the Norwegian Sea or in the Mediterranean Sea) or when the water freezes, expelling most of the salt from the water, so that the salinity and density of the water under the ice increase (e.g. today in the Weddell Sea or Ross Sea, around Antarctica).

—*Planktonic formaminifera.* Thermodynamics requires that both the temperature and sea-water effects be recorded in planktonic foraminifera. Oxygen isotopic variations will be strongly affected by local conditions, latitudes and oceanic circulation patterns.

4.3 Some examples of events allowing correlations

It appears that a certain number of climatic, oceanographic or tectonic events might influence the isotopic record of marine carbonates, such as:

—ice storage on poles and continents and its fluctuations
—bottom and deep-water formation
—fresh-water or saline-water injection events
—changes in the continental biomass
—transgression, regression, etc. . . .

while chemical events:

—marine precipitation of sulphide
—oxydative sulphide weathering
—precipitation of sulphates into evaporites, etc. . . .

influence the isotopic record of marine sulphates. Some of these events have already been identified on sulphur, carbon and oxygen isotopic curves and could already be used as stratigraphic markers.

4.3.1 *Sulphur isotope stratigraphy*

Present-day sulphates have an average $\delta^{34}S$ of +21.0‰ (Langguth and Nielsen, 1980). However, it has been known for some time that the sulphur isotope ratio in the world ocean surface has undergone major excursions during the Phanerozoic that could be documented from evaporite sulphate samples (Thode *et al.*, 1961). The resulting sulphur isotope age curve (Figure 11) for marine sulphates shows that the $\delta^{34}S$ has varied through time between extreme values of +10‰ and +30‰, maximal values being reached during the Cambrian (+30‰) and minimal values during the late Permian (+10‰). The $\delta^{34}S$ increases during Mesozoic times and reaches average values similar to present-day ones, around +20‰. The figure was drawn according to the compilations by Nielsen (Nielsen, 1978; Nathan and Nielsen, 1980), but the time scale used is that proposed in this volume and the mean curve deduced is less smoothed, in order to emphasize the presumed events. Consequently, one should be able to separate locally evaporites of different ages (Rouchy and Pierre, 1979).

But Figure 11 also highlights three rises in $\delta^{34}S$ which take place over a relatively short time, and which were labelled 'catastrophic chemical events' by Holser (1978). These events, which are characterized by sharp rises in $\delta^{34}S$ of the sulphate of the whole world ocean and by greater overshoots locally (this, in fact, constitutes two successive changes), have been identified and named for the formation in which they are most sharply displayed: the *Röt event* in the Early to Middle Triassic (approximately 240 Ma), the *Souris event* in the Late Devonian (approximately 370 Ma) and the *Yudomski event* in the Late Proterozoic (approximately 635 Ma). In addition, two reverse events seem to have occurred in the Late Permian and within the Late Palaeogene. Some of these events (like the Permian one) are documented by hundreds of samples from different locations (Nielsen, 1979) while others, such as the Late Palaeogene event or the Röt event (Nielsen, personal communication, 1981), probably represent only regional changes.

According to the same author, numerous short-term $\delta^{34}S$ excursions exist: an example of such an event has been identified by Langguth and Nielsen (1980) for the Dinantian of Belgium; the rather low $\delta^{34}S$ values (between +12 and +15‰) which mark the Early Carboniferous might also been found in some other countries. The sulphur isotopic ratio from sulphates may certainly be regarded as a possible means of correlation on a regional scale, and also sometimes on a global scale.

4.3.2 *Carbon isotope stratigraphy*

Carbon isotopic shifts have already been identified in deep-sea sediments, and, whatever their origin may be, they can be used as stratigraphic markers. The Cretaceous–Tertiary boundary event is one possible example.

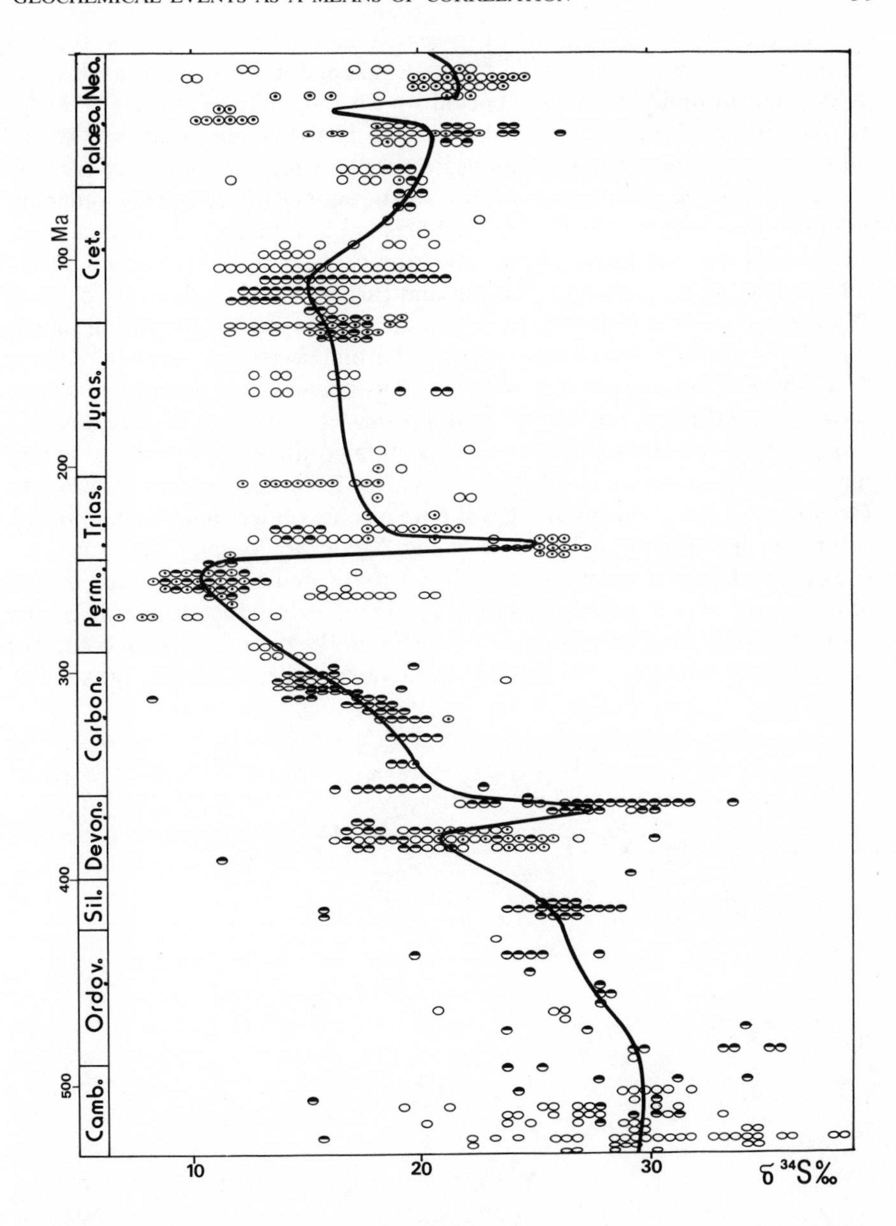

Figure 11. Sulphur isotopic ratios in sulphates collected from Phanerozoic formations. Original dates are according to the compilations by Nielsen, but the time scale is that of this book and the mean curve is less smoothed. Dotted figures are for European samples, mid-black figures are for North American samples, open figures are for samples from other parts of the world.

Recent successes in drilling the deep-sea contact of the Cretaceous–Tertiary boundary have made it possible to study this major event in ocean history. A detailed isotopic record is reported by Thierstein and Berger (1978) for the DSDP site 356, on São Paolo plateau, in the South Atlantic. Oxygen and carbon isotope determinations were performed on the fine fraction (<62 μm) of the sediment, which consists mainly of well-preserved nannoliths; it has been shown that the stable isotopic variations of the nannofossil fraction parallel those of the planktonic foraminifera (Margolis *et al.*, 1975). The latest Cretaceous oxygen and carbon isotope values are close to −2.0‰ and +1.9‰ respectively, which are typical for Maastrichtian plankton elsewhere in low latitudes. At the termination event, both $\delta^{18}O$ and $\delta^{13}C$ values drop sharply within 40 cm. This phenomenon seems to indicate 'warming' and the formation of a strong oxygen minimum (Figure 12).

Thierstein and Berger (1978) have postulated for this Cretaceous 'termination' an *injection* of fresh water coming from the Arctic. The Arctic Ocean was isolated during the latest Cretaceous; on tectonic opening of the Labrador passage, the light water of the Arctic would enter the remaining ocean, producing a layer of fresh water estimated at about 53 m by the authors and which could mix down to 100 m to produce one half of the normal salinity. The salinity could be such that the ocean plankton could not survive, thus explaining the massive extinctions which occurred at that time.

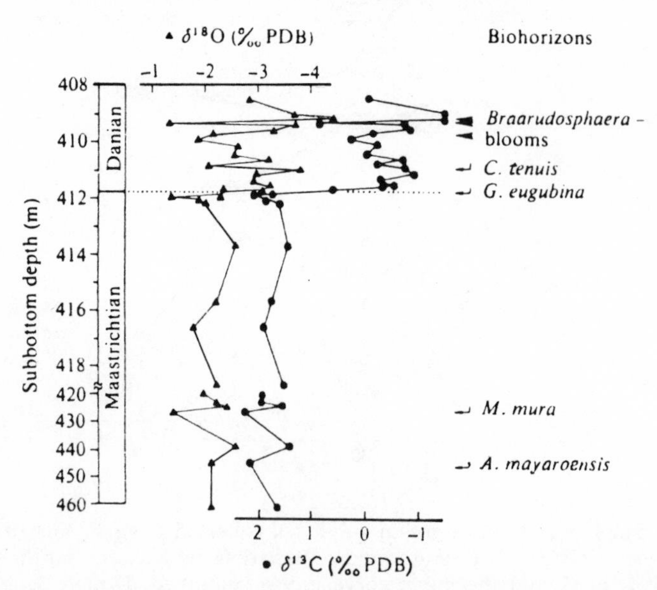

Figure 12. $\delta^{18}O$ and $\delta^{13}C$ curves of the fine fraction carbonate for DSDP site 356, São Paulo plateau, South Atlantic. From Thierstein and Berger (1978) reproduced by permission of Macmillan Journals Ltd.

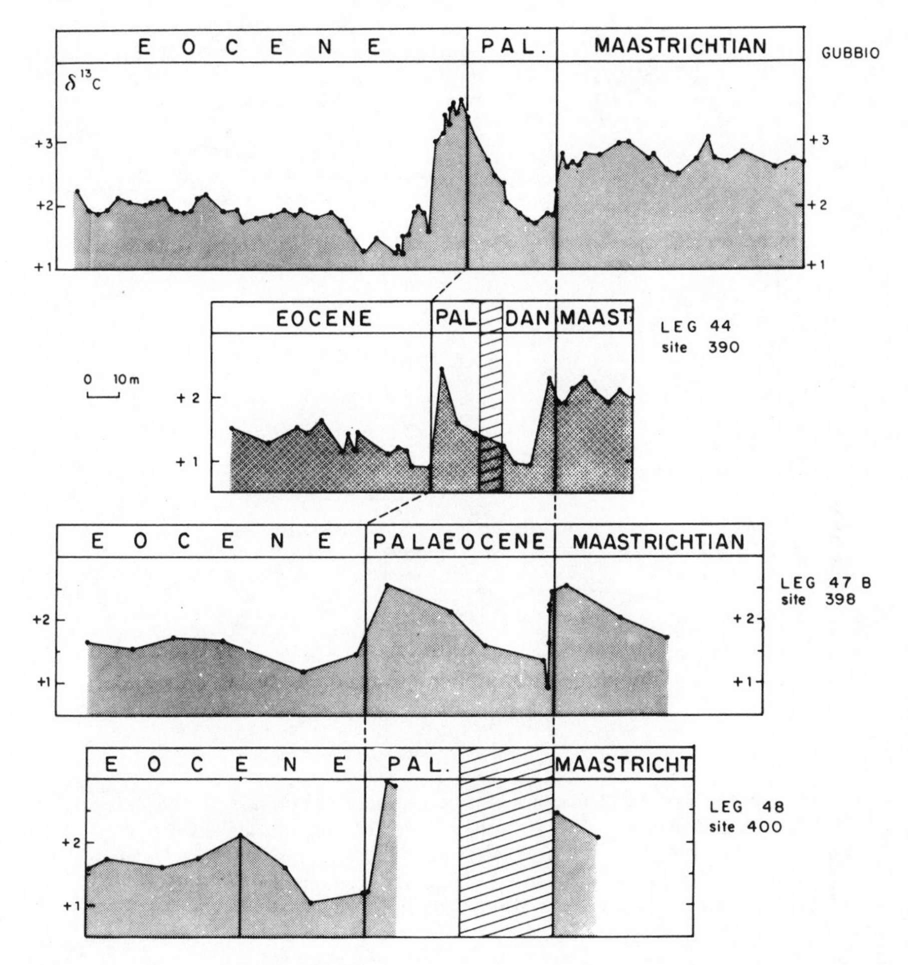

Figure 13. $\delta^{13}C$ curves from Late Cretaceous to Early Eocene time based on bulk carbonate isotopic analysis, in land section (Gubbio, Italy) and deep-sea sediments (DSDP sites 390, 398 and 400 in the North Atlantic). From Létolle and Renard (1980) reproduced by permission of Gauthier Villars et Cie.

This fresh-water lid also explains the sudden and synchronous $\delta^{18}O$ and $\delta^{13}C$ decreases in planktonic carbonates. A similar decrease in $\delta^{13}C$ values has also been reported for emerged series by Létolle and Renard (1980) (Gubbio section) and for deep-sea series at sites 390 (North Atlantic), 398 (North Atlantic, W Portugal) and 400 (North Atlantic, Bay of Biscay). According to the latter authors the $\delta^{13}C$ shift should be related to the marine regression leading to a reduction of the euphotic zone (Figure 13).

A search for $\delta^{13}C$ signals and biostratigraphic events in Deep Sea Drilling

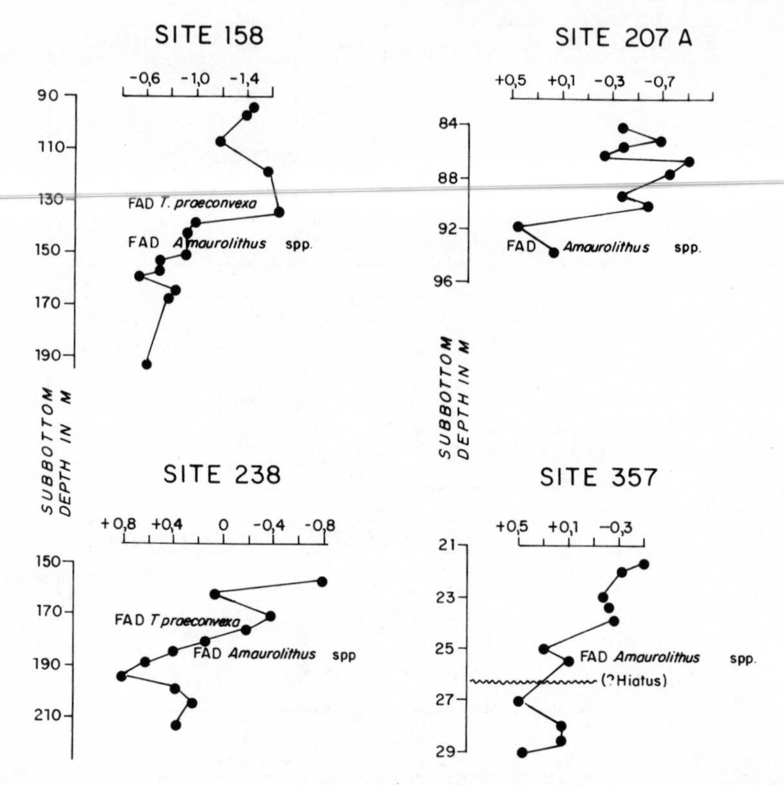

Figure 14. Carbon isotope stratigraphy of benthonic foraminifers of Late Miocene of DSDP sites and occurrence of important phytoplanktonic datum levels: site 158 on *Globocassidulina subglobosa* from North Pacific Ocean; site 207A on *Uvigerina spp.* from southwestern Pacific Ocean; site 238 on *G. subglobosa* from Indian Ocean; and site 357 on *Oridorsalis spp.* from South Atlantic Ocean. After Haq *et al.* (1980).

Project cores to improve chronological resolution led to the recognition of more or less global variations in foraminiferal $\delta^{13}C$:

—in the latest Miocene, between 6.1 and 5.9 Ma (Burkle *et al.*, 1979; Haq *et al.*, 1980), an apparently global decrease in benthonic foraminiferal $\delta^{13}C$ (Figure 14);

—in the Middle Miocene, in the North Atlantic, a rapid increase in benthonic foraminiferal $\delta^{13}C$ (Blanc *et al.*, 1980) interpreted as the onset of Arctic water spreading;

—near the Eocene–Oligocene boundary and at the Palaeocene–Eocene boundary in bulk carbonate $\delta^{13}C$ (Figure 15).

4.3.3 *Oxygen isotope stratigraphy*

Recently a new stratigraphy of the Pleistocene has emerged, classifying deep-sea sediments after the isotopic composition of oxygen in calcareous

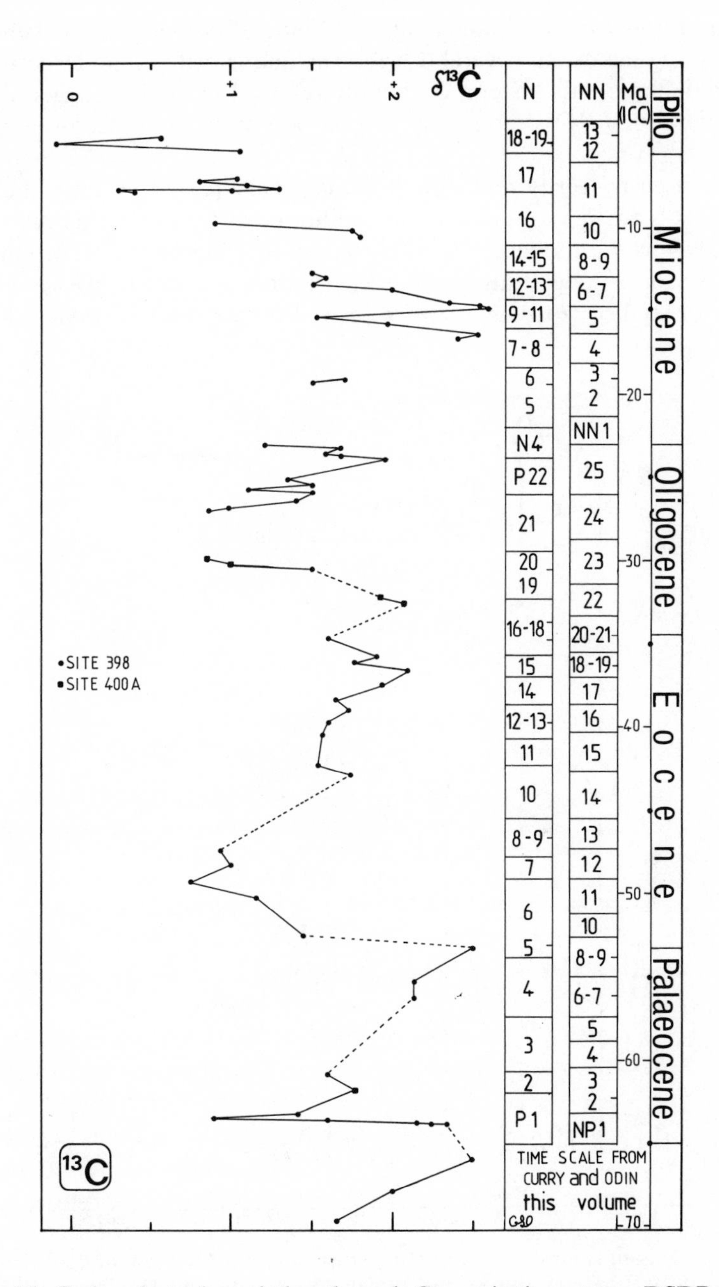

Figure 15. Carbon isotopic variation through Cenozoic time at two DSDP Atlantic sites. Time scale based on foraminiferal zonation. Original data from R. Létolle and M. Renard quoted along the time scale worked out in this volume.

shells of foraminifera. Emiliani first (1955b; 1966) used it to subdivide deep-sea sediments of several Caribbean and North Atlantic cores into numbered stages. His system, commonly referred to as the 'oxygen isotope stratigraphy' or $\delta^{18}O$ stratigraphy, is now widely used by deep-sea researchers.

One can easily verify that, for the Quaternary period at least, the $\delta^{18}O$ variations do answer to the minimal conditions of (1) worldwide reproducibility, (2) short duration and (3) synchronism, listed previously (Section 4.1):

(1) Oxygen isotope stratigraphy is clearly independent of the geographic location of studied cores as has been proved by the study of Indian, Pacific,

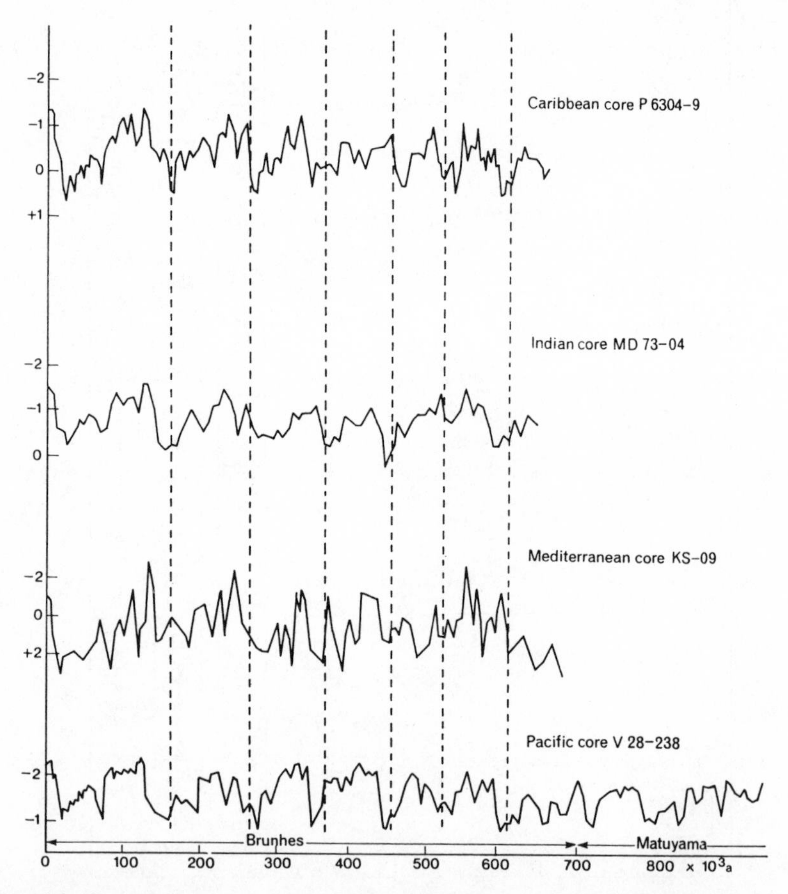

Figure 16. Oxygen isotopic composition of foraminiferal tests in four cores from the Caribbean Sea (Emiliani, 1966), the Indian Ocean (Bé and Duplessy, 1976), the Mediterranean (Cita *et al.*, 1977) and the Pacific (Shackleton and Opdyke, 1976). Time scale in ka. In all figures $\delta^{18}O$ variations are reported in ‰ against the international PDB-1 standard.

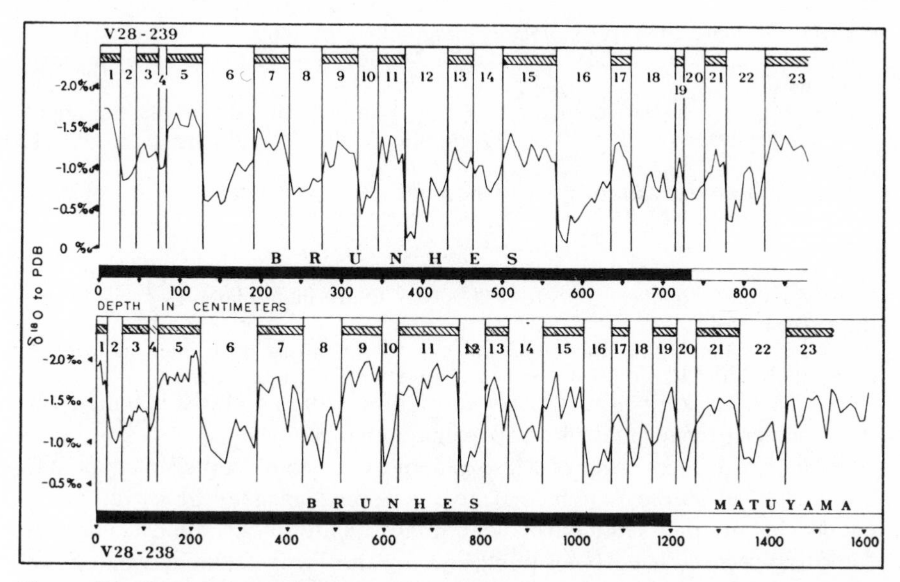

Figure 17. Oxygen isotope and palaeomagnetic record in upper 880 cm of Pacific cores V28-239 (above) and V28-238 (below). From Shackleton and Opdyke (1976) reproduced by permission of Macmillan Journals Ltd.

Atlantic and Mediterranean cores (Figure 16). Cores 28–238 and 28–239 from the equatorial Pacific could conveniently serve as a type locality of the $^{18}O/^{16}O$ record. Isotopic stages 1–23 are clearly identified on these cores, the highest number matching the oldest age (Figure 17).

(2) Boundaries between stages, defined by their depth in the core, are placed near the level of fastest change in the isotopic ratio, mostly halfway between the neighbouring maxima and minima. Boundaries separating especially pronounced isotopic maxima from exceptionally pronounced minima were called '*terminations*' by Broecker and Van Donk (1970). These terminations generally correspond to a duration of less than a thousand years for the Holocene period and less than ten thousand years for older stages of the Quaternary.

(3) The cross-confrontation of different methods complementary to each other has permitted the localization of various isotopic fluctuations, or isotopic stages, on a chronological scale, for instance:

- The reversal of the earth's magnetic field, K–Ar dated in land-based volcanics to about 0.73 Ma (Dalrymple, 1972), was recognized in a number of deep-sea cores (Opdyke, 1972). In core V_{28-238}, it was located at the base of isotopic stage 19. In V_{28-239}, it was found in isotopic stage 20. Undoubtedly, the reversal occurred in the vicinity of the stage boundary 19/20.
- Oxygen isotope variations in corals made it possible to accurately date the ^{18}O peaks of stages 5, 7, 9 in coral reefs at Barbados using the U-Th method (Shackleton and Matthews, 1977).

• LAD *Globoquadrina pseudofoliata* occurs in ^{18}O stage 7 in tropical waters (Thompson and Saïto, 1974) (*c.* 220,000 a ^{18}O age interpolated from the magnetic reversals scale).
• FAD *Emiliana huxleyi* occurs in ^{18}O stage 8 in tropical, subtropical, transitional and subpolar waters (*c.* 275,000 a ^{18}O age interpolated from the magnetic reversals scale).
• LAD *Stylatractus universus* occurs near the base of ^{18}O stage 11 in tropical to subantarctic waters (*c.* 460,000 a ^{18}O age interpolated from the magnetic reversals scale).
• LAD *Pseudoemiliana lacunosa* occurs in ^{18}O stage 12 in tropical, subtropical, transitional and subpolar waters (*c.* 474,000 a ^{18}O age interpolated from the magnetic reversals scale).
It should be noted that the numerical ages of all levels have been assigned by extrapolation and hence are subject to change with additional dates.

The oxygen isotope Quaternary stratigraphy quoted above is linked with the Northern Hemisphere ice-cap volume changes.

Further, the recovering of Mesozoic and Cenozoic deep-sea sedimentary sequences led to the search for stable isotopic signals and biostratigraphic events in marine sediments older than Quaternary. Almost complete Cenozoic records have been published for the Pacific (Douglas and Savin, 1975; Shackleton and Kennett, 1975; Margolis *et al.*, 1977) and for the Atlantic (Figure 18), but in the case of the Mediterranean (Vergnaud Grazzini, 1978) the record does not go further back than −18 Ma. Some examples of isotopic events are studied below; these are sometimes related to stratigraphic boundaries, sometimes not.

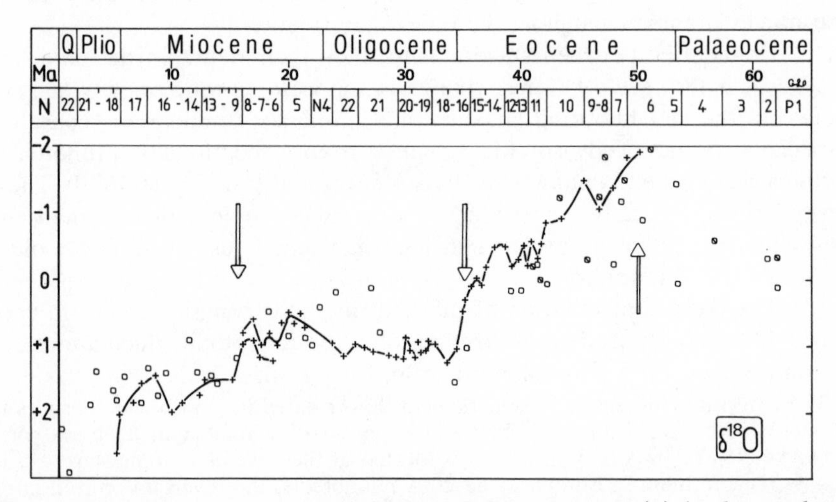

Figure 18. Oxygen isotopic composition of bulk carbonates (□) in the northeast Atlantic (DSDP sites 400A and 401) and planktonic foraminifers (+) in the South Pacific (DSDP sites 277, 279 and 281) after Vergnaud Grazzini *et al.* (1979) and Shackleton and Kennett (1975). Numbers refer to foraminiferal biozonations.

One should distinguish the events which result from a glacial effect (ice storage at the poles), mid-Pliocene or mid-Miocene, from the events which result from a temperature effect (onset of cold bottom water circulation), Early Eocene or approaching the Eocene–Oligocene boundary.

The 3.2 Ma old mid-Pliocene stratigraphic level corresponds to the first ice accumulation on the Northern Hemisphere.

A shift in the oxygen isotopic curves has been observed in an equatorial Pacific core $V_{28\text{-}179}$ and localized through palaeomagnetism at about 3.2 Ma by Shackleton and Opdyke (1977) (Figure 19). An identical shift is also observed in the equatorial Atlantic at DSDP site 366A (Blanc, 1981). The date of 3.2 Ma correlates well with the first appearance of ice-rafted debris at site 116 (DSDP Leg 12) and with the first glacial evidences in Iceland (McDougall and Wensink, 1966).

The mid-Miocene oxygen isotopic shift corresponds to a major ice accumulation on Antarctica.

Generally low δ^{18}O values in the Early Miocene sediments are followed by a general increase. This shift has been interpreted as corresponding to the establishment of the Antarctic ice cap and has been observed in all Miocene deep-sea sediments deposited between foraminiferal zones N8 and N10 (site 277-289, South Pacific, site 366A, equatorial Atlantic, site 116, Rockall plateau, site 400A, north east Atlantic, site 167, equatorial Pacific). This shift can be non-monotonous, but its amplitude is comprised between 1 and 1.5‰. It could serve to divide the Tertiary into a preglacial and glacial epoch according to Shackleton and Kennett (1975) (Figure 18).

Wide variations in the volume of stored ice are also responsible for marked δ^{18}O peaks which may be used for stratigraphy, as, for instance, a strong positive δ^{18}O peak in zone NP 25, in the Late Oligocene (Vergnaud Grazzini and Rabussier-Lointier, 1980b), or another positive δ^{18}O peak in Late Miocene times (Messinian) (Adams *et al.*, 1977).

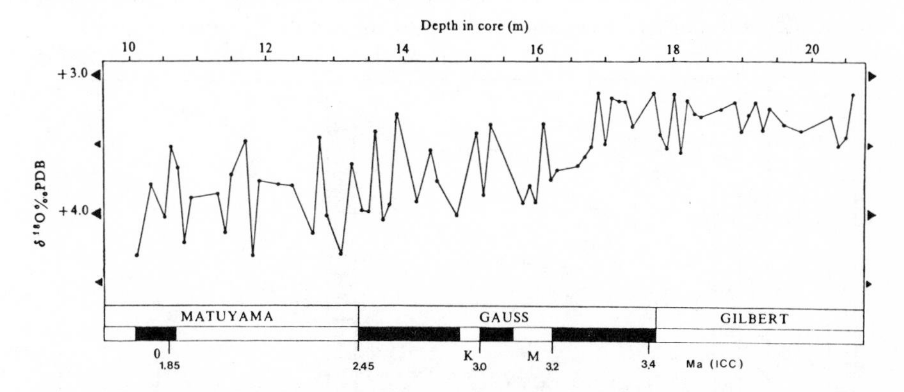

Figure 19. Oxygen isotopic composition of *Globocassidulina subglobosa* in Pacific core V28-179 from 10 to 21 m. *G. subglobosa* living on the sea floor at the coring site today would have an δ^{18}O content of about +3.5‰. Magnetic events are indicated by O (Olduvai), K (Kaena) and M (Mammoth). From Shackleton and Opdyke (1977) reproduced by permission of Macmillan Journals Ltd.

The two events quoted below, the negative $\delta^{18}O$ peak in the Early Eocene and the Late Eocene shift, are considered the result of an onset of bottom-water circulation.

In the Early Eocene, foraminiferal zones P6–P7, isotope composition of the oxygen reaches minimal values in all oceans and sites. These minimal values correlate with the warmest episode of the thermal history of world oceans, when there did not exist any deep cold circulation. The progressive $\delta^{18}O$ increase which occurs after that throughout the whole Tertiary might correspond to the progressive onset of deep cold circulation, progressive cooling and progressive ice-cap building (Figure 18).

The Late Eocene shift first appeared as a 0.75–1.5‰ $\delta^{18}O$ rapid increase throughout zones NP 20–NP 22. It has been observed in the Pacific as well as in the Atlantic, in deep-sea sediments as well as in epicontinental seas (Kennett and Shackleton, 1976; Vergnaud Grazzini, 1979; Vergnaud Grazzini and Rabussier Lointier, 1980a). This shift (Figure 20) was interpreted as the result of first deep cold circulations induced by glaciations at sea-level around the Antarctica (Shackleton and Kennett, 1975); as an 'injection' event from the Arctic Sea (Thierstein and Berger, 1978); or as an ice building at the South Pole (Matthews and Poore, 1980).

In many sedimentary sequences, the duration of an isotopic change may be artificially shortened by the frequent occurrence of hiatuses; in the series where sedimentation was not interrupted (site 366A for instance), or when isotope analyses where carried on monospecific samples (Keigwin, 1980), the isotopic change looks more gradual (Figure 21). In fact, the $\delta^{18}O$ fini-Eocene enrichment is a major event in the Southern Ocean oxygen isotope record, but is considerably less in magnitude than the 2‰ change that occurred gradually from mid-Early Eocene to the Late Eocene. Benthonic foraminiferal and isotopic data (Corliss, 1981) indicate that bottom-water circulation may have developed during the Middle Eocene–Early Oligocene interval, with a 3–4 deg C bottom-water cooling near the Eocene–Oligocene boundary representing only *part of this development.*

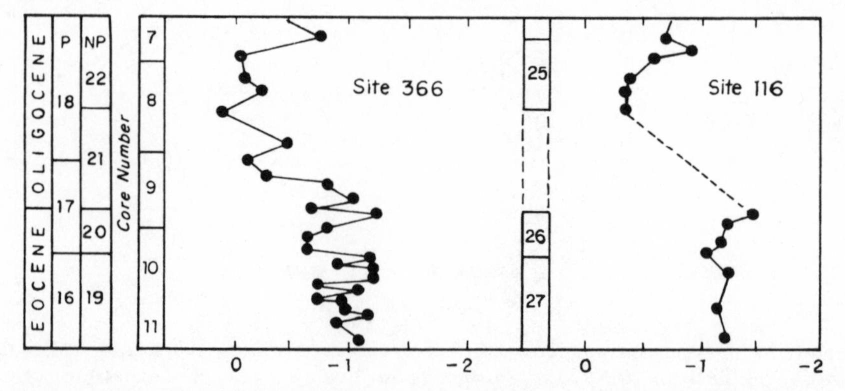

Figure 20. The 1‰ $\delta^{18}O$ variation in bulk carbonates near the Eocene–Oligocene boundary in the equatorial Atlantic (DSDP site 366) and in the North Atlantic (DSDP site 116). Note a marked hiatus at site 116. After Rabussier Lointier (1980).

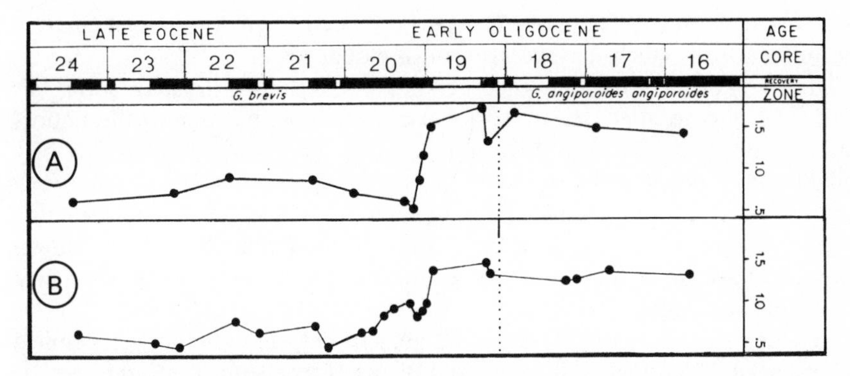

Figure 21. The 1‰ $\delta^{18}O$ variation in benthonic foraminifera in the South Pacific DSDP site 277. Curve A is based on mixed species analyses (after Shackleton and Kennett, 1975); curve B is based on analysis of monospecific samples of *Oridorsalis umbonatus* (after Keigwin, 1980).

4.4 Conclusion

The use of the sulphur isotopic age curve may yield locally interesting information on sulphate age and genesis; the accurate location of the major events already identified in the stratigraphic column needs further investigation. The use of carbon isotopes is recent; the systematic investigation of the major events remains to be done; the first results obtained on Cenozoic formations are promising. The use of oxygen isotopes is now widespread for Quaternary interpolated stratigraphy. This method of correlation may also be applied to sediments older than Quaternary.

The resolution of these methods is limited by a number of factors, some of which are common to those of palaeontological correlations while others are not: varying sedimentation rates, selective dissolution, diagenesis, etc., ...

On the whole, the precision of the methods greatly decreases with the age of the sediments. For the $\delta^{34}S$ curve, the absolute value of isotopic ratios seems to be useful for stratigraphic interpretation, while for the $\delta^{13}C$ and $\delta^{18}O$ curves, only the shape of the curve is informative. The strontium isotope ratio curve studied in the following chapter displays similar characteristics.

5 CONCLUSIONS

Recent developments in the understanding of the actual value of biostratigraphic correlation show that caution must be observed in extrapolating local observation of faunal changes. At the same time, geochemists have shown that some events may be characterized at the scale of a basin or even of an ocean or more. These events may be related to environmental changes the influence of which may be seen on a global scale.

The study of the sedimentological characters of the stratigraphic series has led in the past to approximate regional correlations; but due to the important influence of the local environment, they are now considered poorly reliable on a global scale.

Some examples of climatic changes (ice ages), of tectonic events (opening of oceans) and of biological events (sulphur reduction) have been shown above to have an influence on the elemental and isotopic compositions of the whole sea-water.

Geochemists may now therefore be able to help in defining geochemical events easily identifiable in most sediments. Geochemical stratigraphy is thus a new way to help in geological correlations. Some examples are reported here to show both the great possibilities of this stratigraphy and its limitations, caused especially by possible deformation of the signals registered in the local sedimentary series. Parallel investigations of the palaeontological and geochemical characteristics of the sequences available seem the best way to build up an improved system of stratigraphic correlations. We will see in the two chapters following how other characteristics may complete this knowledge.

Acknowledgements

We thank Dr. H. Nielsen for his kind comments on the geochemistry of the sulphur and improvement of the paragraph relating to this topic.

Résumé

Les données récentes concernant la compréhension de la relativité des corrélations biostratigraphiques montrent que l'extrapolation des coupures de faune adoptées dans une coupe donnée doit être faite avec précaution. Dans le même temps les géochimistes ont montré que certains évènements pouvaient être caractérisés à l'échelle d'un bassin ou même d'un océan voire sur l'ensemble du globe s'ils pouvaient être mis en relation avec un réel changement d'environnement.

L'étude des caractères sédimentologiques de la série stratigraphique a conduit par le passé à des corrélations régionales: faciès des grès rouges, craie, glauconie de base par exemple. Ces corrélations sont aujourd'hui considérées comme peu dignes de confiance à l'échelle du globe par suite de l'influence de l'environnement de dépôt local.

Cependant, il existe des exemples de changements climatiques (âges glaciaires) d'évènements tectoniques (ouverture d'un océan) ou biologiques (réduction bactérienne) qui ont eu une influence instantanée et mesurable sur la composition élémentaire ou isotopique de l'eau marine à l'échelle du globe. Les éléments en traces, les isotopes du soufre, du carbone et de l'oxygène ont spécialement été étudiés et permettent d'établir des coupures dans certains niveaux de l'échelle des temps fossilifères.

La *stratigraphie géochimique* devient alors une voie nouvelle permettant d'aider à la résolution des problèmes de corrélation. La présente contribution donne quelques exemples de ces possibilités en même temps qu'elle essaie d'en définir les limites d'application liées à la déformation possible des signaux enregistrés localement par la série sédimentaire. Il en résulte qu'une investigation des caractéristiques géochimiques d'une séquence disponible se révèle complémentaire de l'investigation paléontologique, les deux recherches développées parallèlement devenant la meilleure façon d'aboutir à un système de corrélation digne de confiance.

(Manuscript received 11-4-81)

Numerical Dating in Stratigraphy
Edited by G. S. Odin

4
The marine-strontium geochronometer

GUNTER FAURE

1 HISTORICAL PERSPECTIVE

In 1948, F. E. Wickman proposed that the isotopic composition of strontium in the oceans should have changed continuously as a result of the radioactive decay of ^{87}Rb to ^{87}Sr in the rocks of the earth's crust. Gast (1955), Gerling and Shukolyukov (1957) and Hedge and Walthall (1963) tested this idea but found that the actual increase of the $^{87}Sr/^{86}Sr$ ratio was far less than predicted. The idea that the $^{87}Sr/^{86}Sr$ ratio of the oceans has increased smoothly with time was revived briefly by Hurley *et al.* (1965), but Brookins *et al.* (1969) demonstrated that the $^{87}Sr/^{86}Sr$ ratios of Permian limestones in eastern Kansas do not fit the 'marine geochron' Hurley and his colleagues had proposed. The turning point in this controversy came with the publication of a paper by Peterman, Hedge and Tourtelot in 1970 which convincingly demonstrated that the $^{87}Sr/^{86}Sr$ ratio of the oceans has fluctuated repeatedly throughout Phanerozoic time.

The study of strontium in marine carbonate and evaporite rocks was accompanied by efforts to measure and to explain the $^{87}Sr/^{86}Sr$ ratio of sea-water. The first measurement on record is by Ewald *et al.* (1956), followed by Herzog *et al.* (1958), Gast (1961), Compston and Pidgeon (1962) and by Hedge and Walthall (1963). The first comprehensive study of strontium in sea-water was made by Faure *et al.* (1965). They reported an average $^{87}Sr/^{86}Sr$ ratio of 0.70893 ± 0.00024 (2σ) for sea-water from the North Atlantic, concluded that strontium in the oceans was isotopically homogeneous, and proposed a model to account for the numerical value of the $^{87}Sr/^{86}Sr$ ratio.

2 STRONTIUM IN SEA-WATER

The number of measurements of the $^{87}Sr/^{86}Sr$ ratio in sea-water has grown to several hundred. The data have so far failed to show any real variation of the $^{87}Sr/^{86}Sr$ ratio in the oceans. Even the strontium in the

Hudson Bay of Canada is isotopically indistinguishable from that in the major oceans (Faure *et al.*, 1967). The reasons for the isotopic homogeneity of strontium in the oceans are: (1) a long residence time of about 5×10^6 years; (2) rapid mixing in the oceans in about 10^3 years; (3) high concentration of strontium of 7676 ± 68 μg/kg (Brass and Turekian, 1974) compared to about 68.5 μg/l in average river water. About 60% of the $^{87}Sr/^{86}Sr$ ratios of sea-water reported since 1965 range from 0.70890 to 0.70920. The average of these determinations is 0.70906 ± 3 (2σ) relative to a value of 0.70800 for the Eimer and Amend $SrCO_3$ isotope standard.

3 MARINE STRONTIUM OF PHANEROZOIC AGE

Strontium is removed from the oceans primarily by coprecipitation with calcium in carbonate minerals. These minerals exclude rubidium so that their $^{87}Sr/^{86}Sr$ ratios remain virtually constant with time. Marine carbonates and other minerals with low Rb/Sr ratios thus preserve a record of the $^{87}Sr/^{86}Sr$ ratio of the sea-water in which they were deposited. Differences in the measured values of this ratio can therefore be interpreted as evidence of time-dependent variation of the $^{87}Sr/^{86}Sr$ ratio in the oceans, provided the following conditions are satisfied: (1) the strontium in the rock or mineral was derived from sea-water and not from volcanic rocks or other non-marine materials that may react with sea-water; (2) the $^{87}Sr/^{86}Sr$ ratio was not altered by *in-situ* decay of ^{87}Rb, diagenesis, dolomitization, recrystallization, regional metamorphism or chemical weathering; (3) the material is known to be of marine origin and strontium in the oceans was isotopically homogeneous; (4) the age is known from palaeontological or isotopic dating; (5) the $^{87}Sr/^{86}Sr$ ratio of the authigenic marine minerals is not altered by the release of strontium from admixed detrital minerals during dissolution in the laboratory.

Unreplaced skeletal calcite and aragonite analysed by Peterman *et al.* (1970), Dasch and Biscaye (1971), Boger *et al.* (1973) and Brass (1976) is probably most reliable but is not always obtainable. Fine-grained limestones and dolomites with low non-carbonate residues are also suitable (Veizer and Compston, 1974; Faure *et al.*, 1978). Care should be taken to avoid specimens containing smectite or illite clays that may release strontium into solution during the dissolution of the carbonate minerals by cold 0.1 N HCl. The $^{87}Sr/^{86}Sr$ ratios of marine dolomites are apparently not changed by dolomitization even though the concentration of strontium is greatly reduced (Faure *et al.*, 1978). Pure marbles formed by regional metamorphism of limestones in Greece were analysed by Tremba *et al.* (1975). The $^{87}Sr/^{86}Sr$ ratios of these marbles do not differ appreciably from those of unrecrystallized limestones of similar age. This conclusion differs from Gittins *et al.* (1969), who found evidence that the $^{87}Sr/^{86}Sr$ ratios of calcite

Table 1. Summary of $^{87}Sr/^{86}Sr$ ratios of marine strontium in 20 million year increments.

Time interval (Ma)		$^{87}Sr/^{86}Sr \pm 2\sigma$	References
Present	(19)	0.70906±3	1–18
0–20	(9)	0.70893±15	18–22
20–40	(1)	0.70813	18
40–60	(5)	0.70783±28	18, 19, 20
60–80	(8)	0.70757±15	17, 18, 20, 23, 24
80–100	(1)	0.70755	19
100–120	(8)	0.70722±23	18, 20, 22, 25, 26
120–140	(4)	0.70720±8	17, 18, 22
140–160	(3)	0.70713	18, 22, 24
160–180	(3)	0.70722	19, 22, 27
180–200	(2)	0.70731	27
200–220	(12)	0.70773±21	18, 22, 24, 27
220–240	(1)	0.70813	18
240–260	(2)	0.70712	18, 22
260–280	(1)	0.70722	22
280–300	(3)	0.70736	18, 28, 29
300–320	(1)	0.70792	18
320–340	(5)	0.70845±39	18, 22, 30
340–360	(4)	0.70811±29	18, 22, 32
360–380	(4)	0.70836±32	18, 31, 32
380–400	(1)	0.70830	22
400–420	(3)	0.70865	31, 32
420–440	(2)	0.70797	18, 32
440–460	(0)	—	—
460–480	(2)	0.70860	22, 32
480–500	(3)	0.70905	22, 32, 33
500–520	(3)	0.70881	22, 28
520–540	(1)	0.70926	22

The sea-water values were restricted to >0.7089 and <0.7092. The $^{87}Sr/^{86}Sr$ ratios were adjusted to 0.70800 for the Eimer and Amend $SrCO_3$, where possible.

1. Gopolan & Wetherill (1968)
2. Wasserburg *et al.* (1969)
3. Papanastassiou & Wasserburg (1969)
4. Murthy & Beiser (1968) (Pacific Ocean)
5. Gray *et al.* (1973)
6. Birck and Allègre (1972)
7. Murthy *et al.* (1971)
8. Papanastassiou *et al.* (1970)
9. Papanastassiou & Wasserburg (1971)
10. Tera *et al.* (1970)
11. Faure *et al.* (1967)
12. Ikpeama *et al.* (1974)
13. Jones & Faure (1967)
14. Faure *et al* (1965)
15. Hamilton (1966)
16. Hildreth & Henderson (1971)
17. Brass (1976)
18. Peterman *et al.* (1970)
19. Starinski *et al.* (1980)
20. Dasch & Biscaye (1971)
21. Clauer (1967a)
22. Veizer & Compston (1976)
23. Jorgensen & Larsen (1979)
24. Tremba *et al.* (1975)
25. Kesler & Jones (1981)
26. Clauer (1981)
27. Faure *et al.* (1978)
28. Faure & Barrett (1973)
29. Brookins *et al.* (1969)
30. Boger *et al.* (1973)
31. Haden (1977)
32. Kessen *et al.* (1981)
33. Hedge & Walthall (1963)

in Grenville marbles on the Canadian Precambrian Shield had been altered by metamorphism. Other kinds of samples that may be suitable under appropriate conditions include biogenic chalk (Jorgensen and Larsen, 1979), gypsum, anhydrite and celestite (Kesler and Jones, 1981), calcium phosphate in conodonts (J. Kovach, personal communication) and groundwater in marine limestone and dolomite aquifers (Starinsky *et al.*, 1980). The suitability of marine zeolite (phillipsite) and neoformed clay minerals is in doubt. Pushkar and Peterson (1967) obtained an average $^{87}Sr/^{86}Sr$ ratio of 0.70893 ± 0.00052 (2σ) for three phillipsites from the Pacific Ocean that is identical to the $^{87}Sr/^{86}Sr$ ratio of sea-water. On the other hand, Hoffert *et al.* (1978) demonstrated that zeolite of Middle Eocene age in two cores taken in the South Pacific contains strontium derived by submarine weathering of basalt, whereas smectites contain a mixture of strontium derived from volcanic sources and sea-water.

Measurements of $^{87}Sr/^{86}Sr$ ratios of marine deposits reported in the literature have been averaged in increments of 20 million years based on the time scale proposed in this volume. Whenever possible, the values were adjusted to 0.70800 for the Eimer and Amend $SrCO_3$ standard. The results are listed in Table 1 and have been plotted *versus* geological time in Figure 1. A curve representing the time-dependent variation of the $^{87}Sr/^{86}Sr$ ratio in the oceans throughout Phanerozoic time was constructed by linear interpolation between data points. The resulting curve is very similar to those previously constructed by Peterman *et al.* (1970), Faure and Powell (1972), Veizer and Compston (1974), Brass (1976), Faure (1977), Hart and Staudigel (1978) and by Brevart and Allègre (1978).

4 POSSIBLE CAUSES FOR THE VARIATION

The isotopic composition of strontium in the oceans is controlled by inputs from three sources: (1) marine carbonate and evaporite rocks primarily of Phanerozoic age ($^{87}Sr/^{86}Sr = 0.708 \pm 0.001$); (2) volcanic rocks on the continents and in the ocean basins ($^{87}Sr/^{86}Sr = 0.704 \pm 0.002$); (3) sialic rocks of Precambrian age and younger detrital sedimentary rocks derived from them ($^{87}Sr/^{86}Sr = 0.720 \pm 0.005$). This model was originally proposed by Faure *et al.* (1965), and was restated by Faure (1977) and Faure *et al.* (1978). It was supported by Peterman *et al.* (1970) and elaborated by Brass (1976), who formulated mass balance equations for strontium entering and leaving the oceans. Strontium released by diagenesis, dolomitization and weathering of marine carbonate rocks greatly dominates the inputs. Faure (1977) and Faure *et al.* (1978) concluded that between 60 and 80% of the input is recycled marine strontium. Brass (1976) derived a value of 75% for this parameter. A model proposed by Spooner (1976) is in error because he failed to include strontium derived from marine carbonate

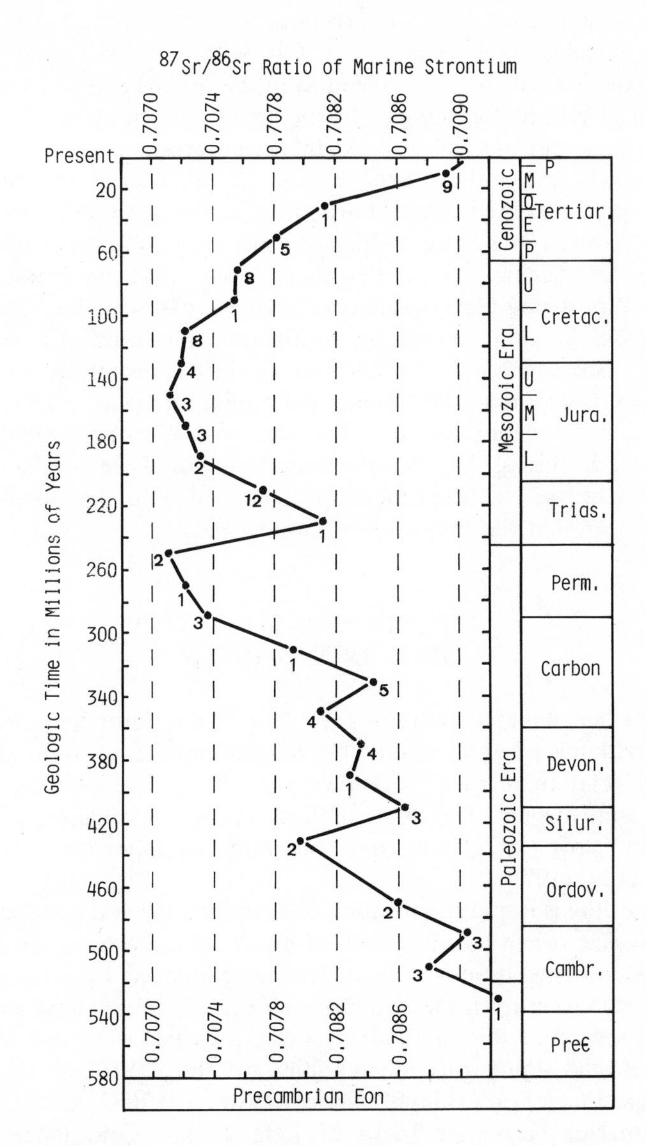

Figure 1. Variation of the ^{87}Sr/^{86}Sr ratio in the oceans throughout Phanerozoic time. The data were adjusted to 0.70800 for the Eimer and Amend $SrCO_3$ standard. Averages were calculated in 20 million year increments. The error bars are two standard deviations of the mean. The numeral beside each data point indicates the number of determinations by different authors. The time scale is based on results presented in this volume. A summary of the data is presented in Table 1.

rocks (Brass and Turekian, 1977). Nevertheless, Spooner drew attention to the interaction between sea-water and hot volcanic rocks on the sea floor and within the oceanic crust. These interactions probably have a greater effect on the chemical and isotopic composition of the volcanic rocks than of sea-water.

The fluctuations of the $^{87}Sr/^{86}Sr$ ratio in the oceans displayed in Figure 1 are primarily caused by changes in the inputs of strontium derived from young volcanic and old sialic rocks. Increased erosion of old rocks on the continents may cause the $^{87}Sr/^{86}Sr$ ratio of the oceans to rise, whereas increased volcanic activity may lower it. The fluctuations of the $^{87}Sr/^{86}Sr$ ratio of sea-water therefore reflect tectonic processes on a global scale. However, the specific causes that shaped the curve have not yet been identified. Armstrong (1971) suggested that increases in the $^{87}Sr/^{86}Sr$ ratio resulted from erosion caused by continental glaciation. Clauer (1976b) noted that increases in the $^{87}Sr/^{86}Sr$ ratio of the oceans tended to follow the pan-African, Caledonian, Hercynian and Alpine orogenic events. This idea was subsequently elaborated quantitatively by Brevart and Allègre (1978). The *marked* decline of the $^{87}Sr/^{86}Sr$ ratio during the Permian period may be related to increased volcanic activity associated with the opening of the Atlantic Ocean and the break-up of Gondwana.

5 THE MARINE-STRONTIUM GEOCHRONOMETER

The time-dependent variation of the $^{87}Sr/^{86}Sr$ ratio of marine strontium permits two kinds of applications: (1) to differentiate between marine and non-marine carbonate rocks of known age; (2) to date carbonate rocks of known marine origin. The first application was originally suggested by Clauer and Tardy (1971) and later elaborated by Jones and Faure (1978) and Neat *et al.* (1979).

The use of the isotopic composition of strontium for dating of sedimentary rocks of marine origin has not received much attention from geochronologists at academic institutions. Our colleagues employed by certain oil companies are said to use this method more widely. Examination of the curve in Figure 1 shows that this method of dating *does not yield unique solutions* because of the numerous fluctuations of the $^{87}Sr/^{86}Sr$ ratio during Phanerozoic time. For example, values between 0.7082 and 0.7086 may occur in marine carbonate rocks of Late Tertiary and Palaeozoic age. However, an $^{87}Sr/^{86}Sr$ ratio greater than 0.7082 indicates that the rock is unlikely to be of Mesozoic age. Perhaps the marine-limestone geochronometer will be most successful in *excluding* possible times of formation. In summary, the variation of the $^{87}Sr/^{86}Sr$ ratio of the oceans during Phanerozoic time is a geological phenomenon of fundamental importance that may, in addition, serve certain practical purposes in studies of the origin and time of deposition of marine carbonate rocks.

Acknowledgements

The financial support of the Division of Polar Programs of the National Science Foundation through Grant DPP-7721505 is gratefully acknowledged. This is Contribution No. 58 of Isotopia, the Laboratory for Isotope Geology and Geochemistry at The Ohio State University.

Résumé du rédacteur

Les minéraux carbonatés d'origine marine ont préservé l'enregistrement des fluctuations du rapport isotopique du strontium au cours des temps Phanérozoïques. Ce rapport isotopique a varié au cours du temps en fonction des apports de strontium aux océans. Les principales sources du strontium sont: les roches carbonatées marines ($^{87}Sr/^{86}Sr$ proche de 0.708), les roches volcaniques (rapport proche de 0.704), les roches sialiques anciennes et les roches sédimentaires dérivées (rapport proche de 0.720). L'accroissement des apports en strontium à rapport élevé a été attribué aux glaciations continentales et aux périodes orogéniques. L'accroissement des apports en strontium à rapport bas peut être lié à l'ouverture de l'Océan Atlantique et à la disjonction du continent du Gondwana au Mésozoïque. Les variations du rapport isotopique du strontium en fonction du temps sont utiles pour distinguer les roches carbonatées marines et non marines d'âge connu ou pour 'dater' des roches carbonatées d'origine marine reconnue.

(Manuscript received 8-10-1980)

Numerical Dating in Stratigraphy
Edited by G. S. Odin

5
Palaeomagnetic stratigraphy as a correlation technique

JAMES E. T. CHANNELL

The two preceding chapters in this volume have shown how biological, sedimentological and geochemical events can be used as means of stratigraphic correlation. This chapter will summarize (1) the processes whereby palaeomagnetic data are recorded in the stratigraphic column, (2) the applications of the palaeomagnetic record as a means of correlation, and (3) the uncertainties associated with the method.

1 THE NATURE OF THE EARTH'S MAGNETIC FIELD

Spherical harmonic analysis of the present Earth's magnetic field indicates that over 99.5% is of internal origin. The geomagnetic field is generated by the motion of highly conducting iron–nickel fluids in the outer part of the Earth's core. The motion is controlled by thermal convection and by the Coriolis force generated by Earth rotation. The Earth's field approximates to an axial geocentric dipole, although non-dipole components generated at the core–mantle interface, and to a much lesser extent ionospheric currents, perturbate the dipolar model. It undergoes variations in strength and direction on time scales from hours to millions of years. As the nature of the longer-term variations becomes better understood, their palaeomagnetic observation in rocks will become a powerful method in stratigraphic correlation.

1.1 Short-term (daily, monthly, yearly) variations

These rarely exceed 0.1% of the average field strength and are due to electrical current variations in the ionosphere. They are only of minor interest to geologists, but it is interesting to note that certain organisms (viz. honeybees) can apparently sense these small variations in field strength (<60 nT) and use them as a chronometer (Lindauer, 1977).

1.2 Secular (slow) variations

Secular variations of the geomagnetic field are of more interest to geologists as they provide a potential tool for relative dating and correlation. In direction, they rarely exceed 30° from the main dipole axis, and tend to average out over time periods of 5000–10,000 years. They are produced essentially by variations of the non-dipole components of the earth's field which drift westwards at an average rate of about 0.15° of longitude per year. Magnetic observatory records date from the end of the sixteenth century. Archaeomagnetic work on ancient kilns and pottery extends the record back several thousand years. However, much of our knowledge of secular variation has come from the study of Recent lake sediments with high deposition rates. Some detailed records are available back to 10,000 years b.p. and master curves for palaeomagnetic field direction and intensity can now be constructed for particular regions and used as archaeological and limnological dating tools.

1.3 Geomagnetic reversals

The dipole (main) component of the Earth's field has reversed its polarity at irregular intervals from Early Precambrian to the present day. Due to the dipole nature of the Earth's field, these *geomagnetic reversals are contemporaneous worldwide phenomena* and when recorded in rocks provide a unique stratigraphic marker. The process of reversal, in the few places where it has been recorded, takes a few thousand years. The last unquestioned reversal of the geomagnetic field occurred about 700,000 years ago. Intervals of 'reversed' or 'normal' (as present-day) polarity in the geological past are commonly referred to either as 'epochs' (duration $>10^5$ years) or as 'events' (duration $\sim10^4$–10^5 years). It has, however, been suggested that the terms 'chron' and 'subchron' should be substituted, as these terms clearly refer to time (Hedberg, 1976). Magnetic stratigraphy in pre-Pleistocene rocks is based on the recognition of patterns of alternating magnetic polarity in stratigraphic section. These patterns can be correlated from one section to another, and if it can be shown that the magnetization was produced during formation (igneous cooling or sedimentation) of the rock, they can be correlated with radiometric ages and with biozonations. This chapter is largely devoted to the uses of magnetic polarity stratigraphy because the polarity signature is the most useful magneto-correlation tool in pre-Pleistocene rocks.

1.4 Geomagnetic excursions

It has been suggested that there have been several so-called geomagnetic 'excursions' (duration less than 10^4 years) during the present Brunhes epoch of normal polarity. A geomagnetic excursion is defined as 'a sequence of virtual geomagnetic poles which extend beyond 45° of latitude from the pole and return to the original polarity after a short period of time' (Thompson, 1977). Such excursions are potentially valuable for detailed stratigraphic studies in the Quaternary and a great deal of effort has been devoted to searching for them. Many candidates have been suggested: e.g. at Mungo Lake, Australia (Barbetti and McElhinny, 1976), at Mono Lake and Lake Erie in North America (Denham, 1976; Creer *et al.*, 1976), in lavas at Laschamp, France (Bonhommet and Zähringer, 1969) and in Swedish varves (Mörner *et al.*, 1971; Noël, 1975).

These Brunhes geomagnetic excursions are a point of controversy. The Mono Lake excursion was not found less than 100 km away at Clear Lake (Verosub, 1977). The Lake Erie excursion has not been found in detailed sampling of lakes in Minnesota (Banerjee *et al.*, 1979). The Laschamp 'excursion' has been questioned by Heller (1980), who suggested that the aberrant directions might be the result of the interaction of two magnetic phases rather than a faithful record of the Earth's magnetic field. The excursions detected in the Swedish varves have been questioned by Thompson (1976), who considered that the aberrant directions could be explained by reorientation of grains by sedimentological and coring processes. Geomagnetic excursions have yet to be shown to be a characteristic feature of the earth's field. None of these possible excursions can be categorically correlated to another and excursions have yet to be detected at different lake sites in the same region. One of the most convincing 'excursions', which may qualify as an 'event', is known as the Blake event and has been documented in several northwest Atlantic sediment cores (Smith and Foster, 1969; Denham, 1976). This event has also been observed in a core from southern Italy, where it is estimated to have ended 90,000 years ago and to have had a duration of up to 50,000 years (Creer *et al.*, 1980).

2 THE PALAEOMAGNETIC RECORD IN ROCKS

2.1 Analytical uncertainties

The development of the astatic magnetometer (Blackett, 1952) paved the way for the modern era of palaeomagnetic research. The innovation allowed the magnetization direction in rock samples to be accurately measured for the first time. The first magnetometers were capable of measuring intensities of about 10^{-2} A/m and were suitable for palaeomagnetic measurements of igneous rocks and highly magnetized red sediments. Modern superconducting magnetometers (Goree and Fuller, 1976) can accurately measure magnetizations of 10^{-5} A/m, and many more sediment varieties can now be studied. Some superconducting magnetometers measure the magnetization simultaneously along three orthogonal axes and hence give the sample magnetization direction without the need to change sample orientation in the magnetometer. Measurement itself is more or less instantaneous and the time is taken up in sample-handling and in lowering the sample into the measurement area. The speed of operation makes it practicable to measure magnetization direction three or more times, determine a mean result and estimate an uncertainty associated with the measurement. The circular standard deviation of three independent measurements of a sample increases sharply at intensities below 10^{-5} A/m (Figure 1).

Rock samples are usually collected in the field using a small gasoline-powered drilling machine. Cores, usually 2.5 cm in diameter, are oriented using a magnetic or sun compass and are cut to 2.5 cm lengths for measurement. The uncertainties in core orientation are less than 2° provided that a well-designed orienting table is used (Tarling, 1971, p. 58). A sun compass must be used for orienting highly magnetized igneous rocks.

The magnetization of a rock sample usually comprises a number of superimposed magnetization components. The processes which give rise to the different magnetization components will be discussed in the next section. The job of the palaeomagnetist is to determine the directions of different components. This is done by progressively demagnetizing the sample using alternating field, thermal or chemical leaching techniques. The sample magnetization is measured between each progressive demagnetizing stage until the sample has been completely demagnetized. If the resistance to one

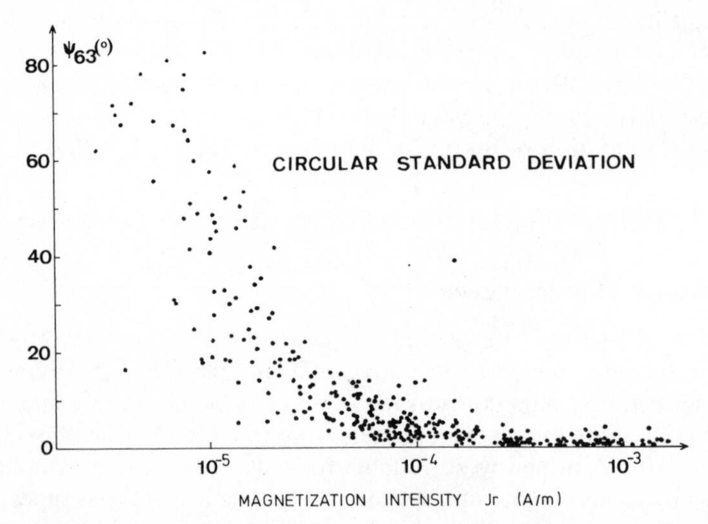

Figure 1. Circular standard deviation for three independent measurements of a sample, plotted against mean intensity (in A/m) of the sample. (Data from SCT cryogenic magnetometer.)

or other of the demagnetizing techniques is different for the different magnetization components, then it is possible to distinguish the components. Different magnetization components can be recognized using orthogonal projections (Zidjerveld, 1967) of demagnetization data (Figure 2). Linear portions on the projections are discrete magnetization components. The component directions can be deduced graphically or by least-squares techniques (e.g. Kirschvink, 1980). Besides orthogonal projections, intersections of great circles on stereographic projections can be used to resolve magnetization components from progressive demagnetization data (Halls, 1978; Hoffman and Day, 1978).

In magnetostratigraphic studies, samples are usually collected at regular stratigraphic intervals in the section. The mean magnetization direction from the whole section gives the mean palaeomagnetic pole. The latitude of the

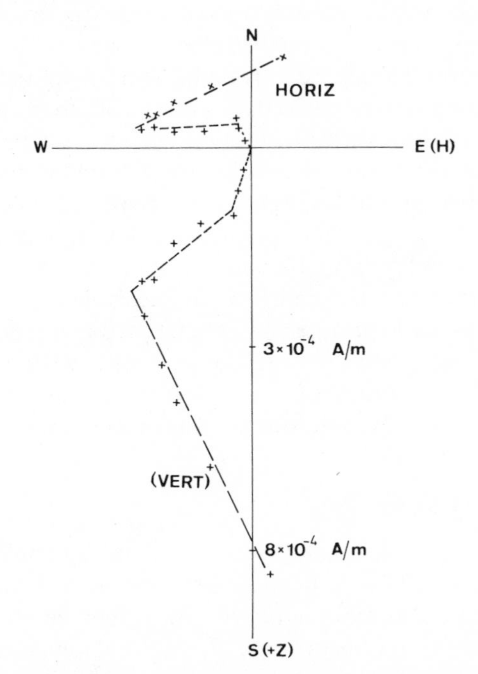

Figure 2. Orthogonal projection (Zidjerveld, 1967) of thermal demagnetization data from a red chert sample. Three magnetization components with distinct blocking temperature spectra are recognized. The crosses represent the end points of the magnetization vector with progressive increase in demagnetization temperature (NRM, 100°C, 150°C, . . . in steps of 50 degC to 600 degC). The vector end points are plotted in the horizontal plane (HORIZ) with respect to orthogonal axes: N (North), S, E, W; and in the vertical plane (VERT) with respect to the horizontal (H) and vertical (Z). Magnetization intensity is given in A/m decreasing to zero at the origin of the plot.

pole position for each sample, relative to the mean palaeomagnetic pole (virtual geomagnetic polar (VGP) latitude), is then plotted against stratigraphic position. VGP latitude is a useful and commonly employed measure of polarity.

The sampling interval depends on the resolution required. For magnetic stratigraphy studies, in the Mesozoic for instance, we may wish to resolve all polarity intervals with duration greater than 10^5 years. An estimate of sedimentation rate, such as 5 m/Ma—typical for pelagic sediments—gives a sampling interval of 0.5 m.

Magnetostratigraphic units are sequences of beds characterized by a common palaeomagnetic polarity signature. These units are referred to either as 'zones' or 'subzones' depending on the amount of time they represent (Subcommission on Magnetostratigraphic Nomenclature, 1973). A zone or subzone should be defined by at least two samples, as a single-sample polarity zone may be a spurious result due to orientation error. When all magnetozones in a certain sampling window have been defined, the magnetostratigraphy may be correlated with other sections or with the ocean floor. Magnetostratigraphic data are essentially a *binary signal*: one reversed or normal zone varies from another only in length. Characteristic patterns of normal and reversed polarity are required to make anambiguous correlation. Reversals in the geological past were very variable in frequency. Short polarity zones may either be not recorded, or not sampled in some sections, making correlation ambiguous. In sedimentary sections a constant sedimentation rate is optimal and facilitates correlation of magnetostratigraphy from one section to another. In practice, sedimentation rates fluctuate; bedding planes, especially in carbonate rocks where they are often dissolution surfaces, may represent considerable amounts of time. Therefore the appearance of polarity patterns will vary from section to section.

2.2 Genetic uncertainties

The use of the palaeomagnetic record as a stratigraphic correlation tool depends on the rocks having been magnetized at a definite time in their history. A rock can become magnetized in a number of different ways at different times during its history and the natural remanent magnetization (NRM) of a sample is often a composite of a number of different magnetization components. When using the palaeomagnetic record as a correlation technique, we are only interested in that magnetization component which records the geomagnetic field direction at the time of formation of the rock. This 'primary' magnetization component is then a stratigraphic marker which can be correlated with a biozonation or with radiometrically determined ages.

2.2.1 *Igneous rocks*

Igneous rocks become magnetized as they cool through the 'magnetic blocking temperatures' of magnetic minerals present in the rock. Titanomagnetites (Fe_2TiO_4–Fe_3O_4 series) are the most common magnetic minerals in igneous rocks. The blocking temperatures depend both on the composition of the titanomagnetite and on grain size. Samples generally exhibit a '*blocking temperature spectrum*' which reflects a range in composition and/or grain size. The blocking temperatures in titanomagnetite-bearing rocks may range up to 578°C, which is the Curie point of pure magnetite. As mentioned above, secular variation averages out over time periods of 5000–10,000 years. However, igneous rocks cool (and become magnetized) much more rapidly and therefore secular variation will generally not be averaged out in igneous rock samples. Most igneous rocks have sufficient concentrations of magnetic minerals to become measurably magnetized during cooling. The more basic are generally more strongly magnetized, as they contain higher concentrations of magnetic minerals such as titanomagnetites. Fine-grained lavas are usually more stably magnetized (higher coercivity, i.e. more difficult to demagnetize with alternating fields) than plutonic rocks because fine (single-domain) grains carry a 'harder' magnetization than larger (multi-domain) grains.

The 'primary' magnetization component in an igneous rock is a thermoremanent magnetization (TRM) produced during cooling. This magnetization direction can be correlated with a radiometrically deduced age. However, in most igneous rocks, the primary magnetization is immersed in the noise of later ('secondary') magnetization. This secondary magnetization may take the form of a viscous remanent magnetization (VRM) or of a chemical remanent magnetization (CRM). VRM is the magnetization which builds up in a rock sample due to the prolonged effect of the present Earth's field. The probability that the remanent magnetization of a grain will reorient itself in line with the present Earth's field is a function of composition, domain state (grain size), temperature and time. VRM built up during the present Brunhes epoch of normal polarity can contribute significantly to the NRM. VRM in titanomagnetites can usually be destroyed by alternating field demagnetization at peak fields of a few tens of millitesla. However, VRM in minerals with higher coercivity (such as haematite) may be very resistant to alternating field demagnetization and is most easily destroyed using thermal demagnetization methods. Chemical remanent magnetization (CRM) is an important source of secondary magnetizations in igneous rocks, for example titanomagnetites tend to oxidize to titanomaghemites. It is sometimes, but not always, possible to separate the magnetization components due to the parent and the oxidation product(s) by careful thermal demagnetization.

2.2.2 Sedimentary rocks

Sedimentary rocks can become magnetized by the mechanical alignment of ferromagnetic grains in the soft sediment, by growth of authigenic minerals during diagenesis or later alteration, or by the build-up of VRM.

Depositional remanent magnetization (DRM) is commonly due to the alignment of (titano)magnetite or specular haematite grains in the direction of the ambient magnetic field. The preferred orientation occurs shortly after deposition, in the water-saturated upper layers of the sediment. We can regard such a DRM as a 'primary' magnetization recording the direction of the geomagnetic field at the time of deposition of the sediment.

Haematite is a magnetic mineral commonly produced during early diagenesis of sediments. the pigmentary haematite in red beds may be due to the alteration of preexisting detrital iron–magnesium minerals, the breakdown of iron-bearing clay minerals, or to the dehydration of oxyhydroxides (such as goethite) deposited from sea-water. An authigenic ferromagnetic mineral becomes magnetized when it grows through a critical volume. The time lag between deposition and growth of authigenic haematite to its critical volume will depend on local diagenetic conditions. Therefore a haematite CRM cannot be considered a 'primary' magnetization.

Ferromagnetic iron sulphides such as pyrrhotite ($Fe_{1-x}S$) and greigite (Fe_3S_4) are other authigenic minerals found in sediments, and their magnetizations may significantly postdate deposition. Sulphides form in sediments with high organic content where sulphate is reduced to sulphide (by bacteria) and the sulphide then combines with aqueous iron in sea-water or with iron deposited as oxyhydroxide.

Ferromagnetic goethite (FeOOH) can be deposited from sea-water and is commonly the precursor to authigenic haematite or iron sulphides. However, goethite can also form as a product of low-temperature oxidation (weathering) of iron-bearing minerals such as oxides and sulphides. Pyrite-bearing sediments often contain goethite as a late alteration product and this goethite dehydrates to haematite under some conditions.

The above discussion indicates that the magnetization of a sediment may be very complex. The best analytical approach is to determine the magnetization components by careful demagnetization, ascertain by rock magnetic study the carriers of each magnetization component, and by association of certain carriers with certain acquisition processes, isolate the primary component. Magnetite (or titanomagnetite) can usually be considered as a detrital or biologically produced mineral in marine sediments, and therefore the magnetization component due to magnetite can be associated with the palaeontological age of the sediment. Haematite (specularite) may also be detrital but it is in addition a common authigenic mineral giving pigment to red beds. In favourable circumstances it may be possible to constrain the age of a diagenetic magnetization by the use of field tests such as the classic 'fold' and 'conglomerate' tests (Graham, 1949). If samples are collected on the limbs of a fold, the magnetization directions will either group or disperse after rotating the fold limbs to the horizontal. An enhanced grouping after unfolding indicates that the magnetization predates folding. The 'conglomerate' test involves sampling pebbles in a conglomerate. If the magnetization directions of the pebbles are random, then the magnetization predates the formation of the conglomerate.

The ideal sediment for magnetic stratigraphy is one that contains magnetite as the only ferromagnetic mineral. A red pigment generally indicates the presence of authigenic haematite but does not necessarily indicate that the haematite contributes to the remanence. Rapid burial of organic matter gives rise to the authigenic formation of ferromagnetic sulphides (pyrrhotite and greigite) or paramagnetic sulphides (pyrite) which often undergo later alteration to goethite and haematite. Therefore these chemically reduced sediments, which may of course be excellent palaeontologically, may be very noisy magnetically.

In order to establish a magnetic polarity stratigraphy for a sedimentary section, it is an advantage to have fairly slow but constant sedimentation rates. Moderately slow sedimentation rates (a few metres per Ma) are advantageous for two reasons. Firstly, if the sedimentation rate is slow, more polarity reversals will be found in a given length of section, enhancing the chance of observing a characteristic pattern. Secondly, slow deposition rates in carbonate rocks allow the concentration of detrital or biological magnetite. Platform limestones with high sedimentation rates are generally too weakly magnetized for precise measurement as the magnetite content is diluted by the rapid calcium carbonate accumulation. Constant sedimentation rates are an advantage as they facilitate the matching of reversal patterns. Hiatuses and large fluctuations in sedimentation rate cause distortion of the reversal pattern and may make correlation difficult from one section to another.

3 SECULAR VARIATION OF THE GEOMAGNETIC FIELD AS A CORRELATION TOOL

The secular variation record can be used as a correlation tool for archaeological artifacts and Recent sediments.

3.1 Archaeomagnetism

The bricks from pottery kilns and ancient fireplaces have a thermoremanent magnetization dating from their last firing. This magnetization direction can often be dated using ^{14}C contents of ashes. The variations in declination and inclination from archaeomagnetic studies in southern England (Aitken, 1970; Tarling, 1975; Noël, 1978) have been combined with London magnetic observatory records to give standard curves (Figure 3) which can be used for dating other kilns and fireplaces in the vicinity. As secular variation is mainly due to changes in the regionally varying non-dipole components of the Earth's field, such a curve does not have global applicability. Contemporaneous variations in magnetic direction may be quite different for places a few thousand kilometres apart. However, long-term variations in intensity of the total geomagnetic field are dominated by the strength of the dipole moment and are believed to be sinusoidal in nature (Figure 4). If this is the case, long-term geomagnetic field variations are essentially a function of latitude and can be correlated worldwide. Moreover, palaeointensity measurement does not require the orientation of the sample to be known, and therefore a wider variety of materials (such as pottery) can be used. Cox (1968) has analysed the palaeointensity data for the past 8000 years (figure 4) and Barbetti (1973) has used this curve to estimate the age of aboriginal fireplaces from palaeointensity measurements. Higher-frequency, low-amplitude fluctuations in palaeointensity such as those given by Bucha *et al.* (1970) and Walton (1979) are mainly due to the non-dipole field, and are useful for precise but local correlation.

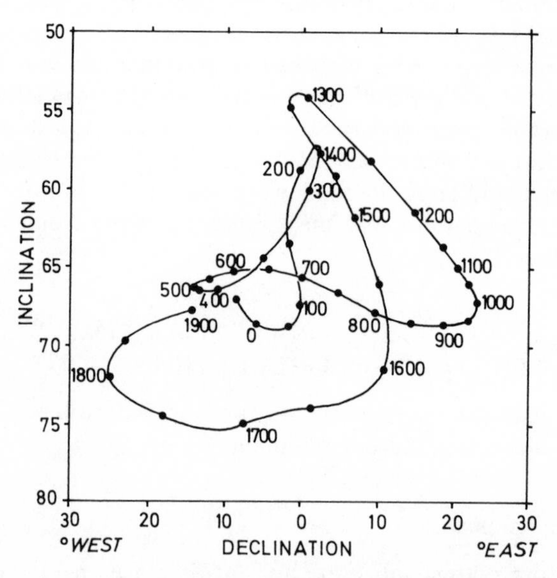

Figure 3. Curve showing changes in direction of the earth's magnetic field since 50 BC. All directions are corrected to Meridan, the traditional centre of England (after Noël, 1978).

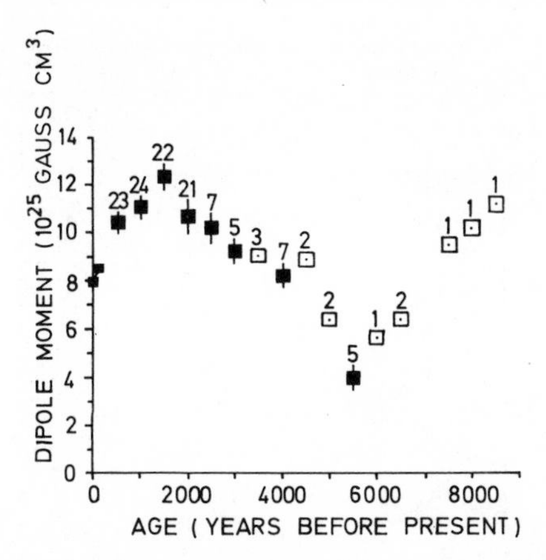

Figure 4. Variations in the dipole moment of the geomagnetic field. Each point is the average of all reduced dipole moments with ages in the indicated 500 years class interval. The number of data in the class interval is given above each point and the standard deviation is shown by a vertical line. Open symbols are intervals with too few data to determine a reliable value of the dipole moment (from Cox, 1968, reproduced by permission of the author and the American Geophysical Union).

3.2 Recent lake sediments

Lava flows can be dated radiometrically and their thermoremanent magnetizations provide spot readings of the magnetic field which can be used to build up a picture of secular variation. However, lavas do not erupt with regularity and gaps occur in the record. The study of wet and dry Recent lake sediments with high deposition rates (1 m/1000 yr) provides a more continuous record of secular variation. The magnetization of lake sediments is usually due to the alignment of magnetite grains either during deposition or within the water-saturated top few centimetres of sediment. Two processes can, and often do, spoil the record in lake sediments: (i) sedimentological factors such as currents and depositional slopes give rise to mechanical reorientation of detrital grains; (ii) post-depositional alteration causes the growth of new magnetic minerals with magnetizations younger than the primary sedimentation.

Several beautiful records have been obtained from lake sediments (Figure 5) in northern Britain. The declination pattern is particularly characteristic and tends to have a periodicity of about 3000 years. Sufficient localities have now been studied palaeomagnetically and dated using pollen and ^{14}C that a master declination curve for northern Britain is available (Thompson, 1977). This curve may be used as a reference curve and as a dating tool but, as mentioned above, is only of local applicability and should not be used as a reference outside NW Europe.

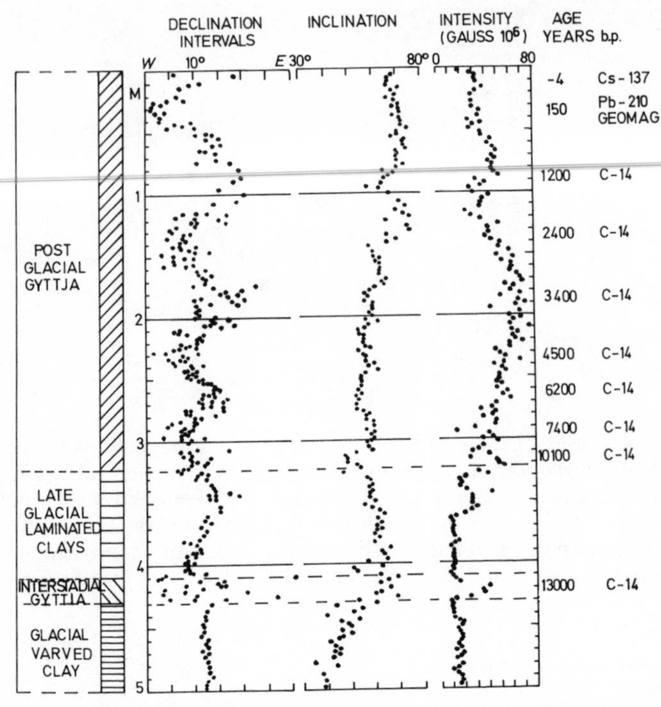

Figure 5. Lake Windermere palaeomagnetic data 14,000 BP to present. Record built up from data from several cores. Dating from ^{14}C, ^{210}Pb, ^{137}Cs, and 1820 westerly maxima in magnetic observatory records (Geomag). (Modified from Thompson, 1977 by permission of Blackwell Scientific Publications Ltd.)

4 THE DEVELOPMENT OF THE GEOMAGNETIC REVERSAL TIME SCALE

In order to use magnetic reversal stratigraphy as a correlation technique, it is necessary to develop a standard reversal time scale. The reversal pattern for the last 5 Ma has been established and calibrated using radiometrically dated lavas on land. Most of our knowledge of the reversal time scale for the Cenozoic and Mesozoic is derived from oceanic magnetic anomalies, but parts of the time scale are now being refined by study of sedimentary sections on land.

4.1 Radiometrically dated lavas on land

Although it had been suggested since the beginning of the century that the earth's field reversed its polarity in the geological past, it was the developments in the K–Ar dating method which paved the way for the confirmation

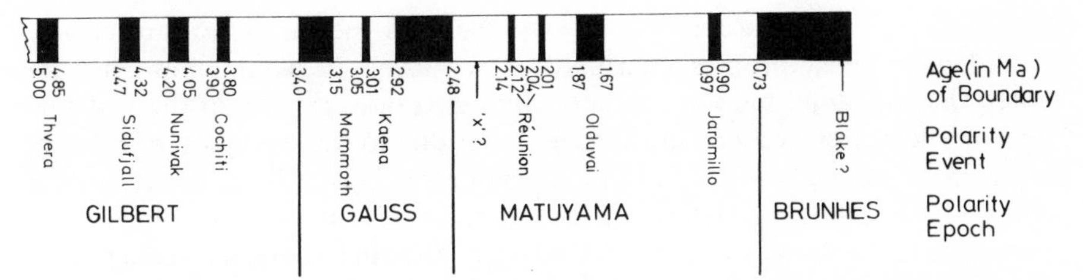

Figure 6. Late Cenozoic polarity time scale. Black corresponds to normal polarity and white to reversed polarity (from Mankinen and Dalrymple, 1979).

and dating of the recent geomagnetic reversals. The first reversal time scale was given by Cox *et al.* (1963) and was based on nine radiometrically dated lava flows erupted during the last three million years. Since then the reversal time scale has been revised and extended. However, the extension of the time scale is limited by the lack of resolution of the K–Ar dating method. For ages greater than 5 Ma, the typical value of ±2% for the precision of a K–Ar age is equivalent to ±0.1 Ma, which is longer than the duration of many of the shorter polarity intervals. The polarity time scale for the last 5 Ma is given in Figure 6 (Mankinen and Dalrymple, 1979). There is still controversy on the age of some events, notably those at the base of the Matuyama epoch (see McDougall, 1979). The polarity time scale has been extended further back in time, to about 7 Ma, using stratigraphically related Icelandic lavas (McDougall *et al.*, 1977). The K–Ar ages were plotted against stratigraphic height above the base of the sequence and reversals dated by interpolation along the least-squares regression line fitting the age data.

4.2 Oceanic magnetic anomalies

The most complete record of the reversal time scale during the Mesozoic and Cenozoic stems from the study of oceanic magnetic anomalies. Using Doell *et al.*'s (1966) date of 3.35 Ma for the beginning of the Gauss normal polarity epoch, Heirtzler *et al.* (1968) constructed a polarity time scale for the Late Cretaceous and Cenozoic based on the assumption of constant spreading rate (1.9 cm/a) in the South Atlantic. The South Atlantic profile, V20, was the best one available at the time and the assumption of constant spreading rate was more or less *ad hoc*. However, the assumption has proved a surprisingly good one. It is only in the last few years that magnetostratigraphic and palaeontological data from land sections and from deep-sea cores have substantially modified the Heirtzler *et al.* (1968) time scale. The oceanic crust is a remarkably efficient recorder of geomagnetic reversals. The resolution of oceanic crust as a recorder increases with spreading rate but it is interesting to note that even the short-lived events with duration of the order of 10^5 years (such as the Jaramillo) are clearly recorded on most profiles measured at sea-level. It should, however, be noted that the amplitude of the anomalies decreases rapidly in the first few tens of kilometres away from the ridge crest due to the oxidation of oceanic layer 2A titanomagnetities to titanomaghemites. This reduces the recording ability of the pillow lavas (layer 2A) and enhances the relative contribution of deeper layers (notably layer 3A and 3B). Although there is a reduction in the recording ability of oceanic crust during the first few million years of ageing, recent studies of land sections in the Cenozoic and Mesozoic have turned up very few reversals which were not previously detected in the oceanic record. The problem with the oceanic record is that *it is very difficult to date.* Palaeontological determinations on the oldest sediments overlying anomalies are available where basement was reached during DSDP. *These determinations are often associated with large uncertainties* and extrapolation to other anomalies, assuming constant spreading rate, is hazardous.

4.3 Land sections and oceanic sediment cores

Land sections and oceanic sediment cores have been extensively used to calibrate the reversal time scale against various biozonations. Conventional piston-coring techniques in the oceans have provided palaeontological control on the reversal sequence as far back as the Early Miocene (Opdyke, 1972; Opdyke *et al.*, 1974; Theyer and Hammond, 1974). The hydraulic piston corer (HPC), which is a relatively new addition to the Glomar Challenger (DSDP vessel), provides increased piston-core penetration and in recent legs much of the Cenozoic has been sampled. This new corer will be an important source of data over the next few years. Conventional rotary-cored DSDP material is not optimal for magnetostratigraphy due to incomplete recovery, core distortion and disturbance of the magnetism of the cores.

Land sections are providing palaeontological calibration of the polarity time scale, especially in the Early Tertiary and Mesozoic. The pelagic limestone section at Gubbio (Central Italy) has yielded a very well defined Late Cretaceous reversal stratigraphy which can be tied to a detailed foraminiferal biostratigraphy (Lowrie and Alvarez, 1977; Alvarez *et al.*, 1977). Middle Cretaceous magnetostratigraphy has been studied in Italian pelagic limestone sections (Channell *et al.*, 1979; Lowrie *et al.*, 1980c) and the extent of the Cretaceous 'long normal' interval estimated (Figure 7). The Cretaceous magnetostratigraphy can be correlated with the reversal stratigraphy derived from oceanic magnetic anomalies (Figure 8) and the land-derived magnetostratigraphy provides palaeontological dating for the oceanic anomalies.

Our knowledge of Jurassic magnetostratigraphy is based almost entirely on the Late Jurassic–Early Cretaceous (M) sequence of oceanic magnetic

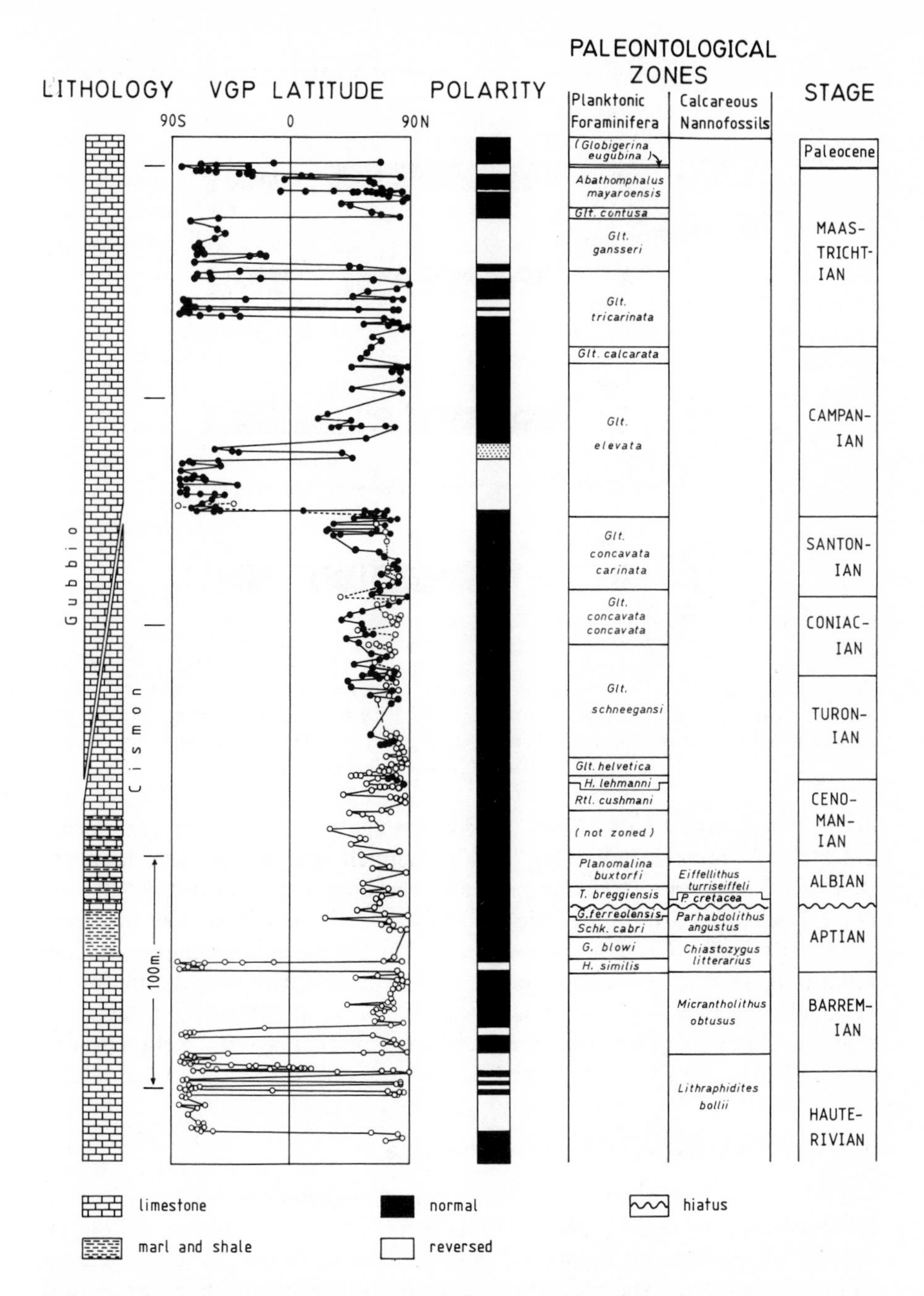

Figure 7. Virtual geomagnetic polar (VGP) latitudes (latitudes relative to the palaeomagnetic pole) are plotted for samples from Gubbio and Cismon (Italy). Those which plot close to 90°N are normal polarity and those which plot close to 90°S are reversed. The polarity stratigraphy is unambiguous and can be correlated with the micropalaeontology (from Lowrie *et al.*, 1980b, copyrighted by the American Geophysical Union).

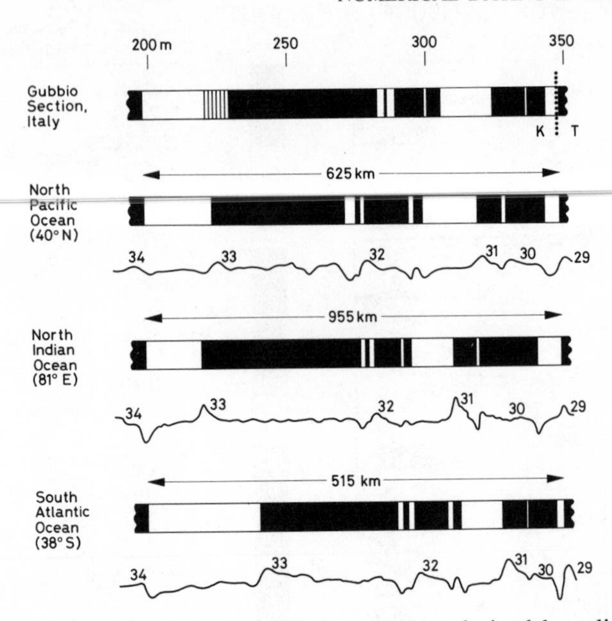

Figure 8. Comparison of the Gubbio polarity sequence derived from limestone land sections with Late Cretaceous geomagnetic polarity sequences derived from oceanic magnetic anomaly profiles (from Lowrie and Alvarez, 1977; reproduced by permission of the Royal Astronomical Society).

anomalies (Larson and Hilde, 1975), although recent studies of pelagic limestone sections in Italy (Ogg, 1980; Channell *et al.*, 1980) have indicated that reversals derived from land sections can be matched with those derived from Jurassic oceanic magnetic anomalies. Although it is now generally accepted that the Cretaceous 'quiet zone' in the oceanic magnetic anomaly record signifies an extended period of normal polarity, the nature of the Jurassic quiet zone is a matter of debate. Cande *et al.* (1978) have suggested that the Jurassic quiet zone signifies a lower geomagnetic field intensity rather than a period of prolonged normal polarity.

4.4 Mesozoic–Cenozoic polarity time scale

The polarity time scale in Figure 9 is an up-to-date summary for the Mesozoic and Cenozoic. The absolute ages of stage boundaries follow the scheme adopted in this book, and the qualification of these dates is given in following chapters. The polarity time scale for the Plio-Pleistocene is from Mankinen and Dalrymple (1979) (Figure 6).

The polarity reversal pattern for the Miocene and Palaeogene is based on the oceanic record from LaBrecque *et al.* (1977). We have expanded and contracted different parts of the sequence to be consistent with (i) the

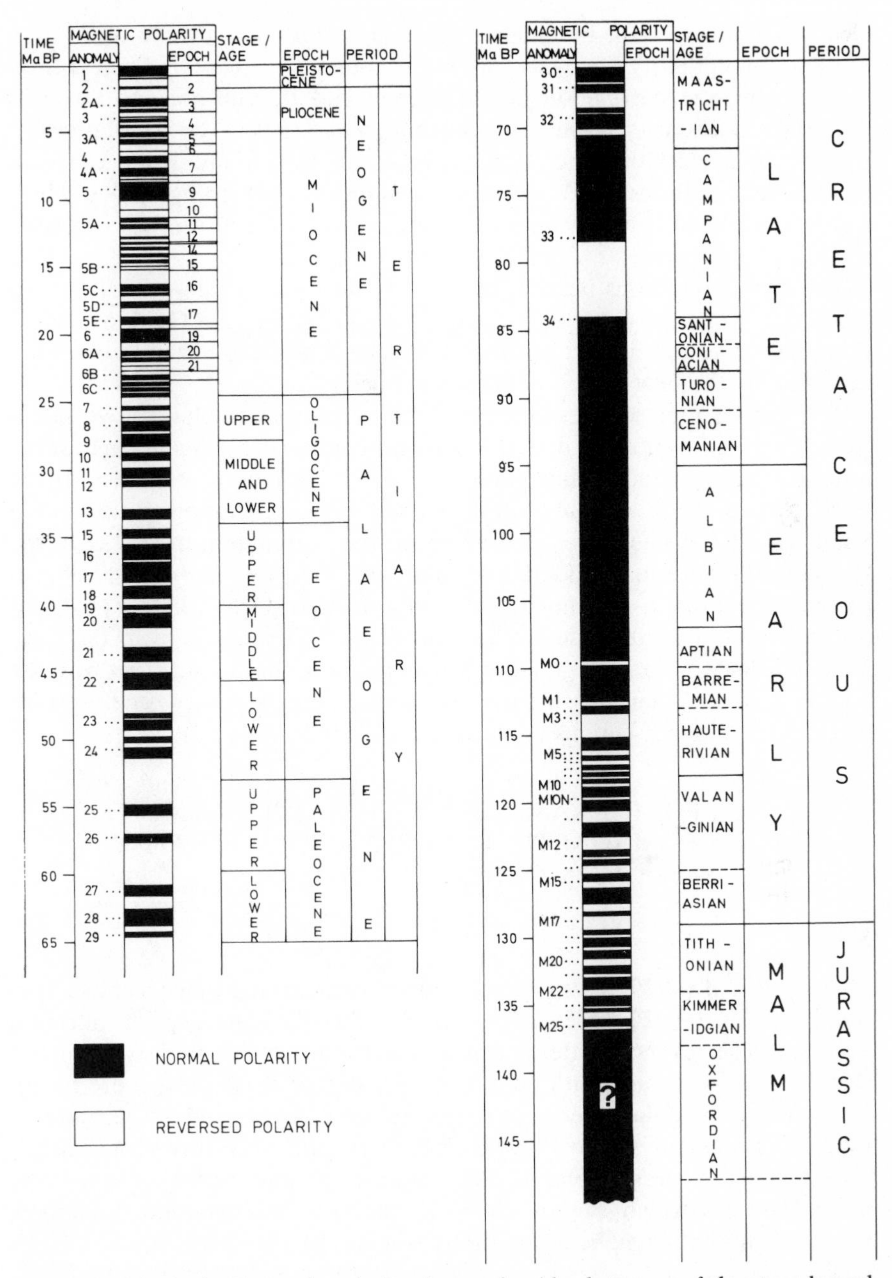

Figure 9. Mesozoic–Cenozoic polarity time scale. Absolute ages of the stage boundaries are qualified elsewhere in the book. The magnetic polarity sequence is mainly from oceanic magnetic anomaly data, with correlation to the stage boundaries mainly from palaeontologically dated land sections (see text).

absolute dating of stage boundaries adopted in this book, (ii) the correlation of Oligocene polarity reversals with stage boundaries given by Lowrie *et al.* (1980a), (iii) the correlation of Palaeocene and Eocene polarity reversals with stage boundaries given by Napoleone *et al.* (1979). The correlations of Lowrie *et al.* (1980a) and Napoleone *et al.* (1979) are from Umbrian (Italian) land sections which give unambiguous reversal stratigraphies, which can be correlated with foraminiferal zones, which in turn can be correlated with stage boundaries.

The Late Cretaceous part of the polarity time scale (Figure 9) is derived from the work of Lowrie and Alvarez (1977) in the Gubbio section (Umbria). The appearance of the reversal sequence is, again, somewhat modified by the new dating of stage boundaries.

The Early Cretaceous–Late Jurassic (M) sequence of polarity reversals is from the oceanic anomaly data (Larson and Hilde, 1975). Again parts of the sequence are expanded and contracted to fit various constraints: (i) the Early Aptian age of anomaly M0 as derived from the Cismon section in the Southern Alps (Channell *et al.*, 1979); (ii) the correlation of M17 with the Berriasian–Tithonian boundary and M22 with the Tithonian–Kimmeridgian boundary from land sections in the Southern Alps (Ogg, 1980). Constraint (ii), assuming a more or less constant oceanic spreading rate in the Late Jurassic, places M25 in the middle Kimmeridgian. This is consistent with the middle Kimmeridgian age for the end of the Jurassic 'quiet zone' derived from Umbrian land sections (Channell *et al.*, 1980).

5 FOUR EXAMPLES OF THE APPLICATION OF GEOMAGNETIC REVERSAL STRATIGRAPHY AS A CORRELATION TOOL

5.1

Opdyke *et al.* (1979) have sampled eight separate stratigraphic sections from sediments of the Plio-Pleistocene Upper Siwalik subgroup of northern Pakistan. The reversal patterns can be correlated unambiguously from one section to another, and with help from radiometric dates on two prominent bentonitized tuffs, the reversal stratigraphy can be correlated to the standard reversal chronology of the Plio-Pleistocene (Figure 6). Hence stage boundaries can be accurately placed. All sections contain important vertebrate fossil localities which can be dated in the time stratigraphic framework provided by the magnetic stratigraphy (Figure 10).

5.2

As mentioned above, the inherent uncertainty in the K–Ar dating method limits the precise absolute dating of reversal boundaries to the last 5–10 Ma.

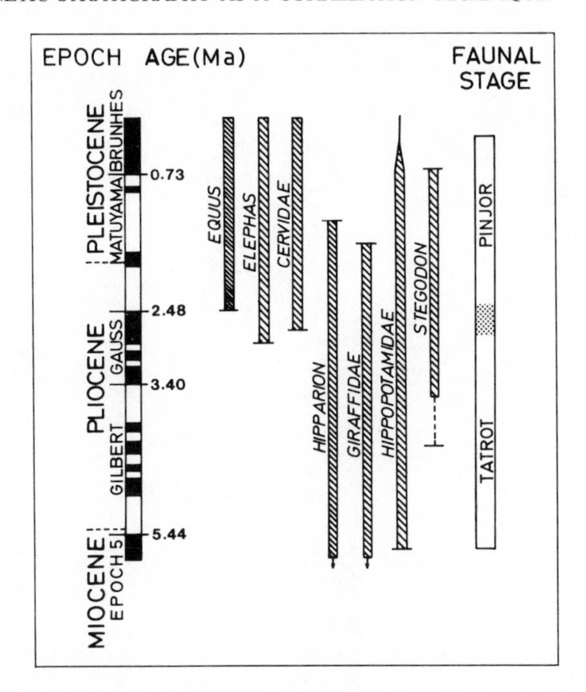

Figure 10. The dating of stratigraphically important taxa in the Pliocene and Pleistocene of the Siwaliks using the magnetostratigraphy (after Opdyke *et al.*, 1979).

However, older reversal sequences on land can be palaeontologically dated, and stage boundaries assigned. If the land-based reversal stratigraphy can be matched to that given by oceanic magnetic anomalies, the oceanic magnetic anomalies can be stratigraphically dated. Lowrie and Alvarez (1977) have been able to date the Late Cretaceous oceanic magnetic anomalies by comparing the interpreted reversal patterns from the oceans with that obtained from a palaeontologically well-dated pelagic limestone section at Gubbio (Figure 8).

5.3

Magnetic stratigraphy is a means of correlating sedimentary sections, and is particularly useful where palaeontological and lithological correlation is difficult. Borehole logs can be correlated using the magnetic stratigraphy provided that the palaeolatitude at the time of deposition was not too low. Borehole logs are not oriented in azimuth, therefore correlations are made only on the basis of the inclination of the magnetization; in the case of low palaeolatitudes the magnetic inclination may not change significantly with changing polarity.

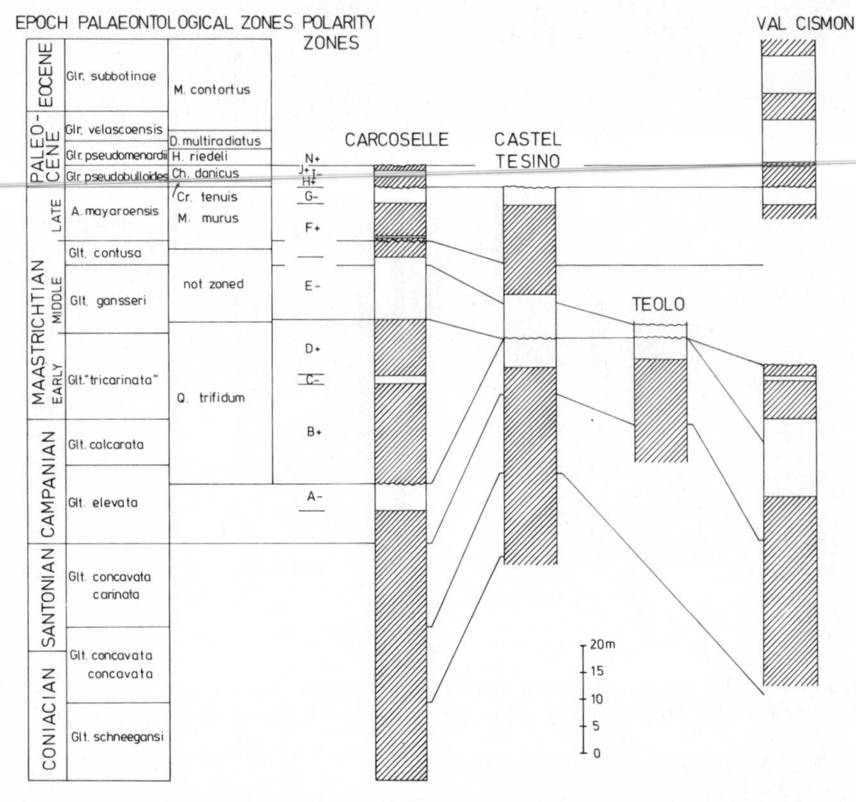

Figure 11. Comparison of polarity zones in four sections from the Belluno basin (Italy). Shaded and unshaded zones represent normal and reversed polarity respectively (after Channell and Medizza, 1980).

Channell and Medizza (1980) have studied four sections of pelagic limestone from the Belluno basin in northern Italy. The magnetic stratigraphy in conjunction with the palaeontology allows the comparison of sedimentation rates in different parts of the basin and gives an estimate of the amount of time represented by hard grounds in the sections (Figure 11). The palaeontology permits the unambiguous correlation of the magnetozones, and the magnetozones themselves can then give a finer-scale correlation between sections.

5.4

Reversals of the geomagnetic field are *synchronous* world-wide phenomena; the process of reversal is thought to take a few thousand years. Therefore, reversal stratigraphy provides a unique method of correlating

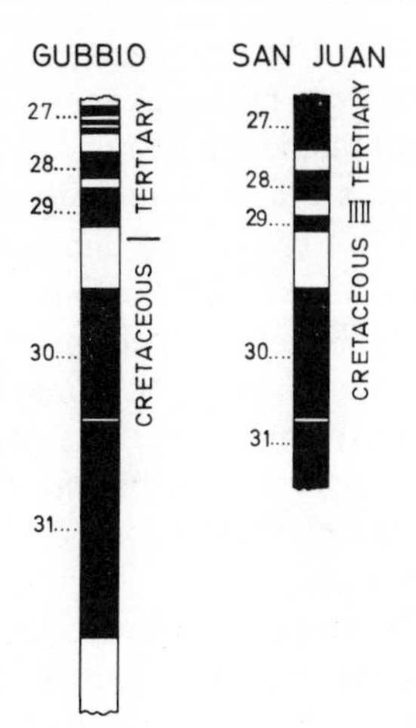

Figure 12. Correlation of magnetozones from pelagic limestones at Gubbio (Lowrie and Alvarez, 1977) and terrestrial sediments in the San Juan Valley. The Cretaceous–Tertiary boundary as given by dinosaurs (San Juan) and foraminifera (Gubbio) appears not to be synchronous (from Butler *et al.*, 1977; reproduced by permission from *Nature*, **267,** p 318. Copyright Macmillan Journals Ltd).

different palaeontological zonations from different environments. One example is given by Butler *et al.* (1977), who compared the magnetic stratigraphy near the Cretaceous–Tertiary boundary for two very different sedimentary environments. The terrestrial sediments of the San Juan basin (New Mexico) are dinosaur-bearing, whereas the pelagic limestones at Gubbio (Lowrie and Alvarez, 1977) are dated using foraminifera. The Cretaceous–Tertiary boundary as defined by the dinosaur remains and that as defined by the foraminifera apparently do not lie in the same magnetozone (Figure 12). Butler *et al.* (1977) considered that the data indicate that the Cretaceous–Tertiary boundary in marine and in terrestrial rocks is not synchronous. If this is the case, catastrophic causes for the extinctions at this boundary (such as meteorite impact) could be eliminated from further consideration. There is, however, evidence for hiatuses in the San Juan sediments which puts the magnetostratigraphic correlation in doubt (Alvarez and Vann, 1979).

6 CONCLUDING REMARKS

Due to the dipole nature of the earth's field, geomagnetic polarity reversals are, on a geological time scale, synchronous worldwide phenomena. They provide a unique stratigraphic marker when recorded in rocks. A wide variety of igneous and sedimentary rocks record the direction of the geomagnetic field at the time of their formation (during sedimentation or igneous cooling). Polarity stratigraphy can be used as a means of correlation not only between sedimentary sections with different biozonations, but also between radiometrically dated lava sequences and palaeontologically dated sedimentary sequences. Magnetic stratigraphy provides a means of correlating different biozonations with each other and with radiometrically determined absolute ages. It is, however, only in the last few years that technological advances in magnetometer design have allowed the potential of the method to be fully realized.

In order to correlate the magnetic polarity of a rock sample with a biozonation or a radiometric age, the 'primary' magnetization component (that produced during deposition of a sediment or cooling of an igneous rock) must be resolved. The 'primary' signal is generally immersed in noise from secondary magnetization processes, such as growth of magnetic minerals during diagensis or weathering. Progressive demagnetization techniques are used to resolve the various magnetization components and isolate the primary component.

Magnetostratigraphic data are essentially a binary signal; a magnetostratigraphic zone may be either 'normal' or 'reversed'. Therefore a unique correlation between two magnetostratigraphies depends on the recognition of a characteristic pattern of normal and reversed zones.

In order to exploit the potential of magnetic polarity stratigraphy as a correlation technique, it is necessary to have a standard polarity time scale (Figure 9) relating polarity reversals to stage boundaries and to absolute ages. Progress has been made in the last few years in calibrating the Cenozoic and Late Cretaceous reversal sequence, and the Early Cretaceous and Jurassic are now a focus of attention.

Résumé du rédacteur

Le champ magnétique terrestre est engendré par le mouvement des fluides hautement conducteurs de la partie externe du noyau de la terre entraînant en gros un dipôle axial. Ce champ magnétique est variable en direction et en intensité à une échelle de durée allant de l'heure au million d'années. C'est l'enregistrement dans les roches, des variations de polarité du champ magnétique, nécessairement synchrone pour tout le globe, qui est utilisé comme méthode de corrélation stratigraphique. Les progrès technologiques récents permettent de mesurer une très faible magnétisation dans une roche.

Les données recueillies par cette méthode sont essentiellement constituées d'un signal binaire: polarisation normale ou inverse, chaque séquence de roche de même sens de polarisation étant nommée magnétozone.

Les incertitudes analytiques et génétiques de la méthode sont analysées et les principes d'application sont abordés.

Grâce à l'accumulation des données, on a pu établir une échelle type des inversions magnétiques dont les coupures se font de plus fines tandis que peu à peu, l'échelle, d'abord établie pour le Cénozoïque supérieur, s'étend aux temps plus anciens. Essentiellement deux types d'enregistrement ont été parallèlement étudiés: (1) la magnétisation des planchers océaniques à partir des dorsales vers les marges continentales (des temps récents aux temps anciens); (2) les sections verticales à terre ou dans des sondages traversant des séries sédimentaires immergées. Les premiers (planchers océaniques) donnent des informations plus approximatives que les seconds en raison de la difficulté de leur calibration stratigraphique. La corrélation entre les deux séries de données reste encore à éprouver localement. Une échelle magnétique du Méso et du Cénozoïque est proposée en fonction des données les plus récentes. Quelques exemples des possibilités d'application de la méthode, en tant que moyen de corrélation très précis, sont proposés.

(Manuscript received 30-10-1980)

Section II

Isotopic dating

Numerical Dating in Stratigraphy
Edited by G. S. Odin

6
The physical decay constants

NOËL H. GALE

1 INTRODUCTION

The preceding section has shown the various possibilities for correlation of stratigraphic levels. The present chapter is part of a Section in which the various methods and techniques of numerical dating of these stratigraphic levels will be discussed in the light of the most recent data.

Central to the task of providing a numerical time calibration of the stratigraphic column is the problem of selecting a method of assigning absolute numerical dates to the emplacement of rocks whose stratigraphic position in the column is well established. This problem reduces in part to that of finding a reliable chronometer with which to measure absolute geological time intervals. Any accurate chronometer must depend on a mechanism which operates at a known rate and which is linked to a recording system in which events marking the beginning and end of the time interval are clearly and sharply recorded. The rate of the mechanism must not be affected by external factors (such as temperature, pressure, etc.) nor by the beginning and ending events themselves, and the recording system must either not record other events occurring within the time interval of interest or its record of the 'start' and 'stop' events must not be disturbed by such other events.

The history of attempts to find a reliable chronometer for the measurement of geological time has been given elsewhere (Holmes, 1937; York and Farquhar, 1972; Harper, 1973; Faure, 1977; and this volume). Suffice it to say here that no reliable chronometer was found until the discovery of natural radioactivity around the turn of the century by Becquerel and the Curies; shortly after that discovery, in 1904–1907, both Rutherford and Boltwood proposed that the ages of uranium minerals could be measured by the amount of the daughter elements helium or lead that had accumulated in them by the radioactive transmutation of uranium over geological time.

Radioactive decay was shown by Rutherford (1900) to follow an exponential law, whereby if by P we denote the number of parent atoms present at the time t, then:

$$\frac{dP}{dT} = -\lambda P. \tag{1}$$

In this equation λ is a constant characteristic of the particular radioactive nuclei concerned, and is called the decay constant or the disintegration constant. Integration of this equation yields

$$P = P_0 \exp(-\lambda t) \tag{2}$$

where P_0 is the initial number of parent atoms at time zero. If we define the half-life, T, as the time at which $0.5P_0$ atoms remain, then

$$\lambda T = \ln 2. \tag{3}$$

If we note that the number D of daughter atoms at the time t is given by $D = P_0 - P$, then equation (2) may be rearranged to give the fundamental equation of geochronology:

$$t = \frac{1}{\lambda} \ln \left| 1 + \frac{D}{P} \right|. \tag{4}$$

This equation demonstrates that, given the value of λ, the determination of a geological age reduces in principle to the determination of the amounts of parent and daughter isotopes present in the rock or mineral being dated.

In practice matters are rarely so simple, but it is not the purpose of this chapter to discuss such matters as the use of Rb–Sr isochrons or the U–Pb concordia curve, etc., which are adequately discussed elsewhere (see, for example, Faure 1977; Gale and Mussett, 1973). Here the emphasis is on the accurate determination of the decay constants λ, but first we must mention that equation (3) was derived assuming that λ is a constant and that the only alterations in the amounts of daughter or parent in the system are due to radioactive decay. Neither assumption is entirely trivial. The assumption that no gain or loss of parent or daughter has occurred other than by radioactive decay is always open to question, and it is part of the objective of subsequent chapters, to discuss how one can detect and overcome lack of closed system behaviour so as to be able to derive reliable dates for various rock types and radioactive decay schemes.

That the decay constant λ for a particular radioactive isotope is indeed a constant, entirely unaffected by changes in the environment of the decaying atoms, has long been accepted and justified by noting that radioactive decay is a property of the atomic nucleus. That is true for decay by emission of an α-particle, a γ-ray or a nuclear β-particle, but both the electron capture and internal conversion modes of decay involve extranuclear electrons and may *in principle* be affected by changes in the electron density in the vicinity of the nucleus. Several studies have shown that the decay constants of ^{7}Be,

^{99m}Tc and ^{131}Ba increase very slightly (a few tenths of a percent) when these atoms are subjected to pressures in excess of about 100 kbar (see, for instance, Hensley *et al.*, 1973). Of the natural radioactivities used in geochronology only ^{40}K decays partly by electron capture, and there is no evidence that potassium now in the earth's crust has been subjected to sufficiently high pressures for long enough to affect the amount of radiogenic ^{40}Ar produced and some evidence that the ^{40}K electron capture process is less affected than ^{7}Be by external conditions (York and Farquhar, 1972). The possibility that the fundamental constants of nature (i.e. the gravitational constant G, the radioactive decay constants, etc.) may be functions of time for cosmological reasons has been seriously considered (Dirac, 1939; Dicke, 1959), but the consensus of opinion is that the largest conceivable variation is too small to produce measurable discrepancies in geological ages based on the hypothesis of constant λs.

Three major criteria have to be satisfied by a radioactive isotope if it is to be useful as a geochronometer in calibrating the stratigraphic column. (1) The half-life of the radioactive isotope must be approximately of the order of the age of the earth, 4.5×10^9 a. (2) The radioactive parent isotope must be reasonably abundant in terrestrial rocks. (3) Significant enrichments of the daughter isotope must occur, i.e. unradiogenic atoms of the daughter isotope must not be overabundantly represented in the chronometer used. In practice this limits the naturally radioactive isotopes useful as geochronometers to ^{40}K, ^{87}Rb, ^{238}U, ^{235}U, ^{232}Th and ^{147}Sm. In the numerical calibration of the stratigraphic column it is necessary to use more than one radioactive decay scheme to provide the geochronometric data. It is therefore vital that the decay constants for each radioactive isotope used are accurately and absolutely known if a consistent and absolute numerical time scale is to be developed. (The fission decay constant of ^{238}U is considered below by D. Storzer and G. Wagner.)

2 DECAY CONSTANTS OF ^{238}U, ^{235}U, ^{232}Th

These isotopes are the basis of the U, Th–Pb isotopic dating method (embracing Th–Pb, U–Pb and Pb–Pb isochrons, the concordia diagram, etc.), the U–He method being little used. The complexities of the decay series which intervene between the decay of the parent atoms and the formation of the stable end products (isotopes of lead) will not be discussed here (see Evans, 1955). Instead the decay series are summarized as:

$$^{238}_{92}\text{U} \rightarrow {}^{206}_{82}\text{Pb} + 8\,^{4}_{2}\text{He} + 6\beta^- + 47.4\ \text{MeV/atom}$$

$$^{235}_{92}\text{U} \rightarrow {}^{207}_{82}\text{Pb} + 7\,^{4}_{2}\text{He} + 4\beta^- + 45.2\ \text{MeV/atom}$$

$$^{232}_{90}\text{Th} \rightarrow {}^{208}_{82}\text{Pb} + 6\,^{4}_{2}\text{He} + 4\beta^- + 39.8\ \text{MeV/atom}$$

Direct physical determinations of decay constants are usually made by measuring the rate of decay (dP/dt) of a known quantity P of the radioactive isotope in question. From equation (1) the decay constant is then given by

$$\lambda = \frac{(dP/dt)}{P} \tag{5}$$

the specific activity of the sample. Such measurements are technically difficult, particularly when accuracies of better than ±1% are required.

In principle, the determination of the decay constants for ^{238}U, ^{235}U and ^{232}Th simply requires the determination of the emission rate of α-particles from a known weight of isotope. In practice, for uranium, determination of the weight requires: (1) chemical analysis of the amount of uranium present, (2) tests for the presence of other interfering elements, (3) mass spectrometric analysis to determine the fraction of uranium present in the form of the desired isotope, and (4) a sample-preparation technique which ensures that each sample contains an accurately known weight. Determining the α-particle emission rate requires that: (1) each sample be counted with a precisely known counting efficiency, (2) each sample be counted long enough to make negligible the overall statistical counting error, and (3) an α-particle energy analysis be made to determine the fraction of the total α-activity derived from the isotope of interest. Propagation of errors requires a very high accuracy indeed in each of these steps if the error in the final decay constant is to be better than ±1%.

Most of the earlier measurements of the decay constants of ^{238}U and ^{235}U were made on samples of natural uranium by α-particle counting. Such measurements are difficult because natural uranium also contains ^{234}U, which contributes about as much α-activity as ^{238}U, whilst the α-activity of ^{235}U in natural uranium is only about 2% of that due to ^{238}U and ^{234}U. For ^{238}U, Kienberger (1949) attempted to avoid these problems, first by combining a measurement of the specific α-activity of enriched ^{234}U with isotopic determinations of the ^{234}U content of natural uranium, and secondly by another experiment in which he determined the α-activity of highly enriched ^{238}U containing very small amounts of ^{234}U and ^{235}U. Most of the earlier measurements were vitiated by uncertain corrections for sample self-absorption and α-particle back-scattering. It is therefore surprising that a tabulation (Table 1) of the available measurements of the half-life of ^{238}U shows a very close agreement between the results of Kovarik and Adams (1932; 1955), Curtiss *et al.* (1941) and Kienberger (1949), a feature which for a long time suggested that this half-life was known with an error less than 1% and caused a value of 4.51×10^9 years to be adopted by geochronologists (Doe, 1970).

A tabulation of the measurements of the half-life of ^{235}U shows the earlier measurements not to be in good agreement, yet, since the ^{235}U

Table 1. Measurements of ^{238}U half-life.

Reference	Counter	Material	Measured specific activity (dis./mn × mg)	Specific activity of ^{238}U (dis./mn × mg)	^{238}U half-life (units of 10^9a)
Kovarik and Adams (1932; 1955)	Grid collimator ion chamber	Natural U	1503 ± 6	739.7 ± 3.0	4.508 ± 0.018
Schiedt (1935)	Intermediate-geometry ion chamber	Natural U	1517 ± 15	747 ± 8	4.46 ± 0.05
Curtiss *et al.* (1941)	2π ion chamber	Natural U	1501 ± 3	738.7 ± 1.5	4.514 ± 0.009
Kienberger (1949)	2π ion chamber	Natural U	1502.0 ± 1.5	739.2 ± 0.7	4.511 ± 0.005
Kienberger (1949)	2π ion chamber	^{238}U	742.7 ± 1.6	742.7 ± 1.6	4.890 ± 0.010
Leachman and Schmitt (1957)	2π ion chamber	^{238}U	728.9 ± 4.8	728.9 ± 4.8	4.57 ± 0.03
Steyn and Strelow (1959)	Liquid scintillator	Natural U	1519.7 ± 2.4	748.1 ± 1.2	4.457 ± 0.007
Jaffey *et al.* (1971)	Intermediate-geometry proportional counter	^{238}U	746.19 ± 0.41	746.19 ± 0.41	4.4683 ± 0.0020

Table 2. Measurements of ^{235}U half-life.

Reference	Method	Material	Measured activity ratio* $\frac{A_{235}}{A_{238}} = R$	Specific activity of ^{235}U†	^{235}U half-life (units of 10^8a)
Nier (1939)	R measured from Pb/U ratios	Natural uranium	0.046 ± 0.001	4800 ± 100	7.04 ± 0.15
Sayag (1951)	R measured by ion chamber energy analysis	Natural uranium	0.0467 ± 0.0017	4870 ± 180	6.94 ± 0.25
Fleming *et al.* (1952)	Specific activity	Enriched ^{235}U		4740 ± 100	7.12 ± 0.16
Knight (1950)	Specific activity	Enriched ^{235}U		4753 ± 100‡	7.10 ± 0.16
Würger *et al.* (1957)	R measured by ion chamber energy analysis	Natural uranium	0.04677 ± 0.00073	4880 ± 80	6.92 ± 0.11
Deruytter *et al.* (1965)	R measured with a Si detector	Natural uranium	0.04645 ± 0.00060	4845 ± 50	6.97 ± 0.07
White *et al.* (1965)	Specific activity	Enriched ^{235}U		4741 ± 60	7.12 ± 0.09
Banks and Silver (1966)	R measured from Pb/U ratios	Natural uranium		4809^{-20}_{+50}	$7.022^{-0.029}_{+0.073}$§
Jaffey *et al.* (1971)	Specific activity	Enriched ^{235}U		4798.1 ± 3.3	7.0381 ± 0.0048

* All activity ratios measured by energy-pulse analyses (i.e. except Nier) have been corrected to the value 87.4% for the fraction of the ^{235}U activity not hidden by ^{235}U or ^{234}U peaks.
† Measured for enriched uranium; calculated for experiments with R measurement.
‡ Corrected as in Fleming *et al.* (1952).
§ Corrected using the value of $T_{1/2}(^{238}U)$ obtained by Jaffey *et al.* (1971).

half-life is shorter, decay has occurred over several half-lives for minerals of great age, and the need for accuracy is greater than for ^{238}U (Table 2).

It was the need to improve the accuracy of our knowledge of the half-life of ^{235}U which prompted Jaffey *et al.* (1971) to make careful measurements both of that and of the half-life of ^{238}U. Their work was characterized by the most painstaking attention to detail. Samples of ^{238}U and ^{235}U isotopes enriched to greater than 99.98% of the respective isotope were molecular-plated onto discs and α-counted in a special intermediate geometry α-particle proportional counter with geometry less than 2π, in order to obviate back-scattering and sample absorption effects. The isotopic composition (^{238}U, ^{236}U, ^{235}U, ^{234}U, ^{233}U) of the enriched isotope samples was determined by high-precision mass spectroscopy and the α-particle energy spectrum was determined for each sample with a silicon surface barrier detector. The concentration of uranium in the solutions used for preparing α-counting samples was determined by a very accurate titration method. Other precautions and methodological details are described at length in their paper (Jaffey *et al.*, 1971), and a critical discussion is given of previous measurements.

For ^{238}U the half-life measured by Jaffey *et al.* (see Table 1) is accurate to ±0.05% but is nearly 1% lower than the previously accepted value. The close agreement between the earlier results of Kovarik and Adams (1932; 1955), Curtiss *et al.* (1941) and Kienberger (1949) seems to be fortuitous, and is reminiscent of the similar close agreement of pre-World War II measurements of the velocity of light, which subsequently proved to be many standard deviations away from later, more accurate measurements. For ^{235}U the measurement of Jaffey *et al.* (see Table 2) is accurate to ±0.07%. There is no doubt that the half-lives measured by Jaffey *et al.* (1971) should be universally adopted in geochronology, and this has been officially recommended by the IUGS Subcommission on Geochronology (Steiger and Jäger, 1977).

Table 3 summarizes the available direct physical measurements of the

Table 3. Measurements of the half-life of ^{232}Th.

Method	$T_{1/2}(\times 10^{10}a)$	Reference
α-counting of natural Th	1.39 ± 0.03	Kovarik and Adams (1938)
Counting of 2.62 MeV γ's, ^{208}Tl	1.42 ± 0.07	Senftle *et al.* (1956)
Nuclear emulsion	1.39 ± 0.03	Picciotto and Wilgain (1956)
	1.401 ± 0.008	Le Roux and Glendenin (1963)

half-life of ^{232}Th. The agreement is good, but the value of Le Roux and Glendenin is recommended since their measurement seems technically superior to the others.

3 HALF-LIFE OF ^{87}Rb

Amongst the half-lives important for geochronology that of ^{87}Rb has for a long time been the most poorly known. Until about 1977 most geochronologists were using a half-life of either 5.0 or 4.7×10^{10} years, which could not be regarded as a very satisfactory situation. The fundamental difficulty in direct physical determination based on counting experiments is that of counting accurately the large number of low-energy β-particles emitted by ^{87}Rb. ^{87}Rb decays to the ground state of ^{87}Sr by a third forbidden β transition (see Evans, 1955, p. 209, for an explanation); this is reflected in the energy spectrum of β-particles, which shows no relative maximum but increases monotonically as the energy decreases. The end-point energy is about 275 keV, but about 20% of the β-particles emitted have energies <15 keV.

The main problem in disintegration rate measurements is the detection of the low-energy β-particles. With scintillation counters they are masked by noise up to some keV; for Rb-doped scintillation crystals, errors can arise when the rubidium is not perfectly dissolved (Neumann and Huster, 1974). Proportional counter tubes can be made sensitive to β-particles of very low energies, but with Rb one has to use solid sources, and self- and mount-absorption and back-scattering effects are major problems.

We may safely ignore the direct physical counting measurements prior to 1959 (they are tabulated by Hamilton, 1965); in that year Flynn and Glendenin published an account of experiments to measure the half-life and β-spectrum of ^{87}Rb using a liquid scintillation counter method. They quoted a value of $(4.70 \pm 0.10) \times 10^{10}$ years for the half-life, a figure adopted by one camp of geochronologists for nearly all the following two decades. In their work the count rates from a liquid scintillator doped with Rb-octate were measured at different discriminator settings. Noise and spurious counts arising below a certain bias level forced extrapolation of the counting curves to zero bias in order to obtain the true activity. However, the extrapolation procedure used is not based on physical principles, and the extrapolation errors given were quite arbitrary. Moreover, the shape of the β-spectrum which they observed is not in accord with the more careful measurements of Rüttenauer and Huster (1973).

A different, indirect approach had been taken earlier by Aldrich *et al.* (1956). Aldrich *et al.* quoted a half-life of $(5.0 \pm 0.2) \times 10^{10}$ years by adjusting the Rb–Sr ages in pegmatite minerals to agree with ^{235}U–^{207}Pb ages from the same rocks, using the half-life for ^{235}U measured by Fleming *et al.*

(1952). This was the basis for the ^{87}Rb half-life used by the other camp of geochronologists for the next 20 years, but there was no doubt that a reliable direct physical measurement of the half-life was much to be desired.

An alternative direct approach to the problem is to allow a sample of strontium-free rubidium to decay for a known time and to measure the amount of accumulated radiogenic ^{87}Sr with a mass spectrometer using the isotope dilution technique. For a decay time less than about 20 years, the decay constant is given as

$$\lambda = |^{87}\mathrm{Sr}|^* \, |^{87}\mathrm{Rb}|^{-1} \, t^{-1}$$

where $|^{87}\mathrm{Sr}|^*$ is the amount of radiogenic strontium accumulated over a time t by a known quantity of rubidium $|^{87}\mathrm{Rb}|$. In 10 years, 20 g of $RbClO_4$ would be expected to contain about 0.33 ng of radiogenic ^{87}Sr, a quantity which it is possible to measure using isotope dilution and ion-counting mass-spectrometric techniques. The major experimental problems are: (1) the preparation of a sufficiently pure sample of rubidium, free of strontium; (2) after the sample has decayed for 10–20 years, the separation of trace amounts of strontium from the rubidium to prevent isobaric interference in the mass spectrum; (3) the quantitative determination of the amounts of rubidium and strontium.

In 1956, McMullen *et al.* (1966) prepared four batches of purified $RbClO_4$ by conversion of about 1 kg of RbCl to the perchlorate by precipitation with perchloric acid, followed by successive recrystallizations. After seven years they measured the amount of radiogenic strontium produced in portions of each batch of $RbClO_4$, finding an average value for the half-life of $(4.718 \pm 0.039) \times 10^{10}$ a. Their work was vitiated by the failure to monitor mass fractionation during the mass-spectrometric measurements, so this value may be subject to systematic error.

It is interesting to note that if one recalculates the old geochronological comparison of Rb–Sr and U–Pb ages of pegmatities made by Aldrich *et al.* (1956), using the new ^{235}U and ^{238}U half-lives measured by Jaffey *et al.* (1971), then the comparison with ^{235}U–^{207}Pb ages yields a mean ^{87}Rb half-life of $(4.87 \pm 0.21) \times 10^{10}$ a, whilst the comparison with ^{238}U–^{206}Pb ages yields $(4.83 \pm 0.22) \times 10^{10}$ a (Table 4). There was already a suggestion here that none of the extant half-lives for ^{87}Rb was satisfactory, but the errors are of course too large for certainty.

A new determination of the ^{87}Rb half-life by absolute counting was made by Neumann and Huster (1972; 1974; 1976), in which back-scattering was removed by counting in 4π geometry and adsorption corrections were minimized by using very thin sources and backings, whilst enhanced absorption in non-uniform layers was avoided by using RbCl sources prepared by vacuum distillation and kept in a dry atmosphere. To avoid counting losses and spurious counts, attention was paid to ensuring sufficient electrical

Table 4. Comparison of U–Pb and Rb–Sr data on pegmatites.*

Pegmatite locality	Mineral	U–Pb Ages† $^{206}Pb/^{238}U$ (Ma)	$^{207}Pb/^{235}U$ (Ma)	Mineral	$[^{87}Sr]^*/Rb$	$T_{1/2}$ ^{87}Rb‡ (10^{10} a)	$T_{1/2}$ ^{87}Rb§ (10^{10} a)
Bikita quarry, S Rhodesia	Monazite	2616	2642	Lepidolite	0.0380	4.86	4.91
Viking lake, Saskatchewan	Uraninite	1773	1804	Biotite	0.0270	4.61	4.69
Bob Ingersoll mine, S. Dakota	Uraninite	1565	1577	Lepidolite	0.0239	4.59	4.63
				Muscovite	0.0244	4.50	4.53
				Microcline	0.0244	4.89	4.93
Cardiff uranium mine, Ontario	Uraninite	1011	1006	Biotite	0.0140	5.03	5.01
Fission mine, Ontario	Uraninite	1030	1035	Biotite	0.0141	5.14	5.16
Chestnut Flat mine, Carolina	Uraninite	371.5	374.7	Muscovite	0.00515	5.01	5.06
				Microcline	0.00535	4.82	4.87

* Primary U–Pb and Rb–Sr data from Aldrich *et al.* (1956)
† U–Pb ages recalculated using the U half-lives measured by Jaffey *et al.* (1971).
‡ ^{87}Rb half-life calculated from the $^{206}Pb/^{238}U$ age.
§ ^{87}Rb half-life calculated from the $^{207}Pb/^{235}U$ age.

conductivity of the backing foils; counting statistics were improved by using large-area sources of enriched ^{87}Rb and effective background reduction. The amount of ^{87}Rb in the sources was determined using a microbalance and checked by isotopic and chemical analysis. Attention was paid to avoiding afterpulsing effects and source contamination by radioactivities other than ^{87}Rb. Their result for the half-life was $(4.88^{+0.06}_{-0.10}) \times 10^{10}$ years. In their 1976 paper, Neumann and Huster critically evaluated all previous direct counting measurements and concluded that the most reliable value for the ^{87}Rb half-life derived from absolute β-counting was their own result using a 4π proportional counter.

In 1977 Davis *et al.* reported a new value for the half-life based upon the direct method of measuring the accumulated radiogenic ^{87}Sr in a sample of a pure rubidium salt. They worked again on the four batches of purified $RbClO_4$ prepared by McMullen *et al.* in 1956. Davis *et al.* (1977) employed careful chemistry to separate the radiogenic ^{87}Sr from the $RbClO_4$ in a form relatively free from rubidium, determined the amount of ^{87}Sr by mass-spectrometric isotopic dilution using NBS SRM-988 99.89% ^{84}Sr spike, measured the strontium isotopic ratios 84/88, 87/88 and 86/88 to enable mass fractionation corrections to be made, monitored ^{85}Rb to correct for isobaric interference at mass 87, and measured ^{87}Rb in the original samples by mass-spectrometric isotope dilution using an enriched ^{87}Rb spike and making corrections for fractionation. Ion beams were detected with a Spiraltron electron multiplier used in an ion counting mode and a mass spectrometer resolution of 600 permitted the resolution of all hydrocarbon peaks in the strontium mass range. The result of their work was a half-life of $(4.89 \pm 0.04) \times 10^{10}$ years. This value is in very satisfactory agreement with the value of $(4.88^{+0.06}_{-0.10}) \times 10^{10}$ years obtained from the direct β-counting work of Neumann and Huster (1974; 1976); both measurements are in accord with the much less accurate geochronological comparison of permatite ages recalculated above (see Table 3) using the Jaffey *et al.* (1971) U-decay constants. These measurements justify the acceptance of the IUGS recommendation to use $\lambda(^{87}\text{Rb}) = 1.42 \times 10^{-11} \text{a}^{-1}$ in geochronology—see Steiger and Jäger (1977).

4 HALF-LIFE OF ^{40}K

The decay scheme (Figure 1) of ^{40}K is complex; it decays by both electron capture and positron emission to ^{40}Ar, and by negatron emission to ^{40}Ca. Approximately 10% of all decays proceed by orbital electron capture to the first excited state of ^{40}Ar, which then falls to the ground state by emitting a 1.46 MeV γ-ray. In the other major branch ^{40}K decays, with the emission of negatrons, to the ground state of ^{40}Ca. In addition there are two weak transitions (one by electron capture, one by positron emission) direct to the ground state of ^{40}Ar.

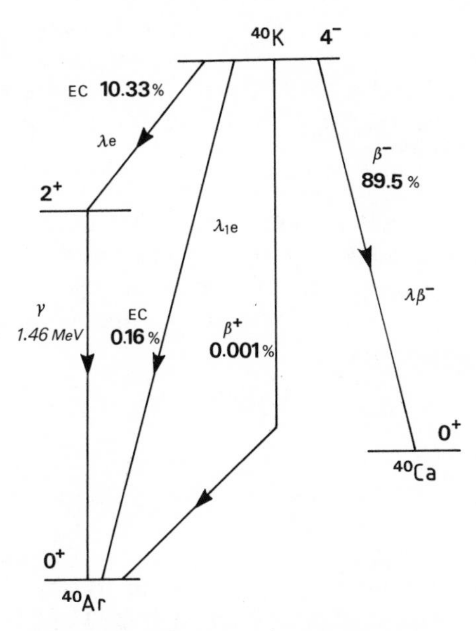

Figure 1. Decay scheme of ^{40}K isotope.

The decay leading to ^{40}Ca is accessible to direct β-counting experiments. The electron capture branch to the first excited state of ^{40}Ar has to be investigated by counting the associated 1.46 MeV γ-rays, but the electron capture branch to the ^{40}Ar ground state, because of the difficulty of counting the directly associated low-energy X-rays or Auger electrons, has to be inferred from the measured positron decay.

In a familiar notation, we may express a particular partial decay constant λ_i of the ^{40}K present in natural potassium in terms of the associated measured activity dn_i/dt (disintegrations per second per gramme: d.p.s.g.$^{-1}$ of natural potassium) by the relation:

$$\lambda_i = \frac{dn_i}{dt} \cdot \frac{A}{f} \cdot \frac{Y}{N} \text{a}^{-1} \tag{6}$$

where f is the atomic abundance of ^{40}K in natural potassium, A is the atomic weight of potassium, N is Avogadro's number $= 6.02252 \times 10^{23}$ atoms per mole on the ^{12}C scale, and Y is the number of seconds in a mean solar year $= 3.15569 \times 10^7$ s (Weast and Selby, 1966).

We may denote by λ_e and λ_{β^-} the partial decay constants for the electron capture branch to the excited state of ^{40}Ar and the ground state of ^{40}Ca respectively. Until recently these decay constants, accessible to direct measurements of the specific γ- and β-activities of natural potassium, were the only ones taken into account in geochronology. Earlier measurements of these specific activities were not in very good agreement, but in 1958 they were reviewed by Aldrich and Wetherill, who recommended values of $\lambda_e = 0.585 \times 10^{-10}$ a^{-1} and $\lambda_{\beta^-} = 4.72 \times 10^{-10}$ a^{-1}. These values remained in general use until a reappraisal of the decay constants was made by Beckinsale and Gale in 1969.

These authors first reviewed the more recent determinations of the specific β- and γ- activities of natural potassium, making an attempt to assess which determinations were likely to be more reliable in the light of modern knowledge of techniques of absolute low-level β- and γ- counting. The weighted mean β- and γ- activities computed from the six selected reliable determinations were:

$$\frac{dn_{\beta^-}}{dt} = 28.27 \pm 0.05 \text{ d.p.s.g.}^{-1} \text{ K}$$

$$\frac{dn_\gamma}{dt} = 3.26 \pm 0.02 \text{ d.p.s.g.}^{-1} \text{ K.}$$

If these activities are inserted into equation (6) together with the latest data for the atomic weight of potassium |39.098304 ± 0.000058| and for the atomic abundance of ^{40}K |(1.167 ± 0.0004) × 10^{-4}|, taken from the work of

Garner *et al.* (1975), there result:

$$\lambda_{\beta^-} = (4.963 \pm 0.009) \times 10^{-10}\,\mathrm{a}^{-1}$$
$$\lambda_e = (0.5723 \pm 0.0035) \times 10^{-10}\,\mathrm{a}^{-1}.$$

For the best value of the ratio of positron decay to negatron decay in the decay of ^{40}K, Beckinsale and Gale took the weighted mean of the three best determinations (Tilley and Madansky, 1959; Engelkemeir *et al.*, 1962; Leutz *et al.*, 1965), yielding

$$\beta^+/\beta^- = (1.15 \pm 0.13) \times 10^{-5}.$$

The corresponding activity of this very weak β^+ branch to the ground state of ^{40}Ar is $(3.25 \pm 0.37) \times 10^{-4}$ d.p.s.g.$^{-1}$ K.

For their estimate of the activity of the electron capture branch direct to the ground state of ^{40}Ar, Beckinsale and Gale used the theoretical calculation of Brosi and Ketelle (see Engelkemier *et al.*, 1962) that it should be 155 times as intense as the positron decay branch for a maximum β^+ energy including rest mass of 1.961 Mc^2. On this basis Beckinsale and Gale approximated the electron capture activity direct to the ground state as $(5 \pm 1) \times 10^{-2}$ d.p.s.g.$^{-1}$ K; with the new data for potassium of Garner *et al.* (1975), this yields a decay constant of $\lambda'_e = (8.78 \pm 1.76) \times 10^{-13}\,\mathrm{a}^{-1}$.

The best estimates of the decay constants are therefore:

$$\lambda_{\beta^-} = (4.963 \pm 0.009) \times 10^{-10}\,\mathrm{a}^{-1}$$
$$\lambda_e + \lambda'_e = (0.581 \pm 0.004) \times 10^{-10}\,\mathrm{a}^{-1}.$$

whilst the total decay constant is $(5.544 \times 0.010) \times 10^{-10}\,\mathrm{a}^{-1}$, corresponding to a half-life of 1.250×10^9 a. Clearly the β^+ transition is so excedingly weak as to be unimportant in geochronology.

5 HALF-LIFE OF ^{147}Sm

^{147}Sm decays by emitting an α-particle of energy 2.314 MeV to ^{143}Nd. Though the half-life of ^{147}Sm, about 10^{11} years, is long, it is sufficiently short to produce small but readily measurable differences in ^{143}Nd abundance over time spans of 10^8 years or more, even though the fractionation between Sm and Nd is rather limited.

Though the use of the ^{147}Sm–^{143}Nd method tends to be laborious in that it requires very high mass-spectrometric precision to measure the small changes in ^{143}Nd abundance, it may have some advantages in that it may be less subject to disturbance by post-crystallization alteration than the Rb–Sr or U–Pb systems. This was early demonstrated for meteorites by the work of Unruh *et al.* (1977) on Pasamonte, and has been demonstrated for terrestrial rocks by the work of Hamilton *et al.* (1978; 1979). The method has not yet

Table 5. Summary of the most precise measurements of the half-life of ^{147}Sm.

Method	$T_{1/2}(\times 10^{11}$ a)	Reference
Liquid scintillation	1.05 ± 0.02	Wright *et al.* (1961)
Liquid scintillation	1.04 ± 0.03	Donhoffer (1964)
Liquid scintillation	1.08 ± 0.02	Valli *et al.* (1965)
Cylindrical ionization chamber	1.06 ± 0.02	Gupta & MacFarlane (1970)
Weighted mean 1.060 ± 0.008 $(1\sigma) \times 10^{11}$ a		
Corresponds with $\lambda = 6.539 \times 10^{-12}$ a^{-1}		

been used much for time-scale work, but one may instance Sm–Nd dates of 508 ± 6 Ma and 501 ± 13 Ma on the Bay of Islands gabbros, Newfoundland (Jacobsen and Wasserburg, 1979).

Table 5 gives a summary of the most precise values of $T_{1/2}$ for ^{147}Sm obtained by direct counting within the last 16 years. The agreement between measurements is very good, and the weighted mean value given in Table 5 is recommended for general use in Sm–Nd dating. It has been used by Lugmair and his colleagues since the inception of Sm–Nd dating.

In order to avoid systematic differences between different laboratories in age interpretations of Sm–Nd isochrons, it is also necessary that standardization should be introduced into the matter of selection of the principal reference values for Sm and Nd isotope ratios used to correct for instrumental mass fractionation during mass spectrometry. There has been a regrettable lack of agreement on this point between the principal laboratories involved. Thus Lugmair at La Jolla (Lugmair *et al.*, 1975; 1976; Lugmair and Marti, 1978) normalizes to the equivalent of ^{142}Nd/^{146}Nd$= 1.5817$, or ^{148}Nd/^{144}Nd$= 0.241572$, O'Nions *et al.* at Lamont-Doherty (Hamilton *et al.*, 1978; 1979) normalize to ^{146}Nd/^{144}Nd$= 0.7219$, whilst Wasserburg *et al.* at the California Institute of Technology (De Paolo and Wasserburg, 1976) normalize to ^{150}Nd/^{142}Nd $= 0.2096$.

The Wasserburg normalization is not equivalent to the others. Our best estimate of the initial ^{143}Nd/^{144}Nd isotopic composition at the time of planet formation comes from the isochron intercept value for the meteorite Juvinas. The Wasserburg normalization results in an initial Nd ratio of 0.50598, whilst the La Jolla and Lamont-Doherty value is 0.50677 (Lugmair *et al.*, 1976).

In the interest of avoiding systematic differences in the Sm–Nd ages reported by different laboratories, we urge the general adoption in terrestrial geochronology of the La Jolla values of:

$$^{148}\text{Nd}/^{144}\text{Nd} = 0.241572$$

$$^{150}\text{Sm}/^{149}\text{Sm} = 0.53406.$$

6 GEOCHRONOLOGICAL COMPARISONS USING THE RECOMMENDED HALF-LIVES FOR URANIUM, RUBIDIUM, POTASSIUM AND SAMARIUM

Table 6 summarizes the recommended values for the chief parameters used in geochronology. It is recommended that nuclidic masses used in geochronology be taken from Wapstra and Gove (1971). The recommended values for U, Th–Pb, Rb–Sr and K–Ar dating conform with the recommendations of the IUGS Subcommission on Geochronology (Steiger and Jäger, 1977); those for Sm–Nd dating conform with majority practice, and it is hoped that they will be universally adopted to avoid confusion in the literature.

Table 6. Recommended constants for use in geochronology

Uranium	
$\lambda(^{238}U) = 1.55125 \times 10^{-10}\ a^{-1}$ (1)	Atomic ratio $^{238}U/^{235}U = 137.88$ (2)
$\lambda(^{235}U) = 9.8485 \times 10^{-10}\ a^{-1}$ (1)	
Thorium	
$\lambda(^{232}Th) = 4.9475 \times 10^{-11}\ a^{-1}$ (3)	
Rubidium	
$\lambda(^{87}Rb) = 1.42 \times 10^{-11}\ a^{-1}$ (4, 5)	Atomic ratio $^{85}Rb/^{87}Rb = 2.59265$ (6)
Strontium	
Atomic ratios $^{86}Sr/^{88}Sr = 0.1194$ (7)	$^{84}Sr/^{86}Sr = 0.056584$ (8)
Potassium	
$\lambda_{\beta-} = 4.962 \times 10^{-10}\ a^{-1}$ (9, 10)	Isotopic abundances (10): $^{39}K = 93.2581$ atom %
$\lambda_e + \lambda'_e = 0.581 \times 10^{-10}\ a^{-1}$ (9, 10)	$^{40}K = 0.01167$ atom %
	$^{41}K = 6.7302$ atom %
Argon	
Atomic ratio $^{40}Ar/^{36}Ar$ atmospheric $= 295.5$ (11)	
Samarium	
$\lambda(^{147}Sm) = 6.539 \times 10^{-12}\ a^{-1}$ (12)	Atomic ratio $^{150}Sm/^{149}Sm = 0.53406$ (13)
Neodymium	
Atomic ratio $^{148}Nd/^{144}Nd = 0.241572$ (13)	

(1) Jaffey *et al.* 1971; (2) Barnes *et al.*, 1972; (3) Le Roux and Glendenin, 1963; (4) Neumann and Huster, 1974; 1976; (5) Davis *et al.*, 1977; (6) Catanzaro *et al.*, 1969; (7) Nier, 1938; (8) Moore *et al.*, 1977; (9) Beckinsale and Gale, 1969; (10) Garner *et al.*, 1975; (11) Nier, 1950; (12) This article; (13) Lugmair and Marti, 1978.

All the dates calculated or recalculated in this volume will use this set of decay constants exclusively, abbreviated (ICC): International Conventional Constants.

In the past, systematic differences have been apparent betwen the U–Pb, Rb–Sr and K–Ar dating methods which are attributable to errors in the decay constants previously adopted. If we take the uranium decay constants as those best founded on absolute direct physical measurement, then it is reassuring to find further arguments in favour of a ^{87}Rb decay constant of 1.42×10^{-11} a^{-1} in the comparison of Rb–Sr and U–Pb ages in meteorites (Wetherill, 1975) and in the geochronological comparisons of Afanass'yev *et al.* (1974). A number of geochronological comparisons (given in Neumann and Huster, 1976) show that concordance to within 1.5% is obtained between Rb–Sr and K–Ar dating if the constants recommended for ^{40}K and ^{87}Rb in Table 6 are adopted. A comparison of K–Ar and Rb–Sr ages in rapidly cooled igneous rocks (Tetley *et al.*, 1976, discussed in Steiger and Jäger, 1977) shows that the ^{87}Rb decay constant estimated on the basis of the ^{40}K constants of Table 6 is $1.42 \pm 0.01 \times 10^{-11}$ a^{-1}, in excellent agreement with the present recommended value. Lastly, a comparison of total fusion ^{40}Ar/^{39}Ar, K–Ar and Rb–Sr dating of biotites from Early Palaeozoic granitoid plutons in New South Wales, Australia, showed excellent agreement when the dates were calculated with the constants of Table 6 (Tetley and McDougall, 1978).

Résumé du rédacteur

La datation numérique des roches utilise la décroissance radioactive des isotopes naturellement instables de l'écorce terrestre. La transmutation de ces isotopes radioactifs se fait à une vitesse constante propre à chaque isotope quelles que soient les conditions en ce qui concerne les désintégrations par émission de particules α ou β d'origine nucléaire et rayonnement γ; et pratiquement constante pour les autres modes.

Les isotopes utiles à la datation des roches considérées dans le présent volume sont ^{40}K, ^{87}Rb, ^{238}U, ^{235}U, ^{232}Th et ^{147}Sm; la décroissance radioactive par fission de ^{238}U est envisagée dans la contribution suivante.

Les valeurs actuellement admises pour ces vitesses de décroissance radioactive des isotopes instables sont rappelées ainsi que les données qui ont contribué à les établir.

La série de constantes proposée ici est celle qui est exclusivement utilisée pour calculer les âges discutés dans ce volume.

(Manuscript received 12-9-1980)

Numerical Dating in Stratigraphy
Edited by G. S. Odin

7

Interlaboratory standards for dating purposes

GILLES S. ODIN and 35 collaborators

This paper was documented with the collaboration of C. J. Adams, R. L. Armstrong, G. P. Bagdasaryan, A. K. Baksi, K. Balogh, I. L. Barnes, N. A. I. M. Boelrijk, F. P. Bonadonna, M. G. Bonhomme, C. Cassignol, L. Chanin, P. Y. Gillot, A. Gledhill, K. Govindaraju, R. Harakal, W. Harre, E. H. Hebeda, J. C. Hunziker, C. O. Ingamells, K. Kawashita, E. Kiss, H. Kreuzer, L. E. Long, I. McDougall, F. McDowell, H. Mehnert, R. Montigny, P. Pasteels, F. Radicati, D. C. Rex, C. C. Rundle, C. Savelli, J. Sonet, E. Welin and J. L. Zimmermann.

1 INTRODUCTION

A large number of samples are used as a means of verification of the calibration of a system of measurement. Practically all countries and sometimes individual laboratories have their own internal standard and this does not facilitate an easy comparison when a very accurate calibration is necessary as for time-scale calibration. The reason is that very few possibilities exist for interlaboratory calibration due to the difficulty in obtaining *homogeneous, abundant* and *adequately verified* material.

The first standards used are now well known. In western countries they are the biotite B 3203 or muscovite P 207, while the sample Asia 1–65 is most frequently used in eastern countries. However, today the question of availability arises (Table 1). The muscovite Bern 4 M has also been available for a long time. Unfortunately its specific behaviour has presented analysts with problems.

More recent standards are now available and these have been distributed after a preliminary study of reproducibility and homogeneity of the major components. Biotite LP-6 was distributed after a careful test of the K and Ar reproducibility and homogeneity; few data are still published. Details on the glauconitic mica GL-O, for which numerous experiments on behaviour during analysis have been done, are presented here.

Table 1. Commonly used reference material.

Name	Mineral	Initial quantity	Origin
Asia 1–65	Effusive rock	?	Igem Moscow (SSSR)
Bern 4 M	Muscovite	2.4 kg flakes	University Berne (Switzerland)
Bern 4 B	Biotite	4.5 kg flakes	University Berne (Switzerland)
B. 3203	Biotite	?	MIT (USA)
GL-O	Glauconite	9.0 kg grains 9.5 kg powder 8.5 kg reserve	CRPG Nancy (France)
LP-6 Bio 40–60#	Biotite	8.0 kg flakes	USGS Menlo Park (USA)
Mica–Fe	Biotite	45.0 kg flakes +powder	CRPG Nancy (France)
Mica–Mg	Phlogopite	20.0 kg flakes 50.0 kg powder	CRPG Nancy (France)
P 207	Muscovite	1.1 kg flakes	Depleted

Other possible standards which are available in exceptional quantity are also discussed.

2 THE FEW AVAILABLE REFERENCE MATERIALS

MIT's biotite B 3203 was one of the first minerals widely used for improvement of the calibration of the K–Ar lines. For reviews of available results see Müller (1966) and Dalrymple and Lanphere (1969). These are shown in Table 2; in spite of the fact that measurements were done when the techniques were not fully refined, the reproducibility is not bad. This is probably linked with the high quantity of radiogenic argon and low atmospheric contamination: there are no atmospheric correction problems. However, these qualities are also a drawback, since no improvement of the line is carried out with regard to the atmospheric argon contamination and this proportion of radiogenic argon is rarely obtained in the majority of samples.

In western countries the most widely used standard was probably the US Geological Survey's muscovite. According to the final compilation by Lanphere and Dalrymple (1976a), the interlaboratory standard deviation was nearly 1.2% of the K and Ar content. For argon, the pooled intralaboratory standard deviation was 1.6%, which indicates an inhomogeneity of the aliquots used. In fact, experience showed that, in spite of careful mixing of

Table 2. K and Ar in biotite B 3203.

Potassium (wt %K)	rad./tot. (%)	rad. Ar (nl/g)
7.56 ± 0.14 (1σ)	>99	388 ± 7 (1σ)

Table 3. Data on P 207: interlaboratory mean values for K and Ar (Lanphere and Dalrymple, 1972).

Potassium (wt %K)	rad./tot. (%)	rad. Ar (nl/g)	Apparent age (Ma, ICC)
8.56 ± 0.10 (1σ)	95	28.2 ± 0.3 (1σ)	82.9 ± 1.0

the bottle content, the subsampling was not always representative of the whole sample. This is linked with incomplete purification of the initial muscovite. Ingamells and Engels (1976b) calculated that, due to the inhomogeneity of the P 207, the minimum wieght needed to ensure a representative subsample at better than 1% for potassium was 1.5 g for a 98% confidence level and 0.170 g if a confidence level of only 68% was accepted. A 'good' standard (homogeneous) is generally obtained for potassium when only 5–50 mg are needed.

Moreover, the initial quantity of available sample was low: 1.1 kg (Lanphere and Dalrymple, 1965). But after six years of active life, this standard is exhausted. The characteristics of P 207 are shown in Table 3 for K and Ar.

Results were also obtained using the Rb–Sr dating method; these are shown in Table 4. The inhomogeneity of the sample and possibilities for contaminations in the laboratory lead to rather variable results. According to Jäger (1979b), the data 'do not allow to distinguish between excess common strontium due to contamination in the laboratory or inhomogeneous distribution of strontium-rich inclusions in the mica' or other minerals.

In the eastern countries, one of the most widely used samples is an effusive crushed and sieved rock: Asia 1–65. It contains 44.41 nl/g of radiogenic argon (Hamor *et al.*, 1979). It was compared with both muscovite Bern 4 M and P 207 (Afanass'yev *et al.*, 1970) and more recently with GL-O under normal conditions of temperature and humidity. K. Balogh (1980, personal communication) found a ratio

$$\frac{\text{Asia 1–65}}{\text{GL-O}} = 1.789 \pm 0.018.$$

Table 4. Rb–Sr data on muscovite P 207 according to Lanphere and Dalrymple (1972).

Rb (10^{-6} mole/g)	Common Sr (10^{-6} mole/g)	$^{87}Sr/^{86}Sr$	$^{87}Rb/^{86}Sr$	Apparent age (Ma, ICC)
9.46 ± 0.35 (1σ)	0.1045 ± 0.013 (1σ)	0.944–1.10	198–336	85.7 ± 2.4 (1σ)

Results obtained on the Berne standards have been summarized in the chapter by M. Flisch (see also NDS 200).

3 BIOTITE LP-6

Since 1974, biotite LP-6 (full designation LP-6 Bio 40–60#) has been proposed as a new mineral standard for K and Ar measurements. About 8000 g were prepared and 600 bottles of 7–8 g each were still available in October 1974. According to Ingamells and Engels (1976b), the biotite is more than 99.9% pure and the flakes 250–420 μm in size. A detailed study of the homogeneity showed that the weight of subsample necessary for obtaining a 1% sampling uncertainty was 5 mg for potassium and nearly 50 mg for argon, i.e. if quantities less than these are taken for measuring K and Ar respectively, there will be actual differences of more than 1% from one aliquot to another (Table 5).

Unlike the case of P 207 muscovite, the problem which arises is not in the purity of the separated mineral but in the unfortunate fact that there is an *inhomogeneity in the biotite itself.* According to Ingamells and Engels, LP-6 appears to be a mixing of two biotites, one with 8.10% potassium (%K) and one with 8.53%. The apparent ages of these two different biotites are of the order of 121 and 138 Ma respectively (ICC recalculated). In order to avoid this problem, these authors recommend the use of 50 mg for potassium measurements and 500 mg for argon measurements, after careful mixing of the bottle to ensure the correct representation of these subsamples. Such quantities were rarely used for argon measurements. The recommended values for this standard are potassium (%K), 8.33±0.03 and argon, 43.26±0.44 nl/g, in the conditions quoted above, i.e. careful mixing and splitting of the bottle content and sufficient quantity weighed. In these conditions, the ±

Table 5. Relative uncertainty in K and Ar measurements using LP-6 (after the system of calculation used by Ingamells and Engels, 1976).

Sample weight (g)	Relative uncertainty (%, 1σ)	Expected range in 100 analyses
Potassium		(%K)
0.1	0.22	8.27–8.38
0.01	0.71	8.15–8.47
0.005	1.0	—
Argon		(nl/g)
0.3	0.40	—
0.1	0.71	42.45–44.07
0.05	1.0	—

Table 6. Results of potassium measurements on LP-6 Bio 40–60# (wt %K) with number of analyses in parentheses (after Ingamells and Engels, 1976b; Kreuzer and Harre, Rex, personal communications). The interlaboratory mean value is 8.33, which becomes 8.37 on eliminating the three values with more than ±2% deviation from the mean; however, according to Ingamells (personal communication), the recommended value of 8.33 should be preferred. This result is the mean obtained by McDowell (Austin) after 8 measurements.

Armstrong	8.43 (3)	Katsura	8.39	Muysson	8.43 (2)
Baksi	8.37 (8)	Harre	8.30(8)	Rex	8.36 (15)
Delaloye	7.78 (4)	Krueger	8.41 (4)	Rijova	8.61
Ferrara	8.00 (2)	Leutwein	8.48 (3)	Schlocker	*8.33 (n)*
Gittins	8.42	MacIntyre	8.29	Smith	8.33
Hebeda	8.37 (6)	McBride	8.40 (10)	Turner	8.32 (*n*)
				Webb	8.34 (8)

given represents the maximum interbottle differences estimated subjectively (C. O. Ingamells, personal communication, 1980).

Some results obtained in different laboratories are shown in Tables 6 for potassium and 7 for argon. Of 19 available results for potassium, seven are at variance with the recommended value by ±1% (Table 6); of 16 results available for argon, six are at variance with the recommended value by ±1% (Table 7). This shows that the recommendations made by Ingamells and Engels were not carefully observed.

The interlaboratory mean for argon (excluding the two extreme values which are poorly reliable) is as low as 42.95 (nl/g), with a standard deviation

Table 7. Results of argon measurements on LP-6 Bio 40–60 (in nl/g): the number of analyses is shown in parentheses when available (after Ingamells and Engels, 1976; Armstrong at Yale, Bonhomme, Hebeda, Kreuzer and Harre, Montigny, Rex, personal communications, 1980; Aronson and Hower, 1976). According to the authors of the standard the recommended value (43.26) must be preferred to the interlaboratory mean of 42.95±0.43 (eliminating the two extreme values), which becomes 43.10±0.23 (±1σ; 12 labs) if we eliminate two more data (Baksi, personal communication, 1980); see further data in NDS 196. After drawing of this table new results were proposed by F. McDowell: mean of 13 measurements on 90–120 mg of samples = 43.73±0.037.

Armstrong	43.24 (2)	Kreuzer RSS9 1977	43.22 (6)	MacIntyre	42.12 (3)
Aronson	40.08 (3)	Kreuzer RSS9 1978	42.69 (2)	McBride	42.78 (2)
Baksi	43.12 (8)	Kreuzer CH4 1978	43.32 (6)	Montigny	42.83 (7)
Bonhomme	42.7 (1)	Kreuzer CH4 1979	43.22 (4)	Radicati	43.03 (1)
Giuliani	42.09 (5)	Kreuzer CH4 1980	42.98 (1)	Rex	43.49 (3)
		mean	43.18 (19)		
Hebeda 1972	43.10 (9)	Krueger	43.31 (4)	Turner	43.22 (4)
Hebeda 1980	43.17 (9+3)	Leutwein	45.38 (2)	Webb.	43.17 (6)

of 0.43 (nl/g) for 14 laboratories. Eliminating two more low values, which deviate from the mean by nearly 2%, the interlaboratory mean becomes 43.10 (nl/g) for 12 laboratories and is nearer the recommended value (see more data in NDS 196). The apparent age calculated using the recommended value is 127.7 ± 1.4 Ma, ICC.

In conclusion, it is necessary to emphasize the fact that, as a result of a complete preliminary study of the sample homogeneity, specific conditions of analysis have been determined; it is absolutely essential that analysts keep in mind these limitations when using standard minerals for calibrating their line or for improving their calibration. Unfortunately, in the present case a large amount of mineral is needed and possibly this will limit the use of this standard, especially for the 39/40 technique, in which small quantities of monitor are commonly used.

4 REFERENCE MATERIAL 'GLAUCONITE GL-O'

4.1 Collection

The reference material GL-O was separated from the basal Cenomanian found in the cliff at Cauville (Normandy). This glauconitic horizon was selected in order to obtain a large quantity of evolved to highly-evolved glaucony of known history and good homogeneity. The aim in establishing this standard was initially to replace the depleted P 207 and to obtain a more homogeneous standard which was clean and could be measured quickly.

The original rock was collected by G. and Ch. Odin in the cliff about 20 m above the beach. Nearly 30 cm of the vertical outcrop were eliminated before collection of 300 kg of a calcareous, highly glauconitic sand (see location in Odin and Hunziker, this volume).

4.2 Purification

The process of preparation is shown in Figure 1. The grains were never heated at more than 65°C and were cleaned in 200 g portions by ultrasonic agitation. Twenty-seven kg of pure glaucony, 160–500 μm in size (non-attractable at an intensity of 0.35 A and attractable at 0.45 A were obtained in this way. Microscopic investigations do not show any common impurity. Six fractions of 20 g were tested for grains heavier than glaucony: only 0–36 thin mica flakes, 2–3 tourmalines, 0–2 staurotides and 1–5 opaques were found. The purity was then estimated as better than 99.95% with the exception of calcite which has been found in cracks in a few pellets.

The initial substrates of glauconitization were themselves probably calcitic debris and were completely replaced by glauconitic minerals. The important thing here is that, in any case, the initial ratios of possibly polluting elements (Sr and Ar) were in equilibrium with the same environment as the glauconitic minerals themselves, i.e. the sea-water.

4.3 Naming

X-ray diffraction showed that the glauconitic minerals were almost micaceous: the mineral was named *glauconite* and the reference sample was then named '*glauconite GL-O*' which, in the new Nancy code, means GLauconite prepared by Odin.

4.4 Conditioning

Mixing and preliminary homogenization were carried out at Nancy by hand mixing and then dividing into six portions with a plastic divider sampler. This was done twice.

Portions 1 and 4 (9 kg of grains) were then automatically divided in the same manner until 442 bags of 20–25 g were obtained: GL-O g.

Portions 2 and 5 (9.5 kg) were finely crushed for a period of less than 50 s in a tungsten carbide vibrocrusher and thus automatically divided to obtain 460 bags of 20 g of powder: GL-O p.

Portions 3 and 6 were kept in reserve (Figure 1).

4.5 Tests for homogeneity

Several tests for homogeneity were done. Ten bags of GL-O g and 10 bags of GL-O p were opened and 0.5 g was analysed from each. In one bag of GL-O g and one bag of GL-O p 10 aliquots of 0.5 g were weighed. The results of the analyses of these 40 aliquots are shown in Table 8. It should be noted that weighings were done without preliminary baking in an oven, the water content being measured by difference in weight between 'normal' conditions and after heating at 1000°C for 15 h. There is a good agreement between the mean values and the standard deviation obtained for the 10 aliquots taken from 10 different bags and the 10 aliquots taken from a single bag. This is commonly considered as an indication of a good homogeneity of the subsample (here 0.5 g) for the elements under consideration. Then, powder on one hand and grains on the other are homogeneous.

However, these initial results seem to show slight systematic differences between grains and powder: more water and potassium and less iron and aluminium in the grains compared with the powder. The question of the water content is an important one, as for all standards and samples which contain adsorbed water. In order to ascertain the behaviour of this water, several specific studies were performed.

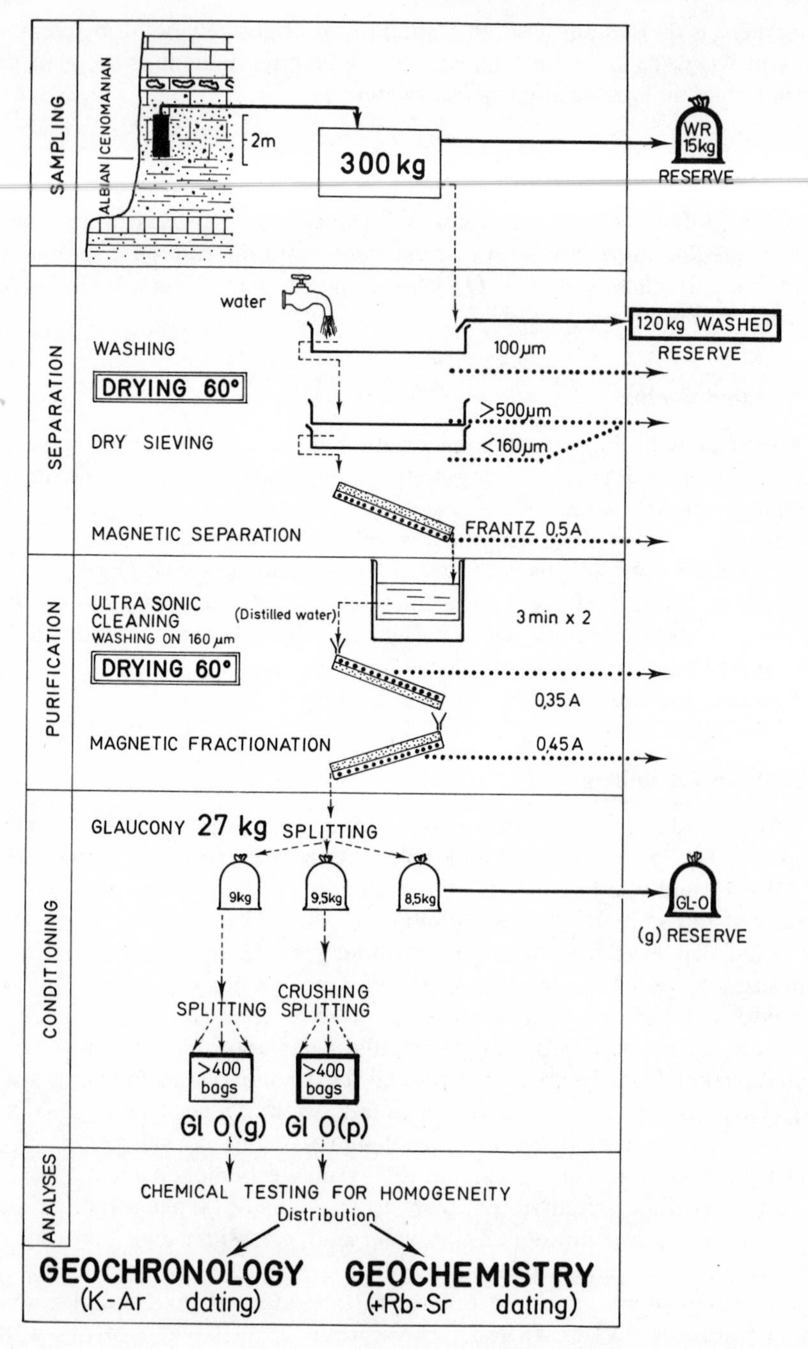

Figure 1. Process of preparation of the reference material 'glauconite GL-O'.

Table 8. Results of analyses of 40 subsamples of GL-O (from de la Roche *et al.*, 1976). The method of analysis is given: AA, atomic absorption; SP, photoelectric emission spectrometry; H, heating at 1000°C. The mean, $\bar{x}$, and standard deviation, s, are calculated.

	Method	10 bags × 1 aliquot		1 bag × 10 aliquots	
		$\bar{x}$	s	$\bar{x}$	s
GL-O g					
SiO_2 %	AA	50.91	0.20	50.82	0.30
Al_2O_3	AA	7.21	0.11	7.18	0.07
Al_2O_3	SP	7.28	0.13	7.35	0.25
Fe_2O_3	AA	19.07	0.21	18.85	0.20
Fe_2O_3	SP	18.92	0.47	18.92	0.40
MnO	SP	0.01	—	0.01	—
MgO	SP	4.33	0.09	4.30	0.09
CaO	SP	0.85	0.05	0.82	0.07
TiO_2	SP	0.05	0.01	0.04	0.01
Na_2O	AA	0.01	—	0.02	—
K_2O	AA	8.12	0.08	8.04	0.09
H_2O	H	8.6	0.12		
GL-O p					
SiO_2 %	AA	51.09	0.20	51.00	0.40
Al_2O_3	AA	7.32	0.15	7.10	0.07
Al_2O_3	SP	7.59	0.18	7.51	0.37
Fe_2O_3	AA	19.41	0.22	19.28	0.20
Fe_2O_3	SP	19.34	0.40	19.11	0.50
MnO	SP	0.01	—	0.01	—
MgO	SP	4.32	0.15	4.35	0.17
CaO	SP	0.85	0.05	0.83	0.06
TiO_2	SP	0.05	0.01	0.04	0.01
Na_2O	AA	0.03	0.01	0.03	0.01
K_2O	AA	7.91	0.05	7.91	0.05
H_2O	H	8.1	0.19		

From 10 bags of GL-O p and 10 bags of GL-O g an aliquot was weighed in normal conditions (20–25°C; humidity 60%). These aliquots were heated at 105°C for 15 h and then at 1000°C for 15 h (a frequently used treatment in chemical analyses for determining the adsorbed and structural water). The results are shown in Table 9. It should be noted that, as in Table 8, the total water content is slightly lower in the powder than in the grains. This does not explain the apparently higher content of potassium displayed in the grains (Table 8). Therefore a final test was done to establish whether an actual difference between grains and powder might be verified when the water was eliminated.

Table 9. Loss of water from 10 aliquots of different bags of GL-O p and 10 aliquots of different bags of GL-O g, percent weight (from de la Roche *et al.*, 1976). The arithmetic mean, $\bar{x}$, and standard deviation, s, are calculated. Elimination of the aberrant value, quoted in parentheses, produces a mean total water content of 8.27% with a standard deviation as low as 0.04.

	GL-O p			Gl-O g		
	105°C	1000°C	Total	105°C	1000°C	Total
	2.60	5.73	8.33	3.31	5.34	8.65
	2.53	5.78	8.31	3.26	5.45	8.71
	2.51	5.80	8.31	3.22	5.44	8.66
	2.52	5.77	8.29	3.07	5.64	8.71
	2.40	5.86	8.26	3.03	5.64	8.67
	2.44	5.81	8.25	2.82	5.89	8.71
	2.48	5.77	8.25	2.77	5.90	8.67
	2.38	5.86	8.24	2.97	5.76	8.73
	2.48	(5.39)	(7.87)	2.99	5.72	8.71
	2.71	5.48	8.19	3.11	5.57	8.68
$\bar{x}$	2.51	5.73	8.23	3.06	5.64	8.69
s	0.10	0.16	0.13	0.18	0.19	0.03

Table 10. Results of analysis of GL-O after heating at 1000°C for 15 hr. $\bar{x}^*$ is the actual mean value of the 20 aliquots measured after heating at 1000°C; $\bar{x}$ is the mean value recalculated according to the loss of weight measured between normal temperature and the heated sample: Table 9 (from de la Roche *et al.*, 1976).

		10 bags × 1 aliquot			1 bag × 10 aliquots	
	Method	$\bar{x}$	s	$\bar{x}^*$	$\bar{x}$	s
GL-O g						
SiO_2	SP	50.84	0.22	55.68	50.74	0.13
Al_2O_3	SP	7.48	0.06	8.19	7.44	0.08
Fe_2O_3	SP	19.29	0.09	21.13	19.32	0.11
MgO	SP	4.31	0.09	4.72	4.28	0.06
CaO	SP	0.87	0.03	0.95	0.92	0.03
K_2O	AA	7.87	0.05	8.62	7.88	0.08
H_2O	H	8.69	0.03		8.70	0.05
GL-O p						
SiO_2	SP	51.00	0.30	55.60	50.98	0.25
Al_2O_3	SP	7.54	0.07	8.22	7.56	0.07
Fe_2O_3	SP	19.39	0.14	21.14	19.35	0.19
MgO	SP	4.27	0.08	4.66	4.31	0.06
CaO	SP	0.91	0.03	0.99	0.92	0.02
K_2O	AA	7.90	0.09	8.61	7.92	0.08
H_2O	H	8.27	0.13		8.30	0.04

Ten aliquots were taken from one bag and one aliquot from 10 bags of the powder and the same was done for the grains. The 40 samples were first heated at 1000°C for 15 h and loss of weight noted. This 'total' loss of weight remains remarkably consistent with that shown during the two preceding sets of experiments. A total of 100 analyses was thus available: 8.27 and 8.30% in Table 10 compared with 8.27% in Table 9 for the powder; 8.69 and 8.70% in Table 10 compared with 8.69% in Table 9 for the grains. The results of analyses of six other elements are also given in Table 10. They show that no systematic difference appears between grains and powder. Therefore the apparent differences seen in Table 8 are not analytically significant. The powder and the grains are homogeneous both separately and together for all elements except water content.

From the above results and a cooperative study done in 1975 by 19 independent chemical laboratories, recommended values for this reference material may be proposed (Table 11). As both analytical processes are commonly used, recommended values of either the sample weighed 'as received' or after having first been dried at 105°C for 15 h before weighing are given. It must be emphasized that for our main purpose a strict adherence to consistent weighing conditions is necessary for all rocks and minerals containing a non-neglectable adsorbed water content, especially for purposes of calibration, as will be demonstrated later. In all cases the analyst must provide, together with his results, the specific weighing conditions which are an integral part of the analysis; otherwise, the data are of less value for possible intercomparison.

Table 11. Recommended values for some major elements of GL-O. Note that preliminary drying of the grains before weighing obviously leads to an increased percentage of components other than H_2O^- in the rest of the sample.

	GL-O p (as received)	GL-O p after heating at 105°C, 15 hr	GL-O g after heating at 105°C, 15 hr	GL-O g (as received)
SiO_2	50.90	52.20	52.20	50.60
Al_2O_3	7.55	7.75	7.76	7.50
Fe_2O_3	17.15	17.61	17.60	17.05
FeO	2.19	2.25	2.25	2.18
MgO	4.46	4.58	4.58	4.43
K_2O	7.95	8.16	8.15	7.90
H_2O^+ (1000°C)	5.58	5.72	5.82	5.64
H_2O^- (105°C)	2.52	—	—	3.06
Total	98.30	98.26	98.36	98.36

4.6 Behaviour

Numerous experiments were carried out in order to determine the behaviour of the sample GL-O. Fairly strong acid leachings of the powder and grains with acetic acid and hydrochloric acid are reported below (Odin and Rex, this volume).

Morton and Long (1980) measured the Rb and Sr content and isotopic ratios on GL-O before and after ammonium acetate and 0.1 N hydrochloric acid leaching. The Rb–Sr apparent age does not change after ammonium acetate treatment. It is perhaps slightly lowered with 0.1 N hydrochloric acid treatment. In fact, Pasteels *et al.* (1976) have presented arguments which seem to show that no detectable apparent age change occurs during leaching with 0.1 N HCl. No data are available on the leachates in the two referenced works and the question remains partly open.

Odin and Rex (this volume) report the behaviour of GL-O g in $MgCl_2$ and NaCl concentrated solutions heated at 90°C, with regard to both potassium and argon contents. The apparent ages appeared unchanged by these treatments although a clear loss of radiogenic argon and potassium was observed. On the other hand, an equivalent experiment at 90°C in deionized water showed an increase in the apparent age of GL-O: radiogenic argon was partly removed, but potassium was extracted in a greater proportion.

With regard to heating procedures, very comprehensive documentation is currently available for the conditions of both dehydration and rehydration and for temperatures of radiogenic argon departure from the structure.

No loss of radiogenic argon from GL-O is detectable between 50 and 180°C; the experiments were done at Moscow by L. Chanin as well as at Berne and Strasbourg (see Odin and Bonhomme, this volume). In the same set of experiments, for which a specific system of heating was tested, the first detectable loss of argon was observed at 250°C.

In an independent series of step-by-step heating experiments, J. L. Zimmermann, at Nancy, has shown that most radiogenic argon departure (nearly 98%) occurs between 200 and 600°C for GL-O. No more radiogenic argon is present above 800°C. In the same time, the adsorbed water is removed by heating under vacuum at 100°C and the structural water essentially removed during the steps at 400 and 500°C (Zimmermann and Odin, 1979, and this volume). These data will greatly help in defining the best method for preparing and measuring the radiogenic argon content in this standard.

Data on the preliminary heating of GL-O under atmospheric pressure before weighing and before analysis are given in Odin (this volume) and Odin and Bonhomme (this volume). The experiments show that heating under atmospheric pressure is not recommended for GL-O as the samples are not effectively cleaned better than during vacuum pumping and, moreover, it appears that the atmospheric argon contamination is enhanced by this process: it seems that air partly replaces the adsorbed water in the structure. The question of adsorbed water will be discussed further in the following section.

The above experiments provide a preliminary knowledge of the standard better than for any previously available standard.

4.7 Adsorbed water

Many authors have already emphasized the importance of the question of adsorbed water in natural rocks or minerals used for chemical analyses. Langmyhr (1969) noted in particular the case of the USGS andesite AGV-1: according to Goldich *et al.* (1967), this sample contains 1.2% of adsorbed water, but after two years of storage in the laboratory only 0.05% of adsorbed water was extracted at 105°C. Similarly, Abbey and Flanagan (1975) have emphasized that different temperature and humidity conditions during storage may have a pronounced effect on the hygroscopic water content. The sample GL-O g contains nearly 3% adsorbed water which is 'easily' exchangeable with the atmosphere (Table 9). By comparison, it should be remembered that a mica biotite or phlogopite only contains 0.3–0.5% water, while a basalt may contain 0.5–1% hygroscopic water. Thus GL-O is especially sensitive to the temperature and humidity of the ambient atmosphere.

The results of a preliminary experiment involving heating at nearly 110°C have shown that about 3.5% of the weight is lost (see Odin, Figure 8, this volume). A more sophisticated recent study has shown that various parameters may greatly change this loss of adsorbed water depending on the weight heated, the temperature of heating, the surface on which the sample is heated and the initial atmospheric humidity. This leads to many difficulties in obtaining reproducible results.

Figure 2 shows the results of heating GL-O g at 105°C for 15 h. The samples were weighed in normal conditions: 20–25°C; atmospheric humidity, 45–60%. Nine samples were heated at different times; only six results are shown in Figure 2. The three others are shown in Figure 3 in order to illustrate the readsorption of water after heating more accurately (on a different time scale). The losses of weight are calculated as percentages of the initial weight. It is seen that the loss of weight is very rapid: after one hour 4–7% of the initial weight is lost. Results of the same order are obtained after 15 h of heating. The important scattering is quite well related to the initial quantity of sample (see the Table in Figure 2). In fact all the samples were put in the oven in the same kind of glass bottle, the basal surface of which was nearly 1 cm^2. It may be assumed that the loss of weight is related to the area of this surface, because in another experiment tubes

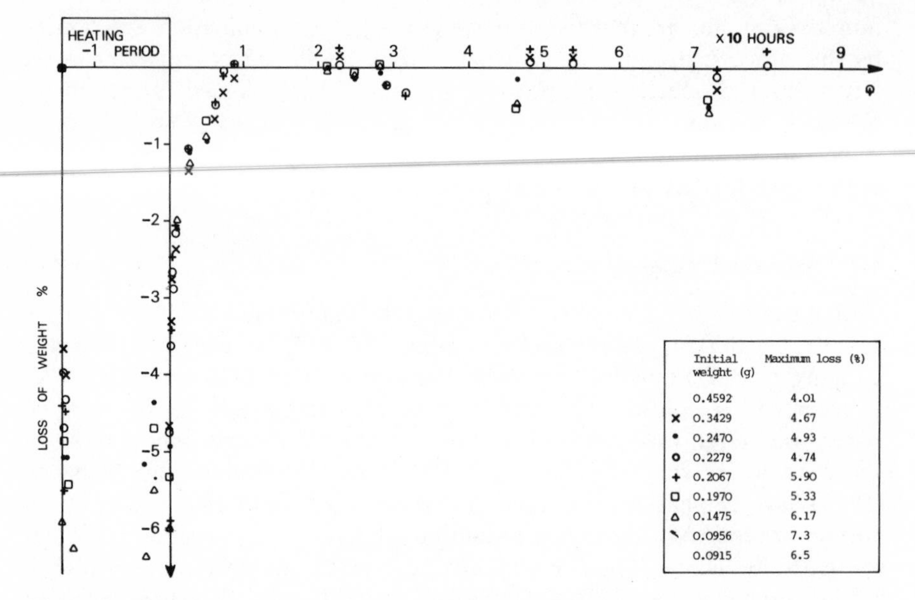

Figure 2. Departure and readsorption of water in six aliquots of GL-O. Heating was done at 105°C. The 'true' weight, during the few minutes after removal from the oven is not practically measurable.

with a smaller diameter (basal surface 0.2 cm^2) were used for 0.1 g of GL-O and only 3.5% of the weight was lost (see Odin, this volume).

The scattering of the results is also related to the fact that the readsorption is so quick that it is very difficult to weigh all the samples at exactly the same moment after removal from the oven. After something like 3–10 h the initial adsorbed water is readsorbed. In the preceding experiment done in Leeds, using long tubes of small diameter, the time necessary for complete readsorption was much longer—several days. Therefore the speed of readsorption of water is also related to the quantity and the surface of the sample. In fact, as for all exchanges of phase, the phenomenon is related to the availability of the water in relation to the reacting surface. The phenomenon is well illustrated in Figure 3, in which the heaviest sample (the smallest surface of exchange compared to the weight) is shown to have lost 4% of its weight, whilst the two smallest samples (the greatest surface of exchange because the grains are in only one thin layer at the bottom of the bottle) lost 7% of their weight. After heating, the readsorption was much quicker in these last two bottles.

Figure 4 shows the results of experiments effected at 65°C. Obviously the loss of weight is lower. As above, the smallest quantities react more quickly both for desorption and readsorption of water.

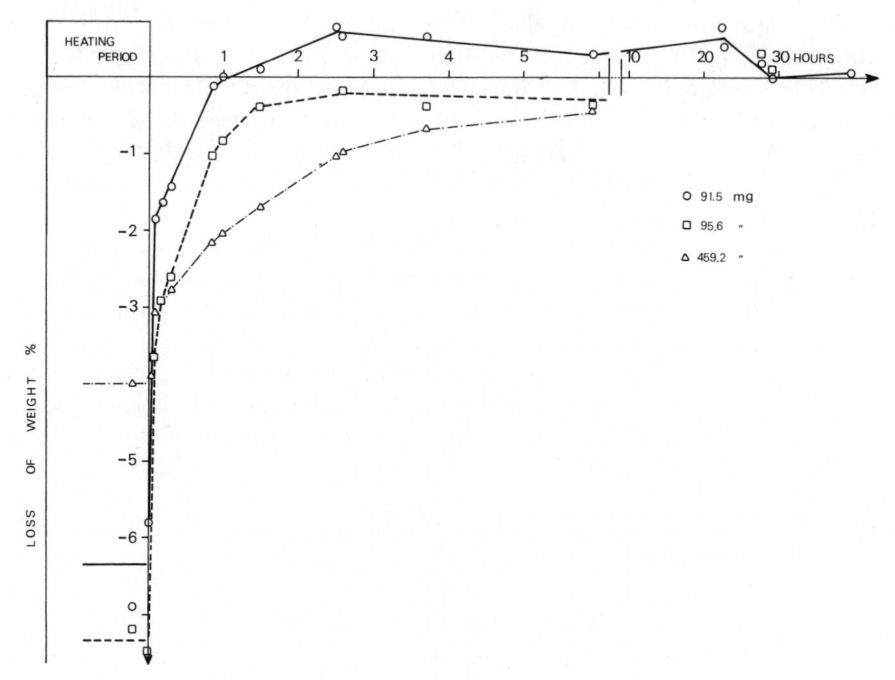

Figure 3. Readsorption of water in three aliquots of GL-O of different weights after heating at 105°C.

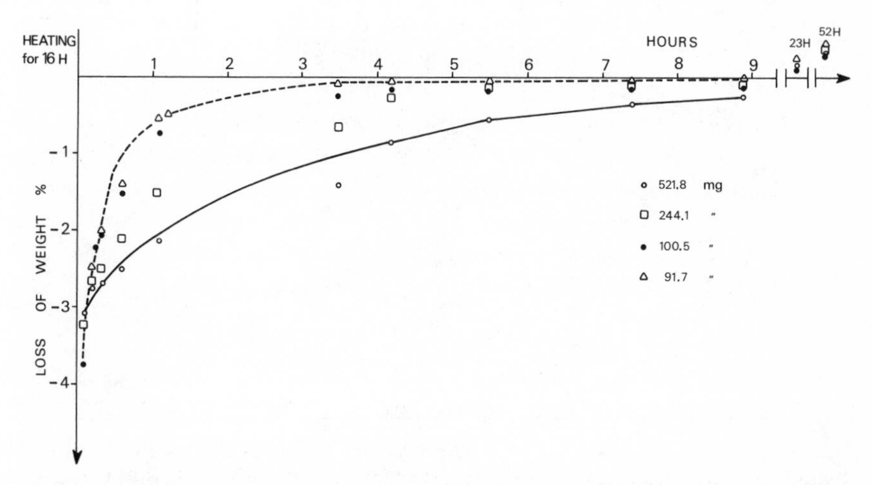

Figure 4. Readsorption of water in four aliquots of GL-O after heating at 65°C.

The main conclusion of these experiments, (confirmed by equivalent experiments not reported here) is that *it is not easy to obtain reproducible results when heating GL-O* g *at* 105°C if conditions are not exactly normalized: quantity of sample heated, temperature of heating, type of bottles to contain the sample, time period between termination of heating and weighing, etc. Therefore it is better not to heat or dry this material preliminarily.

GL-O was heated at higher temperatures in order to establish more reproducible conditions. A temperature of 800°C was chosen because it is clearly higher than the temperature of departure of the structural water (a constant quantity in a given phyllite) according to the thermal curves. As the structural water is removed, the grains change colour from green to red-brown. There is probably oxidation of the ferrous iron of the octahedral layer. X-ray diffraction (Figure 5) shows that the phyllitic structure remains

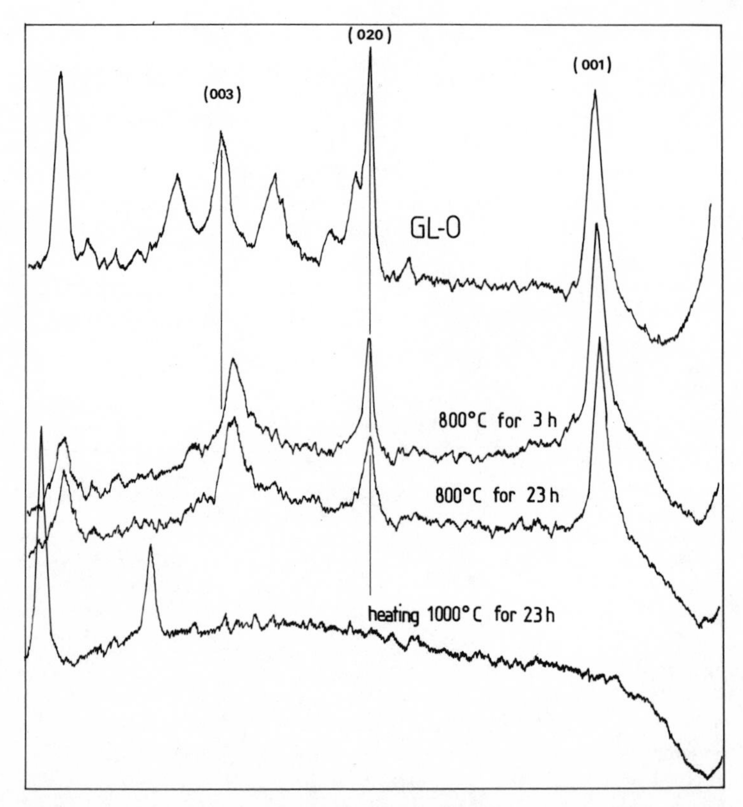

Figure 5. X-ray diffraction on powdered GL-O g: the peaks characteristic of a phyllite disappear after heating at more than 800°C. After heating at 800°C the (003) peak is modified and (hkl) peaks are eliminated except (020).

organized, although considerable disorder is created during heating at 800°C (most (hkl) peaks disappear while the (001) and (020) peaks remain unchanged).

Table 12 gives the results of the various preliminary experiments. Heating at nearly 800°C for a rather short period (experiments 1 and 2, Table 12) leads to an easily reproducible loss of weight whether the initial weight is low or high. Several weighings at different times after leaving the samples

Table 12. Results of heating GL-O g at high temperature. The initial weight is given on the left. Each sample was weighed several times after heating stopped. The interval between termination of heating and weighing is given in parentheses, together with the corresponding loss of weight (as a percentage of the initial weight). Note the good reproducibility of the experiments whatever the initial weight.

Weight (mg)	Loss of weight		
(1)			
85.25	(+0.3 h) = 7.54%		
105.90	(+0.3 h) = 7.30%	(+14 h) = 6.57%	
185.75	(+0.3 h) = 7.47%	(+14 h) = 6.16%	(+24 h) = 6.59%
429.79	(+0.3 h) = 7.53%	(+14 h) = 6.74%	(+37 h) = 6.49%
(2)			
115.58	(+0.3 h) = 7.29%	(+1.7 h) = 6.82%	(+16.5 h) = 6.48%
306.20	(+0.3 h) = 7.5%	(+1.7 h) = 6.85%	(+16.5 h) = 6.51%
(3)			
187.58	(+0.3 h) = 8.38%	(+0.7 h) = 8.07%	(+6.5) = 7.71%
(4)			
134.10	(+0.5 h) = 8.63%	(+0.9 h) = 8.30%	(+3.5 h) = 8.18%
			(+24 h) = 7.69%
156.75	(+0.7 h) = 8.50%	(+0.9 h) = 8.38%	(+3.5 h) = 8.28%
			(+24 h) = 8.06%
322.65	(+0.5 h) = 8.44%	(+0.8 h) = 8.27%	(+24 h) = 8.08%
434.04	(+0.7 h) = 8.46%	(+0.8 h) = 8.31%	(+3.5 h) = 8.15%
			(+24 h) = 7.91%
(5)			
74.34	(+0.4 h) = 9.58%	. . . .	(+23 h) = 9.57%
163.68	(+0.4 h) = 9.53%	. . . .	(+24 h) = 9.5%
214.65	(+0.5 h) = 9.33%	. . . .	(+24 h) = 9.3%
398.39	(+0.3 h) = 9.44%	. . . .	(+25 h) = 9.3%

(1) heating for 3 hr 600–800°C
(2) heating for 6.5 hr 700–800°C
(3) heating for 23 hr 700–800°C
(4) heating for 23 hr 800°C
(5) heating for 20 hr 1000°C.

out of the furnace show a slight readsorption of water, which is much lower than that after heating at 105°C. Longer experiments (Table 12, experiments 3 and 4) show that equilibrium was not attained after 3–6 h. The readsorption of water is also diminished after heating for longer periods.

Heating at 1000°C (experiment 5, Table 12) produces a slightly higher loss of weight compared with heating at 800°C and the grains are then dark chocolate brown. No readsorption of water was observed during the next 24 h. Thus, this loss of weight will be the most easily measurable, whatever the atmospheric conditions, after heating at 1000°C.

X-rays show that the phyllitic structure completely disappeared after this experimental heating at 1000°C.

Having established the best heating conditions, it is possible to envisage the measurement of the 'normal' loss of weight for a GL-O subsample, weighed after equilibrium and with well-defined 'normal' conditions of temperature and hygrometry. This remains to be done with accuracy. However, since one may now foresee the measurement of the adsorbed water content of this sample to an accuracy a few tenths of a percent of the weight, it will be easy to adjust the weighed sample by comparing the total loss of weight of a subsample, weighed under every possible condition, with a total loss of weight of GL-O under normalized conditions. We tentatively propose here to adopt the value obtained in Nancy (total loss of 8.69% ± 0.03) as the *normalized loss of weight at* 1000°*C* for 15 h.

This improvement has not been fully accomplished but is the first good approximation to an answer to the problem.

Using a completely different method, we tried to produce saturation of GL-O by water. For this experiment, two samples were stored in closed containers where the humidity was maintained at 95–100% The results obtained are shown in Figure 6. The water gain was rather slow and was still unstable after three days. The samples reverted to their initial weight one day after having been reintroduced into a normal atmosphere. No serious possibilities for the measurement of initial water content may be foreseen with this method.

We have emphazised the importance of adsorbed water in the use of standards in order to demonstrate that special care must be taken in this regard. In particular, the weighing conditions must be kept in mind in measuring the age of a rock. Standard GL-O is especially sensitive, as are all glauconies (but less than most of them). However, it must be noted that possible inhomogeneity related to adsorbed water content remains of the order of the usual analytical uncertainty, as evidenced by the results obtained below in different laboratories which, to our knowledge, were not especially attentive to this question. In particular it should be kept in mind that the different laboratories participating in the cooperative study below are subject to varying climatic conditions (see, for example, the calculated

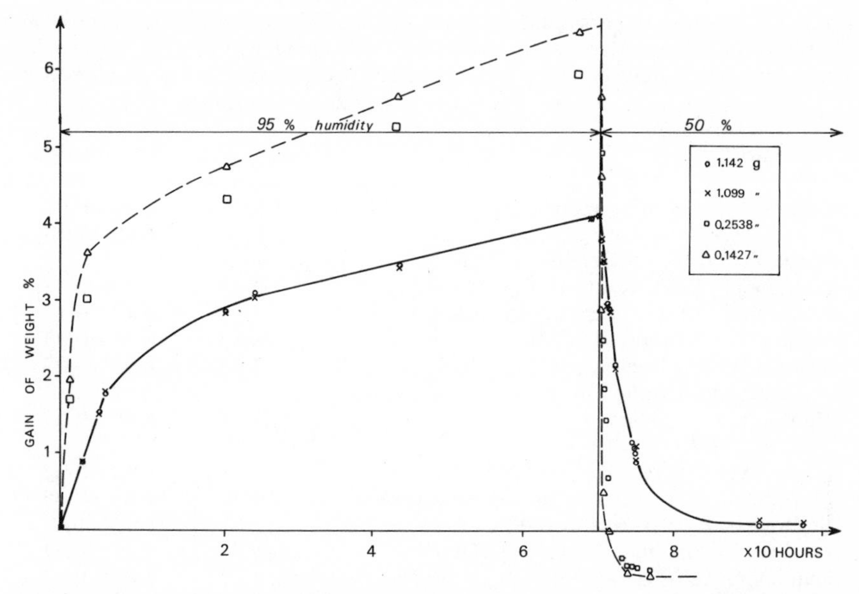

Figure 6. Gain and loss of weight after storage of GL-O in a 95% humid atmosphere. Drying, after 72 hr of humidification, was performed in an atmosphere with 50% humidity (as for weighing) for the two samples of nearly 1 g but at 30% humidity for the two other samples; this explains why their weight becomes lower than initially. Note that the gain and loss of water is quicker for smaller quantities. The same kind of bottle (basal surface 4 cm^2) was used for all samples.

standard deviations by Cassignol, Kreuzer, etc., and the interlaboratory standard deviation for Ar).

4.8 Interlaboratory comparison of K–Ar dating

The reference material *glauconite GL-O* was sent to 55 geochronological laboratories in 24 countries: the United States (11), France (9), Germany, Great Britain, the Soviet Union, Switzerland (3 each), etc. The grains (GL-O g) were used for K–Ar dating, the powder (GL-O p) being recommended for Rb–Sr dating because of its better homogeneity with regard to the common strontium.

The results obtained for *potassium* content by 13 laboratories are listed in Table 13. The interlaboratory mean value is in good agreement with the first data obtained in Nancy, i.e. 6.56% K (Table 13) compared with 6.54% (Table 10), as well as with the recommended value, 6.56% K (Table 11). Hence the recommended value for GL-O g measured 'as received' is confirmed here. The relative interlaboratory standard deviation is less than 1%.

Table 13. Results of potassium analyses on GL-O obtained in geochronological laboratories. The interlaboratory mean was calculated without the lowest value (based on a single determination only). The relative interlaboratory standard deviation is less than 1%.

Laboratory	Measurements	wt %K $\pm 1\sigma$
Adams, C. J.—Lower Hutt (NZ)	8	6.52 ± 0.08
Balogh, K.—Debrecen (Hungary)	5	6.60 ± 0.10
Bagdasaryan, G. P.—Jerevan (USSR)	1	(6.31)
Bonadonna, F. P., Radicati, F.—Pisa (Italy)	2	6.49 ± 0.18
Bonhomme, M. G.—Strasbourg (France)	5	6.47 ± 0.14
Chanin, L.—Moscow (USSR)	8	6.58 ± 0.03
Hebeda, E. H., Boelrijk, N. A.—Amsterdam (Netherlands)	2	6.55 ± 0.07
Hunziker, J. C., Jäger, E.—Berne (Switzerland)	3	6.50 ± 0.04
Harre, W., Kreuzer, H.—Hanover (Germany)	42	6.59 ± 0.02
McDougall, I., Kiss, E.—Canberra (Australia)	10	6.64 ± 0.05
McDowell, F. W.—Austin (US)	3	$6.55 (\pm 0.05)$
Rex, D. C., Gledhill, A.—Leeds (GB)	5	6.52 ± 0.02
Rundle, C. C.—London (GB)	4	6.64 ± 0.03
Savelli, C.—Bologna (Italy)	4	6.59
Interlaboratory mean (11 labs)		6.55_7
1σ		0.05_5

Therefore, one may consider the potassium content of GL-O g weighed 'as received' to be 6.56 ± 0.06 (weight %K).

The *radiogenic argon* content was measured in 17 laboratories; in two of them, GL-O was measured at two different epochs by two different analysts and the mean value used for interlaboratory mean calculation. This mean value of 24.85 (nl/g) was obtained (Table 14) with a standard deviation of 0.24 (nl/g), which is less than 1% of the radiogenic argon content.

No one laboratory obtained a value with a deviation of more than 2% from this mean. If we eliminate the three extreme values, the mean becomes 24.83, and the interlaboratory standard deviation is reduced to 0.18 nl/g for these 14 independent laboratories.

All measurements were done by the isotopic dilution technique; but a study on the reproducibility of GL-O using direct calibration by air pipetting (Kirsten, 1966) was also undertaken by Cassignol and Gillot (this volume). The 39/40 technique has been used but no representative results were obtained, apparently due to diffusion of ^{39}Ar (M. Soroïu, Bucharest, and M. Maluski, Montpellier, personal communications).

The long-term reproducibility (more than 1 year) was tested in several laboratories, such as Hanover where an especially good intralaboratory

Table 14. Radiogenic argon and apparent age of GL-O. Errors given are 1 standard deviation $\left(1\sigma = \left\{\frac{\sum (x_i - \bar{x})^2}{n-1}\right\}^{1/2}\right)$. Details on the data obtained in Hanover are given in the Appendix. The two data quoted from Hanover (two different apparatuses but same intralaboratory calibration) and from Leeds (two different analysts at two different periods) are counted for one laboratory each. The relative standard deviation of the mean radiogenic argon contents is less than 1% and no analytical result deviates more than $\pm 2\sigma$ from the mean value.

Date		rad. Ar ($\pm 1\sigma$) (nl/g)	No. of measurements	Apparent age (Ma, ICC)
(4—1978)	Adams, C. J.	24.53±0.35	11	94.2±1.7
(1980)	Armstrong, R. L., Harakal, R.	24.58±0.	5	—
(6—1980)	Balogh, K.	24.82±0.50	2	94.2±2.4
(1978)	Bagdasaryan, G. P.	25.22±0.64	7	(99.8±5.8)
(6—1974)	Bonadonna, F. P., Radicati, F.	25.07	2	96.6±3.6
(1—1975)	Bonhomme, M. G., Odin, G. S.	24.82±0.11	6	96.1±2.1
(1978)	Cassignol C.—Gif sur Yvette (France)	24.95±0.07	9	—
(9—1974)	Chanin, L.	24.95±0.29	6	95.0±1.2
(8—1980)	Hebeda, E. H.	24.9 (±0.5)	2	95.2±2.2
(3—1977)	Hunziker, J. C., Odin, G. S.	24.75±0.26	14	94.8±1.0
	{Kreuzer, H.	{24.74±0.12	43 (CH4)	{94.1±0.5
	{Kreuzer, H.	{24.65±0.12	23 (RSS9)	{93.8±0.5
(4—1975)	McDougall, I.	24.65±0.08	3	93.1±0.7
(4—1977)	McDowell, F. W.	25.28(±0.33)	3	96.7±1.5
(8—1980)	Mehnert, H.—Denver (US)	24.39(±0.24)	2	—
(8—1980)	Montigny, R.	24.82±0.27	4	—
(6—1974)	{Rex, D. C.	25.21±0.40	2}	95.8±1.5
(10—1979)	{Rex, D. C., Odin, G. S.	24.82±0.21	5}	
(7—1974)	Rundle, C. C.	25.08±0.18	5	94.7±0.8
Interlaboratory mean (17 labs)		$24.85_4 \pm 0.24$	(12 labs)	95.03±1.11

reproducibility was obtained (see Appendix). This is certainly related to peculiarly careful preliminary division of the standard. This was separated 'as received' into small quantities of 100–200 mg, which were then stored in individual small tubes, one tube being weighed and measured when needed.

From the available data in Table 14, thirteen apparent ages can be calculated. Of these, one appears unreliable, as evidenced by the very wide spread of data. From the remaining 12 apparent ages a mean value of 95.03 Ma was calculated, with a standard deviation of a little less than 1.11 Ma.

From the analytical data given, we have calculated (Table 15) the ratio LP-6/GL-O obtained at short-term intervals. A similar short-term interval comparison is available for the ratio P 207/GL-O: UBC, Vancouver, 13 P 207 = 27.93±0.30, 5 GL-O = 24.58±0.07, ratio = *1.136* (R. L. Armstrong, personal communication); Gif-sur-Yvette, preliminary calibration,

Table 15. Ratio of radiogenic argon content LP-6/GL-O in several laboratories. After drawing of this table new results were provided by E. Welin (Stockholm) 2 LP-6/2 GL-O = 1.763 ± 0.050 and by McDowell (Austin) 13 LP-6/3 GL-O = 1.730.

Date	Laboratory	*n of* measurements	Ratio
1979–1980	Amsterdam—Hebeda	3 LP-6 / 2 GL-O	1.743
1978	Bologna—Savelli	Calibration / 4 GL-O	1.764
1979	Hanover—Kreuzer	10 LP-6 / 66 GL-O	1.749
1974–1976	Leeds—Rex	4 LP-6 / 2 GL-O	1.724
October 1979	Leeds—Odin	1 LP-6 / 4 GL-O	1.746
1974–1976	Pisa—Radicati	1 LP-6 / 2 GL-O	1.717
June 1980	Strasbourg IPG—Montigny	7 LP-6 / 4 GL-O	1.726
1980	Strasbourg Univ.— Bonhomme, Odin	2 LP-6 / 6 GL-O	1.702
Interlaboratory mean			1.734
σ			0.020

3 P 207 = *c.* 28.93, 9 GL-O = 24.95, ratio = *1.131* (Ch. Cassignol and P. Y. Gillot, personal communication); Berne, 3 P 207 = 28.22 ± 0.12, 6 GL-O = 24.78 ± 0.10, ratio = *1.139* (personal measurements, 1974).

In summary, one may recommend the following values for GL-O: potassium, *6.56 ± 0.06* (wt %K); this is the initially proposed value (de la Roche *et al.*, 1976, Table 11) which was confirmed later; argon, *24.8 ± 0.2* nl/g; this is the value obtained from the selection of the best-known values from laboratories which have done a sufficient number of analyses with a normal internal reproducibility.

These values were obtained on samples stored under normal temperature (20–25°C) and about 60% humidity conditions.

4.9 Interlaboratory comparison of Rb–Sr dating

The results available from Rb–Sr dating are shown in Table 16. As for P 207 above, there is an inhomogeneity in the data which may be attributed

Table 16. Results of analysis of GL-O g (grains) or p (powder). Apparent ages are recalculated using a conventional strontium isotopic ratio for a Cenomanian sea of 0.708.

Laboratory	Rb (ppm)	Sr (ppm)	$^{87}Rb/^{86}Sr$	$^{87}Sr/^{86}Sr$	T ICC
Austin (US)—Long, L. E.	g 241	21.4	32.4	0.7511	93.7 ± 1.1
Berne (Switz.)—Hunziker, J. C.	g 234.8	18.45	36.84	0.7551 ± 8	89.9 ± 2.7
Berne (Switz.)—Hunziker, J. C.	g 232.7	18.80	35.83	0.7543 ± 8	91.0 ± 2.7
Washington (US)—Barnes, I. L.	g 237.7 ± 0.2 (3)				
Washington (US)—Barnes, I. L.	p 240.1 ± 0.2 (3)				
Amsterdam (Neth.)—Boelrijk, N. A.	g 239.4	18.3	38.0	0.7560	88.8
Amsterdam (Neth.)—Boelrijk, N. A.	g 242.3	19.3	36.5	0.7548	90.2
Amsterdam (Neth.)—Boelrijk, N. A.	g 240.7	—	—	—	—
Amsterdam (Neth.)—Boelrijk, N. A.	g 246.3	19.2	37.3	0.7538	86.3
Amsterdam (Neth.)—Boelrijk, N. A.	p 242.6	19.7	35.8	0.7534	89.2
Amsterdam (Neth.)—Boelrijk, N. A.	p 242.3	20.2	34.9	0.7510	86.7
Amsterdam (Neth.)—Boelrijk, N. A.	p 240.1	18.8	37.1	0.7533	85.9
Amsterdam (Neth.)—Boelrijk, N. A.	p 241.7	17.4	40.4	0.7544	80.9
Brussels (Belg.)—Pasteels, P.	g 235.0	19.52 ± 0.20	(34.9)	0.7518 ± 13	88.2 ± 2.1
Brussels (Belg.)—Pasteels, P.	g 233.3	18.67 ± 0.20	(36.2)	0.7536 ± 7	88.5 ± 2.4
Sao Paulo (Brazil)—Kawashita, K.	g 229.0	19.6	33.97 ± 1.41	0.7533 ± 30 0.7514 ± 38	91.9
Leeds (GB)—Gledhill, A.	g 238.0	17.5	39.5	0.7608 ± 5	94.1
Leeds (GB)—Gledhill, A.	g 237.5	18.5	37.3	0.7540 ± 2	86.8
Leeds (GB)—Gledhill, A.	g 238.7	21.4	32.4	0.7491 ± 1	89.3
Leeds (GB)—Gledhill, A.	g 238.1	19.7	35.1	0.7514 ± 1	87.1
Nancy (France)—Sonet, J.	p 229.7 ± 1.8 (4)	19.8 ± 0.2 (4)	33.58	0.7539 ± 14	96.2
Strasbourg (France)—Bonhomme, M. G.	g 248.2 ± 7.5	19.2 ± 0.4	37.47	0.7550 ± 22	88.3
Strasbourg (France)—Bonhomme, M. G.	g 252.4 ± 5.0	19.4 ± 0.4	37.71	0.7541 ± 3	86.1

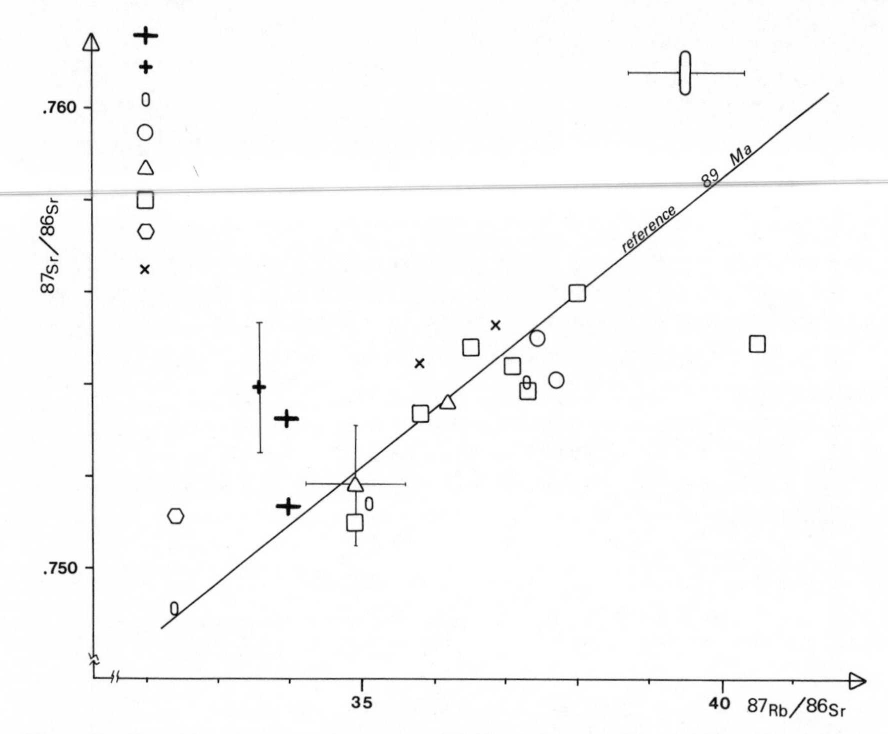

Figure 7. Graphic representation of the Rb–Sr dating on GL-O. The eight different figures represent eight different laboratories.

both to the analytical process and to a possible heterogeneity of the sample. However, the powder was better homogenized than the grains during conditioning and the scatter is equivalent, so it appears that the analytical process may be at fault; however, too few results are currently available to allow more accurate assessment. The scattering of Rb–Sr apparent ages is much greater than that of the K–Ar ages. A graphic presentation of the data shows that the representative points are rather well situated along the reference line 89 Ma (Figure 7).

4.10 Conclusions: How to use GL-O

Glauconite GL-O was carefully prepared with regard to its purity and homogeneity. A significant quantity remains available.

The study of its behaviour under diverse conditions related to isotopic dating was carefully documented. This facilitates the proposal of recommendations for its use.

At the present state of knowledge, it appears better to weigh the sample *without* preliminary heating or drying, whatever kind of analysis is to be undertaken. Further sample must be stored under normal temperature and humidity conditions. As far as possible, the subsamples used for K and Ar must be weighed under equivalent conditions. In this case GL-O will prove an accurate and reproducible sample for testing the whole dating process in a laboratory.

If one needs to calibrate or improve a line for argon only, we propose the following procedure where it is difficult to obtain the normalized weighing conditions, as well as for a more rigorous process of measurement.

(1) Weigh an aliquot for measuring argon and, at the same time, weigh a second aliquot (or two) of 200 or 300 mg in silica capsules.

(2) Put the second (and third) aliquot in a furnace at 1000°C (±50) for 15 h.

After cooling, weigh the sample and calculate the loss of weight.

(3) Normalize the weight of the first aliquot to a normal loss of 8.69%: if the loss has been higher, decrease the measured weight; if the loss has been lower, increase the measured weight of the aliquot.

Provided that division of the sample is done carefully, subsamples of nearly 100 mg show a long-term intralaboratory reproducibility of better than 0.5%. The powder substandard is recommended for Rb–Sr dating, while grains must be used for K–Ar dating: crushing has led to a loss of several percent of radiogenic argon (L. Chanin, personal communication).

Heating of GL-O at temperatures of about 100°C produces very few reproducible weighings for two reasons: the departure of adsorbed water depends on diverse factors other than the easily measurable temperature; the readsorption of water is initially very quick after heating is stopped. Therefore, the actually quantity weighed is less reproducible after heating than under normal conditions.

On the basis of interlaboratory comparisons, this standard appears at least as good as other presently available ones, even when no special care is taken. Several studies aimed at establishing still more easily reproducible conditions for the measurement of water content under all possible climatic conditions are now under way.

After weighing, heating the sample under atmospheric pressure is not recommended to avoid increasing atmospheric argon contamination.

Under vacuum, we recommend preheating at nearly 150°C (see Odin and Bonhomme, this volume).

During the extraction of radiogenic argon the analyst must be careful not to increase the temperature too rapidly in order to avoid sudden breaking of the grains and loss of the sample. The best way is to heat at a temperature lower than 600°C, for about 3–4 min. Most of the gases are extracted before the crucible becomes red (at 800°C). A temperature of 1100°C is sufficient for fusion in our experience; the whole heating period may be as short as 8–10 min. We had no problems in purifying the extracted gases on a titanium furnace or zirconium trap, even if preliminary heating at 150°C was not undertaken.

We still hope to improve the knowledge of this standard in the near future and we agree with Lanphere and Dalrymple's opinion that knowledge about a standard is an ongoing question, which may always be improved.

5 BIOTITE 'MICA–Fe' AND PHLOGOPITE 'MICA–Mg'

These two micas were prepared at Nancy in 1967–1968. They are available in very large quantities as both powder and flakes (Table 1). Their purity and homogeneity were positively tested (Roubault *et al.*, 1968–1970; Ingamells and Engels, 1967a,b).

The last available report concerning the chemical components appeared recently (Govindaraju, 1979). Few analytical data are currently available concerning the geochronology; they are brought together in Table 17.

However, due to the very significant quantity available and the well-recognized homogeneity of the samples, especially of the phlogopite, it would be useful to obtain more analytical data for these two minerals.

Until now no inhomogeneity has been found in these samples and this will probably give useful results using the neutron activation technique. Such studies are under way.

Table 17. Results of K–Ar dating on biotite and pholgopite reference materials. The two samples are available from the following address: CRPG, Dr K. Govindaraju, C.O. no. 1, 54500 Vandeuvre-les-Nancy, France. The Rb–Sr apparent ages obtained in Nancy, i.e. mica–Fe = 315 ± 10, mica–Mg = 495 ± 15, are equivalent to the K–Ar apparent ages calculated with the data given below.

Laboratory	Potassium (%K)	rad. Ar (nl/g)
Mica–Fe		
Nancy, J. L. Zimmermann	(interlaboratory mean) 7.26	98.23
Amsterdam, E. H. Hebeda	7.33 (2 meas.)	96.95 (2 meas.)
Mica–Mg		
Nancy, J. L. Zimmermann	(recommended value) 8.30	195.7
Amsterdam, E. H. Hebeda	8.49 (2 meas.)	188.85 (2 meas.)
Leeds, D. C. Rex, G. S. Odin	—	194.1 (2 meas.)

Acknowledgements

The senior author is greatly indebted to the publishers for improving the English translation. This chapter is a contribution to the IGCP Project 133.

Résumé

L'utilisation d'un minéral de référence interlaboratoire est essentielle lors de la datation d'un niveau stratigraphique destiné à contribuer à la connaissance de l'échelle numérique de la colonne stratigraphique. C'est une partie intégrante du résultat analytique. Cette contribution rassemble les résultats actuellement disponibles aussi bien sur des 'étalons' anciens généralement épuisés, que sur deux étalons récents: LP-6 et GL-O et plus spécialement sur ce dernier.

Les résultats obtenus dans de nombreux laboratoires sont cités. En particulier des centaines d'analyses chimiques totales, de mesures de potassium et d'argon ont été réalisées sur les deux plus récents étalons distribués. Ces mesures ont été réalisées à la fois pour connaître la reproductibilité interlaboratoire et les limites propres à l'étalon lui-même dans un même laboratoire. Cette préoccupation, nouvelle dans la définition des étalons utiles à la géochronologie, a mis en évidence les conditions les meilleures de reproductibilité; elles sont définies ici et il est recommandé aux analystes d'en tenir compte rigoureusement. Moyennant quoi, un étalonnage interlaboratoire à mieux que 1% sur un âge peut être obtenu. Ceci est un progrès évident par rapport aux essais de calibration 'absolue' qui peuvent difficilement atteindre cette précision et sont surtout beaucoup plus délicats à mettre en oeuvre. Les résultats concernent essentiellement les analyses par la méthode K–Ar, les plus couramment utilisées dans le domaine de ce volume.

(Manuscript received 12-12-1980)

APPENDIX: AN EXAMPLE OF INTRALABORATORY REPRODUCIBILITY FOR GLAUCONITE GL-O

W. Harre and H. Kreuzer

GL-O was analysed 61 times for potassium and 66 times for argon on two different lines over six years in Hanover. The detailed results are shown below in order to test the reproducibility of the sample.

In equivalent conditions, the data obtained on the reference biotite LP-6 are as follows: K (wt %): 8 determinations, 1978, 8.30 ± 0.013 (1σ); Ar (nl/g): 19 determinations, 1978–1980, 43.18 ± 0.29 (1σ); sample aliquots 100 mg for potassium analysis, 150 mg for argon analysis.

Table 1. Results of potassium analyses (wt %K) on GL-O g. The weighted mean of the listed values was calculated by limiting the number of determinations to $n \leq 8$ for each set of data. Harre's determinations were done on splits of 100 mg with a digitized automatic processing double-channel EEL 170 flame photometer with internal Li standard. Digital pipettes (Dilutor) were used; calibrations were done against standard solutions. The increased scatter observed in 1980 is attributed to the apparatus, as inferable from the always three replicates on the same solution. This example demonstrates that the proposed standard may lead to adequate reproducibility in spite of the theoretical possible changes related to the adsorbed water content.

Year	No. of measurements	Mean (wt %K)	$\sigma = \left\{\frac{\sum (x_i - \bar{x})^2}{n-1}\right\}^{1/2}$
1974	2	6.587	0.019
1975	4	6.568	0.009
1976	8	6.583	0.018
1977	4	6.597	0.019
1978	7	6.608	0.020
1979	1	6.629	—
1980 (Jan.–Apr.)	9	6.605	0.029
1980 (May–Dec.)	26	6.577	0.027
Weighted mean 1974–1980		6.591	0.015

Table 2. Results of argon analyses on GL-O g. The data obtained on the two different lines are quoted separately. The mean given is the arithmetic mean of all determinations. The quoted values were corrected for observed relative shifts of the result on the intralaboratory standard biotite SV. The corrections were calculated for the individual periods of three–six months from about 10 analyses on the biotite SV each.

Analyst	Year	Line	No. of determinations	Mean (nl/g)	Standard deviation (nl/g)
Kreuzer, H.	1974–75	CH_4	28	24.76	0.13
Kreuzer, H.	1978	CH_4	10	24.67	0.08
Kreuzer, H.	1979	CH_4	5	24.78	0.05
			Mean 43	24.74	0.12
Kreuzer, H.	1974–75	RSS9	13	24.61	0.09
Kreuzer, H.	11–1977	RSS9	4	24.57	0.17
Kreuzer, H.	3–1978	RSS9	2	24.75	0.02
Odin, G. S.	8–1979	RSS9	4	24.80	0.05
			Mean 23	24.65	0.12

Numerical Dating in Stratigraphy
Edited by G. S. Odin

8
Potassium–argon analysis

M. Flisch

1 SPIKE CALIBRATION ON A NEW ARGON LINE

During 1979 and 1980, a new argon extraction, purification and measuring system was set up at the Department of Isotope Geology, University of Berne.

Compared with the system described by Purdy (1972) and Hunziker *in* Odin *et al.*, (1976), improvements were introduced to achieve lower blank levels and a better reproducibility. The design of our Pyrex line is shown in Figure 1.

1.1 Gas extraction

A molybdenum crucible contains the samples, which is extracted in a water cooled quartz-glass furnace. A temperature of 1500–2000°C depending on the sample, is induced by the coils of an RF generator. The extraction process takes about 30 mn.

For each sample a clean crucible and furnace are used to minimize cross-contamination. In a clean furnace the observation of the sample during the extraction is not impeded by the deposition of a metallic mirror, usually found upon the cooled inner surface.

1.2 Gas purification

The purification system incorporates two stages to separate the noble gases. In stage one there are a Ti- and a Cu/CuO-getter (in Figure 1 Cu and Ti I) and a trap which is cooled by dry ice (CT). A Ti-getter (Ti II) is sufficient for the second stage. To pump the gas quantitatively from stage one into stage two a charcoal finger (C II) cooled by liquid nitrogen is used. The whole purification process takes approximately 70 mn.

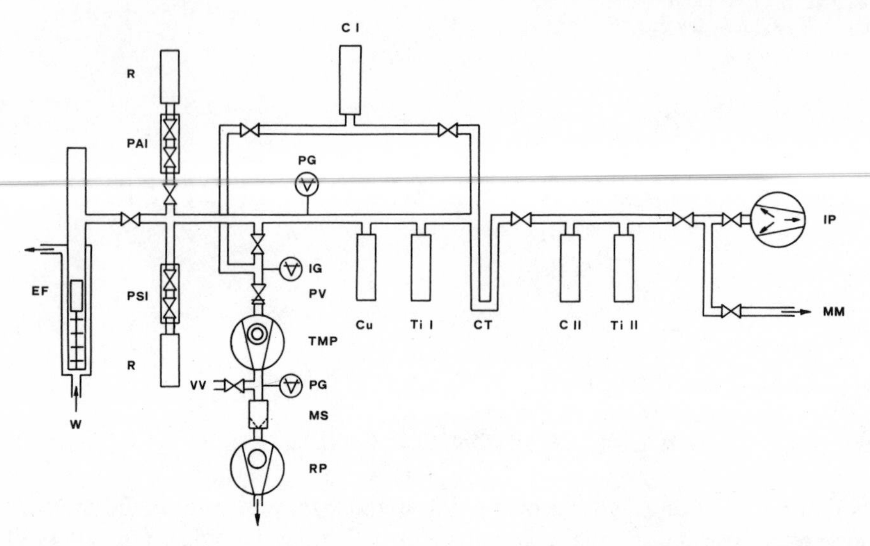

Figure 1. Extraction and purification system. EF, extraction furnace; W, water; R, reservoir; PAI, pipette air inlet; PSI, pipette spite inlet; PG, Pirani gauge; IG, ion gauge; PV, pneumatic valve; VV, ventilation valve; MS, molecular sieve; TMP, turbomolecular pump; RP, rotary pump; IP, ion pump; C, charcoal finger (I and II); Ti, Ti-getter (I and II); Cu, Cu/CuO-getter; CT, cold trap; MM, mass spectrometer.

1.3 Mass spectrometer and data processing

The isotopic abundances are measured in a Micromass 1200 static mass spectrometer. The magnetic sector is of 12 cm radius and 60° deflection. The Faraday cup is connected to a Cary 31 amplifier. The signal is digitized in a Solartron IM 1490 digital voltameter. Via an optical coupling interface, these data are processed in a PDP 8/e minicomputer.

The uncertainty for an air + spike measurement is lower than 0.03% for $^{40}Ar/^{38}Ar$ and better than 0.1% for $^{40}Ar/^{36}Ar$.

Fifteen air measurements of the ratio $^{40}Ar/^{36}Ar$, made over a period of more than one year, give a mean value of 301.0 ± 0.2. A discrimination correction is used according to the following formulae:

$$(^{40}Ar/^{36}Ar)_{SR} = \frac{295.5 \times (^{40}Ar/^{36}Ar)_{SM}}{(^{40}Ar/^{36}Ar)_{AM}}$$

$$(^{38}Ar/^{36}Ar)_{SR} = \frac{295.5 + (^{40}Ar/^{36}Ar)_{AM} \times (^{38}Ar/^{36}Ar)_{SM}}{2 \times (^{40}Ar/^{36}Ar)_{AM}}$$

here $(^{40}Ar/^{36}Ar)$ air real = 295.5; AM = air measured; SM = sample measured; SR = sample real.

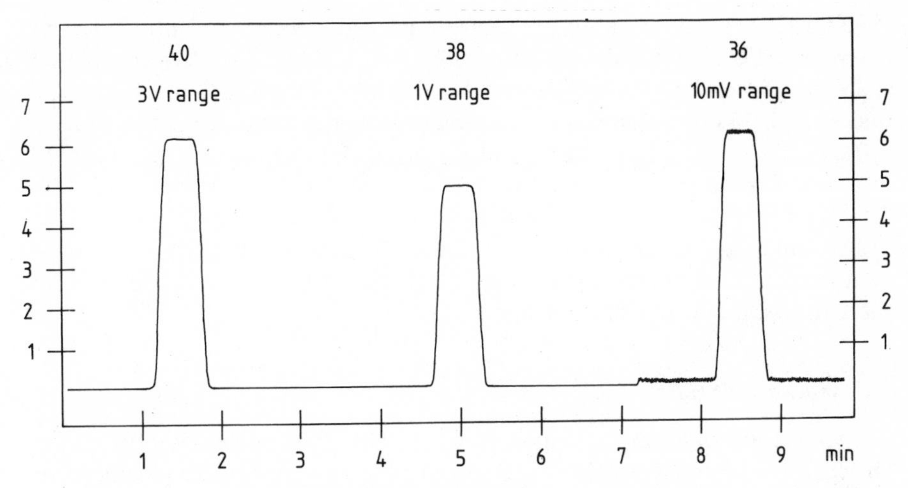

Figure 2. Scan of an air–argon+spike mixture.

All the samples, including air shots and standards, are measured by Hall probe controlled peak switching. In each measurement the following set is measured six times: baseline ^{40}Ar, mass ^{40}Ar, baseline ^{38}Ar, mass ^{38}Ar, baseline ^{36}Ar, mass ^{36}Ar, and once more baseline ^{36}Ar. Every mass and baseline is registered for approximately 30 s, depending on the intensity. After the baseline correction, linear regression lines are calculated with the six peaks of each mass. The positive or negative correlation coefficient is usually better than 0.998. The initial intensities of the masses are determined by the intercepts of the regression lines and the inlet time.

The scan in Figure 2 shows peak shape and abundance sensitivity.

1.4 Vacuum system, line cleaning and blank

A Turbovac 120 (Leybold) backed by a two stage rotary pump ED 50 (Edwards) is provided to pump the purification and extraction system to the ultra high vacuum range. A 25 l/s ion pump (Riber) is connected to the second purification stage.

Following each measurement the extraction and purification lines are baked for 12 h at 175°C. After baking we observe a vacuum better than 5×10^{-8} torr. For cleaning purposes the getters and charcoal fingers are heated up to the following temperatures: Ti, 850°C; Cu/CuO, 530°C; charcoal fingers, 420°C.

Even lower pressure levels are generated by means of an additional charcoal finger (C1), if the residual gases are frozen ahead of the extraction by liquid nitrogen.

Extraction, purification and measuring procedures are standardized. This is important for the blank correction. The blank values for ^{40}Ar are smaller than 3×10^{-8} cm^3 STP. A ^{40}Ar rad blank and a ^{38}Ar spike blank are still undetected. With a significant ^{40}Ar rad blank the ratio $^{40}Ar/^{36}Ar$ for air (301.0 ± 0.2) would have been shifting to higher values for the past 14 months. The absolute sensitivity for a single mass is better than 3×10^{-13} cm^3 STP.

The pumping in the mass spectrometer is done by a 25 l/s Varian ion pump and a 50 l/s Sorbac getter pump. Balzers UVD-8H valves are used for the extraction and purification line.

1.5 Spike calibration

The isotope dilution technique is used. The ^{38}Ar spike 99.9997% was enriched by Schumacher (1975). Reproducible spike shots are produced by a pipette. Two valves in connection with a known volume (approx. 0.1 ml) deliver a calibrated amount of spike. The calibration was performed using two different techniques.

1.5.1 Calibration with a known air volume

An evacuated receptacle of 1013 cm^3 was filled with a known amount of air by using a pipette 64 times. We made corrections for atmospheric pressure and humidity. The same pipette was then used to produce air shots with a known volume. This way we were able to calibrate the ^{38}Ar spike. The results are shown in Table 1.

1.5.2 Calibration using standard materials

The ^{40}Ar rad content of the standard material determines the amount of material used, usually 0.1–0.3 g. This produces a signal in the range of $3 \times 10^{-12} - 2 \times 10^{-11}$ A. For the ^{40}Ar rad contents of the standard materials the following values given in the literature have been used.

B 4 M: $6.31 \times$ nl STP ^{40}Ar rad/g

see Jäger *et al.* (1963), Hunziker (1974), Flisch (this chapter, section 2)

B 4 B: $5.33 \times$ nl STP ^{40}Ar rad/g

see Jäger *et al.*, (1963), Hunziker (1974), Flisch (this chapter, section 3).

LP-6: $43.25 \times$ nl STP ^{40}Ar rad/g

see Ingamells, and Engels (1976), Engels and Ingamells (1977).

Gl-O: $24.80 \times$ nl STP ^{40}Ar rad/g

see Odin *et al.*, (1976), Zimmermann and Odin (1979).
The results are shown in Table 1.

Table 1. Spike calibration.

Calibrated with	No of measurements	$\times$ nl STP ^{38}Ar spike for the first shot	Stand. dev. 1σ(%)	% ^{40}Ar rad.
B 4 M	8	1.333 ± 0.007	± 0.53	62.0 ± 1.4
B 4 B	6	1.327 ± 0.007	± 0.53	79.7 ± 2.3
LP-6 Bio	6	1.336 ± 0.007	± 0.52	97.0 ± 0.7
GL-O	6	1.334 ± 0.009	± 0.67	95.0 ± 1.6
Mean		1.3325 ± 0.0038	± 0.28	
Air	15	1.349 ± 0.003	± 0.22	

The spike value calibrated with the known amount of air argon shows a slight increase of about 1%, most probably caused by the summation of a systematic error when the pipette was actuated 64 times. We are trying to improve this calibration method.

The 15 independent measurements of an accurately known air volume have given a *reproducibility of* ±0.2%, including the purification and measuring system. Higher standard deviations on a sample or a standard which is measured several times are most probably caused by inhomogenities in the sample or by behaviour during the extraction process.

1.6 Comparison of standards B 4 M, B 4 B, LP-6 and GL-O

In connection with the spike calibration we calculated for each standard a corresponding spike value. The ^{40}Ar rad content of each standard was determined in relation to each spike value. The results are shown in Table 2.

Table 2. Calculated values of the various standards based on different spike calibrations (in nl STP ^{40}Ar rad/g).

Spike calibration based on	Calculated values of the various standards B 4 M	B 4 B	LP-6	GL-O
B 4 M		5.354 ± 0.040	43.15 ± 0.32	24.78 ± 0.21
B 4 B	6.282 ± 0.047		42.96 ± 0.32	24.67 ± 0.21
LP-6	6.324 ± 0.047	5.366 ± 0.040		24.84 ± 0.21
GL-O	6.315 ± 0.054	5.358 ± 0.046	43.18 ± 0.37	

2 POTASSIUM AND ARGON RESULTS ON MUSCOVITE BERN 4 M

		% K	*n*	× nl STP ^{40}Ar rad/g	% Ar rad	*n*
Adams	Rice Univ., Houston	8.53	5	6.501		5
Afanass'yev *et al.*	Acad. of Sciences, Moscow	8.59	4	6.46		4
Armstrong	Univ. Yale	8.77	6	6.29		3
Armstrong	Univ Brit. Columbia	8.58	1			
Bonhomme	Univ. Louis Pasteur, Strasbourg			6.22 ± 0.07		8
Bonhomme *et al.*	Univ. Louis Pasteur, Strasbourg			6.23 ± 0.14		11
Cantagrel	Univ. Clermont-Ferrand	8.61 ± 0.06	8	6.67 ± 0.23	45–65	4
Cordani *et al.*	Univ. Sao Paolo			6.225		2
Damon and Livingston	Univ. Arizona, Tucson	8.70	3	6.183		1
Flisch	Abt. fuer Isotopengeol. Univ. Berne			6.324 ± 0.047 Spike cal. with LP-6 6.315 ± 0.054 Spike cal. with GL-O	62.0 ± 1.4	6
Funk	Max Plank Inst., Mainz	8.46	1	6.14		1
Harre	BGR, Hanover	8.664 ± 0.038	6			
Hebeda *et al.*	ZWO Lab. voor Isotopen-Geol., Amsterdam	8.74	2	6.22	63.5	4
Horn *et al.*	Univ. Heidelberg	8.84 ± 0.03	9	6.17	52–69	7
Hunziker	Abt. fuer Isotopengeol. Univ. Berne			6.32 ± 0.07 Spike cal. with P 207	60–64	10
Kreuzer	BGR Hanover			6.368 ± 0.025	64.4 ± 0.6	11
Lanphere and Dalrymple	*J. Geoph. Res.*, **70**	8.616	3	6.42		2
McDougall	Australian Nat. Univ., Canberra	8.718 ± 0.042	7	6.325 ± 0.048	62.3–66.3	4
McDowell	ETH, Zurich	8.73	3			

n Number of measurements.

		%K	n	×nl STP ^{40}Ar rad/g	%Ar rad	n
Mueller	K–Ar Dating, Schaeffer and Zaehringer (66)	8.81/8.84	2			
Ohmoto *et al.*	*Econ. Geol.*, **61**	8.67	1	5.75/6.00 6.11/6.33		4
Purdy and Jäger	Abt. fuer Isotopengeol. Univ. Berne	8.70±0.05	29	6.32±0.07 Spike cal. with air	54.7±3.1	8
Savelli	Univ. Bologna			6.24		1
Signer and McDowell	*Eclogae Geol. Helv.* (1970)			6.30±0.26	51.8–65.4	6
Soroiu	Bucarest Academy			6.376	43.7–52.0	2
Steiger	*J. Geoph. Res.*, **69**	8.635	1	6.18		1
Welin	Naturhistoriska Riksmuseet, Stockholm	8.68±0.13	7	6.16±0.14		7
Zimmermann	Univ. Nancy	8.70±0.06	4	6.27±0.33	45–60	4

3 POTASSIUM AND ARGON RESULTS ON BIOTITE BERN 4 B

		% K	n	×nl STP ^{40}Ar rad/g	% Ar rad	n
Armstrong *et al.*	Phys. Inst., Berne	7.90	2	5.12	16–46	3
Armstrong	Univ. British Columbia	7.85	1			
Armstrong	Univ. Yale	7.83	4	5.33		3
		7.96	5	5.31		4
Bonhomme	Univ. Louis Pasteur, Strasbourg			5.41		1
Bonhomme *et al.*	Univ. Louis Pasteur, Strasbourg			5.37±0.07		11
Cantagrel	Univ. Clemont-Ferrand	8.05±0.12	4	5.40±0.10	83–87	4
Cmerlatti	Univ. Sao Paolo	7.77	2			

n Number of measurements.

		%K	n	×nl STP ^{40}Ar rad/g	%Ar rad	n
Flisch	Abt. fuer Isotopengeol. Univ. Berne			5.354±0.040 Spike cal. with B 4 M	79.7±2.3	6
				5.366±0.040 Spike cal. with LP-6		
				5.358±0.046 Spike cal. with GL-O		
Hebeda *et al.*	ZWO Lab. voor Isotopen-Geol. Amsterdam	7.91	2	5.31	56.6–81.5	3
Hunziker	Abt. fuer Isotopengeol. Univ. Berne			5.29 Spike cal. with P 207	66–71	7
McDougall	Australian Nat. Univ., Canberra	7.968±0.045	3	5.334±0.045	52–83	4
McDowell	ETH, Zurich	7.88	2			
Purdy and Jaeger	Abt fuer Isotopengeol. Univ. Berne	7.90	10	5.33±0.08 Spike cal. with air	67.6	4
Steiger	*J. Geoph. Res.*, **69**	8.04	1	5.01		1
Welin	Naturhistoriska Riksmuseet, Stockholm	7.86±0.13	8	5.24±0.06		8

Acknowledgements

The writer wishes to thank Professor E. Jäger for much helpful advice. Help from other members of the Berne Isotope Geology Laboratory was appreciated. Mr. R. Kraehenbuehl made some argon analyses. Mr. R. Brunner designed the console and selected the auxiliary devices. The research is supported by the Schweizerischer Nationalfonds zur Foerderung der wissenschaftlichen Forschung.

Résumé du rédacteur

Une nouvelle ligne d'extraction de purification et de mesure a été construite et testée en 1979 et 1980 à Berne. Quatre minéraux de référence disponibles actuellement ont pu être comparés à cette occasion: Biotite Bern 4B, Muscovite Bern 4M, Biotite LP-6 Bio et Glauconite GL-O.

L'auteur profite de cette occasion pour faire le point des données parvenues à ce jour concernant les 2 minéraux de référence préparés à Berne.

(Manuscript received 23-1-1981)

Numerical Dating in Stratigraphy
Edited by G. S. Odin

9

Range and effectiveness of unspiked potassium–argon dating: experimental groundwork and applications

CHARLES CASSIGNOL and PIERRE-YVES GILLOT

INTRODUCTION

Ten years ago, the 'Centre des Faibles Radioactivités' decided to set up a laboratory of potassium–argon dating devoted to young rocks: essentially Pleistocene and Holocene volcanic rocks. The first problem to be solved was that of argon isotopic measurement (potassium is dosed by atomic absorption spectrometry; this is the most frequently used method and we shall not discuss it further). For argon isotopic measurements and using the static method in mass spectrometry, we worked out an operating procedure in which the mass spectrometer becomes an instrument of absolute measurement; spiking was thereafter discontinued. The same principle has also been envisaged by other workers: Dalrymple (1969); Baksi (1973); Marty (1979). As our research developed, the advantages of this method were emphasized: substantial progress was exhibited in the measurements of young and very young ages. Examples of these will be given below following the presentation of the procedure. We shall not consider here either the principle of the K–Ar dating method or its customary way of application (see Schaeffer and Zähringer, 1966; Dalrymple and Lanphere, 1969).

1 PRINCIPLE OF MEASUREMENT OF RADIOGENIC ARGON

Radiogenic argon (Ar*), i.e. ^{40}Ar resulting from ^{40}K decay, is measured by mass spectrometry. It is, in fact, mixed with a certain amount of contaminating argon, some from the minerals themselves, the rest from laboratory processes (line blank). In general, we will assume (unless there are contrary indications) that the isotopic composition of the contaminating argon is that of the natural (atmospheric) argon. Consequently, knowing the abundances of ^{40}Ar and ^{36}Ar in atmospheric argon, it is possible, after measuring the ^{36}Ar in the sample, to deduce the radiogenic argon from the difference between total ^{40}Ar and contaminating ^{40}Ar.

The mass spectrometers used in the K–Ar method run often in the 'static' mode. After extracting and purifying the argon from a sample, we introduce it into the cell of the MS, in which we have previously created a vacuum. We then measure the signals, meanwhile the pumping is halted.

Under these conditions, an important decreasing in the peaks becomes apparent, e.g. >10% within half an hour (Dalrymple and Lanphere, 1969), which makes it impossible to establish any link between the height of these peaks and the partial pressures of the three isotopes. The use of a 'spike' of ^{38}Ar, the peak of which changes in the same way, provides a mobile quantitative reference and makes it possible to discount the drift. It also makes it possible to discount partial transfers of the gases.

Leaving out the spike has been made possible by an observation made by one of us (Cassignol, 1973). If we continue to purify argon while measuring it, by inserting a getter in the cell, the signals virtually stop drifting. The evolution of the ^{40}Ar peak may not exceed one per thousand in one hour. If we insert a new sample, identical to the former one, in the cell after clearing it, the heights of the signals remain unchanged. If we introduce a sample of argon extracted from the atmosphere with a convenient pipette (see below), the processing makes it possible to detect the daily variations of the atmospheric pressure. The heights of the peaks then depend directly on the magnitude of the introduced sample. Since it is possible to calibrate the MS in a separate experiment, the quantitative reference provided by the spike is no longer necessary. In addition, as we shall see, the transfers of the gases over the whole process can be rendered as quantitative as we need. Thus, abandoning the isotopic dilution offers no major drawback. On the contrary, we shall now emphasize the advantages of this.

(a) The suppression of the drift of the signal improves the accuracy of the measurements. In order to take full advantage of it, we must discontinue continuous scanning of the spectrum and operate with stable parameters: i.e. by peak-switching or simultaneous collection.

(b) Omitting the spike frees us from the necessity to know its actual isotopic composition.

(c) A direct comparison between ^{38}Ar and ^{40}Ar peaks, where spiking is used, has some meaning only if the response of the mass spectrometer is directly proportional to the partial pressures of the two isotopes; which appear not to be the case, at least for our set-up. In our measurements, we take strictly into account the unproportionality of the response of the spectrometer.

(d) Omitting the spike allows us to measure mineral ^{38}Ar. If the ratios of ^{36}Ar and ^{38}Ar in the sample are the same as in atmospheric argon, we may consider natural argon to be the contaminating agent. In a few rare cases we have encountered and worked out some anomalies in the ratios of the rare isotopes.

(e) Last but not least, the suppression of the spike makes possible the use and language of metrology. For instance, it is possible to define a single working point for the MS where we can carry out the three measurements of argon necessary to find the percentage of radiogenic argon in a mineral sample, as we shall see later. This working point is determined by the working parameters of the ion source (intensity of electronic current, accelerating voltage of electrons), by the environmental data, such as temperature, by the nature of the spectrometer background (among other factors, the residual peaks) and by the amount of introduced argon (the load). This load in practice consists of the amount of ^{40}Ar, the other isotopes being very rare. On the contrary, when using a ^{38}Ar spike, the load corresponds to the sum of the ^{38}Ar and ^{40}Ar amounts.

Let us note that the approach to measuring a peak of ^{36}Ar in a mineral sample and in a similar atmospheric argon sample (i.e. the same peak of ^{40}Ar) is equivalent to the Borda double weighing method (weighing right with a false balance). The reliability of the measurement of radiogenic argon depends solely on the proportional response of the detector.

Let us make it clear from this point on that the spectrometer background has to be cleared of the helium introduced together with the mineral argon samples before every measurement. Coming from the atmosphere, this helium crossed the wall of a silica tube heated to 900°C and containing titanium sponge for purification purposes. If helium were present in the mass spectrometer while the argon was being measured, the working point of the set-up would be altered in an uncontrollable way and the results of the measurements would be unreliable. We had the opportunity to check this point, and from now on we shall consider that helium has been cleared off, through a process that we shall describe later.

In addition, we must keep the residual gases to a constant level that should be as low as possible. This is where the getter comes in. At the outset of our research we observed that the drift of the argon peaks was simultaneous with the rise in residual gases. The use of a getter eliminates the drift together with the rise. Not any kind of getter can be used. Titanium sponge, in fact, does not trap hydrogen at the temperature (900°C) where it efficiently traps active gases such as O_2, N_2 and CO_2. Notably, water is decomposed and hydrogen evolves. The getters we use are industrial zirconium–aluminium getters (manufactured by the Milan firm SAES), which trap any active gas efficiently, including hydrogen and hydrocarbons, at a temperature not exceeding 400°C.

The device we use is a cylindrical assembly of metal plates, covered with the Zr–Al alloy, radially arranged and heated by a filament radiating from the axis of the cylinder. The whole device is contained in a cylindrical coaxial bulb, which has electric leads for heating the filament and is connected to the line. Under normal working conditions, the supply voltage for the filament is 12 volts. Under activated conditions (30 volts), the temperature rises so as to evolve any previously absorbed hydrogen, which is then eliminated by pumping.

2 PRACTICE

2.1 Description of the argon measuring system

We shall limit ourselves to a succinct description here. The essential information is given in Figure 1. We emphasize the simplification of the pumping system and the omission of ionic pumps (they caused some trouble because of their bad performance for argon and helium). The transfer of the gas in two steps through the traps C_1 and C_2 is also essential. The efficiency of this mode of transfer is close to one part in ten thousand, and the overall process does not last more than 15 minutes.

The line is made up of stainless steel. It, as well as the mass spectrometer cell, can be baked after the magnet has been removed. Note also that there is no ionization manometer, the operation of which would enhance memory effects.

2.2 Operating mode

After preparing the mineral samples on a separate set-up, i.e.:

—pre-degassing of the mineral at room temperature or at any temperature up to 300°C;

—vacuum melting in an induction furnace;

—purification of the evolved gases through hot titanium sponge, excluding any other chemical absorbent or cold trap (we ensure the quantitative recovery of argon by using a large cooled active carbon trap while melting and keeping it at 400°C during the purification stage);

the mineral argon samples are enclosed in Pyrex glass bulbs $M_1, M_2, \ldots$ (see Figure 1) arranged in an admission unit P and the gas is admitted into the line by a magnetic hammer acting on a break-off seal.

Argon is then attracted in the part of the line close to the cell by traps C_1 and C_2 working in cascade. Helium is eliminated by opening the valve V_{12} between the line and the pumping system. V_6 is then shut and C_1 and C_2 brought to room temperature. The cell is evacuated by opening V_1. After closing V_1, V_2 is opened and argon allowed to expand into the cell; after

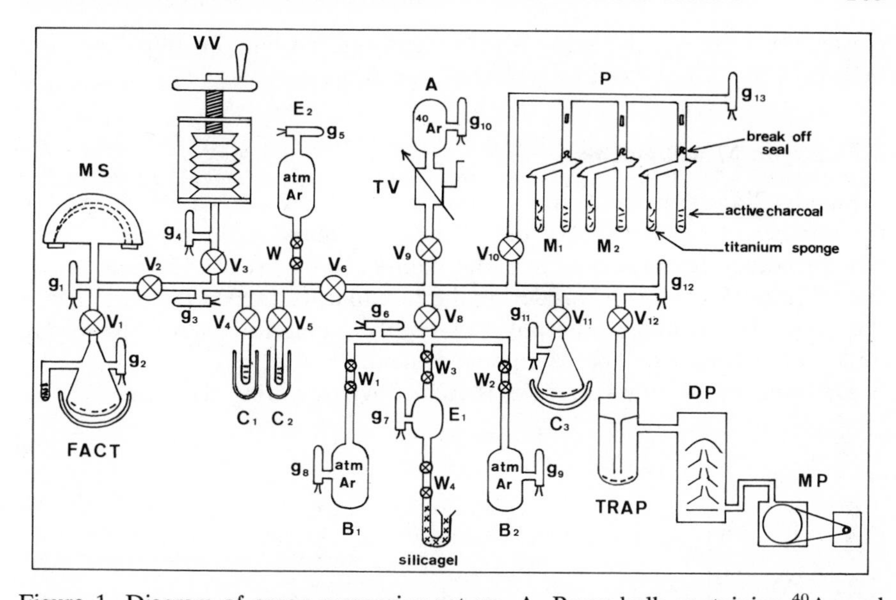

Figure 1. Diagram of argon measuring set-up. A. Pyrex bulb containing ^{40}Ar and connected to the main line through a throttle valve (TV). B_1, B_2. Pyrex bulbs containing atmospheric argon. They are connected to the line through double expansion valves (W_1, W_2), making it possible to pipette into the line discrete amounts of gas, different for the two bulbs, without any isotopic fractionation. E_2. Secondary gas standard, a bulb similar to B_1 and B_2 except that it is located next to the mass spectrometer, containing atmospheric argon. It obviates daily use of the main standard E_1 and is periodically calibrated according to the latter. C_1, C_2. Active charcoal traps. These are used to attract mineral samples, as well as the amounts of air argon pipetted from E_1, in the compartment of the line next to the cell. They contain very little charcoal and run alternatively, the corresponding valve being shut when the trap is not running. C_3. Active charcoal trap. This is used only to improve the vacuum in the line after the operation of the DP. E_1. Main gas standard. This device permits a constant quantity of atmospheric argon to be admitted into the line. It is made up of two double expansion valves (W_3, W_4) separated by an auxiliary chamber volume in which the dry air admitted undergoes a purification step. An aliquot is then sampled through the second valve and purification is achieved in an annexe of the line. FACT. Fast active charcoal trap, applied to the MS cell. G_1–G_{13}. Getters. MS. 180° mass spectrometer, fitted with a permanent magnet giving a uniform induction of 35 tesla in the gap. The source is of the Nier type, with no electron trap or ion repeller and with two half-plates adjusted to the optimal ^{40}Ar signal peak. The collector has a simple slit (peak switching). The ions are collected by means of a Faraday cup (without electron multiplier). The radius of the ionic trajectory is 6 cm and the accelerating voltage for ^{40}Ar is 516 V. Pumping system including DP.: silicon oil diffusion pump, separated from the line by an active charcoal trap working at room temperature, backed by a two-stage mechanical pump MP. V_1, V_2. All metal valves (diam. 10 or 20 mm). VV. Variable-volume stainless-steel bellows container operated by a wheel-driven screw. This makes it possible to continually vary the volume of the line, and hence the argon pressure, without any isotopic fractionation. $W_1, W_2, \ldots$ double expansion valves (pipettes).

closing the valve V_2, the argon is then measured. During this step, while the bellows VV are not working, valve V_3 remains closed.

2.3 Argon Measurements

Figures 2 and 3 bring us to the operating mode. What we have to find is the number of atoms of ^{40}Ar and ^{36}Ar in a mineral argon sample (^{38}Ar is not considered for the moment). The number of atoms of ^{40}Ar is given by the diagram (Figure 2). Considering the fact that we exceed the accuracy of the graph by parabolic interpolation on the numerical values, we easily obtain an accuracy of one part per thousand.

The number of atoms of ^{36}Ar is found by comparing the height of the

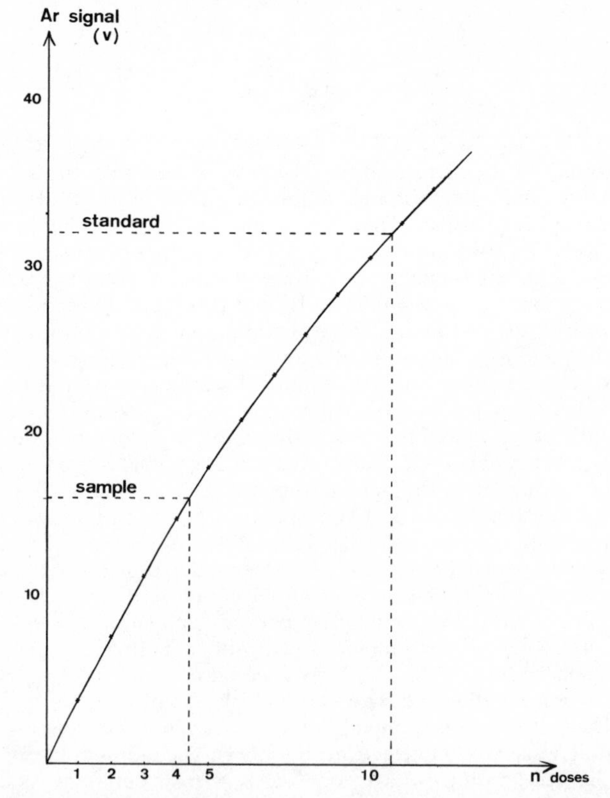

Figure 2. Diagram of the MS response as a function of argon content. Abscissae show successive amounts of argon pipetted from B_1, ordinates heights of the ^{40}Ar peaks referred to a supposed resistor of 10^{11} ohm (the actual resistor is about 200 times lower). On the diagram we have indicated the heights corresponding to a mineral sample and a standard. The ratio between the corresponding amounts of ^{40}Ar is equal to that of the abscissae.

corresponding peak to that of the same isotope in atmospheric argon under the same working conditions—i.e. the mass spectrometer giving the same 40 peak. We thus determine the 'instrumental ratio' $^{40}Ar/^{36}Ar$ (uncorrected) for the sample (RE) and for atmospheric argon (RA). The procedure is as follows. When the measurements of mineral argon are completed, the cell is cleared through the charcoal trap 'FACT', then filled with atmospheric argon from B_1 and/or B_2 up to a point near the former filling point. During this process we keep V_3 open, and operating the bellows allows us to fit the height of the ^{40}Ar peak to a few parts in ten thousands. Under these conditions, the proportion of Ar* in the sample is (RE−RA)/RE and the amount of radiogenic argon is $Ar^* = Ar(RE-RA)/RE$, Ar being the total ^{40}Ar.

We emphasize that neither the actual isotopic ratio of the sample nor that of the atmospheric argon (295.5) are used in this expression.

In practice, each measurement on mineral argon, following a measurement on a standard, is followed by the corresponding measurement on atmospheric argon and by the drawing out of the response diagram.

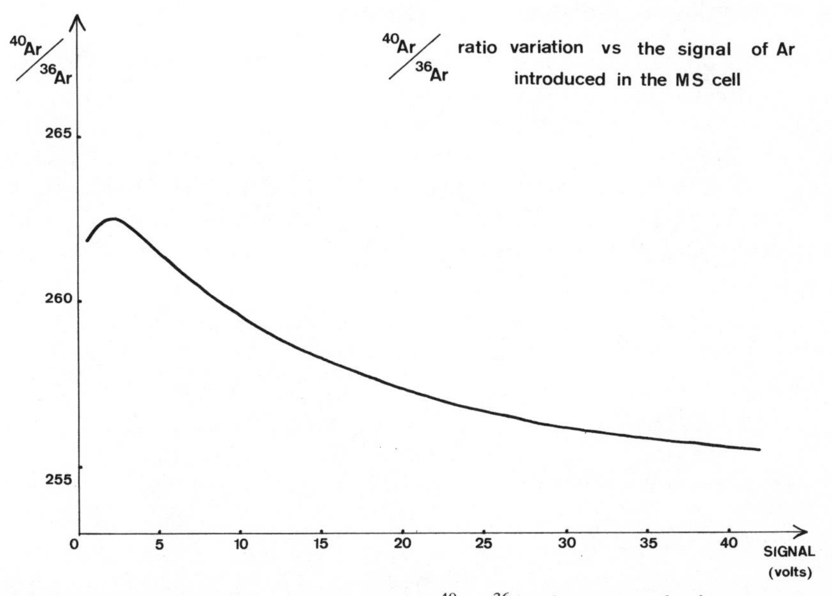

Figure 3. Value of the 'instrumental ratio' $^{40}Ar/^{36}Ar$ for atmospheric argon, as a function of the ^{40}Ar peak. It should be noted that this ratio depends on argon pressure (i.e. on the working point, as defined earlier), and that its average value differs from the actual ratio (295.5). Consequently the correction for the contamination cannot be carried out by using a constant ratio for atmospheric argon, but by using the ratio corresponding to a filling up of the mass spectrometer with atmospheric argon, giving the same ^{40}Ar peak as the mineral sample.

We recently improved this operating mode by a third measurement upon pure ^{40}Ar with the same 40 signal. The purpose is to work out the spectrometer background, and more particularly the height of the residual 36 signal. We operate for convenience in the dynamic mode. We fit the height of the 40 peak by operating the throttle valve and V_{12} (leading to the diffusion pump). The height of the 36 signal resulting from this measurement is subtracted from those given by each of the first ones.

We measure ionic currents by peak-switching or simultaneous collection. In the former case, accuracy in measuring the 36 peaks reaches one part per thousand, as well as that obtained for the ratios RE and RA. Simultaneous collection gives an accuracy at least twice as good.

However, the reliability of these results implies that particular care has been given to the quality of the supply equipment. For instance, the electric power for the pilot and measuring circuits is provided by a motor-driven alternator, which ensures a very regular voltage supply completely free from any industrial interference. The laboratory is also equipped with a fairly sophisticated air-conditioning system.

2.4 Calibration of the atmospheric samples

The amounts of the standard (atmospheric argon) delivered by the bulb E_1 are attracted by C_1 and C_2 and generally processed as a mineral argon sample. As we do not know the constructional volumes of E_1 accurately, we calibrate this standard by comparison with a mineral standard that we process as an ordinary sample. This mineral standard (GL-O) was chosen after world supplies of former international standards (Bern 4 M and P 207) grew rare. Several laboratories, including ours, simultaneously operated

Table 1. Analytical data on the standard glauconite GL-O. Sample weights around 0.3 g; mean value: 24.95±0.15 (2σ) and ±0.50 for standard uncertainty (Cassignol *et al.*, 1977); authors' value: 24.82 nl/g (Odin and coll., 1976).

Date of measurement	Atmospheric argon (%)	Radiogenic ^{40}Ar (nl/g)
30 05 75	28.9	25.07
30 05 75	17.8	24.85
02 06 75	19.9	24.94
19 11 75	31.1	25.01
20 11 75	15.0	24.90
16 02 76	8.9	24.87
17 06 76	11.9	24.98
21 06 76	8.5	24.91
10 08 76	4.9	25.04

the comparison of this new standard with the latter. The average result of these measurements is 24.82 nl of Ar* per gramme of GL-O. This will be our reference value.

Our own measurements led to an average value of 24.95 nl/g within an accuracy better than 0.5% (2σ) (Cassignol *et al.*, 1977). The measurements extended over one year (Table 1). For routine measurements we use the secondary standard E_2, which is easier to operate. Its value decreases as amounts of argon are pipetted from the bulb. The decrement, which is very small (about 3/10,000), is periodically checked with the main standard.

2.5 Analytical control and uncertainty

Except in the case of highly contaminated samples, mainly coming from very young minerals, the accuracy of the ages is that of potassium measurements. So the method can measure quite variable ^{40}Ar signals, such as those obtained for glaucony samples by G. S. Odin, and with apparent ages ranging from less than 1 Ma for a glaucony dredged off California (G.585 A) to 130 Ma (Portlandian) (Table 2). However, the most original field of application corresponds to the young ages and analyses in which the amount of contaminating argon is preponderant.

Table 3 represents the impact on age following the use of an incorrect value of the instrumental atmospheric ratio RA. Such errors would occur if we were to use an average value instead of measuring the RA at every dating process . . . . Observing the form of the plot, we see that the impact of this error is negligible and even not apparent in the case of little contaminated samples. On the other hand, the knowledge of the RA to within an

Table 2. Argon measurements of glaucony samples, a comparison with previous data; analytical uncertainties quoted to 2σ.

Sample	rad. ^{40}Ar (%)	Radiogenic ^{40}Ar $\pm 2\sigma$ (nl/g) Present study	Previous data	Laboratory
G.585 A*	0.0	0.05 ± 0.02	0.0	Leeds
ML210 A*	9–12	1.08 ± 0.03	1.02 ± 0.15	Berne
ML279 A*	22–26	1.66 ± 0.03	1.50 ± 0.08	Berne
G.150 A NDS4	56–83	9.69 ± 0.15	9.67 ± 0.22	Berne
			9.74 ± 0.20	Strasbourg
G.101 A NDS13	45–65	10.06 ± 0.15	10.04 ± 0.32	Berne
			10.01 ± 0.25	Strasbourg
G.5132 A NDS30	78–85	10.02 ± 0.15	10.28 ± 0.18	Berne
G.440 A NDS37	63–72	11.50 ± 0.17	11.58 ± 0.17	Strasbourg
G.348 A NDS66	79–90	25.50 ± 0.38	25.58 ± 0.30	Berne
G.482 A USNDS76	84–85	36.12 ± 0.54	35.93 ± 0.79	Berne

* See Odin and Dodson, this volume, Chapter 14 and Part II of this book.

Table 3. Plot of the errors for the apparent age as a function of contamination and relative error for the $^{40}Ar/^{36}Ar$ ratio. The plotted values are numerical applications of the expression (dRA/RA)($c/1-c$) to the sign.

dRA/RA (%) \ Contamination (%)	10	20	50	80	90	99
10	1	2.5	10	40		
5	0.5	1.2	5	20		
1		0.25	1	4	9	
0.3			0.3	1.2	3	30
0.1			0.1	0.4	0.9	10

accuracy of one per thousand allows us to determine an age to about 10% when the sample is contaminated to 99%.

We might observe that the presence in the cell of a foreign gas such as helium, if we omit to eliminate it, or a spike of ^{38}Ar alters the working point of the mass spectrometer, and particularly the ratio $^{40}Ar/^{36}Ar$ for the sample (RE). Under these conditions it is impossible to fix the same working point for the RA, and hence to proceed to a reliable dating. Unless the RA of a spectrometer is independent from the working point (i.e. in the first place from the filling point), the isotopic dilution method is not reliable as far as young ages are concerned.

Comparisons between $^{40}Ar/^{36}Ar$ ratios obtained for different measurements of atmospheric argon at the same working point of the mass spectrometer reveal a $\pm 0.2\%$ (2σ) accuracy for the value of this ratio (e.g. 260 ± 0.5 for the RA at one point of the curve in Figure 3). This accuracy typically represents an age value limit of $\pm 10{,}000$ years for a normally contaminated (0.2 nl of Ar per gramme) basalt (1% potassic) and less than 1000 years for sanidines.

3 REFINEMENTS IN PROCESSING MATERIALS

When considering the accuracy of argon and potassium measurements, the large spread and meaninglessness of some ages reveal the existence of disturbing factors the origin of which may be searched for in the rocks themselves. We shall limit our consideration here to the Recent effusive volcanic rocks. Such rocks can contain extraneous ^{40}Ar (Damon *et al.*, 1967; McDougall *et al.*, 1969; Dalrymple, 1969), magmatic inherited argon enclosed in early crystallizing mineral phases, or chips of older material coming from basement. This extraneous argon is responsible for the lack of consistency of some datings. In fact, recurrent datings on a basalt of one of the oldest flows of the Chaîne des Puys, at Egaules (Massif Central, France), demonstrate this in evidence. When whole-rock (size 1–3 mm) was analysed the apparent ages ranged between 110,000 and 380,000 a. Dating measurements carried out on the groundmass (after removing phenocrysts of magnetite, pyroxene and olivine, denser than the groundmass, and phenocrysts of plagioclase and chips of older granitic material, less dense) gave an apparent age of 75,000 ± 8000 a, and the age result dispersion became consistent with that of Ar measurements.

The basalt of the Puy de la Vache flow (Massif Central, France), sampled at Saint Saturnin, processed in whole-rock, was dated to 30,000 ± 6000 a (Table 4). Previously dated by radiocarbon on charcoal sampled under the lava flow, its age is 7800 ± 300 a (Pelletier *et al.*, 1959). This basalt is virtually free from xenolites, but contains a noticeable amount of phenocrysts of augite (about 5%). Removing them lowers the apparent K–Ar age to 12,000 ± 6000 a. The amount of excess $^{40}Ar^*$ observed in the whole-rock analyses was approximately found in the pyroxenes (2×10^{-3} nl/g) after atmospheric correction.

Following these observations, we now systematically operate by sorting the phases, which enables us to process a more homogeneous and more age-representative material. The technique we use is gravity separation. The selected fraction of the crushed rock (generally 250–500 μm) is processed in bromoform–ethanol or di-iodo-methane–benzene mixtures. This operation does not affect the argon measurements by enhancing unwanted residuals in the spectrometer.

An observation of the rock in thin section makes it possible to determine an approximate density for each component of the lava to be dated (among which is the groundmass).

Table 4 presents other dating results on lavas of diverse chemical composition from different volcanic provinces, either Recent or dated by other methods. It tends to demonstrate that the K–Ar method is reliable under 0.1 Ma within an accuracy better than ±2000 a in favourable cases.

4 APPLICATIONS

4.1 Tephrochronology and reconstitution of the late volcanic activity in the Gulf of Napoli

We shall base our investigations on a few results collected by working upon the volcanoes of the Campanian plain and the Gulf of Napoli. This particular area offers for easy consideration the products of the activity of the volcanoes of the Phlegrean Fields and of the Somma Vesuvius, as well as of the volcanic islands of Procida and Ischia (Figure 4). Each one of these

Table 4. Analytical data on volcanic rocks of known ages. 1, basaltic flow of Puy de la Vache sampled at Saint-Saturnin. a, Pelletier *et al.* (1959); 2, andesitic flow of Volvic, sampled at La Nugère. b, Guérin, personal communication; 3, trachytic ashflow deposit sampled on Puy Lacroix. c, Brousse *et al.* (1966); 4, trachyandesitic flow of Laschamp; 5, basaltic flow of Olby; 6, basaltic flow sampled at Egaules. d, Brousse and Rudel (1973), e, Guérin and Valladas (1980); 7, olivine-rich oceanic flow erupted from Piton de la Fournaise in 1961; 8, 1971 basaltic flow sampled in the flow core; 9, Nicolosi basaltic flow; 10, shoshonitic flow from the base of the modern series of Stromboli sampled at Punta Frontone; 11, obsidian flow of Rocche Rosse, northeastern end of the island. f, Pichler (1961–1965); 12, trachytic flow erupted from l'Arso in 1302; 13, leucitic flow erupted from Vesuvius in 1944; 14, trachytic dome of Caprara connected with Astroni crater. g, Alessio *et al.* (1971).

Material K %	rad. ^{40}Ar (%)	rad. ^{40}Ar (10^{-3} nl/g)	Apparent age 10^3 a $\pm 2\sigma$	Age (a)
Chaîne des Puys (Massif Central, France)				
1 Whole-rock	0.95	1.3	27.0 ± 6.0	7800 ± 300 (^{14}C)[a]
1.29	0.75	1.7	32.0 ± 8.0	
	1.50	1.9	35.0 ± 6.0	
Groundmass	0.60	0.7	13.0 ± 4.0	
	0.35	0.5	10.0 ± 5.0	
	0.60	0.8	14.0 ± 5.0	
2 Whole-rock	0.30	0.4	8.0 ± 5.0	11,000 ± 3000 (TL)[b]
2.45	0.20	0.3	5.0 ± 5.0	
3 Whole-rock	−0.90	—	—	8000 ± 300 (^{14}C)[c]
4.30	−1.15	—	—	
4 Whole-rock	1.80	3.1	44.0 ± 5.0	See Figure 5
1.77				
Groundmass	1.40	2.7	38.0 ± 6.0	
1.83	2.85	2.6	36.0 ± 3.0	
5 Whole-rock	2.10	3.2	49.0 ± 5.0	See Figure 5
1.67				
Groundmass	1.70	3.1	46.0 ± 6.0	
1.75	1.40	2.9	42.0 ± 6.0	
6 Whole-rock	1.50	10.4	230.0 ± 30.0	35,000 (^{14}C)[d]
1.25	1.00	4.9	110.0 ± 20.0	
	1.60	11.2	250.0 ± 30.0	70,000 ± 6500 (TL)[e]
	1.80	17.1	380.0 ± 40.0	
Groundmass	2.20	4.1	70.0 ± 11.0	
1.49	3.40	5.2	85.0 ± 7.0	
	3.00	4.5	75.0 ± 7.0	
Island of Reunion (Indian Ocean)				
7 Whole-rock	0.0	<0.9	<50.0	1961 PC flow
0.50	0.1	<1.0	<70.0	
Mount Etna (Sicily)				
8 Whole-rock	0.1	<0.6	<10.0	1971 PC flow
1.45	0.1	<0.4	<7.0	

Table 4. *Continued.*

	Material K %	rad. ^{40}Ar (%)	rad. ^{40}Ar (10^{-3} nl/g)	Apparent age 10^3 a $\pm 2\sigma$	Age (a)
9	Groundmass	0.0	<0.0	<5.0	16th century PC
Eolian Islands (Italy)					
10	Groundmass	0.20	0.8	5.0± 5.0	Subactual
	4.03	0.25	1.0	6.5± 5.0	
11	Glass	0.0	<0.2	<2.0	Late Roman decade,
	5.0	0.1	<0.3	<3.0	about 1400 BP[f]
Campania (Italy)					
12	Whole-rock	0.3	1.5	8.0± 5.0	1302 PC
	5.0	0.2	1.0	5.0± 5.0	
	Groundmass	0.1	<1.2	<5.0	
	4.5	0.0	<1.0	<4.0	
	Sanidine	0.4	0.8	2.0± 1.0	
	9.0	0.2	0.4	1.0± 1.0	
		0.0	<0.4	<1.0	
13	Whole-rock	0.2	<1.8	<8.0	1944 PC
	5.5				
	Leucite	0.0	<3.0	<5.0	
14	Sanidine	0.9	1.4	3.3± 0.8	3700±200 (^{14}C)[g]
	10.55	1.2	1.8	4.2± 0.8	

volcanoes has displayed activity at sometime within the historical era: eruption of Monte Nuovo in the Phlegrean Fields in 1538, the almost continuous activity of the Vesuvius from 1631 to 1944, and the Arso flow in 1302 in the island of Ischia.

4.1.1 *Phlegrean Fields*

The volcanic activity of the Phlegrean Fields has traditionally been separated into three main periods (Rittmann, 1950), and Table 5 gives the results of measurements carried out by the K–Ar method upon the sanidine, the mesostase or the glass of trachytic lavas belonging to those three phases.

The first period includes polygenic breccias (Breccia Museo), the Piperno facies (trachytic ashes deposited when hot, hence indurated and shaped into glassy fiammes), as well as trachytic domes still outcropping in the north-western part (e.g. Cuma) which represent, with the Monte di Procida, the heights left after the collapse that occurred in the Gulf of Napoli and in the central part of the Phlegrean Fields. The eruption of the Campanian ignimbrite can be related to that first period; this formation outcrops on a surface

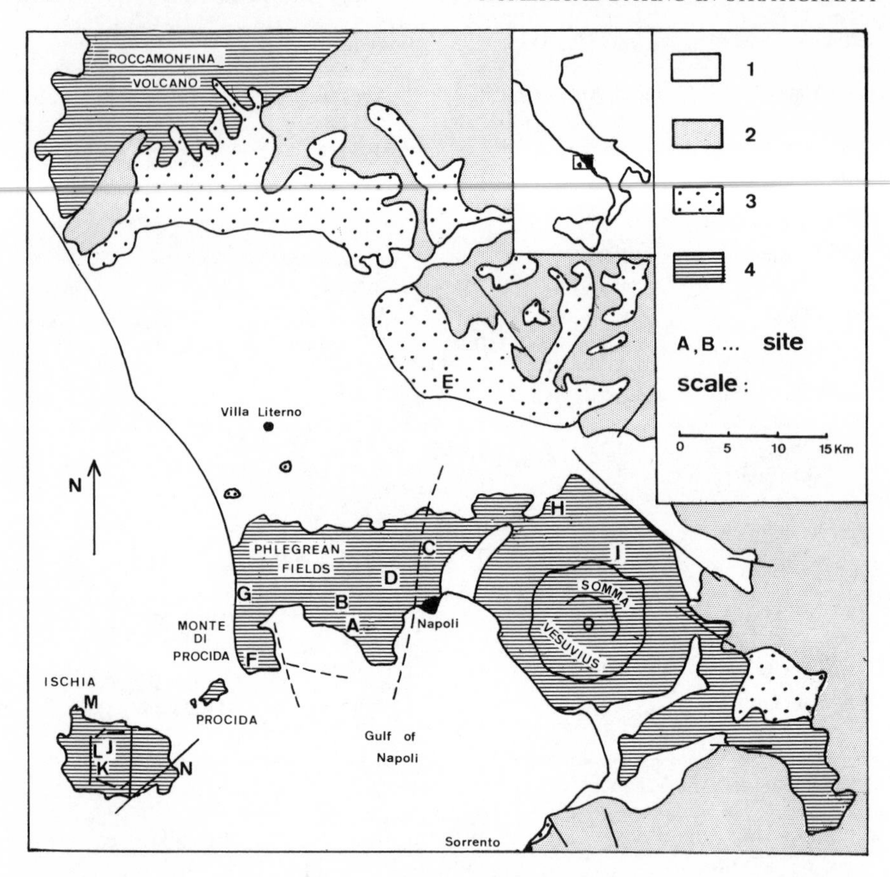

Figure 4. Geological sketch map of the Campania volcanic area. 1, Alluvium and lacustrine deposits; 2, Apenninic formations; 3, Campanian ignimbrite; 4, volcanic complexes. A, B, . . . correspond to samples reported in Tables 5, 6 and 7.

of 500 km^2. This monotonous ashflow deposit also covered a surface fifteen times larger, extending from the Roccamonfina in the north to the Sorrento Peninsula in the south; its boundary (shown on the map in Figure 4) towards the east exceeds the Somma Vesuvius area. This pyroclastic flow could be correlated to the level Y_5 observed in numerous cores in the eastern Mediterranean (Thunell *et al.*, 1979). ^{14}C data allowed its age to be located at between 28,000 and 35,000 years (Di Girolamo and Keller, 1972). The datings we worked out mean that the Piperno might be just a peculiar facies of the Campanian ignimbrite, close to its emission point, which is probably located between Napoli and Villa Literno, along an 'Appenninic' NW–SE fault (Barberi *et al.*, 1978).

Table 5. Age data on lavas from the Phlegrean Fields. Sample weights: 4–8 g. A, Monte Olibano trachytic dome; B, Caprara trachytic dome; C, pumice level connected with Neapolitan yellow tuff; D, Piperno sampled at Pianura; E, Campanian ignimbrite sampled at Santa Agata dei Goti; F, Monte di Procida, welded scoriae from cliff of the Marina di Vita Fumo; G, Cuma trachytic dome.

Site	Material K %	rad.^{40}Ar (%)	rad.^{40}Ar (10^{-3} nl/g)	Apparent age $\pm 2\sigma$ 10^3 a (ICC)
A	Sanidine	1.4	1.6	3.8±0.6
	10.50			
	Groundmass	1.0	1.2	4.0±1.0
	7.50			
B	Sanidine	0.9	1.4	3.3±0.8
	10.55	1.2	1.8	4.2±0.8
C	Sanidine	9.9	6.4	15.4±0.6
	10.54	8.8	5.8	13.8±0.6
		3.6	6.3	15.2±0.9
D	Sanidine	1.8	9.8	30.5±3.5
	8.17			
	Groundmass	1.7	7.7	33.0±4.0
		1.8	7.8	33.5±4.0
E	Sanidine	9.7	14.8	33.5±1.0
	11.19	8.9	15.2	34.5±1.0
		8.8	14.0	32.0±1.0
F	Sanidine	4.5	13.3	38.2±1.7
	8.85	4.2	13.4	38.5±1.8
G	Groundmass	6.8	8.1	36.5±1.2
	5.60	7.2	8.2	37.0±1.2

The second period is characterized by the eruption of the 'Neapolitan yellow tuff', pyroclastic deposit outcropping between Napoli and the Phlegrean Fields. Measured by means of ^{14}C data, the age of this formation would be located between 11,000 and 13,000 years (Alessio *et al.*, 1971); the dating we present here has been calculated from the sanidines belonging to a level of trachytic pumices connected with this ignimbritic ashflow. The Neapolitan tuff, in fact, owes its colour and induration to alteration, and its facies is not suitable for K–Ar dating.

Lastly, the third period includes the latest materials produced by the different craters of the Phlegrean Fields up to the last eruption of Monte Nuovo in 1538. This time also, the results we obtained for the trachytic domes of Caprara and Monte Olibano, connected with the Astroni crater, are totally compatible with the ^{14}C dating of the Astroni projections back to 3700 years (Delibrias *et al.*, 1969).

4.1.2 Vesuvius

Besides the coherence of age measurements with stratigraphic observation (when possible), or with radiocarbon data, we foresee the testing of the internal coherence of the K–Ar measurements by analysing separated mineral phases (when the facies of the rock allows) such as sanidine or leucite and a microlitic phase (groundmass). An example of this type of test is given by the measurements carried out on Vesuvian flow (Figure 4 and Table 6): the flow of leucititic tephrite of Castello di Cisterna corresponds to the upper part of the mass of flows resulting from the first period of activity of the Somma Vesuvius. The southern part of that unit is covered by the pyroclastic products of the Somma, underneath which a palaeosoil was ^{14}C dated back to 17,000 years (Delibrias *et al.*, 1969). After comparing the results obtained with the leucite phenocrystals and the groundmass, separated from the 250–500 μm fraction part of the crushed rock, we extracted from mesostase, recrushed down to 80–125 μm, the microlites of leucite and of plagioclase. All the ages, corresponding to a potassium percentage varying between 2.5 and 16.5 and various amounts of contamination, range between 19,000 and 22,000 years; this is also an indirect test for the precision of the correction for the contamination.

Another flow of this unit has been dated: the PFSV 153 sample corresponds to a drilling at Case Trapolino. It makes it possible to date back 30,000 years the beginning of the eruptive activity of the Somma Vesuvius. The whole activity of the Vesuvius has thus covered the last 30,000 years and its products would lie on the Campanian ignimbrite (Barberi *et al.*,

Table 6. Age data on lavas from Vesuvius. Sample weights: 3–8 g. H, tephritic lava flow of Castello di Cisterna; I, tephritic lava from the first cycle of activity of Somma Vesuvius, sampled from a drilling at Case Trapolino.

Site	Material K %	rad.^{40}Ar (%)	rad.^{40}Ar (10^{-3} nl/g)	Apparent age $\pm 2\sigma$ 10^3 a (ICC)
H	Leucite	1.55	12.5	19.0±2.5
	Phenocryst	1.50	12.6	19.5±2.5
	16.57			
	Groundmass	1.40	3.4	21.2±3.0
	4.07	1.25	3.1	19.7±3.0
	Plagioclase			
	microlites	0.70	2.0	20.2±5.8
	2.54			
	Leucite			
	microlites	1.00	10.7	20.3±4.0
	13.32			
I	Groundmass	1.80	5.9	32.0±3.5
	4.65	1.70	5.3	29.0±3.5

1978), whereas older measures (Civetta *et al.*, 1970) gave an age of 250,000 ± 80,000 years to one of these flows drilled at Scaffati.

4.1.3 *Island of Ischia*

Finally, let us describe a few recent analyses carried out on the volcanic formations of Ischia, which permit a reconsideration of its volcano-tectonic context. Ischia, in fact, appears to be a volcano-tectonic horst, composed of pumice flow, volcanic and volcano-sedimentary deposits: the so-called 'green tuff'; beach levels as well as the presence of diffused glauconites testify to the fact that this subaerial formation was first completely immersed before its surrection up to 780 m (Monte Epomeo, the highest point of the island). This horst would be the result of the intrusion of a shallow trachytic magmatic chamber, which originated small volcanoes distributed along the horst faults and conjugate directions (Rittmann, 1930; 1948).

The sanidine of a sample of green tuff taken from the top of Monte Epomeo has been dated back to 840,000 years (Gasparini and Adams, 1969) and 640,000 years (Cappaldi *et al.*, 1976); Evernden and Curtis (1965) had worked out an age of 83,000 years from the sanidine of a sample of green tuff from Forio (this green tuff had been removed from the western flank of Epomeo). The figures we present in Table 7—95,000 years for a sample taken near the base of the green tuff at the flank of the Bocca di Serra (southwest of Epomeo), 55,000 years for a sample from the top of Monte Epomeo, agreeing with the age of 110,000 years obtained on a trachytic block included in the green tuff—show that the whole history of the island would be much more recent than thought previously, and the green tuff would result from the activity and the dismantling of a first volcanic complex still outcropping in the southeastern part of the island, where some volcanic formations have been estimated at around 120,000 years (e.g. a flow above Campagnano), and in the north at Monte Vico dated back to 75,000 years. In this case, the surrection of the horst would be younger than 55,000 years, which would explain the intense tectonic and volcanic activity that has been identified in some areas of the island (the total volcanic activity in the northeastern part is younger than 8000 years), as well as the freshness of the relief of Monte Epomeo, which is very readily erodable, being made out of tuff. These new values corroborate the first datings of Evernden and Curtis (1965).

4.2 Age of Laschamp palaeomagnetic excursion

In 1967, Bonhommet and Babkine discovered two lava flows with inverse thermoremanent magnetism at Laschamp and Olby, in the Recent volcanic province of La Chaîne des Puys, Massif Central, France. They reveal the

Table 7. Age data on volcanic formations of island of Ischia. Sample weights: 3–8 g. J, green tuff sampled at the top of Monte Epomeo (780 m); K, green tuff sampled from the flank of the Bocca di Serra (370 m); L, trachytic block embedded in green tuff on the flank of the Pietra del Acqua (520 m); M, Monte Vico trachytic dome; N, trachytic flow above Campagnano.

Location and rock type	Site	Material K %	rad.^{40}Ar (%)	rad.^{40}Ar (10^{-3} nl/g)	Apparent age $\pm 2\sigma$ 10^3 a (ICC)
Monte Epomeo	J	Sanidine	3.3	21.2	56.0± 3.0
top (780 m)		9.7	3.5	20.5	53.5± 3.0
Green tuff					
Bocca di Serra	K	Sanidine	6.3	37.6	95.0± 3.0
flank (370 m)		10.0	6.5	37.2	94.5± 3.0
Green tuff					
Pietra del Acqua	L	Groundmass	0.6	23.4	100.0±35.0
flank (520 m)		6.03	0.7	26.8	113.0±35.0
Trachytic block in green tuff					
Monte Vico	M	Sanidine	15.6	16.0	72.0± 2.0
Trachytic dome		5.70	16.5	16.7	75.0± 2.0
		Groundmass	16.4	15.6	75.0± 2.0
		5.19			
Campagnano	N	Groundmass	13.8	26.4	129.0± 5.0
Trachytic flow		5.15	13.0	27.2	133.0± 5.0

occurrence of magnetic field excursions in very recent time: previous chronological studies have shown that a major part of the volcanic activity in this province was around 10,000 years (Brousse *et al.*, 1969). A range of ages between 8000 (^{14}C dating of a palaeosoil below a trachytic ash deposit partly covering the sites) (Brousse *et al.*, 1966) and 20,000 years (Bonhommet and Zähringer, 1969) was proposed. Failure to find indisputable evidence of this reversed magnetic event in the same age range in other regions of the world in spite of numerous investigations (Denham and Cox, 1971; Verosub and Banerjee, 1977) attests the necessity for new age investigations on the Laschamp and Olby lava flows which record the excursion.

Table 4 presents analyses (whole-rock and separated groundmass) of samples of the two lava flows. The results on whole-rock, at 44,000±5000 and 49,000±5000 years (Gillot and Cassignol, 1977; Gillot *et al.*, 1979), are slightly larger than the one obtained on the separated groundmass. This indicates the presence of extraneous argon in the early crystallized minerals and perhaps xenolites as shown by the scattering of the results observed for some samples of the Laschamp flow (Bonhommet and Zähringer, 1969; Hall

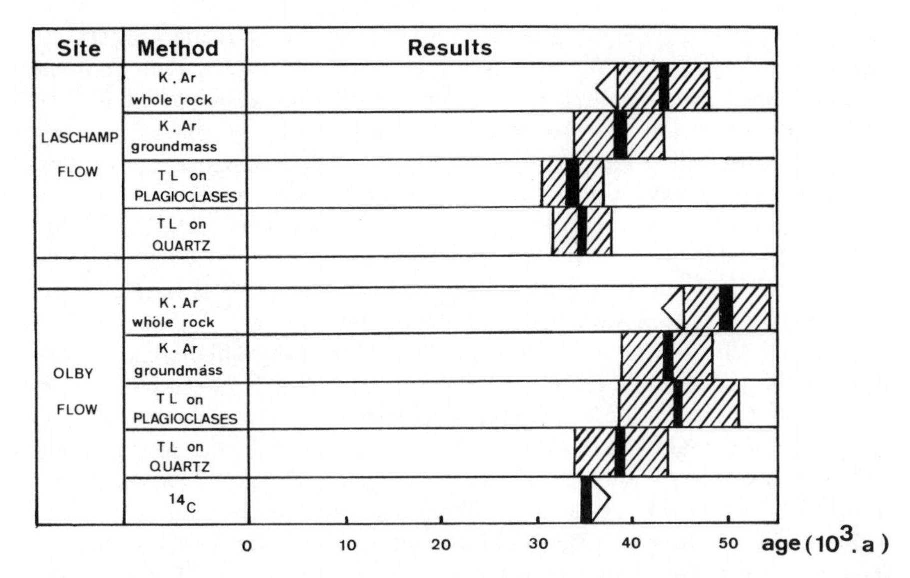

Figure 5. Age results on Laschamp and Olby lava flows; comparison between different methods (Gillot *et al.*, 1979). Uncertainties correspond to 2σ. New K–Ar data on groundmass are reported in Table 4.

and York, 1978). The most reliable ages are 37,000 ± 5000 and 44,000 ± 5000 years. We also discovered a palaeosoil under the Olby flow, and radiocarbon data indicate that the age of this flow is ⩾36,000 years. An attempt to date with thermoluminescence (TL) on the quartz of a granitic enclave in the Laschamp flow and quartz pebbles in the palaeosoil under the Olby flow and TL dating of the plagioclases of these two flows gave other age indications of around 35,000 years for Laschamp and 40,000 years for Olby, with about a 10% accuracy (Figure 5) (Gillot *et al.*, 1979). Conventional K–Ar and $^{39}Ar/^{40}Ar$ data obtained by Hall and York on these lavas (whole-rock) (1978) indicate weighted age values of 45,200 ± 2500 years (K–Ar) and 48,400 ± 7900 years ($^{39}Ar/^{40}Ar$). Thus, the apparent age values obtained are between 30,000 and 50,000 years and the range of ages for this magnetic event appears to be much greater than previously estimated. In any event, further research on materials of anomalous TRM in that range is still necessary.

5 MEASURING MINERAL ^{38}Ar AS A CONCLUSION

We shall discuss only briefly the systematic measurement, made possible by discontinuing spiking, of the ^{38}Ar of the mineral samples, our studies in this field being only at a beginning.

The success of our overall procedure rests upon the correctness of several assumptions, which are: on the one hand, contaminating argon, from the minerals themselves together with the line blank, has the isotopic composition of natural argon; on the other hand, during the pre-degassing step, there is no noticeable loss of radiogenic argon, nor any noticeable isotopic discrimination accompanying the reduction in the contamination.

In practice, the correctness of these assumptions is very frequently fulfilled, which results in age independence of the pre-degassing parameters (temperature and duration) within certain limits. This fact, as shown above, enabled us to measure very highly contaminated samples.

However, all cases do not present themselves so favourably. Some materials show themselves difficult, if not impossible, to date. For example, we met isotopic discrimination during the pre-degassing step. This phenomenon had been already pointed out (Baksi, 1973; Gourinard, 1975; McDougall *et al.*, 1976). It involves, in spiked conditions, an apparent excess of radiogenic argon. The measured age is higher than the actual one because ^{36}Ar was preferentially desorbed during baking, and its value depends on the pre-degassing parameters. In unspiked conditions, one can also observe an alteration in the $^{36}Ar/^{38}Ar$ ratio, which is in principle half of that of the $^{36}Ar/^{40}Ar$ ratio, which enables the correction of contamination. However, the procedure lacks accuracy.

We also met the other type of anomaly: e.g. a microlitic phase dated to 2.3 and 1.7 Ma when pre-degassed respectively at room temperature and 200°C. The ratios $^{36}Ar/^{38}Ar$ being in both cases those of natural argon, the phenomenon is shown to be a loss of radiogenic argon.

Although the deductions we can derive from examination of the preceding cases are far from satisfactory, they emphasize the potentialities of the unspiked method, which in fact extends the scope of the potassium–argon method.

It appears that these new problems, if they could be examined, cannot be solved with the help of our existing mass spectrometer. We are now designing a new and more powerful apparatus, with a higher radius of curvature for ionic trajectories (12 instead of 6 cm) and a higher accelerating voltage for ions (3000 instead of 500 V), with the help of which we hope to improve on our present performances and also to cast some light on some still unsolved questions.

Acknowledgements

We are greatly indebted to Ian McDougall for helpful discussions on the subject developed here; and to Hans J. Lippolt for thorough comments on and criticisms of an earlier version of this manuscript. Data on the Phlegrean Fields, Vesuvius and the island of Ischia were obtained through the efficient collaboration of, respectively: Lucio Lirer (University Napoli), Roberto Santacroce (University Pisa) and Sergio Chiesa, Franco Forcella, Giorgio Pasquaré and Luigina Vezzoli (University Milano).

Résumé du rédacteur

Cette contribution est la description d'une technique permettant l'analyse de l'argon des roches sans utilisation du procédé de la dilution isotopique mis en oeuvre à ce jour par la quasi-totalité des laboratoires, et la présentation de quelques résultats significatifs.

Le procédé consiste à obtenir du spectromètre de masse une stabilité et une reproductibilité des signaux isotopiques (durant toute la mesure) aussi parfaites que possible, alliées à la connaissance complète de la réponse de l'appareil, discrimination isotopique comprise. La *stabilité du signal* est obtenue par une *purification très complète* du gaz introduit, poursuivie pendant toute la mesure par un getter logé dans la cellule du spectromètre, ce qui assure également la *reproductibilité*, et donc autorise la suppression du traceur. La *discrimination isotopique* est déterminée à chaque mesure d'argon minéral en traitant une prise de même grandeur d'argon atmosphérique, la pression étant ajustée très finement au moyen d'un *appareil à soufflet*. On compare enfin la réponse obtenue pour l'argon 40 extrait de la roche et pour des doses successives, connues, d'argon atmosphérique permettant de tracer une courbe d'étalonnage.

Les tests de la technique (sur des roches d'âge connu) montrent que la qualité des datations n'est pas inférieure à celle obtenue par la dilution isotopique pour des échantillons peu pollués par de l'argon atmosphérique. Mais l'intérêt du procédé réside en la possibilité de dater avec une plus grande confiance des échantillons très pollués. En particulier, des âges récents significatifs ont été obtenus sur des échantillons contenant moins de 1% d'argon radiogénique dans l'argon total extrait. Il est alors possible de recouper l'échelle des âges du ^{14}C. Enfin une ouverture intéressante sur l'avenir est la possibilité d'utiliser les isotopes de masse 36 et 38 de l'argon comme un repère susceptible de déceler les éventuels fractionnements durant l'histoire des roches ou les irrégularités de la composition isotopique de l'argon.

A performances équivalentes d'un spectromètre de masse donné, cette technique se révèle fondamentalement plus favorable que la technique utilisant la dilution isotopique pour ces deux possibilités d'application (échantillons contenant une faible proportion d'argon radiogénique, reconnaissance du fractionnement des isotopes de l'argon).

Les applications géochronologiques proprement dites concernent le volcanisme récent de l'Italie méridionale.

(Manuscript received 17-2-1981)

Numerical Dating in Stratigraphy
Edited by G. S. Odin

10
The $^{39}Ar/^{40}Ar$ technique of dating

FRANCIS ALBARÈDE

1 INTRODUCTION

The $^{39}Ar/^{40}Ar$ technique of dating is a modified version of the classical $^{40}K/^{40}Ar$ technique and is presented from the need to solve some of the latter's major shortcomings:

(1) The necessity for measuring potassium and argon on two aliquots of the sample whence difficulties arise due to possible sample inhomogeneities;

(2) Samples which have experienced a complex thermal history and argon loss are generally useless for K–Ar dating: a technique discriminating 'leaky' and retentive parts of datable systems will be of considerable interest under these circumstances;

(3) Inherited 'excess' argon sometimes hampers age determination: it might be hoped that excess argon has crystallochemical properties differing from those of radiogenic argon and a method to characterize these two components is then necessary.

In 1962, Sigurgeisson extended to the K–Ar method a technique of dating neutron irradiated samples previously devised by Jeffery and Reynolds (1961) for the iodine–xenon dating of meteorites (extinct radioactivities). In his original technique, experimentally set up by Merrihue (1965), the sample to be dated is irradiated by a fast neutron flux which induces the reaction ^{39}K (n, p) ^{39}Ar, thus enabling the separate measurement of K and ^{40}Ar on two aliquots and by a different technique to be replaced by a single isotopic ratio measurement. Merrihue and Turner (1966) pursued the analogy with the I–Xe method of dating and showed that stepwise degassing of argon at increasing temperatures was able to resolve some difficulties for samples partially outgassed during a complex thermal history.

The major concepts of the $^{39}Ar/^{40}Ar$ technique have been laid down within a few years by now classical papers concerning both theoretical aspects (simulation of the degassing, synthetic age spectra) (Turner, 1968;

Fitch *et al.*, 1969) and the methodological approach (mass interferences, recoil phenomena, . . .) (Mitchell, 1968; Turner, 1970b; 1971; Turner and Cadogan, 1974).

The popularity of the $^{39}Ar/^{40}Ar$ technique coincided with the return of Apollo missions. It was rapidly realized that it provided a technique really competitive with other conventional methods (Rb–Sr, U–Pb) hampered by the size and chemical composition of available lunar samples and furthermore gave unique information about geochemical parameters of the planetary environment (trapped gases, solar wind, spallation reactions). The method obtained more and more spectacular success and benefited from significant improvements of the experimental techniques.

Applied to terrestrial problems, the technique did not keep all the promise of its application to the lunar samples. Insufficient account of excess argon or loss by alteration in aerial and submarine samples, the irreducible complexity of data in the case of a complex thermal history and meaningless plateau ages, disturbing effects of particle recoil during irradiations, all seemed to converge to ruin the hoped-for ability of this technique to solve the complex situations for which it was devised. In fact, the real possibilities of the technique probably lie between those promoted by the Apollo optimistic view and those resulting from the rather disappointing results of the first studies on earth samples. The complexity of dating these latter samples, often involved in a hydrous environment, with complex tectonic as well as thermic histories, requires extensive studies before the right information may be extracted.

2 FUNDAMENTAL ASPECTS OF THE $^{39}Ar/^{40}Ar$ TECHNIQUE

The basic equation of the $^{40}K^{40}Ar$ method (e.g. Dalrymple and Lanphere, 1969) for a t_s old sample

$$^{40}\mathrm{Ar}_{\mathrm{rad}} = \frac{\lambda_\varepsilon}{\lambda_\varepsilon + \lambda_\beta}\, {}^{40}\mathrm{K}\{e^{(\lambda_\varepsilon+\lambda_\beta)t_s} - 1\}$$

where $\lambda_\varepsilon = 0.581 \times 10^{-10}\ \mathrm{yr}^{-1}$ and $\lambda_\beta = 4.962 \times 10^{-10}\ \mathrm{yr}^{-1}$ (Steiger and Jäger, 1977), and rad stands for radiogenic, requires two distinct analyses for Ar and K concentrations.

Irradiation of the sample with fast ($E > 1$ MeV) neutrons induces the reaction ^{39}K (n, p) ^{39}Ar and noting $R = (^{39}\mathrm{Ar}/^{39}\mathrm{K})$ the yield of this reaction, the above equation may be rewritten:

$$\frac{^{40}\mathrm{Ar}_{\mathrm{rad}}}{^{39}\mathrm{Ar}} = \frac{\lambda_\varepsilon}{\lambda_\varepsilon + \lambda_\beta} \frac{^{40}\mathrm{K}}{^{39}\mathrm{K}} \times \frac{1}{R}\{e^{(\lambda_\varepsilon+\lambda_\beta)t_s} - 1\}.$$

Inclusion in the irradiation can of a sample called a monitor whose age t_M has been previously determined by conventional K–Ar dating technique ensures that the reaction yield is reasonably similar for both samples and enables the unknown t_s to be determined from the equation:

$$\frac{(^{40}Ar_{rad}/^{39}Ar)_{\bar{S}}}{(^{40}Ar_{rad}/^{39}Ar)_M} = \frac{e^{(\lambda_\varepsilon+\lambda_\beta)t_s}-1}{e^{(\lambda_\varepsilon+\lambda_\beta)t_M}-1}$$

where $(^{40}Ar_{rad}/^{39}Ar)_M$ is the corresponding ratio determined on the monitor according to this procedure.

It must be emphasized that dating of a sample only requires the determination of the argon isotopic composition of sample and monitor and knowledge of the K–Ar age of the monitor. Any absolute concentration in K or Ar is basically useless and uncertainties due to sample aliquoting are removed.

In fact, this procedure is rarely used in its simple form, i.e. the total outgassing of the sample. The experimentalist will try to extract more information from irradiated samples by releasing the argon in discrete steps at increasingly higher temperatures in order to characterize sample domains with different K–Ar properties. In this case, data are usually represented by plotting the $^{40}Ar_{rad}/^{39}Ar$ ratio or apparent age of each individual gas fraction against the cumulated fractional amount of ^{39}Ar released by the sample at successively higher temperatures: this is the conventional release pattern or age spectrum representation (Turner *et al.*, 1966), which gives a meaningful 'weight' to individual fraction ages.

As suggested by Turner (1968) and Fitch *et al.* (1969) from two different theoretical approaches, stepwise argon release is a promising tool for dating disturbed samples. Turner (1968) considers a sample for which the argon clock is set at T_0 (Figure 1) and remains closed till the time T_1. At T_1, the sample undergoes a thermal disturbance in which radiogenic argon is lost by *volume diffusion*, so depleting the outermost part of the grains. From T_1 to the present, argon is again accumulated in the sample working as a closed system. In this case numerical simulation of age spectra (Figure 1) enabled Turner to show that memory of either T_1 or T_0 (rarely both) may be expected from the data depending on the intensity of the disturbance at T_1. The importance of this model rests on the evidence that most datable minerals (with the notable exception of biotite) have been shown to lose their argon by volume diffusion at temperatures <800–1000°C: K-feldspars (Foland, 1974), plagioclase (Turner *et al.*, 1973), phlogopite (Giletti, 1974), and for many samples considered by the author, muscovite and amphibole (see below).

Alternatively, Fitch *et al.* (1969) consider that a system may present sites of different intrinsic retentivity, an assumption which may be paralleled with the kinetic loss model of Hart (1964) or Musset (1969). High retentivity sites are the normal crystallographic sites; low retentivity sites may be

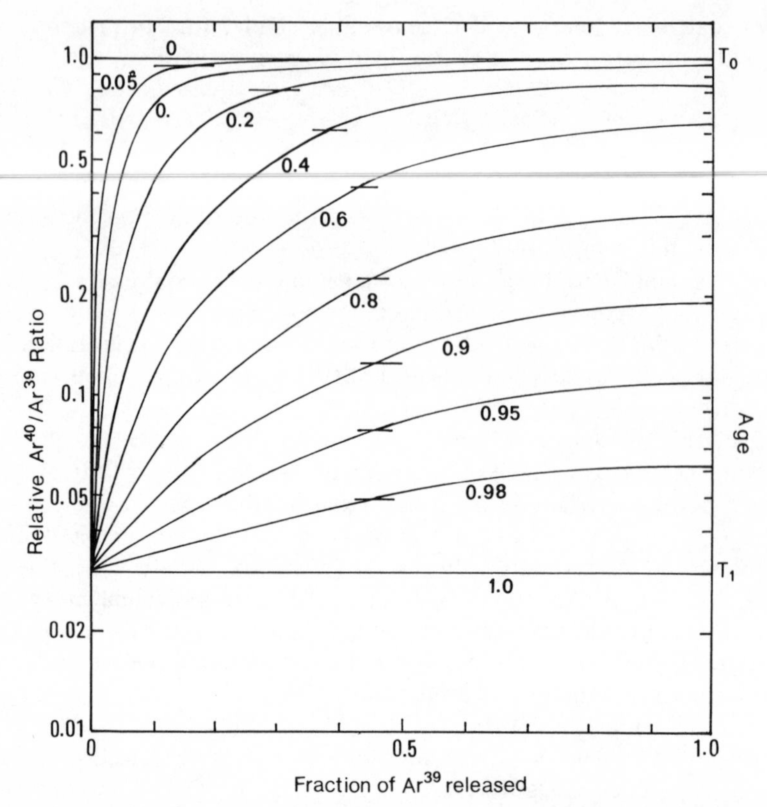

Figure 1. Theoretical continuous age spectra for a spherical mineral of age $T_0(^{40}Ar/^{39}Ar = 1)$ which suffered argon loss by volume diffusion of different intensities at $T_1(^{40}Ar/^{39}Ar = 0.3)$. The numbers on the curves are the fraction of ^{40}Ar lost at T_1 and the horizontal bars indicate the average $^{40}Ar/^{39}Ar$ ratio (or the corresponding K–Ar age) of the sample. From Turner (1968) reproduced by permission of Pergamon Press, Oxford.

associated with surfaces, cracks or deformed parts of the lattice. Fitch *et al.* (1969), adopting a history similar to that of Turner's model, emphasize the fact that low retentivity sites will lose their radiogenic argon during the T_1 event, leaving the high retentivity sites more or less in their original state. Age spectra expected from this model are similar to those depicted in Figure 1 from Turner's model. In both cases, the commonly adopted test for the T_0 (crystallization) and T_1 (disturbance) age significance is the fact that several consecutive steps give consistent ages within the error bars, thus defining respectively either a high-temperature plateau age or a low-temperature plateau age.

Geological tests of these approaches have had a diverse success. Disturbed lunar samples have shown the ability of the $^{39}Ar/^{40}Ar$ technique to keep a memory of a crystallization age in their high-temperature plateaux (Turner, 1977). Classical examples of contact metamorphism at Eldora, Colorado (Hart, 1964), or Duluth (Hanson and Gast, 1967) have been reinvestigated with this technique respectively by Berger (1975) and Hanson *et al.* (1975).

Samples which are either far from the thermal anomaly and left little disturbed or very close to the intrusive and profoundly reset at the time of intrusion give reliable dating information, but not the partially overprinted samples. Similar conclusions are arrived at by Dallmeyer (1975; 1979) and Albarède *et al.* (1978), who worked on regionally remetamorphosed terrains. More serious is the contention of several authors that meaningless plateau ages may be obtained on disturbed samples due to experimental artifacts (Hanson *et al.*, 1975; Fleck *et al.*, 1977; Albarède *et al.*, 1978) or impure mineral mixtures. In all cases, it is dangerous to rely on a limited number of data; even with reasonably good plateau ages and the analysis of minerals with different retentivity properties in the same area, it is necessary

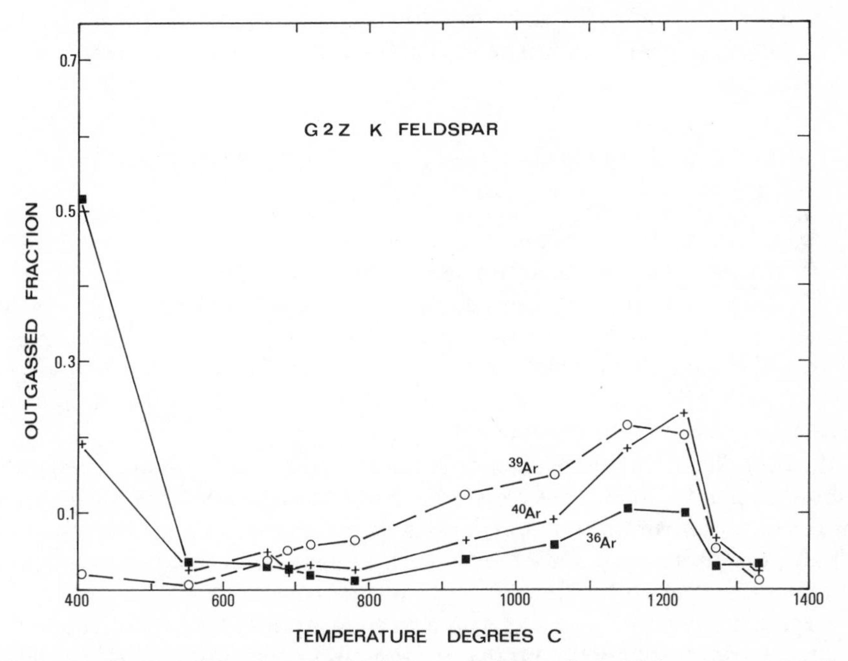

Figure 2. Comparative release of Ar isotopes at increasingly higher temperature for a K-feldspar sample. Atmospheric ^{36}Ar is released at low temperature (<800°C) while trapped (dissolved) ^{36}Ar is released at high temperature in approximate proportion with $^{40}Ar_{rad}$ and K-produced ^{39}Ar. From Albarède (1976).

to unravel complicated histories (Albarède *et al.*, 1978). Analysis of data from polytectonic environments further requires the disturbing effects of mechanical overpressure related to rock deformation to be taken into account, as they have been shown both on natural samples (Maluski, 1978) and in experiments (Ozima *et al.*, 1979) to severely affect the age spectra.

Another purpose of the stepwise heating technique is to detect possible excess of argon; this is of prime importance, especially in the dating of mafic rocks. U-shaped age spectra have been suggested by Kaneoka (1974) and Lanphere and Dalrymple (1976) to provide a diagnostic for ^{40}Ar excess but the technique provides no correction of this excess and has been shown to depend on irradiation conditions (Stettler and Bochsler, 1979). Trapped argon cannot be resolved from $^{40}Ar_{rad}$ due to evidence of non-atmospheric dissolved ^{36}Ar released at high temperature together with the other Ar isotopes (Figure 2) (Kaneoka, 1975; Albarède, 1976) or geologically absurd plateau ages, reaching 5000 Ma for a Greenland biotite (Pankhurst *et al.*, 1973).

In both cases (thermal disturbances or excess argon), the $^{39}Ar/^{40}Ar$ technique provides an interesting hint on the detection of these conventional traps of the K–Ar method, but in the case of terrestrial samples does not offer a systematically successful tool to correct these effects if considerable attention is not paid to experimental details and suitability of materials to be dated.

3 EXPERIMENTAL DIFFICULTIES INHERENT IN THE $^{39}Ar/^{40}Ar$ TECHNIQUE OF DATING

If neutron irradiation enables some major shortcomings of conventional K–Ar dating to be overcome, it is also a source of characteristic difficulties which, up to now, have been more or less satisfactorily solved. Major effects are due to (1) inhomogeneous neutron irradiations; (2) interference reactions which alter the 'normal' argon composition; (3) recoil effects of target nuclei; and (4) in the case of stepwise heating, changes in the Ar release pattern relative to non-irradiated samples.

Neutron fluence inhomogeneities in the range of 5–10% on the irradiation vessel scale are common but are readily corrected through a flux mapping by pieces of nickel wire inserted at various places in the can and counted after irradiation. In this way, the uncertainty level may be lowered to well below 1% and disregarded for the remainder of the calculations.

Beside the central $^{39}K(n, p)^{39}Ar$ reaction, neutron irradiation of K and Ca contained in the samples produces argon isotopes (mass 36–40) typical values of which are given in Table 1, taken from Turner (1977). If for an irradiation these K and Ca derived argon isotopic compositions are known (by irradiating K and Ca pure salts), precise corrections can be made as long

Table 1. Argon isotopes present in neutron-irradiated samples (from Turner, 1977, reproduced by permission of Pergamon Press).

Source	Typical relative abundance				
	^{36}Ar	^{37}Ar	^{38}Ar	^{39}Ar	^{40}Ar
Atmospheric argon	$\equiv 1.00$		0.190		295.5
Radiogenic argon					$\equiv 1.00$
Neutron interactions on K	$<1\times10^{-4}$	$<1\times10^{-4}$	$(1.0–1.5)\times10^{-2}$	$\equiv 1.00$	$(1–6)\times10^{-2}$
Neutron interactions on Ca	$(1–3)\times10^{-4}$	$\equiv 1.00$	$(1–5)\times10^{-4}$	$(6.4–7.3)\times10^{-4}$	$<6\times10^{-4}$
Neutron interactions on Cl			$\equiv 1.00$		

as the Ca/K ratio of the sample is not too high and its age not too young (Turner, 1971). Basically, Ca products major interferences on ^{36}Ar and ^{39}Ar which can be corrected by a measurement of ^{37}Ar (radioactive with a half-life of 35.1 days), K an interference on ^{40}Ar which is corrected by the K derived ^{39}Ar. A detailed discussion of interference correction may be found in Mitchell (1968), Brereton (1970) or Dallmeyer (1979), while error estimation and minimization are discussed by Turner (1971) and Dalrymple and Lanphere (1971). In general, interference corrections do not introduce significant errors in the age of K-rich minerals (micas, K-feldspars), while errors in ages of Ca-rich minerals (plagioclase, amphibole) vary from several percent for young (less than 100 Ma) samples to a fraction of a percent for Precambrian samples. Care must be taken in that such an error evaluation has to be made for each irradiation, and of course corrections have also to be applied to the monitor.

Nuclear reactions cause the daughter nuclei (^{39}Ar, ^{37}Ar) to recoil out of their original site. Recoil energies estimated at a few hundredths of a keV are sufficient to produce displacements of the order of 0.1 μm. If no K-rich phase is adjacent to the analysed grain, it could be expected (Brereton, 1972; Turner and Cadogan, 1974) that a *c.* 0.1 μm thick rim depleted in ^{39}Ar and ^{37}Ar would exist around it. Convincing experimental evidence of recoil effect for ^{39}Ar has been provided by Huneke and Smith (1976), who found ^{39}Ar impoverished low-temperature fractions on K-rich/K-poor artificial mixtures, but a surprising lack of this effect has been observed for ^{37}Ar. Recoil has been thought to be responsible for anomalously high $^{40}Ar_{rad}/^{39}Ar$ ratios (or age) in various geological systems composed of grain mixtures of different K content components, such as devitrified basalts (Fleck *et al.*, 1977), K-feldspars (Albarède *et al.*, 1978) or argillaceous minerals (Halliday, 1978). Recoil effects can be minimized by avoiding measurements of fine-grained heterogeneous samples, but some cases such as those quoted above lead to insoluble difficulties due to the scale (<1 μm) of the chemical inhomogeneities.

Changing the neutron doses can change the argon release pattern of some samples. Horn *et al.* (1975) found that high fluence increased the retentivity of argon in whole-rock lunar samples and decreased plateau ages by some 2%. Stettler and Bochsler (1979) have also shown that the Ar release pattern of glassy basalts may be modified by the irradiation, which induces changes in retentivity and redistribution of excess argon. Finally, attention must be drawn to temperature elevation in the reactor can by nuclear reactions, especially when cadmium shielding is used to cut off thermal neutrons and to lower the sample activity. If temperatures higher than 150–200°C are reached some argon may be lost and low-temperature steps become insignificant.

A conclusion of all authors is that the lowest neutron fluence permissible by mass spectrometer sensitivity and blank levels should be preferred anyway to minimize the aforementioned artifacts, an absolute upper limit of 10^{19} neutrons/cm^2 being a reasonable estimate (Horn *et al.*, 1975).

4 WHAT CAN BE DATED?

In most cases, dating of terrestrial samples may make use of high K/Ca samples, although lunar studies have shown that this condition is not always necessary. Basically, any system with moderate (>10 μm) grain size and homogeneous K and Ca distribution can be dated provided (1) its K/Ca ratio is not unreasonably low (>1/100) or its age too young (in practice, a few Ma), (2) no excess argon is present and (3) the sample did not suffer argon loss subsequent to the time of its crystallization. The $^{39}Ar/^{40}Ar$ technique with SWH is more restrictive than the K–Ar conventional technique as far as condition 1 is concerned but has the advantage of indicating the cases for which conditions 2 and 3 are not fulfilled. Moreover, use of Turner's (1968) or Fitch *et al.*'s (1969) models enables the age of some samples which underwent argon loss (condition 3) to be restored. We will consider whole-rocks and different minerals in turn.

4.1 Whole-rocks

Despite some recent success in dating basaltic rocks from Iceland (Musset *et al.*, 1980), the use of whole-rocks is not in general recommended as they are usually inhomogeneous mixtures of chemically different phases with different grain sizes and argon retentivity. If any phase introduces a complication (excess argon, loss of radiogenic argon, recoil, . . .) this will necessarily introduce a bias on the whole-rock data.

There are unfortunately some cases where the choice is inescapable, especially when rocks do not enable mineral phases to be separated. Volcanic rocks are a prime example, particularly important for basalts, for which the K–Ar method is the only way to provide reliable results, or fine-grained mafic rocks (dolerites, gabbros). $^{39}Ar/^{40}Ar$ results are often disappointing (Lanphere and Dalrymple, 1971; Brereton, 1972; Ozima and Saito, 1973; Stukas and Reynolds, 1975; Bottomley and York, 1976; Kaneoka, 1980) owing to excessively complex age spectra. In most cases the presence of excess argon may be established, but no convincing procedure to extract the 'true' age has been found. Fleck *et al.* (1977) and Stettler and Bochsler (1979) have shown that devitrification and irradiation artifacts may further increase this complexity. In turn, $^{39}Ar/^{40}Ar$ results strongly suggest that K–Ar ages on whole-rocks must be handled with the utmost care.

Shales are also unsuitable for mineral separation, but data from Miller (in Brereton, 1972) or Halliday (1978) and results on phyllites obtained by Dallmeyer (personal communication) suggest that they may occasionally provide more reliable results, although small-scale inhomogeneities and fine grain size magnify recoil effects and argon loss during the irradiation (Halliday, 1978).

4.2 Micas

Muscovite and biotite are able to provide reliable plateau ages. When addressing thermally imprinted minerals muscovites (Figure 3) present age spectra fairly similar to those predicted by Turner's (1968) volume diffusion model (e.g. Hanson *et al.*, 1975; Miller, in Brereton, 1972; Albarède *et al.*, 1978) (Figure 3), while disturbed biotites have more complex spectra. An important controversy has arisen between authors about the significance of biotite plateau ages. For Hanson *et al.* (1975) or Albarède *et al.* (1978), biotite experimental outgassing controlled by dehydration may induce meaningless plateau ages even on severely disturbed minerals (Figure 3). For others (Berger, 1975; Dallmeyer, 1975), irregularities on the age spectrum differentiate disturbed from undisturbed minerals. Besides thermal overprinting, excess argon may also affect the age spectra of biotites: a sample from Greenland containing proved excess argon (ages >5000 Ma) (Pankhurst *et al.*, 1973) presents a nicely defined plateau age with none of the characteristic shown by other systems (U-shaped spectra, Lanphere and Dalrymple, 1976). All of these observations suggest that biotite outgassing is probably a unique and poorly understood process, a point which will be returned to later. So, interpretation of biotite age plateaux in terms of crystallization or cooling age must be delayed until some independent evidence (age of a different mineral, for instance) has been found. Attention must be paid when interpreting progressive variations of biotite plateau ages on a regional scale in terms of progressive cooling to the possibility that they

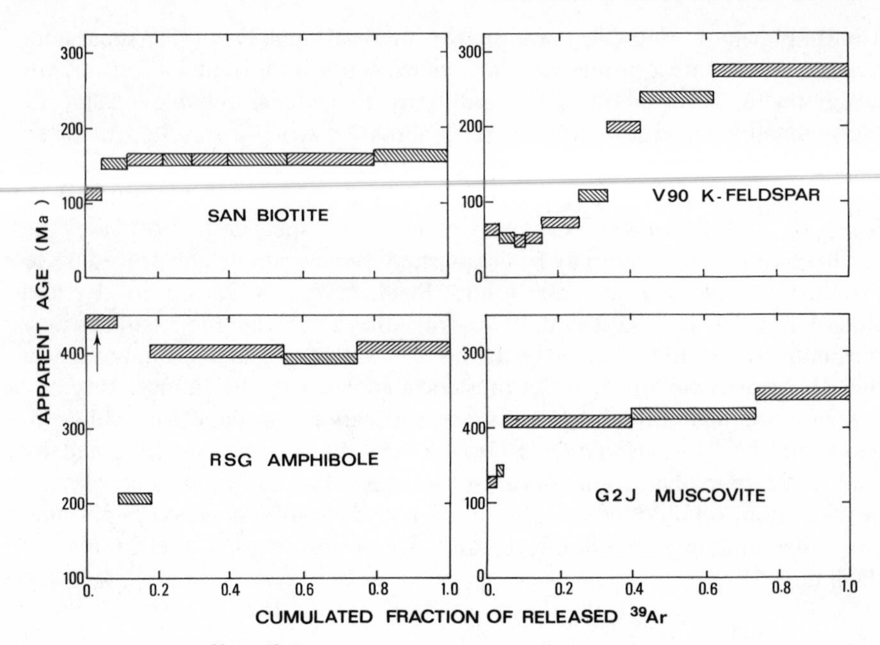

Figure 3. Typical $^{39}Ar/^{40}Ar$ age spectra of common rock-forming minerals which have experienced reheating subsequent to their crystallization. Biotite age spectra indicative of a thermal disturbance may present an age 'plateau' (this case) or not. Simplified from Albarède *et al.* (1978).

could conceivably result from a slight thermal disturbance blurred by the argon extraction procedure (Albarède *et al.*, 1978). Related minerals: glauconitic minerals have been shown (Brereton *et al.*, 1976) to be hardly suitable for dating as being sensitive to argon loss and recoil during irradiation, probably due to the small size of the homogeneous individual grains. In such a case, conventional K–Ar ages should probably be preferred (see NDS131, 132).

4.3 Amphibole

As for the K–Ar dating conventional technique, amphibole is a reliable chronometer which, despite some occurrences of excess argon, has a fairly good retentivity (Hanson *et al.*, 1975; Dallmeyer, 1975; Lanphere *et al.*, 1977; NDS128). Inconsistently high ages may be observed on low-temperature fractions (Bryhni *et al.*, 1971; Hanson *et al.*, 1975; Berger, 1975; Albarède and Michard-Vitrac, 1978) but it is unclear whether they reflect excess argon or recoiled ^{39}Ar. Partially imprinted amphiboles often present age spectra similar to those calculated for volume diffusion (Turner,

1968) (Figure 3). However, as for the conventional technique, one must bear in mind that even minute amounts of intermingled biotite can significantly bias the results if both minerals have been differentially affected by a thermal disturbance (Berger, 1975).

4.4 Feldspars

There is a considerable discrepancy between the ability of plagioclase to produce reliable results in terrestrial and extraterrestrial environments. Lunar samples (e.g. Turner, 1977) have proved that this mineral dates crystallization events on the moon with an accuracy comparable to that of the Rb–Sr method (of the order of 1%). Data on terrestrial plagioclases are few (Lanphere and Dalrymple, 1971, 1976; Albarède *et al.*, 1978; Maluski, 1978) and have turned out to be disappointing. Their U-shaped age spectra cannot be easily interpreted in terms of a single phenomenon like excess argon, ^{39}Ar recoil or partial resetting. A precise interpretation of these spectra is required for dating. Lanphere and Dalrymple (1976) in fact suggest that the minimum age is the closest to the actual one, but if one considers that U-shaped spectra may partially result from isotope redistribution during irradiation as suggested by Stettler and Bochsler (1977) for basaltic glasses, the total fusion age has to be preferred and there is no benefit to be derived from the SWH $^{39}Ar/^{40}Ar$ procedure of dating.

Low-temperature K-feldspar has long been known as a poor argon retentive mineral (e.g. Lanphere and Dalrymple, 1969). A typical $^{39}Ar/^{40}Ar$ age spectrum of K-feldspar presents an age trough between 10 and 30% of ^{39}Ar released and a gently dipping high-temperature zone (Figure 3) (Berger, 1975; Albarède *et al.*, 1978; Maluski, 1978). This mineral, suspected to be intrinsically leaky to radiogenic argon, has been shown by Foland (1974) to have 'normal' Ar diffusion rates. This is supported by several examples of $^{39}Ar/^{40}Ar$ plateau ages on K-feldspar from totally undisturbed environments (Albarède, 1976; Albarède *et al.*, 1978). Tectonic shattering of submicroscopically exsolved feldspars has been shown to account for both their easy argon loss and high-temperature anomalously high ages (Albarède *et al.*, 1978). The unsuitability of K-feldspars for $^{39}Ar/^{40}Ar$ dating is balanced by their sensitivity to subtle thermomechanical disturbances, thus providing a check for the reliability of ages obtained for other minerals (e.g. biotite).

5 THE DATING OF DISTURBED MINERALS

If a system (rock or mineral) to be dated undergoes a thermal or mechanical disturbance subsequent to its crystallization, it can be hoped that, in favourable circumstances, Ar isotope degassing data can help unravel such a complex history. In the above discussion, it has been shown that case studies in the terrestrial environment have produced ambiguous results due to various difficulties:

— high-temperature fractions do not necessarily retain crystallization ages even in the case of moderate disturbances;

— false plateau ages resulting from different Ar loss mechanisms in natural and experimental conditions may be overlooked, especially for biotite and to a lesser extent for amphibole.

It can be checked that plateau ages do record a meaningful geological event if different minerals from the same sample, *including K-feldspar*, give consistent data (Albarède *et al.*, 1978). In other cases, the Ar loss mechanisms must be assessed before any model of age correction can be set up. For instance, data on hydroxylated minerals should be analysed to determine if experimental Ar degassing is controlled by volume diffusion, dehydration or a combination of both. It can be safely assumed that for many minerals, potassium hence ^{39}Ar—or better calcium hence ^{37}Ar—are homogeneously distributed. In that case, ^{39}Ar or ^{37}Ar release *vs.* temperature may be used through a classical diffusion analysis to check whether these isotopes are released by volume diffusion assuming a spherical, cylindrical or sheet-like general geometry of the grains (Turner *et al.*, 1973).

Let us consider, for example, a muscovite and a biotite from the metamorphic aureole around the Duluth gabbro, Minnesota (Hanson *et al.*, 1975), and an albite from the Llano Uplift, Texas (Lanphere and Dalrymple, 1971). If D is the diffusion coefficient at temperature T, Δt the duration of the temperature step and a the grain size, the quantity $D\,\Delta t/a^2$ may be determined from the data (e.g. Albarède, 1978) if one assumes that the sample is roughly isogranular. A plot of $\log(D\,\Delta t/a^2)$ *vs.* $(1/T)$ should be a straight line if loss is governed by volume diffusion, its slope being proportional to the so-called activation energy of the diffusion process. Inspection of Figure 4 indicates that this is the case at $t<650°C$ for albite, $t<1000°C$ for muscovite, a conclusion strongly supported by the comparison of the local slope of the corresponding segments on Figure 4 with Ar diffusion activation energies independently determined for similar material (Foland, 1974; Giletti, 1974). At higher temperatures, other loss mechanisms (dehydroxylation, structural transitions . . .) must be considered to account for the departure from linearity. This has the important consequence that none of Turner's (1968) or Fitch *et al.*'s (1969) models accounts for high-temperature steps of the age spectra of these minerals, and their interpretation in terms of 'age' is still questionable. Similarly, argon extraction from biotites does not follow the volume diffusion behaviour at any temperature, a common case being a double-peaked ^{39}Ar release expressed as a bump at around 950°C on the Arrhenius plot.

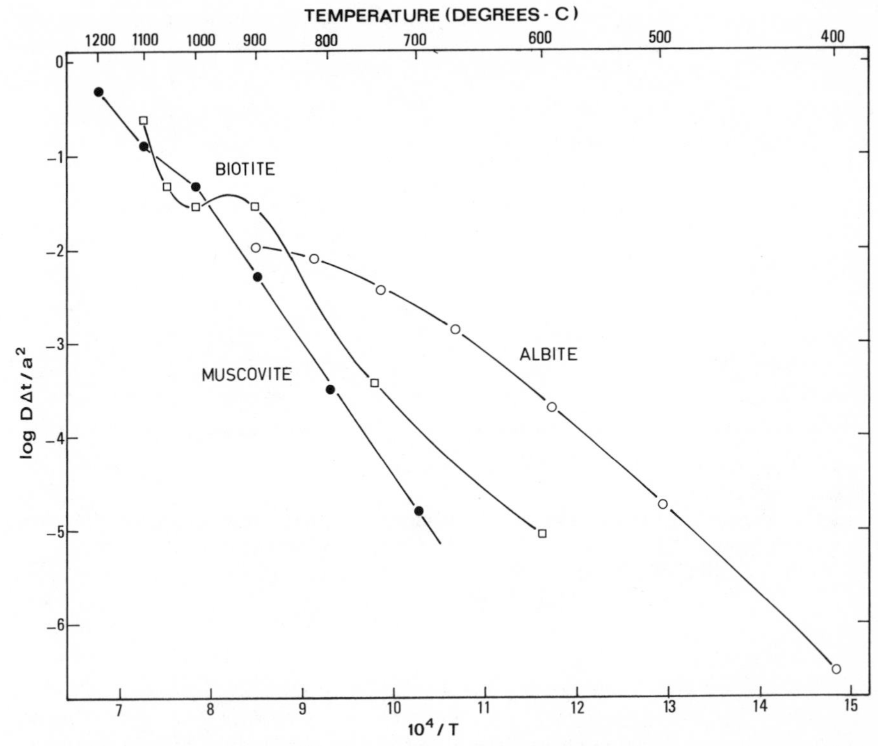

Figure 4. Arrhenius plot of ^{39}Ar release *vs.* reciprocal absolute temperature. Diffusion parameters $D\,\Delta t/a^2$ are obtained from volume diffusion models with either spherical (albite) or sheet-like (muscovite and biotite) geometry. Data from Hanson *et al.* (1975) for micas, from Lanphere and Dalrymple (1971) for albite.

Conversely, the area of volume diffusion behaviour contains valuable information which can be used to estimate the intensity of any perturbation affecting the system to be dated (Turner *et al.*, 1973). Application of the linear inverse theory can restore the spatial isotope distribution in the grains (Albarède, 1978), such as the one depicted in Figure 5 for a lunar plagioclase. In the latter case, application of the volume diffusion laws to low-temperature steps has been shown to provide an age correction in fair agreement with what is given by other methods (Albarède, 1978). This example shows that although most authors incline to 'trust' higher temperature ages for disturbed samples, they probably represent mixed ages difficult to interpret. Attention must also be given to low-temperature results, which, in spite of their poorer experimental quality, may provide useful information for complex chronometries.

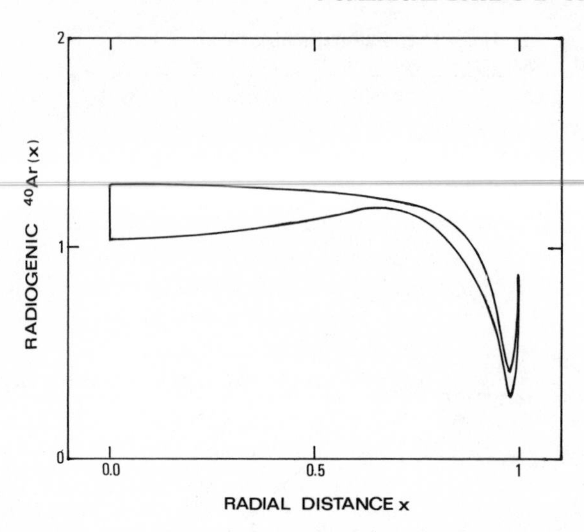

Figure 5. Example of spatial isotope distribution restored from Ar release data: the lunar anorthosite 15415. Spherical geometry of the mineral grains is assumed (0 is the centre, 1 the edge). From Albarède (1978), reproduced by permission of Elsevier.

6 CALCULATING AGES

$^{39}Ar/^{40}Ar$ ages are usually calculated following two main procedures, from either the age spectrum or the $^{40}Ar/^{36}Ar$ *vs.* $^{39}Ar/^{36}Ar$ isochron diagram.

In the age spectrum representation, in which $^{40}Ar/^{39}Ar$ ratios are corrected for atmospheric ($[^{40}Ar/^{36}Ar]_{atm} = 295.5$) contamination, only medium- and high-temperature steps currently provide plateau ages, i.e. ages which are similar to each other. The first difficulty is to define the low- and high-temperature limits of the plateau, since they are obviously dependent on many parameters including grain size, irradiation dose or heating schedule. It is usually agreed to accept for a plateau only adjacent steps (>3) which represent a large (>50%) cumulated fraction of the outgassed ^{39}Ar and whose ages differ by no more than a few percent, depending on the analytical uncertainty of each individual age determination. The plateau age is taken as the ^{39}Ar fraction and/or individual step error weighted average of the chosen individual step ages. There is little doubt that much subjectivity enters this choice and that the calculated analytical uncertainty assigned to the plateau age is somewhat arbitrary. Little work has been published to date which could show the reproducibility of plateau parameters from one experiment to another. However, common examples of age plateaux defined at better than 2% may be found in lunar and terrestrial literature. Care must be taken not to arbitrarily multiply the steps to decrease the apparent accuracy on the ages, especially when different consecutive steps give ages which scatter much beyond the analytical uncertainties.

The isochron plot $^{40}Ar/^{36}Ar$ *vs.* $^{39}Ar/^{36}Ar$ (Merrihue and Turner, 1966; Brereton, 1972) theoretically makes it possible to take into account contaminant argon whose $^{40}Ar/^{36}Ar$ ratio is different from 295.5 and to handle excess argon problems. Points representative of each step should be aligned on a straight line having ($^{40}Ar_{rad}/^{39}Ar$) for slope and the $^{40}Ar/^{36}Ar$ ratio of the contaminant for intercept (Figure 6). This representation suffers from four shortcomings: (1) points have similar weight irrespective of absolute amounts of argon released at the corresponding step; (2) argon loss may not be modelled as easily as by the age spectrum; (3) position of the points in the diagram is largely controlled by the isotope ^{36}Ar, which has little informative value; and (4) assumption is made that the non-radiogenic component is unique.

Point 3 is critical, as tremendous variations in the ^{36}Ar release during the heating schedule stretch the point along a line going through the origin (Dalrymple and Lanphere, 1974) and give a misleading impression of good alignments when $^{40}Ar/^{36}Ar$ variations are large.

Point 4 has been demonstrated to be often erroneous, as atmospheric contamination dominates the non-radiogenic argon at low temperature while a component dissolved in the lattice is released at high temperature together with $^{40}Ar_{rad}$ and ^{39}Ar (Kaneoka, 1975; Albarède, 1976) (Figure 6).

If one takes these restrictions into account, ages can be calculated in the isochron diagram by conventional fitting techniques. Points will be weighted by errors on $^{40}Ar/^{36}Ar$ and $^{39}Ar/^{36}Ar$ ratios, but the strong covariance of these ratios introduced by the measurement of the $^{40}Ar/^{39}Ar$ ratio must be calculated for example, from:

$$\operatorname{cov}\left(\frac{^{40}Ar}{^{36}Ar},\frac{^{39}Ar}{^{36}Ar}\right)=\frac{1}{2}\frac{^{40}Ar}{^{36}Ar}\times\frac{^{39}Ar}{^{36}Ar}\left\{\frac{\operatorname{var}(^{40}Ar/^{36}Ar)}{(^{40}Ar/^{36}Ar)^2}+\frac{\operatorname{var}(^{39}Ar/^{36}Ar)}{(^{39}Ar/^{36}Ar)^2}-\frac{\operatorname{var}(^{40}Ar/^{39}Ar)}{(^{40}Ar/^{39}Ar)^2}\right\}.$$

Using this, parameters of the best fit straight line may be calculated—after steps indicative of argon loss have been discarded—according to any procedure such as those proposed by York (1969), Williamson (1968) or Minster *et al.* (1979) for correlated errors. Again, it must be emphasized that nothing guarantees that ultimate results (age, intercept, errors) are independent of the heating schedule. Thus, care will be taken in applying statistical tests to the best fit results (chi-square tests of hypothesis).

Finally, the error made in the determination of the monitor age t_M by the conventional K–Ar technique will be propagated on the $^{39}Ar/^{40}Ar$ results.

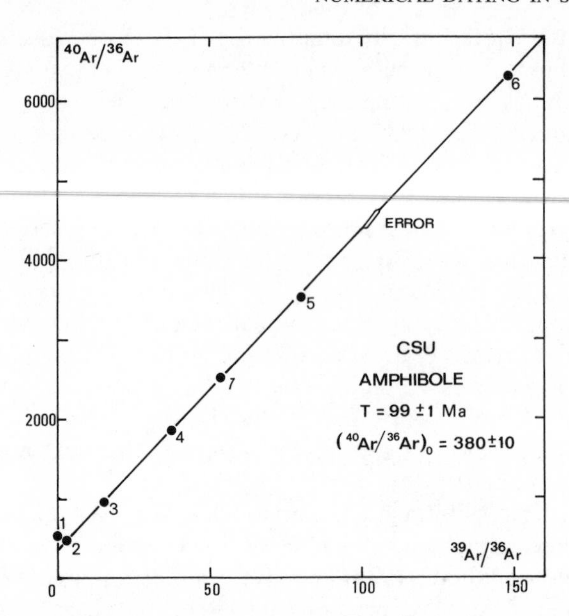

Figure 6. Isochron diagram commonly used in $^{39}Ar/^{40}Ar$ dating. The example is an amphibole in a lherzolite from the French Pyrénées. Slope is indicative of the $^{40}Ar_{rad}/^{39}Ar$ ratio (whence age is calculated), intercept yields the $^{40}Ar/^{36}Ar$ ratio of the non-radiogenic component supposed to be unique. From Albarède and Michard-Vitrac (1978), reproduced by permission of Elsevier.

Amplification of this error will be minimized if the age of the monitor is reasonably close to that expected for the samples to be dated. A major problem remains: the availability of reliable monitors presenting a good chemical homogeneity and carefully calibrated for K and Ar by different laboratories.

CONCLUSIONS

The stepwise heating technique of $^{39}Ar/^{40}Ar$ dating which proved to be a reliable chronometer for lunar material has turned out to be less powerful in the terrestrial environment. However, in contrast to the conventional K–Ar technique, it is able to detect excess argon and argon loss although no general procedure is available to correct these effects. Physical modelling of argon loss and use of plateau ages are often disappointing due to the variety of loss mechanisms in nature (volume diffusion, dehydration, . . .) and systematic use of plateau ages may turn out to be dangerous, especially for hydroxylated minerals (mica, clays, glauconitic minerals). Attempts to use degassing data with temperature might offer a powerful analysis of partially imprinted mineral ages. It is emphasized that multimineral analysis drastically improves the quality of the chronometric information even from poorly retentive minerals (feldspars). Experimental difficulties may be satisfactorily accounted for (fluence inhomogeneities, interferences) except in the case of ^{39}Ar loss by recoil, which is scarcely avoidable with small grain size and inhomogeneous samples.

Acknowledgements

Thanks are due to T. Staudacher, R. D. Dallmeyer and G. S. Odin for comments on the manuscript.

Résumé du rédacteur

La technique de datation $^{39}Ar/^{40}Ar$ consiste à irradier un échantillon afin de transmuter l'isotope de masse 39 du potassium en argon de même masse. Ceci permet de déduire d'une seule prise d'essai les deux quantités d'isotope mises en jeu dans la méthode de datation K–Ar.

Le flux d'irradiation est difficilement mesurable avec précision et est d'homogénéité imparfaite; on irradie alors à la fois l'échantillon et un témoin d'âge supposé connu. Ceci est une incertitude importante de la méthode. Le chauffage par paliers de l'échantillon permet de mesurer des âges apparents à différentes températures d'après les gaz extraits successivement de différents sites cristallographiques: ceci révèle parfois des irrégularités fort instructives car elles mettent en évidence de possibles altérations de l'échantillon ou un apport isotopique initial anormal; sinon, on obtient un 'âge plateau'.

Les incertitudes analytiques propres à cette technique sont analysées; elles sont liées à l'irradiation qui peut entraîner des perturbations dans l'échantillon.

La technique peut être appliquée avec succès aux micas et amphiboles, aux roches totales à grain fin ou aux feldspaths mais parfois sans succès. Les minéraux sédimentaires sont peu favorables.

(Manuscript revised 29-9-1980)

Numerical Dating in Stratigraphy
Edited by G. S. Odin

11
The application of fission track dating in stratigraphy: a critical review

DIETER STORZER and GÜNTHER A. WAGNER

1 INTRODUCTION

The past decade has seen numerous applications of fission track dating related to problems of the stratigraphic column. Substantial contributions to this subject have been claimed. Rather than review the individual contributions the present article attempts to discuss the methodical potentials and limitations of fission track dating for stratigraphic purposes. This will enable the reader to assess critically the published fission track age data. The principle of fission track dating is treated only briefly, since it has been comprehensively reviewed recently by Fleischer *et al.* (1975). The emphasis lies on the discussion of the parameters and constants which influence the precision and accuracy of fission track ages and, hence, the usefulness of fission track dating in stratigraphy.

2 PRINCIPLE

Natural uranium consists of 99.3% uranium-238 and 0.7% uranium-235. Both isotopes decay by the emission of α-particles. Apart from its natural α-decay, uranium-238 also undergoes *spontaneous fission*, but at a much slower rate. Its fission half-life is about 10^{16} years and, thus, more than one million times longer than its α-decay half-life. The energy released during this fission process is about 200 MeV and appears essentially as kinetic energy of the two fission fragments. The highly ionized fragments rip through the material in opposite directions and release their kinetic energy by disordering—mostly ionizing—the target atoms. In non-conducting solids, such as most minerals and glasses, the fragments leave along their path disordered channels, the 'fission tracks'. These solids are called 'solid state nuclear track detectors'. The latent fission tracks, which have a diameter of about 100 Å, can be enlarged by suitable etching until they become visible under an optical microscope. In minerals the length of the fission tracks (both fragments) is typically 20 μm or less. The width depends on the etching conditions and is a few micrometres.

Many glasses and minerals contain some uranium as a trace element. During their geological history they steadily register and store latent fission tracks, the more the older they are and the richer in uranium. If one polishes an internal face of such a track detector and subjects it to an appropriate etching procedure the spontaneous fission tracks become visible under a microscope (Figures 1 and 2). The number of fission tracks per unit area ('track density') is proportional to the age of the sample. In order to calculate the age, one has to know the uranium content. The uranium content is implicitly determined by irradiating the sample with thermal neutrons in a nuclear reactor. Thermal neutrons induce fission of uranium-235. The fragments of this fission also leave—analogous to those of the spontaneous fission of uranium-238—etchable tracks of practically the same physical properties. The number of *induced fission* tracks is proportional to the uranium content. The fission track age t is calculated according to the following equation:

$$t = \frac{1}{\lambda_d} \ln \left(\frac{\lambda_d \cdot P_s \cdot I \cdot n \cdot \sigma_f}{\lambda_f \cdot P_i} + 1 \right)$$

where λ_d = total decay constant of uranium-238, λ_f = spontaneous fission

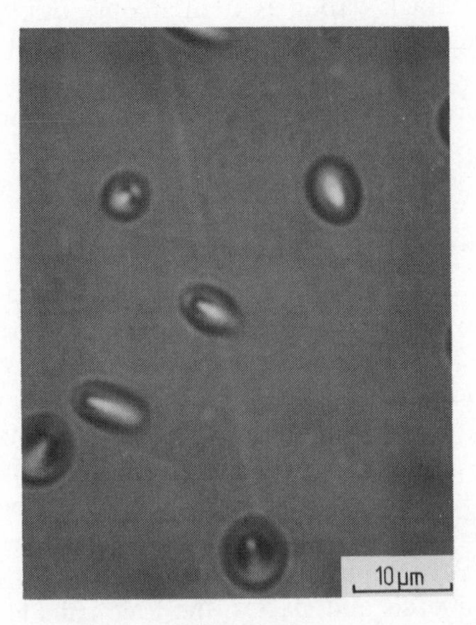

Figure 1. Induced fission tracks in tektite glass (Australite) etched 15 sec at 23°C in 48% HF.

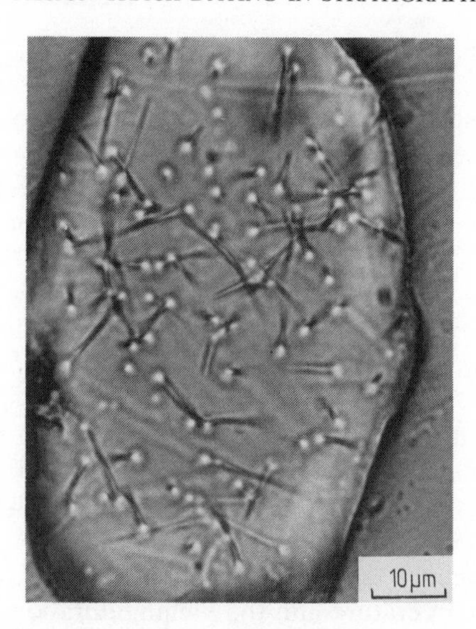

Figure 2. Induced fission tracks in an apatite crystal (Odenwald) etched 35 sec at 23°C in 5% HNO_3.

decay constant of uranium-238, P_s = number of spontaneous fission tracks per unit area, P_i = number of induced fission tracks per unit area, I = isotopic ratio uranium-235/uranium-238, n = integrated thermal neutron flux, and σ_f = cross-section of thermal neutron induced fission for uranium-235.

The age derived in this way dates the time of the beginning of fission track record in the mineral or glass. However, this presumes that since that time all fission tracks were accumulated and that the spontaneous fission tracks are revealed for counting with the same efficiency as the induced fission tracks. Ideally this age should correspond to the time of mineral formation or of another geologically significant event. Unfortunately, in reality the above assumptions are rarely fulfilled, as discussions in the next section and the section on track fading will show.

3 APPLICATION

3.1 Experimental complications

Serious difficulties in fission track dating arise from the fact that spontaneous fission tracks are often revealed with an efficiency different to that of induced tracks. Different efficiencies for revealing both types of tracks necessitate appropriate corrections in the age equation. The following factors influence the efficiency of track revealing:

(a) the induced tracks are often not recorded in the sample itself but in an external detector such as mica which is adjacent to the sample during neutron bombardment (Price and Walker, 1963). Since different materials have different track registration and track etching efficiencies, it is necessary to calibrate the efficiency ratio for each sample–external detector combination (Reimer *et al.*, 1970; Gleadow and Lovering, 1977).

(b) Fission tracks can be etched on faces in an internal (4π-geometry) or external (2π-geometry) position during track registration. The use of faces with different registration geometry for spontaneous and for induced tracks requires calibration (Reimer *et al.*, 1970).

(c) In minerals the crystallographic orientation of the etched face influences the efficiency (Green and Durrani, 1977). This effect can be circumvented by revealing spontaneous and induced fission tracks on faces of identical crystallographic orientation.

(d) The track etching conditions, such as type of etching reagent, its concentration, its temperature and the etching duration, have a pronounced effect on the efficiency (Storzer, 1972). Identical conditions for spontaneous and induced fission tracks are obtained by simultaneous etching of both types of tracks.

(e) Thermal treatment of samples changes the efficiency. The effect is mainly due to partial fading of tracks which results in lower track etching efficiency, but also physical changes of the detector material as a consequence of the heating may contribute to efficiency changes. The annealing of spontaneous fission tracks may be due to natural heating (see section on track fading) or to inappropriate sample preparation. It is a common practice to erase the spontaneous fission tracks from the sample by strong heating before neutron irradiation. This treatment may change the efficiency with which the induced fission tracks are revealed.

(f) The accumulation of radiation damage due to natural α-decay may change the efficiency of minerals such as zircon and sphene. According to Gleadow (1978), the 'spontaneous tracks in sphene are therefore not directly comparable in their etching characteristics to tracks induced in the sphene after thermal annealing' of the α-radiation damage.

(g) *Subjective criteria* are used for track identification under the microscope. This source of error is sometimes underestimated when spontaneous and induced fission tracks are not alike, for example due to track fading and when there is a large time lapse between the counting of the spontaneous fission tracks and of the induced ones.

Another complication arises if the uranium is not homogeneously distributed within the sample. One way of solving this difficulty is to count the spontaneous and induced fission tracks on identical or nearby faces of the sample. External track detectors are also used for this purpose. A different approach is based on taking a sufficiently large area or statistically adequate number of crystal grains for the counting of both types of tracks in order to obtain representative uranium contents of the total sample.

3.2 Dating techniques

The various difficulties listed in the previous section have led to the development of several techniques of fission track dating. The choice of which of the techniques is the optimal, depends on the type of material, its grain size, the spontaneous-fission track density, the degree of track fading and the homogeneity of uranium distribution. In the stratigraphic application of fission track-dating the population technique for small glassy shards and apatite grains and the external track detector technique for single zircon and sphene grains are the ones mainly used.

3.2.1 Population technique

The population technique (Naeser, 1967; Wagner, 1968; Storzer and Gentner, 1970) is used for sufficiently large populations of small (typically between 100 and 200 μm) mineral grains or glass shards which are separated from their host rocks. The population is split into two subpopulations, one of which is used for the counting of spontaneous fission tracks and the other—after neutron irradiation—for counting the induced ones. The grains or shards are embedded into a matrix of epoxy and are polished and etched. The tracks per unit field are counted in a sufficiently large number (several hundred) of single grains to have statistical validity. The population technique is based on the assumption that the subpopulations represent the same average uranium content. This assumption can be checked by track counting statistics (Johnson *et al.*, 1979; Naeser *et al.*, 1979). The main advantage of the population technique is its experimental simplicity, the registration of spontaneous fission tracks in the sample material under the same 4π-geometry, the identical etching conditions of spontaneous and induced tracks and the statistical elimination of difficulties due to crystallographic orientation and inhomogeneously distributed uranium. The subpopulation which is used for the induced fission tracks is usually heated before neutron irradiation in order to erase the spontaneous fission tracks. Otherwise the known spontaneous fission track density has to be subtracted after neutron irradiation ('subtraction technique'). The population technique is disadvantageous for minerals with very high and very inhomogeneous uranium contents, as is often the case with zircon and sphene concentrates. It is commonly used for apatite grains and for glass shards. A similar procedure is applied when large crystals and pieces of glass are available. The samples are cut or cleaved into two adjacent halves. One half is irradiated. The two halves are polished, etched simultaneously and the corresponding track densities measured in a statistical manner using a unit field.

3.2.2 *Reetching*

The reetching of an existing external or newly polished (internal) face allows the determination of the spontaneous and induced fission tracks on the same face. After etching the spontaneous fission tracks the sample is irradiated (without pre-annealing for spontaneous-track erasure) and etched again. In case an internal face is used for the spontaneous fission tracks, this technique has the disadvantages of different registration geometries, because the induced tracks would be on an external face. The presence of large spontaneous-track etch pits, might make counting of induced tracks difficult.

3.2.3 *Repolishing*

Repolishing (by polishing down 20 μm or more) of an internal face which has been used for spontaneous track counting recreates an internal face with 4π-geometry for track registration. The reetching and repolishing techniques are useful for samples with inhomogeneously distributed uranium because spontaneous and induced fission tracks are counted on nearly identical faces. Also they do not present the problem of different crystallographic orientation for both types of track counts. These techniques allow the dating of small, single grains with sufficiently high uranium content. They are used for micas, large crystals and pieces of glass.

3.2.4 *External track detector method*

In the external track detector method a track detector material is placed in close contact with the sample face which contains the spontaneous fission tracks and this sandwich is irradiated with neutrons. Induced fission fragments leave the sample and are recorded in the external detector, where they are revealed by etching (Price and Walker, 1963). Muscovite plates or plastic foils which must be sufficiently clean of uranium, are used as detector materials. In practice, one embeds small crystal grains in a matrix of epoxy or Teflon plastic and polishes and etches them to reveal the spontaneous fission tracks on an internal surface. The section is then covered with the external detector and irradiated. The erasure of the spontaneous fission tracks in the sample by annealing is not necessary. This method is therefore specially suitable for minerals with high α-radiation damage. It allows the dating of small single grains with sufficiently high uranium content. It is thus commonly applied to sphene and zircons. The disadvantage of this technique lies in the necessity to correct for the different efficiencies due to material and geometry differences of the sample/detector pair in the specific etching conditions used. Also, size comparisons between spontaneous and induced tracks for recognizing thermal fading of tracks cannot be made with this technique. The density measurement of spontaneous and induced fission tracks for single grains with varying uranium contents allows the application of an isochron age calculation (Gleadow and Lovering, 1975).

3.3 Fission track fading and age correction

Latent fission tracks are, like any radiation damage, very sensitive to elevated temperatures (Fleischer *et al.*, 1965a). Thermal energy reactivates the radiation damage accumulated along latent tracks, so that a latent track heals up gradually and becomes shorter with increasing annealing temperature. At the same time, the etching velocity along this latent track is reduced. The consequence of these two effects is that increasingly fewer and shorter fission tracks cross an etched reference surface. In the case of micas, this phenomenon was realized at the very beginning of fission track dating (Maurette *et al.*, 1964; Fleischer *et al.*, 1964; Bigazzi, 1967) and was also discussed as a possible reason for anomalously low fission track ages. Certainly, the extent of partial fading of fossil fission tracks at a given ambient temperature and, hence, the degree of thermal age-lowering, depends to a large extent, on the track retention properties of the specific material to be dated. These retention properties are characteristically different among the various detector materials. As a first approximation, track retention temperatures are lower for glasses than for minerals. Also, for a given material the tracks do not disappear at one single temperature but over a wide temperature range, due to the fact that the thermal energy (activation energy) required for further track fading increases with the degree of track fading. This is shown in Figure 3, taking a North American tektite as an example. The Arrhenius plot represents results of an annealing experiment obtained for a set of tektite discs with 'fresh' reactor-induced tracks which were tempered at different temperatures and for various times (Storzer and Wagner, 1971). Within the measured field of this representation all points with identical degree of track density reduction are aligned on straight lines. On the assumption that the slopes (i.e. activation energies) of these lines are constant, track retention properties at low ambient temperatures prevailing for geological times can be estimated by extrapolating laboratory annealing data.

These estimated track retention temperatures are compiled for the materials most commonly used in fission track dating in Table 1. For glasses, especially for sideromelanes, it seems evident that thermally lowered fission track ages are to be expected if the geological ages exceed some million years even if the samples never experienced temperatures higher than

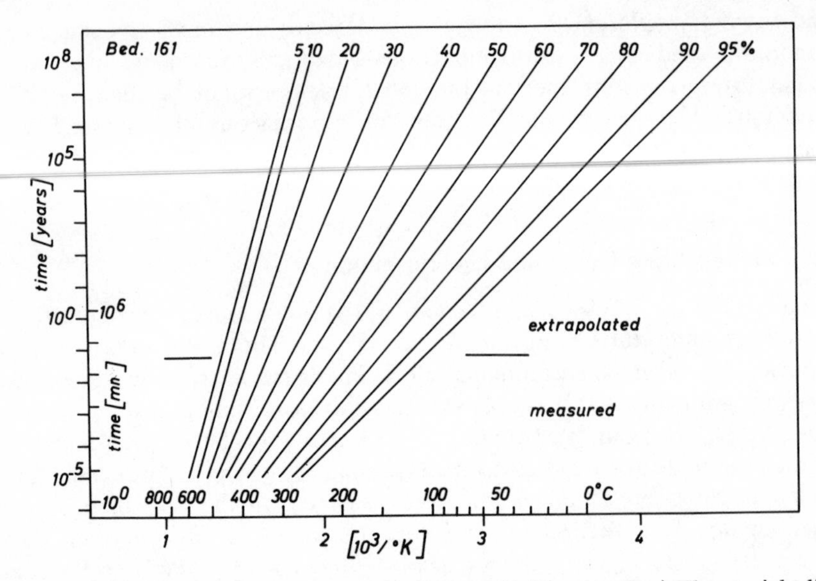

Figure 3. Annealing characteristics of bediasite 161 (Fayette Cty.) The straight lines represent the reduction to a residual percentage from a given track density (p/p_0) after annealing (etched 25 sec at 23°C in 48% HF). (From Storzer and Wagner, 1971, reproduced by permission of Elsevier Publ. Cy.)

Table 1. Fission track retention temperatures for some materials of stratigraphic interest.

Material	Degree of track loss	Range of retention temperatures (°C) for different annealing times					Reference
		1 hr	1 a	10^2 a	10^4 a	10^6 a	
Glass							
Rhyolitic	Beginning	190–240	45– 85	5– 40	(−30)– 5	—	1, 2, 3
	Half	345–450	205–260	175–210	130–160	90–115	
	Complete	490–660	390–500	330–440	285–390	240–330	
Basaltic	Beginning	110–130	(−16)– 13	—	—	—	3, 4, 11
	Half	170–235	72–124	37– 84	8 51	(−16)– 24	
	Complete	230–345	175–246	144–207	116–173	92–143	
Tektite	Beginning	105–330	65–155	25– 95	(−5)– 50	(−30)– 20	3, 5, 6, 7
	Half	300–460	195–280	140–220	100–180	65–140	
	Complete	495–580	390–450	340–400	290–360	250–310	
Apatite	Beginning	270–280	150–165	110–125	70– 90	40– 60	3, 8, 9, 10
	Half	320–330	215–240	175–200	140–165	110–140	
	Complete	360–370	260–285	225–250	190–220	160–200	

1, Storzer (1970); 2, Suzuki (1973); 3, Storzer, unpublished results; 4, MacDougall (1976); 5, Storzer and Wagner (1969); 6, Durrani and Khan (1970); 7, Storzer and Wagner (1971); 8, Wagner (1968); 9, Naeser and Faul (1969); 10, Wagner and Reimer (1972); 11, Selo unpublished results.

5–10°C. Actually, numerous examples of anomalously low glass fission track ages are known (e.g. Fleischer *et al.*, 1965b; Storzer and Wagner, 1969; Storzer, 1970; Suzuki, 1973; MacDougall, 1976; Wagner *et al.*, 1976; Arias *et al.*, 1981).

Also, in the case of apatite, fossil fission tracks are expected to fade partially at ambient temperatures within some million years (Table 1). Recent apatite data from drill holes suggest that the track retention temperatures for apatite might be somewhat higher than those extrapolated from laboratory annealing experiments (Gleadow and Duddy, 1981; Naeser, 1979; 1981). According to the new data, tracks will start to fade at 50°C (Gleadow and Duddy, 1981) or 75°C (Naeser, 1981) within a hundred million years. These temperatures have to be compared with about 25°C for 10^8a derived from the annealing experiments.

The relatively low track retention temperatures for apatite, agree well with the observation that fossil tracks in this mineral are frequently shorter in length than induced tracks (Wagner and Storzer, 1970; Green, 1980). There is however, at present, considerable debate as to the effect of this shortening on fission track ages. It has been argued (e.g. Fleischer *et al.*, 1975) that shortening of fossil tracks up to 20% relative to induced tracks has no effect on the density of fossil tracks and, hence, on the age.

It is not by chance that these authors also use the 20% lower value for the spontaneous-fission decay constant (see following section). It is clear that this choice of the decay constant and a fossil-track length reduction may not be unconnected. In addition, this supposition is probably based on muscovite (which might be a special case) annealing experiments by Fleischer *et al.* (1964). They observed that 'tracks appear to shorten before their number per unit area decreases'. In principle, such a phenomenon is typical for annealing experiments on californium-252 fission tracks. As discussed in detail previously (Storzer and Pellas, 1977), the use of external ^{252}Cf tracks for studying the annealing behaviour of internal uranium tracks leads to aberrant results. On the basis of extensive annealing experiments on glasses and apatites using 'fresh' reactor induced tracks it could be shown that there exists from the very beginning of fading a definite relationship between track length (or size in the case of glasses) and track density reduction for both, glasses (Storzer and Wagner, 1969) and apatite (Wagner and Storzer, 1970). This relationship is shown for an apatite in Figure 4. Storzer and Wagner (1969) took advantage of this characteristic fading behaviour in correcting thermally lowered fission track ages.

In the '*track-size*' correction technique the lengths of fossil tracks (etch-pit sizes in the case of glasses) are compared with those of the induced tracks. In this way, thermally lowered track ages can be realized by the reduced length ratio (l/l_0) of fossil to induced tracks. The proper correction factor for the partial fossil-track fading, is then obtained by means of a

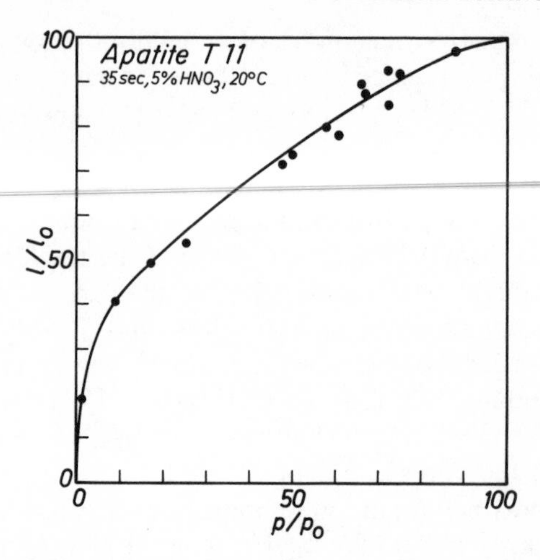

Figure 4. Average fission track length (l) *versus* track density (p) of partially annealed tracks in apatite, normalized to the average length (l_0) and density (p_0) of thermally unaffected fission tracks. The measurements were performed on a population of unoriented apatite crystals (from Wagner, 1973).

correction curve, as shown in Figure 4. Corrected fission track ages are concordant with K–Ar ages obtained for the same materials or with geological ages (Storzer and Wagner 1969; 1971; Storzer, 1970; Arias *et al.*, 1981). In fact, correction curves always have to be calibrated for the material to be dated as well as for the etching conditions used to develop fossil and induced tracks (Storzer, 1972). The 'track-size' correction technique is especially interesting for the detection of fossil thermal events, or more generally for retracing the thermal history of rocks (Storzer, 1970; Wagner and Storzer, 1975).

A further correction procedure for thermally lowered fission track ages of glasses and apatites is the '*plateau-annealing*' technique (Storzer and Poupeau, 1973). This technique is based on the observation that the thermal energy required for further track fading increases exponentially with the degree of track fading. Because of this, the stepwise annealing of a sample containing both partially faded fossil tracks and thermally unaffected induced tracks, results initially in greatly different rates of fading of these two kinds of tracks. At some point in the annealing procedure, depending on the degree of fossil-track fading, the two rates of fading become equal. This is shown in Figure 5 for a North American tektite, for a tachylitic glass and for an obsidian from Colombia. Also shown in Figure 5, for the three glass types, are the increase in the apparent fission track ages with increasing degree of annealing, until a plateau is reached where the age remains virtually constant. The fission track plateau ages are concordant with the

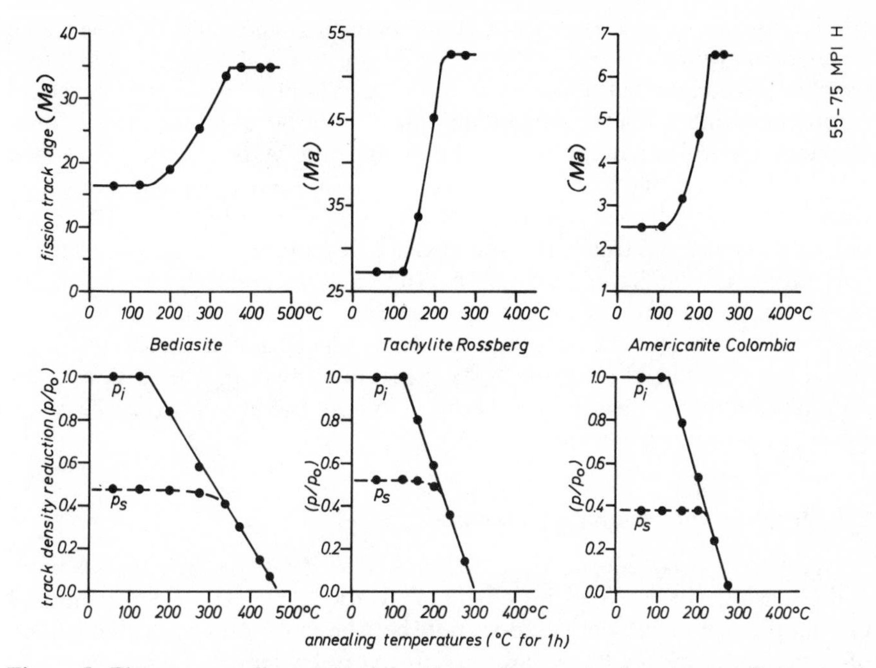

Figure 5. Plateau-annealing correction. Above, increase in the apparent fission track ages with increasing temperature of annealing up to the plateau age. Below, with increasing degree of annealing the two rates of fading for fossil and induced tracks become equal.

corresponding K–Ar and/or 'track-size' corrected fission track ages (Storzer and Poupeau, 1973; Wagner *et al.*, 1976; Arias *et al.*, 1981).

Although the 'plateau-annealing' technique is unable to detect and to separate mixed ages due to a fossil thermal event (Storzer, 1970), its application to thermally lowered fission track ages provides a more simplified correction procedure compared with the 'track-size' correction technique. The precision of the latter lies between 5 and about 40% depending on the degree to which the fossil fission tracks have been faded. The precision of fission track plateau ages is normally of the order of 5% (1σ).

4 UNCERTAINTIES OF EXPERIMENTAL PARAMETERS AND PHYSICAL CONSTANTS

4.1 Fission track counting

In order to minimize experimental errors which may affect the density ratio p_s/p_i of spontaneous to induced fission tracks, one should prepare the unirradiated and irradiated aliquots of a sample under strictly the same conditions of, for instance, mounting, polishing and etching. The same etching conditions are best achieved by immersing both aliquots simultaneously in the etching bath. The spontaneous and induced fission tracks should be counted with the same microscopic magnification using identical criteria for track identification. In optimal cases the ratio p_s/p_i can be determined with a precision of about ±3 to ±5% (1σ). Such cases presume sufficiently high track counting statistics, homogeneous uranium distribution, the absence of partially faded fossil tracks and the absence of features such as dislocations, microlites and small bubbles which after etching might be mistaken for fission tracks. In practice, one or more of these disturbing influences commonly causes larger uncertainties. It should be noted that most errors which are quoted for fission track ages published in the fission track literature represent the *precision* of the ratio p_s/p_i and not the *accuracy* of the fission track age.

4.2 Isotopic composition of uranium

Apart from very special circumstances (e.g. the Oklo 'reactor', Naudet and Renson, 1975), the isotopic composition I ($^{235}U/^{238}U$) of natural uranium in terrestrial samples, does not show large variations which might be of any significance for the accuracy of fission track ages. Even in meteorites and lunar rocks no deviations from this isotopic ratio were detected (Rosholt and Tatsumoto, 1971; Chen and Wasserburg, 1981). The value recommended by Steiger and Jäger (1977) for geochronological use is $I = 7.253 \times 10^{-3}$ (Shields, 1960; Cowan and Adler, 1976). Its accuracy is probably around ±1‰.

4.3 Spontaneous fission rate

Compared to the ^{238}U α-decay constant, which is sufficiently well known (value recommended by Steiger and Jäger, 1977: $\lambda_\alpha^{238} = 1.55125 \times 10^{-10}\ a^{-1}$, Jaffey *et al.*, 1971), the value for the decay constant of ^{238}U spontaneous fission is still under debate (e.g. Wagner *et al.*, 1975). This leads to the largest systematic uncertainty affecting the accuracy of fission track ages. In fact two values for the fission decay constant are currently in use: $\lambda_f^{238} = 7.03 \times 10^{-17}\ a^{-1}$ (Roberts *et al.*, 1968) (or $6.85 \times 10^{-17}\ a^{-1}$, Fleischer and Price, 1964) and $\lambda_f^{238} = 8.46 \times 10^{-17}\ a^{-1}$ (Galliker *et al.*, 1970). There is strong evidence (Gentner *et al.*, 1972) that the $8.46 \times 10^{-17}\ a^{-1}$ value gives the more plausible age results whenever partial fading of fossil fission tracks is considered and corrected. This evidence is presented in Figure 6. On the other hand, Naeser *et al.* (1977) reported best agreement between zircon fission track ages (obtained by means of the external detector method) and K–Ar ages of coexisting minerals from volcanic rocks where track fading

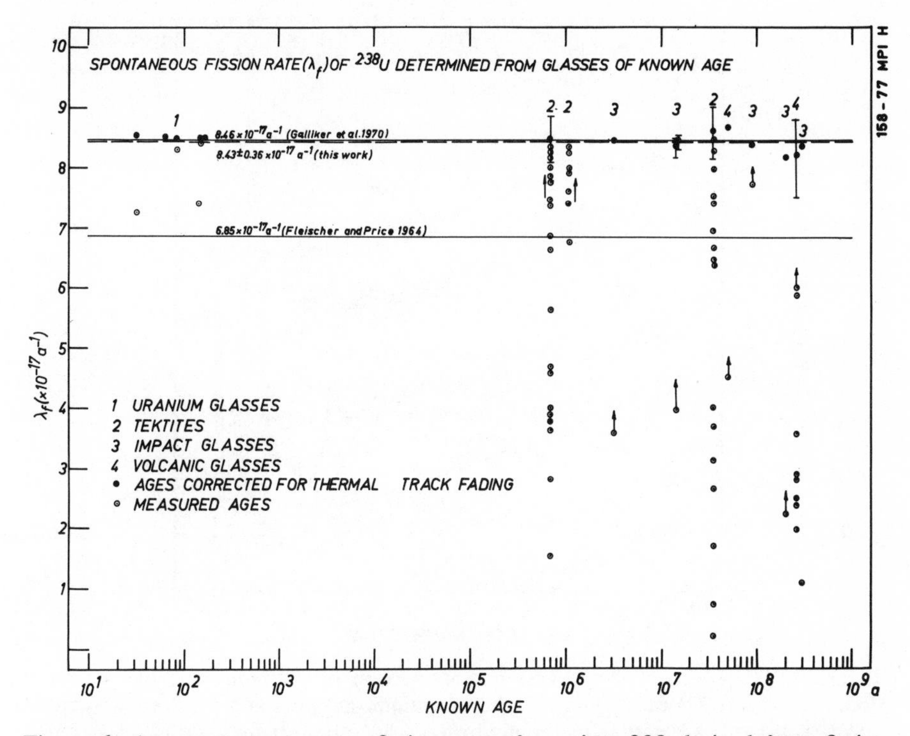

Figure 6. Apparent spontaneous fission rate of uranium-238 derived from fission track dating of glasses with well-documented age *versus* the sample age. Only after correction of partial thermal fading of fossil fission tracks do the results come close to the value $8.46 \times 10^{-17}\ a^{-1}$ (Galliker *et al.*, 1970). For comparison the result of Fleischer and Price (1964) is also shown; they used the same approach but did not take into account the effect of partial fossil-track fading.

should be minimal for a λ_f value of $6.9 \times 10^{-17}\ a^{-1}$. In our opinion this result has to be regarded with some reservation because these authors did not take into account uncertainties due to geometry correction and neutron dosimetry. However, we do not wish to discuss here which λ_f value should be recommended, but to draw the reader's attention to the fact that two constants differing by about 20% are still used for age calculation.

4.4 Fission cross-section and neutron dosimetry

The uranium-235 fission cross-section σ_f^{235} for thermal neutron absorption is strongly energy-dependent (Figure 7). Below the energy of 0.2 eV this cross-section, analogous to the activation cross-sections of copper, cobalt, gold and indium, follows a '$1/v$- law', with v being the neutron

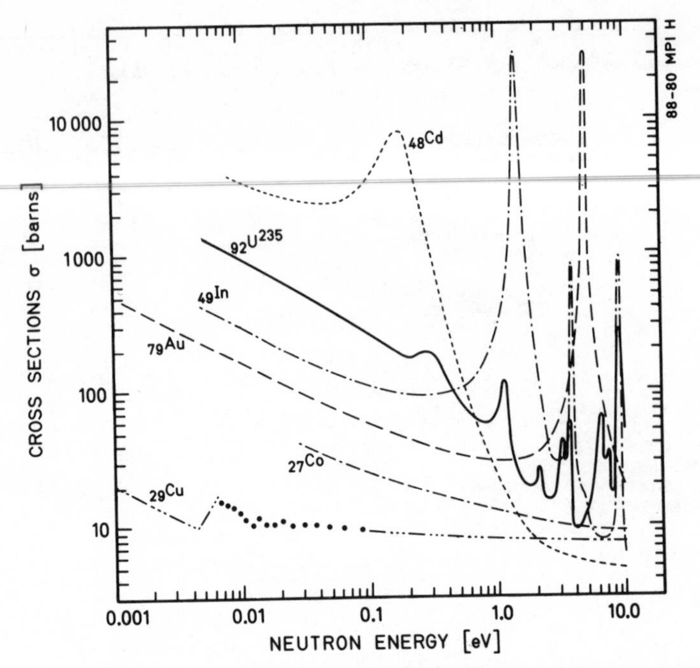

Figure 7. Dependence of the fission cross-section of uranium-235, the activation cross-sections of copper, cobalt, gold and indium monitors and the total absorption cross-section of cadmium from the neutron energy (after Hughes and Harvey, 1955).

velocity (Fermi, 1947). At energies exceeding 0.2 eV, the fission cross-section displays resonance behaviour. Below the conventional cadmium cut-off at 0.5 eV, a neutron flux is considered as thermal, between 0.5 eV and 0.5 MeV as epithermal and above 0.5 MeV as fast. In order to exclude epithermal neutron fission of uranium-235 or even fast neutron fission of uranium-238 and thorium-232, thermal reactors should have thermal to epithermal flux ratios exceeding about 100 (Beckurts and Wirtz, 1964). However, even for reactors with thermalized neutron fluxes, the thermal neutron energy spectra, and thus the thermal fission cross-section σ_f, may vary due to different working temperatures. Therefore the cross-section is conventionally reduced to the velocity $v_0 = 2200$ m/sec ($E = 0.0253$ eV) of a highly thermalized neutron flux. This reduced thermal fission cross-section for uranium-235 $\sigma_f = 580.2 \pm 1.8$ barn (Hanna *et al.*, 1969) is used for fission track age calculation. This reduction is allowable as long as the neutron fluxes are well thermalized and obey Maxwellian energy distribution, since the fission cross-section as well as the activation cross-section of the flux monitors closely follow the '$1/v$- law'. However, due to deviations from strict '$1/v$- behaviour' and to some contributions of epithermal neutrons in practice, the value $\sigma_f = 580.2$ barn has about ±5% uncertainty for thermal neutron reactor facilities.

The time-integrated thermal neutron flux is obtained by measuring the neutron induced activity of the monitor used (Cu-, Co-, Au-, In-foils). It should be pointed out again that for flux determination with 1/v- monitors it is not necessary to know the neutron spectrum exactly as long as it is thermal, and that deviations from the '$1/v$- law can be corrected for if the neutron spectrum is Maxwellian (Beckurts and Wirtz, 1964, De Soete *et al.*, 1972). As both monitor and U-235 in the sample follow the $1/v$- law, the activation of U-235 or the density of induced fission tracks in an irradiated sample can, in principle, be precisely monitored with metal activation systems in the thermal region of the neutron spectrum. However, if the neutrons are not well thermalized significant errors can be introduced into the calibration of the flux monitored with gold or indium due to high resonance peaks of the activation cross-sections in the epithermal region (Figure 7). Copper as monitor remains questionable because of resonance peaks in the thermal region. Whenever resonance peaks affect the dosimetry the value 'n' for the integrated flux will be overestimated, whereas the density of the induced fission tracks 'p_i' will be too low, which increases the apparent fission track age.

Additional uncertainties regarding flux calibrations are introduced by flux perturbations during irradiation with thermal neutrons. Whenever a sample is placed in the diffusing neutron medium of a thermal reactor for neutron activation or in order to induce fission of the U-235 nuclei, some neutrons will be absorbed, causing flux perturbations inside and outside this sample. Diffusing into the sample, neutrons are absorbed so that the flux decreases towards the sample's interior. Due to this self-absorption effect, steep density gradients of induced fission tracks may occur inside irradiated samples. On the other hand, as the neutrons absorbed within the sample are prevented from back-scattering into the moderator, the flux cannot remain constant in the vicinity of the absorber, which leads to a flux decrease towards this sample. This flux-depression effect is one of the principal error sources in the calibration of thermal neutron fluxes because neutron dosimeter and sample are not at the same location. Both effects together are the reason for steep lateral and axial flux gradients (up to several percent per centimetre) which occur within an irradiation container. The extent of this flux perturbation, depends strongly on the sample's dimension and its concentration of elements with high cross-sections for neutron absorption (e.g. boron, cadmium, rare earths), as well as on the nature of the crystal-mounting medium (e.g. epoxy resin can cause severe flux depression). It can be minimized by irradiating only small amounts of samples in well thermalized reactor positions, preferentially in a thermal column. If important self-absorption and, hence, flux depression effects are expected for samples with, for example, higher rare earth contents, these samples may each be irradiated individually in close contact with a neutron dosimeter and the induced tracks may then be studied as a function of sample depth. The effects of flux perturbation introduce the need for correction factors which, unfortunately, in real cases are difficult to calculate (Beckurts and Wirtz, 1964; De Soete *et al.*, 1972). Equally difficult, if not even more complex, is the thermal–neutron flux calibration itself. Important uncertainties are introduced by the efficiency calibration of metal activation systems (De Soete *et al.*, 1972), for example due to neutron attenuation and self-shielding effects in gold or the incomplete β^+ annihilation in copper monitors, as well as by the calibration of the counting system itself. It is difficult to estimate the contribution of the different effects inherent in irradiation with thermal neutrons to arrive at an overall uncertainty for the absolute, integrated neutron flux. Relative measurements may probably be reproduced with a precision of about 3% using the same thermal reactor facility with its given parameters. As to the absolute value for an integrated flux, it seems, however, realistic to assume that even in the best case (quantitatively thermalized neutron spectrum, small-dimensioned sample in close contact with an unproblematic neutron monitor) the accuracy will not be better than ±10%. An example concerning the problem of inter-reactor thermal–neutron flux calibration is given by Wagner *et al.* (1975), where for the same sample the results vary up to 30% among the different thermal reactors. A further example is the preirradiated NBS glasses SRM 961–964 which were distributed for calibration of thermal neutron fluxes (Carpenter and Reimer, 1974). For these reference glasses, differences up to 19% exist between the copper and the gold calibration, whereby the copper monitors always give the lower values although both monitors were activated simultaneously in the same reactor facility.

In conclusion, it is evident that important uncertainties exist regarding the fission track ratio (p_s/p_i), the decay constant (λ_f), the fission cross-section (σ_f) and the integrated neutron flux (n) calibration. This leads to an overall absolute uncertainty of about 15–20% (1σ) for fission track ages. Therefore it is emphasized that the fission track method in the present state, even in the best case, hardly meets the accuracy required in stratigraphy. The only way to escape this dilemma would be the introduction of a *common age-standard* in fission track dating as similarly used routinely in $^{39}Ar/^{40}Ar$ dating. By irradiating sample and age-standard simultaneously all uncertainties with the exception of the error of p_s/p_i disappear. Therefore, the major challenge for the future will be the search for an age-standard which can satisfy the following restrictions: geologically well documented, well dated by other radiometric methods, no indication of thermal fission track age-lowering, easy to handle for a simple fission track dating technique, and abundant in material. However, there are already age-standards such as the 'Fish Canyon' tuff (Naeser *et al.*, 1981) or a moldavite (Storzer *et al.*, 1973) used in some fission track dating laboratories. Unfortunately these age standards do not meet all the mentioned criteria. When these presently available standards are used, the accuracy of the resulting fission track age may already be as good as about 10% (1σ).

5 GEOLOGICAL MATERIALS

5.1 Volcanic glasses and tektites

The geological materials most commonly used in the stratigraphic application of fission track dating are glass shards separated from volcanic ash beds. Ash layers which are intercalated in sedimentary series seem to be ideally suited for stratigraphic purposes. Air-borne ashes are quickly transported from the eruption to the deposition site. Thus, the age of their volcanic formation, which is determined by fission track dating, and the age of deposition are practically identical. However, serious problems may arise if glass shards of different generations are incorporated within the same bed due to reworking of the layer during its geological past.

Glassy particles can be easily handled for fission track dating as long as they are not smaller than about 100 μm. They are usually dated with the 'population technique'. Occasionally microlites and gas bubbles within the glass matrix can be confused with the fission track etch pits. Volcanic glasses contain typically few ppm uranium, which allows also the dating of very young samples down to several thousand years. It is therefore not surprising that volcanic glasses have been used in order to solve stratigraphic problems mainly of the Pleistocene period, which is difficult to date with other radiometric methods. However, there are also numerous studies on Tertiary ash beds. In geographical terms, bentonite glasses from Germany (Storzer and Gentner, 1970) and glass shards from ash layers in Italy (e.g. Selli *et al.*, 1977), New Zealand (e.g. Seward, 1976), the central and western United States (e.g. Boellstorff, 1976), Japan (Suzuki and Yamanoi, 1970), Morocco (Arias *et al.*, 1976) and East Africa (Fleischer *et al.*, 1965c; Hurford, 1974) have been dated with fission tracks. Also glass shards from volcanic ash layers in deep-sea sediments have been investigated (MacDougall, 1971). In the same way, microtektites which occur in deep-sea sediments are useful for stratigraphic application of fission track dating, as has been demonstrated for microtektites from the Caribbean Sea (Glass *et al.*, 1973) and from the Indian and Atlantic Oceans (Gentner *et al.*, 1970). In the latter case a good agreement with the palaeomagnetic and palaeontological stratigraphy of the deep-sea cores was obtained. Sometimes obsidian also is suited for stratigraphic studies. Stratigraphically significant fission track ages were determined for Mediterranean obsidian flows and xenolithic obsidian inclusions within pumice (Wagner *et al.*, 1976; Seward *et al.*, 1980).

Although natural glasses do not present any particular experimental problems in fission track dating, they generally exhibit the phenomenon of partial fading of fossil tracks. The fading may be caused by secondary thermal events, especially in volcanic areas, but also the *normal ambient temperatures* are sufficient for partial track fading in most glasses. Even the mentioned microtektites from the Caribbean deep-sea sediments had suffered a 20% loss of fossil tracks, although they had spent all their geological past at the low temperatures of the deep-sea bottom. Since partial fading of fossil tracks results in lowered fission track ages of the glasses, such ages are obviously useless for stratigraphic applications. Undoubtedly all serious fission track researchers would probably agree on this point. Therefore, it is odd that individual authors simply ignore the fading effect and interpret such 'raw' fission track ages on glass shards as stratigraphically significant formation ages (Boellstorff and Steineck, 1975). Fortunately, *partial fading of fossil tracks* can be independently recognized by size comparison of fossil fission tracks with induced ones in the same sample. It is therefore essential to carry out size measurements of fission tracks when dating glasses. Once partial fading of tracks has been recognized, one of the

Table 2. Comparison between 'apparent' and 'size' and 'plateau' corrected fission track ages of stratigraphically well documented glasses (from Arias *et al.*, 1981).

Sample site	Type of occurrence	Palaeontological age	'Apparent' age (Ma)*	'Size' and 'plateau' corrected age (Ma)*	
Monte Amiata (Tuscany)	Lava flow	Pleistocene	0.33	0.42±0.05	0.36±0.04
Valle Ricca (Latium)	Glass shards in sandy clays	Neogene–Quaternary boundary	1.38	2.13±0.27	2.03±0.26
Marco Simone (Latium)	Glass shards in sandy clays	Middle Pliocene (Globorotalia *crassaformis* zone)	2.24	3.32±0.30	3.18±0.31
Roccastrada (Tuscany)	Lava flow	Upper Pliocene	1.98	3.23±0.23	3.20±0.17
S. Vincenzo (Tuscany)	Lava flow	Pliocene	3.84	4.96±0.37	5.23±0.31
Monte Arci (Sardinia)	Perlite	Lower Pliocene	2.73	4.96±0.69	4.67±0.45
S. Maria in Carpineto (Marche)	Glass shards in silt	Messinian	2.77	6.16±0.84	6.12±0.62
Izarorene (Morocco)	Glass shards in clay	Messinian	2.04	6.45±0.46	6.80±1.10
Torrente Tarugo (Marche)	Glass shards in silty clays	Chattian–Aquitanian boundary	6.25	22.31±2.41	19.73±2.18
Tripoli di Contignago (Emilia)	Glass shards in tripoli	Chattian–Aquitanian boundary	10.66	24.22±1.54	22.65±1.25

* Calculated with $\lambda_f = 6.85 \times 10^{-17}$ a^{-1}.

two correction techniques described above has to be applied. One has clearly to distinguish between the 'raw', uncorrected and the corrected fission track age of glasses. Only the latter is generally significant in stratigraphic studies. This has been recently demonstrated by Arias *et al.* (1981). These authors dated numerous Pleistocene and Tertiary volcanic glasses from Italy by using both correction techniques simultaneously (see Table 2).

5.2 Zircon

The most important mineral for stratigraphic applications of fission track dating is zircon. Zircon occurs commonly as accessory mineral in volcanic tephra and tuff layers. The relatively high uranium content in zircon allows the age determination of single grains. The experimental techniques of handling single grains down to 100 μm in size for fission track dating have been largely elaborated (Naeser, 1969; Gleadow *et al.*, 1976). The dating of individual grains has the advantage of distinguishing between *primary* and *detrital zircon* grains provided their fission track ages are sufficiently different from one another. Detrital zircons have been recognized in this way by Gleadow (1980). Older zircon populations may also derive from basement rocks disrupted during volcanic eruption (Seward *et al.*, 1980). Whatever the source of such older zircons, they contaminate the fission track age of the primary zircons, which is the relevant one for stratigraphic interpretation. Unlike glass, fission tracks in zircon are not sensitive to ambient temperatures which are typical for rocks near the surface of the earth. Therefore, the danger of lowered fission track ages due to fossil-track fading is much less for zircons than for glass shards from tephra layers. This may not be valid for zircons from much older stratigraphic units which, at some time during their long geological past, were exposed to higher temperatures due to subsidence. Studies on borehole rocks indicate that temperatures of *c.* 70°C prevailing for 10^8 years can significantly lower the fission track age of zircon (Naeser 1979; 1981). Also secondary thermal events, such as occur in volcanic areas, may be sufficiently strong partially to anneal fission tracks in zircon.

On the other hand, there are some experimental difficulties in the fission track dating of zircons as has been recently described by Gleadow (1980). These difficulties are caused by changing etching efficiencies according to the level of alpha radiation damage and the crystallographic orientation of the etched faces and tracks. Another complication may be introduced by the presence of acicular inclusions within the zircon which after etching might be mistaken for tracks.

As a dating technique for zircon the 'external detector technique' is usually applied. Since the induced fission tracks are recorded in a different detector material (mostly muscovite) and under different detection geometry, *correction factors* are necessary. Unfortunately, careful studies on the correlation between geometry correction and α-radiation damage as shown by Gleadow (1981) for sphene do not yet exist for zircon. Such correction factors are not trivial because they have to take into account the various etching efficiencies of the single zircon grains. They present a serious source of error.

Stratigraphic studies using zircon have so far been carried out on Pleistocene and Tertiary sedimentary and volcanic series, mainly in the western United States (e.g. Izett and Naeser, 1976), in East Africa (e.g. Hurford *et al.*, 1976; Aronson *et al.*, 1977), Japan (Tamanyu, 1975) and Greece (Seward *et al.*, 1980). Also much older stratigraphic units have been dated with zircons. Ross *et al.* (1978) reported the fission track dating of zircons from Ordovician and Silurian bentonites and other volcanically derived rocks in Great Britain. Unfortunately, these latter authors were apparently unaware of the efficiency problems in the external detector technique, and they did not consider the possibility of fossil-track fading in zircon. This impedes the unreserved acceptance of their data for stratigraphy.

5.3 Apatite

There are only very few stratigraphic applications using apatite fission track ages. Like zircon, apatite occurs as accessory mineral in tephra layers. It is relatively rich in uranium, commonly between 1 and 100 ppm. Apatite grains, down to 100 μm in size, are well suited for fission track dating. Usually, the population technique is applied. In order to detect detrital apatite grains the external detector technique for single-grain dating is useful (Gleadow, 1981). A serious drawback of apatite in its application to stratigraphy is its low thermal stability for tracks. In this respect, apatite behaves in a similar way to glass. Even ambient temperatures typical of near-surface rocks may cause considerable fading of tracks in apatite (Wagner and Storzer, 1970). This has recently been proven with length measurements on fossil and induced fission tracks in apatite from Tertiary tuffs (Green, 1980), including apatite from the Fish Canyon tuff which is used by some workers as an age-standard. Since the fading of fossil tracks very probably lowers the fission track age, a correction procedure analogous to that for glass may also be appropriate for apatite fission track ages (Wagner and Storzer, 1970; 1972). Therefore, 'raw' uncorrected apatite fission track ages of young volcanic horizons should not be used for stratigraphic studies. This statement applies even more to the apatite fission track ages on Early Palaeozoic stratotypes in Great Britain of Ross *et al.* (1978), who ignored the general phenomenon of track fading in apatite.

6 CONCLUSIONS

Although fission track dating has proved to be a valuable tool in geochronology, especially for young (e.g. Quaternary) samples which are difficult to date by other radiometric methods, its stratigraphic application is still limited. Due to relatively large uncertainties in the decay constant for spontaneous U-238 fission and in the neutron dosimetry, fission track dating at present cannot contribute much in establishing a time framework for stratigraphy. With regard to these uncertainties, it is surprising that for some stratigraphic studies fission track ages with errors as small as ±2% are given.

Another drawback of fission track ages for stratigraphic work is the fading of tracks. By measuring track sizes and by dating coexisting minerals track fading can be detected. Once fading has been recognized, appropriate age correction techniques can be applied.

If *accuracies* of about 10% are sufficient for a stratigraphic problem, fission track dating may be a useful method. However, for age comparison of different stratigraphic layers a much higher *precision* as good as ±3% (1σ) may be obtained provided the samples are dated in the same way and there is no track fading. For anyone who has illusions about the present state of stratigraphic utility of fission track dating we recommend the reading, as a

Table 3. Controversial potassium–argon and fission track dating of markerbed KBS tuff in the hominid-bearing Koobi Fora formation (east of Lake Turkona, northern Kenya).

Fitch and Miller (1970)	40/39 Ar on feldspars	2.61 ± 0.26 Ma
Hurford (1974)	Fission tracks in glass shards ($\lambda_f = 8.42 \times 10^{-17}$ a^{-1})	1.8 Ma (1)
Fitch *et al.* (1974)	40/39 Ar on feldspars	2.61 ± 0.26 Ma (2)
Curtis *et al.* (1975)	K/Ar on feldspars and glasses	1.82 ± 0.04 Ma 1.60 ± 0.05 Ma (3)
Hurford *et al.* (1976)	Fission tracks in zircons ($\lambda_f = 6.85 \times 10^{-17}$ a^{-1})	2.44 ± 0.08 Ma
Fitch *et al.* (1976)	40/39 Ar on feldspars	2.42 ± 0.01 Ma
Wagner (1977)	Fission tracks	~1.98 Ma (4)
Naeser *et al.* (1977)	Fission tracks	2.44 ± 0.08 Ma (5)
Drake *et al.* (1980)	K/Ar on feldspars and glasses	1.83 ± 0.06 Ma
Gleadow (1980)	Fission tracks in zircons ($\lambda_f = 6.9 \times 10^{-17}$ a^{-1})	1.87 ± 0.04 Ma
McDougall *et al.* (1980)	K/Ar on feldspars	1.89 ± 0.01 Ma

(1) Age interpreted as time of thermal overprint. (2) Age scatter: 0.52–2.64 Ma; 2.6 Ma preferred formation age; ~1.8 Ma time of thermal overprint. (3) Ages from two different areas. (4) Results of Hurford *et al.* (1976) recalculated with $\lambda_f = 8.46 \times 10^{-17}$ a^{-1}. (5) Reply to Wagner: proof of the correctness of the 2.4 Ma value from best agreement of 34 zircon fission track ages with K–Ar ages of coexisting minerals.

kind of illustrating example, of the contributions by Hurford (1974), Hurford *et al.* (1976), Wagner (1977), Naeser *et al.* (1977) and Gleadow (1980) to the controversy over the age of the hominid-bearing KBS tuff in East Africa (for illustration, see Table 3).

Finally, it should be pointed out that the utility of the fission track method for stratigraphy can be improved in the future. This would require diminishing the uncertainties about the *decay constant* and *neutron dosimetry* and introducing appropriate *age-standards* in fission track dating.

Résumé du rédacteur

L'uranium naturel est constitué de 99.3% d'uranium 238 et 0.7% d'uranium 235. La méthode de datation par traces de fission est basée sur la fission spontanée de l'uranium 238 environ 2×10^6 fois plus lente que la transmutation α. Les fragments de fission sont éjectés, au cours de la réaction spontanée, en direction opposée et créent, le long de leur trajectoire dans des solides diélectriques, une zone de dommage ou 'trace latente'. Cette trace latente, en forme de cylindre d'environ 100 Å de diamètre, peut être agrandie par une attaque chimique préférentielle jusqu'à des dimensions observables en microscopie optique. Dans les minéraux et les verres, la longueur des traces ainsi 'révélées' peut atteindre 20 μm.

La densité des traces de fission est fonction de l'âge de la roche et de la concentration en uranium. La concentration est déterminée en induisant la fission de l'isotope 235 dans un flux de neutrons thermiques. Connaissant le flux utilisé, le rapport: densité de traces spontanées sur densité de traces induites, permet le calcul de l'âge apparent d'une roche.

Les sources d'incertitude analytique de la méthode sont nombreuses. Elles comprennent *l'efficacité de la révélation* des traces, variable selon de nombreux facteurs assez bien dominés aujourd'hui par quelques spécialistes; la *répartition de l'uranium* dans le solide, souvent hétérogène; *l'identification des traces* et leur comptage au microscope. L'utilisation de diverses techniques de révélation permet d'opérer des recoupements de résultats et d'espérer une reproductibilité du rapport densité de traces spontanées/densité de traces induites (p_s/p_i) de 3 à 5% (1σ). Un second problème est l'actuelle absence d'agrément sur la valeur de la *constante de désintégration* à laquelle s'ajoute la difficulté de l'estimation précise de la *valeur du flux intégré* de neutrons utilisé dans le dosage de l'uranium. Au total les auteurs estiment qu'une incertitude absolue de 15 à 20% (1σ) sur l'âge apparent obtenu doit être prise en compte.

Ces dernières incertitudes (constante et flux) peuvent cependant être en partie éliminées par l'utilisation d'un minéral de référence d'âge interlaboratoire auquel on compare l'échantillon analysé. Compte tenu du fait qu'à l'heure actuelle un tel témoin de référence de qualité et commun n'existe pas, l'incertitude totale ne peut être ramenée qu'autour de 10% de l'âge mesuré. Cependant, si l'on dispose d'une série d'échantillons de comportement voisin, analysés dans des conditions strictement identiques, ils pourront être comparés avec seulement l'incertitude liée à la détermination du rapport (p_s/p_i).

En dehors des questions analytiques, les auteurs ont observé l'existence d'un phénomène *d'effacement des traces* latentes sous l'effet de facteurs thermiques. L'instabilité thermique des traces varie selon le solide considéré: l'effacement est beaucoup plus rapide dans une apatite que dans un zircon par exemple. Cet effacement a pu être estimé et des courbes de correction établies à partir d'observations expérimentales. Si l'effacement a été partiel, on peut alors calculer des *âges corrigés.*

En résumé, les âges par traces de fission sont fondés sur un principe aisé à comprendre et ne nécessitent pas d'équipement personnel coûteux si l'on a accès à un réacteur thermonucléaire. Ceci a conduit de nombreux géologues à commencer des recherches sans connaître en détail les difficultés de mise en oeuvre de la méthode. Après 18 années, la méthode est cependant bien maîtrisée dans certains laboratoires où la précision de 10% est atteinte grâce à l'utilisation d'une référence d'âge interne. Si cette précision est suffisante pour les âges récents, ou lorsque d'autres méthodes ne peuvent être utilisées, pour les âges plus anciens il faut encore attendre qu'un meilleur agrément intervienne notamment sur la question d'un matérial de référence commun.

(Manuscript received 23-2-1981)

Section III

Utilization of sediments as chronometers

Numerical Dating in Stratigraphy
Edited by G. S. Odin

12
A comparison of rubidium–strontium and potassium–argon apparent ages on glauconies

EDDY KEPPENS and PAUL PASTEELS

1 INTRODUCTION

Upon the recognition of the decay of ^{87}Rb to ^{87}Sr and of ^{40}K to ^{40}Ar as dating tools for minerals and rocks, glaucony was considered as a useful material for wide applicability in the absolute dating of sediments. As in the other chapters of this volume, the word *glaucony* (ies) will be used as a *facies* name in order to avoid confusion with a peculiar *mineralogical* composition. Understanding of the geochemical behaviour of the green pellets greatly depends on this facies/composition distinction. A first survey of the feasibility of glaucony dating, using both K–Ar and Rb–Sr methods, was made at the Massachusetts Institute of Technology, Cambridge, Massachusetts (USA) in the late 1950s. Later, much less attention was paid to the Rb–Sr method than to the K–Ar method, use of the latter becoming widespread in glaucony age-measuring.

The use of glaucony appears to be limited by a number of factors dealt with in detail in other parts of this book. The evaluation and occasional corrections of glaucony apparent ages are based on our knowledge of:

(1) the origin of the mineral (i.e. the behaviour of the relevant nuclides during the constitution or first closure of the system) (= genetic uncertainties);

(2) the phenomena that influence the apparent age (i.e. the behaviour of the relevant nuclides during the time elapsed since this first closure) = historical uncertainties).

Early or late diagenetic processes operating at low temperature, burial diagenesis, tectonization and alteration due to weathering may thus influence the glaucony ages, as shown elsewhere in this volume. In this connection, several factors are still unknown and/or subject to controversy.

The comparison of K–Ar and Rb–Sr ages on the same sample may help to elucidate the problem of glaucony age-measuring. On a first consideration, it can reasonably be expected that, in a number of cases, K and Rb will behave more similarly than will Ar and Sr. This, however, does not exclude the fact that a different behaviour of K and Rb may be expected under different circumstances.

Reported cases where comparisons of K–Ar and Rb–Sr apparent ages on the same sample are available will be considered here in turn, in the light of what is known on the behaviour of mother and daughter nuclides. Some additional information on the same point is expected as a result of the comparison itself.

In the present short review, Precambrian cases will be considered as well as those relating to the Phanerozoic, since the former may illuminate interpretations of the latter.

2 PRECAMBRIAN AND PALAEOZOIC GLAUCONIES

2.1

Nine Palaeozoic and a single Cenozoic (Early Eocene) sample from a large variety of locations were analyzed by Hurley *et al.* (1960). Results are given in Figure 1.

Three of the Palaeozoic samples came from folded or deeply buried rocks. Certain other Palaeozoic samples are supposed to have had 'histories of deep burial' also. In the Early Eocene sample, K–Ar and Rb–Sr ages are concordant but 10–20% higher than the estimated age for Early Eocene deposition.

In the Palaeozoic samples, three different cases are observed:

(1) In five samples, K–Ar and Rb–Sr ages are concordant but about 5–25% lower than the estimated age of deposition, the latter being a subject for discussion on its own.

(2) In two other samples, K–Ar ages are about 15% higher than Rb–Sr ages. The former are still about 10–25% lower than the estimated age of deposition.

(3) In two samples, Rb–Sr ages are about 25% higher than K–Ar ages. In both cases, Rb–Sr ages are close to the probable age of sedimentation. K–Ar ages are 20 and 35% younger than the time of deposition. All in all, radiometric ages are 10–35% lower than expected, two Rb–Sr ages excepted.

In the authors' opinion (Hurley *et al.*, 1960) the low radiometric ages can be explained by a slow post-depositional uptake of K (and Rb) in the expandable layers of the glauconies. In addition, in some samples Ar losses from the expandable layers, due to deep burial, cause excessively low K–Ar ages.

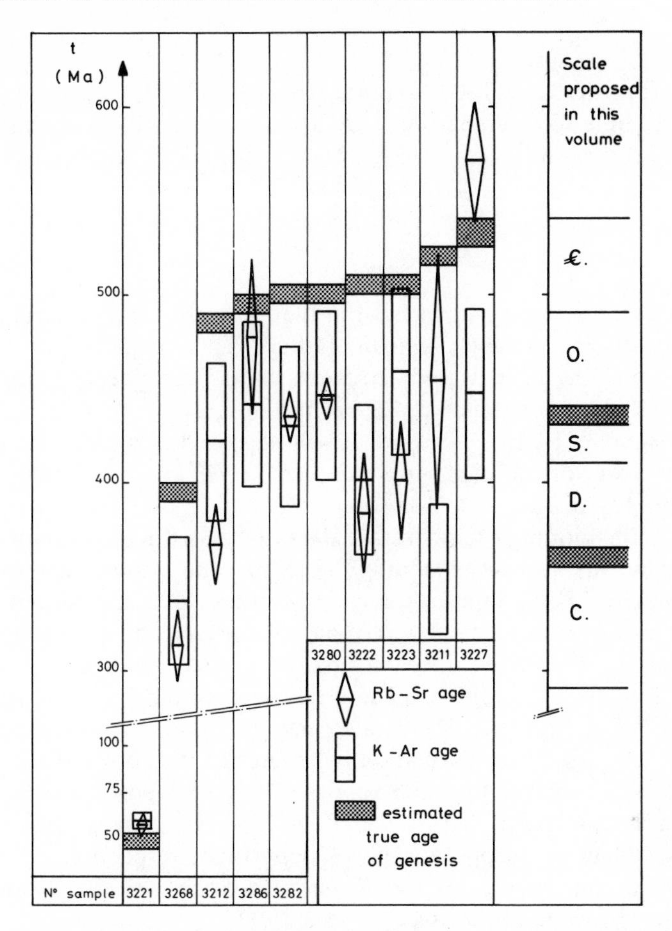

Figure 1. K–Ar and Rb–Sr apparent ages from Palaeozoic glauconies (data from Hurley *et al.*, 1960).

The late, slow K uptake interpretation (late diagenetic evolution) in glauconies has been challenged by other investigators (Obradovich, 1964, pp. 25–27; Velde and Odin, 1975). However, it is observed that all high Rb–Sr age/low K–Ar age glauconies have the highest K/Rb ratios, which apparently is in keeping with the original interpretation. The differences between K–Ar and Rb–Sr ages would then result from differences between the K *versus* Rb uptake rate.

Other interpretations, however, remain possible. In this respect the probable burial diagenesis and/or tectonization of the majority of the sampled deposits must be kept in mind.

2.2

Concordant K–Ar (1100 Ma) and Rb–Sr (1086 Ma) ages on a single glaucony sample (containing 6.94% K and 294 ppm Rb) from the Precambrian belt series (Montana, USA) were reported by Gulbrandsen *et al.* (1963).

2.3

Eleven Precambrian glauconies from the Australian Carpentaria province were analysed by McDougall *et al.* (1965). The sampled area has been tectonized and the investigated sediments have been deeply buried (up to 12,000 m in places). Results are given in Figure 2.

In six samples both K–Ar and Rb–Sr ages agree within the limits of analytical error. In the other five samples Rb–Sr ages are about 10–15% higher than K–Ar ages.

Comparison with high-temperature data on related igneous units suggests that the glaucony ages are generally lower than the assumed age of deposition, the highest being interpreted by the authors as minimum ages. In the Tawallah group, the K–Ar ≃ Rb–Sr apparent ages may be close to the age of sedimentation. The Dook Creek formation sample yields a K–Ar ≃ Rb–Sr age 10–20% lower than the estimated true age. In the Crawford formation, the Rb–Sr ages and one K–Ar age may be more or less close to the estimated true age of the formation, whereas the four other K–Ar ages are definitely 5–20% lower than the supposed age of deposition. The Mullera formation sample yields Rb–Sr and K–Ar ages that are probably 20 and 25% lower than the supposed age of deposition, respectively. Finally, an evaluation of the validity of the K–Ar = Rb–Sr age of the Wessel group sample is seen to be impossible.

The fact that most radiometric ages quoted above are too low cannot be explained by the model proposed by Hurley *et al.* (1960). Contrary to the case reported by the former contributors, the samples for which the following sequence is observed.

$$t \text{ (true age, estimated)} \geqslant t \text{ Rb–Sr} > t \text{ K–Ar}$$

are not characterized by a low Rb/K ratio. The reverse situation is observed in most instances. Taking into account the great depth of burial of the sediments, escape of radiogenic nuclides seems a more likely explanation than does K or Rb uptake. Ar loss is generally considered as being strongly temperature-dependent (Amirkhanoff *et al.*, 1961; Polevaya *et al.*, 1961; Odin *et al.*, 1977), while radiogenic Sr loss may be more dependent on compositional changes accompanying structural changes.

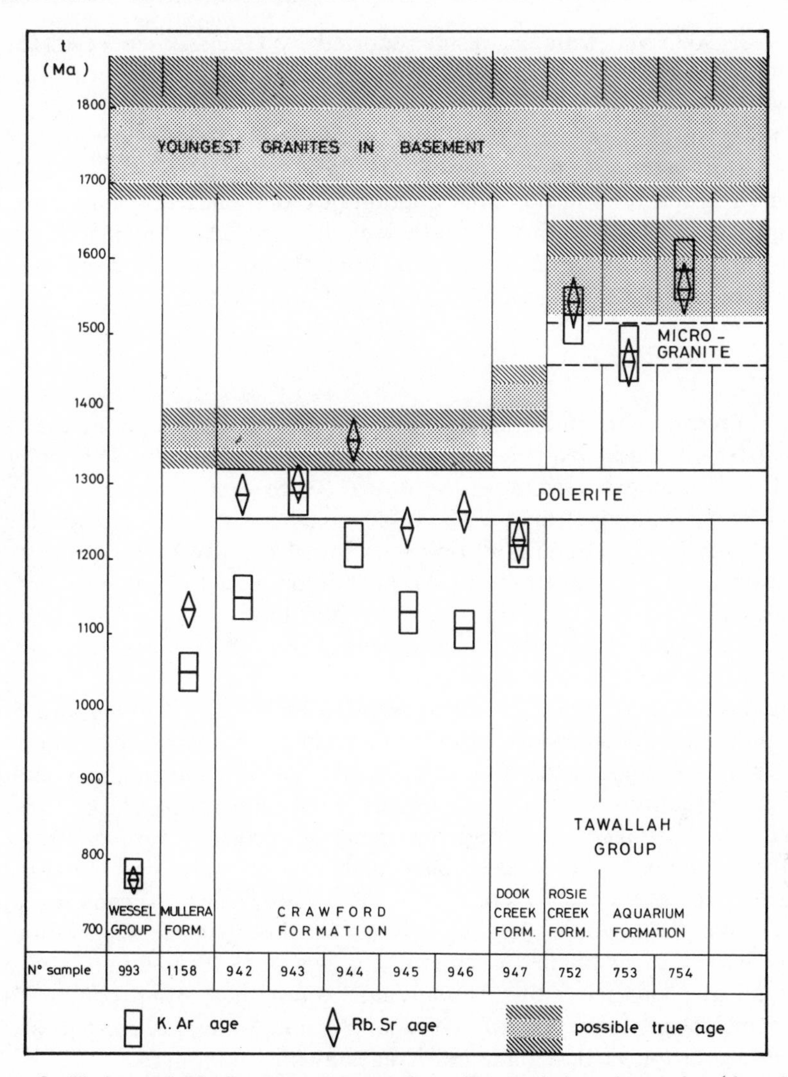

Figure 2. K–Ar and Rb–Sr apparent ages from Precambrian glauconies (data from McDougall *et al.*, 1965).

2.4

Seven samples of the Precambrian belt series (Montana, USA) yield concordant K–Ar (1050–1140 Ma) ages, Rb–Sr (1050–1100 Ma) conventional ages and a (1070 Ma) Rb–Sr isochron age (Obradovich and Peterman, 1968). In addition, these results are consistent with the whole-rock isochron figure of 1075 Ma, with the result reported by Gulbrandsen *et al.* (1963) (quoted above, Section 2.2) and with K–Ar ages on associated intrusive and eruptive rocks.

The consistency of these data must undoubtedly be considered in close association with the presence of the low temperature–pressure 1 Md micapolymorph in both the illitic fraction of the argillaceous sediments and the glauconies. Indeed, this shows that the considered sediments are essentially unaffected by burial diagenesis, contrary to most Precambrian and Palaeozoic deposits outcropping in different parts of the world.

2.5

Five Precambrian and a single Early Palaeozoic sample from the Australian Ngalia Basin (Northern Territory) were analysed by Cooper *et al.* (1970). The authors calculate conventional Rb–Sr ages introducing different initial $^{87}Sr/^{86}Sr$ ratios (from 0.70 to 0.75). In all cases Rb–Sr ages are higher than K–Ar ages. Assuming an open-sea initial ratio of 0.707 (Veizer and Compston, 1976), the Rb–Sr/K–Ar age differences for the Precambrian samples become 6–13%. Due to an internal discrepancy in the reported Rb–Sr data for the Early Palaeozoic sample, these data will not be discussed.

The investigated region has been affected by folding, faulting and overthrusting; the sediments may have been deeply buried. Diagenesis is obvious (the authors observe 'a reaction relationship between the glaucony and the poorly crystallized material', 'complete recrystallization of the quartz', 'patches and stringers of ferruginous material occur(ing) throughout (most of) the sample(s)'). All samples have high to extremely high Rb contents (350 to over 600 ppm) as compared to low to normal K contents (4.0–6.8%). We shall ignore the implications of this unusual observation on glaucony genesis or further evolution. The fact deserves mention, however, since all glauconies (about 50) analysed by our own group contain 200–300 ppm Rb only. The authors consider Ar and Sr loss as most probable, and interpret the highest radiometric ages as minimum ages.

2.6 Discussion of Palaeozoic and Precambrian cases

From the observed data on Precambrian and Palaeozoic deposits, it appears that glaucony ages are only significant in the cases where it can be demonstrated that *no diagenetic changes*, implying a significant temperature rise, have occurred. In other cases where burial diagenesis or even incipient metamorphism are probable, glaucony ages appear to be rejuvengated. In general, Rb–Sr ages seem to be less affected; in some cases they are concordant with or close to the time of deposition. This at first sight would suggest that Ar loss by any process (including diffusion and/or recrystallization) at relatively high temperature occurred, radiogenic Sr being more strongly held under the same conditions. However, if a lowering of glaucony K–Ar ages in rocks having undergone a metamorphism is a demonstrated fact (Frey *et al.*, 1973), its mechanism may nevertheless not be evident in some cases.

It has been shown (Polevaya *et al.*, 1961; Zimmermann and Odin, this volume) that Ar loss from glauconies accompanies the removal of chemically combined water, causing changes in the lattice structure, at temperatures above 200 or 300°C. In this model one can expect that radiogenic Sr also, at least to a certain extent, will be removed with temperature rise. In other mica-type minerals, it has been repeatedly demonstrated that Rb–Sr ages are more resistant to thermal changes than K–Ar ages on the same mineral (Hart, 1964; Aldrich *et al.*, 1965).

On the other hand, exchange reactions due to circulating ground-water should more effectively alter Rb–Sr ages than K–Ar ages. In certain experiments the latter are not affected by removal of up to 50% of the K initially present (Kulp and Engels, 1963; Odin and Rex, this volume).

In some cases, post-depositional uptake of K and/or Rb (aggradation) cannot be ruled out on the base of the K/Rb ratio of the sample. Whether this uptake is purely accidental or accompanies a normal evolution to an ordered micaceous structure in the case of a disordered one is a point which remains obscure.

3 MESOZOIC AND CENOZOIC GLAUCONIES

Before considering these cases, it should be borne in mind that here the problematics in glaucony dating are very different from those previously discussed. A 10% Ar or Sr loss will represent a 2 Ma lowering for a 20 Ma old sample. This may remain unnoticed because of stratigraphical or analytical uncertainty. On the other hand, a 30 Ma age may be measured on a reworked sample deposited 15 Ma ago. This represents a 100% error on the deposition time, if the radiometric age measured on the glaucony is supposed to correspond to it. Such a problem becomes non-existent for the Precambrian if the absolute uncertainty on the age remains the same, i.e. 15 Ma. This is indeed the analytical error on an accurate measurement performed on a 1000 Ma sample. Finally, sediments free from diagenetic alterations are much more common in Mesozoic and Cenozoic terrains.

3.1

Seven samples of the Oligocene–Miocene glauconitic sand complex of the northern part of the Belgian Basin were analysed by Odin *et al.* (1974).

Results (including unpublished Rb–Sr results) are given in Figure 3. Rb–Sr figures given are the weighed mean of results of two independent analyses made on aliquots that underwent different ultrasonic treatments: one without, the other including hydrochloric acid leaching, following a method proposed in Pasteels *et al.* (1976).

A single sample from the Diest formation yields near concordant K–Ar and Rb–Sr ages of about 15 Ma. This formation is considered as belonging to the Late Miocene. However, sedimentological evidence for the reworking of the glaucony is strong (Gullentops, 1957). The underlying Zonderschot member (of the Berchem formation), reported to be of late Middle Miocene age, contains glauconies that yield a K–Ar age of 15.5 ± 1.0 Ma. It appears that the Diest formation glauconies were reworked from the underlying sands and that this reworking did not substantially affect its K–Ar or Rb–Sr ages.

From the Antwerpen member (of the Berchem formation), three samples were collected. There is a good agreement between K–Ar and Rb–Sr ages in one case. Two additional samples yield Rb–Sr ages which also agree rather well with the ages formerly mentioned. These radiometric ages are, however, a few million years higher than the admitted figure for the Middle Miocene, which would correspond to the time of deposition of the Antwerpen member. This is a case of uncertain interpretation, for reworking may be invoked yet is not demonstrated.

As to the Kiel and Edegem samples, it appears that one sample from the Edegem member (Antwerpen formation) must be considered separately from the remaining four. The latter yield internally consistent Rb–Sr ages close to 30 Ma and somewhat less consistent K–Ar ages of about 24 Ma. According to recent palaeontological research on the Edegem member (the Kiel member is decalcified and contains no fossils), it would be either of Middle Oligocene (P 20) age, based on its planktonic foraminifera content (Hooyberghs and De Meuter, 1972), or of earliest Miocene age (zone NN3), based on calcareous nannoplankton (Martini and Müller, 1973).

Though the data (biostratigraphic and radiometric) are ambiguous, a reworking of both foraminifera and glaucony may be invoked. Nannoplankton, being too small and brittle, cannot be reworked. Greensands containing 30 Ma old glaucony and Late Oligocene fauna have been recognized in boreholes (Zand van Voort). The glaucony from the Kiel and Edegem members is slightly oxidized. It should be noted that if the discrepancy between Rb–Sr and K–Ar apparent ages is linked to the postulated reworking, then more Ar has been lost than Sr, or more Rb gained than K.

From this it would appear that reworking may occur without alteration of the radiometric ages on glauconies. One case is noted, however, where reworking is probable and where the systematic discrepancy between K–Ar and Rb–Sr ages on the same sample and the scatter of Rb–Sr and K–Ar ages

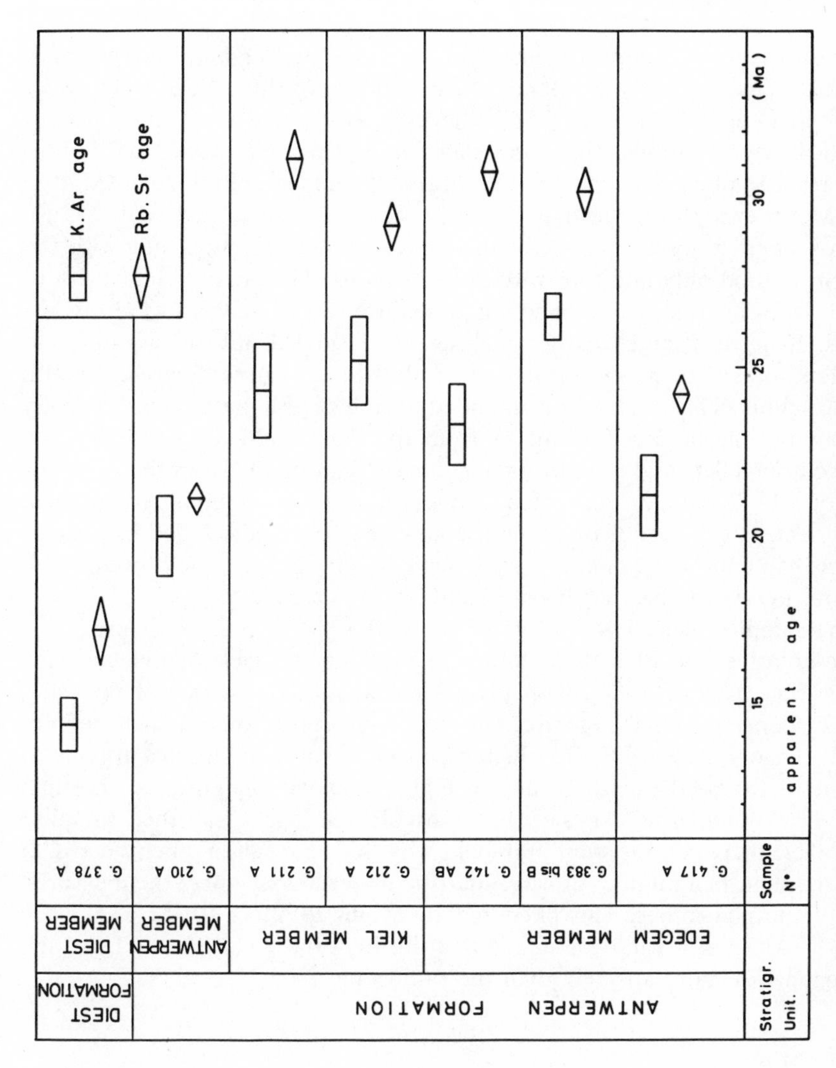

Figure 3. K–Ar and Rb–Sr apparent ages from Oligocene–Miocene glauconies from the Belgian Basin (data from Odin *et al.*, 1974).

of different samples from the same formation are unusually large for Mesozoic and Cenozoic samples.

In cases where reworking is obvious, radiometric data may be used to trace the source of the reworked material.

3.2

A series of 10 samples from eight localities in Belgium and northern France, representing the 'Bande Noire' (base of the Asse clay), were analysed (Keppens *et al.*, 1978a; NDS 84). According to recent palaeontological investigations, the Asse clay (or at least the power part of it) displays assemblages of the NP 15 calcareous nannoplankton zone (Martini and Moorkens, 1969; Martini, 1969a).

Two samples were analysed with the K–Ar method only, four with the Rb–Sr method only and four with both methods. The possible influences of grain size, mineralogy and handling procedure before analysis were investigated. In most samples Rb–Sr analyses were carried out on two different aliquots; one of them was submitted to ultrasonic treatment combined with acid leaching. This treatment is intended to free clayish material occasionally containing inherited radiogenic Sr from the cracks and the surface of the glauconitic pellets and to reduce the common Sr content. From the comparison of Rb–Sr data on untreated aliquots and aliquots submitted to sonic treatment, Rb–Sr results on certain samples have been discarded because of the probable presence of excess radiogenic Sr, i.e. the cleaning procedure, in general necessary, has not been effective in all cases.

The samples yield a K–Ar age of 40.9 ± 0.5 Ma. This is considered as a representative age of sedimentation in view of the lack of evidence for reworking, diagenetic or other alteration and the consistency of the data. This evidence is in part negative, but the Rb–Sr ages of the retained samples yield an age of 41.8 ± 0.8 Ma, which is considered as an additional argument.

One of the retained Rb–Sr ages is higher (beyond analytical errors) than the corresponding K–Ar age and most Rb–Sr ages on other samples. However, there is no *a priori* argument why this figure should be rejected. It appears that, in a number of cases, detritic material may affect the measured age of the glauconies. This phenomenon seems to affect Rb–Sr ages more than K–Ar ages. It can be assumed that this material is clayish and holds the radiogenic Sr more strongly than the radiogenic Ar.

3.3

Seven samples of the Early Cenomanian from several localities in the Paris Basin are analysed by both the K–Ar and Rb–Sr method by Odin and Hunziker (this volume). Results are given in Figure 4. A few additional samples were measured by either the K–Ar or the Rb–Sr method.

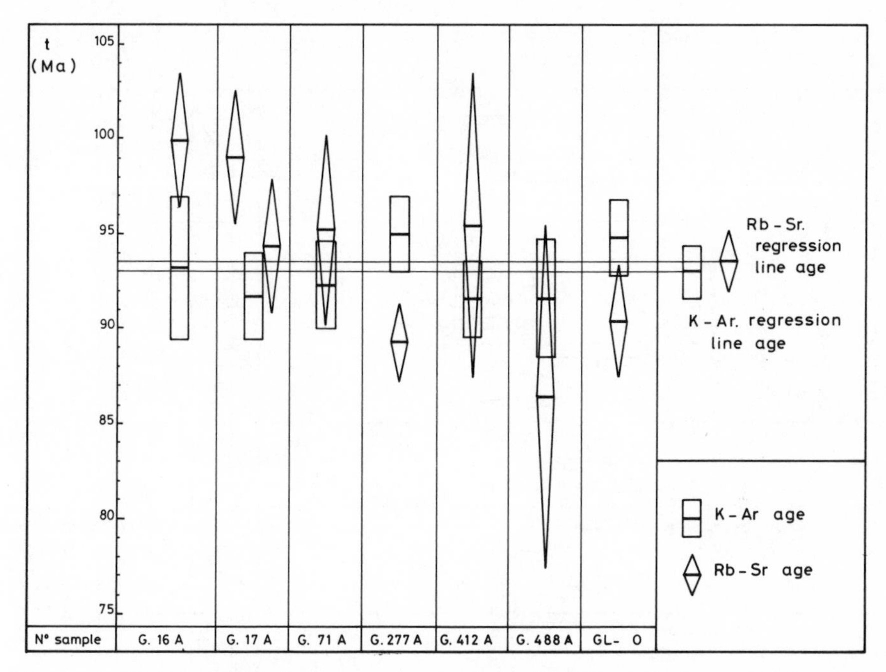

Figure 4. K–Ar and Rb–Sr apparent ages from Cenomanian glauconies from the Paris Basin (data from Odin and Hunziker, this volume).

In some samples differences up to 6% were observed between K–Ar and Rb–Sr ages; these discrepancies occur in both directions. Some of these discrepancies are analytical and due to the high common Sr content of the samples.

K–Ar and Rb–Sr regression age calculations are in excellent agreement, yielding respectively 93.0 ± 1.4 Ma and 93.5 ± 1.6 Ma. As mentioned elsewhere in this volume, this age is considered to be a good estimate of the age of closure of the glauconies.

3.4

Eleven samples from several stratigraphically well located levels of the Middle and Late Cretaceous of the Boulonnais (northern France) and the Mons Basin (Belgium) were analysed by the K–Ar and the Rb–Sr method (Keppens *et al.*, 1978b; NDS 77–83).

On several samples different grain size fractions were analysed. For all samples Rb–Sr measurement were made on aliquots that were submitted to ultrasonic treatment and acid leaching. From a few samples control untreated aliquots were analysed. In all but one sample (Late Aptian) a good

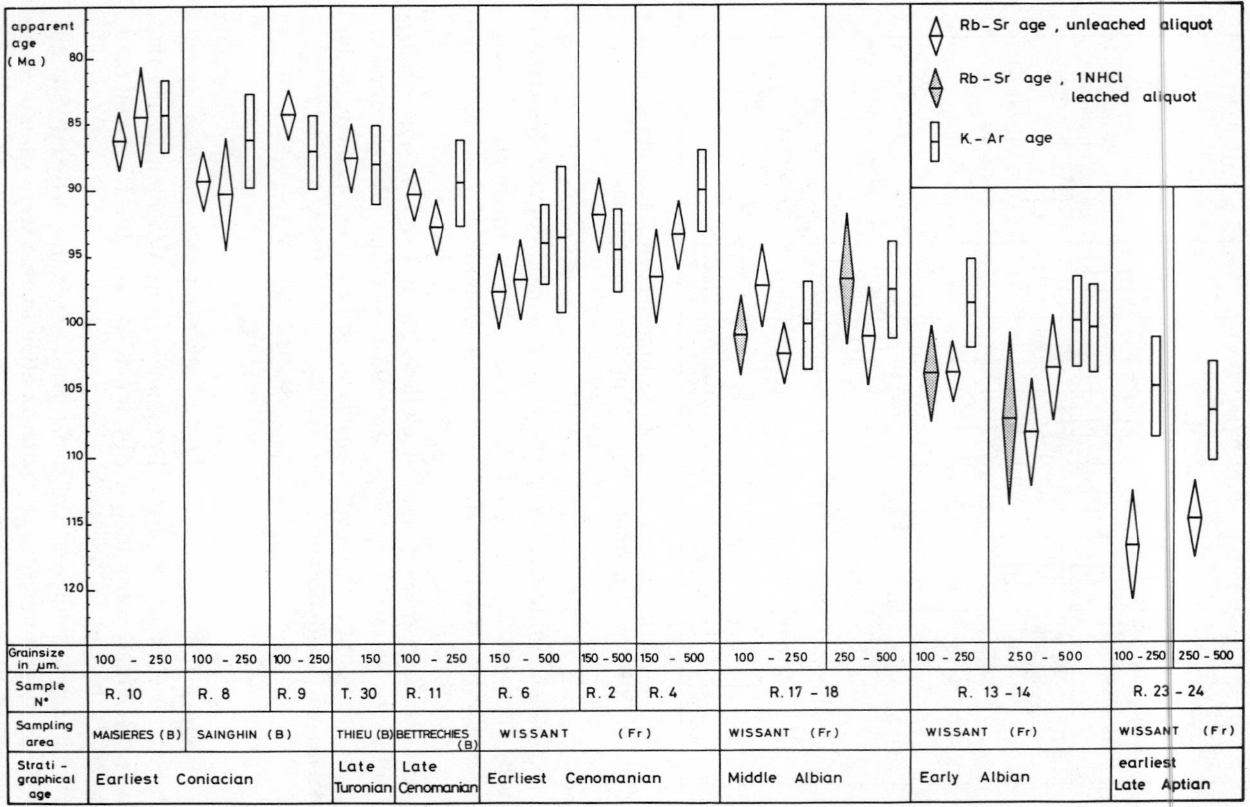

Grainsize in µm.	100 - 250	100 - 250	100 - 250	150	100 - 250	150 - 500	150 - 500	150 - 500	100 - 250	250 - 500	100 - 250	250 - 500	100 - 250	250 - 500
Sample N°	R. 10	R. 8	R. 9	T. 30	R. 11	R. 6	R. 2	R. 4	R. 17 - 18		R. 13 - 14		R. 23 - 24	
Sampling area	MAISIERES (B)	SAINGHIN (B)		THIEU (B)	BETTRECHIES (B)	WISSANT (Fr)			WISSANT (Fr)		WISSANT (Fr)		WISSANT (Fr)	
Stratigraphical age	Earliest Coniacian			Late Turonian	Late Cenomanian	Earliest Cenomanian			Middle Albian		Early Albian		earliest Late Aptian	

Errata: Sainghain and Bettrechies are in France; figures for leached and unleached aliquots are reversed.

Figure 5. K–Ar and Rb–Sr apparent ages from Aptian to Coniacian glauconies (data from Keppens *et al.*, 1978b).

agreement between K–Ar and Rb–Sr figures was observed (Figure 5). All radiometric ages except those of the Aptian sample are considered as very close to sedimentation times or closure of the chronometers.

In the Late Aptian sample (NDS 77), for both grain size fractions (100–250 μm and 250–500 μm), the Rb–Sr age is higher than the K–Ar age (by 10 and 7% respectively). Reworking of the glaucony in this sample probably did not occur (see NDS 77), but the presence of inherited radiogenic Sr (and occasionally also radiogenic Ar) cannot be excluded on the base of the available data. It must be mentioned, however, that this sample definitely has the lowest K and Rb content of the whole series (respectively 5.25 and 5.20% versus 6.18–7.40% K and 209 and 220 ppm versus 248–292 ppm Rb). In addition, X-ray powder diffraction patterns showed a disordered, relatively open lattice indicating that about 20% expandable layers were present. In this study *a clear relationship is observed between the discrepancies of K–Ar versus Rb–Sr ages on the one hand and the amount of expandable layers on the other.*

3.5 Discussion of Mesozoic and Cenzoic cases

In those cases where deep burial and very high temperature influence can be excluded, processes that alter the radiometric ages are thought to be mainly attributable to leaching by, exchange with or diagenesis induced by ground-waters, to alteration by weathering due to percolating fresh water, and to sea-water or run-off water action. It must be borne in mind that in a number of Mesozoic and Cenozoic sediments never deeply buried and outcropping at present time the formerly mentioned processes have not remained the same due to changes in sea-level and weathering regimes.

Of importance in relation to these processes, is the position of the radiometric nuclides in the glauconitic mineral and in the associated components that may be included in the analysis. These associated components are either:

(1) carbonates, containing at first approximation only Sr with an isotopic composition which is very close to the initial composition assumed for the glauconitic material;

(2) detritic clayish material containing excessive radiogenic Sr and radiogenic Ar; or

(3) both.

Regarding the glauconitic material itself, approximately all the nuclides involved are located in the interlayer and to some extent on the edges of the 2:1 sheets. Their behaviour can at least to a certain extent be explained in terms of cation exchange reactions, and thus vary according to a number of mineralogical features of the glauconitic material, among which is the amount of expandable layers.

Understandably, Ar atoms, having no charge, do not intervene as directly as the K^+-, Rb^+- and Sr^{2+} ions in these chemically governed processes. This fact has been experimentally demonstrated by Polevaya *et al.* (1961).

Some information about certain systematics in the behaviours is given by studies on the influence on radiometric glaucony ages of natural and artificial alterations such as washing, leaching, weathering, cleansing techniques, etc. As for the K–Ar method, these problems are discussed in detail by Odin and Rex (this volume).

Our group had studied the influence of different cleansing techniques of glauconies, including ultrasonic treatment and acid leaching, on their Rb–Sr age. Preliminary results of this investigation were presented by Keppens *et al.* (1973) and discussed by Pasteels *et al.* (1976); additional results were reported by Keppens *et al.* (1978a, b). It had been demonstrated that:

(1) the Rb content is not noticeably changed upon acid leaching and/or ultrasonic treatment, i.e. not more than an occasional slight (less than a few percent) increase due to the removal of low-Rb non-glauconitic components. Morton and Long (1980) found a similar behaviour in five out of six Palaeozoic glauconies submitted to acid leaching.

(2) the Sr content is reduced on acid leaching by a factor of 2 to 10, depending upon the acid used, ultrasonic treatment having only a minor effect and bringing about a decrease of not more than 20%. A reduction by a factor up to 100 was found in six Palaeozoic glauconies by Morton and Long (1980), using hydrochloric acid.

(3) in many cases neither acid leaching nor ultrasonic treatment changed the apparent age of the sample. Radiogenic Sr appeared to behave as Rb and was not removed in spite of the drastic decrease of the total Sr content. A similar radiogenic Sr *versus* common Sr behaviour has been observed by Laskowski *et al.* (1980) in five (0.1 N) hydrochloric acid leached Palaeozoic glauconies. In all cases the leaches yielded a $^{87}Sr/^{86}Sr$ ratio near 0.709, indicating that only common Sr was leached out.

In some other cases where glauconitic samples appeared to contain radiogenic Sr in excess, the latter was found to be removed by ultrasonic treatment, whether acid leached or not, and to a much lesser extent by acid leaching without ultrasonic treatment. These results suggest that in a number of cases glauconitic samples are contaminated by a component containing inherited radiogenic Sr. The latter is (in some cases partially) removed by mechanical cleansing.

The influence of treatment of glauconitic samples with NH_4OAc, HOAc and Na-EDTA on their Rb–Sr results has been investigated by Morton and Long (1980). In nine glauconies all containing less than 20% expandable layers, the treatments affect the common Sr much more than the Rb content as well as the radiogenic Sr content of the samples. The effects on the apparent ages, however, seem to differ rather widely depending upon the sample: different models are proposed by the authors. It appears that further research on the physicochemical properties of glauconies is needed in order to obtain a more conclusive understanding of the radiometric age altering processes at low temperature in relatively undisturbed deposits.

4 CONCLUSIONS

(1) Most investigated Palaeozoic and Precambrian deposits have been affected by tectonization and/or deep burial. Whereas the mechanisms through which these phenomena influence the radiometric glaucony ages are not yet fully understood, it appears from the available data that in most cases the glaucony clocks are set back, and that this setting back is more substantial in the K–Ar than in the Rb–Sr systems. On the other hand, perfectly concordant data are obtained on Precambrian sediments that were demonstrated never to have been affected by high temperature/pressure involving processes.

(2) Unfortunately, these early and in most cases discouraging results seem to have been generalized, saddling glauconies with a bad reputation as a geochronometer.

(3) It appears, however, that in undisturbed sediments, mostly to be found in Mesozoic and Cenozoic terrains, a majority of conveniently selected glauconies yield K–Ar and Rb–Sr ages in good agreement.

(4) Reworking of glauconies having a more important influence in young sediments necessitates sedimentological studies to enable correct interpretation of the radiometric results.

(5) The presence of detritic contaminants containing excessive radiogenic Sr may in a number of cases affect the corresponding radiometric age. This problem can in most cases be solved by applying appropriate cleansing techniques to the analysed samples.

(6) Radiometric age altering processes on glauconies in low temperature/pressure conditions are not yet fully understood (especially in the Rb–Sr system). From the present date it appears, however, that in cases where reworking or the presence of detritic contaminants is not observed, K–Ar and Rb–Sr age discrepancies can be related to open lattice glaucony samples. Whether these open lattices are the result of alteration or indicate an uncompleted glauconitization may in many cases not be clear. This problem, however, is of minor importance provided that only closed lattice glauconies are selected for dating purposes.

(7) Finally, an additional problem that may be of certain importance in (especially young) undisturbed sediments is the initial radiometric glaucony age. As far as the K–Ar system is concerned, this problem is discussed by Odin and Dodson (this volume). As for the Rb–Sr initial glaucony age, present knowledge does not enable us to reach any conclusive understand-

ing. The initial Sr–isotopic ratio in forming glauconies, the influence of inherited radiogenic ^{87}Sr in cases where the glauconies are formed from degraded mica-type minerals, the behaviour of Rb and Sr during early diagenesis, whether this process is considered systematically to be part of the glauconitization process or not, etc., . . . are questions that remain open to debate. In any case, it would seem that different Rb–Sr and K–Ar initial glaucony ages may account for differences in Rb–Sr and K–Ar ages of young glauconies.

Résumé du rédacteur

La comparaison des âges apparents K–Ar et Rb–Sr d'une quarantaine de glauconies anté-Mésozoïque montre que les âges Rb–Sr aussi bien que K–Ar peuvent être rajeunis. Cependant ce rajeunissement peut pratiquement toujours être expliqué par une diagenèse profonde. Les âges Rb–Sr sont plus sensibles à une circulation aqueuse, les âges K–Ar plus sensibles à un échauffement.

Pour les glauconies plus récentes, la discordance entre âges Rb–Sr et K–Ar a été observée en liaison avec une forte proportion de feuillets expansibles dans les minéraux glauconitiques des grains verts. Ces glauconies, constituées de minéraux non fermés, se révèlent dans bien des cas, de pauvres géochronomètres, pour l'une ou l'autre méthode. Des études plus poussées seraient nécessaires en vue de déterminer leur possibilités d'utilisation éventuelle en géochronologie.

Des âges plus anciens que supposés sont liés à des phénomènes d'héritage soit dûs au remaniement des grains, soit dûs à des composants non authigènes mêlés dans les grains verts.

L'attaque acide permet de diminuer de 2 à 10 fois les teneurs en strontium des glauconies sans altérer sensiblement leur âge apparent. Ceci permet d'étendre le champ d'application de la méthode Rb–Sr aux glauconies du Néogène. Cependant, cette technique, liée à un traitement aux ultrasons, permet dans certains cas d'éliminer les composants hérités qui élèvent l'âge apparent des graines et d'obtenir ainsi un âge proche de celui de la formation.

(Manuscript received 6-1-1981)

APPENDIX: ANALYTICAL FEASIBILITY OF RUBIDIUM–STRONTIUM DATING OF YOUNG GLAUCONIES

Eddy Keppens

It is common knowledge that the precision of Rb–Sr ages, obtained under given technical conditions, depends greatly on the radiogenic Sr to common Sr ratio of the sample.

In many post-Mesozoic bulk glauconies, the $^{87}Sr^{*}/^{87}Sr$ ratio is such that Rb–Sr analyses do not provide good precision (the asterisk designates radiogenic component). Obradovich (1968) dated two Pliocene glauconies using the Rb–Sr method, obtaining (95% confidence level) analytical errors of 20 and 84%.

The subject of this short contribution is a rough evaluation of the possibility of applying the Rb–Sr dating method to young glauconies, from the strictly analytical point of view. More precisely, we are considering the calculation of conventional ages, based on an initial ratio $(^{87}Sr/^{86}Sr)_i$ and the case where both $^{87}Sr/^{86}Sr$ ratio and ^{86}Sr amount are provided by a single mass-spectrometric analysis using the isotope dilution technique with a pure ^{84}Sr spike. Furthermore, we shall consider that a 2σ error of 2% on the $^{87}Sr^*$ determination is allowable. This figure will provide an error on the apparent age comparable to those obtained by the K–Ar method on the same material.

If $^{87}Sr^*/^{87}Sr$ is not too high (less than about 0.1), i.e. in cases where

$$E_a{}^{86}Sr\{(^{87}Sr/^{86}Sr)-(^{87}Sr/^{86}Sr)_i\} \ll {}^{86}Sr E_a\{(^{87}Sr/^{86}Sr)-(^{87}Sr/^{86}Sr)_i\}$$

and where

$$(^{87}Sr/^{86}Sr) \simeq (^{87}Sr/^{86}Sr)_i$$

it can be easily demonstrated that

$$E_r{}^{87}Sr^* = \frac{\{E_r^2(^{87}Sr/^{86}Sr) + E_r^2(^{87}Sr/^{86}Sr)_i\}^{1/2}}{^{87}Sr^*/^{87}Sr}$$

where E_a and E_r represent random error and relative error respectively. In fact, with an increasing $^{87}Sr^*/^{87}Sr$ ratio these approximations become less possible and the proposed model deviates further from reality.

Let us assume that $^{87}Sr/^{86}Sr$ and $(^{87}Sr/^{86}Sr)_i$ have a 2σ error of ± 0.0002 ($\simeq 0.03\%$ in low radiogenic samples), then

$$E_r{}^{87}Sr^* \simeq \frac{0.04\%}{^{87}Sr^*/^{87}Sr}.$$

The introduction of a random error on $(^{87}Sr/^{86}Sr)_i$ derives from the following consideration.

When a conventional $(^{87}Sr/^{86}Sr)_i$ value is used in glaucony dating, it is assumed that the initial ratio of the glaucony was that of its first closure, i.e. the sea-water, and that this ratio was also recorded in carbonates formed during the same period. This value is then provided by $^{87}Sr/^{86}Sr$ measurements on carbonates, whether associated or not with the glaucony. This determination, apart from having geochemical uncertainties, has at the least an analytical uncertainty given by the 2σ error on the $^{87}Sr/^{86}Sr$ measurement(s).

To provide a $^{87}Sr^*$ value with a precision not worse than 2%, the sample should contain not less than 0.02 $^{87}Sr^*/^{87}Sr$. Knowing that $^{87}Sr^* \simeq 1.4 \times 10^{-11}\,a^{-1} t\,^{87}Rb$, and that $^{87}Sr \simeq$ (Total Sr/14), the relationship between Sr content, Rb content, age and $^{87}Sr^*/^{87}Sr$ ratio can be derived, as has been done in Figure A.1. In this figure, the age–Sr content conditions are given

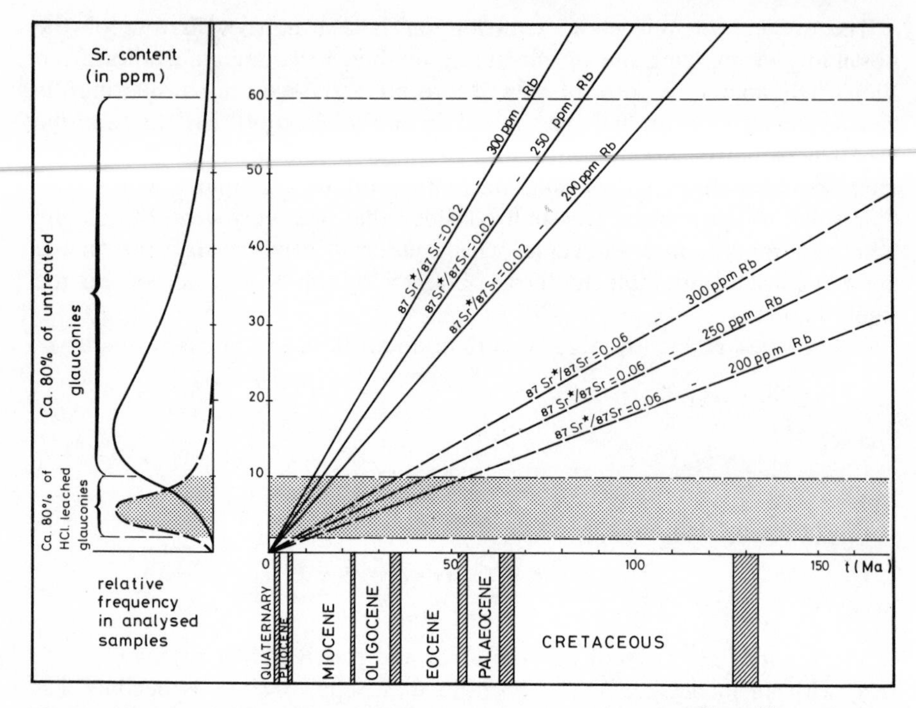

Figure A.1. Age–Sr content conditions for glauconies to yield $^{87}Sr^*/^{87}Sr$ ratios of 0.02 (continuous lines) and 0.06 (broken lines) for different Rb contents. Left from ordinate, Sr-content distribution is given as observed in untreated bulk glauconies (continuous line) and in HCl-leached glauconies (broken line). It appears that leaching of glauconies with HCl greatly increases their suitability for Rb–Sr age measuring, from the strictly analytical point of view at least.

for glauconies containing 200 ppm, 250 ppm and 300 ppm Rb, to yield a $^{87}Sr^*/^{87}Sr$ ratio of 0.02 (continuous lines) and 0.06 (broken lines). The latter ratio is required to obtain a 2% precision on the $^{87}Sr^*$ determination provided by a $^{87}Sr/^{86}Sr$ measurement with an error of *c.* ±0.001.

It appears that for the Cenozoic era, even with a 2σ error of ±0.0002 on the $^{87}Sr/^{86}Sr$ value, a number of bulk glauconies (see frequency–distribution curve on the left along the ordinate in Figure A.1—continuous line) will not provide precise age measurements.

It has been demonstrated, however, that a treatment, including acid leaching, applied to the glauconies reduces their common Sr content without altering their $^{87}Sr^*/^{87}Rb$ ratio (Hurley *et al.*, 1960; Keppens *et al.*, 1973; Pasteels *et al.*, 1976; Keppens *et al.*, 1978a, b; Morton and Long, 1980). Our group has been applying this technique using 1–2.5 N hydrochloric acid, reducing the Sr content by a factor 5 to 10. On these aliquots, containing in most cases between 1 and 15 ppm Sr (see frequency–distribution on the left along the ordinate in Figure A.1—broken line), precise dating can be made even on Miocene samples. However, it must be taken into account that isotopic ratio measurements on very low Sr glauconies will be less precise than on glauconies containing 'normal' amounts of Sr. Using samples selected for their low Sr content, Pliocene–Pleistocene glauconies could yield precise data, i.e. apparent ages.

It must be borne in mind, however, that the time needed for the chronometer to form as a closed system must be neglectable compared to the age measured. In glauconies, this closing time is not instantaneous and may amount to several hundred thousands or even millions of years, producing a substantial uncertainty in the dating of post-Oligocene material (Odin, 1975).

It can be concluded that the application of the Rb–Sr dating method to glauconies is not limited (upwards) by the analytical procedure, but by the genetic uncertainties related to the chronometer itself.

Numerical Dating in Stratigraphy
Edited by G. S. Odin

13
The rubidium–strontium method applied to sediments: certitudes and uncertainties

NORBERT CLAUER

A critical examination of the Rb–Sr dating method applied to sediments for stratigraphic purposes emphasizes uncertainties which are (see the introductory chapter of this volume): (1) of *stratigraphic* type and concern the geologist; (2) of *genetic* type and concern the mineralogist; (3) of *historical* type and concern the sedimentologist; and (4) of *analytical* type and concern the geochronologist. Therefore, it is obvious that this dating technique needs a completely fresh approach which is necessarily different from the classical geochronology. However, Rb–Sr dating of sediments is now possible using whole-rocks or separated authigenic (e.g. clay) minerals. For that, defined selection criteria should be systematically used and the recent progress in clay mineralogy and sedimentology better integrated into interpretations.

1 HISTORICAL BACKGROUND

Wickman (1948) was the first to suggest that ancient marine limestones should be studied to test the possibility of an increase in the $^{87}Sr/^{86}Sr$ ratio of sea-water with time, owing to the accumulation of ^{87}Sr from the decay of ^{87}Rb. However, Herzog *et al.* (1953) and Gast (1955), among others, found abnormal values for palaeontologically known material. Later, Peterman *et al.* (1970), with improved analytical capabilities, could report an irregular increase of the $^{87}Sr/^{86}Sr$ *vs.* time due to temporal fluctuations.

With Cormier's first Rb–Sr determinations on 'glauconites' (1956) began the study of authigenic minerals that are enriched in Rb and depleted in Sr. The results were encouraging, as the ages were *reasonable* compared to the suggested time scale (Cormier *et al.*, 1956; Herzog *et al.*, 1958). However, the calculated ages of so-called glauconites were often considered to be 10–20% lower than the admitted ages (Hurley *et al.*, 1960; Hurley, 1961). This discrepancy was attributed to a preferential leaching of radiogenic ^{87}Sr from minerals having various proportions of expandable layers with weak structural bonds. Since the proportion of these layers appeared to have been negatively correlated with age, the authors assumed that glauconitic fractions alter during post-depositional burial. This information led them to introduce a 'correction' factor even for clay minerals having concordant Rb–Sr and K–Ar ages (Hurley *et al.*, 1962), and the practice of this empirical calculation was continued even in recent times (Sanford and Mosher, 1975). The low ages were then systematically related to late closures of the Rb–Sr systematics in response to diagenetic events, whereas the high values were attributed to detrital origins of the clays (Hurley *et al.*, 1959; 1961; 1963a,b; Hower *et al.*, 1963).

Bonhomme *et al.* (1966) emphasized the need for integrating the knowledge of clay mineralogy and sedimentology into age interpretations, since clays may contain components *formed* or *modified* in sedimentary environments (Lucas, 1962; Millot, 1964). These authors assumed that *in situ* formation involves a Sr isotopic equilibrium with the environment, whereas modification may lead to this equilibrium depending on its degree. Obviously the proposition of Bonhomme *et al.* underscores the need for a good knowledge of the genetic mechanisms of the clay minerals and of the behaviour of their Rb–Sr systematics. Evidence for Sr isotopic equilibrations was reported in a number of studies, which found isotopic ages in good agreement with the stratigraphic ages (Obradovich and Peterman, 1968; Bonhomme *et al.*, 1970; Clauer, 1973; Harris and Bottino, 1974a,b; Harris, 1976; Harris and Baum, 1977; Keppens *et al.*, 1978a,b). The clues to isotopic equilibration were sought from different perspectives, such as the effects of weathering (Odin *et al.*, 1974), late diagenesis (Bonhomme *et al.*, 1969; Perry and Turekian, 1974; Thomas *Filho*, 1976; Stanley and Faure, 1979), slight metamorphism (Bonhomme and Prévôt, 1968; Clauer and Bonhomme, 1970a,b; Clauer, 1974; Clauer and Kröner, 1979) and laboratory treatments (Chaudhuri and Brookins, 1969; 1979; Chaudhuri *et al.*, 1970; Pasteels *et al.*, 1976; Morton and Long, 1980). A comprehensive study of these different aspects was attempted by Clauer (1976b), special attention being given to the relations between origin, nature and Sr isotopic composition of the clay minerals. Chaudhuri (1976) also emphasized the necessity to relate origin, nature and Sr isotopic composition of the clays. More recently, Clauer (1979a) expounded on criteria for a successful application of the Rb–Sr method for dating of sediments.

Since 1962 (Compston and Pidgeon, 1962), whole-rock analysis has been the most common method of dating sedimentary rocks. A large degree of analytical scatters prevented any logical interpretation of sediment history

(Compston *et al.*, 1966; Bofinger *et al.*, 1968; Brookins *et al.*, 1970; Chaudhuri and Brookins, 1970; Kawashita, 1972; Clauer, 1973; 1974; 1976b; Spears, 1974; Thomas *Filho*, 1976). With less degree of data scatter, sedimentation times or metamorphic events could sometimes be dated (Bofinger and Compston, 1967; Chaudhuri and Faure, 1967; Bonhomme and Prévôt, 1968; Moorbath, 1969; Bofinger *et al.*, 1970; Gebauer and Grünenfelder, 1974; Clauer *et al.*, 1980). Hofman (1971) and Hofman *et al.* (1974), in splitting up whole-rocks and in analysing separately the different size fractions, emphasized the relationship between isotopic heterogeneity and relative amount of different detrital components. Therefore, in spite of a recent proposition of Cordani *et al.* (1978), no general satisfactory guidelines can be established for a successful dating of sediments by the whole-rock Rb–Sr method.

Owing to the many conflicting and unexplained results that have emerged from the Rb–Sr isotopic investigation of sedimentary rocks, recent investigators have often sought to address the issues of proper selection of samples and of appropriate laboratory preparation prior to isotopic analyses, as well as of the interrelationship between mineralogy, sedimentary environment and Rb–Sr systematics. The most important factor is no longer the analytical result, but its geological significance. A detailed discussion of the state-of-the-art of the Rb–Sr method applied to sediments and of criteria that have the potential to provide meaningful results is proposed here.

2 THE PRESENT Rb–Sr DATING POSSIBILITIES OF SEDIMENTS

Two methods currently exist in Rb–Sr dating of sediments: the study of whole-rocks and the study of separated authigenic minerals.

2.1 Dating of whole-rocks

The whole-rock dating proposal of Cordani *et al.* (1978) rests on the assumption that sediments, especially shales, can be well enough mixed in a homogeneous environment to represent a *uniform* mixture datable for stratigraphic purposes. This may be true in some cases, but it fails to explain the analytical scatters in many whole-rock studies which have been recognized in the above report and elsewhere. Additional information is necessary: (1) to explain the dispersions; and (2) to obtain a required reproducible application. The scatters could result from sudden and large variations in the proportions of the components or from diverse inherited Sr isotopic compositions for identical components.

An explanation for the effects of mineral variations on whole-rock Sr isotopic compositions can be given, based on a model adapted from Faure *et*

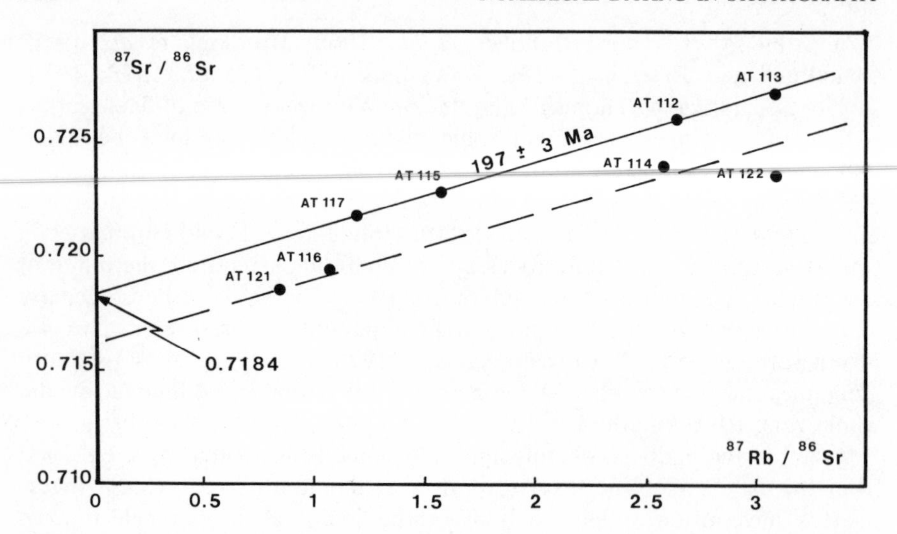

Figure 1. Isochron diagram of the rocks from the Botucatu formation (from Cordani *et al.*, 1978, reproduced by permission of the American Association of Petroleum Geologists).

al. (1965) and Faure (1977) with the formula:

$$({}^{87}\mathrm{Sr}/{}^{86}\mathrm{Sr})_R = ({}^{87}\mathrm{Sr}/{}^{86}\mathrm{Sr})_{M^m} + ({}^{87}\mathrm{Sr}/{}^{86}\mathrm{Sr})_{F^f} + ({}^{87}\mathrm{Sr}/{}^{86}\mathrm{Sr})_{C^c}$$

where R = whole-rock, M = micas, F = feldspars and C = clay minerals and where $m+f+c=100\%$. If we consider now that the rocks of the Botucatu formation in Brasil (Cordani *et al.*, 1978, p. 103) contain only these three components and that the mean $^{87}Sr/^{86}Sr$ ratios of these minerals are 0.740 ± 0.002, 0.720 ± 0.002 and 0.710 ± 0.002 for respectively the micas, the feldspars and the clay minerals during sedimentation, the formula becomes:

$$({}^{87}\mathrm{Sr}/{}^{86}\mathrm{Sr})_R = (0.740 \times m) + (0.720 \times f) + (0.710 \times c).$$

The extremest possible ratios for the rocks will then be 0.710 if they are exclusively clayey ($c=100$; $m=0$; $f=0$) and 0.740 if they are only micaceous ($c=0$; $m=100$; $f=0$). In the study, the whole-rock isochron has an intercept at 0.7184, which characterizes the $^{87}Sr/^{86}Sr$ value of the samples during deposition (Figure 1). Plotted on a diagram which outlines the field of the model (Figure 2), the value at 0.7184 shows that to lie on the isochron the rocks must contain less than 25% micas, less than 85% feldspars and between 15 and about 75% clay minerals. The points located below the isochron have an initial $^{87}Sr/^{86}Sr$ ratio at about 0.716, which requires an

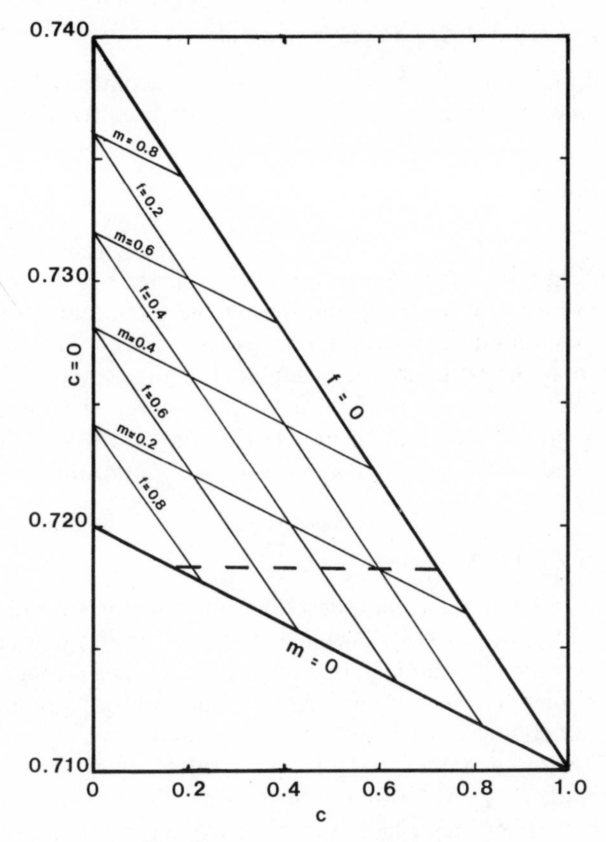

Figure 2. Model of the $^{87}Sr/^{86}Sr$ ratio in a whole-rock considered as a three-component mixture (adapted from Faure, 1977). The dashed line represents the initial value of the isochron.

abnormal increase of clay minerals with correlative decreases of feldspars and micas.

Different weathering states of components, especially when enriched in Rb-like biotites, may induce analytical scatters of the whole-rocks. A leaching of about 60% K during natural weathering of biotites can reduce their $^{87}Sr/^{86}Sr$ ratio by about half: 1.589 to 0.765 (Clauer, 1978).

The present state of the Rb–Sr dating of whole-rocks may be summarized by some lines extracted from the publication of Cordani *et al.* (1978): '... More work is necessary to demonstrate the effectiveness of the proposed mechanism of uniform dispersion of detrital material in the depositional basins ...'.

2.2 Dating of separated authigenic minerals

Among all the authigenic minerals, carbonates, zeolites, siliceous concentrations and clay minerals constitute components used for dating investigations.

2.2.1 *Use of the $^{87}Sr/^{86}Sr$ ratio from carbonates*

The $^{87}Sr/^{86}Sr$ ratio of marine carbonates is primarily controlled by the degree of continental weathering and the intensity of ocean-floor spreading. The former continuously provides the sea-water with Sr enriched in 87 isotope, whereas the latter source is depleted in this isotope. The results of Peterman *et al.* (1970) and of others allowed, however, the construction of a 'scale' *versus* time. Faure and Barrett (1973) proposed using it as a reference to date the carbonates by comparison (see Faure, this volume).

2.2.2 *Rb–Sr dating of zeolites*

Zeolites have been principally described in rocks of volcanic origin (see Mumpton and Ormsby, 1976). They may form: (1) from pyroclastic material in closed evaporitic basins; (2) in open lakes and ground-water tables; (3) in marine environments; (4) during slight metamorphism; (5) in zones with hydrothermal activities; (6) in alkali soils from volcanic particles or clay minerals; (7) without precise relations with volcanic precursors (Mumpton, 1973; Sheppard, 1973; Munson and Sheppard, 1974). About fifteen groups of zeolites have been described; the more common are clinoptilolite and phillipsite with analcime, chabazite, laumontite and mordenite.

No Rb–Sr or even K–Ar dating studies of zeolites seem to be available in the literature. Such an attempt is in progress (Clauer, in preparation) on phillipsites from a Middle Eocene volcano-sedimentary level drilled in the South Pacific (Hoffert *et al.*, 1978). Formed from volcanic material, they give a Miocene isochron age (Figure 3). A provisional interpretation of this discrepancy is that phillipsites form very slowly, almost over several million years. The age of 15 Ma has, therefore, no relation to sedimentation time given by the radiolarians, but is more typical for a late closure of the Rb–Sr system of the zeolites. The occurrence of three points between the isochron and the stratigraphic reference line could mean that the lowered ages are the consequence of a more or less continuous migration of radiogenic ^{87}Sr from the newly formed minerals. In fact, the three samples contain small amounts of Mn micronodules with high Sr concentrations and typical sea-water $^{87}Sr/^{86}Sr$ ratios; their occurrence in phillipsitic fractions induces an artificial upward displacement of the points (Figure 3).

Zeolites seem to be datable by the Rb–Sr method. Further attempts are

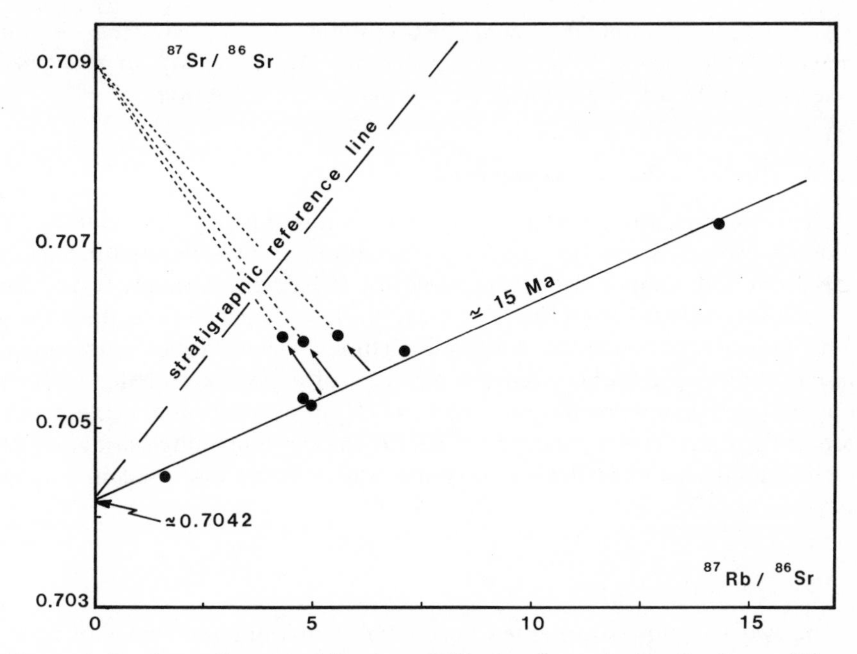

Figure 3. Isochron diagram of Tertiary phillipsites from the Pacific Ocean (Clauer, unpublished).

necessary to interpret the analytical results, but they should be made on different types of zeolites and take into account the nature of the parent material and of the environment.

2.2.3 Rb–Sr dating of siliceous neoformations

The siliceous neoformations are often represented in bentonites, shales and cherts by cristobalite which can precipitate from biogenic opal, volcanic glass and sometimes from quartz (Peterson and Von der Bosch, 1965; Wise *et al.*, 1972; Wise and Weaver, 1974; Wilding *et al.*, 1977). The possibility of a mixed volcanic and biogenic origin has also been considered for cherts (Gibson and Towe, 1971; Castellarini and Sartori, 1978).

Brueckner and Snyder (1979) briefly reported a dating attempt on cherts, but it seems that the material was contaminated by external Sr during its formation or during late modifications. A study on quartz and opal Ct from Deep Sea Drilling Project Leg 62 confirmed that these minerals evolve easily as open systems during recrystallizations after dissolution of the biogenic siliceous phase (Karpoff *et al.*, in press). As they contain only low Sr amounts—less than 7 ppm—it is obvious that even minor external contaminations of Sr easily modify the Sr isotopic balances.

The siliceous minerals seem also to be datable by the Rb–Sr method. Nevertheless, complementary studies are necessary to recognize especially those which are unaffected by late recrystallizations.

2.2.4 Rb–Sr dating of clay minerals

Clay minerals are common components of sediments. The glauconitic minerals especially can be easily recognized and easily separated mechanically from bulk samples for dating purposes. These were certainly two of the reasons why geochronologists tried early to date them. Despite the apparently ambiguous results of studies by Hurley *et al.* (1960), subsequent investigations have emphasized the necessity of a good knowledge of their mineralogical compositions and genetic mechanisms for a proper interpretation of the data. The significance of geological and mineralogical factors, as applied to the isotopic data of clay minerals, will be discussed in a later section.

2.2.5 Conclusion

The two techniques presented here involve two different types of problems. The dating of whole-rocks avoids the problems of sample preparation and mineral separation but raises problems of appropriate sampling and isotopic homogeneity or equilibration. The dating of separated minerals needs care in fraction separation, representative sampling, and understanding of the relationship between isotopic homogeneity and genesis mechanism.

3 SOME FUNDAMENTALS ABOUT CLAY MINERALS

Before any discussion of the structural and chemical aspects of clay minerals and a short presentation of their genetic modes, the precise meaning of some commonly used words is outlined.

3.1 Definitions

The terms *clay* and *glaucony* are used by sedimentologists to describe *facies.* They should only be used for that purpose to avoid confusion with mineral descriptions. In this latter case, the terms clay mineral or clay fractions should be used for pure minerals or mixtures of several clayey components. A preference should also be given to *glauconitic minerals* for mineral characterizations of 'green pellets'.

The term *glauconite* is also ambiguous in the sense used by many English-speaking authors to describe *all* the iron-rich micaceous clay minerals occurring as green pellets in sediments. Odin (1975), among some mineralogists, only uses the term glauconite 's.s.' or *glauconitic mica* for the K-rich end-member of the *family* of glauconitic minerals or protoglauconites (Porrenga, 1967); the other end-member being a smectite. It is at least important to consider that the best adapted nomenclature (Odin and Létolle, 1980) is that which takes into account the evolution of the minerals, and we will see that this estimation is important for dating attempts of clay minerals.

3.2 Separation techniques

Most of the clay minerals are *dispersed* in the sediments; only the glauconitic minerals occur as rounded or subrounded grains or pellets. They are normally concentrated in the smaller than 2 μm grain fraction which is easily extracted from the rocks. The samples are gently crushed, after removal of the weathered parts, in an agate mortar. The crushed material is then mixed and dispersed in distilled water. The upper part of the suspension, as a consequence of Stokes' law, contains the clay minerals. It must be noted here that the smaller than 2 μm fraction is *not* a mineralogical fraction. It may also contain various amounts of accessory minerals like quartz, calcite or feldspars (Figure 4). These are often of detrital origin in sediments and their occurrence, therefore, poses additional problems in

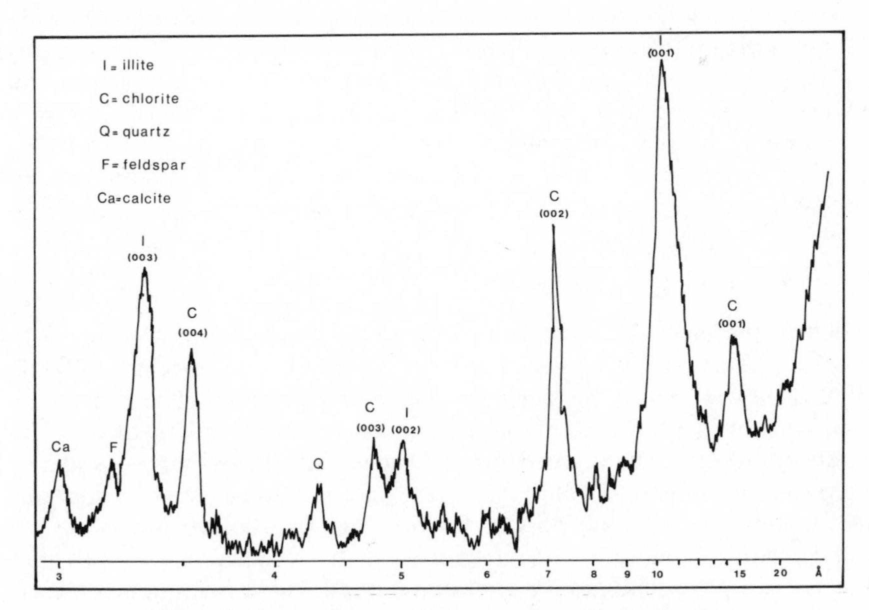

Figure 4. X-ray diagram of a 'contaminated' clay fraction.

isotopic studies. A systematic X-ray control is then necessary, not only to determine the mineralogical composition of the fractions, but also to check their purity and thus to exclude the 'contaminated' ones.

Hydrochloric or hydrogen peroxide treatments, among others, are sometimes necessary to extract the clay minerals from limestones, marls or rocks enriched in organic material. Preliminary attempts have shown that such treatments, if *uncontrolled*, may induce preferential leachings of radiogenic ^{87}Sr (Clauer, 1976b; 1979a). It is recommended, as long as no systematic investigations are available, to avoid the isotopic study of clay fractions extracted by these techniques.

The extraction technique is different for glauconitic pellets. The rocks are sieved after disaggregation by washing. The pellets are enriched by magnetic separation, cleaned by ultrasonic treatment and eventually purified by hand-picking. Sometimes a leaching with diluted acid, acetic acid for preference, cleans the pellets of calcitic impurities (for further details see for example Odin, 1969, and this volume). The purity and crystallographic quality of the grains must also be systematically checked by X-rays.

3.3 Determination technique

Because of their very small size, clay minerals can only be recognized routinely by X-ray diffractometry. Their classification has often to be modified depending on the progressive understanding of these 'special' minerals. A very full compilation has been proposed by Pedro (1965) with complements by Brindley and Pedro (1973).

The clay minerals are formally identified through their reactions to treatments like solvation with vapours of ethylene–glycol and hydrazine–monohydrate or heating at 490°C. The details of these reactions may be found in Thorez (1975), who provided a detailed outline of the procedures for X-ray identification of the different groups of clay minerals.

3.3.1 The group of micas

Illite is the most abundant sedimentary mineral of the mica group; it is also the most common clay mineral. It has a basic reticular distance (001) of 10 Å (Figure 4) which characterizes the thickness of the elementary unit composed by a primarily Al octahedral layer surrounded by two Si–Al tetrahedral layers. These layers are held together largely by K ions at the interlayer site (Figure 5). Illites are generally insensitive to solvation and heat treatments commonly applied for their identification (Figure 6).

If Fe becomes more abundant than Al in the octahedral positions, the mineral is a glauconite 's.s' or glauconitic mica. There is no mineralogical

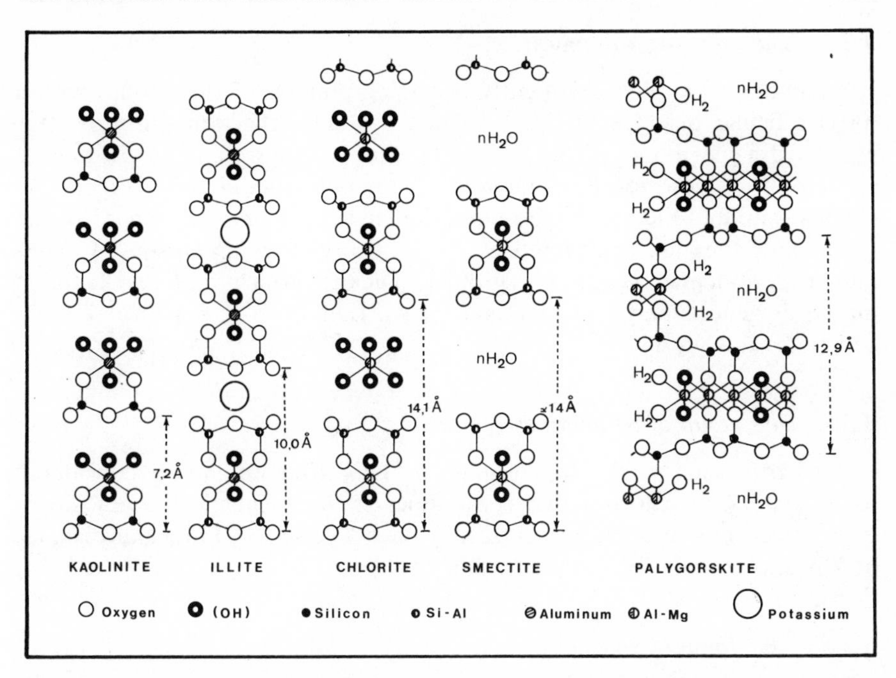

Figure 5. Schematic presentation of the structure of clay minerals (from Millot, 1964, after Brindley, 1951, reproduced by permission of Masson et Cie).

and/or chemical continuity between it and illite (Velde and Odin, 1975). Sabatier (1949) found disorders in the piling up of the glauconitic layers and Burst (1958) described various structures assigned to the same word '*glauconite*'. Odin (1975) emphasized that the glauconitic mica is only one possible component of the facies glaucony.

Pyrophyllite and talc, which may be added to this group, occur also in sedimentary rocks. They characterize either hydrothermal or slight metamorphic environments.

3.3.2 The group of chlorites

Chlorites are less common in sediments than ilites. Insensitive to ethylene–glycol and hydrazine–monohydrate vapours, they react to heating by a change in the heights of the peaks (001) and (002) (Figure 6). The basic (001) reticular distance is 14 Å (Figures 4 and 5).

Chlorites are complex minerals because of numerous octahedral and brucitic substitution possibilities (Foster, 1962).

3.3.3 *The group of kaolinites*

Kaolinite is the most frequently occurring mineral of this group, with a (001) reticular distance of 7.2 Å (Figure 5). This reflection shifts to 10 Å after hydrazine–monohydrate solvation and is removed after heating to about 500°C; these reactions allow its distinction from chlorite (Figure 6).

Halloysite and metahalloysite have similar behaviours with regard to the treatments; they are considered by Novikoff (1974) as precursors of kaolinite in continental weathering profiles. Dickite, which replaces kaolinite during diagenesis and slight metamorphism (Dunoyer de Segonzac, 1969), completes this group.

3.3.4 *The group of berthierines*

Berthierines are similar to kaolinites in their structure and are considered to be similar to chlorites and serpentines in their chemical composition. Insensitive to the usual treatments, they have a basic (001) reticular distance of 7 Å and contain Fe instead of Al.

3.3.5 *The group of smectites*

Also called 'expandable minerals', the smectites have a mica-like structure (Figure 5) with weak bonds between the layers. This allows the introduction of supplementary water in variable amounts into the interlayer. The

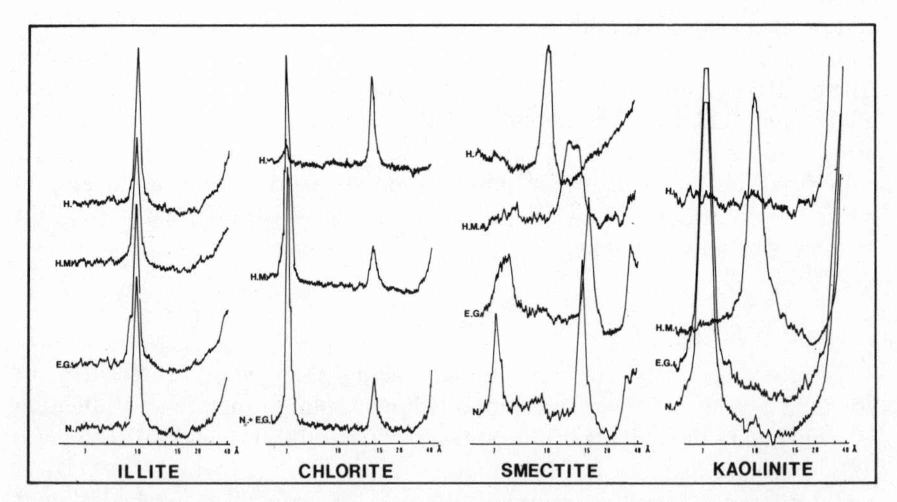

Figure 6. X-ray diagrams of different clay minerals before and after treatments. N, normal; E.G., ethylene-glycol solvation; H.M., hydrazine-monohydrate solvation; H, heated.

periodicity of the basic (001) reticular distance is, therefore, variable around 14 Å. The replacement of this water by ethylene–glycol produces a swelling of the minerals towards 17 Å, whereas the reticular distance is about 12 Å after hydrazine–monohydrate solvation. During heating, the water is definitively extracted from the interlayer and the basic reflection shifts to 10 Å (Figure 6).

The smectite group includes numerous *series* depending on the tetrahedral Si-substitutions by Al, the valence and the nature of the octahedral cations and the nature of the compensator cations of the interlayer. The dioctahedral series, for instance, progresses through intermediate components from montmorillonite without tetrahedral Si-substitutions to beidellite with substitutions:

$$Na_{0.33}(Al_{1.67}Mg_{0.33})\,Si_4O_{10}(OH)_2 \longleftrightarrow Na_{0.36}(Al_{1.46}Fe^{3+}_{0.50}Mg_{0.04})(Si_{3.64}Al_{0.36})O_{10}(OH)_2.$$

Stevensite and saponite are the end-members of the trioctahedral series.

3.3.6 The group of mixed-layered minerals

Clay minerals poorly known or difficult to classify because of their intermediate characteristics are often classed in this group. They present regular or random alternations of layers which may be illite–chlorite–smectite and even kaolinite-like (Lucas, 1962).

3.3.7 The group of fibrous minerals

In this group are minerals which have a fibrous appearance and not a sheet-like structure (Figure 5). Moreover, they are always insensitive to treatments. Palygorskites, also called attapulgites by the French authors, have a basic reticular distance of 10.5 Å; that of the sepiolites lies at 12.1 Å.

3.3.8 Conclusion

The above classification is relatively rigid and formal. It does not correspond to strict reality because the groups are not synonymous with mineralogical types. Thus, the mixed layers probably represent intermediate and/or temporary forms between more or less evolved mineral types. This may be illustrated by the two following series or families:

smectite ⟷ illite–smectite mixed layers ⟷ illite

smectite ⟷ chlorite–smectite mixed layers ⟷ chlorite

One can also consider that the dioctahedral or trioctahedral series of smectites, through their intermediate compositions, set up steps of mineral evolutions. Odin (1975) in the same way, proposes a relationship between the glauconitic smectite and the glauconitic mica. If this is true, the term *mineral families* should be used.

3.4 Origin and evolution

Three different types of clay mineral geneses can be formally distinguished (broader discussion in Clauer and Millot, 1978). Their 'inheritance', which is the most common type, results from deposition of detrital particles brought from the nearby continents. Their 'neoformation' is the result of an ionic chemical precipitation from the environment. Their 'transformation' is a modification of detrital clay minerals into other clay minerals by ion exchanges with the environment. This transformation can occur with or without a modification of the structures. This nomenclature is broad, because it neither conveys completely the true sequential steps of the genesis of the minerals nor their formation or modification period. These aspects are of major significance for stratigraphic dating purposes: it is worthwhile to know the moment of the isotopic equilibrium during mineral evolution.

The knowledge of the origin of a clay mineral may depend on the scale of the analytical approach. Thus, Odin (1972a), Aubry and Odin (1973), Lamboy (1976) and Hoffert (1980), for instance, could observe tiny neoformations of glauconitic minerals or of smectites on detrital minerals by scanning electron microscope. These neoformations are not detectable by X-ray diffraction, which integrates the whole population of minerals and therefore takes very little account of the lesser minerals in small amounts. Other recent studies showed neoformations of smectites from volcanic material (Honnorez, 1972; Melson and Thompson, 1973; Seyfried *et al.*, 1976; Andrews, 1977; Scarfe and Smith, 1977; Hoffert *et al.*, 1978; Noack, 1979) or near-active ridges (Dasch *et al.*, 1971; Hoffert, Perseil *et al.*, 1978; Corliss *et al.*, 1978).

The mineral formation or modification modalities are also important. They depend, especially for glauconitic minerals, on the physical properties of the potential substrate like coprolites, foraminifera tests, carbonate chips (Odin, 1969) or weathered detrital minerals (Odin, 1972a). These substrates must represent semi-confined microenvironments allowing diffusion-like exchanges with the external marine environment (Odin, 1976b). An iron-rich smectite is formed first by crystalline growth; then it is progressively modified into glauconitic minerals more and more enriched in K. This geochemical evolution is also accompanied by a physical modification of the pellets, which become darker green (Odin and Giresse, 1972; Giresse *et al.*, 1980).

The concept of mineral family with intermediate stages which represent more and more evolved minerals poses the problem of the intimate evolution from original end-member, detrital or not, into a new mineral. If this evolution is progressive, the intermediate minerals form probably more or less continuous 'solid solutions'. If a part of the original material recrystallizes or if crystalline overgrowth occurs on original minerals, the population will rather be a more or less homogenized mixture. The establishment of the precise mode depends again on the analytical scale.

It must be remembered that clay minerals have not an unique origin and that they may result from different formation mechanisms. The sedimentary smectites have already been discussed. This may also be illustrated by the illites, which are often of detrital origin owing to their formation from degradation of previous micas, but which can also be authigenic, especially in older sediments. Their authigenesis in sediments may be supported by their synthesis at low temperature during laboratory experiments (Harder, 1974). Contrary to many 2.1 type minerals, kaolinites never form in marine environments (Paquet, 1970). This information is also of value to geochronologists, as it indicates that, either (1) kaolinite is secondary in the rocks and was formed during continental weathering, or (2) kaolinite is detrital.

It is also important for stratigraphic datings to establish the start and the duration of a clay mineral formation or modification. The genesis of glauconitic minerals is *early* during sedimentation and their formation precedes burial (Odin and Giresse, 1972; Odin, 1973a). This estimation is, on the other hand, more difficult for the other clay minerals, except for the smectites which fill up the clefts of basalts after effusion (Hart and Staudigel, 1978). For the smectites from deep-sea red clays, for instance, it is as yet difficult to work out any evolutionary time scale because these clays are often a reworked material (Hoffert, 1980) and tend to have had ionic exchanges with interstitial waters (Clauer *et al.*, 1975; Clauer, 1979b). Migrations of radiogenic ^{87}Sr from clays to water, which may happen during these ionic exchanges, are suggestive of mineral modifications. Diagenetic recrystallizations of clay minerals also occur late relative to sedimentation time; they may be *influenced* by migrating fluids or may occur during burial of the sedimentary series. Sometimes these diagenetic minerals may be easily recognized because they have lath-like and not sheet-like forms (Sommer, 1978).

In summary, neoformations of clay minerals may have different forms depending on the environment. The substrates in the neoformations and the original material in the modifications may supply either no cations or a part of the cations necessary to the neoformed minerals. The completed modifications are more or less dependent on the nature of the environment and of the potential substrates. The formation of minerals depends on their nature and may be rapid or slow during sedimentation time or late during diagenesis.

3.5 Characterization of evolution stages

The illite *crystallinity index* and the determination of their *polymorphic forms* are based on X-ray diffraction. They often allow the characterization of the origin and the evolution steps of this mineral. One can also evaluate the nature of the environment in studying the carbonates associated with the clastic rocks.

The illite crystallinity index can be measured with a simple, rapid and precise technique proposed by Kubler (1966). It allows differentiation between sedimentary or diagenetic domains and zeolite or greenschist facies. To do this, it is only necessary to X-ray oriented smear-slides of clay minerals, to measure the width of the (001) peak at half-height, and compare it to standard values (Figure 7).

The determination of the polymorphic forms of illites or glauconitic minerals is made with the help of desoriented powder diagrams. Illites occur in sediments as 2 M or 1 M polymorphs (Figure 8). The 1 M type crystallizes only at low temperature and can, therefore, only have an early or late authigenic origin in sedimentary rocks. The 2 M form, on the other hand, crystallizes only at high temperature, above 250°C. Its occurrence in a sediment may, then, be due: (1) to a metamorphism of the rock; or (2) to a detrital origin of the mineral.

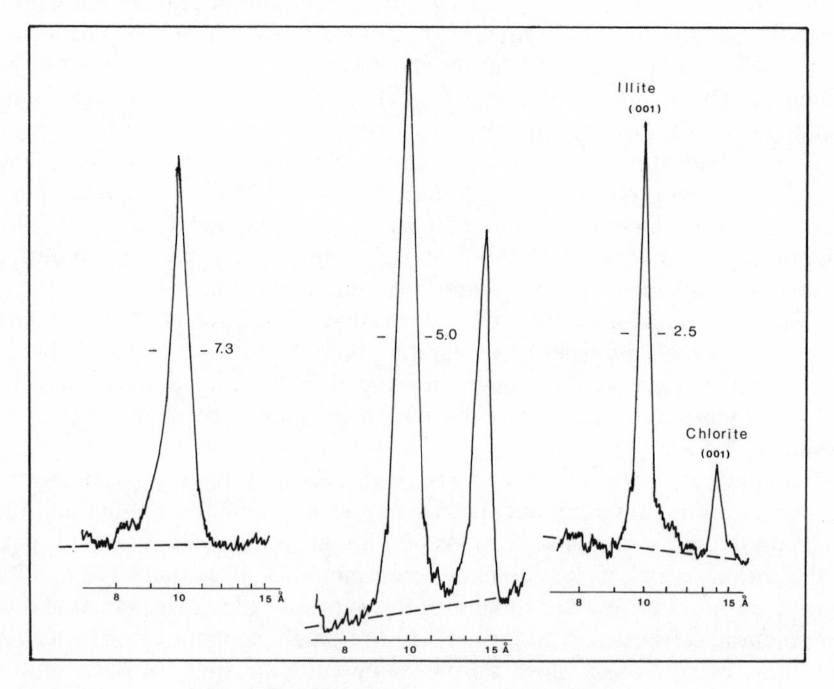

Figure 7. Different illite crystallinity indices.

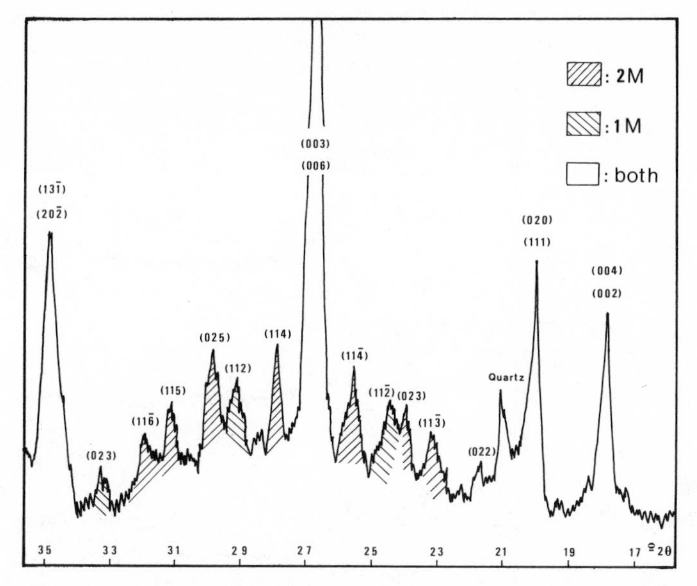

Figure 8. X-ray diagram of a powdered illite containing both 2 M and 1 M polymorphs.

The crystallographic characters of glauconitic minerals occurring in pellets may also be estimated on X-ray diffractograms with the help of the height of the two peaks (112) and (11$\bar{2}$) located on both sides of the (003) one (Odin *et al.*, 1975). The higher they are, compared to the (003) peak, the more the structure is 1 M well-ordered, the smaller they are, the more the structure is 1 Md disordered (Figure 9).

Numerous studies have shown that each recrystallization of carbonates causes a decrease in their Sr content (compilation available in Clauer, 1974) and frequently also an increase in their $^{87}Sr/^{86}Sr$ ratios (Tremba *et al.*, 1975; Veizer and Compston, 1976). Recently, Brand and Veizer (1980) also found a negative correlation between Mn contents and Sr/Ca ratios of carbonates. Veizer (1979) integrated all these results in an unique tri-dimensional diagram (Figure 10) allowing an easy detection of eventual late recrystallizations.

3.6 Conclusion: the choice criteria before isotopic analysis

The above discussion shows that the only type of clay mineral which as yet yields reliable stratigraphic datings is the glauconitic material. There are two reasons for this: (1) an easy standardization of sample separation and preparation due to the nature and colour of the pellets; and (2) a good knowledge of its mechanism and time of formation.

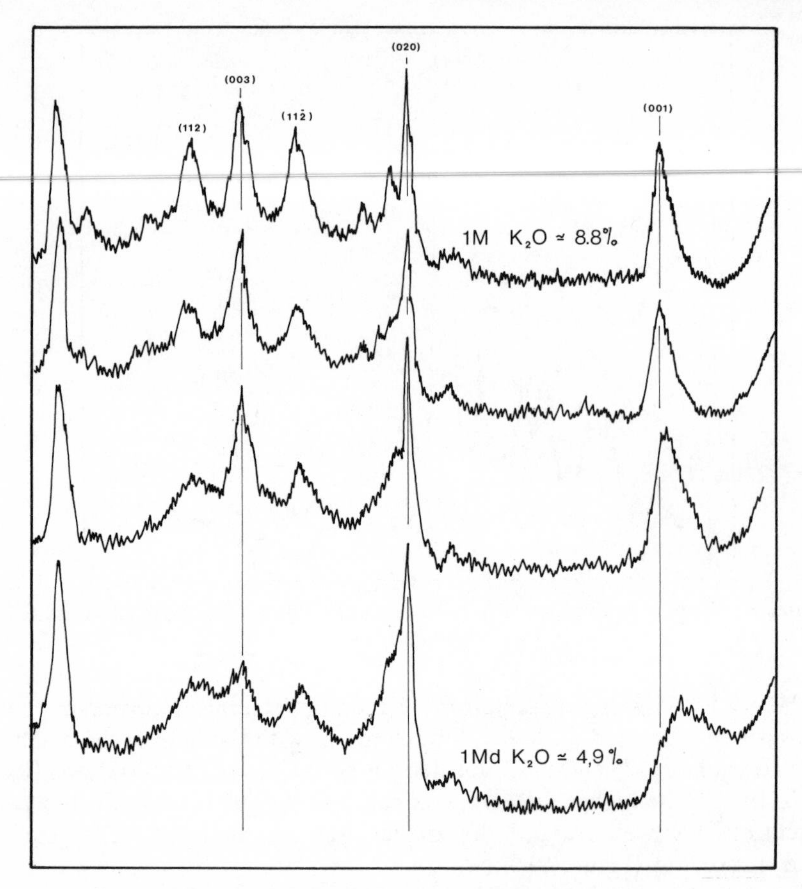

Figure 9. X-ray diagrams of powdered glauconitic minerals showing the relationships between peak heights and K_2O contents.

The problem remains complicated if studies are made on dispersed clay minerals. However, some selected criteria for a successful and meaningful dating of sedimentary rocks may be proposed.

Success is dependent on the preparation of the clay fractions. Every fraction which needs a special treatment, e.g. acid washing for extraction from a limestone or a marl, should be systematically avoided. In contrast, as shown later, mineral leachings with diluted acid may be used if carefully controlled. The clay fractions contaminated by accessory minerals, feldspars especially, must also be eliminated.

The reliability of the procedure is dependent on the nature of the clay fractions. Some minerals are *indicative* either of detrital origin like kaolinite or 2 M illite or of diagenetic or metamorphic recrystallizations like dickite,

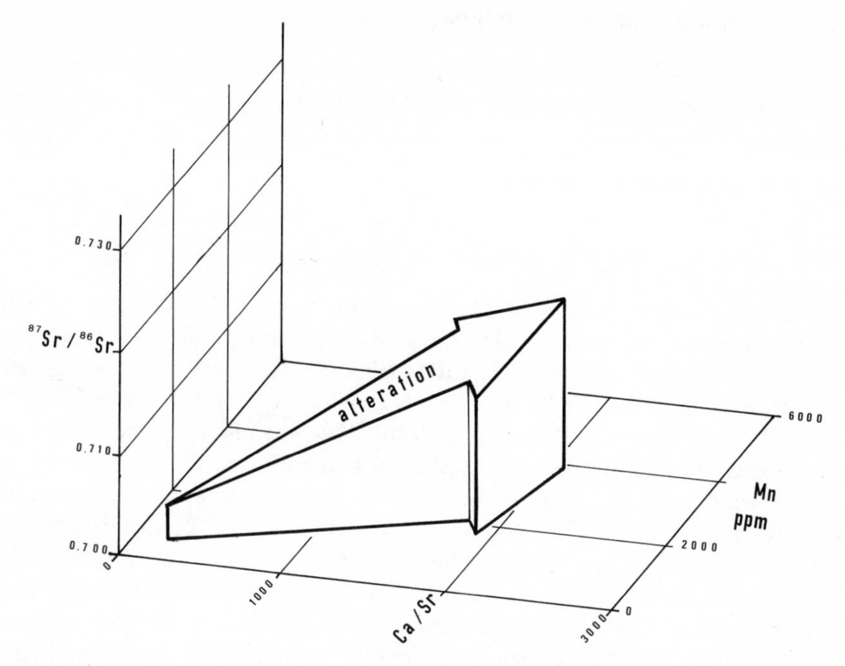

Figure 10. Tri-dimensional diagram for carbonate recrystallization test (after Veizer, 1979). Ca/Sr ratios given in molar weights.

pyrophyllite and again 2 M illite. The fractions containing these minerals must be excluded.

Success is also dependent on the evolutionary stage of the clay fractions and of their environment. It is necessary to determine the evolutionary sequence of the mineral families in order to select only the most evolved fractions, that is to say those which could equilibrate the isotopic composition of their Sr. For that a dense sampling of one or of several shaly horizons from one or several equivalent outcrops is necessary. It is also recommended to control the state of the carbonates associated with the clays, when available.

4 THE UNCERTAINTIES OF THE METHOD

Odin (Chapter 1) proposes classifying the uncertainties occurring in stratigraphic datings of sedimentary rocks by isotopic methods into stratigraphical, genetic, historical and analytical uncertainties. On the basis of this proposition the *actual* possibilities of the Rb–Sr dating of clay minerals will be itemized below into useful results as well as still existing problems.

4.1 The stratigraphical uncertainties

Discussed in Chapters 1 and 2, they are simply recalled here to indicate that they are generally small in marine sediments.

4.2 The genetic uncertainties

4.2.1 Inheritance and Sr isotopic balance

Dasch (1969) showed that the sedimentary material of the Atlántic Ocean, mainly composed of feldspars and clay minerals after decarbonatation, has heterogeneous $^{87}Sr/^{86}Sr$ ratios (Figure 11). These ratios are generally higher than 0.7091 ± 0.004 (2σ), the value compiled for the $^{87}Sr/^{86}Sr$ ratio of the sea-water (Clauer, 1976b). This means that the deposited minerals are detrital and contain an excess of radiogenic ^{87}Sr over that of the sea-water. All the studies made later on the same type of material (Biscaye and Dasch, 1971; Church, 1971; Biscaye, 1972; Dymond *et al.*, 1973; Cooper *et al.*, 1974; Ikpeama *et al.*, 1974; Boger and Faure, 1976; Shaffer and Faure, 1976; Kovach and Faure, 1977; 1978) confirmed this Sr isotopic heterogeneity and inheritance memory. Studies on old clay material showed that these characteristics are obviously maintained after compaction (Chaudhuri, 1976; Clauer, 1976b) (Figure 12).

Giblin (1979), splitting Palaeocene detrital smectites into two different grain size fractions of <0.2 μm and 0.2–0.4 μm, found that the smaller fraction had a significantly lower apparent age than the coarser one: about 168 Ma and about 187 Ma respectively (Figure 13). These values have no

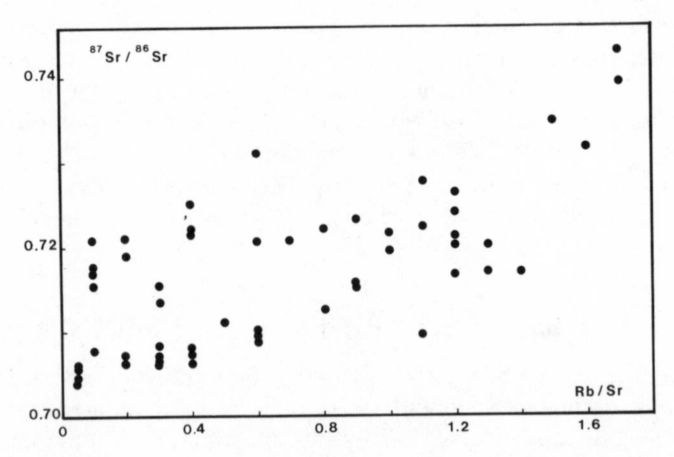

Figure 11. $^{87}Sr/^{86}Sr$ ratios *vs.* Rb/Sr ratios of the Recent sediment from the Atlantic Ocean (Dasch, 1969).

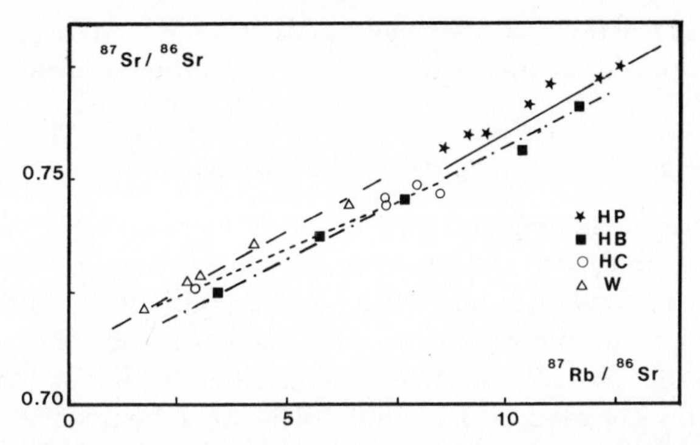

Figure 12. $^{87}Sr/^{86}Sr$ ratios *vs.* $^{87}Rb/^{86}Sr$ ratios of mainly detrital clay minerals (from Chaudhuri, 1976, reproduced by permission of Springer-Verlag).

geological significance; they just confirm the isotopic inheritance memory. Nevertheless the discrepancy is remarkable and could be the result of either: (1) a preferential leaching of radiogenic ^{87}Sr due to the smaller size; (2) isotopic exchanges with sea-water or interstitial fluids favoured also by the smaller size; or (3) neoformations of very tiny smectites. Yeh and Savin (1976) and Yeh and Epstein (1978) referred to similar results with oxygen and hydrogen isotopes. The smaller the clay fractions, the closer their isotopic balances to sea-water values.

Signs of isotopic evolutions are detectable in detrital clay minerals with a tendency towards the sea-water values (or the values of the interstitial fluids)

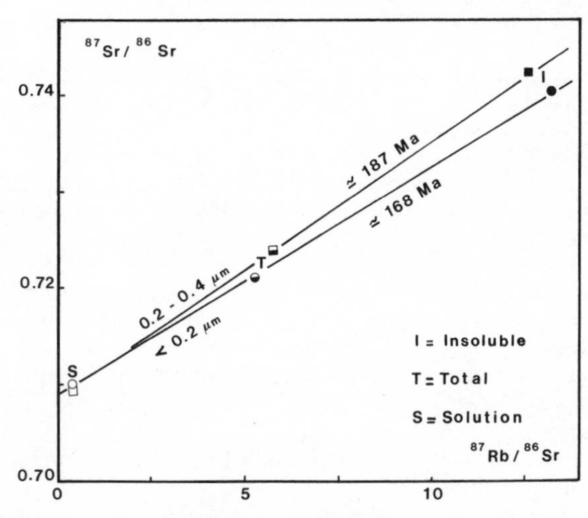

Figure 13. $^{87}Sr/^{86}Sr$ ratios *vs.* $^{87}Rb/^{86}Sr$ ratios of two size fractions from a detrital Palaeocene smectite (Giblin, 1979).

in the smallest fractions. These fine materials can, therefore, no longer be considered as absolutely inert with respect to strontium isotopic exchanges.

4.2.2 Sr isotopic homogenization or equilibration

The Sr isotopic homogenization, which is a necessary requirement for isotopic dating, must be critically examined in sedimentary geochronology. It is obviously important to know the $^{87}Sr/^{86}Sr$ ratio of the minerals at their closure, because the interpretation of the data, whether obtained from the isochron method or calculated from assuming the initial $^{87}Sr/^{86}Sr$ ratio, depends on this knowledge. The mode of isotopic homogeneity has still to be described relative to the size, the nature and the evolution of the clay minerals and relative to the specific micro-environments of the minerals.

Recent smectites extracted from the uppermost 2 cm of deep-sea red clays from the Pacific Ocean have $^{87}Sr/^{86}Sr$ ratios almost similar to that of the sea-water value (Hoffert *et al.*, 1978; Clauer, 1979b). Iron chlorites from the western Spanish off-shore also contain Sr with an $^{87}Sr/^{86}Sr$ ratio similar to the sea-water value (Table 1). A Sr isotopic identity can, therefore, exist between sea-water and clay minerals, even if these minerals are detrital in origin like the Fe chlorites.

This identity has been documented by Odin and Hunziker (this volume) on Cenomanian glauconies from Normandy (France). Eleven glauconitic fractions were measured, as well as two smectites and one leachate. The glauconies lie along an isochron giving an age of 93.5 ± 1.6 Ma (2σ) and an

Table 1. $^{87}Sr/^{86}Sr$ ratios of Recent smectites and Fe chlorites compared to sea-water.

Samples	$^{87}Sr/^{86}Sr \pm 2\sigma$	References
Sea-water	0.70910 ± 0.00035	Compilation in Clauer (1976b)
Smectite (1) (Pacific surface red clay)	0.70917 ± 0.00019	Clauer (unpublished)
Smectite (2) (Pacific surface red clay)	0.70873 ± 0.00050	Clauer (1976b)
Smectite (Pacific surface red clay)	0.70870 ± 0.00030	Clauer (unpublished)
Fe chlorite (Spanish off-shore)	0.7090 ± 0.0015	Clauer (unpublished)

Smectites (1) and (2) represent the same sample with the difference that (1) was measured without treatment and (2) after a 1 N HCl leaching.

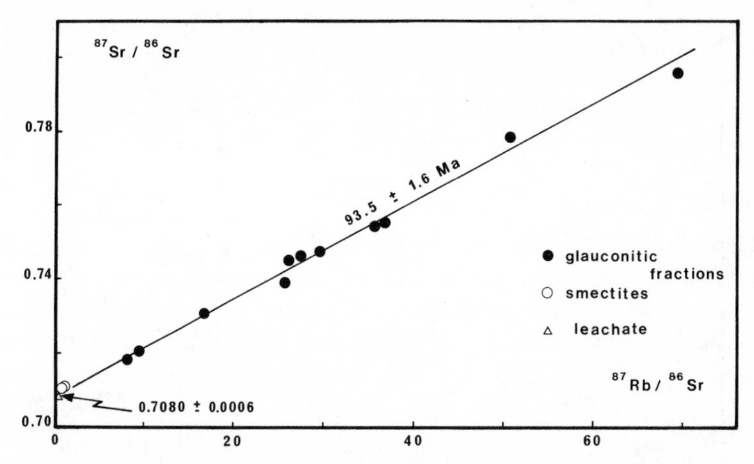

Figure 14. Isochron diagram of the glauconies, smectites and carbonate from the Cenomanian, France (Odin and Hunziker, this volume).

initial $^{87}Sr/^{86}Sr$ at 0.7080 ± 0.0006 (2σ) (Figure 14). The age is similar to that of the K–Ar 'isochron': 93.0 ± 1.4 Ma (2σ) and is compatible with the limits set at 95 and 90 Ma for the Cenomanian (Odin and Hunziker, 1978). The presence of the leachate and the smectites almost on the isochron of the glauconitic fractions suggests that all the minerals had the same $^{87}Sr/^{86}Sr$ ratios during sedimentation time. These ratios were certainly identical to that of their environment, since those of carbonates and contemporaneous water are always identical. In fact, it must be emphasized that this general identity depends also on the dating technique used: the conventional method, for instance, in assuming an initial $^{87}Sr/^{86}Sr$ ratio at 0.7080, would give high apparent ages at 208 and 165 Ma for the smectites (see Odin and Hunziker, this volume) without any reasonable justification other than the analytical precision.

However, the similarity in Sr isotopic composition between clay minerals and their contemporary environment has not always been recognized. A case is the sedimentary sequence in the Mormoiron Basin in the southeastern part of France. An azoic horizon contains smectites and palygorskites immediately covered by Eocene sediments (Triat, 1969). These minerals result from the modification of previous detrital smectites (Trauth, 1974). Nine clay fractions were analysed: two detrital smectites, two modified smectites, two mixtures of smectites and palygorskites and three pure palygorskites. A carbonate phase was also leached from a marl in order to control the isotopic composition of the environmental Sr (Clauer, 1976b). All the clay minerals lie along an isochron at 61.2 ± 1.5 Ma (2σ) and have an intercept at 0.70887 ± 0.00036 (Figure 15). The age is in good agreement

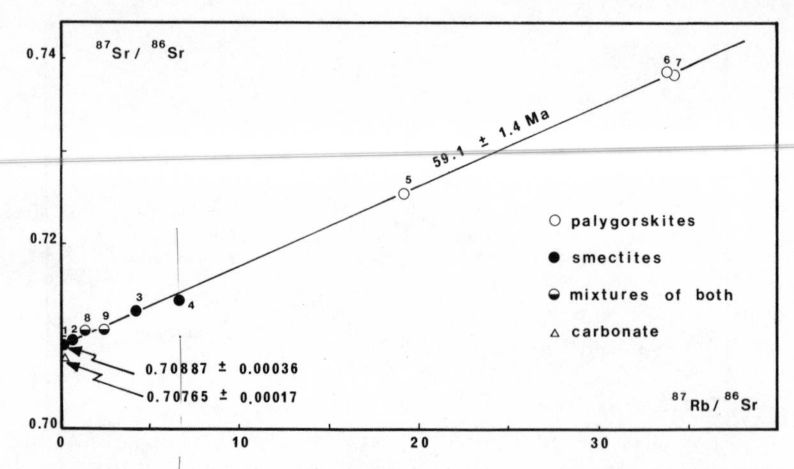

Figure 15. Isochron diagram of the different Tertiary clay minerals from the Mormoiron Basin, France (Clauer, 1976b).

with the stratigraphy, since the Palaeocene is bracketed between 65 and 53 Ma (Chapter 34). The intercept of 0.70887 ± 0.00036 (2σ) for the isochron is significantly different from the $^{87}Sr/^{86}Sr$ ratio of 0.70765 ± 0.00017 (2σ) for the carbonate, and thus from the environmental water. Sr isotopic homogeneity between minerals and environmental water did not occur during sedimentation time, and does not seem to be necessary for dating of authigenic clay minerals. The presence of detrital and modified fractions on the same isochron suggests that mineral evolution is not always accompanied by Sr isotopic equilibration with their environment. This observation and the relative scattering of the smectites around the isochron, which is certainly due to a virtually completed evolution, again indicate the

Table 2. Isochron and conventional ages for the different types of clay minerals from the Mormoiron Basin (France).

Samples		Isochron age ($Ma \pm 2\sigma$)	Conventional ages ($Ma \pm 2\sigma$)
Detrital smectites	1		308 ± 88
	2		196 ± 36
'Transformed' smectites	3		89 ± 3
	4	61.2 ± 1.5	68 ± 5
Palygorskites	5	with	65 ± 3
	6	0.70887 ± 0.00036	63 ± 3
	7		64 ± 2
Mixtures Sm + Pal.	8		145 ± 17
	9		93 ± 11

problem of the conventional dating technique. Thus, calculated individually with an assumed initial $^{87}Sr/^{86}Sr$ at 0.70765, the ages of the fractions would be scattered between 308 and 63 Ma (Table 2), the high values corresponding normally to the samples with low Rb/Sr ratios and those close to the isochron age to the palygorskites with high Rb/Sr ratios.

4.2.3 Timing and mechanism of the Sr isotopic evolution

The study of the variation of ^{87}Sr with depth for smectites and the interstitial waters of red clays from the Pacific Ocean (Clauer *et al.*, 1975; Clauer, 1979b) showed: (1) a gain of ^{87}Sr relative to ^{86}Sr in the fluids; and (2) a loss of ^{87}Sr relative to ^{86}Sr in the clays (Figure 16). This behaviour is interpreted as a transfer of radiogenic ^{87}Sr from clay minerals to the waters. Related to ionic exchanges, it seems to be a sign of slow and continuous evolution going on in the sediment up to several metres depth.

The leachings of clay minerals with diluted acids (Chaudhuri *et al.*, 1970; Clauer, 1976b; Hoffert *et al.*, 1978) are helpful for understanding the

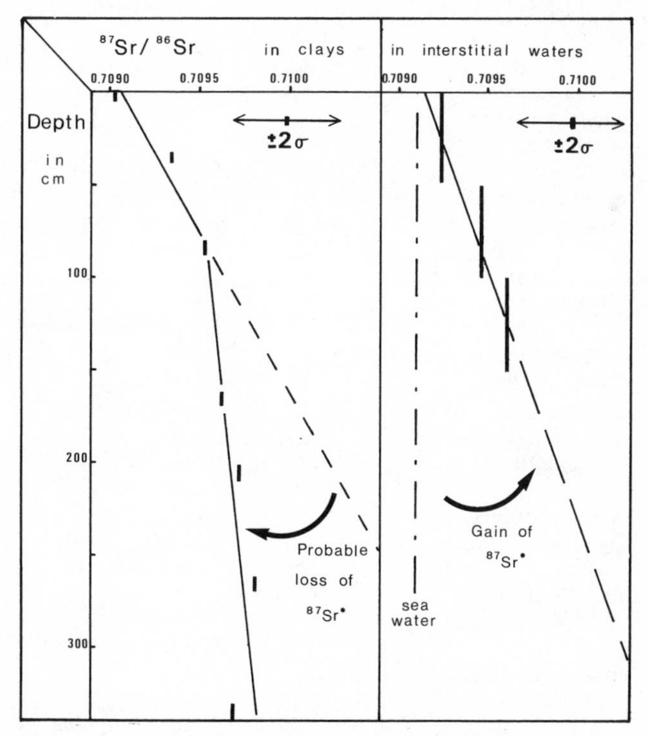

Figure 16. $^{87}Sr/^{86}Sr$ ratios *vs.* depth of clays and interstitial waters from a borehole (Clauer *et al.*, 1975; Clauer, unpublished).

mechanism, because they allow one to analyse separately the Sr trapped in the minerals during their formation and the Sr adsorbed after crystallization or deposition. The laboratory experiments carried out on smectites, which have high cation exchange capacities, confirmed that they are able to adsorb important quantities of marine Sr (Table 3). Moreover, Clauer (1979b) remarked that the $^{87}Sr/^{86}Sr$ ratios of Recent oceanic smectites depend on their origin: (1) they are high, above the sea-water value, when detrital (Dymond *et al.*, 1973; Clauer, 1976b); (2) they are low, significantly below the sea-water value, when formed from basic volcanic rocks (Dymond *et al.*, 1973; Dasch *et al.*, 1971; Clauer, 1976b; Hoffert *et al.*, 1978); and (3) they are slightly below the sea-water value when of hydrothermal origin (Table 4). On the other hand, Hoffert (1980) now interprets the deep-sea red clays as reworked material from different origins, volcanic, detrital and even hydrothermal, and their smectites contain Sr with an $^{87}Sr/^{86}Sr$ ratio which is almost identical to that of sea-water and probably integrates the different origins. The problem is now to study the mixture and find the intermediate

Table 3. Amounts of leachable Sr from smectites.

Samples		Sr (ppm)	References
Detrital smectite			
(Atlantic)	U	51.2	
	L	22.7	Clauer (1976b)
Smectite from red clays			
(Pacific)	U	666	
	L	183.8	Clauer (1976b)
Smectite from red clays			
(Pacific)	U	491	
	L	155.6	Clauer (unpublished)
Smectite from nodule			
core (Pacific)	U	86.9	
	L	39.3	Clauer (1976b)
Smectite from shale			
(Wyoming)	U	248	Chaudhuri and
	L	64.5	Brookins (1979)
Smectite from sands			
(California)	U	117.4	Chaudhuri and
	L	50.4	Brookins (1979)
Smectite from E.P.R.			
(Pacific)	U	338	
	L	164.7	Clauer (unpublished)
Smectite from E.P.R.			
(Pacific)	U	195.2	
	L	20.2	Clauer (unpublished)

U, untreated; L, leached with cold 1 N HCl.

Table 4. $^{87}Sr/^{86}Sr$ ratios of Recent oceanic smectites having different origins.

Samples	$^{87}Sr/^{86}Sr$ ($\pm 2\sigma$)	References
Detrital clay enriched in smectite (Pacific)	0.7175 ± 0.0008	Dymond *et al.* (1973)
Detrital smectite (Atlantic)	0.7212 ± 0.0020	Clauer (1976b)
Hydrothermal smectite (Pacific)	0.70769 ± 0.00013	Clauer (unpublished)
Hydrothermal smectite (Pacific)	0.70830 ± 0.00015	Clauer (unpublished)
Smectite from basalt weathering (Pacific)	$0.70450 \pm$?	Dasch *et al.* (1971)
Smectite of volcanic Eocene origin (Pacific)	0.70550 ± 0.00034	Hoffert *et al.* (1978)

steps between the different end-members. One aspect is almost clarified: there is an adsorption of marine Sr either during deposition or during or after crystallization.

4.3 The historical uncertainties

The Rb–Sr systematics of clay minerals must remain close since the time of crystallization to obtain isotopic data having a stratigraphic significance. Different events such as diagenesis, metamorphism or weathering may, nevertheless, later modify the isotopic balances of these minerals.

The Precambrian group of Bambui (São Francisco craton, Brazil) is mainly composed of siltstones and limestones with interbedded shales. These shales contain illite as the most common clay mineral. These illites have crystallinity indices compatible with the sedimentary or diagenetic domains in the Lages do Batata region. The carbonates as chemical phase are replaced by bedded cherts containing less than 150 ppm Sr with high $^{87}Sr/^{86}Sr$ ratios relative to the contemporaneous 1 Ga old sea-water: 0.7385 ± 0.0020 and about 0.7068, respectively (Veizer and Compston, 1976). These results suggest a diagenetic recrystallization. Therefore, the age of the isochron (Figure 17), 886 ± 11 Ma (2σ), is related rather to the end of late diagenesis than to the sedimentation time, particularly since its intercept is abnormally high at 0.7304 (Bonhomme, 1976).

The Estancia formation in the northeast of Brazil also belongs to the Precambrian. Illite is again the main clay mineral of the shales and siltstones with a crystallinity index located around the boundary between anchimetamorphic (zeolite facies) and epimetamorphic domain (greenschist facies). A recrystallization of the minerals is, therefore, expected. The whole

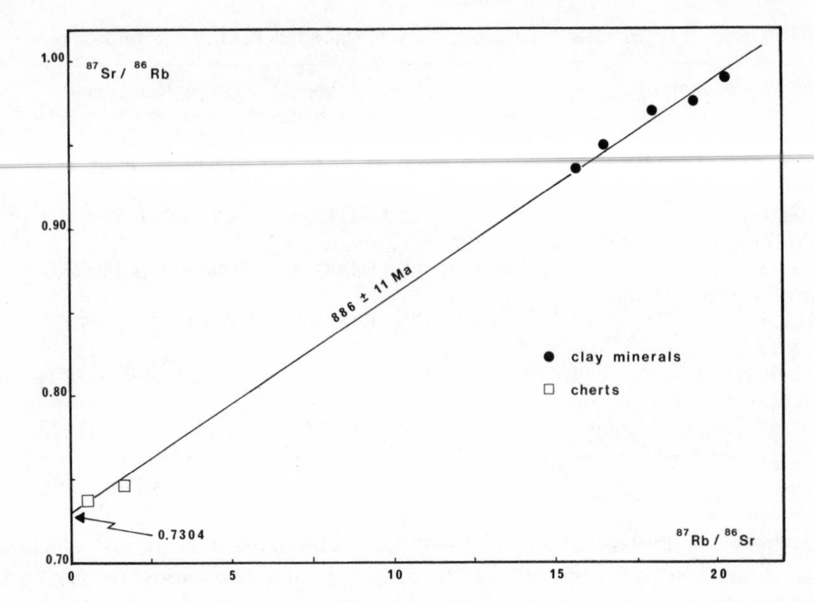

Figure 17. Isochron diagram of the clay minerals and cherts from the Precambrian Bambui group (Bonhomme, 1976).

rocks and fine fractions were effectively rehomogenized during a metamorphic event at 459 ± 15 Ma (2σ). All the points lie on an isochron (Figure 18) with an intercept at 0.7161 (Cordani *et al.*, 1978). This value is again higher than the Upper Precambrian marine $^{87}Sr/^{86}Sr$ ratios, which lie between about 0.7068 at 1.0 Ga ago and 0.7094 at the beginning of the Phanerozoic (Veizer and Compston, 1976).

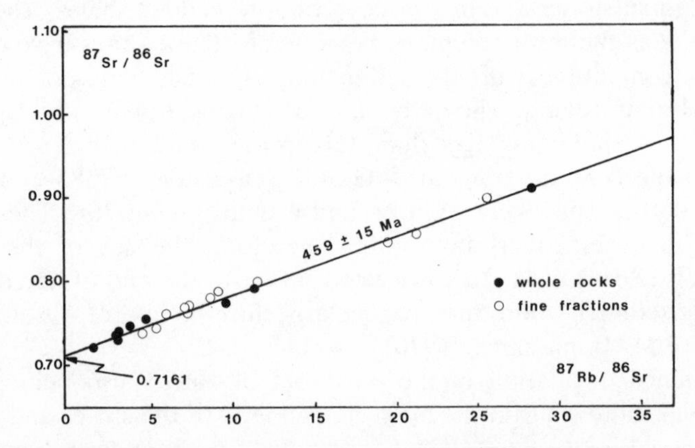

Figure 18. Isochron diagram of the whole-rocks and fine fractions from the Precambrian Estancia formation (from Cordani *et al.*, 1976, reproduced by permission of the American Association of Petroleum Geologists).

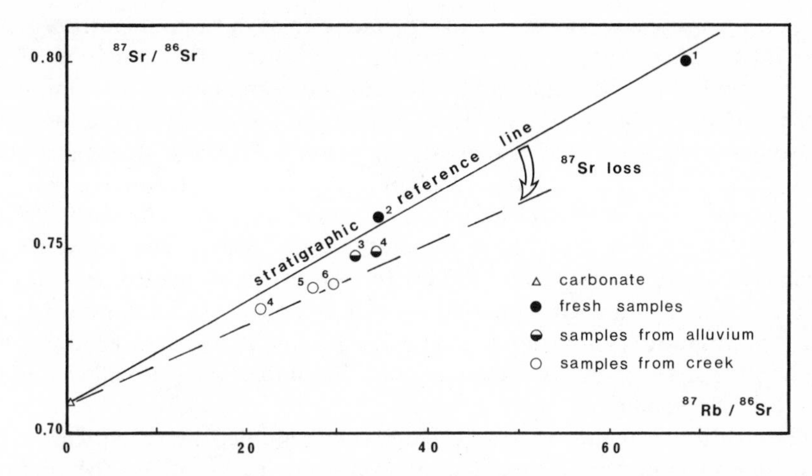

Figure 19. Effect of natural weathering on the Rb–Sr balance of Albian-Cenomanian glauconitic minerals from Normandy, France (Clauer, 1976b).

The effect of weathering on the Rb–Sr systematics of clay minerals was studied on Cenomanian glauconitic minerals from Normandy, France. Two samples (1 and 2 on Figure 19) were collected on a fresh outcrop and from a core. Two others (3 and 4) were extracted from the nearby Quaternary alluvium and three from the mud of a small creek (5–7) (Odin and Hunziker, 1974). The two fresh fractions lie together with the associated carbonate on an isochron representative of the sedimentation time. The weathered samples lie below this line with lower $^{87}Sr/^{86}Sr$ ratios. Weathering induces an increase in the Sr content by addition of external Sr and a leaching of radiogenic ^{87}Sr (Clauer, 1976b). The result is here a decrease in the apparent age of the minerals from about 92 Ma to 78 Ma.

4.4 The analytical uncertainties

The analytical uncertainties begin with the sampling and representativity of the material. This problem is more acute for sediments than for plutonic or volcanic rocks, because the Sr isotopic homogeneity of the sedimentary minerals is not an automatically established fact. Moreover, it is necessary to select the shaly horizons in order to choose the most suitable clay fractions. Therefore, a sampling standardization is difficult because the quality of the outcropping rocks or the availability of drill cores, for instance, may introduce bias in sampling procedures. In any case, it is recommended to collect many samples for preliminary mineralogical studies.

The analytical uncertainties continue with the preparation of the samples and the separation of the granulometric fractions. Engels and Ingamells

(1970), for instance, evaluated the possible sources of sampling errors. They also exist during isotopic measurements, especially in the estimation of the reproducibility of the equipment. This estimation should be systematically given in publications, together with the values of the standards and also with the types of error ($1\sigma, 2\sigma, 2\sigma/\sqrt{N}, \ldots$). The use of the age calculation techniques is also an analytical problem. The isochron technique requires the measurement of several minerals, but gives automatically the $^{87}Sr/^{86}Sr$ ratio of the Sr trapped during mineral crystallization. The conventional technique, proposed by Odin (1973b) for the dating of glauconies, may be used individually on each separated fraction but needs the assumption of the $^{87}Sr/^{86}Sr$ ratio at crystallization time. This technique has another disadvantage: the analysis of an insufficient number of samples does not allow control of their representativity.

The artifical leaching of clay minerals with diluted acid, similar to the technique proposed by Bofinger *et al.* (1968) for rocks, allows the removal of the adsorbed Sr and of the Sr from calcite impurities. This removal increases the Rb/Sr ratio of the sheet silicates (Table 5) and correspondingly

Table 5. Effect of artificial 1 N HCl leaching on the Rb/Sr and $^{87}Sr/^{86}Sr$ ratios, as well as on the conventional age of different clay minerals and a zeolite.

Samples		Rb/Sr	$^{87}Sr/^{86}Sr$ ($\pm 2\sigma$)	Conventional age calculation (Ma $\pm 2\sigma$)	References
Tertiary glaucony					
(Belgium)	U	7.74	0.7151 ± 0.0004	20.1 ± 1.1	Pasteels *et al.*
	L	43.26	0·7436 ± 0.0005	19.6 ± 0.4	(1976)
Tertiary glaucony					
(Belgium)	U	6.44	0.7152 ± 0.0003	24.5 ± 1.4	Pasteels *et al.*
	L	50.79	0.7482 ± 0.0004	18.7 ± 0.4	(1976)
Tertiary detrital					
illite (France)	U	1.41	0.7254 ± 0.0020	131 ± 38	Clauer (1976b)
	L	5.47	0.7396 ± 0.0020	97 ± 11	
Silurian illite					
(Oklahoma)	U	6.46	0.8200 ± 0.0003	427 ± 10	Chaudhuri and
	L	7.83	0.8545 ± 0.0003	461 ± 10	Brookins (1979)
Recent smectite					
(Pacific)	U	0.10	0.70902 ± 0.00019	n.d.	Clauer
	L	0.33	0.70858 ± 0.00050	n.d.	(unpublished)
Tertiary phillip-					
site (Pacific)	U	0.47	0.70770 ± 0.00029	18.2 ± 1.9	Hoffert *et al.*
	L	1.68	0.70538 ± 0.00029	17.1 ± 1.2	(1978)

The different ages were recalculated with the same $\lambda\,^{87}Rb = 1.42 \times 10^{-11} \times a^{-1}$. The two values for the Recent smectite were not calculated, a zero value being assumed.

decreases the error in their ages. To make sure that no preferential leaching of radiogenic ^{87}Sr from the minerals occurs during these experiments, as happened during clay separation from limestones (Clauer, 1979a), it is necessary to measure systematically the $^{87}Sr/^{86}Sr$ ratio of the leachates and to compare it to the value of the contemporary marine Sr. The non-analysis of leachates is an experimental uncertainty easy to eliminate.

4.5 Conclusion

Many uncertainties regarding genetic information from clay minerals still exist. However, a significant number of the other uncertainties can be reduced or even eliminated through: (1) careful control of the analytical process; and (2) integration of the *historical* information with the geochemical and mineralogical data.

5 CONCLUSIONS

Three points should be emphasized at present as provisional conclusions.

(1) Rb–Sr dating of sediments is possible through the use of whole-rocks or separated authigenic minerals such as zeolites, siliceous concentrations and especially clay minerals. Complementary studies are nevertheless necessary for stratigraphic purposes.

(2) The recent progress in clay mineralogy, sedimentology and the geochemistry of their Rb–Sr systematics makes it possible to date sediments after a careful consideration of *preliminary* criteria. However, uncertainties reside in particular in misunderstanding of the genetic mechanisms.

(3) Rb–Sr dating of sediments requires a *fresh* approach in which it is necessary to *adapt* mineralogical and geochemical results to the type of materials dated, because it is they which determine the final interpretation.

Acknowledgements

I would like particularly to thank S. Chaudhuri (Kansas State University, USA) for reading the manuscript and for kindly improving the translation.

Résumé du rédacteur

Un examen critique de l'application de la méthode de datation Rb–Sr aux sédiments a permis de faire le point sur les certitudes et incertitudes actuelles. Celles-ci peuvent être groupées en quatre rubriques ainsi qu'il est proposé dans l'article introductif de ce volume: (1) les incertitudes *stratigraphiques* qui nécessitent une réflexion de géologue, (2) les incertitudes *génétiques* qui nécessitent une réponse minéralogique, (3) les incertitudes *historiques* qui posent des problèmes sédimentologiques, et (4) les incertitudes *analytiques* qui concernent le géochronologiste. Il est donc clair que cette datation nécessite une toute nouvelle approche qui tient compte de ces aspects spécifiques et qui est nécessairement différente de la géochronologie du matériel plutonique ou métamorphique.

Il est montré que la datation Rb–Sr de sédiments est d'ores et déjà possible par l'utilisation des roches totales ou des minéraux authigènes tels que minéraux argileux. Pour ceci, des critères préliminaires devraient être systématiquement utilisés et les récents progrès de la minéralogie des argiles et de la sédimentologie mieux intégrés dans les interprétations.

(Manuscript received 6-11-1980)

Numerical Dating in Stratigraphy
Edited by G. S. Odin

14
Zero isotopic age of glauconies

Gilles S. Odin and Martin H. Dodson

1 INTRODUCTION

The aim of this chapter is to present evidence concerning genetic uncertainties surrounding the use of *glaucony* as a geochronometer. The word glaucony (plural glauconies) will be used here as a facies name to designate the green pellets. This will avoid confusion with a peculiar mineralogical composition of the pellets which may vary (Odin and Létolle, 1980; Odin and Matter, 1981). The understanding of the geochemical behaviour of the green pellets greatly depends on this distinction. Detailed sedimentological studies lead to the conclusion that the chemical composition of glauconitized substrates evolves slowly. There is a complex equilibrium between (1) the sea-water above, (2) the horizon of '*verdissement*' and (3) the underlying sediment. The knowledge of these phenomena leads to the question, what is the isotopic ratio of radiogenic isotopes of the dated glauconies at the moment of their initial closure? We will consider here only the argon isotopes.

Thanks to various examples of glauconitization in progress in Recent sediments, we have been able to distinguish successive stages of evolution of glauconies. We indicate below how these evolutionary stages may be recognized. Their argon content from the deposition of the initial substrate up to a relatively evolved stage is analysed. In carrying out these researches we have identified and tried to answer two questions:

What is the initial isotopic ratio?
What is the precise moment of closure of the chronometer relative to sediment accumulation and the palaeontologic sequence?

The observations and measurements reported in this chapter may be used for correct interpretation of an apparent K–Ar/glaucony age provided that a careful sedimentological study of the dated horizon and sample has been made. An example will be given where a series of measurements, apparently without significance as a whole, may be understood in terms of the experience obtained here so that a useful conclusion can be drawn.

2 THE GENESIS OF GLAUCONY

2.1 Ancient theories and new observations

In the last twenty years a lot of new observations have been collected on the process of glauconitization thanks to the study of Recent stable continental margins. In particular, both sides of the Atlantic Ocean have been studied in detail (Ehlmann *et al.*, 1963; Porrenga, 1967; Giresse, 1969; Lamboy, 1976). A synthesis of these observations, with many new ones, is presented in Odin (1975), from which are taken the schemes presented below.

Various theories have been proposed to explain glauconitization. An early theory involved coprecipitation of Mg, Fe, Al and Si gels and subsequent absorption of K on the sea bottom (Murray and Renard, 1891; Takahashi and Yagi, 1929). During the last twenty years the theory of transformation of degraded layer lattice silicate has been widely accepted (Burst, 1958). In the last few years, however, numerous new results and many difficulties in applying this 'layer lattice theory' have led to a different interpretation which takes into account all the known occurrences of newly forming glaucony. The new scheme is based solely upon detailed observations confirmed in several cases in order to eliminate peculiarities of local environments. Firstly, the study of Recent outcrops when glaucony is forming has shown that an *inherited degraded TOT lattice is very frequently absent.* The second general observation is that in most cases glauconitization occurs *in grains and not in the whole sediment.* We will restrict our discussion to this granular facies, which is the facies used for dating. The third main observation is that glauconitization may *go to completion in close contact with the sea-water.* This means that exchanges with the sea-water must be possible during the whole process; such exchanges can only occur in the first centimetres of a mud, in the first decimetres of a sandy mud and the first few metres of pure coarse sand.

2.2 The 'verdissement' of the grains

The reason why glaucony so commonly occurs in a granular facies is that favourable initial substrates are themselves granular. Observations made on granular glaucony from both ancient and Recent formations suggest that four main kinds of substrate exist:

(1) *Internal moulds* such as foraminifers, ostracods, larval or young gasteropods, pteropods and algae are present in small amounts in most samples

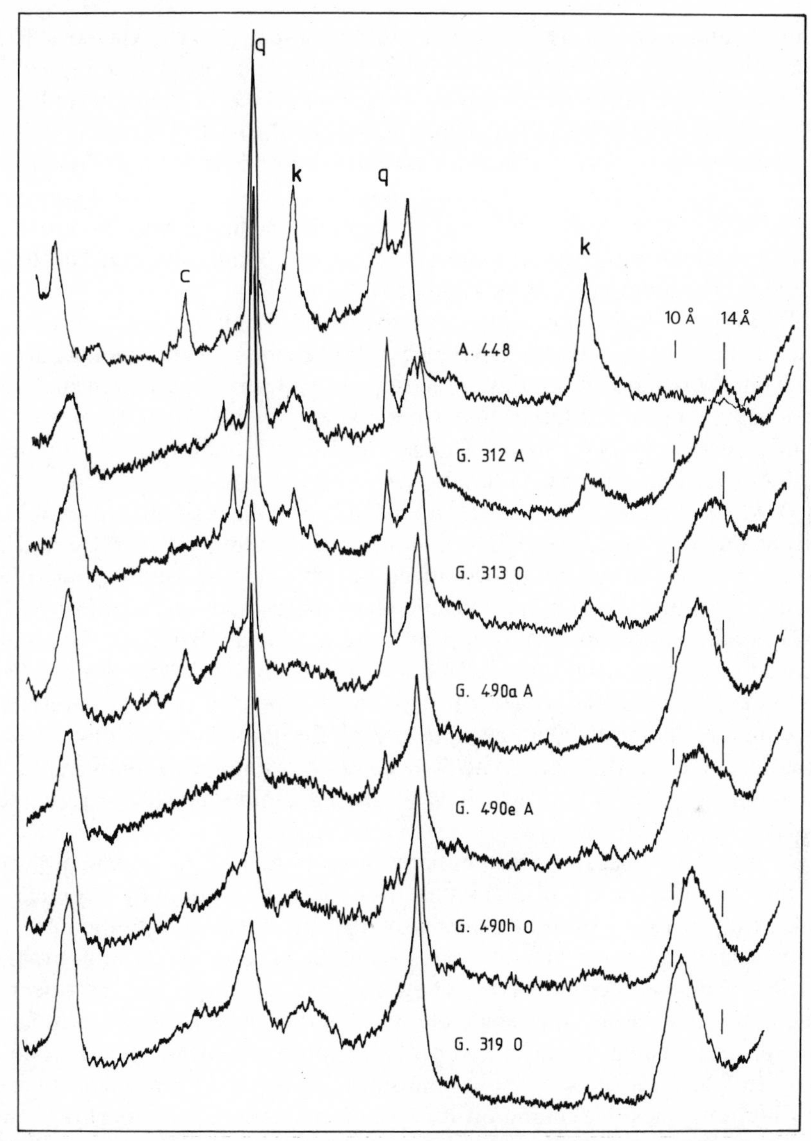

Figure 1. X-ray diffractograms on powder mounts of sediments from the Gulf of Guinea. A, fine fraction; G, glaucony (grains). The samples are arranged from top to bottom in order of increasing potassium content. The evolution is shown: (1) in the change of position of the (001) diffraction peak from 14 Å towards 10 Å; (2) in the disappearance of the initial substrate of glauconitization: the kaolinic (k) fine fraction with a little quartz (q) and traces of calcite (c). The most evolved grains (fraction O: 0.500–1 mm in size of G.319) show no definite traces of the initial substrate. According to the diffractograms no traces of a possible potassium-rich reservoir of radiogenic argon are visible (neither mica nor feldspar).

(Murray and Renard, 1891; Collet, 1908; Caspari, 1910; Cayeux, 1916; Wermund, 1961; Ehlmann *et al.*, 1963; Bjerkli and Östmo-Saeter, 1973).

(2) *Coprolites* occur frequently in great abundance (Takahashi and Yagi, 1929; Moore, 1939; Porrenga, 1967; Tooms *et al.*, 1970; Giresse and Odin, 1973; Boyer *et al.*, 1977). They are especially abundant on the African shelf from the Senegal river mouth to the mouth of the Congo. These coprolites are formed by various mud-eaters (annelids, echinoderms) and contain initially various mineral components: clay, carbonates, quartz. In Recent outcrops the dominant clay is kaolinite.

(3) Biogenic *carbonate debris* formed by disarticulation after disintegration of the organic tissue (sponges, echinoderms) or by biological and mechanical breakdown (molluscs), is frequently observed as a main substrate, for example in the Lutetian and Cenomanian of the Paris Basin and the relict glaucony of NW Spain (Dangeard, 1928; Houbolt, 1957; Lamboy, 1968; 1976; Lamboy and Odin, 1975).

(4) *Mineral grains* and rock fragments are sometimes dominant in old and Recent outcrops (Cayeux, 1916; Galliher, 1935; Odin, 1972a; Hein *et al.*, 1974). We have observed glauconitized quartz, feldspar, biotite, muscovite, calcite, dolomite, phosphate, volcanic glass shards and chert grains.

Carbonate source materials appear to represent an especially favourable substrate (Cayeux, 1932; 1938; Millot, 1964; Lamboy, 1976) together with coprolites. One or the other of these four types of parent materials is undoubtedly the source of the majority of the granular glaucony encountered in ancient formations. This conclusion has important implications for the initial isotopic ratios, which vary greatly amongst these various substrates.

All granular substrates share initially the common characteristic of porosity due to either their primary constitution or alteration by chemical or biological processes. In these pores grow the first green glauconitic minerals, distinguished by their crystal form, seen under the SEM, and by their specific chemical composition. They are iron-rich glauconitic smectites which develop in all the available space (the pores). These authigenic minerals grow using the ions, and probably more complex structures too, in the pore fluids. At the same time, the substrate is also altered. It will persist for different times depending on its ease of alteration in the marine bottom environment. For example, carbonate substrates are altered more easily and are therefore more favourable to glauconitization than are silicate substrates, which are more resistant. The disappearance of the substrate has been observed by SEM and X-ray diffraction, but two questions remain; what is the initial isotopic ratio of the parent material, and how long does that ratio persist?

Figure 1 shows X-ray diffractograms on powder mounts of the glaucony of the Gulf of Guinea at various stages of evolution (Giresse and Odin, 1973),

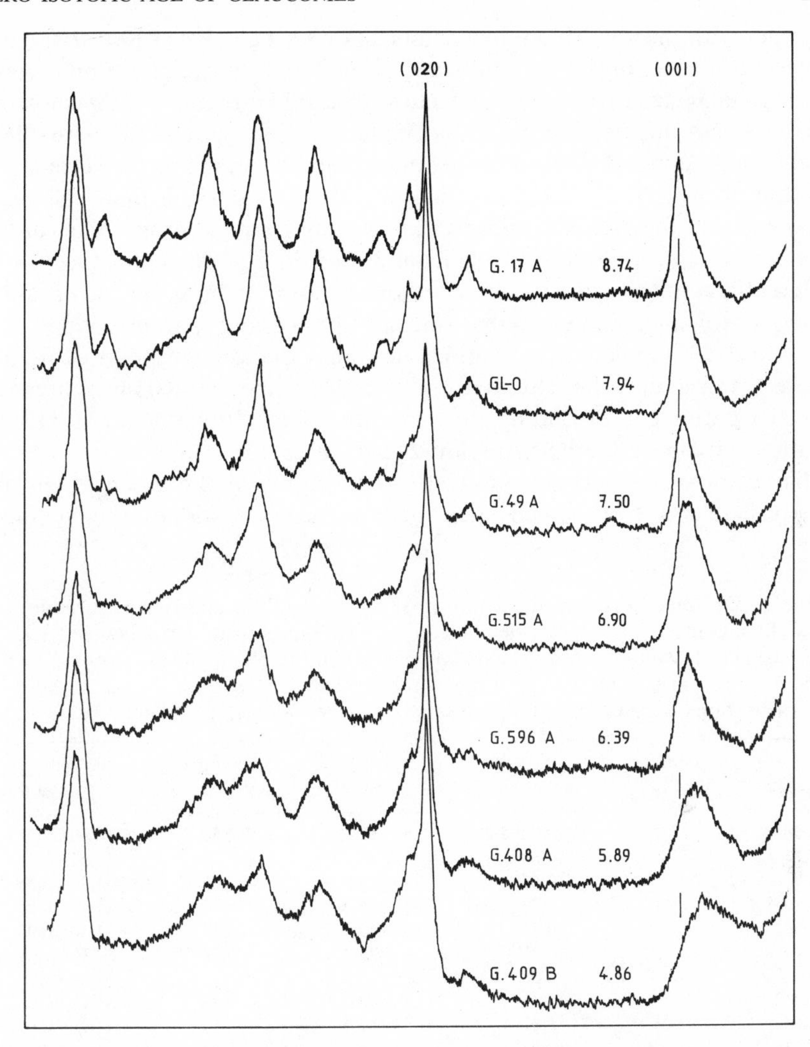

Figure 2. X-ray powder diffractograms of pure glauconies with various potassium contents. All samples are from the Cretaceous and the Palaeogene; the potassium content is given (K_2O wt %). Note the constancy of the (020) peak, whose position may be taken as an internal reference for each diagram. The position and form of the (001) peak is related to the potassium content as well as the form of the other peaks. This system of comparison of X-ray diffractograms permits the choice, before chemical and isotopical analyses, of the fraction or sample of glaucony most rich in potassium. An easy method to obtain a number characteristic of the diagram and of the approximate potassium content is to measure carefully the distance between the middle of the diffraction peak (020) (stable) and the middle of the diffraction peak (001) (function of K_2O content—see Table 1 and Odin, this volume, Chapter 10). Samples with less than 4.86% K_2O are shown in Figure 1.

together with the probably initial composition of the pellets (the fine fraction collected in a nearby area). The stage of evolution of the glauconitic grains may be recognized from various factors. The initial texture of the substrate becomes less and less recognizable under the lens as well as with SEM. Little cracks appear, while protuberances may be formed at the surface due to rapid local growth of glauconitic minerals. At the same time, the paramagnetism of the grains is enhanced, providing a simple property according to which the various steps of glauconitization in a single sediment may be differentiated. The green colour becomes more intense, giving a useful criterion for indicating whether or not the various paramagnetic, densitometric or granulometric fractions represent the same stage of evolution. These criteria may be checked by measuring the potassium content of selected grains corresponding to a specific X-ray diffractogram (a quicker and less expensive method): Figure 2 and Table 1.

The comparison between the stage of evolution of the green grains and the probable age at which their genesis began leads to the synthesis

Table 1. Distance between diffraction peaks (020) (stable) and (001) (variable) and potassium content of a series of glauconies. The samples are arranged in order of decreasing potassium content; the distance between the two peaks increases. Samples from the present continental shelf measured here for argon were included. All these diffractograms were prepared on the same day (see Odin, this volume, Chapter 20).

Sample	Potassium (K_2O wt %)	(001)–(020) (distance in cm)	Sample	Potassium (K_2O wt %)	(001)–(020) (distance in cm)
G.563 A	8.77	10.78±0.05	G.513–2 A	7.24	10.90±0.05
G.17 A	8.74	10.83			10.91±0.05
		10.82	G.534 A	7.02	11.00±0.05
G.16 A	8.49	10.79	G.515 A	6.90	11.12±0.05
G.582 A	8.32	10.81	G.440 A	6.85	11.10±0.05
G.299	8.19	10.83	G.398 A	6.73	11.16±0.05
G.347	8.07	10.88			11.24±0.05
G.412	8.04	10.94	G.440 O	6.64	11.15±0.05
GL-O	7.94	10.87	G.319 A	6.50	11.29±0.05
		10.83	G.596 A	6.39	11.24±0.05
		10.88	G.486 A	6.37	11.28±0.10
G.467 A	7.92	10.88	G.503–22 A	6.15	11.15±0.10
G.468 A	7.83	10.83	G.405 A	6.08	11.50±0.10
G.348 A	7.76	10.99	G.408 A	5.89	11.54±0.10
G.488 A	7.75	10.96	ML.210 A	5.44	11.56±0.10
G.273 A	7.73	10.89	G.409 B	4.86	11.69±0.10
G.288 A	7.72	10.98	G.490 e A	4.20	11.75±0.20
G.286 A	7.71	10.97	G.600–360 A	3.65	12.00±0.20
G.466 A	7.69	10.92	G.356 A	2.80	12.84±0.20
G.585 A	7.51	10.95	G.312 A	2.30	12.75±0.25
G.49 A	7.50	10.93			
G.461	7.31	11.08			
		11.04			

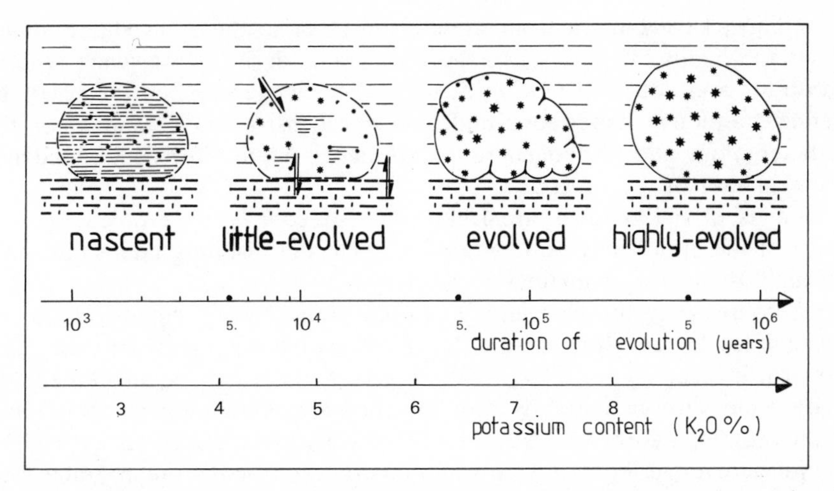

Figure 3. The evolution of an initial substrate during glauconitization. Four points in a continuum of evolution are chosen. The *nascent* glaucony (2–4% K_2O) is a porous substrate in which glauconitic smectites originate by crystal growth. The *slightly-evolved* glaucony contains 4–6% K_2O and traces of initial substrate remain according to X-ray diffractograms and SEM; exchanges between sea-water above and sediment below permit a replenishment of ions at the level of the pores of the substrate. These grains are the connecting link between sea-water and sediment, a unique special environment with a large surface area favourable to authigenesis. The *evolved* glaucony displays 6–8% potassium (K_2O). This stage is characterized by a recrystallization of the authigenic minerals in an environment probably more closed than before. Recrystallization and crystal growth lead to an expansion of the initial substrate, so the grain becomes cracked. No trace of initial substrate remains. The *highly-evolved* glaucony (more than 8% potassium, K_2O) is rounded by external deposition of less evolved new glauconitic minerals in the cracks. At this time, no trace of initial substrate, including possible isotopic inheritance remains. The potassium-rich glauconitic mica becomes slowly closed with regard to sea-water; the chronometer is launched even in the absence of burial. Burial may halt its evolution at any moment, which explains the existence of glauconies at various stages of evolution in the sediments.

proposed in Figure 3. The time scale is indicative; the length of the evolution is a function of local conditions and especially of the nature of the initial substrate, as stated above. The evolutionary process has been artificially divided at four points, but it is in fact *continuous*. In nature the process may be stopped at any moment by sudden burial which isolates the developing glaucony from sea-water.

Finally, we must emphasize that most highly-evolved glauconies present a nondescript appearance of more or less cracked, globular, generally dark green grains which is very difficult to relate to any particular original substrate. However, as shown by Giresse *et al.* (1980), who studied the main possibilities of evolution from the nascent stage to the evolved and sometimes the highly-evolved stage, the final grains derived from each kind of substrate are not statistically identical. A few general criteria may be defined, although experience and a comprehensive understanding of the sedimentology and palaeogeography are necessary to infer with confidence the character of the substrate.

Moulds of microfaunas are frequently not completely evolved, so that some of the grains may be identified as such; the resulting grains are small (around 100 μm in diameter).

Coprolites may produce highly-evolved grains; no specific form can be recognized; the resulting grains (the most evolved of them) may be 200–500 μm in size.

Biogenic carbonate debris probably provide the dominant initial substrates for highly-evolved glauconies; some such initial substrates may be identified even after a long evolution, especially from the characteristic zebra structure arising from the replacement of shell debris of bivalves (Odin, 1969); the resulting grains are certainly the biggest and are frequently of plate-like form; grains of more than one millimetre are not rare and some may reach 3–4 mm, a size not so far observed for other substrates.

Mineral grains and rock fragments give various possibilities depending on the physical properties of the initial mineral and its ease of alteration. The best characterized are grains derived from the evolution of micas, whether muscovite or biotite, which are accordion-like, elongated and easily broken across (Galliher, 1935; Odin, 1972a).

We will examine here two evolutionary series in particular: the coprolites of the Gulf of Guinea (Giresse and Odin, 1973) and the biogenic-carbonate debris of NW Spain (Lamboy and Odin, 1975).

3 Ar DATA AS A FUNCTION OF K_2O CONTENT OF GLAUCONIES

3.1 Results available for ancient series

Various individual published results show that so-called glauconies which are very poor in potassium often give too high argon-40 content for their probable age (see especially Owens and Sohl, 1973). However, the available data do not show that true glauconies are definitely concerned in all cases: we know that various green grains may have the facies of a glaucony without consisting of glauconitic minerals formed by the process shown above in the marine environment (Odin and Létolle, 1978). Ghosh (1972) measured the argon content in four glauconies with various K contents from the Weches formation of the Middle Eocene of the Gulf Coastal Province (Table 2). The author remarks that glauconitization started only after the pellet form

Table 2. Apparent K–Ar ages of glaucony grains from the Middle Eocene of Texas: according to the composition of the grains given by the author (Ghosh, 1972), only the two samples richest in K are typical of true glaucony. The correct age is probably near to 43 Ma; the lower the K content, the higher is the argon inheritance.

Composition (after Ghosh, 1972)	Potassium (wt % K)	Apparent age (Ma, ICC)	Colour
Chlorite + kaolinite	0.77	136	Earthy green
Chlorite + kaolinite	0.88	128	Earthy green
Illite + smectite + interlayered	2.77	53	Green
Interlayered: illite–smectite	4.55	46.5	Green

was reached. These coprolites were probably formed initially with a 'chlorite–kaolinite type of clay' according to Ghosh. The probable age, 40–45 Ma, is only approached by the most K-rich sample which has only reached a slightly-evolved stage. The lower the K content, the higher is the apparent age.

3.2 Results obtained in Recent series

3.2.1 Recent glauconitization in the Gulf of Guinea

The literature gives very few analytical results on Recent glauconies. Cullen (1967) has published a Pliocene–Quaternary K–Ar apparent age on an evolved to highly-evolved glaucony of the sediments of the Chatham Rise (New Zealand). Logvinenko (1976) presented in the IGC of Sydney a communication on glauconies lying in the Pacific Ocean, which gave apparent ages from Cretaceous to Pliocene–Quaternary varying with potassium content in the same way as the Weches glaucony.

Table 3 presents the results obtained for sediments of the Gulf of Guinea collected by P. Giresse and G. Moguedet between the Ogooué river mouth and the Congo river mouth. The fine fraction from which the coprolites originate has been analysed from four sites: A 448; A 593; A 594; A 595 (1). A 595 (2) is a coarser fraction (50–10 μm) from the same sample as A 595 (1). The X-ray diffractograms of the fine fraction (the parent material) show essentially kaolinite with less than 10% and sometimes only traces of open illite and poorly characterized smectite (Figure 1). Quartz is always present and is especially abundant in A 595 (2). In spite of this the potassium content is far from negligible, which is astonishing since potassium does not usually contribute to the lattice of 7 Å clays like kaolinite. This kaolinite is a clay typical of the intertropical zone and is known to be newformed from dissolved ions during alteration (kaolinization of feldspars) and in acid weathered soils.

Bearing this in mind, it is quite surprising to see (Table 3) the very high radiogenic argon content of these newformed clays which have been transported in the sea. If we ignore the probably mode of formation of most of the fine fraction, the apparent age near 500 Ma is compatible with the great age of the crust drained by the Congo river. The Congo craton gives apparent ages of 1000, 1600 and up to 2100 Ma (Cahen, 1961; Clifford, 1968). Even assuming that 10% of the fine fraction is formed with pure closed mica (not present according to X-ray diffraction), with a potassium content of 10% the quantity of argon measured implies that radiogenic argon must be present in the other 90% of the fine fraction, because otherwise the mica would have to give a 5000 Ma age. Thus the first

Table 3. Results of isotopic analysis on fine fractions and recently glauconitized coprolites from the Gulf of Guinea. Potassium measurements are from Miss M. Lenoble, Paris; argon measurements are from GSO in Berne (B), Hanover (H), Leeds (L) and Strasbourg (S); analytical uncertainties 95% confidence level. The sample A 595 (2) twice partly blew out of the crucible. All clays are dominantly kaolinite, with probably a tenth of interlayered illite–smectite and illite; note the homogeneity of the fine fraction in the different samples and the high K and Ar content for a fraction mostly composed of kaolinite after X-ray diffractograms. For the green paramagnetic grains the apparent age diminishes when the potassium content increases, but the zero age is not reached. The genesis began less than 10^5 years ago.

Sample		Potassium (wt % K_2O)	Radiogenic argon (%)	Radiogenic argon (nl/g)	Apparent age (Ma, ICC)
Fine fractions					
A 593 (H)		1.70±0.09	82.3	29.60±0.37	473±26
A 595 (1) (H)		1.55±0.08	87.6	29.93±0.47	517±28
A 595 (2) (H)		1.10±0.06	92.0	>19.21±0.23	>475
			92.2	19.30±0.21	
A 448–2 (B)		1.47±0.07	89.8	29.38±0.55	519±26
		1.56±0.08			
A 594 (L)		1.58±0.08	53.9	30.4±0.6	516±28
Grains					
G.313	(B)	3.00	74.0	15.12±0.40	149±11
			74.2	14.94±0.40	
G.490 h	(B)	3.48	50.2	8.86±0.35	79±7
		3.34			
G.490 e	(L)	4.19	22.4	7.15	52.0±4.5
G.490 a	(B)	4.25	45.6	7.12±0.30	50.3±4.5
	(S)		42.0	6.51±0.17	46.1±3.0
G.319	(B)	6.60	31.9	2.64±0.17	12.3±1.2

conclusion of this study is that radiogenic argon occurs either in kaolinite seen on X-ray diffractograms or in a highly altered continental component which does not diffract X-rays. In both cases, the fine, mainly clay fraction *inherits argon from old magmatic rocks* in spite of their accepted process of genesis (crystal growth from ionic solutions). At present, we do not know where this radiogenic argon is and how it can be retained along the whole geochemical process of alteration, erosion, transportation and deposition in sea-water.

Let us come now to our main objective. Six glaucony fractions have been chosen representing various steps of evolution from nascent to evolved. All samples apparently come from coprolites (Figure 4). The grains are clearly glauconitized: that is, their colour is green, they are definitely paramagnetic and their X-ray diffractograms are dominated by glauconitic minerals. Nascent glaucony is lighter in colour and contains traces of substrate component (seen by X-rays and SEM), while the coprolites retain their shape; evolved grains are darker, all traces of the initial substrate have disappeared, and the initially ellipsoidal form is commonly broken and cracked like popcorn (Boyer *et al.*, 1977), though in places relics of coprolites may be divined.

According to X-ray powder diffractograms, the grains are mainly composed of authigenic marine glauconitic minerals. These are probably formed by crystal growth from ions from pore fluids. Genesis of the less evolved grains, found at the shallowest depth, cannot have begun more than 18,000 years ago (Figure 5). At this time the sea-level was nearly 100–120 m lower

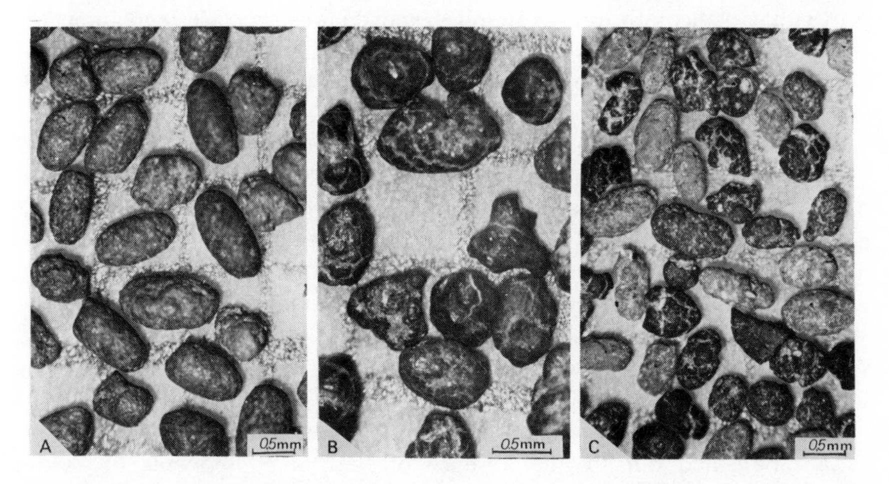

Figure 4. Coprolites from the African shelf at various stages of evolution. G.3130 (A) and G.3190 (B) from the Gulf of Guinea; G.600–030 (C) off Senegal.

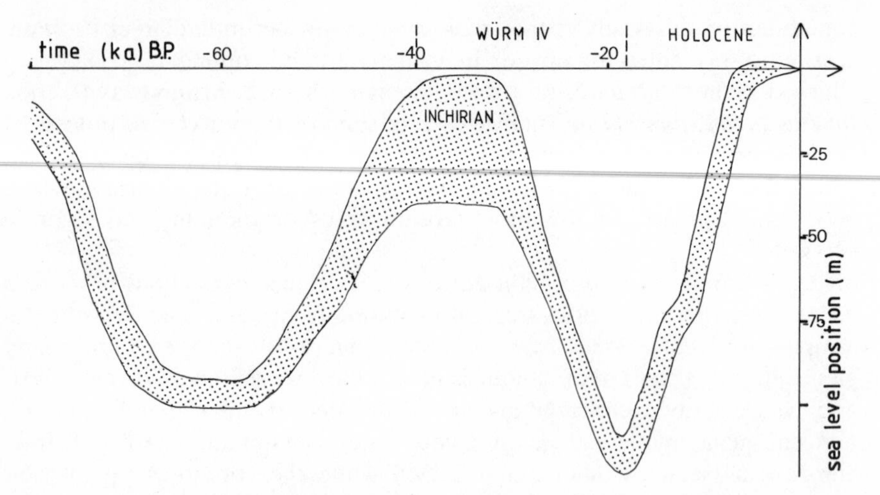

Figure 5. Sea-level changes during the last 70,000 years. Scheme according to Faure and Elouard (1967: Senegal, Mauritania); Martin (1973, p. 292: Ivory Coast); Delibrias *et al.* (1973: Congo); and H. Faure, personal communication (1980). All previously formed glauconies are destroyed during the emergence of the shelf.

than today. The low K glaucony (G.313; G.490) was formed after the Holocene transgression which followed the maximum of regression; the deeper the sample taken, the older may be the beginning of the glauconitization (Giresse and Odin, 1973). Concerning the high K glaucony (G.319), its situation, less than 300 m deep today, leads to the conclusion that it may be older than 18,000 years. Red coloured pellets at about 110 m depth prove that glauconitization certainly began before the last regression, for they are oxidized glaucony grains. The possible age of this material is perhaps Würmian and probably post-Inchirian, 24,000 years according to P. Giresse (Odin and Giresse, 1976).

Despite their Recent age and the mode of genesis discussed above, all the green grains contain a significant quantity of radiogenic argon. This argon must come from the substrate components although, according to X-rays, quartz and only traces of kaolinite remain. When K_2O is 3%, half of the initial radiogenic argon content is still present (Table 2). This shows that the argon retention is on the whole much better than the stability of the structure of inherited kaolinite. When K_2O becomes 4% or more not the smallest trace of substrate is visible either on powder diffractograms or under SEM.

The first fact is: inheritance of radiogenic argon from an initial substrate cannot be ruled out by the absence of detrital components on X-ray diffractograms; radiogenic argon must therefore either reside in a component which is poorly detectable by X-rays, or be trapped in the glauconitic mineral structure when forming (see 4.1 below). The second fact is: even when K_2O is as high as 6.6% there may still be a noticeable quantity of inherited radiogenic argon; in our opinion this leads to the second conclusion, that unless the initial substrate is recognized and carefully investigated, the apparent age of a glaucony must *a priori* be considered as suspect due to the *possibility of argon inheritance at the time zero.*

3.2.2 Glauconitization of various initial substrates

Other data have been obtained in Recent sediments off Senegal (N°600); off Guiana (N°503); in the Aegean Sea (N°486); and off California (N°585) (Table 3). Only the glaucony G.600 has been definitely formed mostly in coprolites, and the fine fraction (A 600) has been measured. Once again, radiogenic argon appears to be inherited by the clay matrix of the sediment. Argon is in lower concentrations than in the Gulf of Guinea but the potassium content is also low. The glaucony, at a nascent stage, also contains inherited radiogenic argon.

The three other samples, of which the substrate is predominantly moulds of microfauna, are representative of an evolved stage. They are all younger than 10^5 years. It is noticeable that, bearing in mind the analytical uncertainties, the radiogenic argon content tends to zero as the potassium content increases.

The glauconitization off NW Spain is, for various reasons, a special case. Its evolution is probably more complex, as no deposits occur in this continental shelf since the Miocene. The present sediments are therefore the result of a long and complex history due to the numerous changes in sea-level, together with large changes in temperature, during the last million years. According to sedimentological studies, the process of glauconitization may have begun in the Late Miocene. Reworking is not excluded, although only a few examples have been recognized, for example we have found in some samples Eocene nummulites, operculina and orbitoides with glaucony inside. However, according to Lamboy and Odin (1975), the glauconitization of these substrates postdates the reworking. At the same time palaeontological studies have shown that glaucony may be present in Quaternary foraminifers further to the east. Glauconitization may have been active in different places at various times since the end of the Miocene. The clearest evidence of the origin of the main glauconitization is that the 90% of the substrates consists of biogenic carbonate debris: echinoderms, molluscs. Argon has not been analysed in this original substrate—it is assumed to be of normal atmospheric isotopic ratio. Two glauconies have been analysed (Table 4). One is slightly-evolved and the other highly-evolved. It should be noted first that the highly-evolved and slightly-evolved glauconies give the same apparent age: they may have been formed and *closed* together.

Table 4. Results of isotopic analysis on samples from various present continental shelves. Analysis carried out as for Table 2. The general trend seen in Table 2 is also well demonstrated here for glaucony coming from the verdissement of coprolites or moulds. When K_2O becomes higher than 7%, zero isotopic age (equilibrium with sea-water for argon isotopes) is reached. In the case of the relict glaucony from NW Spain, the apparent age is independent of the potassium content, due to the absence of radiogenic argon from the recognized initial substrates. (* See also Cassignol and Gillot, Chapter 9.)

	Samples		Potassium	Radiogenic argon		Apparent age
No.	Clay	Substrates	(wt % $K_2O \pm 3\%$)	(% rad./tot.)	(nl/g STP)	(Ma, ICC)
Shelf off Senegal						
A 600–240(H)	Kaol. 5; Sm. 4.5	Coprolites (+ moulds)	0.84–0.79	78.8	13.66±0.24	457±19
G.600–360(H)			3.65	35.5	3.47±0.18	30.0±1.1
				30.1	3.58±0.08	
Shelf off Guiana						
G.503–22 (H)	Kaol. 4; Ill. 3; Ill.–Sm. 3	Moulds (+ various)	6.15	16.8	0.99±0.06	5.0±0.2
Shelf off W Mediterranean						
G.486 (L)	Sm. 7; Ill. + Chl. + Kaol. 3	Moulds + coprolites?	6.37	3.3	0.56±0.20	2.5±1.0
*Shelf off California**						
G.585 (L)	n.d.	Moulds + mica	7.50–7.53	0	0	0
*Shelf off NW Spain**						
ML 210 (B)			5.45	12.6	1.017±0.15	5.8±1.7
ML 279 (B)		Biogenic carbonate debris	8.00	37.7	1.504±0.08	5.8±0.6

Secondly, it is interesting that the apparent age is far from zero: according to the preceding observations on glauconies of various origin, at least as far as the highly-evolved glaucony is concerned, the grains are probably relict; this will be discussed in the following section.

4 ZERO AGE OF GLAUCONIES

4.1 Initial argon isotopic ratio

The figures which follow illustrate the various observations presented above. Figure 6 is based upon the results of Ghosh on Middle Eocene glauconies from the Weches formation. It shows the radiogenic argon content measured and the content which should normally be present if the apparent ages correspond to the probable age of these samples (probably near 43 Ma). We may conclude that argon is inherited from the initial substrate in the glaucony, and this inheritance decreases as the potassium content of glaucony increases.

Another interesting example of radiogenic argon inheritance linked with an early stage of evolution of glaucony has been quoted by Adams (1975).

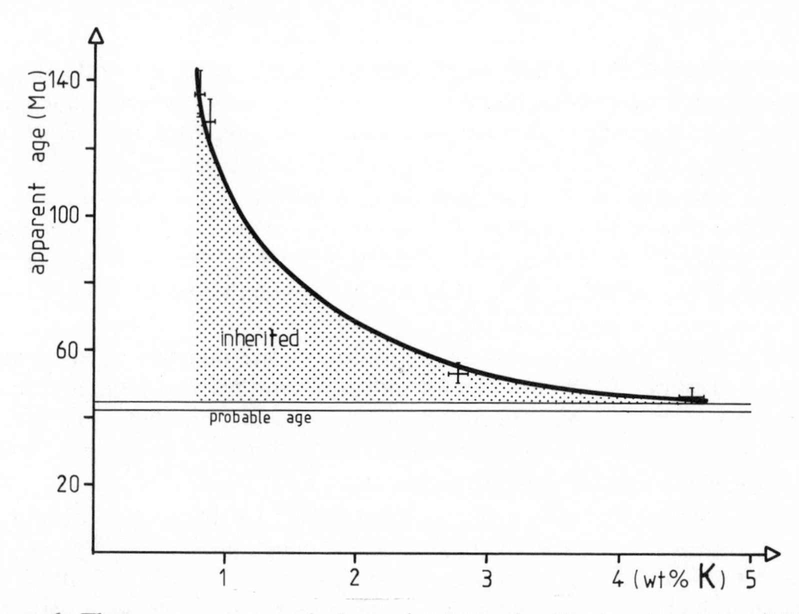

Figure 6. The apparent ages of glauconies from the Weches formation (Middle Eocene of Texas) according to data from Ghosh (1972). The potassium enrichment (glauconitization) leads to slow removal of the inherited radiogenic argon of the initial substrate (probably dominantly mica flakes). Analytical uncertainties are shown.

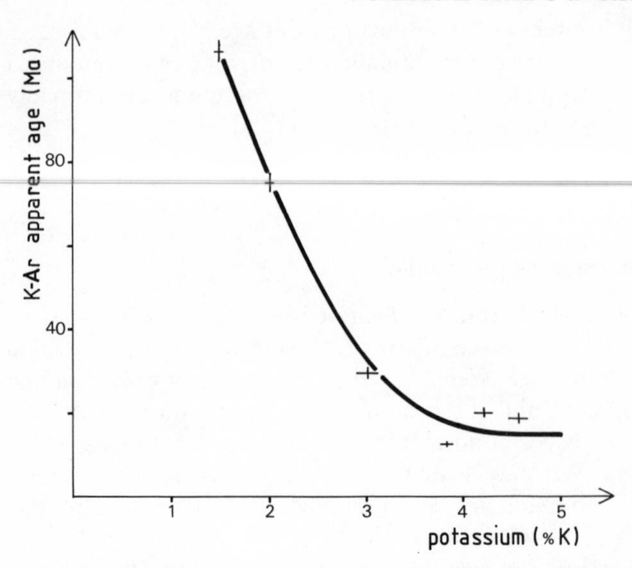

Figure 7. The apparent ages of New Zealand Early Miocene glauconies according to data from Adams (1975): comments as for Figure 6.

In New Zealand a drill-hole near Oamaru has allowed the collecting of Early Miocene samples containing nascent to slightly-evolved glauconies. Figure 7 uses the analytical results obtained by Adams, who infers a process of glauconitization of inherited chlorite in the Miocene sequence sampled. According to Adams, the original chlorite may be derived from a mid-Cretaceous biotite tuff or from the underlying schists which have yielded radiometric ages of 120–140 Ma. The sample with 3.8% K giving the youngest age is the oldest in the drill-hole; although the sample is not deeply buried, it seems that a process of argon loss (NZ is very tectonized and glauconies are K-poor) occurs and complicates the question of the initial zero age. In spite of this, the time of sedimentation may be estimated to be about 20 Ma, as may be expected from the stratigraphic data. The results obtained on Late Oligocene glauconies from Germany (Obradovich, 1964; see NDS 89) show a similar phenomenon.

Figure 8 shows the general trend demonstrated by the glauconitization in the Gulf of Guinea. Although the geochemical process is not exactly contemporaneous for all samples, the quantities of argon measured are much too high to be influenced by the precise moment at which the glauconitization of each sample began. According to this figure, and taking into account the data on samples from Guiana, California and Senegal, it may be stated that even for a glaucony developed from a ^{40}Ar-rich substrate *the initial ratio becomes normal when the potassium content is equal to or* *greater than 7% K_2O.* This value is thus the criterion according to which the possibility of a positive initial apparent age can always be eliminated.

As noted above, the inherited radiogenic argon either remains in structures poorly detectable by X-rays or may be trapped in the growing lattice of glauconitic minerals. This duality is not satisfactory, so we have attempted to put forward arguments for preferring one of these alternatives.

The first point which significantly tilts the balance concerns the K content of the clays probably used for the building of Recent coprolites. In spite of its kaolinic nature, the clay of the Gulf of Guinea is richer in potassium than the clay of Senegal. Moreover, the content of radiogenic argon is rather higher in glauconitized coprolites from the Gulf of Guinea (Figure 8). This indicates that part of the potassium, and inherited argon, is present in inherited structures not revealed by X-rays.

The results of two kinds of experiments reinforce this opinion. Considering the experiments on the leaching of various glauconies with concentrated

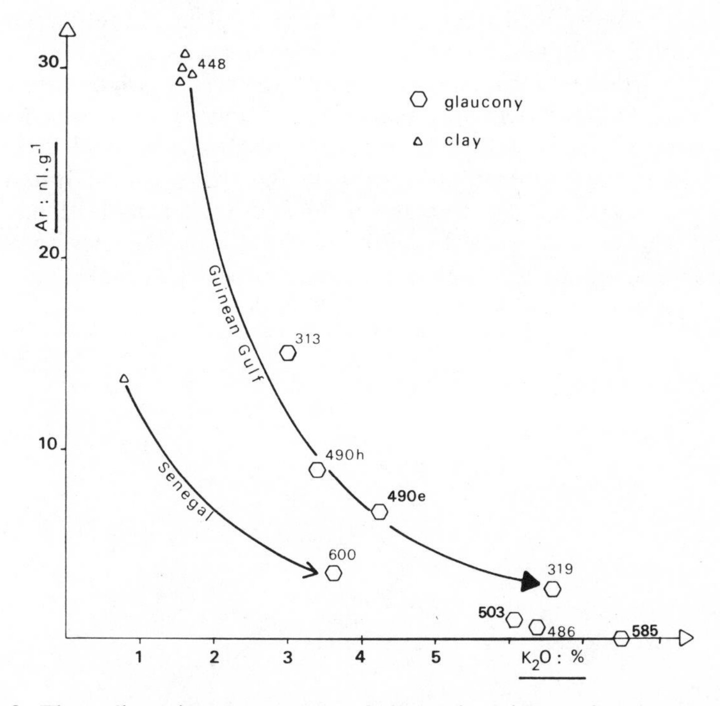

Figure 8. The radiogenic argon content of glauconies lying on the present continental platform. The evolution of coprolites from the Gulf of Guinea is well established, progressing from the fine fraction probably used by mud-eaters to an evolved glaucony. The possible parent material and green coprolites off Senegal show a similar trend, with less initial potassium and argon than in the Gulf of Guinea.

saline solutions (Odin and Rex, Chapter 19) one notices that leaching with a 2 N NaCl solution does not remove as much argon from a Recent glaucony as from ancient glauconies richer in potassium. This suggests that the radiogenic argon of the Recent glaucony is not in a glauconitic-type structure but *in a more solid one.*

Experiments on the heating of various glauconies under vacuum give strong support on this point (Odin and Bonhomme, Chapter 17). Heating at 250°C of various glauconies with 5.7–8% K_2O leads to the loss of a few percent of radiogenic argon by diffusion. However, on a glaucony from the Gulf of Guinea the same treatment has no detectable effect. For us, there are strong arguments to suggest that radiogenic argon of Recent glauconies with a low K content is contained in remnants of old minerals which, though disorganized, are more retentive for argon than are glauconitic minerals.

If we consider now the case of NW Spain, the situation is completely different: the initial argon isotopic ratio of the substrate may be assumed normal. The two results obtained for two different stages of evolution are in line with this assumption. By this means a second kind of isotopic behaviour (Figure 9) may be defined where all the stages of glauconitization have had the same initial isotopic ratio (normal) at the same time. However, bearing in mind the various possible lines of evolution of this Spanish occurrence, we do not claim to have definitely demonstrated this process, which simply remains a logical probability. In the particular case of NW Spain, the analytical data (the straight line given by the three points) lead to the tentative conclusion that the true age of the end of glauconitization, i.e. the closure of glauconitic minerals, occurred 5–6 Ma ago. This closure may be due to various changes of conditions. The analysed glauconies have been

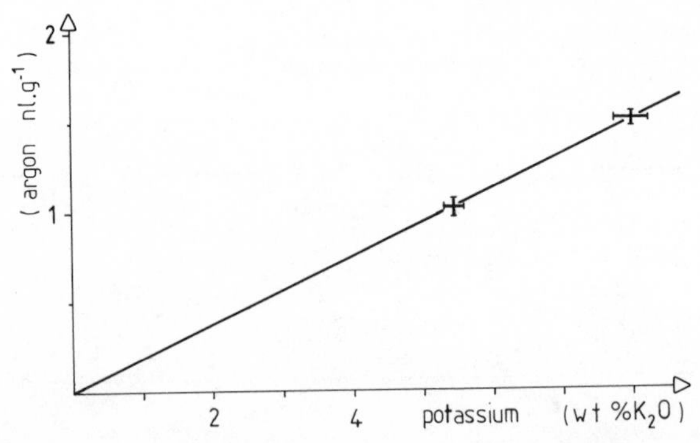

Figure 9. The radiogenic argon content of glauconies off NW Spain. The absence of initial inheritance is due to the fact that the parent material is argon-free (carbonate debris). The positive age, independent of the potassium content, is due to the fact that these glauconies appear relictual (although uncovered since the beginning of the Pliocene).

'dead' since the Miocene–Pliocene limit; although always in the marine environment, they are relicts.

4.2 Closure of a glaucony

The moment of closure of a glaucony and the duration of genesis of a glaucony are different aspects of its origin which must be considered separately.

4.2.1 Duration of genesis of a glaucony

Various arguments exist. The Recent submarine occurrence in the Gulf of Guinea is a very instructive example. Table 5 gives the potassium content of the best evolved grains of various samples taken at different depths off the Congo, Gaboon and Cabinda. The correlation is obvious. The time during which the coprolites are in the sea is estimated from Figure 5 (duration of immersion). But in fact the favourable minimum depth of glauconitization is at 50 m and more. The possible true duration of evolution is the time *after* this depth was reached—the 'maximum duration of evolution' in Table 5. The highly-evolved but oxidized glauconies from 120 m depth in this area must be older: their oxidation occurred during the pre-Holocene regression, 18,000 years ago, when they would have been at very shallow depths. From their highly-evolved state it appears that they were then already 5–10,000

Table 5. Depth of occurrence and 'true age' of the glaucony from the Gulf of Guinea. The duration of immersion may be estimated from a knowledge of recent variations in sea-level. We assume that coprolites are produced by animals living at the bottom of the sea at 0–10 m depth. But the verdissement can only begin at nearly 50–60 m depth, so the maximum possible duration of evolution is lower than the duration of immersion.

Glaucony	Present depth (m ± 10%)	Potassium (% K_2O ± 3%)	Duration of immersion (years ± 10%)	Maximum duration of evolution (years ± 10%)
G.312	85	2.3	15,000	9000
G.313	95	3.0	16,000	9500
G.356	100	2.8	16,000	10,000
G.357	110	3.1	18,000	10,000
G.490	120	3.4–4.3	20,000	12,000
G.363	180	5.0–6.3	Always	32,000 if all
G.318	225	5.1	Always	glaucony is
G.319	300	6.6	Always	actually post-
G.362	300	6.6	Always	Inchirian

years old, giving a total duration of evolution of about 25,000 years. The most evolved glauconies are found today at even greater depths, so they are believed to have required at least that time for their evolution.

In the Aegean Sea two levels of glauconitization giving nascent to slightly-evolved pellets (4.5% K_2O at most) lie in the Würmian, which occurred 80,000 years ago (Robert and Odin, 1975). The duration of their presence in the sea may be estimated as above to be about 10,000 years. They are respectively at 1.2 and 3.0 m below the sea-water–sediment interface; in spite of this very shallow burial it is clear that their evolution must have ceased as soon as the overlying clay was deposited, otherwise they would be more evolved than the coprolites of the Gulf of Guinea.

The evolved to highly-evolved glauconies of the sediments presently available on the continental slope are not very accurately placed in the stratigraphic column. According to Pratt (1963), glaucony has been formed off southern California partly in the interval from Late Pliocene time to Recent near Santa Barbara, and also before, and it may be forming today. According to the data given in Table 4, glaucony G.585 A is not closed for argon isotopes; it is still 'living', but it cannot be stated since when.

As far as we know, all glauconies of the present platforms with near to 8% K_2O may be related to more or less relictual Pliocene–Quaternary sediments.

The ancient sediments may only give general information which must be treated with caution: the exact palaeo-environmental history is very difficult to reconstruct, as the time and environmental conditions are not completely recorded in the deposits. However, investigation of the Middle Cretaceous glauconitizations of the Paris Basin leads to the conclusion that, in the Pays de Caux, an evolved to highly-evolved glaucony was formed between the Albian and the Cenomanian, and probably within Cenomanian time, without any apparent break in the faunal succession. We calculated the mean duration of ammonite zones in this basin to be 0.5 Ma (Odin and Hunziker, this volume). If we assume that the ammonite zones around the Albian–Cenomanian boundary have a normal duration, that there is no general stratigraphic break over the whole NW European basins, and that for less than half of the time of the last ammonite zone of the Albian and of the first of the Cenomanian deposition was normal, without glauconitization, then the maximum duration of the direct exchange between sea-water and glauconitized grains is 0.5 Ma for a glaucony with about 8% K_2O.

By this kind of reasoning, but with fewer examples than are now available, we proposed a few years ago a possible timing for glauconitization (Odin, 1975, p. 21). A nascent glaucony would represent some 10^3–10^4 years of close contact (active exchange) between sea-water and grains, while a highly-evolved glaucony would represent 10^5–10^6 years of evolution. In the absence of contradictory evidence we always adopt these values.

From these estimates, together with the fact that it is preferable to use evolved to highly-evolved glaucony for dating, we conclude that *the moment represented by the glauconitization is not instantaneous, even on the scale of geological time.*

4.2.2 Moment of closure of a glaucony

From the comparison between old glauconies and Recent unburied ones, it appears that the high Ar/K ratio of potassium-poor glauconies is a common factor. This argument suggests that the inherited isotopic ratio is not modified by burial: the evolution of that ratio is, in fact, halted. Although not finally established by sufficient observations, this conclusion will be tentatively accepted after the examination of some more arguments. In particular, we shall see below what happens with potassium. In the Gulf of Guinea various short cores, a few metres high, have been examined for possible variations of crystallization of glauconitic minerals as a function of depth. In no case has a clear variation been observed by X-ray investigations of green pellets. At one point (core 490) three samples have been analysed for potassium content. From the sea bottom (G.490 a) to 80 cm below (G.490 h), the potassium content changes from 4.25 to 3.41% (Table 2). This confirms that the evolution of the pellets is not due principally to the effect of early burial, but that probably burial halted the evolution towards potassium-rich glauconitic minerals. In the same way, glauconies from the cores off Senegal show no particular trend from the interface to 3.90 m depth. In the Aegean Sea, the core studied displays a more evolved glaucony at 3.0 m depth than at 1.25 m. We do not think that burial is the cause of the difference. Finally, the observations made in the Late Ypresian of the Monts des Flandres (Odin, 1971a–1972b) show a much more complete evolution for the more recent levels than for the older ones in a Late Ypresian glauconitic sandy clay series of 55 m thickness. There is a continuous trend from nascent or slightly-evolved at the bottom to an evolved state at the top of the succession. This variation has been related to the speed of deposition, and shows that processes after burial, if they occur, are much less important in the evolution of glaucony than all the mineralogical changes which occur due to interaction with sea-water.

As a final argument, we know from various Recent occurrences (NW Spain, Chatham Rise of New Zealand, southern California) glauconies which are completely evolved without having undergone any burial whatsoever. On the NW Spain platform and the New Zealand Rise the glauconies with 8% or more K_2O seem closed and argon is accumulating in the geochronometer. Cullen (1967) has reported a K–Ar apparent age of near 3 Ma.

In conclusion, we emphasize that isotopic equilibrium, together with the mineralogical evolution, is *achieved mostly or entirely in contact with sea-water* before burial. In the case of the examples of highly-evolved glauconies presently on the sea bottom, we know that near 8% K_2O the geochronometer becomes more or less closed and argon begins to accumulate without any burial. We are not sure, however, that we can generalize from these cases: we know glauconitic minerals with more than 8% K_2O and up to 9%. We have no data to show exactly why not only the two glauconies referred to above are closed, but also the second one from NW Spain which contains only 5.4% K_2O. But as a general rule, the facts given above show that burial *effectively stops the geochemical evolution* of a glaucony. This burial represents, then, the moment of closure of a glaucony whether the initial apparent age of the geochronometer is zero or not. But burial seems to be not the only method to stop the evolution of a glaucony and then to start the chronometer.

5 INITIAL APPARENT AGE OF GLAUCONIES

We shall now describe the recent application of the preceding observations to apparently unreliable results obtained in North Africa; a synthesis will then be proposed which makes practical use of the data assembled here; finally, we shall review what actual progress has been made in the possible use of glauconies thanks to the data presented here.

5.1 Inheritance of argon in Miocene glauconies from North Africa

Five glauconitic samples collected by G. Choubert and A. Faure-Muret in Morocco were studied by D. Tisserant a few years ago together with another one from Algeria collected by L. Hottinger. The five samples from Morocco were related to the Late Miocene (Late Tortonian), while the other was related to probably Messinian time. No sedimentological studies were made to help the interpretation of the planned isotopic measurements; the separated glauconies were directly processed without mineralogical investigations. The analytical data are reassembled in Table 6. Because of the scatter of the apparent ages, the results were not considered useable. However, a recent plot of apparent ages against potassium content (Figure 10) leads to a curve similar to that shown for the glauconies of the Gulf of Guinea. We may therefore conclude that the initial substrates of these glauconies were not free of radiogenic argon—depending on the locality, glauconitization was halted at different stages of evolution—and the most probable age of all these glauconies is that given by the horizontal asymptote of the curve defined in Figure 10 (Tisserant and Odin, 1979). According to the time scale proposed for the Late Cenozoic, the asymptotic value near 8 Ma corresponds to the Tortonian–Messinian limit. Of course this method cannot be regarded

Table 6. Results of isotopic analysis on six Late Miocene glauconies from Morocco (data obtained in the Geochron laboratories after Tisserant and Odin, 1979). Argon content is in inverse relation to potassium content; the apparent ages obtained appear unreliable. The apparent age nearest to the actual one is italicized (see Figure 10).

Sample	Potassium (K%)	rad. Ar (nl/g)	rad./tot. (%)	Apparent age (Ma, ICC)
373	2.47 2.55	3.473 3.809	60.3 28.3	36.9±2.5
376	3.71 3.72	3.025 3.249	48.5 60.6	21.4±1.4
411	4.06 4.06	1.849 2.185	15.8 22.3	12.8±1.5
371	5.06 5.11	2.039 2.050	71.0 68.8	10.4±0.5
422	5.43 5.73	2.409 2.263	55.0 46.5	10.8±0.7
372	*6.16* *6.19*	1.949 1.921	44.2 50.0	*8.1±0.4*

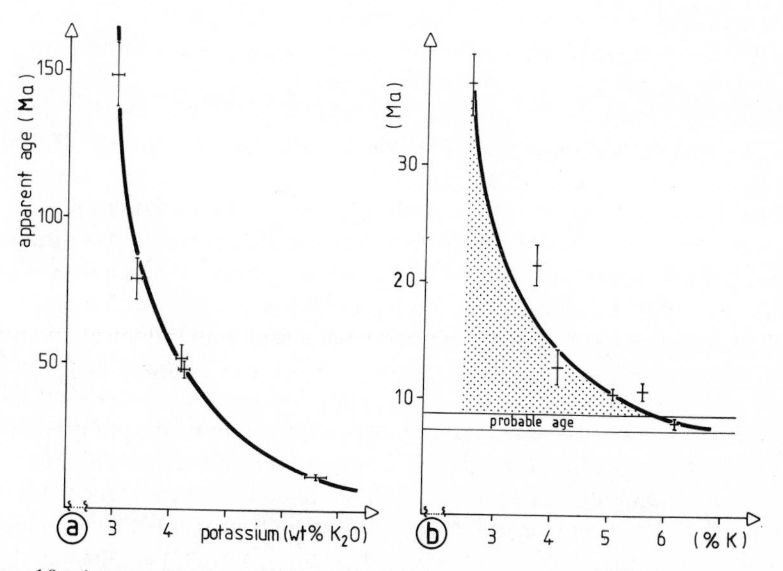

Figure 10. Apparent ages of glauconies from the Miocene of North Africa (b) compared with the evolution of the apparent ages of Recent glauconies from the Gulf of Guinea (a). An asymptotic age is obtained. In both cases, this age is compatible with the known stratigraphic age but it remains inaccurate in the absence of highly-evolved glaucony. Analytical uncertainties are given by the figures.

as a general method for accurate dating of a horizon, especially for the Miocene, but in the absence of a better chronometer, and accepting a wide uncertainty, K–Ar isotopic data may sometimes be used in this manner.

5.2 Correlation between isotopic age and faunal assemblage

The precise correlation between the apparent age given by isotopic measurements and biostratigraphy depends on two factors. The first is the nature of the substrate, and the second the moment of closure of the chronometer.

Figure 11 represents the case of an initial substrate free of radiogenic argon. The y-axis shows the apparent age on an arbitrary scale such that, when the chronometer is closed, the scale becomes identical to that of the x-axis. The x-axis shows on a logarithmic scale the time after the beginning of evolution of the substrate. $F_1, F_2, \ldots$ are hypothetical biozones deposited during the glauconitization process, while F_0 symbolizes a fossil remanié. Zero on the x-axis represents the pause in deposition during which substrates acquire the right character for alteration before the beginning of their evolution towards nascent, slightly-evolved, evolved and possibly highly-evolved glaucony. Four lines of evolution are drawn following four sedimentological events numbered along the x-axis).

Event 1 may be an early burial; closure of the glaucony with a zero apparent age will occur just before deposition of the zone F_2 but fossils F_1 and F_0 may be present in the glauconitic level.

Events 2 and 3 may also be due to burial. In the first case the slightly-evolved glaucony will date a moment just before the deposition of F_4; in the second case the evolved glaucony will date a moment just before the deposition of F_5. In all cases it is better to note that glaucony *predates the next non-glauconitic level*, because we can be sure that the fossils present in the glauconitic level are deposited *before* the closure of glaucony. The closure time must be regarded as the end of the evolution, and therefore as the time of deposition of the next level with the fossils included. In some cases, perhaps, the deposition of a few centimetres of sediments cannot be regarded as an efficient burial; it depends on whether or not exchange occurred between sea-water and evolving grains. Event 4 is a burial which occurs a very long time after the beginning of the glauconitization. In this case the chronometer was restarted *before burial* and the apparent age will be older than the fossils F_6 which are found both mixed with the glaucony and also above the glauconitic level. Such a case is well demonstrated in NW Spain today, as well as in the Early Miocene of the Aquitaine Basin (Odin *et al.*, 1975). In the latter area the apparent age of the glaucony is a maximum age for the most recent fossils mixed with the glaucony. As probably shown in the NW Spanish continental slope, even without apparent burial the

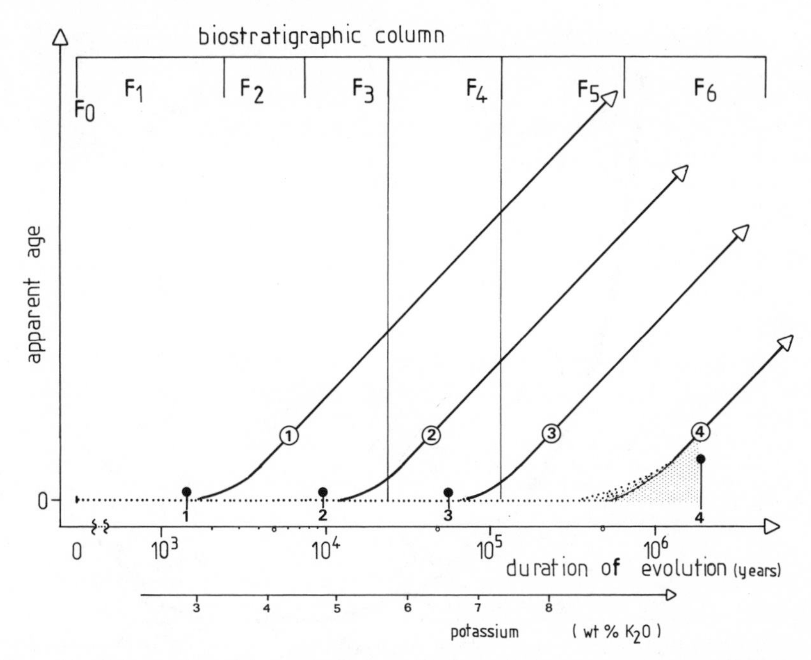

Figure 11. Schematic guide to the interpretation of the apparent ages of glauconies without initial inheritance. The genetic uncertainty alone is taken into account. Four possibilities of burial are shown (bottom). A sequence of six biozones is shown (top, F_1–F_6), with possibility of reworking at the base (F_0). Burial case 1: the glauconitic level may contain fossils F_0 together with F_1 but the moment of closure corresponds to the base of deposit of fossils F_2. The initial apparent age is zero. Burial cases 2 and 3: the slightly-evolved case 2 and the evolved case 3 glauconies are closed at the beginning of deposition of zones F_4 and F_5 respectively, although these fossils are not present with the glauconies but above. Burial case 4: this highly-evolved glaucony accumulates argon before the moment of burial (4); the measured age will be a maximum age for the fossils deposited together with the glaucony.

evolution of glaucony may cease before the highly-evolved step is reached. Probably the very great changes in sea-level during the Quaternary are the causes of these particular phenomena; we consider that such material may be recognized from the glossy appearance of the surface of the grain, which indicates a long evolution of the outer parts without evolution (recrystallization) inside the grain. This characteristic appearance has been observed practically each time that we might infer termination of evolution before burial.

Figure 12 represents the analogous case of a radiogenic argon-rich initial substrate. As in Figure 11, the y-axis scale is quite arbitrary and indicates the apparent age of the grains (or their radiogenic argon content) for various events shown on the x-axis. As before, we have chosen four possibilities.

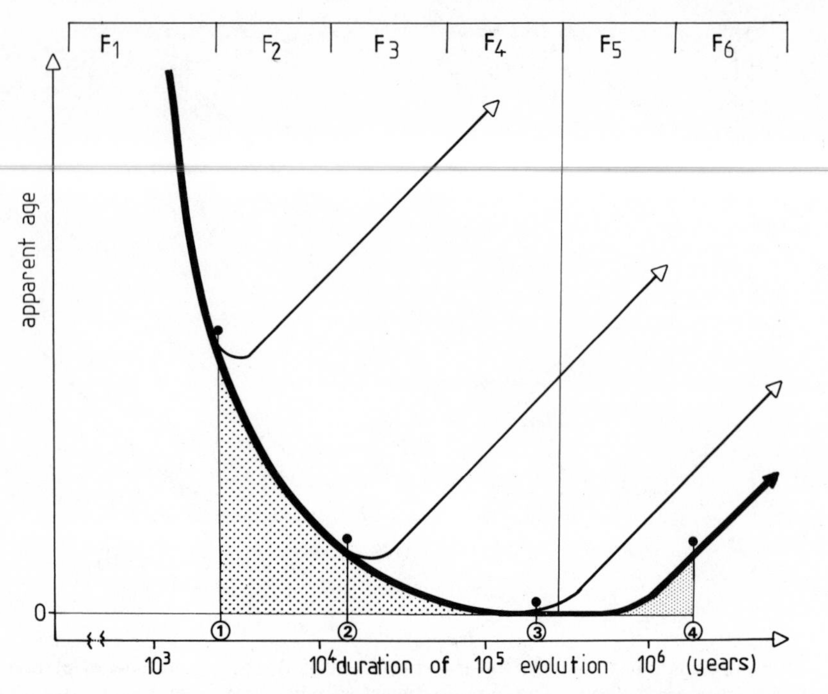

Figure 12. Schematic guide to a discussion of the genetic uncertainties related to glauconies growing on radiogenic argon-rich substrates. Same presentation as Figure 11. The heavy line shows the evolution of the apparent age if neither burial nor change of conditions stops the glauconitization process. If burial occurs when glaucony is nascent (1), inheritance will be very high and easily seen. If burial occurs when glaucony is slightly-evolved (2), the inheritance will be lower and sometimes not obvious; correct estimation of the apparent age will be impossible. The choice of a sufficiently evolved glaucony (7–8% K_2O) will lead to the lowest possible genetic uncertainty. Note that in the case of closure during the deposition of fossils of biozone F_4 together with glaucony, the actual moment of closure will in fact occur during the deposition of the next fossils. Burial case 4 leads to relict glaucony with an apparent age higher than the moment of burial.

Case 1 concerns a nascent glaucony buried by the deposition of sediments only a few thousand years after the beginning of alteration of the initial substrates. Fossils F_2 found together with the glaucony will be predated: the apparent age will certainly be much older than the time of deposition of the fossils. The difference between the presumed age and the measured age will be easily detected in most cases. Such samples must be rejected for dating a rock unless the origin of the green pellets is known.

Case 2, a slightly-evolved glaucony, is the most unreliable if we have no idea of the genetic history of the isotopic chronometer. The small inheritance leads to apparent ages which are not obviously older than the correct age (F_3–F_2 boundary in this example). This may lead to errors in time-scale calibration if the numerical scale was not accurately known before. For this reason a slightly-evolved glaucony must be used for time-scale work only if there is independent evidence that it has not inherited argon in its substrate. Otherwise such materials *should be strictly avoided.*

Case 3, an evolved to highly-evolved glaucony, is the most reliable as far as the normal initial isotopic ratio of argon is concerned. As for the preceding type of substrate, it must be emphasized that the moment zero is the burial by the next sediment. In the figure the moment dated will thus be the base of the zone of deposition of fossil F_5, rather than the moment of deposition of fossils found with the glaucony which include fossils F_0–F_4.

Case 4 is similar to case 4 of Figure 11: the isotopic apparent age predates the fossils F_6. The length of the time interval between closure and burial must be found by study of the biozone succession (a similar reasoning may apply to the perigenic glauconies, whose apparent age may be slightly older than contemporaneous fossils). We should point out that the interpretation may depend on whether the zonal fossils used are nannofaunas or macrofaunas. It is well known that nannofaunas are quickly destroyed in sea-water although macrotests will persist. In other words, there is little chance of finding in a glauconitic bed nannofossils much older than the time of burial, whereas with macrofossils it is better to know the fauna above the glauconitic level than within it.

6 SUMMARY AND CONCLUSIONS

The genesis of a glaucony can rarely be considered to be an instantaneous phenomenon. The initial isotopic composition of the green grains has been shown to be abnormal in various examples in the literature. An attempt has been made above to understand these uncertainties in the sedimentary chronometer utilizing the K–Ar method.

Most of the answers to these questions are contained in the *correct understanding of the genesis of green grains*, the facies used for dating. Granular glaucony is the result of marine authigenesis of the glauconitic minerals inside a necessary initial substrate. This initial substrate may be formed from various chemical components including quartz, silica, mica, feldspars, clay minerals, phosphates, aragonite, calcite, dolomite, and so on, provided that they are susceptible to alteration and may thus protect the growth of glauconitic minerals in their pores. These pores are the passageway for ions from various sources, particularly sea-water for potassium, magnesium and a proportion of the other ions. The initial isotopic ratio of argon may be normal or enriched in radiogenic argon, depending on the initial substrate, as shown by the various examples of glauconitization studied from present-day occurrences on the continental shelf. If the initial substrate is free from radiogenic argon, for example a carbonate, its initial apparent age is zero whether the evolution has reached a nascent, a slightly-evolved or an evolved stage. For an initial substrate rich in radiogenic argon, such as coprolites or an assemblage including moulds of microtests, some of that initial argon remains in a nascent or slightly-evolved glaucony, and even in glaucony with more than 6% K_2O.

For glauconies resulting from the evolution of argon-free initial substrates, as well as for evolved to highly-evolved glauconies of other substrates, the time zero (the closure of the chronometer) may generally be accurately assigned to the moment when exchanges with sea-water are halted by burial. In relation to biozonation, this moment is closer to the *moment of deposition of faunas in the horizon immediately above* the glaucony than to that of the faunas deposited with the glaucony. For a few highly-evolved glauconies, particularly those presenting a rather smooth appearance, it is possible that the closure occurs before burial by the next deposit.

In any case, a very careful and detailed sedimentological and palaeontological study of the glauconitic series appears necessary for correct interpretation of the analytical data in relation to the genetic uncertainties. It is easier to study glauconies with at least 7% K_2O: samples with less potassium may give, depending on the initial substrate, an apparent age which is not identical to the age of end of genesis of glaucony. This possibility may be explored either by microscopic examination or X-ray diffraction, or by isotopic analysis of a less evolved stage of glauconitization which may sometimes be found elsewhere at the same horizon in the same basin. However, isotopic analyses have shown that radiogenic inheritance persists long after all traces of initial substrates have been eliminated according to microscopic or X-ray observations. Various experiments have shown that inheritance probably occurs in almost undetectable structures more resistant to argon diffusion than glauconitic minerals.

Acknowledgements

Samples were given by G. H. Curtis, J. C. Faugères, P. Giresse, M. Lamboy and G. Moguedet. X-ray analysis were performed with the help and permission of M. Leikine and A. Person. Potassium was measured by M. Lenoble. Argon measurements were performed with the help and permission of M. G. Bonhomme and R. Winkler in Strasbourg; J. C. Hunziker and E. Jäger in Berne; H. Kreuzer in Hanover; D. C. Rex in Leeds. To each of them our thanks are due. This chapter is a contribution to IGCP Project 133.

Résumé

L'étude sédimentologique de glauconies formées récemment ou en cours de formation sur les marges océaniques actuelles a permis de reconstituer leur processus de genèse. Il s'agit fondamentalement d'une *néoformation de minéraux spécifiques:* TOT ferrifères et potassiques, suivie de recristallisations dans un micro-environnement semi-confiné. Ce milieu est créé dans des débris minéraux divers, souvent granulaires, en voie d'altération sur le plancher marin. *Les minéraux néoformés se substituent peu à peu au support de verdissement.*

Notre contribution rassemble des données obtenues par la méthode potassium–argon qui permettent de vérifier et de préciser ce mécanisme. L'âge apparent initial des grains verts et le moment de fermeture (le déclenchement du chronomètre glauconie) sont ainsi observés et compris pour la première fois au moment même où les phénomènes se déroulent dans les sédiments récents.

Deux modèles d'évolution isotopique sont établis en fonction du support de verdissement initial. Si le support initial ne contient pas d'argon radiogénique: cas de la glauconitisation de débris carbonatés, l'âge apparent initial des grains verts est nul dès le début de l'évolution. Cependant, dans de nombreux autres cas: glauconitisation de coprolites ou de remplissages de Foraminifères, on mesure dans les graines en cours de verdissement des âges apparents parfois très élevés qui diminuent au fur et à mesure du développement des minéraux glauconitiques. L'expérience montre que seuls les grains évolués formés de minéraux glauconitiques contenant plus 6% (K_2O) de potassium peuvent être considérés comme sûrement dépourvus d'argon hérité au moment de l'enfouissement.

On montre enfin que la fermeture du chronomètre vis-à-vis du milieu extérieur se fait généralement *de façon tardive* durant l'enfouissement. Ainsi le moment zéro (déclenchement du chronomètre glauconie) caractérise plutôt le dépôt du sédiment situé *au-dessus du niveau glauconieux* que le moment du dépôt des fossiles qui accompagnent les grains verts eux–mêmes. L'application de ces données à l'interprétation des âges mesurés dans les glauconies anciennes nécessite une étude sédimentologique détaillée de la série dont on souhaite connaître l'âge de dépôt.

(Manuscript received 10-3-1980)

Numerical Dating in Stratigraphy
Edited by G. S. Odin

15
Effect of pressure and temperature on clay mineral potassium–argon ages

GILLES S. ODIN

Knowledge of the geochemical behaviour of potassium–argon system during the burial of clay minerals is only in its preliminary stage. This chapter reviews the available data on argon and potassium behaviour in clay minerals which have been naturally or experimentally subjected to heating under pressure. The glauconitic minerals and illite mineral group will be considered separately.

The word *glaucony* will be used here as a facies name for green pellets of which there are several mineralogical components as defined by Velde and Odin (1975). This hopefully will avoid the constant confusion linked with the word 'glauconite', which unfortunately also designates one *particular* mineral component (rarely encountered) of the green pellets.

1 NATURAL EXAMPLES OF BURIAL OF GLAUCONIES

Generally, deeply buried glauconies are found in tectonically unstable areas which permit major rapid subsidence. Under these circumstances it becomes very problematic to distinguish between the effect of deep burial and effects related to the pressure and temperature developed during or after burial by tectonism. Radiometric age rejuvenation related to orogenic phenomena is well documented today. Evernden *et al.* (1961) have shown an example in the Liassic from Italy; Frey *et al.* (1973) and Conard *et al.* (this volume) have observed relative loss of argon in the Cretaceous and Tertiary from the Glaris Alps and Provence. Radiometric age lowering may be related to a recrystallization process well illustrated in the extreme case of the Glaris Alps, where incipient metamorphism is observed. Folinsbee *et al.* (1960) stressed that on the North American platform, Cambrian glauconies with 7.3–7.5% K, buried at 10,000 m, showed an apparent age characteristic of the Caledonian orogeny; Jurassic glauconies with 4.5–5.8% K showed an apparent age related to a mid-Cretaceous orogeny, while Albian glauconies with only 4% K showed an apparent age which might be related to the Laramide orogeny (latest Cretaceous–earliest Palaeocene). These observations seem to show that during tectonism the isotopic clock is reset to zero; but no other examples are available and we think it is difficult to imagine the possibility of dating tectonic events assuming a complete rejuvenation of a glaucony during tectonism.

Obradovich (1964) reported analytical results obtained from glauconies buried under varying depths of cover (Table 1). Rejuvenation is more effective at depths greater than 1750 m, but the samples studied by Obradovich were from California where Alpine orogeny was strongly developed. Thus tectonism may have helped the burial effect in this rejuvenation.

Apparent ages on three glauconitic levels taken from cores of holes from the ÖMV Company collected with M. F. Steininger and Mr Brix (NE Austria) are reported below. Sample 506-13, collected from a hole at Porrau, is stratigraphically dated as either basal Miocene or Turonian (similar glauconitic horizons exist in both formations in this basin). Sample 506-18, collected from a hole at Staatz, is regarded as Turonian. Sample G.517 (see NDS 86), collected from an outcrop, was first regarded as Turonian but is now recognized as basal Coniacian. All three samples were tectonized.

The analytical results obtained are shown in Table 2. The conclusions are (i) that the sample from Porrau cannot be of Miocene age; (ii) that the samples from the holes are clearly rejuvenated compared with sample from the surface outcrop. This rejuvenation may be related either to burial or to tectonic processes (especially higher temperature at more than 1000 m depth). In this example the combination of burial at more than 1500 m and

Table 1. K–Ar apparent ages of buried glauconies from California (after Obradovich, 1964). The sample JDO 5 collected from 1750 m shows no rejuvenation while below, at nearly 2500 m depth, the rejuvenation is significant. The different samples were collected from different boreholes.

Sample	Depth (m)	Potassium (% K)	Stratigraphy (Ma)	Apparent age (Ma, ICC)
JDO 5	1750	5.09	24–22	22.8
JDO 25	2470	5.15	65–53	36.0
JDO 26	2470	5.23	65–53	47.6
JDO 27	2490	3.70	65–53	46.0

Table 2. Apparent ages of glauconies from the Vienna Basin; the two samples from cores were questionably attributed to the Turonian or Early Miocene of identical facies. For G.517 A, see NDS 86.

Sample	Depth (m)	Potassium (% K)	Argon (nl/g)	Stratigraphy (Ma)	Apparent age (Ma, ICC)
G.517 A	Outcrop	5.95	20.57	87	86.8±3.3
G.506–13 A	1350	6.52	18.69	90–87 or 24–22	72.3±3.6
G.506–18 A	1950	6.65	21.80	90–87	82.4±4.1

of tectonism leads to a greater argon loss in the deeply buried sediment than in the presently exposed sample, which has, however, also suffered folding.

In New Zealand, an area folded during the Late Alpine times, Early Palaeogene glauconies were studied by Field and Odin (1981). Only the apparent ages obtained from the less deeply buried samples are compatible with the palaeontological data, suggesting that the loss of argon is more

Table 3. Apparents ages of glaucony from the Middle–Late Eocene from California. Except for KA 153 (Domengine fm, late Middle Eocene), all samples come from the basal Kreyenhagen level of middle Late Eocene age. Original dates (1) according to Evernden *et al.* (1960); (2) according to Evernden *et al.* (1961); (3) according to Obradovich (1964); (4) according to new measurements (italicized) on ultrasonically cleaned grains (Obradovich, 1964). All dates given as in the original paper. Differences between (1), (2) and (3) are only due to various approximations in calculation and are not significant; the main difference is between (1) and (2) on the one hand and (4) on the other. This shows that undetected analytical uncertainties occurred in the first study: cleaning of glaucony and measurement techniques. The resulting trend linked with burial is not as evident in column T (4) as in the previous obsolete values: T (1).

Sample KA	Depth (m)	T (1) (Ma)	K (2) %	T (2) (Ma)	K (3) (%)	T (3) (Ma)	K (4) (%)	T (4) (Ma)
189	Road cut	45	5.42	43	5.49	42.5	5.57	*34.3*
153	1400	44	5.82	42	5.82	41.1	5.81	*35.6*
285	1900	36	4.57	33	4.57	32.9	5.26	*34.9*
266	1900	—	—	—	4.96	*27.7*	5.09	*27.5*
1357	2050	—	—	—	4.87	*28.7*	—	—
154	2450	38	5.00	36	5.00	35.7	5.26	*34.2*
147	2500	—	5.44	31.5	5.44	31.4	6.27	*33.6*
265	2800	29	5.16	27	—	—	—	—
160	3300	26	5.83	25	—	—	—	—
267	3700	20	5.23	18	—	—	—	—

related to burial than to tectonism. This loss is closely linked to the crystallography of the glaucony. The two most rejuvenated samples contain 5.8 and 5.6% K. Burial of about 2000 m seems sufficient to lead to an important loss of argon (Field and Odin, 1981).

A frequently quoted example of lowering of apparent K–Ar glaucony ages as a result of burial is that of the Middle to Late Eocene Kreyenhagen formation of California (Evernden *et al.*, 1960). This example must be rediscussed in the light of more recent data obtained on the same samples. These new data were obtained by Obradovich (1964), using more careful preparations. Obradovich cleaned the previous samples of glaucony and obtained the results shown in Table 3. The ultrasonically treated glauconies are richer in potassium compared with untreated ones. The newly obtained apparent ages, quoted here in their original form (old decay constants), are clearly 'younger'. We conclude that the *first set of measurements was not representative of the correct apparent ages of the green grains.* The very clear trend shown by the earlier data is greatly modified by the more careful new measurements. Down to 2500 m depth there is *no* analytical difference between a sample taken from the surface outcrop and in the borehole.

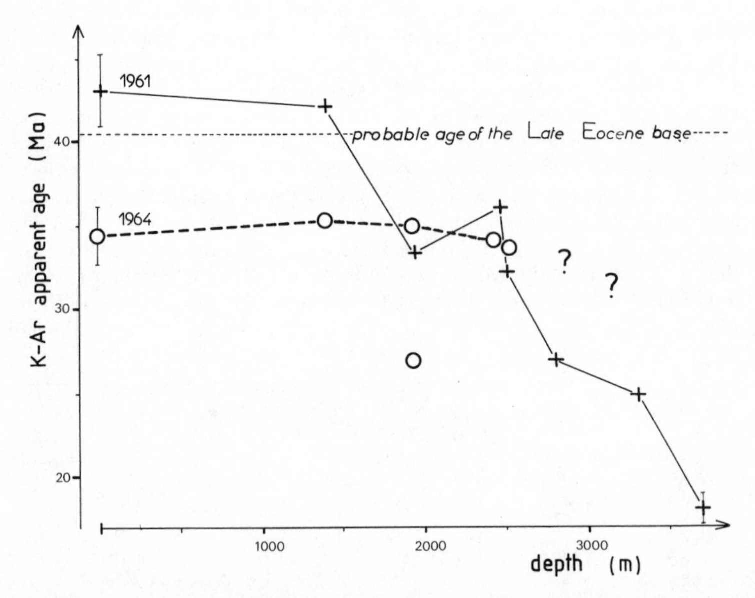

Figure 1. Apparent ages of glauconies from the Kreyenhagen fm according to Evernden *et al.* (1961) (+) and to Obradovich (1964) (○). The frequently cited 'demonstration' of Evernden *et al.* (1961), loss of argon from glauconies with depth of burial, is based on questionable analytical results. The data by Obradovich are considered more reliable. However, at depths greater than 2500 m it may be accepted that argon loss does occur.

Results on the deepest samples have not been redetermined, but we may assume that the very significant rejuvenation observed at 3700 m by Evernden *et al.* (1960) does exist. Figure 1 shows this situation; crosses are for the earlier published data, open circles for the corrected apparent ages.

According to the data of Obradovich, the apparent age of the green grains (35 Ma) is much younger than the accepted age of the Middle–Late Eocene (nearly 40 Ma). This apparent age remains more or less unchanged down to 2500 m and probably decreases at greater depth, perhaps as a result of more severe heating effects. It is concluded that this classical example of rejuvenation due to burial is a much less effective demonstration of the effect than claimed by the original authors.

In summary, few unequivocal examples are available concerning a rejuvenation of K–Ar ages as a result of burial alone. The *phenomenon of rejuvenation exists*, but we do not know at the moment if it is essentially due to burial, tectonic or thermodynamic effects.

2 NATURAL EXAMPLES OF BURIAL OF THE ILLITE–SMECTITE MINERAL SERIES

It should first be recalled that although the glauconitic mineral series and illite–smectite mineral series are of similar crystallographic nature, we do not know if the behaviours of these two series are identical as far as pressure–temperature stabilities are concerned. Velde and Odin (1975) and Odin *et al.* (1977) showed that the two mineral groups are probably quite different in pressure–temperature response. This may easily be understood when one remembers that glauconitic minerals are formed at low temperature and pressure. Illitization of a smectite, on the other hand, is the result of diagenesis under thousands of metres of sediment. From their respective genetic processes, the two mineral series are in equilibrium in completely different pressure and temperature environments when they form.

Perry (1974) studied the evolution of the illite–smectite mineral series in a deep hole cored in very thick, probably Miocene, sediments of the Gulf Coast Province. The smallest grain size fraction (less than 0.5 μm) will be considered here because it is probably the most sensitive to diagenetic changes. This fraction consists of interlayered illite–smectite. Perry estimated the proportion of expandable layers in 15 samples collected from nearly 1800–5800 m depth (Figure 2). The dioctahedral illite–smectite minerals are considered as recrystallized with enrichment in potassium indicated by the lowering of the proportion of expandable layers at greater depth. From Perry's samples, four were selected for argon isotopic measurements: Table 4. It can be seen that, in spite of a high proportion of illitic layers, the potassium content remains low. The apparent age calculated is an 'inherited age', clearly higher than the age of deposition estimated at

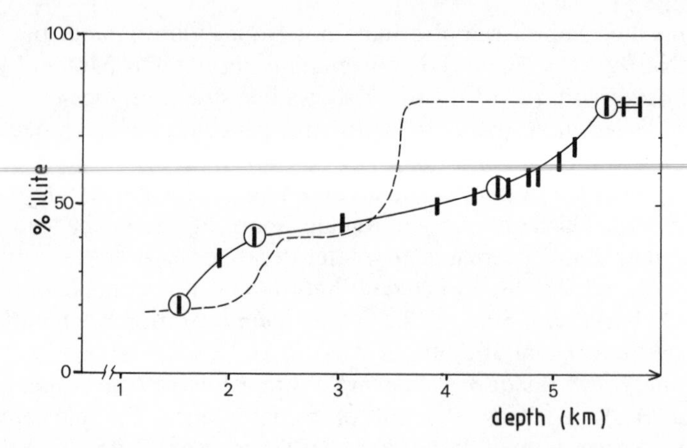

Figure 2. Example of diagenetic change of smectite–illite minerals investigated by Perry (1974). The K–Ar analysed samples are shown in open circles. The broken line shows the equivalent results obtained in another hole by Aronson and Hower.

10–20 Ma; the clay minerals are detrital. This apparent age is clearly rejuvenated at greater depth of burial. The author also observed a rejuvenation of the other grain-size fractions of the same samples and concluded that loss of argon took place during recrystallization, while the original potassium is held in the new mineral.

Aronson and Hower (1976) published a more convincing example. Their data were obtained from a 5500 m deep hole cored in Texas, as in the preceding case. The cored formations are 15–30 Ma old and the size fraction less than 0.1 μm contains more than 85% of interlayered illite–smectite. The potassium content increases from 1.7 to 4.2 (% K), while the proportion of expandable layers decreases from 80 to 20% in the fraction less than 0.1 μm (Figure 3). The illite-rich clay appears at nearly 3700 m depth and no further change occurs at greater depth. The argon content *increases also* with depth, but at a lower rate than the potassium content; as a result, the apparent age of the clay fraction diminishes with depth. Figure 3 shows that

Table 4. Apparent age of illite–smectite clays buried at various depths (recalculated after Perry, 1974).

Depth (m)	% illite layers	Potassium (% K)	Apparent age (Ma, ICC)
1580	20	2.22	168 ± 4
2230	40	2.66	166 ± 4
4420	55	2.38	140 ± 3
5530	80	3.10	111 ± 3

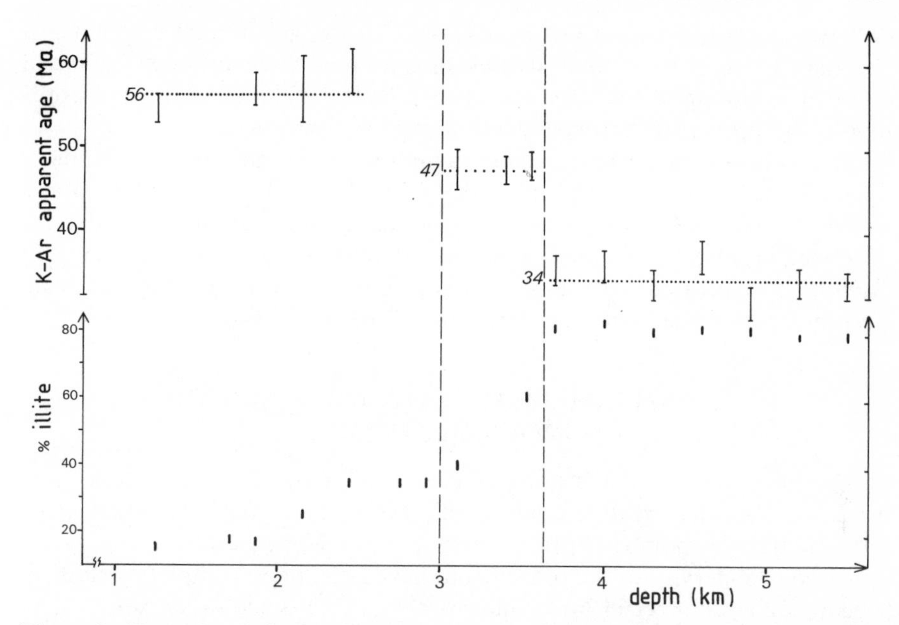

Figure 3. Example of diagenetic change of smectite–illite minerals investigated by Aronson and Hower (1976). The mineralogical changes (below) may be correlated with reequilibrium of the Ar content at two different moments of the evolution (2500–3000 and 3600 m depth). These breaks are more obvious than in Figure 2.

the change in age is *discontinuous*; one may identify three plateaux. The first one, distinct, shows an apparent age of 56 Ma (ICC) for the samples at less than 2450 m; the lowest one, very distinct, shows an apparent age of 34 Ma (ICC recalculated) for seven samples collected from 3700 to 5500 m depth; the intermediate one includes three samples collected from 3100 to 3550 m depth, and shows an apparent age of 47 Ma (ICC).

These data show that a series of steps in the diagenetic isotopic evolution better describes the pattern than a continous trend. There is good correlation between the mineralogical changes and the changes in apparent ages. These *evolutionary steps* should be contrasted with the *continuous change* in temperature and pressure with depth. In my view, the presence of these steps is proof of the association of argon loss with a recrystallization process. This process only occurs when the new thermodynamic conditions become incompatible with the pressure–temperature conditions of equilibrium of the clay minerals crystallized at higher levels. This contrasts markedly with the hypothesis of a progressive diffusion mechanism. At depths greater than 3700 m and down to 5500 m the newly formed minerals remain stable.

Aronson and Hower (1976) insist on a very clear discontinuity situated between 3550 and 3700 m depth (which corresponds to a pressure of

900 bar and a temperature of nearly 140°C if the geothermal gradient was normal). It seems to me that another discontinuity exists between 2450 and 3100 m according to the apparent ages, between 2800 and 3100 m according to the proportion of expandable layers.

In conclusion, the above examples indicate that various potassic dioctahedral clay minerals subjected to deep burial may show loss of argon at a depth of 2500–3000 m, though no argon loss is noted at shallower depths. The temperatures corresponding to these depths of burial have not yet been accurately estimated. The loss of argon appears to be more linked with *recrystallization* than with a continuous diffusion process.

3 EXPERIMENTAL DATA OBTAINED ON VARIOUS GLAUCONIES

The first experimental heating of glauconitic minerals under water pressure was reported by Evernden *et al.* (1960); more recently, Odin *et al.* (1977) have reported further experiments. Various high-temperature micaceous minerals, on the other hand, have been more completely studied: Fechtig and Kalbitzer (1966), Giletti (1974), Giletti and Anderson (1975), Purdy and Jäger (1976). These last studies may be helpful insofar as they sometimes concern the mineral phlogopite which, although formed in completely different conditions of pressure and temperature, curiously enough has shown behaviour similar to that of glauconitic minerals (Evernden *et al.*, 1960; Newman, 1969). In all the experiments, the loss of argon under water pressure has been observed to be less than in the experiments under vacuum at equivalent temperatures.

Concerning glauconitic minerals, experiments were performed in conventional hydrothermal equipment (Velde, 1969) over a one-month period using pelletal glauconies and distilled water sealed in gold capsules. Three glauconies (−86 m, −92 m and −97 m) from the Ypresian cored at Mont Cassel, France, were heated at 2 kbar pressure together with the Cambrian glaucony of the Franconia fm (Hower, 1961; Velde and Odin, 1975). In Figure 4, the potassium content of the green grains used is plotted against the temperature at which the experiments were run. The mineralogical results quoted in Table 5 were used to construct the phase diagram proposed in Figure 4.

The index of refraction measured (Table 5) is a function of the iron content of the observed minerals (Toler and Hower, 1959; Cimbalnikova, 1970; Velde, 1972); the (001) peak and its variation after glycolation indicate the nature of the X-rayed layers: smectite, interlayered, mica; the position of the (060) peak indicates if the structure is dioctahedral (two trivalent ions per unit cell) or trioctahedral (three bivalent ions per unit cell). According to these criteria, the initially dioctahedral glauconitic minerals are

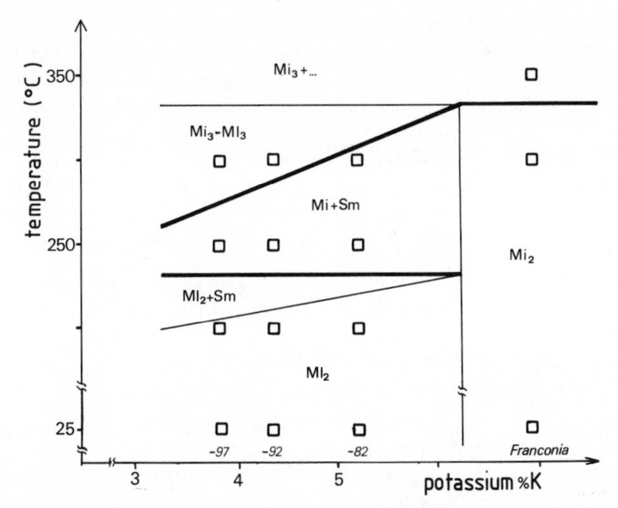

Figure 4. Mineralogical changes as a function of temperature: an experiment on four glauconies of varying mineralogical composition (data from Velde and Odin, 1975, reproduced by permission of the Clay Minerals Society).

enriched in Fe^{++} when temperature increases. The experiments reported must be taken only as an indication of the assemblages which will be produced in nature when the glauconitic minerals of the green pellets experience the initial phase of metamorphism.

In conclusion, the experiments show that the glauconitic minerals pass through different crystallographic stages when they are subjected to hydrothermal activity. A second series of experiments on other glauconies has shown the effect on argon retention (Odin *et al.*, 1977).

Five samples of pelletal glauconies from the Albo-Aptian of the Paris Basin were used with the standard mineral *glauconite GL-O*. The range of K contents was from 6.6 to 4.1% K; the glauconitic minerals were micaceous to 30% expandable in their mixed layered component. The main hydrothermal experiments were performed in rod bomb type autoclaves at 2 kbar pressure as above for a duration of one month. Other durations were also tested. The argon proportion remaining in the solid phase after each experiment is shown in Figure 5.

In Figure 5, essentially five regions are pertinent: that where glauconitic mica exists (Mi), that where various amounts of mixed-layered glauconitic minerals exist (Ml_2), that where glauconitic mica coexists with a smectitic trioctahedral phase ($Mi+Sm_3$), that where glauconitic mica and a mixed-layered chlorite–saponite phase coexist ($Mi+Ml_3$), and finally that where biotite, feldspar, quartz and oxides coexist in the region above glauconitic mica stability. This last region represents total recrystallization and reconstitution of the mineral assemblage used as starting materials. The sample

Table 5. Results of experiments on four glauconies (after Velde and Odin, 1975). The green pellets were heated under 2 kbar water pressure and the solid phase analysed after one month (see interpretation and other data in Figure 4); data from Velde and Odin (1975) reproduced by permission of the Clay Minerals Society.

Temperature (°C)	*n*	(060) Å	(001) Å air-dried	(001) Å glycolated
97 m				
25	1.590	1.514	11.5	10
200	1.598	1.513	10.6+14 (?)	10+18?
250	1.604	1.517	10.6+14 (?)	10
300	1.607	1.519	10.5	9.8+11.4
92 m				
25	1.600	1.515	11.0	10+18
200	1.602	1.514	10.6	10+18?
250	1.606	1.516	10.6	10+slope
300	1.612	1.518	10.5	10+11.4
82 m				
25	1.602	1.519	11.0	10
200	1.604	1.519	10.8	10+slope
250	1.606	1.515	10.8	10+slope
300	1.611	1.521	10.4+14.9	10+16.7
Franconia				
25	1.624	1.517	10	10
250	1.624	1.517	10	10
300	1.626	1.519	10	10
350	No green mica	1.541	10	10

GL-O, composed of the glauconitic mica, can be used to indicate the Ar loss when the mineral subjected to hydrothermal treatment does not recrystallize. We see that a difference of 100°C in the experiments decreases the Ar content of this sample by 5%. As a first approximation, one can attribute most of this Ar loss to diffusion. Further, in the experiments at 200°C, all of the samples show about the same Ar loss when they show no apparent recrystallization, i.e. no new phases appear in X-ray diagrams or under the microscope. Again one can most probably attribute the 2–4% Ar loss in the samples to diffusion. We might reasonably claim that the reproducibility of Ar measurements is better than 2%; the given numbers are thus significant as a whole. Comparison of little recrystallized glauconitic mica (GL-O) and less micaceous (lower potassium content) glauconitic minerals for run temperatures of 250 and 300°C indicates that the appearance of new phases decreases the amount of ^{40}Ar in the solids to a much greater extent than ^{40}Ar loss through diffusion or other processes.

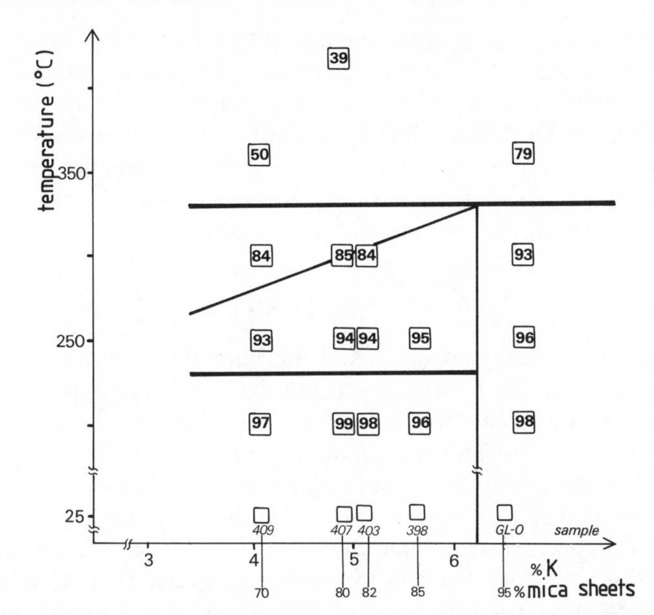

Figure 5. Argon loss from mineralogically diverse glauconies as a function of temperature. The percentage of remaining argon is shown for five temperatures (200, 250, 300, 350, 410°C) for 1 month at 2 kb pressure. The numbers are to be considered with an analytical uncertainty of ±2%.

When glauconitic mica is no longer stable (>300°C), the amount of Ar retained in the solids is much less than in the other experiments. It also appears that the amount of ^{40}Ar present is a function of the bulk composition of the charge. However, a surprising amount of ^{40}Ar is still present in the new phases, suggesting some sort of partitioning between solids and fluids.

It is possible to explain the variation in Ar content of the solids which are mineralogically reconstituted under the experimental conditions in two ways.

(1) Ar diffuses from the solids. Recrystallization is effected by ionic migration of certain elements in the phyllosilicate structure. These mineral segregations produce new phases in individual grains which are seen under the microscope (Velde and Odin, 1975). Ar is then mobile, as are other ions—notably K, Fe and Si. There is, however, no relation between ^{40}Ar retained and the amount of mica present.

(2) Ar is involved in an equilibrium between solid and liquid when minerals are precipitated from solution. During a recrystallization process of the type: mixed-layered (mica–smectite)→mica + expandable saponitic mineral, all elements pass into ionic complexes in solution and precipitate new

phyllosilicates. Oxygen isotope studies (O'Neil and Kharaka, 1976; B. Velde, unpublished data) indicate that mineral recrystallization under hydrothermal treatment is effected by transfer of polymeric silicate units which are smaller than the original phyllosilicates but larger than monomeric silica units. There is no way to determine whether Ar remains in these units during the process or whether it enters the fluid phase. The latter possibility is more probable in the present author's opinion.

4 CONCLUSION

The minerals of the glauconies and the illite–smectite mineral series are crystallographically similar but genetically different. They show a rejuvenation of their K–Ar apparent ages when naturally or experimentally subjected to conditions of burial. This rejuvenation is not just a diffusion process. The available data show that the main phenomenon is a step-by-step argon loss. These steps may be related to recrystallization processes as shown by mineralogical analyses. Although these recrystallizations sometimes proceed to completion, this does not lead to a complete evacuation of the radiogenic argon (a complete reset to zero of the K–Ar clock) except in extreme conditions. Even at more than 400°C, even at present depths of more than 5000 m, the newly formed minerals always retain some radiogenic argon. The experiments also show that the *glauconitic minerals with more than 6% (K) are more stable* and retain their argon better than those poor in potassium. However, if they have been subjected to conditions involving a temperature higher than 200°C these potassium-rich glauconitic minerals will lose argon; their apparent age will become a minimum for the time of their formation.

Acknowledgements

The comments of B. Velde on the first draft of this chapter were helpful. I am indebted to the publishers for the improvement of the English version. This is a contribution to the IGCP Project 133.

Résumé

Les minéraux glauconitiques et la famille minérale diagénétique: smectite–illite sont cristallographiquement semblables mais génétiquement et minéralogiquement distincts. Ces deux familles montrent un comportement voisin lorsqu'elles sont naturellement ou artificiellement soumises à des conditions de température et de pression reproduisant l'enfouissement profond: les âges apparents K–Ar sont rajeunis. Ce rajeunissement n'est pas lié à un simple phénomène de diffusion continu; au contraire des *étapes d'évolution* sont clairement mises en évidence. Les départs d'argon sont liés à des *réorganisations cristallines* déclenchées au-delà des seuils de stabilité des minéraux préexistant. Au cours des recristallisations complètes qui ont pu être observées, l'argon radiogénique n'est pas extrait dans sa totalité: le chronomètre n'est pas remis à zéro mais est seulement rajeuni. Les expériences montrent encore que les minéraux glauconitiques les plus évolués sont plus stables et retiennent mieux leur argon à température élevée que les minéraux moins évolués plus pauvres en potassium. Ces derniers sont donc moins dignes de confiance pour évaluer l'âge d'une roche.

(Manuscript received 1-10-1980)

Numerical Dating in Stratigraphy
Edited by G. S. Odin

16
Potassium–argon dating of tectonized glauconies

MONIQUE CONARD, HANS KREUZER and
GILLES S. ODIN

1 INTRODUCTION

The rejuvenation of K–Ar apparent ages has frequently been assigned to thermodynamic events which were thought to have caused an argon loss. The burial factor is the most frequently quoted. In epicontinental basins of Europe, sedimentary series of sufficient thickness are unusual; on the other hand, the tectonization of glauconitic horizons is not uncommon.

Evernden *et al.* (1961) quoted the example of a sample of Late Liassic of North Italy. This area has been deformed during the Alpine orogeny. The potassium content of the dated fraction is 5.4(% K); the apparent age (ICC recalculated), 93 Ma, 'appears to date the orogeny rather than the time of deposition' of glaucony according to these authors. (The word *glaucony* was defined by Odin and Matter (1981) as a facies name.) As far as we are concerned, in the absence of accurate data on the tectonic history of this particular area, it is more prudent to conclude that noticeable rejuvenation has occurred.

Frey *et al.* (1973) investigated the behaviour of glaucony-bearing formations of Cretaceous and Tertiary age from the Glaris Alps (West Switzerland) by K–Ar dating and petrographic studies. They showed that the apparent ages of glauconies regularly decrease when approaching the zone of metamorphic reactions which transform glauconitic minerals to stilpnomelane. Only in this case can any idea of the probable age of tectonic or metamorphic events be deduced.

To illustrate in more detail the behaviour of glauconies during restricted deformation processes we studied glauconies collected from the subAlpine chains in Haute-Provence (SE France). This area had been investigated in detail for stratigraphic, sedimentological and tectonic purposes in recent years by, among others, Cotillon (1968) and one of us (M.C.). The series is characterized by alternating marls and limestones with several glauconitic horizons generally

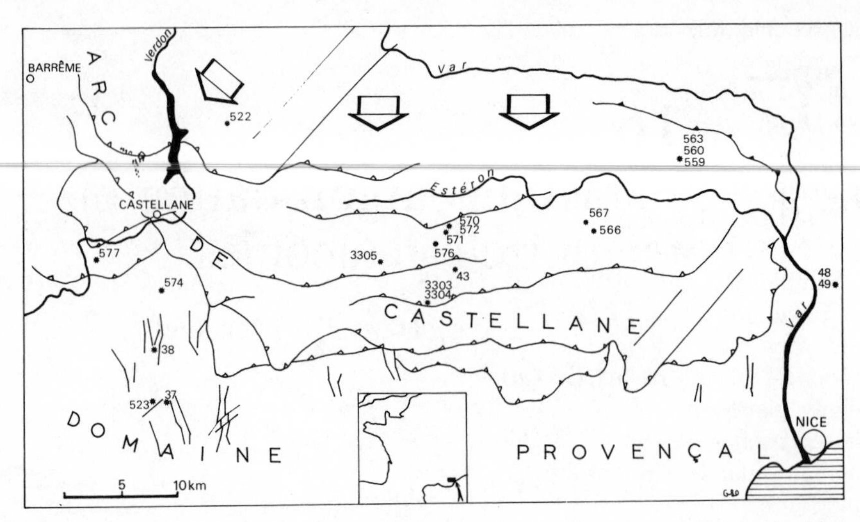

Figure 1. Situation of the dated samples. An important thrust due to the Alpine orogenesis led to EW foldings, overthrusts to the south and transverse faults.

well marked in relation to the biostratigraphic sequence. Thirty-five samples were collected in the field, of which 13 were subjected to K–Ar investigations. Nine come from the Estéron river valley and four from the area near Castellane (Figure 1). Among other interesting aspects from the geological point of view, this area is well correlated to the type section of the Barremian stage at Barrême. In addition, glauconitic samples from this area had already been investigated for dating purposes by Hurley *et al.* (1960), Obradovich (1964), Bodelle *et al.* (1969) and Bonhomme *et al.* (1969).

2 STRATIGRAPHIC DATA

The macrofauna (ammonites, inocerams) or the microfauna (ostracods, foraminifers) collected in the Cretaceous strata allow a regional biostratigraphic scale to be established. Ammonites are not uncommon in the Early Cretaceous (Cotillon, 1968), while globotruncanids were used for the Late Cretaceous (Conard, 1978; 1979). The glauconitic levels were dated as follows.

Early Cretaceous samples were investigated and correlated according to the work of Cotillon (1968), where most of the outcrops are described. Sample 523 is correlated with the Middle Hauterivian by *Pseudothurmannia cruasensis*. With sample 577 the collector (P. Péneaud) found numerous echinoids: *Toxaster lorioli, T. aff. gibbus.* Belemnites (*Duvalia*) are not uncommon. This corresponds with level 13 of Cotillon (1968, p. 95), also

Middle Hauterivian. Sample 522 gave ammonites characteristic of *Pervinquiera inflata* and *Stoliczkaia dispar* zones of the Late Albian. Sample 566 is situated at the base of the Late Cenomanian as defined by the appearance of *Rotalipora deeckeï* and *Whiteinella baltica* at this level. Sample 567 is above a horizon with *Praeglobotruncana helvetica* of Early Turonian age and below a formation with *Dicarinella imbricata, D. primitiva* and *Marginotruncana renzi* of Late Turonian age. In a Turonian series, the sample 572 lies below marls with *D. primitiva* and *D. concavata*, both of Coniacian age.

Sample 570 and, 30 m above, sample 571, come from marls with fauna characteristic of the *D. concavata* zone, Late Coniacian in age. Sample 559 and, 5 m above, 560, are bracketed by samples in which some globotruncanids of the *D. asymetrica* zone, Santonian in age, were discovered. The highly glauconitic sample 563, rich in brachiopods and molluscs (*Neithea quadricostata*), lies above biodetrital limestone which can be correlated with the Santonian because of the occurrence of *D. asymetrica*, but a hard ground separates this recognized Santonian from our glauconitic horizon. The sediments lying above are detrital and do not contain useful fossils. 563 is post-Santonian and pre-Palaeogene.

The biozonation established for the Tethyan area, parallel with the

Table 1. Location and stratigraphy of the selected glauconitic samples. 522 and 523 collected by P. Cotillon, 577 by P. Péneaud, the others by the authors (M.C. and G.S.O). The three problems studied are separated into groups: top, stratigraphical problem; middle, problem of argon diffusion due to tectonics; bottom, problem of calibration of the Early Cretaceous time scale.

Sample	Locality	Lambert coordinates x	Lambert coordinates y	Stratigraphy
563	Vieux-Pierrefeu	981.2	186.1	Pre-Palaeogene, post-Santonian
560	Vieux-Pierrefeu	981.4	186.1	Santonian
559	Vieux-Pierrefeu	981.4	186.1	Santonian
571	Col de Pinpinnier	960.7	180.9	Late Coniacian
570	Col de Pinpinnier	960.8	180.9	Late Coniacian
572	Col de Pinpinnier	960.2	180.4	Late Turonian
567	Le Pous	973.8	182.8	Late Turonian
566	Vallon du Blayoux	976.4	182.4	Late Cenomanian
522	Vergons	942.2	188.2	Late Albian
574	Le Bourguet	936.4	174.4	Late Barremian
576	Le Pré d'Oule	957.7	179.0	Late Barremian
523	Col de la Craou	935.8	166.8	Middle Hauterivian
577	Taloire	930.5	187.8	Middle Hauterivian

biozonation of ammonites of the same area (Porthault, 1974), was correlated with the boreal biostratigraphy (collaborational, 1979). The locality and the stratigraphical age are given in Table 1.

3 GENETIC UNCERTAINTIES

All the samples are from a shelf environment with a more neritic facies to the south passing progressively to open marine facies to the north.

The Early Cretaceous samples are from the neritic province according to P. Cotillon. Berriasian levels are essentially carbonates. The Valanginian consists of marls with benthonic faunas and bioclastic limestones. During Hauterivian times, blue-grey marls were deposited, sometimes very glauconitic, which became progressively more calcareous. Barremian sediments are limestones, often rich in belemnites, with thin horizons of limited glaucony content. The undifferentiated Aptian–Albian is generally marly, with here and there enrichment in detrital grains or glaucony.

The Late Cretaceous often begins in the Estéron valley with glauconitic marls, Cenomanian in age. Limestone banks with detrital grains or glaucony are common in Late Cenomanian deposits. Turonian deposits are calcareous, with thin interlayers of marl and sometimes abundant macrofauna: bivalves (*Exogyra*) and gastropods. Coniacian and Santonian deposits are characterized by alternating marls and limestones with some very glaucony-rich layers. The Late Cretaceous is not recognizable and the Senonian series is overlapped by Palaeogene formations.

In the most neritic province (to the south), some glauconitic horizons are associated with interruptions of sedimentation: for example, sample 523 of the Middle Hauterivian lies on Valanginian bioclastic limestones; samples 574 and 576 are taken from minor discontinuities of deposition in Barremian limestones. The highly glauconitic marls (sample 522) of the Albian are deposited after the major post-Barremian discontinuity. No interruption of deposition can be seen in the Late Turonian (samples 567–572) nor in the Late Cenomanian (566), but these glauconitic samples seem to be linked with the deepening of the sea. The glauconitic accumulation of Coniacian and Santonian (570, 571, 559, 560) would be better explained by dynamic processes: they are probably perigenic.

Many of the glauconies sampled had sufficient time to evolve during the interruptions of deposition mentioned above. The mineralogical studies on powder diffractograms confirm that all the grains contain evolved to highly evolved glauconitic minerals (Figure 2). The glaucony G.522 is the least evolved, but inheritance from the initial substrate cannot be assumed. Due to the evolution it has been impossible to recognize with certainty any of the initial substrates for glauconitization.

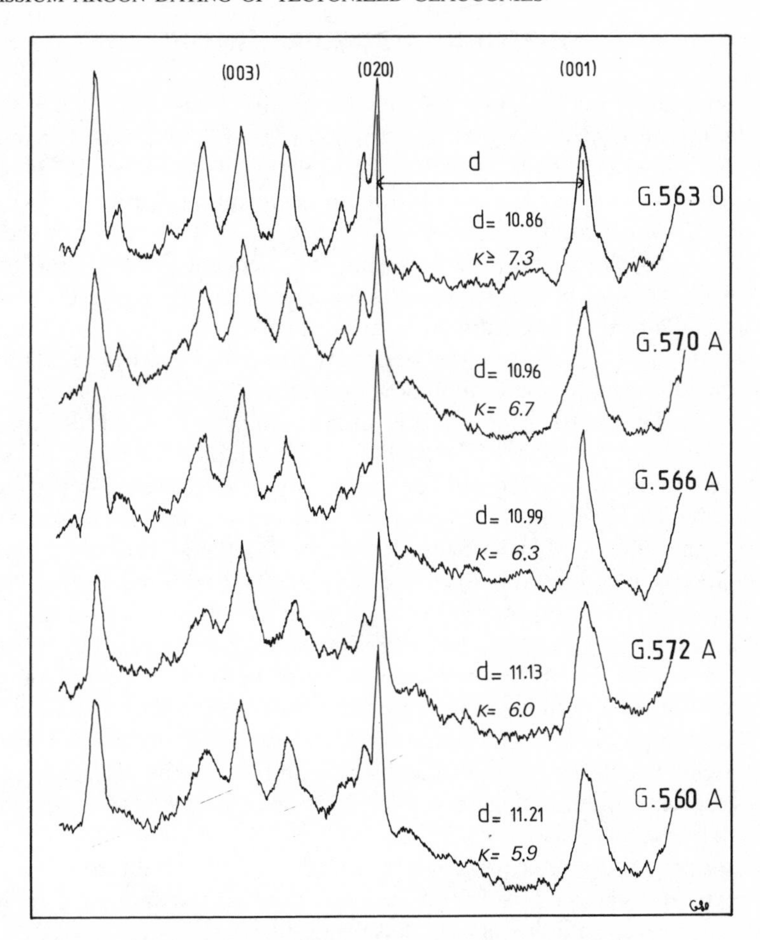

Figure 2. X-ray diffractograms on powdered glauconies from SE France: the potassium content (K%) is quoted together with the characteristic relative distance in centimetres between the mobile (001) and the stable (020) diffraction peaks. The glaucony G.563 O is not the one which was dated but a coarser size fraction (1.0–0.5 mm), which gave the best diffractogram we know for Mesozoic and Tertiary glauconies.

From observation of the grains, no sample should be considered as reworked. The only one which shows clear evidence of transport is G.563. Like various others collected from the same horizon, this glaucony seems to have had a very long genetic history and had probably suffered a period of low sea-level during which some grains were oxidized before burial. These very dense, very dark highly-evolved grains display the most perfect crystallographic structure we have ever seen in glauconies from the Mesozoic and Tertiary.

4 HISTORICAL UNCERTAINTIES

The sampled area belongs to the subAlpine Meridional Chains, the external, less tectonized zone of the Alpine orogeny. This area, corresponding to the eastern part of the Arc de Castellane, is characterized by E–W folds and overthrusts to the south in association with dextral and sinistral transverse faults. These events are the result of N–S compressive movements, developed essentially during the Neogene. In the area to the south of Castellane, bordering the Provençal region, the Cretaceous series (Berriasian–Turonian) is cut by N–S faults. There is no significant burial of all these deposits, which are not generally very thick and do not display recrystallization related to a metamorphic activity.

The glauconitic horizons, especially in limestones, are sometimes tectonically preferred slide zones. We tried to take samples far away from the recognized faults or sliding and no grains were identified as crushed by tectonic effects; however, all grains have suffered tectonic pressures.

Concerning the weathering of the outcrops, the Late Cretaceous samples were collected after elimination of the superfically weathered surface. However, traces of oxidation may be seen in some separated glauconies, especially G.563, where this is due to specific conditions of genesis. As far as the samples from the Early Cretaceous are concerned, we collected the best available glaucony, often trapped betwen two very indurated limestone beds. The sample 574 was obtained more or less thanks to the pedogenesis of the rocks, which dissolved the carbonate and led to the release of green grains from the indurated glauconitic limestone. Indeed, this pedogenesis altered the glaucony also.

In summary, the numerous glauconitic levels of the Cretaceous series of SE France offer various possibilities of application of the dating methods for considering the specific history of these outcrops.

The first is the testing of the possibility of isotopic exchanges due to tectonization processes: the samples 522, 566, 567, 572, 570, 571, 559, 560 are well related to the stratigraphic column and we know their probable numerical ages thanks to a good calibration of this part of the time scale.

The second possibility is regional and concerns the problem of the exact age of the post-Santonian pre-Palaeogene sediments (sample 563) which had already been studied (Bodelle *et al.*, 1969a).

The third is related to the calibration of the Early Cretaceous time scale in the area where type sections are located: samples 577, 523, 576, 574. Several earlier studies or analyses have already been devoted to this question (Hurley *et al.*, 1960; Obradovich, 1964; Dodson *et al.*, 1964; Bonhomme *et al.*, 1969) but taking into account the more detailed geological study of this area available today, we will report better documented results.

5 ANALYTICAL UNCERTAINTIES

The samples were washed and sieved. The glauconies from the 160–500 μm fractions (except sample 577:100–500 μm) were separated and then fractionated using a magnetic separator to select the most evolved part. All glauconies were ultrasonically cleaned twice, first for ten minutes in 0.1 N acetic acid solution then in deionized water. After complete washing on a 160 μm sieve, the samples were dried overnight at 60°C and then stored in polyethylene bags.

The potassium content was measured by means of flame photometry and argon by isotope dilution. The analytical results are shown in Table 2. The argon analyses were performed in different laboratories. An intercalibration was done with the reference material 'glauconite GL-O'. The three lines used gave mean results diverging by less than 3‰ from the recommended value, 24.8 nl/g of radiogenic argon, so no correction was applied.

The 2σ uncertainties are given for analytical data as well as for the calculated apparent age. The constants used are those recommended since

Table 2. Analytical results of radiometric dating of glauconies of SE France; 2σ uncertainties are given. Potassium data are from Miss M. Lenoble (Paris); samples measured twice are indicated by an asterisk. Argon data from G.S.O. in BGR Hanover (H), G.M.I. Berne (B) and the University of Leeds (L).

Dated glaucony	Potassium (K%)	Argon (% rad)	Argon rad. Ar(nl/g)	Apparent age (Ma, ICC)
G.563 A	*7.28±0.10	89.3 H	20.50±0.26	71.0±1.2
		89.2 H	20.45±0.30	
G.560 A	*5.93±0.12	85.2 H	18.85±0.27	79.9±1.8
		84.6 H	18.80±0.24	
G.559 A	6.06±0.18	83.9 H	18.75±0.23	77.9±2.5
G.571 A	*6.03±0.10	85.3 H	19.09±0.24	79.7±1.7
G.570 A	6.74±0.18	88.8 H	21.37±0.26	79.6±2.2
		89.1 H	21.29±0.26	
G.572 A	*6.03±0.15	36.0 L	19.79±0.50	82.5±2.9
G.567 A	6.05±0.15	89.2 H	20.47±0.26	85.0±2.2
		88.8 H	20.45±0.25	
G.566 A	6.31±0.19	85.0 H	21.04±0.26	83.8±2.6
		87.2 H	21.05±0.43	
G.522 A	5.72±0.17	84.1 B	19.43±0.50	85.4±3.3
G.574 A	6.45±0.19	89.8 H	26.59±0.32	103.4±3.2
		90.4 H	26.78±0.32	
G.576 A	6.33±0.19	91.4 H	28.08±0.32	110.7±3.2
G.523 A	*6.96±0.14	91.3 B	31.74±0.66	113.7±3.3
G.577 AB	*7.02±0.14	92.4 H	32.96±0.41	117.0±2.5
		92.3 H	32.99±0.39	

the Congress of Sydney by the IUGS Subcommission of Geochronology (abbrev.: ICC).

6 DISCUSSION AND CONCLUSIONS

Concerning the first problem, loss of argon due to tectonization, we may compare the apparent ages obtained here with the similar results obtained on glauconies from quiet epicontinental basins from Europe. The Albian–Cenomanian boundary may be dated at 95 Ma according to Juignet *et al.* (1975). The Cenomanian–Turonian boundary was estimated near 90 Ma by Odin (1978a), the Turonian–Coniacian boundary may be accepted at 87–88 Ma according to Keppens *et al.* (1978b), and the Campanian–Maastrichtian boundary is probably near 72 Ma if we use the data and discussions by Priem *et al.* (1975b; 1976) and Odin (1976b).

No data are published on the age of horizons between the Middle Coniacian and Middle Campanian in Europe, but most authors agree in estimating a short duration of the Coniacian and Santonian stages. We accept here that the Santonian stage is between 83 and 86 Ma. Table 3 gives the results of the calculation of argon loss for the eight tectonized glauconies G.522 A–G.560 A. Individually, these results must be quoted with caution, but in general it is clear that all glauconies are rejuvenated.

The mean rejuvenation is of the order of $7.5 \pm 2\%$. It is perhaps not by chance that the glaucony poorest in K, sample G.522 A, shows the largest decrease of apparent age. But this sample is also one of the closest to the main Alpine deformation zone.

Concluding, we have demonstrated that in the absence of deep burial, and in the absence of faulting or overthrusting in the immediate vicinity (but where the formations are folded), *tectonic activity leads to a noticeable rejuvenation of the K–Ar apparent ages of glauconies.* The probable loss of argon remains lower than 10% of the argon present, so that the glaucony cannot be used for the dating of the tectonic event.

Concerning the probable age of deposition of the glauconitic levels above the Santonian of Vieux-Pierrefeu, our sample 563 probably corresponds to the sample number 9 in the paper by Bodelle *et al.* (1969a). The ICC recalculated age of sample 9 is 71.6 ± 3.0, identical to our own result. Taking into account a mean loss of argon of 7% for the eight samples of the Late Cretaceous as well as for the two other samples of the same outcrop, we may calculate with some confidence a 'corrected' age of 76.6 for the post-Santonian glauconitization. By this means, we could accept that the glaucony G.563 A completed its evolution nearly 76.5 ± 1.5 Ma ago. From the long time required for the evolution of this glaucony and its marine alteration after genesis, we may conclude that the sea was present for a minimum of 1–2 Ma during the Late Campanian in this area.

Table 3. Tentative estimates of argon losses due to the history of the glauconies (comment as for Table 1). The estimated ages are according to the time scale proposed in Kennedy and Odin, this volume. $A \pm \Delta A$ is the apparent age; $E \pm \Delta E$ is the estimated age; $\Delta\%$ argon loss is $\{(\Delta E/E)^2 + (\Delta D/D)^2\}^{1/2} \times 100$.

Glaucony	Known history	Estimated age	Apparent age	% argon loss
G.563 A	Vertical dip Marine oxidation	Less than 84	69.8–72.2	
G.560 A	Vertical dip	85±2	78.1–81.7	6.0±3.2
G.559 A	Vertical dip	85±2	75.4–80.4	8.4±4.0
G.571 A	Friction and slip	86±1	78.0–81.4	7.3±2.4
G.570 A	Strong dip	86±1	77.4–81.8	7.4±3.0
G.572 A	Strong dip	88±1	79.6–85.4	6.3±4.0
G.567 A	Strong dip	88±1	82.8–87.2	3.4±2.8
G.566 A	Low dip	91±1	81.2–85.4	7.9±3.3
	Weathering		;	
G.522 A	Folding	97±2	82.1–88.7	12.0±4.4
G.574 A	Friction–folding			
	Pedogenesis	112±2	100.2–106.6	7.7±3.6
G.576 A	Friction		106.1–114.4	
	Weathering			
G.523 A	Cambering		110.4–117.0	
G.577 AB	Flexure		115.5–119.5	

Concerning the Early Cretaceous, we shall examine the different uncertainties which affect the apparent ages calculated. The conditions of genesis appear favourable for a good correspondence of the Ar/K ratio with the exact time of deposition: all glauconitic minerals are evolved to highly-evolved and the isotopic zero age has been reached before burial. The cessation of isotopic exchanges with sea-water occurred in all cases at the beginning of the burial, or perhaps a little time before for the highly-evolved glaucony G.577 AB.

The uncertainties arising from the subsequent history are less favourable. We have demonstrated a mean argon loss of 7% for tectonized samples of the Late Cretaceous. This loss may be less important for the Early Cretaceous: the mean potassium content is higher in the oldest glauconies, 6.7% K instead of 6.1% K (Table 2); the Early Cretaceous sampled outcrops are situated further from the centre of the tectonized area (Figure 1). According to these data we will add to the analytical uncertainty an historical uncertainty of +5%, which represents a possible but not a certain rejuvenation. Concerning the Late Barremian outcrops, G.567 A was weathered,

Table 4. Radiometric ages of glauconies from SE France in the literature.

Sample	Stratigraphy	Potassium (K%)	Apparent age (Ma, ICC)	Authors
3305	Late Aptian–Early Cenomanian	5.44	89	Hurley *et al.* (1960)
3303	Valanginian–Hauterivian (?)	5.71	90	Hurley *et al.* (1960)
3304	Valanginian–Hauterivian (?)	5.85	107±8	Hurley *et al.* (1960)
49	Hauterivian	(N) 6.65	112.5±5.5	Obradovich (1964)
48	Hauterivian	(US) 6.96	116±6	Obradovich (1964)
	1 m above 49	(N) 6.83	117±6	Obradovich (1964)
43	Vraconian (Late Albian)	No datum	86.3±4.5	Bonhomme *et al.* (1969)
38	Barremian	No datum	115.0±6	Bonhomme *et al.* (1969)
37	Hauterivian	No datum	104.8±5	Bonhomme *et al.* (1969)

G.574 A was a soil; in spite of a high potassium content, the apparent age of the last sample seems to have been altered during this last stage of its history Retrospectively, the apparent age of G.574 A appears rejuvenated: it is much younger than G.576 A of the same age; it is much younger than a basal Barremian glaucony from England (s. 8011 of Dodson *et al.*, 1964: ICC recalculated 112 Ma); it is younger than glauconies from the Early Albian of the type area in France (Odin, 1979).

The present results may be compared with results in the literature obtained in the same area. Hurley *et al.* (1960) measured three samples, which appear fairly badly stratigraphically allocated (Table 4). All their glauconies are potassium-poor compared with our own ones, and all samples are probably rejuvenated more than the ones presented above. Obradovich (1964) also measured two glauconies of the Hauterivian. His results obtained on the bulk glaucony (N) or after ultrasonic treatment (US) are equivalent to our own, and the strata also appear to be only little tectonized (less than the Late Cretaceous samples). Bonhomme *et al.* (1969) reported Rb–Sr data on three samples equivalent to our own ones. These authors gave a complex interpretation of the results and concluded that there is the possibility of isotopic exchanges of Rb and Sr during Cretaceous times after burial. However, if we calculate conventional apparent ages using new constants and the probable isotopic strontium ratio of the sea-water during the Early Cretaceous, two apparent ages (glauconies 38 and 43) are equivalent to those of our own G.576 A. G.522 A, the Albian sample, is rejuvenated as our Albian one and the Barremian result is not far out. The Hauterivian Rb–Sr age is younger than the K–Ar ages obtained here. As a whole, the various results above agree well with the conclusions obtained from our samples.

7 SUMMARY

The comparison between K–Ar apparent ages obtained on Late Cretaceous glauconies from a tectonized area of SE France and glauconies of quiet sedimentary basins of NW Europe shows that the deformations of Alpine origin lead to isotopic rejuvenation. The mean rejuvenation reaches 7% in this external part of the Alpine area for a thrusting which essentially occurred between probably 30 and 15 Ma ago but was terminated only 5 Ma ago in some parts of the sampled unit, the Arc de Castellane. The last glauconitic and marine deposits of the Late Cretaceous initially assigned to a post-Santonian sea may be dated as Late Campanian, 76.5 ± 1.5 Ma ago.

The Early Cretaceous glauconitic horizons provide welcome age data to add to the few so far available on this section of the time scale. However, the apparent ages obtained must be used with caution, as minimum values. Maximum values may be obtained by taking into account a possible rejuvenation of 5% due to tectonic phenomena. One Late Barremian glaucony yielded an age of deposition of $111^{+9}_{-3.5}$ Ma, two Middle Hauterivian glauconies an age of 115^{+9}_{-3} Ma, which agrees well with results obtained 15 years ago in the Mont Chauve d'Aspremont, north of Nice.

Acknowledgements

The authors are indebted to the publishers for improvement of the English version of this chapter. The last author of the chapter is greatly indebted to Professor Wendt, who gave him permission for a long stay in his laboratory. The recommendations of all colleagues in the BGR were also greatly appreciated. Grateful thanks are also due to Hans Kreuzer's family for their unlimited hospitality. This chapter is a joint contribution to IGCP Projects 89/133.

Résumé

La comparaison entre les âges apparents K–Ar de glauconies du Crétacé récent de Provence et ceux des glauconies récoltées dans les provinces non tectonisées du NO de l'Europe montre que *la tectonique alpine a provoqué un rajeunissement* de l'ordre de 7% de l'âge actuel. Ceci a été observé même pour les échantillons éloignés d'un charriage ou d'une zone faillée mais dans des séries évidemment légèrement déformées. La glauconie la moins évoluée paraît la plus affectée.

On montre que la dernière incursion marine ayant permis la formation de glauconie à la fin du Crétacé peut être rapportée au Campanien récent dans cette région.

Les données obtenues sur le Crétacé ancien de Provence sont parmi les rares disponibles actuellement pour illustrer cette période de temps; mais l'existence des déformations alpines ne permet pas de considérer ces âges apparents comme sûrement représentatifs de l'âge de dépôt. Les âges obtenus sur le Barrémien et l'Hauterivien confirment analytiquement les âges proposés voici 15 ans par Obradovich sur des glauconies récoltées au Nord de Nice.

(Manuscript received 30-5-1980)

Numerical Dating in Stratigraphy
Edited by G. S. Odin

17
Argon behaviour in clays and glauconies during preheating experiments

GILLES S. ODIN and MICHEL G. BONHOMME

1 INTRODUCTION

Argon diffusion due to heating is a process frequently referred to when the measured age, in dating layer silicates, is significantly younger than the assumed one. On reading the literature about sedimentary dating (clays, glauconies), one is led to an unprecise image concerning the exact conditions of radiogenic argon retention–diffusion.

Evernden *et al.* (1960) observed that departure of radiogenic argon occurs between 200 and 600°C in the case of the glaucony sample K–A 153 containing 5.8% K. Evernden *et al.* (1961) showed that an apparent age of 57 Ma (calculated with the constants defined by Smith, 1964) obtained after an overnight bakeout under vacuum at 85°C becomes 49 Ma after heating at 190°C and 34 Ma after heating at 260°C in the case of the glaucony K–A 274 from the Thanetian of England containing 5.0% K (see NDS 16).

Amirkhanov *et al.* (1961) reported an important loss of argon after heating at temperatures as low as 100°C. On the other hand, Polevaya *et al.* (1961) showed that no argon loss may be detected before the temperature of 200°C is reached. Maximum loss is observed at 400°C. This has been confirmed by Sardarov (1963), who reported that argon diffuses essentially after 400°C under vacuum. The mineralogical nature of the analysed glauconies is not defined. Ghosh (1972) and Odin (1975) showed that radiogenic argon loss under high vacuum, begins between 200 and 300°C from glauconies containing evolved (K-rich) glauconitic minerals. The whole spectrum of radiogenic argon loss under vacuum from glauconies was also investigated by Ivanovskaya *et al.* (1973), Aprub and Levsky (1976) and Zimmermann and Odin (1979). No detectable diffusion of argon before 200°C was observed by these authors. In spite of this fact, some authors, however, still assume that radiogenic argon is lost at less than 150°C: for instance McDougall (1978, p. 120).

Few investigations are available for glaucony heated under atmospheric pressure. Hurley *et al.* (1960) heated a Cambrian potassium-rich glaucony at 110°C for 24 h without detecting any lowering of the apparent age.

In order to come closer to the natural conditions, glauconies have been heated under hot-water pressure. A few experiments are reported by Evernden *et al.* (1960), who heated a glaucony in the middle temperature range (300–600°C) under middle to high pressure range (70–700 bar). They concluded that argon diffusion in glaucony and phlogopite is greatly increased in the period during which constitution water is extracted from the lattice. If glaucony is held under confining water pressures sufficient to prevent loss of water, argon diffusion is greatly delayed. New experiments have recently been made at higher pressures, lower temperatures and for a longer duration to come closer to the natural conditions (Odin *et al.*, 1977). The results showed the influence of the initial mineralogical composition of the green granules. The diffusion of argon is related to *recrystallization processes* which are more easily accomplished in a glaucony not much evolved (K-poor glauconitic minerals) than in a highly-evolved one such as GL-O.

The aims of the present chapter are:

(1) to present new analytical results of experiments made in order to estimate the lowest temperature at which radiogenic argon loss begins;

(2) to evaluate the influence of the mineralogical nature of glauconitic components in the behaviour of radiogenic argon;

(3) to try to use these results to put forward documented recommendations about preheating of the glaucony before measuring argon.

2 THE SAMPLES

Four glauconies have been selected. GL-O is the grain preparation of the reference material. It contains nearly 8% of K_2O. The granules are formed with a pure mica-like mineral: the glauconitic mica. They have been prepared from a Cenomanian sample (NDS 62).

G.583 A is the 160–500 μm fraction of a glaucony whose glauconitic mineral contains 7.5% of K_2O. It comes from the Early Lutetian of the Paris Basin, cf. NDS 32.

G.440 A is a glaucony of the same grain size with nearly 6.8% of K_2O. It has been extracted from a sample collected in the Argiles de Varengeville, the Ypresian of the Paris Basin, cf. NDS 37.

G.490 aA is a glaucony collected from a core drilled off the Congo Republic (cf. Odin and Dodson, this volume.). The selected coprolites, 160–500 μm fraction size, contain 4.2% of K_2O. In spite of a careful purification, a very few traces of non-glauconitic minerals always remain detectable on X-ray diffractograms. These minerals are inherited from the initial substrate of glauconitization: a kaolinitic clay mixed with a few other minerals—quartz, illite, etc.

Two clays were selected from the already analysed sample collection of Strasbourg. The first one is an inherited smectitic structure, which contains nearly 1% of K_2O. The second one, of diagenetic illitic type, contains more than 7% of K_2O.

3 BAKEOUT UNDER HIGH VACUUM

3.1 Experimental

The samples are conserved in normal temperature, pressure and hygrometry conditions. They are weighed under these conditions and introduced in the high vacuum system. In Berne, the whole system, the line and furnace containing the sample, is baked at the chosen experimental temperature during 6–12 hr under pressure of 10^{-6} torr. In Strasbourg, the samples are stored in an appendix separately baked at the temperature required for the experiment. The heating system was first gauged with a mercury thermometer and then controlled with a Pt/Pt–Rh thermocouple during the experiments. The temperature of the samples may be assumed to be correctly determined at ±10°C due to variations in the voltage supply during the night. Preheating duration is between 10 and 15 hr.

3.2 Results

Table 1 gives the results on GL-O. A first set of experiments was done in Moscow by L. Shanin on GL-O weighed in dry air. It should be noted that there is no loss of argon at 170°C as at 50°C. The general mean value

Table 1. Argon content (nl/g STP) in the reference material GL-O grains (K% ~ 6.6) after bakeout under vacuum at selected temperatures. The radiogenic *vs.* total argon ratio is given in parentheses.

Moscow	50°C				170°C		
0.75 hr (1)	24.76				24.92	24.93	24.83
					25.11	25.12	
Berne	80°C	90°C	110°C	120°C	140°C		180°C
6–12 hr (2)	24.99	25.02	24.78	24.68	24.71		24.72
			24.75		24.87		
Strasbourg	90°C	110°C	120°C	155°C	170°C	210°C	250°C
10–12 hr (3)	24.73	24.80	24.87	24.72	24.85	24.81	24.40
	(90.9%)	(94.4%)	24.82	(91.6%)	(91.8%)		(90.8%)

(1) From L. Chanin in Moscow (1974), sample weighed in dry air; (2) from G. S. Odin in Berne (1974–1976), calibration with P 207–Bern 4 M.; (3) from G. S. Odin in Strasbourg (1979), calibration with GL-O.

Table 2. *Argon content (nl/g) in three glauconies after bakeout under vacuum at different temperatures.* Data from G. S. Odin in Strasbourg (1979) after calibration of 38 argon spike using GL-O = 24.8 (nl/g). Argon extraction occurs only at temperatures higher than 200°C. The radiogenic *vs.* total argon ratio is given in parentheses. The behaviour of sample 490 aA seems abnormal.

	90°C	110–120°C	155°C	170°C	210°C	250°C
G.583 A		(10.61)				
K% = 6.15	11.16	11.33	11.12	11.06	11.22	10.63
Ar = 11.18 ± 0.12	(87.7%)	(84–87%)	(86.7%)	(85.0%)	(80.1%)	(85.1%)
G.440 A						
K% = 5.58	11.66	11.54	11.55	11.44	11.46	11.18
Ar = 11.58 ± 0.17	(72.2%)	(58.2%)	(74.6%)	(71.6%)	(68.8%)	(69.5%)
G.490 aA						
K% = 3.52		6.46	6.52		6.53	6.54
Ar ≃ 6.51		(41.6%)	(43.0%)		(42.4%)	(41.1%)

obtained is slightly higher than the accepted one. This may be due to the weighing being performed after drying. In the two other sets of experiments in Berne and in Strasbourg, only a single experiment showed a slightly lower content of radiogenic argon. At 250°C, GL-O seems to have lost 1–2% of radiogenic argon. Between 90 and 210°C, no loss can be detected.

The three other samples of glaucony were successively heated at different temperatures during 10–24 hr in order to detect the first noticeable loss of radiogenic argon. The reproducibility of the results may be estimated at ±1.5%. The results are shown in Table 2. The reference value for G.583 A, 11.18 nl/g, is lowered only after heating at 250°C. The reference value for G.440 A is also lowered at this temperature. In both cases no loss can be detected at 210°C; the loss is quite high compared with GL-O. There is no argon diffusion in the Recent glaucony.

In all cases the radiogenic *vs.* total argon ratio remains constant whether the temperature is low or high.

Table 3. Argon content (nl/g) in clays after bakeout under vacuum at selected temperatures. Data from M. G. Bonhomme in Strasbourg. The radiogenic *vs.* total argon ratio is given in parentheses.

	150°C (10 hr)	200°C (11 hr)	250°C (12 hr)
Smectite	5.16	5.13	5.17
K% = 1	(48.3%)	(51.4%)	(56.4%)
Illite	39.9	39.0	40.9
K% = 6	(89.9%)	(90.3%)	(90.2%)

Table 3 gives the results obtained on the two clays. No difference is observed between the three different treatments done before analysis, at 150, 200 and 250°C. This behaviour is similar to that of the Recent glaucony.

4 BAKEOUT UNDER ATMOSPHERIC PRESSURE

4.1 Experimental

The samples have been weighed in Strasbourg in the same conditions as above. The granules are packed in thin molybdenum sheet. The samples are then put into an oven and heated under atmospheric pressure at a regulated temperature during 100 hr before conservation in a silica gel dried desiccator and introduction in the high vacuum line.

A first set of experiments was done after heating at 180°C. As a surprising apparent loss of argon appeared, other experiments were then made at 160°C. The lack of reproducibility between these experiments then led to a second set of measurements with heating at 150 and 210°C, done one year after the preceding experiments.

4.2 Results

All results are shown in Table 4.

Heating under atmospheric pressure at 150 and 210°C for 100 hr does not significantly change the radiogenic argon content. It remains equivalent to

Table 4. Radiogenic argon content after heating for 100 hr of four glauconies at selected temperatures under atmospheric pressure. Argon in nl/g, with the respective radiogenic *vs.* total ratios in percent (data from M. G. Bonhomme and G. S. Odin in Strasbourg). The left-hand column gives the normal radiogenic argon content and rad./tot. Ar normal ratio.

Sample	100 hr 180°C	100 hr 160°C	100 hr 150°C	100 hr 210°C
GL-O	(23.65)	24.57	24.87	24.99
24.8 (90%)	(64.1%)	(87.2%)	(72.6%)	(69.2%)
G.583 A	11.02	11.34	11.32	11.00
11.18 (85%)	(56.7%)	(68.5%)	(62.2%)	(57.2%)
G.440 A	11.05	11.22	11.42	11.42
11.58 (70%)	(37.9%)	(47.3%)	(39.7%)	(32.6%)
G.490 aA	6.51	6.81	5.37	5.93
6.51 (42%)	(28.8%)	(23.6%)	(26.1%)	(27.6%)

that of the untreated samples. The argon behaviour of the Recent glaucony G.490 aA cannot be easily interpreted. Exceptionally stable after treatment under vacuum, the argon content becomes erratic in these experiments. An heterogeneity of the sample is not excluded while the whole granules look homogeneous. It remains difficult to explain such large variations in aliquots larger than 150 mg.

The interesting fact is the *systematic adsorption of atmospheric argon* shown by all samples. Even at 110°C, adsorption of atmospheric argon is already evident. Results are presented in Figure 1.

To determine if atmospheric argon adsorption was not relative to the molybdenum sheet itself, an empty molybdenum ball was heated in air at 210°C during 100 hr. Analysing the argon content, we found the best blank ever obtained for the line: 0.6×10^{-7} cm^3, instead of the normal blank of nearly 1×10^{-7} cm^3.

These experiments were carried out in the same series as untreated

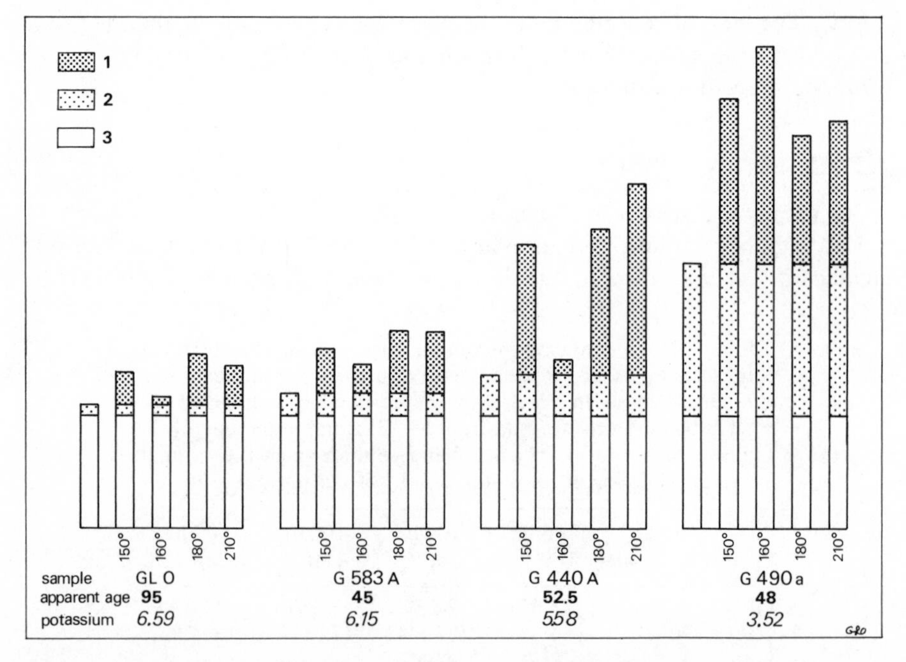

Figure 1. Atmospheric argon contamination in selected glauconies before and after heating under atmospheric pressure. All the data are recalculated for the same quantity of radiogenic Ar (3). For each sample, the normal air argon contamination (2) is given on the left-hand side. The additional atmospheric argon (1) is introduced during or just after heating under atmosphere at the temperature shown at the base of each column. The two kinds of contaminating atmospheric argon seem related to the potassium content of the glauconitic component of the pellets.

Table 5. Adsorption of atmospheric argon after heating of two clays under atmospheric pressure.

	Treatment	^{40}Ar rad. (nl/g)	% Ar. rad.	Additional atm. Ar (nl/g)
Illite 2M	90°C/vacuum	790	99.6	—
K% = 6.7	200°C/atmosphere (240 hr)	809	99.0	+4.9
Illite 1M	90°C/vacuum	904	99.6	—
K% = 7.4	200°C/atmosphere (240 hr)	913	99.1	+4.6

samples. This demonstrates that atmospheric argon is adsorbed by the mineral itself and we may estimate the quantity of atmospheric argon adsorbed on each sample. No relation is detectable between the weight of sample and the quantity of atmospheric argon adsorbed on the different samples. The capacity of adsorption of atmospheric argon increases as potassium content decreases. There is no relation between adsorbed atmospheric argon and temperature at heating temperatures higher than 150°C. There is an approximate inverse relation between the proportion of radiogenic argon present in the sample and the newly adsorbed atmospheric argon.

This increase of atmospheric argon during heating in air has also been determined on the two clays, as shown in Table 5.

5 DISCUSSION

Fifty-six experiments made in three laboratories on four glauconies in order to better define the behaviour of argon during preheating of samples usually done before radiogenic argon extraction and determination are reported here. Ten others were carried out on clays.

5.1 Preheating experiments under vacuum

5.1.1 Evolved to highly-evolved glauconies

The results of all the experiments demonstrate that the loss of argon caused by sample preheating carried out between 100 and 180°C, sometimes used to explain 'too young apparent ages', *does not exist.* Diffusion of radiogenic argon remains less important than may be detected considering the reproducibility of the analytical system, as long as the preheating temperature does not exceed 210°C. The beginning of radiogenic argon loss may be situated between 210 and 250°C for glauconies formed by authigenic minerals containing more than 5.5% K. At 250°C, it appears that the K-richest glauconitic minerals are more closed than the poorest ones. But all samples lost argon after a few hours of heating. This defines a 'blocking temperature' lower than 250°C.

The three evolved glauconies support this conclusion. But we cannot reject the possibility of a slight loss of argon at 200°C in exceptional cases, especially for glauconies with lower potassium content. Considering the only example reported in the literature and extrapolating our results obtained at 250°C, minerals which contain less than 5.5% K are probably more sensitive to heating. It is better not to place complete confidence in the apparent ages obtained with these chronometers.

Compared with some ages obtained very early, it is established by the above experiments that the 'younger ages' recently obtained cannot be explained by radiogenic argon loss during preheating.

Different explanations might be propounded for various 'rejuvenations':

(1) the better definition and deduction of the atmospheric argon contamination;

(2) the possible radiogenic argon loss from the smectitic component of badly evolved glauconies due to alteration;

(3) the age lowering caused by later diagenetic effects, as frequently happens with clays (Clauer, this volume; Bonhomme *et al.*, 1980). The aim of preheating is to remove the adsorbed gases. The clay minerals and those contained in glauconies display thermal curves with the departure of adsorbed water between 100 and 140°C under atmospheric pressure. Thus 120°C appears to be a minimum preheating temperature if one intends to desorb water, organic molecules, air, and consequently atmospheric argon. Outgassing these fluids also protects the titanium furnace. Outgassing does not seem to be better if preheating temperatures as high as 200°C are used. Consequently we recommend a temperature of nearly 150°C for a good removal of adsorbed fluids if maintained for one to two hours. No loss of radiogenic argon can be feared in these conditions if evolved glauconies are analysed.

In the case of samples put in the general bakeout system, the selection of glauconies containing more than 6% K permits a general bakeout temperature as high as 180°C, as demonstrated by the present data, with the guarantee of a complete retention of the radiogenic argon. If the potassium content is less than 5.0%, one might expect a slight argon loss. When the potassium content is between 5.0 and 6.0%, one might expect the same behaviour as for K-rich glauconies, but we did not make any experiments within that range.

5.1.2 Illite and smectites

We have not observed any loss of radiogenic argon in the two clays analysed, but some authigenic marine and lacustrine clays are probably more fragile and further experiments are needed.

The illite has been demonstrated to be diagenetic; the diagenesis is attributed to a regional event (Bonhomme *et al.*, 1980). It has been sufficiently transformed to appear resistant to radiogenic argon loss under vacuum at temperatures as high as 250°C.

The behaviour of the smectite was completely unforeseen. As the crystal lattices of the smectites have a low bound energy, they are expected to display bad argon retention. The experiments reported here demonstrate the contrary. The surprisingly good argon retention may be attributed to the fact that these smectites are inherited and contain a few mica layers not yet isotopically disturbed through weathering.

5.1.3 The Recent glaucony

The special behaviour of the Recent K-poor glaucony, G.490 aA appears unforeseeable, as in the case of the smectites. We might relate it to the fact that the radiogenic argon of this sample is not the product of ^{40}K in an interlayer position as with the other glauconitic minerals studied here. This glaucony is Recent and cannot have accumulated the measured quantity of radiogenic argon in its structure. The radiogenic argon in excess must be trapped in inherited mineral structures which cannot be characterized in X-ray diffractograms. These mineral structures are probably of a different kind, and this would explain the complex behaviour of the radiogenic argon during these preheating experiments.

More experiments are needed to define this behaviour and recognize inherited argon. At the present time, we cannot distinguish between the radiogenic argon inherited from the initial substrate of glauconitization and the argon normally produced within the authigenic glauconitic minerals after their genesis. Step-by-step heating would probably be a better method of detection.

5.2 Preheating under atmosphere

Preheating under atmospheric pressure increases the quantity of adsorbed atmospheric argon. The increase in the total amount of atmospheric argon does not favour the accuracy of radiogenic argon measurement. Consequently, treating the samples in air at temperatures between 150 and 210° is not recommended.

However, we have observed that after such a treatment, the vacuum in the line improves more quickly than without any preliminary heating, especially if a large quantity of sample is introduced. Heating the samples in an oven under atmospheric pressure probably removes the water adsorbed at the surface and inside the layer silicates which compose glauconies and clays. Adsorption of air argon is induced by this treatment. This might be due to the replacement of water by air molecules, among which would be argon, in electrically active surface sites.

6 CONCLUSIONS

(1) Bakeout of evolved glaucony and clay samples under high vacuum does not produce any radiogenic argon loss under 200°C.

(2) Argon loss becomes measurable only for a few percent at temperatures as high as 250°C. It increases as potassium content decreases.

(3) Bakeout of glaucony and clay samples under atmospheric pressure at temperatures between 150 and 210°C considerably increases the atmospheric argon contamination.

(4) Consequently, bakeout of sedimentary phyllitic minerals must be done under vacuum at temperatures up to 150°C to avoid any danger either of radiogenic argon loss or of atmospheric argon contamination.

Acknowledgements

The first author of this chapter is greatly indebted to colleagues in the geochronology unit of Strasbourg, especially R. Winkler who helped him during his two stays in their laboratory. This chapter is a contribution to Project 133 of the International Geological Correlation Programme.

Résumé

Ce travail rassemble les données de 56 expériences de chauffage sous vide ou sous atmosphère, effectuées sur quatre glauconies constituées de différents minéraux glauconitiques ainsi que 10 autres expériences sur des argiles.

On démontre ainsi que les argiles utilisées ou les glauconies évoluées, chauffées sous haut vide ne montrent *aucun départ d'argon* jusqu'à une température strictement contrôlée de 200°C. On ne décèle une fuite d'argon radiogénique (un rajeunissement) qu'à une température de 250°C. L'importance de la fuite est assez directement *liée à la teneur en feuillets expansibles* et à la faible teneur en potassium. Ceci démontre qu'en aucun cas des âges apparents sur glauconies ne peuvent être considérés comme analytiquement trop faibles sous prétexte que la température d'étuvage de 150°C utilisée aurait été trop élevée.

Les expériences de chauffage sous pression atmosphérique démontrent qu'à des températures de 150°C et 210°C on peut provoquer une importante *adsorption d'air* atmosphérique dans les feuillets. Ce préchauffage sous atmosphère est donc déconseillé car il augmente la pollution de l'échantillon de phyllite en argon atmosphérique.

Une température d'étuvage sous vide à 150°C est donc seule conseillée avant les mesures. A cette température, les phyllites sont débarrassées de leur eau d'adsorption sans aucun danger de modification de l'âge apparent.

(Manuscript received 13-6-1980)

Numerical Dating in Stratigraphy
Edited by G. S. Odin

18
Kinetics of the release of argon and fluids from glauconies

JEAN-LOUIS ZIMMERMANN and GILLES S. ODIN

1 INTRODUCTION

In hydrated and hydroxylated minerals, such as micas, argon release cannot be dissociated from dehydration processes (Zimmermann, 1970). Glauconitic minerals from glauconies are components much like other phyllites (e.g. micas); they are hydrated to varying extents. As in the other chapters of this book, the word glaucony(ies) will be used in order to avoid confusion with the mineral, the glauconitic mica, which is hardly ever encountered in nature. Polevaya *et al.* (1961) shows that the release of radiogenic argon from glauconies is controlled by the destruction of the lattice of their mineralogical components.

A first set of analyses has been carried out (Zimmermann and Odin, 1979) on the reference material GL-O proposed as a geochemical standard for both chemical analysis (GL-Op, de la Roche *et al.*, 1976) and geochronological purposes (GL-Og, Odin and coll., 1976). This work, concerning a much evolved and potassium-rich mineral, is complemented here with original results on glauconies with diverse mineralogical compositions (Table 1).

Argon is analysed by means of mass spectrometry on a 180° deflection angle and 5 cm radius THN 205 E instrument. Measurements are made in the static mode; argon is extracted by heating the sample in a resistance furnace.

For the reference material GL-Og, stepwise heating by temperature increments of 100°C has been used; the first step at 200°C is preceded by an overnight outgassing of the sample at 100°C. For other samples, temperature increments of 200°C were applied instead.

Temperature is kept constant until the pressure of gases has reached an apparent stability, generally after two hours. Current step duration is three hours. A known quantity of spike ^{38}Ar is added to the extracted gases.

Water and other fluids (CO_2, CO, N_2, CH_4) are analysed on a separate mass spectrometer. A further set of experiments with linear heating has

Table 1. Mineralogical characteristics of the investigated glauconies. Early glauconies can be considered as pure glauconitic minerals. G.490 hA contains components inherited from the substrate of glauconitization.

Sample	K%	$^{40}Ar^*$ (nl/g)	Expandable layers content (%)	Origin
G.582 A	6.91	145.20	0–5 (mica)	Ordovician, Estonia
GL-O g	6.56	24.80	5–10	Cenomanian, Pays de Caux (France)
G.461 A	6.07	24.88	10–15	Early Albian Aube (France)
G.440 A	5.58	11.58	15–20	Argile de Varengeville, Normandy (France)
G.490 hA	2.83	8.90	50–70	Recent glaucony, Gulf of Guinea

allowed a continuous monitoring of the release for: water (mass 18), carbon dioxide (mass 44), carbon monoxide (mass 28 and mass 12), nitrogen (mass 28 and mass 14), methane (mass 16 and mass 15) and, incidentally, other hydrocarbons (C_2H_2: mass 26; C_2H_6: mass 30).

2 THE THERMAL RELEASE OF ARGON

2.1 Experimental results

Table 2 gives the amounts of radiogenic argon extracted at each step. In all cases the total amount of radiogenic argon released during these experiments (Q_{total}) is less than the released quantity determined by conventional K–Ar dating ($Q_{initial}$). This fact can be explained by slight argon loss during pumping of the extraction line between consecutive steps. The loss does not exceed 10% of the total nor significantly affect the results of the kinetic study derived from these experiments.

If one excepts the young glaucony sample, G.490 hA, from which radiogenic argon is still released at quite high temperatures, 97–98% of argon is extracted between 200 and 600°C. Preliminary work on GL-Og (Zimmermann and Odin, 1979) had shown that most of its argon content is released from 300 to 500°C (85%).

Below 200°C, radiogenic argon is hardly extracted at all (Odin and Bonhomme, this volume). Nevertheless, present results suggest that slight amounts of argon may be extracted from little-evolved glauconies in this temperature range (Table 2): for instance, for the sample G.440 A, argon loss amounts to about 5.6%; this single experiment needs confirmation.

On the other hand, an abundant release of atmospheric argon is observed

Table 2. Release of argon by step heating analysis (step duration: 3 hr). ^{40}Ar, argon extracted during a step (uncertainty ±4%). % extracted, percentage of radiogenic argon extracted during a step relative to the total quantity. $Q_{initial}$, amount measured by total fusion in K–Ar dating. D/a^2 s^{-1}, calculated from the diffusion model (Reichenberg's table).

	G.582 A			GL-O			G.461 A		
T(°C)	^{40}Ar rad.	% extract.	D/a^2 s^{-1}	^{40}Ar rad.	% extract.	D/a^2 s^{-1}	^{40}Ar rad.	% extract.	D/a^2 s^{-1}
200	0.63	1.1	1.7×10^{-9}	0.08	0.35	4.2×10^{-10}	0.27	1.6	3.5×10^{-9}
300				1.6	7.35	7.4×10^{-8}			
400	58.57	44.1	3.3×10^{-6}	5.5	24.9	1.6×10^{-6}	7.07	40.8	2.9×10^{-6}
500				13.2	59.8	2.9×10^{-5}			
600	71.17	53.6	5.6×10^{-5}	1.3	5.8	5.1×10^{-5}	9.74	56.2	5.4×10^{-5}
700				0.09	0.4	5.4×10^{-5}			
800	1.45	1.1					0.15	0.9	
900				0.19	0.9				
1000	0.15	0.1		0.06	0.3		0.09	0.5	
1100				0.05	0.2				
Q_{total} (nl/g)	131.97			22.07	100		17.32	100	
$Q_{initial}$ (nl/g)	145.2			24.80			24.88		

	G.440 A			G.490 hA		
T(°C)	^{40}Ar rad.	% extract.	D/a^2 s^{-1}	^{40}Ar rad.	% extract.	D/a^2 s^{-1}
200	0.55	5.6	3.94×10^{-8}	0.09	1.2	2.1×10^{-9}
300	3.73	37.3	2.94×10^{-6}			
400	2.28	22.8	8.6×10^{-6}	0.74	9.6	1.5×10^{-7}
500						
600	3.24	32.4	4.9×10^{-5}	4.57	59.3	1.04×10^{-5}
700						
800	0.16	1.6		1.41	18.3	2.33×10^{-5}
900						
1000	0.03	3.3		0.89	11.6	
1100						
Q_{total} (nl/g)	9.99	100		7.7	100	
$Q_{initial}$ (nl/g)	11.58			8.9		

during the 200°C step (77–97% of argon is of atmospheric origin) which previous baking at 100°C was unable to extract. At 600°C, radiogenic argon is almost completely released.

Histograms (Figure 1) of fractional amounts of argon released at each step do not differ significantly from those published by Polevaya *et al.* (1961), Ivanovskaya *et al.* (1973) and Aprub and Levsky (1976) nor those compiled by Ghosh (1972). The latter author points out a correlation between the maximum argon release at more than 400°C and the onset of the second endothermic peak observed by differential thermic analyses of glauconies (Kazakov, 1963); this peak corresponds to the destruction of the crystalline frame by release of structural water.

2.2 Kinetics of argon release

The interpretation of the present results by volume diffusion models (Zimmermann, 1972; Zimmermann and Odin, 1979) is greatly dependent

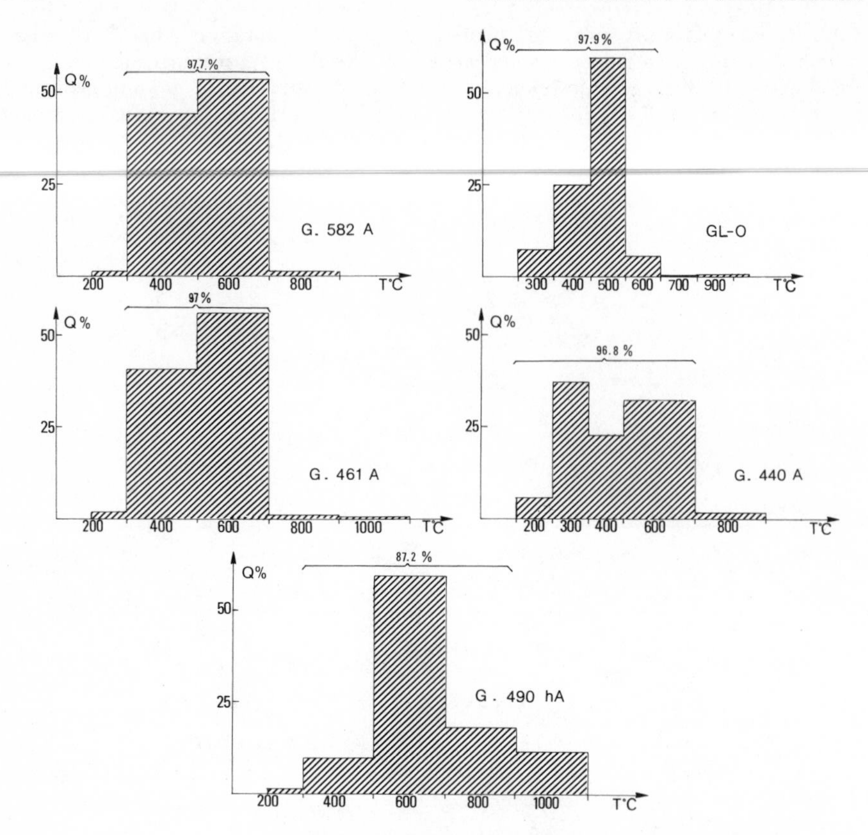

Figure 1. Histograms of the release of radiogenic argon from different glauconies; step heating analysis.

on physical assumptions which may be not wholly appropriate because of the diversity of natural materials and phenomena. Important approximations arise from simplifying assumptions, among them single spherical geometry and uniform size of grains or isotopic behaviour of diffusion laws.

Actual crystallites in glaucony have diameters ranging from 0.1 to 5 μm (Odin, 1975) and assume flattened shapes. Furthermore, the chosen models only take diffusion into account. This process is not likely to be the most important according to Odin *et al.* (1977, see also Odin, Chapter 17).

From the present experimental results (Table 2), and keeping in mind the above-mentioned restrictions, we have calculated the diffusion characteristics of argon loss.

These characteristics are given by the 'Arrhenius equation':

$$D = D_0 \exp(-E/RT)$$

where E is the activation energy in kcal/mole, R is the gas constant, T is the absolute temperature (°K), and D_0 is the characteristic coefficient of diffusion (cm^2/s). The mineral particles are considered as equisized spherical solids and the appropriate solution (e.g. Carslaw and Jaeger, 1959) may be written as:

$$F = 1 - \frac{6}{\pi^2} \sum_{n=1}^{\infty} \frac{e^{-n^2(Bt)}}{n^2}$$

where F is the fraction of total argon lost at time t and t is the time of heating, and $B = \pi^2(D_i/a^2)$ where a is the radius of the hypothetical sphere. Various approximations are discussed by Fechtig and Kalbitzer (1966).

From 200 to 600°C, where more than 95% of radiogenic argon is expelled, the function log D/a^2 *vs.* ($1000/T$) is approximately linear (Figure 2A), which suggests that the kinetics of the process is controlled by volume diffusion. For each sample, the slope of the straight line gives the activation energy of the release of argon (Table 3).

It appears that *the activation energy decreases appreciably with the potassium content,* which suggests that the mineralogical composition of a glaucony must be taken into account for K–Ar dating.

Above 600°C (except for G.490 hA), the diffusion coefficients are no longer significant because most radiogenic argon has been released during the preceding steps.

At 200°C, the amount of outgassed radiogenic argon is very small (close to the experimental uncertainty $\simeq$1.5%) and not much confidence must be placed in extrapolations of the straight lines log D/a^2 *vs.* ($1000/T$) at room temperature. There is probably a change in the slope: the release of argon is more sluggish due to the fact that, at these temperatures, the crystalline lattice of phyllites is in *stable equilibrium.* In particular, octahedral hydroxyl ions are not expelled and the closure of the crystalline framework is at its best.

However, if we extrapolate these straight lines down to room temperature (Figure 2b), we determine diffusion constants from 10^{-18} to 10^{-25} cm^2/s with particles from 0.1 to 5 μm. The smaller these constants are, the higher the potassium content, which would indicate that argon diffusion is more difficult in potassium-rich samples.

Ghosh's compilation (1972) suggests values ranging from 3×10^{-22} to 3×10^{-26} cm^2/s for under vacuum experiments in different laboratories.

Table 3. Activation energy of the release of argon between 200 and 600°C.

	Sample				
	G.582 A	GL-Og	G.461 A	G.440 A	G.490 hA
K%	6.91	6.56	6.07	5.58	2.83
E (kcal/mole)	24 ± 3	28.8 ± 3.2	20 ± 2	17.7 ± 2.3	16.7 ± 2.3

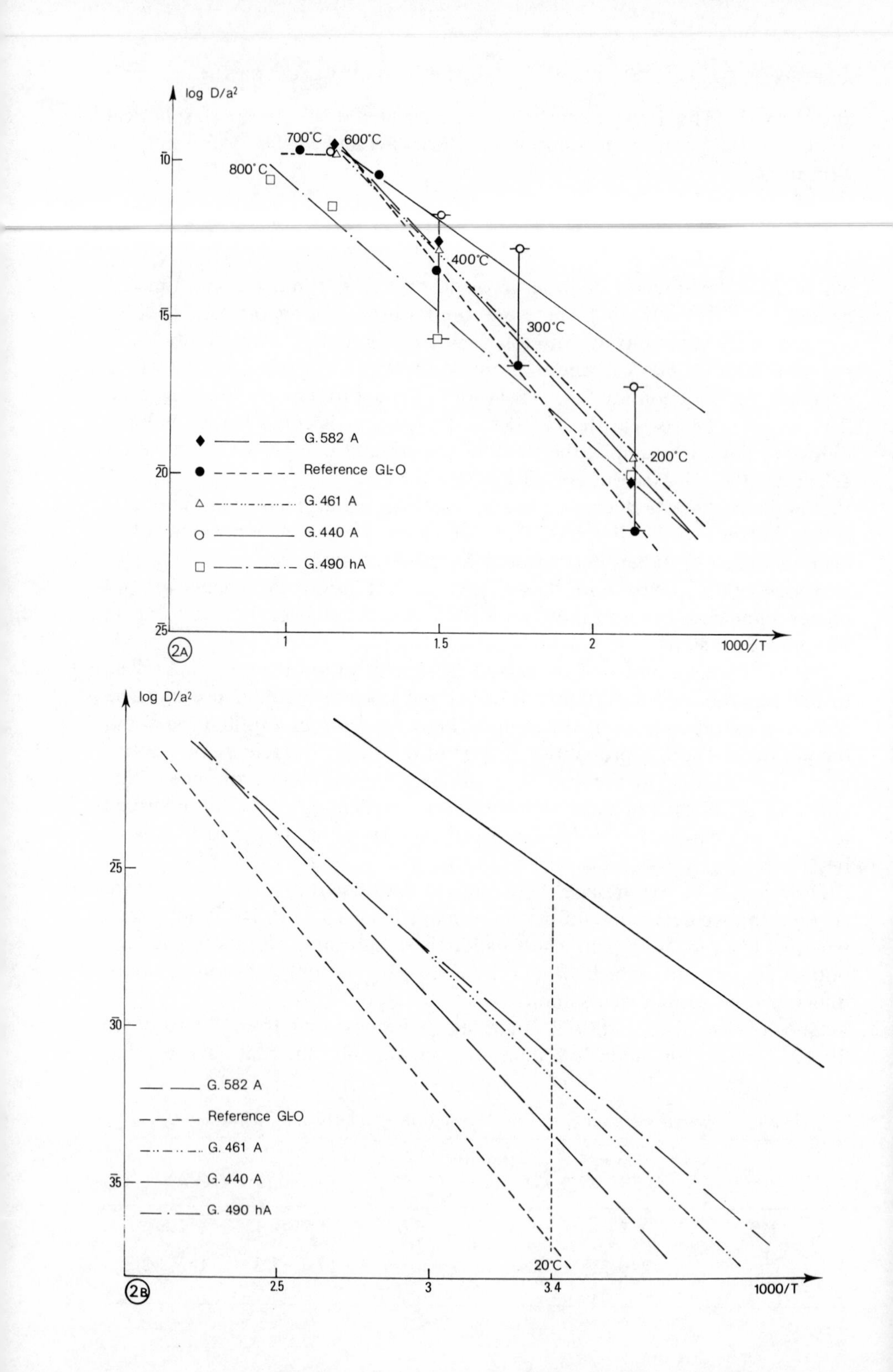

log D/a²
700°C
600°C
800°C
400°C
300°C
200°C
G. 582 A
Reference GL-O
G. 461 A
G. 440 A
G. 490 hA
1000/T
2A
log D/a²
G. 582 A
Reference GL-O
G. 461 A
G. 440 A
G. 490 hA
20°C
1000/T
2B

Evernden *et al.* (1960) also calculate diffusion constants under vacuum and under water pressure with a glaucony of which the mineralogical composition is not known. They determine an activation energy of 28 kcal/mole and a diffusion coefficient at 20°C of 3.22×10^{-22} cm^2/s under vacuum and of 10^{-29}–10^{-30} cm^2/s under water pressure. However, these extrapolated values do not carry much significance because they are extrapolations of phenomena which only occur at higher temperatures, such as crystalline rearrangements.

3 THE RELEASE OF WATER AND CARBONIC FLUIDS

3.1 Experimental results

The step heating analysis allows evaluation of the approximate composition of the fluid phase extracted at each temperature step.

With GL-O, after a preliminary heating under vacuum at 100°C, the water released from 200 to 1100°C amounts to about 4.8% of the sample weight, less than 0.05% for methane, 0.8% for CO_2 and *ca.* 0.4% for carbonic compounds (CO+organic C). The average loss of water between 105 and 1100°C as determined by several laboratories (de la Roche *et al.*, 1976) is 5.6% of the sample weight (Table 4). The difference probably results from a slight loss of water due to the pumping between consecutive steps. Water extracted at 105°C (3.06 wt% according to the above-mentioned study) has been eliminated during preliminary heating at 100°C. The other samples have not been exposed to preliminary heating, but

Table 4. Water loss: below 105°C (H_2O^-), from 105 to 1100°C (H_2O^+), by step heating (H_2O_s). For GL-O, the numbers represent an average from several laboratories (de la Roche *et al.*, 1976). For the other samples, CRPG analyses (M. Vernet, analyst).

	Sample				
	G.582 A	GL-Og	G.461 A	G.440 A	G.490 hA
H_2O^- (% wt)	2.11	3.06	3.73	4.24	6.44
H_2O^+ (% wt)	4.57	5.60	4.60	4.83	5.14
H_2O_s (% wt)	5.20	4.80	5.11	5.24	6.90

Figure 2. A. Argon diffusion characteristics calculated from the diffusion model. The slope of the straight lines gives the activation energies. B. Extrapolation of these straight lines down to room temperature.

Table 5. Percentages of different gases measured at each temperature step as a function of the total quantity of these gases (analysed by mass spectrometry).

	G.582 A			GL-O			G.461 A			G.440 A			G.490 hA		
T(°C)	H_2O	CO_2	N_2+CO	H_2O	CO_2	N_2+CO	H_2O	CO_2	N_2+CO	H_2O	CO_2	N_2+CO	H_2O	CO_2	N_2+CO
100	0.6	—	0.1				0.9	0.8	1.5	0.3	0.3	0.1	2	0.5	0.5
200	1.7	1.8	1.3	2.4	3.3	1.1	2.6	3.7	5.3	0.3	3.3	3.4	5.1	2	2.8
300	4.8	1.8	2	5.4	6.6	4.1	5.9	3.3	11.9	9.3	4.4	5.2	10.7	4.1	9.5
400	17.1	3.5	12.8	41.8	8.1	12.6	25.8	8.6	7.4	38	10.2	10.3	31.7	7.2	9.2
500	56.9	11.4	9.9	38.4	7.3	12.7	47.9	32.8	23.1	40.3	24.5	21.7	36.9	26.9	15.9
600	12.9	10.5	13.1	10.3	33.7	21.9	12.5	15.2	17.8	7.8	17.1	18.2	6.7	8.4	10.3
700	3.8	33.3	38.3	0.8	9.2	5.7	2.9	22.5	24	2.5	23.5	22	5	38	35.7
800	1.6	32.5	7.7	0.4	19.1	8.7	1.1	11.5	6.8	1.2	11.2	10.7	2.1	12	14.3
900	0.3	1.8	5.1	0.2	4.8	9.3	0.2	0.4	0.6	0.1	1.2	1.4	0.1	0.5	0.8
1000	0.2	1.8	4.2	0.2	2.7	8.4	0.1	0.4	0.6	0.1	1.9	3	0.1	0.2	0.5
1100	0.2	1.7	5.5	0.2	5.2	15.5	0.1	0.8	0.9	0.1	2.3	3.9	0.1	0.2	0.5
Total	100	100	100	100	100	100	100	100	100	100	100	100	100	100	100

pumping at room temperature for about 15 hr was used instead; the measured total water amounts are less than those of structural water (measured by heating the sample from 105 to 1100°C).

Step heating experiments also allow the fraction of each fluid extracted at each step to be analysed and compared with total fusion data obtained by mass spectrometry (Table 5).

At 600°C, more than 90% of fluids are released; essentially they contain water (≃96% H_2O), 1–2% CO_2 and traces of CO, N_2 and organic compounds. Above 600°C, i.e. for the remaining 10% of the fluid phase, the water content drops to 40–50% while the CO_2 content increases to about 15–20% and that of CO to about 35–40%. On the whole, the total water content varies from 90 to 96%, and that of CO_2 from 1 to 6%, while methane does not exceed 0.1% and $CO+N_2$ (mass 28) 2–7%.

It is important to keep in mind that ionized gases do not necessarily correspond to the compounds initially present in the minerals (Zimmermann, 1972). Indeed, these green grains are heated at high temperatures and molecules can react in the furnace or undergo cracking in the mass spectrometer source. Water is effectively present at mass 18 but mass 15 (methyl radical) indicates the presence of both methane and dissociation products of heavy organic compounds.

Similarly, carbon dioxide gives not only its molecular mass 44 but also mass 28 (CO), 16 (O) and 12 (C).

So, results must be very carefully interpreted. Figure 3 shows the histograms of release for water and carbonic fluids from the sample GL-O. Carbon (mass 12) has an histogram virtually identical to that of carbon dioxide and probably originated from carbonic compounds.

Figures 4 and 5 correspond to the release of water and carbon dioxide from the other four samples.

The linear heating analysis clarifies some of the above results. The

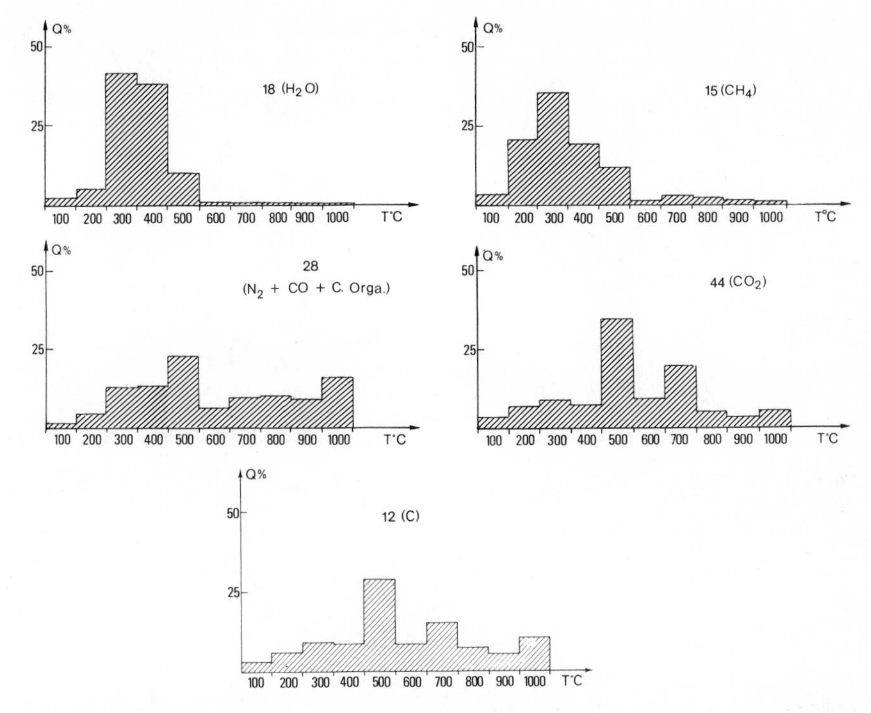

Figure 3. Results of step heating analysis on the reference material GL-O. Histograms of H_2O, CH_4, mass 28 (N_2+CO+C org.) and C release. C is produced from CO_2, CO and organic compound dissociation.

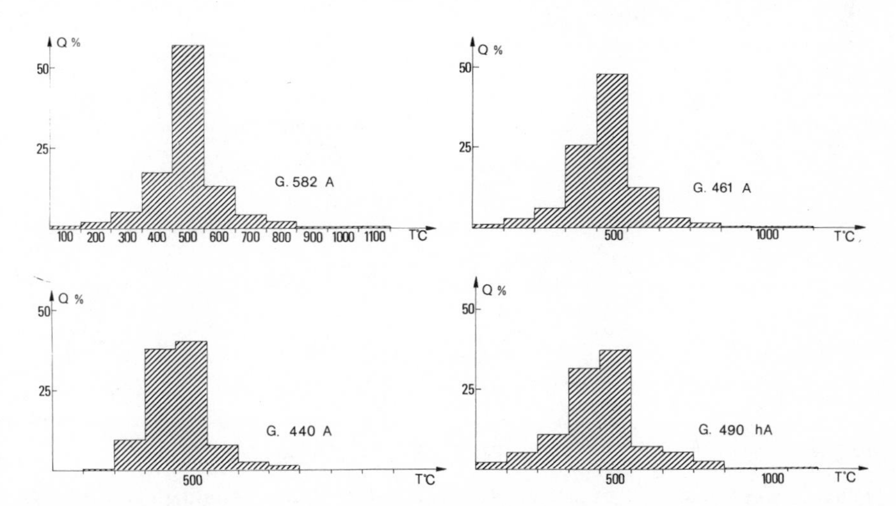

Figure 4. Histograms of the release of water from the four other glauconies. Step heating analysis.

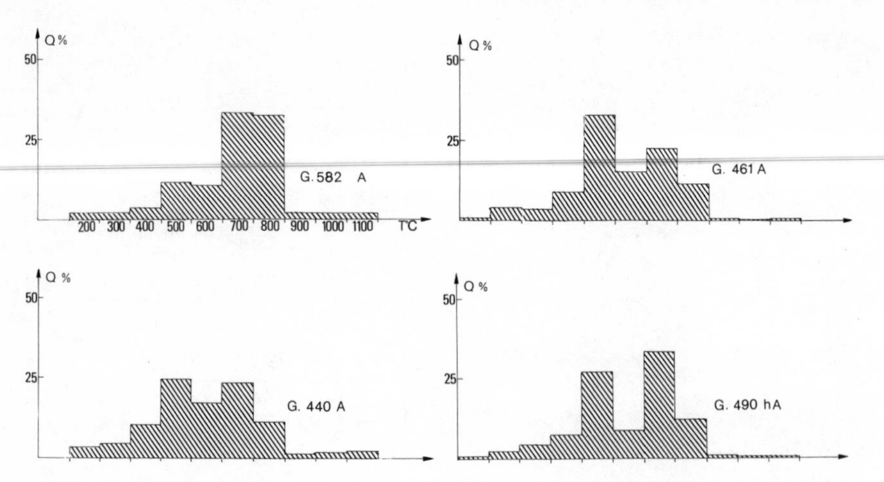

Figure 5. Histograms of the release of CO_2 from the four other glauconies.

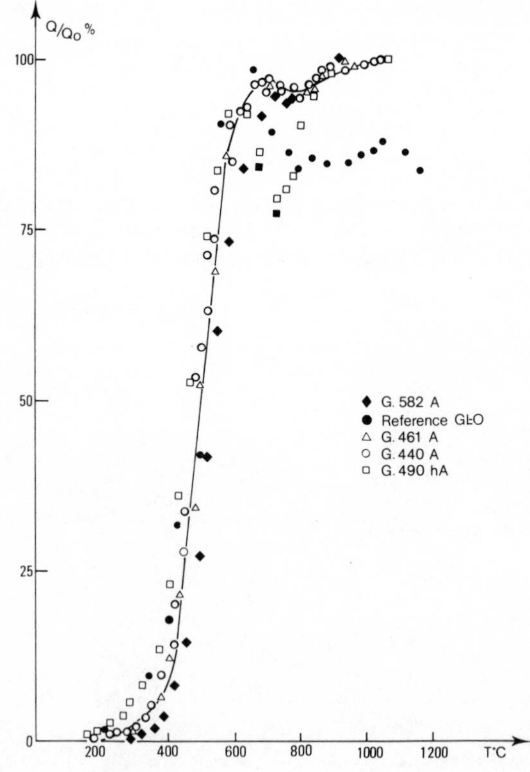

Figure 6. Continuous heating rate release of water. Rate of heating: 10°C/mn. Q is the instantaneous measured quantity, Q_0 the total quantity released during the experiment.

continuous water release curve (Figure 6) has a prominent maximum between 550 and 600°C, then decreases up to 700°C; two other small peaks are apparent one between 750 and 850°C, the other between 1000 and 1100°C. Taking into account the results of step heating experiments, we can consider that the important release of water starting above 200°C (around 300°C) and decreasing above 650°C corresponds to the constitution water, as for differential thermal analyses (McRae and Lambert, 1968) and thermogravimetric analyses (Douillet and Odin, 1968).

With carbon dioxide (Figure 7A), a maximum is observed between 750 and 850°C, depending on the samples; then above 1000°C the curves increase again slowly, probably because of incipient carbonate dissociation. On the methane curves, the maximum lies between 950 and 1000°C (Zimmermann and Odin, 1979). For mass 28 (CO, N_2 and possibly a dissociation mass of heavier organic compounds) a first maximum occurs at about 250°C, corresponding to the release of atmospheric nitrogen. Two other peaks at 750–850°C then at 1000°C (Figure 7B) are also evident. These latter maxima coincide with those of carbon dioxide, indicating that CO results from CO_2 reduction and carbonate dissociation.

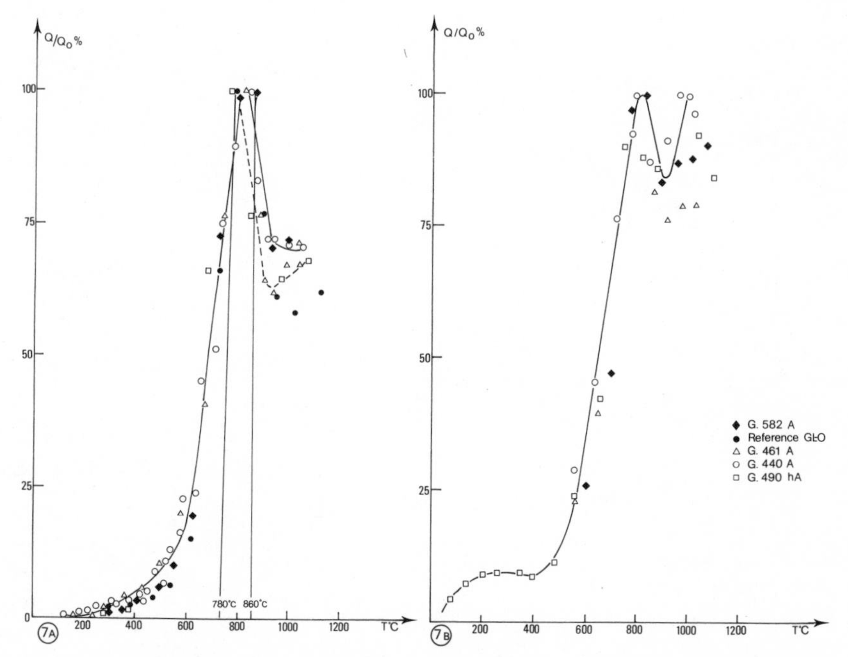

Figure 7. A. continuous release of CO_2. B. continuous release of mass 28 ($CO+N_2+C$ org.). Q is the instantaneous measured quantity, Q_0 the total quantity released during the experiment.

3.2 Kinetics of the release of water

The release of water can be considered as either a diffusion process or a dissociation chemical reaction.

In the first case, from data of step heating analysis, we can use the diffusion model as described above.

From 200 to 600°C, the calculated diffusion parameters D/a^2 (Table 6) give an almost linear relationship when plotted in an Arrhenius diagram (Figure 8). Allowing that the kinetics of the release of water follows a diffusion law, we can also turn to continuous analyses with variable heating rates, using the so-called linear annealing theory (Gerling *et al.*, 1966).

Activation energy can be calculated from temperatures corresponding to the maxima of the derivative functions (or inflexion points of continuous curves: Figure 6) for at least two different heating rates:

$$E = R\left(\text{Log}\,\frac{\beta_1}{\beta_2} + 2\,\text{Log}\,\frac{T_2}{T_1}\right)\frac{T_1T_2}{T_1 - T_2},\; T_1 > T_2$$

where β_1, β_2 is the heating rates in °C/mn, T_1, T_2 is the temperature of the maxima, and R is the universal gas constant. The uncertainties are basically dependent upon the difference $T_1 - T_2$; with the heating rates used in this

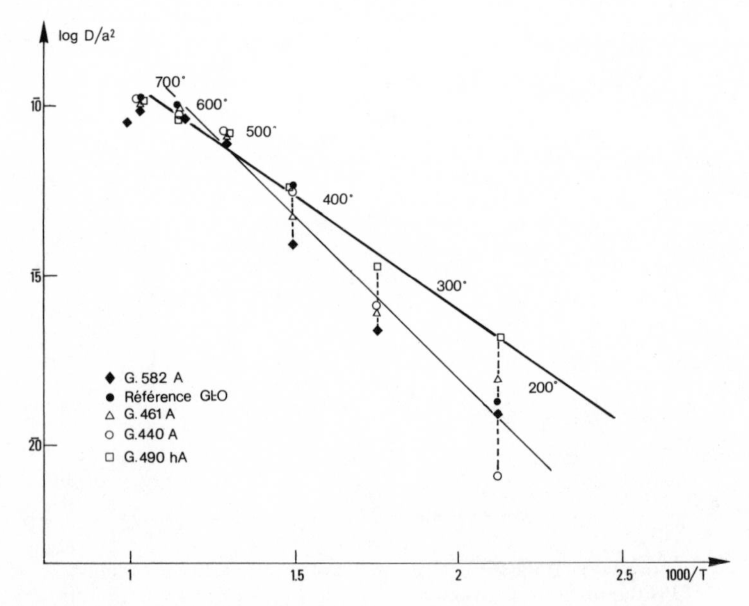

Figure 8. Water diffusion characteristics calculated from the diffusion model.

Table 6. Characteristics of the release of water considered as a diffusion process. $F\%$, fractional part of water released at each step. $D/a^2\ s^{-1}$, calculated from the diffusion model (Reichenberg's table).

Sample	G.582 A		GL-O		G.461 A		G.440 A		G.490 hA	
T(°C)	$F\%$	$D/a^2\ s^{-1}$	$F\%$	$D/a^2\ s^{-1}$	$F\%$	$D/a^2\ s^{-1}$	$F\%$	$D/a^2\ s^{-1}$	$F\%$	$D/a^2\ s^{-1}$
100	0.6	8.4×10^{-10}			0.9	1.1×10^{-9}	0.3	4.2×10^{-10}	2	5.1×10^{-7}
200	2.3	6.8×10^{-9}	2.4	7.3×10^{-9}	3.5	1.5×10^{-8}	0.6	8.4×10^{-10}	7.1	6.3×10^{-8}
300	7.1	6.3×10^{-8}	7.8	7.6×10^{-8}	9.4	1.1×10^{-7}	9.9	1.3×10^{-7}	17.8	4.2×10^{-7}
400	24.2	8.2×10^{-7}	49.6	4.2×10^{-6}	35.2	1.9×10^{-6}	47.9	3.8×10^{-6}	49	4.0×10^{-6}
500	81.1	1.7×10^{-5}	88	2.3×10^{-5}	83.1	1.8×10^{-5}	88.2	2.3×10^{-5}	85.9	2.1×10^{-5}
600	93.9	3.2×10^{-5}	98.3	5.1×10^{-5}	95.6	3.7×10^{-5}	96	3.8×10^{-5}	92.6	3.0×10^{-5}
700	97.7	4.6×10^{-5}	99.1	5.8×10^{-5}	98.5	5.3×10^{-5}	98.5	5.3×10^{-5}	97.6	4.6×10^{-5}
E (kcal/mole)	18 ± 3		20 ± 3		16 ± 3		20.5 ± 3.0		12.6 ± 2.0	

study (5–10°C/mn) this difference is quite small, thus magnifying the expected error. For the reference material GL-O, the maxima of the derivative are located at 490 and 470°C, with heating rates of respectively 10 and 6.66°C/mn; giving an activation energy of 20 ± 3 kcal/mole.

Conversely, if the kinetics of dehydration is controlled by a reaction of dissociation, curves of continuous release of water can be considered as thermogravimetric curves. Also the Horowitz and Metzger (1963) method can be used (Zimmermann and Odin, 1979).

For a first-order reaction, the expression Log Log $\left(\frac{1}{1-Q/Q_0}\right)$ should be linear in temperature and becomes zero at a temperature T_s for which the dehydration rate is maximum. At T_s, water molecules are in equilibrium with water dissolved in the crystalline structure. The activation energy is proportional to the slope of the straight line:

$$\log\log\left(\frac{1}{1-Q/Q_0}\right)=\frac{E\theta}{RT_s^2},\quad \text{with}\quad \theta=T-T_s.$$

From our data (Figure 6), we get three different straight lines which are almost parallel (Figure 9); temperatures T_s are respectively 495°C (GL-O, G.490 hA), 520°C (G.440 A, G.461 A) and 560°C (G.582 A) and the activation energies vary from 17 ± 3 to 22 ± 3 kcal/mole. These results give parameters in the same range as those calculated from diffusion models (Table 6). The activation energies for the release of water are about 4–8 kcal/mole smaller than the activation energies for the release of argon (Tables 3 and 6).

3.3 Kinetics of the release of carbon dioxide

We have applied the same models as those used for water outgassing studies. However, attention must be paid to the small amounts of involved

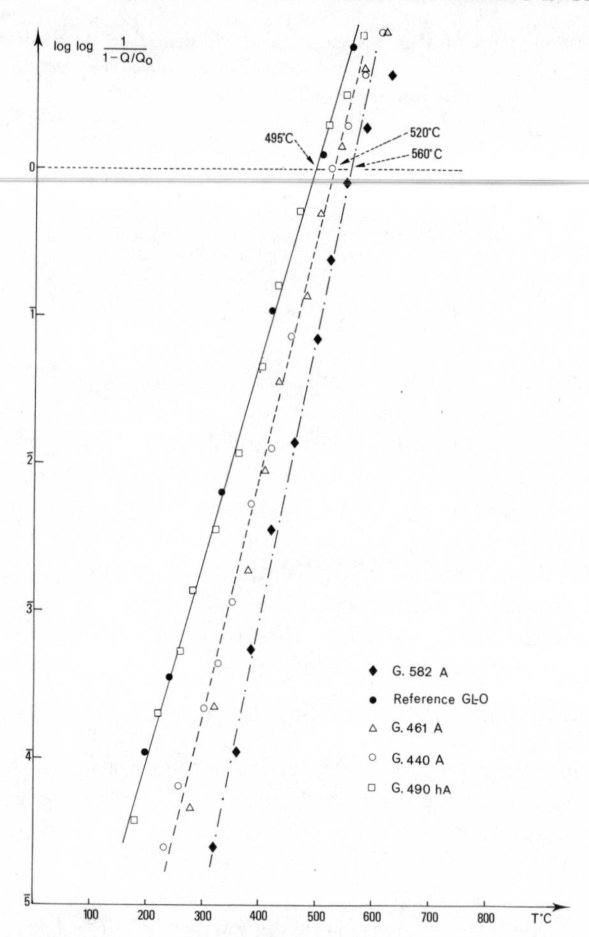

Figure 9. Continuous heating rate analysis. Activation energies for the release of water are calculated from the Horowitz and Metzger model. These energies are proportional to the slope of these straight lines. When the quantity $\log \log \left(\frac{1}{1-Q/Q_0}\right)$ vanishes the temperature corresponds to the maximum rate of release.

CO_2, which are usually less than 5% of the whole fluid and for the determination of which analytical accuracy is rather poor. Also, CO_2 is easily dissociated in both the extraction furnace and the ionization cell.

The diffusion model gives different activation energies for the different samples (Figure 10); these values range from 13 to 19 kcal/mole. With continuous analysis (Figure 7A), the dissociation model (Figure 11) gives, between 300 and 750°C, an activation energy of 20 ± 3 kcal/mole and a T_s

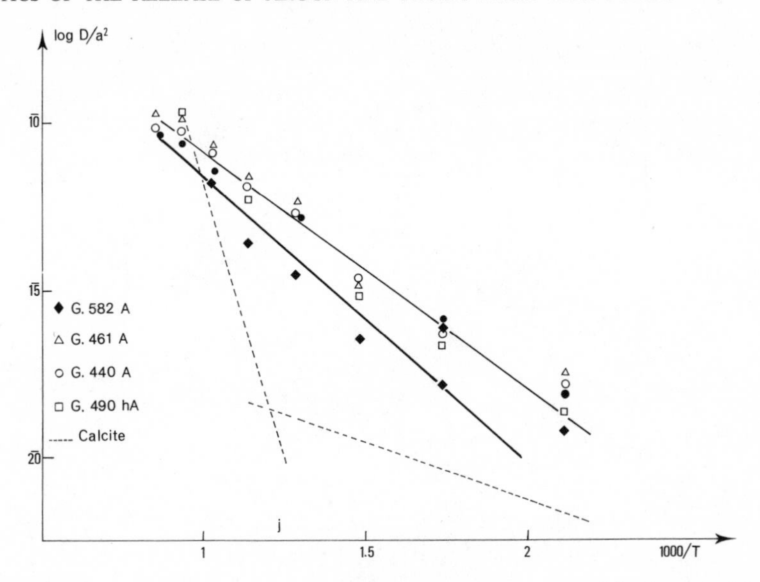

Figure 10. CO_2 diffusion characteristics calculated from the diffusion model. Step heating analysis.

of about 720°C. This 'activation energy' is close to those which have been calculated for water liberation. On the other hand, the results on a pure calcite (Figures 10, 11) are different, suggesting that extracted CO_2 is not produced by carbonate dissociation.

4 DISCUSSION

The preliminary baking of the sample GL-O at 100°C during 15 hr and the pumping out of other samples eliminate most adsorbed water. The onset of radiogenic argon outgassing coincides with that of the constitution water, in the 300–600°C range.

Former investigations have shown that from 80 to 180°C no significant radiogenic argon departure occurs and generally does not accompany the release of adsorbed water (Hurley *et al.*, 1960; Polevaya *et al.*, 1961; Yukhnevich *et al.*, 1969; Odin, 1975).

This adsorbed water, which is eliminated before spectrometric analysis, amounts to 32–56% of the fluid phase (Table 4); a large part of it is adsorbed on and between the sheets. Since very little argon, if any, is released below 200°C, this suggests that radiogenic argon does not occupy the same structural site. For other types of micas, we can assume that most of the radiogenic argon is also not present between the sheets (Brandt and Voronovskij, 1964; Zimmermann, 1970; 1972). The apparent atomic radius

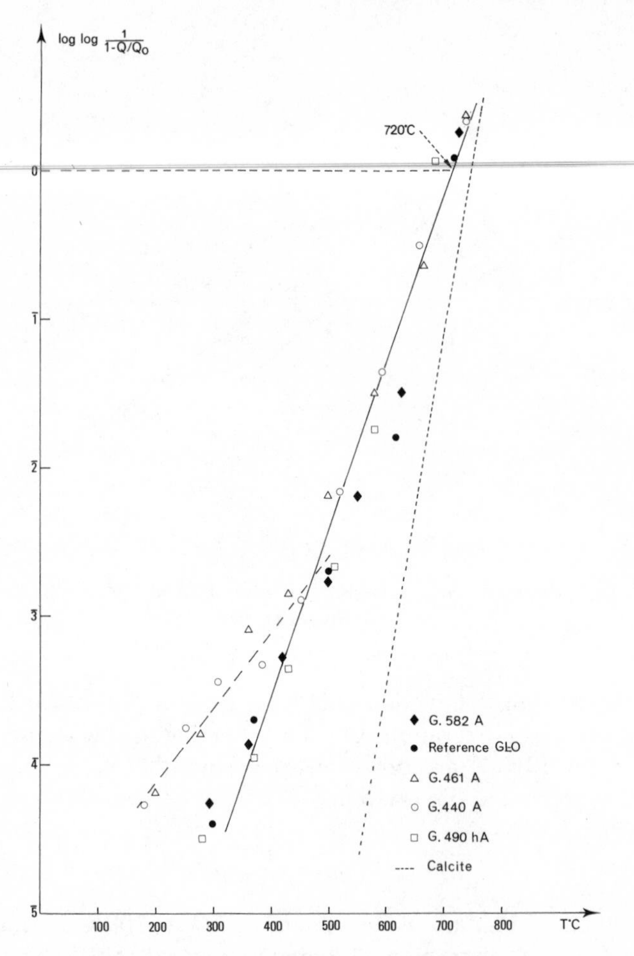

Figure 11. CO_2 continuous analysis. Activation energies of the release of CO_2 calculated from Horowitz and Metzger's model.

of argon, 1.91 Å, is 20% greater than the molecular radius of water; but this fact alone does not account for the better argon retention, particularly at temperatures lower than 200°C. This electrically neutral argon atom is loosely bound in the lattice, unlike the potassium ion.

For glauconitic minerals, activation energies are higher for the release of argon than for the release of constitution water (Table 7). The Recent glaucony G.490 hA is a specific case; relict structures may be observed in it which probably contain radiogenic argon.

In the glauconitic minerals, the radiogenic argon, which is more tightly held than constitution water, is partly trapped in the sheets themselves.

Table 7. Activation energies for the release of argon and constitution water (diffusion model).

Sample	G.582 A	GL-Og	G.461 A	G.440 A	G.490 hA
K%	6.91	6.59	6.07	5.58	2.83
E_{argon} (kcal/mole)	24	28.8	20	17.7	16.7
E_{H_2O} (kcal/mole)	18	20	16	20.5	12.6

According to Brandt and Voronovskij (1964), the recoil energy which is transmitted to the argon atom at the time of its formation from ^{40}K is higher than the Si–O bond energy. Then, ^{40}Ar atom can overcome the oxygen frame binding energy and enter the lattice where, on account of its diameter, it cannot be easily released by volume diffusion until dehydration reactions and concomitant lattice rearrangement become effective at a higher temperature.

Moreover, more potassic glauconies, of which the mineralogical components are better closed, retain more tightly their radiogenic argon. This indicates that a part of the argon remains in potassium sites which interconnect the sheets. These sheets with a high potassium content are poor in interlayer water. The duality between interlayer water (early departure) and radiogenic argon (late departure) can thus be explained.

For the release of argon, a comparison of these results with those on different varieties of micas (Zimmermann, 1970; 1972) shows that the activation energies in more potassic glauconitic minerals (E from 28.8 to 20 kcal/mole) are close to those of phlogopite ($E \simeq 30 \pm 5$ kcal/mole) and biotite ($E \simeq 25 \pm 5$ kcal/mole); however, they are significantly smaller than those of muscovites ($E \simeq 55$ kcal/mole). Evernden *et al.* (1960) calculated a 28 kcal/mole activation energy for both biotite and glaucony (of which the mineralogical compositions are not specified). At 200°C, the diffusion coefficients (D/a^2) for biotite, phlogopite and different glauconitic minerals range from 1.5×10^{-9} to 4×10^{-9} s^{-1}.

In addition, the dehydroxylation activation energies which were evaluated for these micas (biotites, phlogopites, muscovites) are also close to those calculated for the release of the radiogenic argon. In all of these, including the glauconitic minerals, the release of radiogenic argon starts when the lattice begins to be disturbed by hydroxyl ion loss. Kinetics studies of the release of water appear, then, to be a powerful investigative tool of reliability in K–Ar dating.

Acknowledgements

We thank D. Dautel and G. Guyetand for technical assistance and we are grateful to F. Albarède for his comments and improvement of the English of this article. This is a contribution to Project 133 of the International Geological Correlation Programme.

Résumé

Cinq glauconies de composition minéralogique variée ont été soumises à un chauffage sous vide par étapes de 100 en 100°C ou de 200 en 200°C entre 200 et 1100°C. Pour chaque étape on a analysé les masses 40 et 36 pour l'argon; 12 (C); 15 (CH_4); 18 (H_2O); 28 (N_2+CO+divers) et 44 (CO_2). De même, on a effectué un chauffage continu et analysé les masses 18, 28 et 44 des gaz extraits. La cinétique de la libération de ces gaz du réseau des divers minéraux glauconitiques a ainsi été établie pour la première fois en détail.

Une fuite d'argon supérieure à l'incertitude analytique a été décelée dans un échantillon chauffé à 200°C; cette mesure mérite vérification. Le départ d'argon se fait entre 400 et 600°C essentiellement; à 800°C tout l'argon est pratiquement extrait des glauconies anciennes. Au contraire, dans la glauconie naissante d'âge récent, de l'argon est retenu jusqu'à 1000°C. Ceci est la preuve qu'il est piégé dans une structure minérale héritée, plus solide que les minéraux glauconitiques, bien que peu décelable par diffractométrie X. Ce phénomène pourrait être utilisé pour *déceler un âge initial non nul dans une glauconie ancienne* peu évoluée.

L'énergie d'activation calculée pour l'eau de constitution (OH^-) est inférieure à celle calculée pour l'argon qui est ainsi plus solidement retenu que les ions oxhydryles. On est alors amené à supposer que l'argon n'est pas simplement emprisonné entre les feuillets mais réellement imbriqué dans des sites solides du réseau. Cet argon radiogénique n'est *libéré que lors de la désorganisation du réseau phylliteux* marquée par la sortie de l'eau de constitution. Ce n'est donc pas une simple diffusion.

Les glauconies dont les composants minéraux sont plus riches en potassium retiennent plus énergiquement leur argon radiogénique; ceci indique que la rétention de l'argon est quand même liée à la présence de potassium dans les interfeuillets.

La rétention de l'argon dans les minéraux glauconitiques évolués est équivalente à celle des micas phlogopite et biotite mais plus faible que dans une muscovite.

(Manuscript received 10-7-1980)

Numerical Dating in Stratigraphy
Edited by G. S. Odin

19

Potassium–argon dating of washed, leached, weathered and reworked glauconies

GILLES S. ODIN and DAVID C. REX

Glaucony can be altered by surface weathering as well as by the various solutions used in its preparation for analysis. The present chapter reviews data taken from experimental and field samples with a view to obtaining more information on (1) possible ways of processing the chronometer before analyses, (2) the selection of samples, and also (3) the interpretation of the analytical results as a function of field outcrop.

We have especially studied the ultrasonic treatments used to clean the green glaucony grains, the acid treatments used to remove the rock matrix and adsorbed cations, the effect of highly saline solutions intended to quickly reproduce the possible slight natural cation exchange processes, and finally, natural examples of reworking in a continental environment.

1 ULTRASONIC TREATMENT AND K–Ar APPARENT AGES OF GLAUCONIES

In his study on the dating of glauconies, Obradovich (1964) has tested the effect of *moderate ultrasonic treatment* on the potassium and argon content. In all cases the treatment increases the potassium content as shown in Figure 1. The increase in potassium content appears statistically greater in potassium-poor grains. Considering the related argon content, Obradovich did not see any systematic change. However, in a few cases an anomalously high Ar/K ratio (apparent age) is lowered by the treatment. We conclude that ultrasonic treatment frequently removes a 'neutral' non-glauconitic fraction and also in a few cases an argon-rich inherited (polluting) fraction.

The general absence of change of age after moderate ultrasonic treatment is confirmed by Ghosh (1972, pp. 34, 42–43), at least for potassium-rich glauconies. However, he shows that a potassium-poor (3.70% K) glaucony, Middle Eocene in age, became richer in K (4.45% K) after ultrasonic

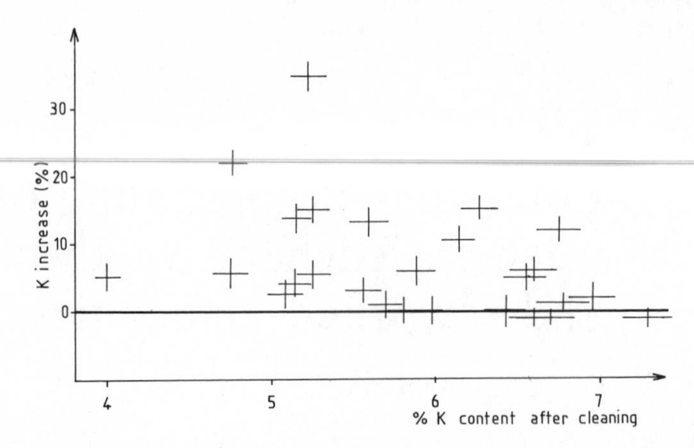

Figure 1. Increase in potassium content after moderate ultrasonic treatment of glauconies. Most data taken from Obradovich (1964); the 2σ uncertainty is given by the figure. There is a tendency for higher enrichment for potassium-poor glauconies.

treatment while its anomalously high initial apparent age (80.1 Ma, ICC) became more credible (46.7 Ma, ICC). The absence of systematic ultrasonic treatment may therefore lead to too high apparent ages. This partly explains the sometimes younger ages found by Obradovich (1964) on remeasuring after ultrasonic cleaning the samples dated by Evernden *et al.* (1961).

We have tested three potassium-rich samples. The samples were ultrasonically agitated in normal water two or three times for 5–10 mn with a careful washing between each agitation. They were then resieved to their original grain size, 160–500 μm, and reseparated with a magnetic separator. The potassium was then measured in two different laboratories and the argon measured once in Berne on both untreated (N) and treated (US) samples. The results are shown in Table 1.

Table 1. Potassium and argon content in three potassium-rich glauconies before (N) and after ultrasonic (US) treatment. We generally prefer to use the US treated samples because this treatment removes possible impurities from the cracks of the grains and the less evolved outer part of the grains when initially present.

Sample	Potassium K(% $\pm 2\sigma$)	Argon (nl/g $\pm 2\sigma$)	K(%)/Ar (nl/g) $\pm 2\sigma$
G.487 A(N)	6.78±0.13	24.93±0.57	3.68±0.11
G.467 A(US)	6.57±0.14	25.40±0.58	3.87±0.12
G.468 A(N)	6.70±0.20	25.41±0.56	3.79±0.14
G.468 A(US)	6.50±0.11	25.26±0.57	3.89±0.11
G.482 A(N)	6.62±0.15	34.46±0.76	5.21±0.16
G.482 A(US)	6.70±0.15	35.93±0.79	5.36±0.16

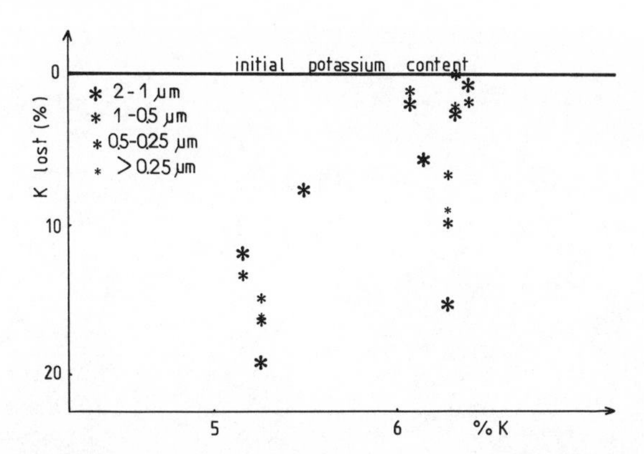

Figure 2. Potassium content on various fractions of ultrasonically disaggregated glauconies. Very strong ultrasonic treatment leads to the break-up of glauconitic crystallites. The proportion of potassium lost is seen to be greater with low-potassium samples. (Results interpreted from Obradovich, 1964.)

The effect of a *long ultrasonic treatment* with a more or less complete disaggregation of the grains was tested on ten samples by Obradovich (1964) and on two samples by Ghosh (1972). The different granulometric fractions obtained were then measured by the first author both for argon and potassium. His results are shown in Figure 2, where the y-axis gives the percentage of loss of the whole potassium content in various size fractions of disaggregated glaucony and the x-axis gives the K content of the untreated sample (grains). In general, it seems that the lower the potassium content, the higher is the *loss of potassium after ultrasonic treatment.* There is no systematic difference between the different sizes of particles chosen. The apparent ages calculated by Obradovich on the different fractions do not show lowering of the apparent age and the author concludes, from this and from electronmicrographs, that his ultrasonic treatment does not disaggregate the microcrystals, but rather fractures the mineral particles. Finally, we should note that, except for the minerals most rich in potassium, the fracturing of the glauconitic minerals leads to a noticeable loss of potassium (accompanied by the argon of the same sites). Scanning electron microprobes show that glauconitic particles are of different size depending on the evolution of the glauconitization. In evolved to highly-evolved grains they are generally larger than 2 μm and may go up to 10 μm (Odin, 1971b). Little-evolved grains, on the other hand, are formed with very small, less coherent mineral particles, which may be easily disaggregated (Figure 3).

In conclusion, we may emphasize that moderate ultrasonic treatment systematically applied to glauconies leads to greater certainty about the significance of the ages obtained. Only the evolved authigenic minerals are dated while the less evolved ones, often peripheral to the grains, may be

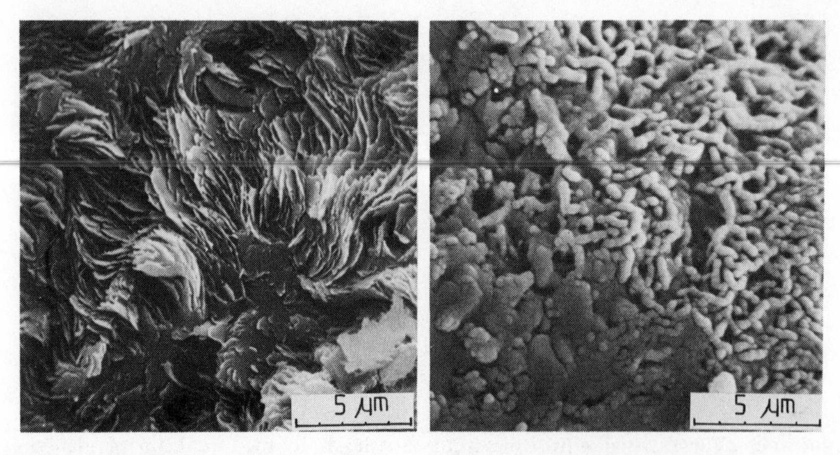

Figure 3. SEM aspects of glauconitic minerals. Left: evolved minerals in the centre of a grain of G.16, a potassium-rich Cenomanian glaucony. Right: little-evolved minerals of the outer part of a grain of GL-O. Fractionation of the small particles of disaggregated glaucony may lead to the choice of the less evolved glauconitic particles of the grains.

removed. But too long a treatment (one hour or more) may lead to the fracturing of the glauconitic particles and is not recommended.

2 ACID TREATMENT OF GLAUCONIES

An acid solution is often used to clean glaucony. This section reviews data on the possibility of altering the chronometer by this treatment.

Hutton and Seelye (1941) show that, as soon as the concentration of hydrochloric acid is higher than 1% (nearly 0.3 N), the glaucony is noticeably altered. Analyses of the supernatant solution by these authors show that iron is one of the first mobilized ions. This is confirmed by Delvaux *et al.* (1966), for whom, although potassium was not the first mobilized ion from the glauconitic minerals structure using hydrochloric acid, in all cases a large proportion of potassium was present in the solutions, showing the possibility of apparent age changes.

Polevaya *et al.* (1961) have tested hydrochloric acid in removing possible adsorbed potassium. They try to replace this removed potassium by putting the acid-leached samples in potassium and calcium chloride solutions. The acid used was 0.05–0.5 N HCl. Their measurements show that the radiogenic argon content does not change and they deduce that no perceptible amount of potassium is superficially adsorbed on the glauconies used.

Thompson and Hower (1973) have tested the loss of potassium as a function of time during alteration of various glauconies in 0.5 N hydrochloric acid solutions periodically replaced. For a more efficient reaction, the grains were systematically disaggregated to less than 2 μm fraction. This leads to a fracturing of the crystallites as shown above for evolved glauconitic minerals; according to us, the choice of the less than 2 μm fraction leads to analysis of the less evolved minerals of a grain. In any case, the experiments referred to lead the authors to the conclusion that potassium is in three kinds of crystallographic sites, one of which does not retain the potassium ions as effectively as the others. The same kind of behaviour is reported below for experiments with saline solutions. With regard to the mineralogical nature of the components of the glauconies used, the authors conclude that with an increasing proportion of the expandable layers (this is linked with the step of evolution of the grains, see Odin and Dodson, this volume), there is an increasing proportion of the total potassium in this fragile site. The proportion of potassium more easily mobilized increases from 2 to 25% of the total potassium as the proportion of expandable layers increases from 0 to 65%. Unfortunately this trend is not very systematic and no one set of argon data illustrates this fully.

From these various experiments we will keep in mind that *hydrochloric acid alters glauconitic minerals at low concentrations* of 0.3 or 0.5 N. However, no proof of a concomitant change of age has been given. If it is true that a significant portion of the potassium may be in a relatively 'open' site, we must try to select glaucony where this proportion is as low as possible, i.e. the most evolved grains where the glauconitic minerals contain only a very small quantity of these open sites. The possible diffusion of argon will then be diminished.

We have tried to find conditions of acid leaching efficient for removing impurities and less aggressive for glauconitic minerals. The results discussed below are taken from Pomerol and Odin (1974). Acid leaching has been carried out on the reference material GL-O, which is formed with potassium-rich minerals. GL-O was tested under its grain facies as well as after crushing it in a tungsten steel rings mortar. All experiments quoted here were done with 2 g of glaucony agitated in 250 cm^3 of solution. Table 2

Table 2. Results of analyses of supernatant solutions after leaching of 2 g of GL-O at 50°C using rapid filters.

		Elements in μg cm^3					
		Al	Fe	Mg	K	Ca	Na
HCl	Grains	13.3	33	16.5	20	125	4.0
	Crushed	26	150	27	52	190	7.8
CH_3COOH	Grains	2.1	1.1	4.2	5.6	64	3.0
	Crushed	7.8	8.1	8.3	30	122	5.3
H_2O	Grains	0.2	0.9	0.4	3.1	5.4	1.6
	Crushed	(120)	7.2	3.9	22	11.1	2.8

shows the results of analyses obtained by B. Pomerol after leaching with normal hydrochloric acid, normal acetic acid and deionized water. After leaching, the solutions were removed by leaching on a very rapid first filter and then a second rapid filter. The filtrate was evaporated and then taken up in 10 N hydrochloric acid solution for analysis by atomic absorption methods. From these first results it is clear that, at equivalent concentrations, *hydrochloric acid is much more aggressive than acetic*, especially for the trivalent ions in layer sites (Al, Fe). For potassium and the two other ions in interlayer sites, the difference of mobilization is less important, and the behaviour of magnesium is intermediate. As a result, we strongly recommend the use of acetic rather than hydrochloric acid for removing impurities (carbonates).

From the difference betwen crushed samples and grains, we first thought that part of the powder had probably passed through both filters because of the important quantities of ions measured in deionized water. A second set of experiments was thus carried out with slow filters; the results of these are shown in Table 3. In order to check for mobilization of new ions during the dissolution of the filtrate, two experiments were done without 10 N HCl remobilization. These blanks show no difference between experiments redissolving with HCl and with water. The quantities measured are thus *certainly extracted by the leaching*. After this check it is clear that far more ions are removed from crushed samples by various solutions than from granular samples. This crushing appears to us very dangerous during acid leaching although there is no evidence for a concomitant change of apparent age. It appears that acetic acid is more sensitive to the preliminary crushing, but at equivalent concentrations attack by acetic acid on crushed samples is less important than the attack by hydrochloric acid on grains for the ions from layer sites and is of the same order for the interlayer cations.

Considering next the results obtained with deionized waters on crushed

Table 3. Results of analyses of solutions after leaching of 2 g of GL-O using slow filters. (1) with a 10 N HCl retake of the evaporate; (2) without a 10 N HCl retake of the evaporate in order not to leach powder which possibly passed through the filters.

		Elements in $\mu g\ cm^3$				
		Fe	Mg	K	Ca	Na
CH_3COOH	(1) Grains	1.8	2.35	2.85	19	1.7
(50°C)	(1) Crushed	7.9	4.4	18.5	53	1.5
50°C	(1) Crushed	0.2	0.5	10.8	4.6	
H_2O 50°C	(2) Crushed	0.1	0.4	10.6	2.5	
20°C	(1) Crushed	0.2	0.5	10.4	4.0	
20°C	(2) Crushed	0.1	0.4	8.8	2.3	

glauconies, it seems that only a very small proportion of the interlayer cations may be mobilized. Again it appears better not to crush the grains.

To sum up, we *recommend the use of acetic acid instead of hydrochloric acid* in order to remove the carbonates possibly present in the cracks of the green grains. Ultrasonic powdering of the grains before this treatment leads to noticeably greater alteration of the crystallites and is best avoided. Insufficient data are available to show definitely that a change in K–Ar apparent age of the authigenic minerals may be due to acid leaching. We must, however, remember that the removal of inherited components by this treatment is efficient and often desirable (Rb–Sr method). If the hypothesis of multisite potassium position is accepted the experimental acid leaching shows that the most evolved grains will be less subject to loss of argon by diffusion. A combined ultrasonic and acid treatment will be the most efficient and less time-consuming.

3 WEATHERING IN SALINE SOLUTIONS

3.1 Introduction

The potassium used as radioactive parent for dating is situated in the interlayer sites of the crystal lattice of the glauconitic minerals. These interlayers of sheet silicates may be submitted to cation exchanges by saline solution in nature. We will test here the possibility of change of age related to this alteration.

Glauconitic minerals may be more or less rich in expandable layers; they will be more expandable when the potassium content is lower (Figure 4). Manghnani and Hower (1964) first studied and established the effect of the proportion of expandable layers on the cation exchange capacity of glauconitic minerals. The cation exchange capacity may be measured by shaking powdered samples of glaucony in a concentrated solution of calcium chloride as proposed by Cimbalnikova (1971). Part of the calcium is sorbed by the minerals, and the quantity sorbed may be determined chemically in milliequivalents per 100 grams (meq/100 g). Both these papers propose a cation exchange capacity of the order of 15–30 meq/100 g for the glauconitic mica. This value is equivalent to that obtained on high-temperature micas. The c.e.c. increases with the percentage of expandable layers in glauconitic minerals. At the other end of the mineralogical family, a glauconitic smectite has never been tested but equivalent minerals generally give values of roughly 100 meq/100 g. These data show that cation exchange will be related to the mineralogical nature of glauconitic components. Because of this, we use here glauconies with different mineralogical compositions to achieve a more complete understanding. Cimbalnikova (1971) indicates that these physicochemical properties are very difficult to reproduce, although

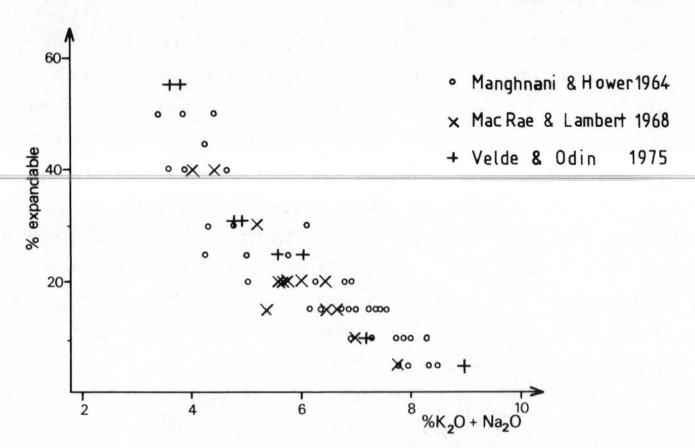

Figure 4. Ratio of expandable interlayers from glauconies *vs.* interlayer cation content. The potassium-poor glauconitic minerals with more expandable interlayers will permit easier exchanges.

the general trend, a higher c.e.c. for glauconies with less potassium, is always apparent if a sufficient number of samples is examined; we, however, can only present data on a small number of samples for potassium and argon measurements.

The most useful and complete information on which to base this project is reviewed in the work by Newman (1969). This author shows that measurable amounts of potassium are released when a mica is shaken with a salt solution for several hours; more potassium is released when the supernatant solution is removed and replaced with fresh. Sodium chloride is more effective than magnesium, calcium or lithium chlorides in exchanging potassium. Newman uses normal (molar) solutions in his experiments; he also tries different temperatures and shows that at 60°C the release is faster but the final concentration the same as at 25°C, and notes that between 0 and 10% of the total exchange the amount of potassium in solution decreases sharply (Figure 5). Between 20 and 80%, exchange curves are almost flat: the potassium in solution is independent of the percentage of potassium exchanged. This observation also shows that potassium is probably in two different sites of the crystallites. After changing the solution every three days some experiments required six months for a complete removal of potassium. In the case of glaucony the amount extracted each time is constant after three months (Figure 5B). For each step, at 60°C, the extraction is complete after three days in the case of phlogopite. Many experiments in various fields have shown that phlogopite and some glauconitic minerals display similar behaviour. Newman's work shows that 'glauconite is much less sensitive to exchange than biotite, as sensitive as phlogopite while muscovite is less sensitive than glauconite'. Unfortunately, the author

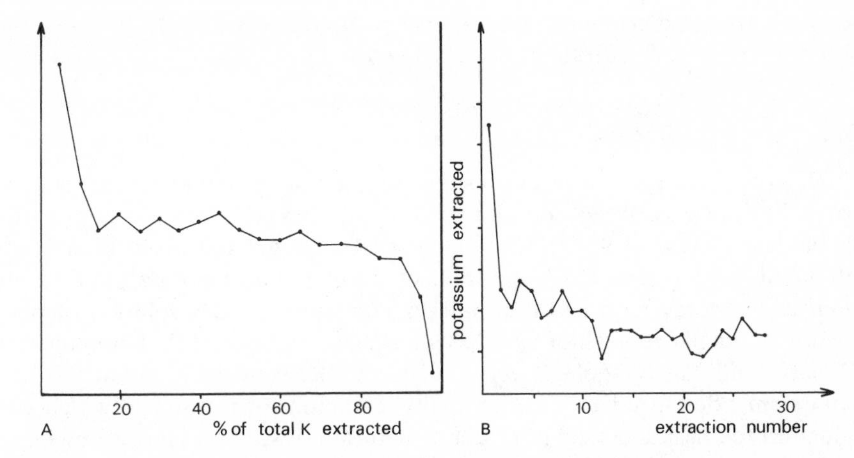

Figure 5. Departure of potassium in mica leached by 1 N chloride solution (after Newman, 1969). A. Sample of curve obtained for phlogopite; y-axis shows the quantity of potassium extracted during each period of three days. B. Curve obtained on a glaucony; after three months of periodic changes of solution the extracted quantity remains stable; the extraction is not complete.

does not give the precise mineralogical composition of his 'glauconite' and only one sample was studied.

Robert (1972–1973) has measured potassium extraction in glauconitic minerals and various micas. He generally uses normal sodium and barium chloride solutions at 60°C, renewing them after four days. He notes that, whereas extraction proceeds from trioctahedral micas (such as biotite), dioctahedral ones (such as glauconitic mica) require the use of very pure salts and a low ratio, mineral *vs.* solution, for an efficient exchange of ions. (Namely, 30–50 mg of mineral with 1 l of solution.) No argon was measured in any of these materials.

Kulp and Bassett (1961) have carried out cationic exchange on biotites using 2 N magnesium chloride solutions continuously stirred and heated at 100°C. The results obtained are shown in Table 4. The argon is preferentially leached. One should note, the very important proportion of potassium

Table 4. Cation exchange on biotite in two normal magnesium chloride solutions: data by Kulp and Bassett (1961).

Potassium (K%)		rad.^{40}Ar/^{40}K	
Before leaching	After leaching	Before	After
7.24	1.64	0.0590	0.0481
7.24	6.93	0.0590	0.0526
5.69	4.36	0.1380	0.1082

leached after experiments during only 1–4 weeks without changing the solution compared with data given below on glauconies.

3.2 New experiments

We now present some recent results obtained by us. All experiments were carried out in glass flasks containing 1 l of solution and 4.0 g of sample. The whole was heated at 90–95°C in a temperature-regulated oven. Four kinds of liquids were used: a 2 N solution of pure sodium chloride (117 g/l *pro analysi* of Merck with less than 0.01% potassium); a 2 N solution of pure hydrated magnesium chloride (203 g/l $MgCl_2$, 6 H_2O R.P. Normapur of Prolabo and $MgCl_2$, 6 H_2O *pro analysi* of Merck with less than 0.001% potassium); deionized water; and finally, common town water to which was added in the flask a crystal of calcite to saturate it with calcium carbonate so that the condition could be compared with glaucony occurring in the chalk and for comparison with the low pH of the deionized water.

Two periods of time were tried first, 10 days with two changes of solution and 33 days with one change of solution; the results showed little loss of either potassium or argon. Hence we did a second set of experiments of 77 days' duration with six changes of solution. The suffixes used for designating the samples are as follows: 2 Na^+ for solutions of NaCl; Mg^{++} for solutions of $MgCl_2$; d.w. for deionized water; Ca for water saturated in $CaCO_3$; 10 d (days), 33 d or 77 d for the duration of experiments and the relevant number of changes of solution.

The samples used were 4 g. The fractions were 160–500 μm in size. The reference material *glauconite GL-O* (grains), a highly-evolved component, was the most completely tested; an evolved Albian glaucony (near 6.1% K); and a slightly less evolved (5.7% K) Ypresian glaucony. A Recent nascent glaucony with almost 2.8% K was also studied.

A posteriori, we note that the extraction of potassium was clearly linked with the change of colour seen on the grains. Some of them become very pale green. Most grains slightly change colour from dark green to mid-green; a few of them completely change colour to yellow-green. This suggests that weathering is not homogeneous, but affects some grains more than others apparently identical in character. This must be remembered in the field and may indicate possible weathering of an outcrop. We have frequently noticed that glaucony from slightly weathered outcrops contained two kinds of grains, different in colour. This criterion may be an index of alteration. We also observed that concentrated solutions provoked alteration in the glass of the flasks, transparent thin films swimming in the liquid. Finally, the calcium carbonate tends to encrust the grains in the bottle with the calcite crystal. Before analysis of these grains they were washed with 0.1 N acetic acid and then sieved, dried at 60°C and purified by magnetic separation.

Table 5. Results of weathering of GL-O in various solutions heated at 95°C. Calculation of the loss is given in percent of the total quantity initially present. It cannot be given to an accuracy better than ±2.0% when only one datum is available. All potassium data by Miss M. Lenoble in Paris; argon measured by Odin in Leeds (L), Strasbourg (S) or Berne (B).

Sample and treatment	Argon (% rad./tot.)	(nl/g)	Lab.	Potassium (K%)	Loss K(%)	Ar(%)
GL-O	93	24.80		6.56±0.07	—	—
$GL\text{-}O_{d.w.24d}$	64.9	24.92	L	6.37–6.51	2.0	0
$GL\text{-}O_{d.w.77d}$	92.7	24.74	S	6.32–6.46	3.0	0
$GL\text{-}O_{Ca\ 77d}$	89.4	24.62	S	6.39	3.0	0.5
$GL\text{-}O_{Mg^{++}10d}$	74.1	(25.73)	L	6.63	0	0
$GL\text{-}O_{Mg^{++}10d}$	36.8	24.72	L	6.63		
$GL\text{-}O_{Mg^{++}33d}$	70.7	24.06	L	6.35	3.5	3.5
$GL\text{-}O_{2Na^{+}33d}$	60.8	23.29	L	6.15 }	6.5	5.5
$GL\text{-}O_{2Na^{+}33d}$	89.7	23.67	S	6.15 }		
$GL\text{-}O_{2Na^{+}77d}$	89.2	21.27	S	5.60	15.0	14.0

The results of the most complete experimentation on GL-O are shown in Table 5; those obtained on the other three samples are shown in Table 6.

The potassium measurements were all obtained in the same laboratory at Paris. The argon data were mostly obtained in Leeds (9.1979), with a few from Strasbourg (4.1980) and Berne (before 1977). The intercalibration between different laboratories was tested with at least four measurements on the reference *glauconite GL-O* (grains) for which we assumed a value of 24.8 nl/g. In Strasbourg the calibration of the spike was done with GL-O, while in Leeds the mean value obtained on five analyses of GL-O was 24.82 and no correction was applied. The higher values of atmospheric components obtained in this laboratory were due to a small leak which slightly reduced the normal reproducibility and accuracy.

3.3 Discussion and conclusions

The sodium chloride solution is more aggressive than the magnesium chloride one on GL-O, in agreement with the observations of Newman (1969). After 10 days and two changes, the last solution did not produce any alteration. This superior efficiency of the sodium chloride solution is also well evident for G.461, but is less clear for G.440 and does not hold for G.490; this sample will be interpreted later.

As well noted by Robert (1972), the results obtained on GL-O and two other glauconies clearly show that weathering by concentrated solutions removes far fewer interlayer cations from glauconitic minerals than from the biotite studied by Kulp and Basset (1961).

Table 6. Results of weathering of three glauconies in various solutions heated at 95°C (legend as for Table 5).

Sample and treatment	Argon (% rad./tot)	Argon (nl/g)	Lab.	Potassium (K%)	Loss K(%)	Loss Ar(%)
G.461	88	24.88	B	6.07 ± 0.17	—	—
$G.461_{d.w.24d}$	83.5	24.62	L	5.87	3.5	1.0
$G.461_{Mg^{++}33d}$	85.9	22.55	L	5.64	7.0	9.5
$G.461_{2Na^{+}33d}$	69.3	21.84	L	5.45	10.0	11.0
	81.9	22.60	L			
G.440 A	70	11.58	S	5.58 ± 0.35	—	—
$G.440\ A_{d.w.24d}$	54.9	11.13	L			
	18.3	11.19	L	5.27	5.5	3.5
$G.440\ A_{d.w.77d}$	81.4	11.10	S	5.12–5.15	8.0	4.0
$G.440\ A_{Mg^{++}33d}$	47.5	11.19	L	5.24	6.0	3.5
	66.7	11.11	L			
$G.440\ A_{2Na^{+}33d}$	54.4	10.81	L	5.20	7.0	6.5
$G.440\ A_{2Na^{+}77d}$	76.9	9.64	S	4.47	20.0	17.0
G.490 hA	50	8.90	B	2.83	—	—
$G.490\ hA_{Mg^{++}33d}$	52.9	8.09	L	1.97–2.04	30	9.0
$G.490\ hA_{2Na^{+}33d}$	50.4	8.82	L	2.10	25	3.0
	24.1	8.33	L			

Looking at the general scheme of extraction shown by Newman (see Figure 5A), we conclude that, for the sodium chloride solution on GL-O, the first leaching removed nearly 4% of the total potassium and all the six next ones nearly 1.8%. If this is correct, we may predict that more than 50 changes of solution would be needed for a complete removal. In the same manner we obtain for G.440 A an initial loss of potassium of approximately 6% of the total with the first extraction before the first change of solution and nearly 2.5% at each of the next six changes of solution. If this is correct, only 35 changes of solution would be sufficient for a complete removal. These conclusions could be tested in the future in more complete experiments.

Concerning our 'blanks' done with water, we observe that deionized water at 90°C is not neutral with regard to the glauconitic minerals. As for experiments with concentrated solutions, we changed the supernatant from time to time. As a result, looking at samples GL-O and G.440 A, it appears clear that more potassium is extracted after 77 days and six changes than after 24 days and one change. One result would not have been conclusive due to the analytical uncertainty, but the four results obtained on these two samples and the one result on G.461 are very consistent: deionized water provokes the extraction of potassium from the glauconitic minerals.

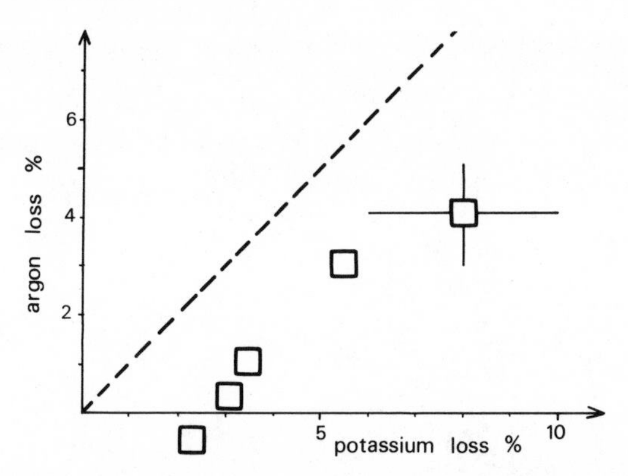

Figure 6. Potassium and argon loss observed after leaching of glauconies at 95°C in deionized water. The 2σ uncertainty is given for one sample; apparent ages appear increased.

The argon extracted is systematically less abundant than the potassium (see Figure 6). This may be interpreted as a different site localization for the exchanged potassium and the radiogenic argon, the first being more open (see Zimmermann and Odin, this volume, for the site of argon). There is no significant difference between treatment with deionized water and with water saturated in calcium carbonate.

Now comparing the weathering obtained on GL-O and G.461 and G.440, it appears that GL-O is less leached by deionized water, 2 Na^+ or Mg^{++} solutions than the two other samples. It may be surmised that *the most closed minerals* (*with more potassium*) *are more stable* with regard to the various weathering effects.

With the exception of the treatment with deionized water, in which there is a systematic tendency to an increase in the age, it may be stated that, at least for a moderate departure of potassium, the *apparent ages of the glauconies have not been changed* by these treatments. This is shown in Figure 7, where potassium and argon are plotted as a function of the loss calculated from Tables 5 and 6. Practically all the samples are on or near the line of slope unity if we assume a minimum uncertainty of $\pm 1\%$ for the change in argon content and of $\pm 2\%$ for the change in potassium content. As a result, it seems that during leaching by concentrated solutions the sheets are progressively completely opened, freeing Ar and K equally wherever they are in the sheets, while other sheets remains completely closed. This result parallels the observation of the colour of the grains: some grains are very altered, others not.

Now, as far as the nascent glaucony G.490 hA is concerned, its behaviour can hardly be understood in the same way as above.

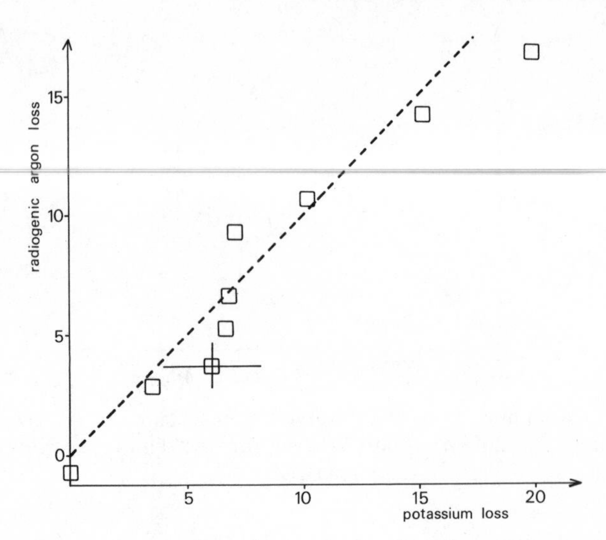

Figure 7. Potassium and argon loss observed after leaching of glauconies at 95°C in concentrated solution. The 2σ uncertainty is given for one sample. The line of slope 1 is drawn; apparent age is not changed but sample G.490 hA is not shown.

The very important loss of potassium may be understood as due to the very open crystal structure of nascent glauconitic minerals. However, the relatively low loss of argon after Mg^{++} treatment and very low loss of argon (both relative and absolute) after the 2 Na^{+} treatment leads us to look for an original interpretation. We suggest that although no traces of potassium and argon-rich minerals are seen either by X-ray diffraction or by scanning electron microprobe, radiogenic argon exists trapped in mineral structures different from those of normal glauconitic minerals. This argon-rich mineral is probably a very altered and disorganized one since its structure does not reflect X-rays, although it is probably abundant in the matrix used by mud-eaters. Then, in fact, the coprolites used are composed of three kinds of minerals: (1) the authigenic ones: low-K glauconitic smectites; (2) the inherited ones: essentially kaolinite, as in the whole Gulf of Guinea, where the kaolinitic mud is deposited from the soils of the tropical continent; and (3) other little-organized mineral structures where radiogenic argon is trapped. These may be some groups of Si tetrahedra originating from the alteration of mica or feldspars of the very old African shield.

In conclusion, we should remember that moderate artificial weathering with saline solutions does not affect the apparent age of glauconies. This apparent age will be much less affected by intense weathering if the glaucony is a highly-evolved one. Treatment of glauconies with deionized water at 95°C causes a loss of potassium. This loss is essentially for less-evolved glaucony but is negligible for normal periods and temperatures of washing.

4 NATURAL LEACHING AND REWORKING OF GLAUCONIES

This section reviews available data on the effect of the reworking of glauconies on their apparent potassium–argon ages. Although rarely investigated, this possibility must be systematically considered in calibrating the stratigraphic column. At least in Europe, examples of reworking, discussed below, have been observed in sediments being deposited today as well as in Quaternary, Palaeogene and Lower Cretaceous ones.

4.1 Recent weathering

The present and Quaternary reworking of glauconitic horizons has been investigated by Odin and Hunziker (1974) using the potassium–argon method. Their data, completed analytically since that time, are summarized here.

In Normandy (France), the glaucony-bearing base of the Cenomanian, together with the Late Albian, has a wide distribution in horizontal outcrops in river section or in cliffs (Figure 8 and Odin and Hunziker, this volume). Glaucony is subjected to weathering and slight fluvial transport following separation from the glaucony-bearing bedrock by streams and creeks which dissolve the carbonate fraction. From numerous samples collected (partly with the help of P. Juignet) in the Pays d'Auge, 10 samples representative of various degrees and possibilities of weathering have been selected by means of field data and X-ray diffraction studies. Diffractograms show that the non-weathered grains are composed of tightly closed glauconitic minerals (the interlayers are very thin, due to an abundance of potassium which collapses the successive layers). In these reworked glauconies the alteration may be detected by the slight opening of the interlayers shown on powder diffractograms.

Figure 9 especially shows the change in the appearance of the (001) diffraction peaks which indicate a mean initial thickness of the 'layer+interlayer' close to 10 Å, as for all micas. After weathering the (001) peak shows that this 'layer+interlayer' thickness increases to 11 Å or more.

All samples have been prepared in the same manner. We have not tried to analyse the most oxidized grains from the weathered samples collected in alluvium. On the contrary, we systematically selected the best available grains in order to have exactly the same conditions of dating as for a normal sample. Three samples, G.466, G.467 and G.468, are from an outcrop SE of Lisieux (see NDS 67); they are themselves partly reworked from Late Albian to basal Cenomanian. G.347, of Albian age, comes from the west

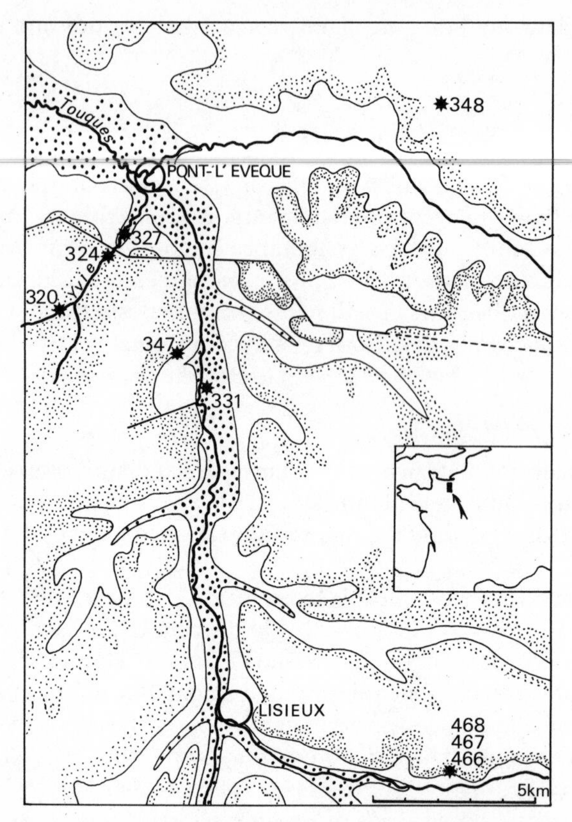

Figure 8. Geological sketch of the Pays d'Auge (Normandy) with sample localities. 335 is equivalent to 331 in a more eastern valley (Risle river). Quaternary alluvium is heavy stippled, Albo-Cenomanian formation light stippled; between these there is Jurassic limestones and above the Cenomanian an alteration formation occurs.

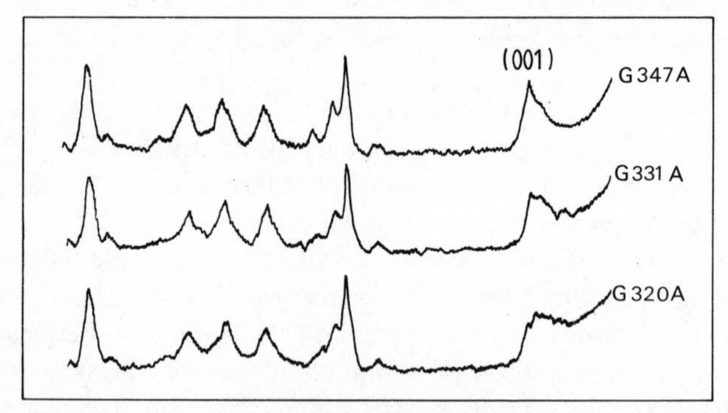

Figure 9. X-ray diffractograms of powdered glauconies. The evolution of the (001) diffraction peak shows a progressive opening (swelling) of the interlayers related to potassium extraction.

bank of the Touques river (Juignet *et al.*, 1967), which probably directly supplied the Quaternary alluvium of this river (G.331), and G.348 comes from a borehole east of Pont-L'Évêque. G.331 and G.335 come from Quaternary sediments of the low terrace of the Touques and the Risle rivers, now covered 1 m above with meadows. G.320, G.324 and G.327 are glauconies collected in a streamlet which cuts the glauconitic horizons before reaching Jurassic strata. They are sediments accompanied by mud; the water is saturated with carbonate, leading sometimes to incrustations of roots.

The analytical results are shown in Table 7. Compared with published data, the new potassium and argon results do not change the general significance of the conclusions. All argon measurements have been obtained in Berne. The results quoted here are recalculated using GL-O as reference with an assumed content of 24.8 nl/g, slightly higher than the mean value obtained previously with this equipment.

Figure 10 presents a view of the situation of the glauconies with regard to two reference isochrons: 98.5 Ma represents the mean apparent age (ICC)

Table 7. Results of isotopic measurements on glauconies taken in the bedrock and in alluvium coming from their natural disaggregation. Ar/K ratio in the same units as in the preceding columns. 2σ uncertainties are given. (m) indicates the mean of two or more independent analyses. Data completed and recalculated after Odin and Hunziker (1974).

Sample	Outcrop	Potassium (K%)	Argon rad.Ar (nl/g)	Argon (% rad./tot.)	Ar/K
G.468 A(US)	*In situ* outcrop	6.50±0.11(m)	25.26±0.57	88.7	3.89±0.11
G.467 A(US)	*In situ* outcrop	6.57±0.14(m)	25.40±0.58	87.8	3.87±0.12
G.466 A	*In situ* outcrop	6.38±0.26(m)	25.42±0.76	66.9	3.98±0.20
G.347 A	*In situ* outcrop	6.70±0.20(m)	26.53±0.88(m)	89(m)	3.96±0.18
G.348 A	*In situ* hole	6.44±0.19(m)	25.58±0.30	87(m)	3.97±0.13
G.331 A	IVry alluvium	6.35±0.19	24.74±0.53	87.7	3.90±0.14
G.335 A	IVry alluvium	5.98±0.18	23.05±0.54	78.0	3.85±0.15
G.320 A	Present river	6.08±0.24(m)	22.15±0.30(m)	85(m)	3.64±0.15
G.324 A	Present river	6.07±0.18(m)	22.70±0.86(m)	79(m)	3.74±0.18
G.327 A	Present river	5.98±0.18	22.64±0.50	86.9	3.79±0.14

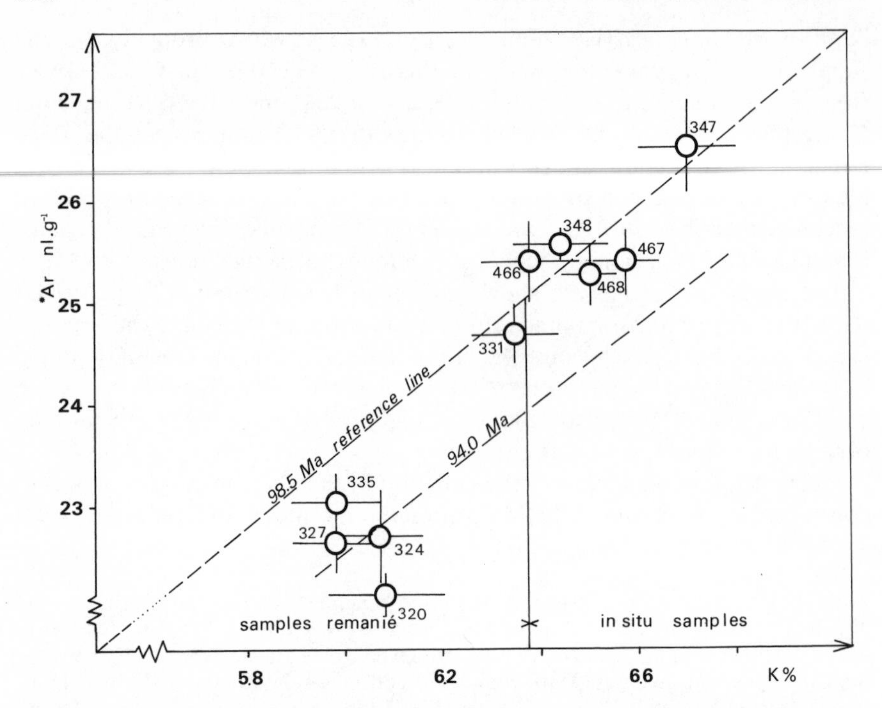

Figure 10. Argon content as a function of potassium content of *in situ* and remanié glauconies from Normandy (Odin and Hunziker, 1974). 1σ analytical uncertainty is shown; the apparent age of samples remanié is not very different from the other ones except for G.320A, which appears low.

of *in situ* samples, and 94 Ma is the initial apparent age of possibly Early Cenomanian glauconies. The figure shows only 1σ uncertainties on some samples. Analyses of only five weathered and five *in situ* samples are insufficient to produce statistically significant conclusions. However, the potassium contents of the glauconies reworked in fresh water, under temperate climate, are clearly below those of the *in situ* samples; up to 10% of initial K is lost, which corroborates the diffractograms. Argon contents are also lowered. The Ar/K ratios are not far from those of probable source glauconies. There is one sample, G.320, which is outside the 2σ uncertainty applied to the reference line 98.5 Ma. This sample is topographically the highest (see Figure 8) and it is probable that it comes from a younger glauconitic horizon than the mean of the others.

In another area two samples taken in beach sands below the cliffs in the Pays de Caux have shown potassium, argon and Ar/K ratios equivalent to those of glauconies collected in the cliffs themselves (Odin and Hunziker, 1974). No weathering is apparent; it appears that sea-water is less aggressive than fresh water ... which is conceivable, as glauconies are formed in sea-water.

In summary, the isotopic study of the *in situ* glauconies collected in Normandy and the reworked samples from Quaternary and present alluvium provides the first definitive evidence of natural alteration. From highly-evolved grains, the reworking may lead to the extraction of 10–15% of the potassium content. An equivalent departure of radiogenic argon is observed. At the worst, the data suggest a slight rejuvenation of the apparent age which does not pass beyond the 2σ uncertainty. According to these observations it appears that reworking of a glaucony from one horizon to the next in a continental environment with temperate climate will be easily detected by isotopic analysis if the weathering is moderate and the difference between the new deposited level and the reworked one is greater than the analytical uncertainty.

4.2 Ancient reworking in the Cretaceous

Allen *et al.* (1964) have proposed a new example of reworking of glauconies from Late Jurassic to Early Cretaceous. Two glauconitic horizons were sampled in the Wealden (Early Cretaceous) sediments of southern England. Wealden sediments are non-marine facies in which detrital components suggest Early Volgian sources (that is, Late Kimmeridgian and Early Portlandian stages). Around and beneath the Weald area the Early Volgian is sandy and glauconitic. Three samples of *in situ* Early Volgian were collected in a pit, a new excavation and a borehole (samples S.8003, S.8004, S.8005). Two samples (S.5501 and S.3501) were collected in the Wealden sediments. The results are shown in Figure 11. Argon was measured with preliminary outgassing and once without; no change in radiogenic argon concentration was observed.

The authors conclude that the apparent ages are remarkably consistent. A possible Palaeozoic origin has been invoked but the results show that the glauconies must, if detrital, be Jurassic. The detrital origin of glauconies is recognized on sedimentological grounds: the presence of detrital elements characteristic of a Jurassic source, the oxidation state and the low potassium content of Wealden glauconies.

A different hypothesis may also be put forward, if a certain number of preliminary assertions are changed. If marine horizons exist in Wealden series, and useful initial substrates were present, it is conceivable that instead of being weathered the low-potassium glauconies are 'little-evolved'. Hence they would always contain inherited radiogenic argon. These hypothetical authigenic Wealden glauconies would then appear older than the Cretaceous glauconitization, and by chance might seem coeval with the Early Volgian. The present authors do not support such a coincidence.

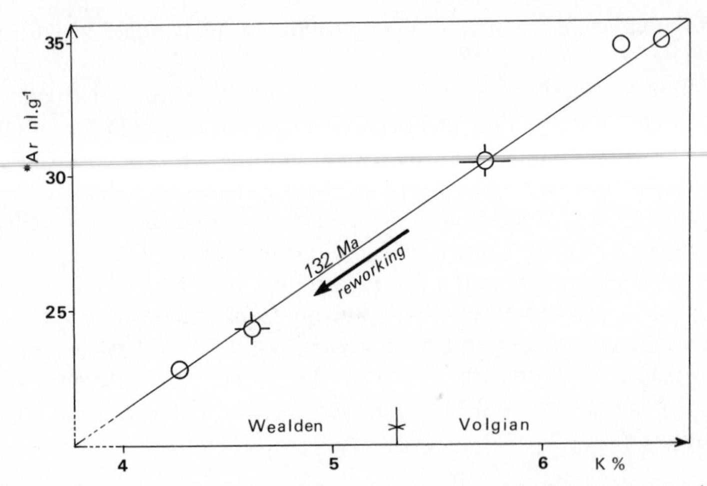

Figure 11. Argon content as a function of potassium content of Wealden (probably remanié) and Early Volgian glauconies from the Weald area (data from Dodson *et al.*, 1964). 2σ uncertainty is shown; no change of apparent age is seen in spite of a strong potassium extraction—if the hypothesis of reworking is correct.

4.3 Ancient reworking in the Palaeogene

Another kind of application of the slight change of initial age during a reworking was tested in SE France by Triat *et al.* (1976). In this area the epicontinental marine sedimentation of the Albian and Early Cenomanian are characterized by the genesis of abundant glaucony. Marine sedimentation seems to have ceased from the Middle Cenomanian until the Burdigalian transgression in the Rhone Basin. During the Late Cretaceous a general tropical alteration is developed in the area. Continental deposits occur during the Middle Eocene and Oligocene. However, some of the Palaeogene levels contain glaucony which has been interpreted by some workers as the trace of a marine incursion. For others the presence of strata containing glauconitic pebbles and pieces of Cenomanian fossils shows the possible reworking of Cretaceous glaucony. These samples of Cretaceous rocks (altered more or less *in situ*) and two Palaeogene samples were chosen after separation of glaucony. The characteristics, stratigraphic age and apparent age are shown in Table 8. The analytical results obtained in Berne and Paris before 1976 were completed later in the sàme laboratories.

The potassium constant of Middle Cretaceous glauconies becomes very low due to *Late Cretaceous tropical alteration.* In some cases, glaucony has been completely kaolinized. We may estimate the minimum potassium loss in Cretaceous glauconies as 10–15% and the maximum as 20–25%. In spite of this moderate loss, of the same order as those observed in the two preceding

Table 8. Analytical results of dating of altered and reworked glauconies from SE France. Stratigraphical data by J. M. Triat and G. Truc; potassium measurements by Miss M. Lenoble in Paris; argon data by G. S. Odin in Berne, calibration with GL-O grains assumed to be 24.8 nl/g radiogenic argon. Ages calculated using ICC values, completed and recalculated after Triat *et al.* (1976).

Sample	Stratigraphic level	Potassium (K%)	Radiogenic argon (nl/g)	Apparent age (Ma ± 2σ)
5	Sannoisian	6.07	19.87	82.4 ± 4.0
1	Sannoisian/Ludian	5.71	21.39	93.9 ± 4.5
3	Early Cenomanian	5.35	19.39	91.1 ± 3.0
7	Albian or Cenomanian	5.30	17.53	83.3 ± 4.0
2	Albian	5.76	18.76	82.1 ± 4.0

studies, the apparent ages have been reduced by probably very little (sample 3) up to almost 15% (samples 2 and 5). The samples redeposited in the Palaeogene sediments are not more altered than the *in situ* ones: both their potassium contents and their apparent ages exceed the lowest values for the weathered material. They indicate a typical Cretaceous age which excludes the possibility of a marine episode during the Palaeogene in the area studied.

To sum up, the continental tropical alteration of the Middle Cretaceous levels leads to a potassium loss of less than 15%, and at the same time the apparent age may be lowered by up to 20%. This shows that the tropical alteration, leading to very weathered and oxidized horizons, is much more aggressive than a temperate climate one for a sheet silicate chronometer. The glauconies reworked from these levels in Palaeogene continental clayey sediments are no more altered than the original ones.

5 SUMMARY

The data summarized above on natural and experimental leaching of glaucony has led to the following conclusions.

In the field, it must be noted if the possible weathering of the glauconitic horizon occurred under a moderate temperate climate or as an effect of a hot oxidizing climate. The presence of oxidized rusty sand layers in the outcrop will indicate which type of weathering took place.

In the laboratory, the microscopic examination may show the presence of a few rusted or yellow-green grains in a normal green glaucony. This may indicate a weathering effect. If weathering was moderate and the glaucony evolved to a highly-evolved state, the K–Ar apparent age may be assumed to be unaltered. But, if the glauconitic minerals present in the grains are little-evolved and K-poor, then a slight lowering of the K–Ar apparent age probably occurs. If weathering took place under a tropical climate, the lowering of the initial apparent age is practically certain.

The preliminary acid leaching of glaucony may lead to an effective exchange of potassium, especially using the hydrochloric acid at concentrations as low as 0.3 N. The cationic exchanges will be higher for a less-evolved glaucony than for a highly-evolved one as they are clearly related to the presence of a greater proportion of expandable (K-poor) layers. In order to diminish the leaching effect three precautions may be taken: (1) elimination of the less-evolved glauconies for K–Ar dating; (2) use of less concentrated hydrochloric acid, or the use of acetic acid which is much less aggressive than hydrochloric acid; (3) removal of the less-evolved grains, or part grains, by ultrasonic cleaning.

However, the use of ultrasonic treatment must be limited in order not to break the most evolved crystallites. A complete disaggregation of grains lowers the potassium content; these disaggregated grains are not representative of the initial mineralogical nature of the glaucony sample.

The best glauconitic fraction, the one which will have more chance of being representative of the age of closure of the chronometer with regard to K–Ar dating, may be obtained in eliminating the less-evolved grains (paramagnetic fractionation) and by cleaning the grains by ultrasonic agitation for a few minutes in a 0.1 N acetic acid solution.

If no evolved to highly-evolved glaucony is available, the measured K–Ar age will probably have been lowered by weathering.

Acknowledgements

The first author is greatly indebted to Professor M. Dodson, who gave him permission for a very instructive stay in his laboratory, and to other students and colleagues of his laboratory for their efficient help even when problems occurred over weekends and at night. J. C. Hunziker, in Berne, and M. G. Bonhomme, in Strasbourg, also helped him to complete the data; they are greatly thanked, as are other helpful colleagues of these two laboratories. Madeleine Lenoble carried out the potassium measurements with accuracy and efficiency; P. Juignet and J. M. Triat helped with this research both in the field and with their own samples. To all of them our thanks are due. This is a contribution to IGCP Project 133.

Résumé

Cette contribution rassemble des données récentes et souvent inédites permettant d'évaluer l'action de diverses solutions sur l'âge apparent K–Ar de glauconies de composition variée.

Sur le terrain les conditions climatiques d'un éventuel rinçage des grains verts est un facteur important. La présence de niveaux de couleur rouille dans la formation glauconieuse indique un rinçage énergique sous climat chaud. Cette oxydation est néfaste au maintien de l'âge apparent.

Au laboratoire, un rinçage plus discret peut être supposé lorsqu'on observe de rares grains vert pâle ou rouille. Si ce rinçage est resté modéré et que la glauconie qui l'a subi est évoluée à très évoluée, on montre que *l'âge apparent K–Ar n'est pas modifié.* Mais pour une glauconie peu évoluée, l'âge apparent sera légèrement abaissé. Sous climat chaud, une altération oxydante abaisse sûrement l'âge apparent des glauconies même si l'on prend le soin de sélectionner les grains restés verts.

Le lavage à l'acide, utilisé pour nettoyer les grains, conduit à un échange de potassium spécialement si l'on utilise l'acide chlorhydrique même à des concentrations aussi faibles que 0.3 N. On peut aisément diminuer l'effet de ce rinçage en sélectionnant des glauconies évoluées, beaucoup moins sensibles à ces échanges cationiques—en utilisant des acides peu concentrés mais surtout—en *préférant systématiquement l'acide acétique à l'acide chlorhydrique*; ce dernier est beaucoup plus agressif, à concentation identique, et n'est pas beaucoup plus efficace pour éliminer les carbonates qui peuvent coller des minéraux hérités dans les fissures des grains verts.

L'utilisation des ultra-sons doit rester limitée; la partie désagrégée des grains devant être rejetée en tant que chronomètre, car il s'agit de cristallites cassés; souvent moins évolués que dans le reste du grain, ils ont subi une agression certaine durant ce traitement.

Des extractions de potassium en milieu concentré (NaCl ou $MgCl_2$) à chaud entraînent une décoloration des grains: un bon critère de rinçage. Les minéraux glauconitiques sont moins affectés par cette altération que les biotites. Les minéraux plus évolués sont plus résistants que ceux des glauconies moins évoluées. Les âges apparents sont modifiés vers un âge *plus élevé* dans les expériences utilisant l'eau désionisée ce qui pourrait laisser supposer qu'une partie de l'argon n'est pas dans les mêmes sites que le potassium ainsi qu'il a été déjà proposé lors des expériences de chauffage par étape. Dans les solutions salines, au contraire, les âges apparents ne sont pas modifiés sensiblement tout au moins jusqu'à l'extraction de 15% du potassium initialement présent.

Le comportement tout à fait particulier de la glauconie récente, dont on sait que l'âge est un âge hérité du support de verdissement, est intéressant. Les solutions salines extraient une grande proportion du potassium initialement présent (dans les minéraux glauconitiques) mais très peu d'argon (dans les minéraux hérités). Ce comportement pourrait être utilisé pour identifier un âge initial non nul dans une glauconie ancienne peu évoluée.

(Manuscript received 28-5-1980)

Numerical Dating in Stratigraphy
Edited by G. S. Odin

20
How to measure glaucony ages

GILLES S. ODIN

This chapter summarizes the methods and special precautions needed for sample collection, preparation and analysis of glaucony, and the interpretation of analytical data obtained from this sedimentary chronometer for use in the calibration of the time scale.

1 INTRODUCTION

We will confine the discussion here to the granular form of the green pigment defined as *glaucony* by Odin and Matter (1981), which is the easiest form to use in sediments. Glaucony occurs as part of a rock in which not all the components are in geochemical equilibrium. Such a rock is a mixture of *detrital components* and *components of authigenic marine origin*. The granular facies of glaucony is the result of glauconitization of granular substrates. This '*verdissement*' mainly occurs within four kinds of preliminary substrates (see Odin and Dodson, this volume): coprolites, infillings, biogenic carbonate debris and detrital minerals. From these considerations, *radiogenic isotope inheritance is clearly the normal expectation* for the development of glaucony in some substrates. This inheritance will survive until the whole substrate is eliminated. As isotopes of the substrate itself may be incorporated in growing glauconitic minerals, these new minerals require sufficient time for complete equilibration with the environment in which they develop. Glauconitization occurs in contact with the open sea-water, with some restrictions. In this environment, temperature is low, 5–15°C, reactions are very slow and equilibrium will only be attained gradually. If the initial substrate is itself neutral or in equilibrium with the sea-water with respect to the isotopes used, no inheritance occurs. However, if the substrate is enriched in radiogenic isotopes and is only slowly altered (mica) or not altered (inherited glaucony), true equilibrium, an initial zero age, will never be obtained.

Chemical exchanges with sea-water must occur to permit development of glaucony. As long as chemical exchanges occur, even after burial, glaucony development is continued inside the sediment. According to Odin and Matter (1981), the components of glaucony have fundamentally two origins: the substrate minerals and the authigenic minerals. The initial substrate is an integral part of the glaucony and must be distinguished from impurities (Figure 1). Several minerals may make up the substrate: calcite, aragonite, micas, all clay minerals, quartz, silica, etc., were observed. The important considerations here are that they may be enriched in radiogenic isotopes and that they are more or less difficult to eliminate during glauconitization of the particle.

Authigenic minerals are part of a mineralogical family ranging from a smectite component, an open structure, to a micaceous component, a closed structure, with all intermediate possibilities. The properties of a glaucony as geochronometer are directly dependent on the variety of the glauconitic minerals. The variety will itself be dependent on the stage of evolution of

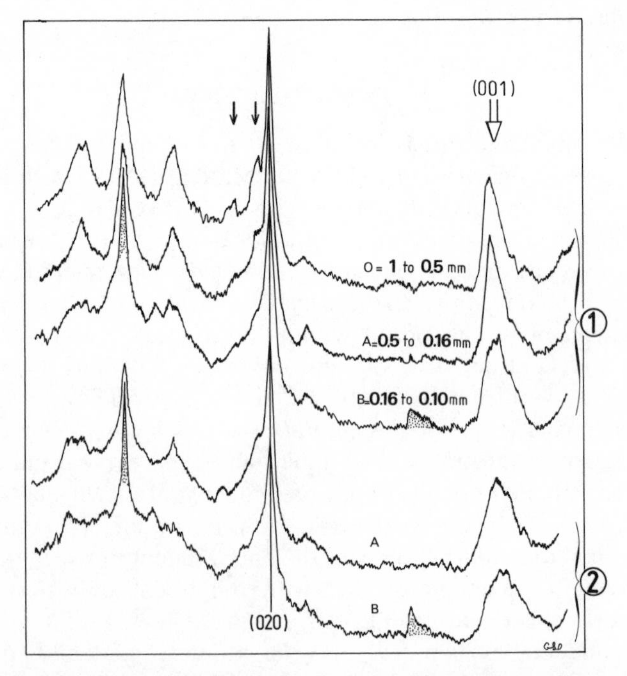

Figure 1. Evolution of glauconies as a function of their grain size. The glaucony (1) was separated into three size fractions: O, A, B; the glaucony (2) into two fractions: A, B. The diffractograms show a clear change from the lower size fractions (B) to the higher size fractions: the inherited substrate disappears (here: quartz and kaolinite, dotted areas); the (001) peak becomes higher together with other (*hkl*) peaks. This permits *selection* of the best grains and also identification of the possible *original substrates* for glauconitization. The (020) peak is a constant.

the grains; the minerals with a closed structure in the most evolved grains will have:
—a high paramagnetic susceptibility
—a low cation exchange capacity
—a high potassium content
—a low content of adsorbed water
—a low content of expandable layers (see Odin and Rex, this volume)
—a small principal reticular space.

These physicochemical properties will directly influence the reliability of the grains as chronometers. During the history of the sediments, the isotopic equilibria will be influenced by two factors: (1) the fact that completely different components (with their own isotopic compositions) are present side by side; (2) the fact that a deep burial will completely change the thermodynamic factors affecting the equilibria compared to those which occur on the sea-floor. The pressure for glaucony genesis is between 10 and 30 bar (100–300 m water); each additional 100 m of accumulated sediments will add a pressure of nearly 25 bar. Diagenetic mineralogical changes are well known for diffuse clay minerals at more than 2000 m: 500 bar and 60–80°C higher. These possible diagenetic changes as well as the superficial alteration will be more important in a more open structured (i.e. less evolved) and K-poor glauconitic mineral.

2 FIELD SAMPLING

The dating process begins with geological observations in the field. Core samples do not permit these observations, which are especially useful in tectonized areas (salt tectonics in North Germany) or in series with several glauconitic horizons (facies of different ages may look very similar). Other factors such as the possibility of contamination during coring, and the limited quantity of samples available, also make this sampling process unsatisfactory in numerous cases. An accurate *biochronological study* may help in estimating the duration of the depositional break which is required for the complete evolution of glaucony (see chapter by Odin and Hunziker, this volume). No fossils occur in horizons with evolved glaucony, except for very thick shells or reworked moulds, which give poor biochronological control. Hence a palaeontological study done below and above the glaucony sample is necessary.

The *genetic uncertainties* defined above may be partly solved in the field by examination of the vertical sequence (rate of deposition) and of lateral environmental variations (palaeogeography) in order to discern whether or not the glaucony has developed *in situ*.

The arguments concerning the *historical uncertainties* can be considered in the field; they include an estimate of the depth of burial, the tectonic events, and possibilities of weathering. With regard to the last question, a knowledge of the position and seasonal changes of the underground water level and the presence of impermeable levels which stop the vertical water circulation is useful.

The environmental data in the field should preferably be obtained with the help of a specialist on the local geology and after a preliminary study of some test samples at the laboratory. The sampling can then be done with three main aims: (1) to collect the superficially least altered samples; (2) to collect a sufficient quantity of each sample to satisfy criteria for dating; (3) to collect a sufficient number of samples for biochronological and palaeogeographical vertical and lateral studies. Preparation of an outcrop needs no specific comment except that it is rarely sufficient to remove the outermost few centimetres to get unaltered sediment: experience shows that metres may have to be removed in sands. As far as possible, a very glauconitic sampling cut will be better situated facing north in the northern hemisphere, because the sun may heat an outcrop fiercely due to the dark colour of glaucony, which traps radiant heat.

One hundred grammes of 'total' glaucony is a suitable quantity permitting a convenient division into size fractions. Fifteen to 25 g is the minimum adequate to represent the level. Depending on the sediment, this would require from one to several tens of kilogrammes of total rock, but it is well known that only the very glauconitic sediments have some chance of yielding an evolved to highly-evolved glaucony, with the possible exception of some chalk. Usually glaucony-poor sediments require a lot of work to recover either a little-evolved glaucony or an inherited one. Five to ten samples collected from a vertical series each metre or each two metres are useful for the elucidation of the glauconitization process of a glaucony-rich horizon. To complete the sampling, several sections may be sampled laterally at several kilometres or tens of kilometres distance, according to an itinerary perpendicular to the palaeogeographic zones.

Finally, if one intends to use the Rb–Sr isotopic clocks it is necessary to collect a *carbonate phase* which may be representative of the sea-water Sr isotopic composition at the time and place where the glaucony formed. This presents difficulties. Coastal shells are sometimes present together with glaucony. These shells cannot be representative of the environment of genesis of glaucony (50–500 m depth): they have been formed *before* the glaucony, and show that glauconitization often follows a deepening of the sea (Odin, 1980). The measured strontium isotopic ratio of this shell material is thus not contemporaneous with the glauconitization but anterior and characteristic of another environment. According to Renard (1975a), there may be a large lowering in strontium content between original aragonite from a shell and the calcite in which it is often replaced during early diagenesis. However, Peterman *et al.* (1970) present some comparisons where no isotopic differences appear after recrystallization. We now have 20–100 collected samples which need to be studied in the laboratory.

3 SEDIMENTOLOGICAL STUDY IN THE LABORATORY

The sedimentological study requires that whole-rock-samples are available for study. Three stages of study may be distinguished: (1) the splitting of the sample in various phases; (2) the identification of the obtained fractions; (3) the choice and cleaning of selected glauconies.

3.1 Fractionation of the sediment

A disaggregation with refined petroleum after drying at 60–70°C is preferable to crushing when the first treatment is sufficient. For a porous limestone or sandstone the use of liquid nitrogen may be helpful. The sample is arranged in an empty glass vacuum desiccator; then water with a few drops of detergent is introduced. After a few days the sample is put successively in liquid nitrogen and in water. This process breaks the sample. Washing on a 50 μm sieve and dry sieving are the most convenient possibilities for beginning the size fractionation. The clay fraction is kept, after

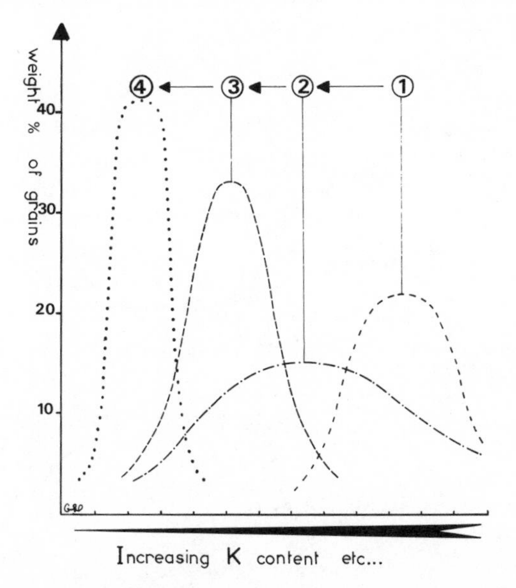

Figure 2. Histograms of distribution of the green grains of various glauconies: (1) *nascent* glaucony; (2) *little-evolved* glaucony; (3) *evolved* glaucony; (4) *highly-evolved* glaucony. With increasing K content, the paramagnetism, density, index of refraction, etc., increase, but expandable layer content, cation exchange capacity, etc., decrease.

decantation of silts, for diffractometric analyses. Experience shows that the coarser glaucony is generally more evolved than the finest one (Figure 1, see also NDS 123), the easiest to separate being the 150–500 μm grains. The sample is sieved before magnetic treatment, which helps to achieve a better separation. The Frantz isodynamic separator is generally used—though it would be less expensive, easier to repair and less noisy to use a system made by a craftsman employing oscillating electromagnets as vibrators. We usually separate all the attractable grains at a high magnetic intensity (0.6 A) with a lateral inclination of 16° and a longitudinal one of 15–25°. After two or three passes the whole glaucony is separated at various intensities (two–three passes each) and the different fractions are weighed.

The form of the histogram of distribution of the grains (Figure 2) as well as its position on the x-axis is informative. Glaucony evolution changes a large flat histogram to a high narrow figure at lower intensity of supply of

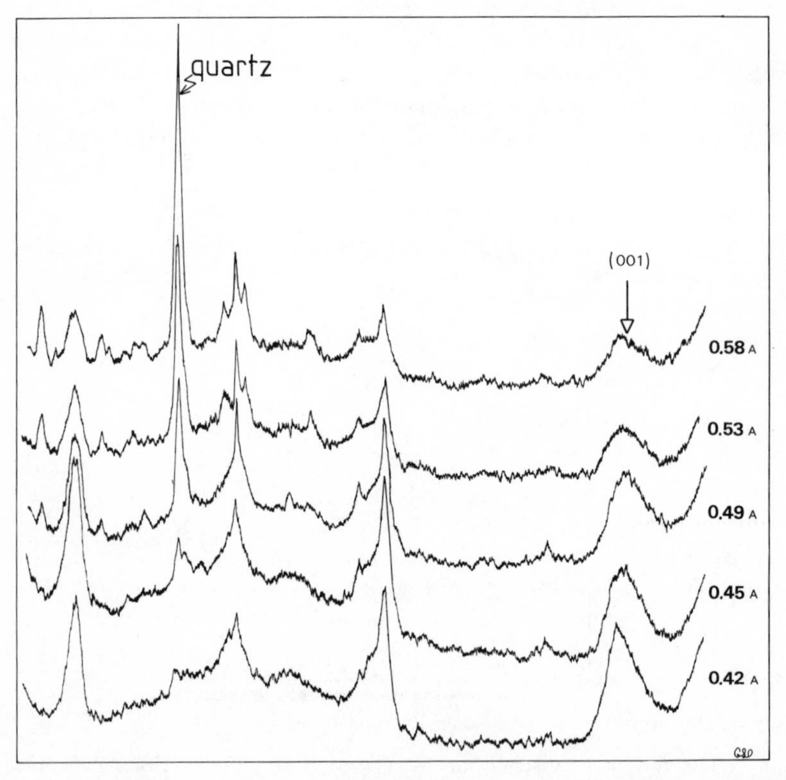

Figure 3. X-ray diffractograms of various paramagnetic fractions of the glaucony G.486 A (Würmian of the Mediterranean Sea). Intensity of current supply to the magnetic separator is indicated (on right). The most paramagnetic grains (more evolved according to the (001) peak and more pure) are attracted at lower intensity.

the magnet. The 'best glaucony' is obtained by eliminating the tails of the histogram and choosing the paramagnetically most susceptible grains (Figure 3). Observations on the various size fractions may make possible the recognition of the initial substrate of glauconitization (Giresse *et al.*, 1980).

Allogenic grains show traces of their redeposition: rounded grains, sometimes brown when reworked in the sea at less than 30 m depth. But rounded grains may also be obtained after a very long *in situ* evolution. The brown grains may easily be removed using bromoform as they are more dense than the unoxidized ones.

3.2 Analysis of mineralogical components

The study of the clay fraction may be of some use in recognizing a parent material. For example, if we assume that the initial substrate is coprolite, the less evolved of the grains must have a composition showing a resemblance to the clay fraction which was the food for the mud-eaters which excreted the coprolite material.

The most useful and necessary apparatus for aiding the interpretation of glaucony apparent ages is the X-ray diffractometer. The quickest and most informative technique is the use of a *non-oriented powder.* Ten milligrammes are sufficient for a powder preparation: after crushing of the grains in an agate mortar, a few drops of acetone are added to obtain a liquid mud which is spread out with a spatula onto a convenient substrate (generally a glass slide). As soon as the mud dries, the preparation may be used in the apparatus. A middle inertia of integration together with a not too thin entry slit and a cobalt anode (high Fe content) are recommended. Examples of the results are shown in various chapters in this volume (see Odin and Dodson, Odin and Hunziker). We have demonstrated that by this method it is possible to measure the potassium content.

The results of two series of measurements based on 42 experiments are shown in Figure 4. The distance between the (001) and the (020) peaks of the diffractogram is the criterion used. The position of the (001) peak is measured according to Figure 5 and the deduced K_2O content is correct at $\pm5\%$. A good calibration will need standards of glaucony with 8.5, 8, 7.5, 7, etc., % K_2O.

The origin of green grains may also be deduced from a series of diffractograms of the various paramagnetic and size fractions tested. From coarser grains to smaller ones as well as from more paramagnetic to less paramagnetic ones, we may sometimes detect the increase of peaks characteristic of an initial substrate, when present. A general law we systematically follow is to select the best fraction (K-richest) of a glaucony and try to improve the quality of this fraction, so as to have more chance of selecting an undisturbed chronometer.

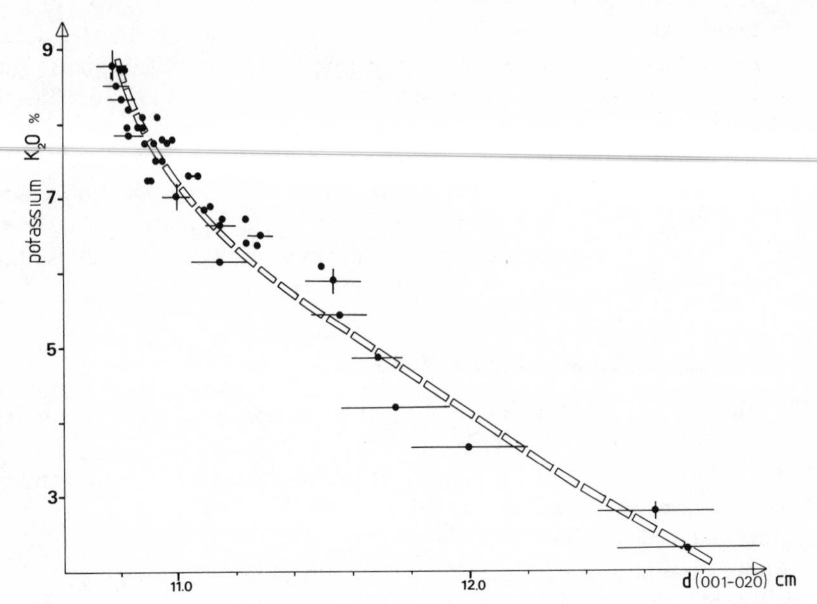

Figure 4. Distance between peaks (001) and (020) as a function of the potassium content of glauconies: the position of the (020) peak is constant on the diffractograms, while the position of the (001) peak is a function of the potassium content. Analytical uncertainties (2σ) on the potassium content and the distance d between (001) and (020) peaks are given for several samples.

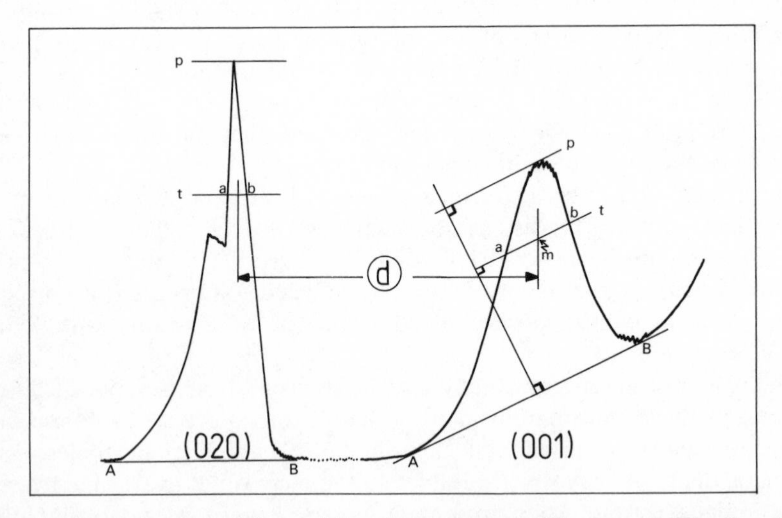

Figure 5. Determination of the position of a peak on a diffractogram. The line AB is constructed at the base of the peak, then the parallel at the top of the peak, then the parallel t a third below. This line cuts the diffractogram at a and b. The middle of ab:m gives the position of the peak.

3.3 Cleaning of glaucony before dating

Three techniques are used for cleaning the fractions of grains selected according to X-ray analyses with a view to dating: bromoform flotation, electrostatic table, ultrasonic cleaning.

Bromoform flotation will help in removing pyrite, ilmenite, haematite, goethite, staurolite, pyroxenes, amphiboles, some biotites and chlorites or mixed grains. These minerals either contain iron, which sometimes renders difficult Rb–Sr separation on the column, or possibly radiogenic isotopes.

Electrostatic fractionation may use a special vibrating table or a sheet of thick glossy paper; this helps in removing the frequently observed flakes of partly glauconitized micas.

Ultrasonic treatment is routinely used for cleaning the cracks developed in grains during the growth of glauconitic minerals. The highly-evolved glauconies sometimes show a thin external film covering the grains. This film is composed of less-evolved glauconitic minerals (Lamboy and Odin, 1975) and is easily removed by ultrasonic treatment. Routinely we use two agitations of 5–10 mn duration with wet sieving between and after to retain only the largest grains. The use of the Rb–Sr isotopic clock needs a special ultrasonic treatment for removing strontium adsorbed in glauconitic minerals; this permits a better accuracy (Pasteels *et al.*, 1976). Some authors use ultrasonic cleaning 0.1–1 N hydrochloric acid. We recommend (see Odin and Rex, this volume) the use of a milder 0.1 N acetic acid solution. Whatever acid is used, acetic or not, it is highly recommended to *keep the leachate* for a test of the strontium isotopic composition.

After cleaning, the sieved sample may be separated a last time with the magnetic separator. A rest of one week at normal temperature and humidity is useful to permit a *hygrometric equilibration* of the grain after drying at less than 70°C and before storing in waterproof bags.

4 ISOTOPIC ANALYSES

With regard to the analytical process, two special problems should be mentioned in connection with glaucony: the conditions of moisture during weighing and the effect of preheating. As noted by Langmyhr (1969), the moisture conditions are important when weighing rocks and minerals due to the adsorbed water present in most of them. As for all other properties, glaucony presents a variability for this one too. Figure 6 shows the adsorbed water content as a function of the potassium content in more than 50 pure glauconies analysed by Odin and Matter (1981). Adsorbed water (H_2O^-) was measured by weighing before and after heating overnight at 110°C. As previously quoted by Odin (1975), the total $K_2O+H_2O^-$ is generally between 10 and 11%, and these two equitotal lines are shown in the figure.

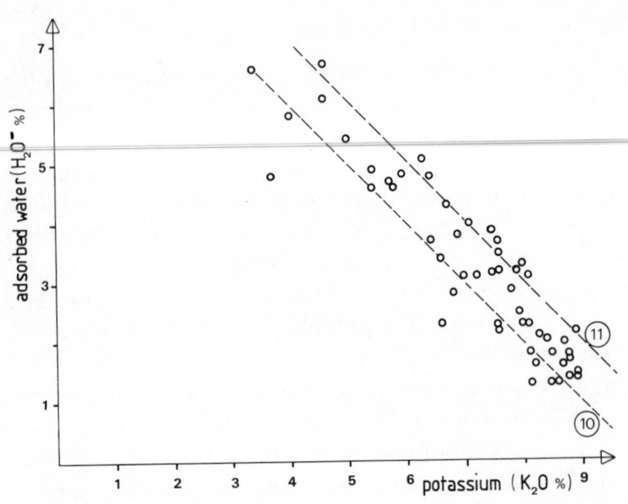

Figure 6. Adsorbed water content as a function of potassium content in glauconies.

Thus, the K-poor glauconies will be more subject to dehydration–rehydration processes and weight changes than those richest in K.

The loss of weight after drying at 100°C (evaporation) was studied in various glauconies by E. Elewaut (1977, unpublished data), and some of his results are given in Figure 7. The loss of water in two samples appears much higher than in the preceding figure, which leads us to a new experiment (see below).

Readsorption is initially very quick, but after one hour it becomes much slower and the initial weight is not recovered after five hours. It is, however, clear that the three potassium-richest glauconies have lost much less water. Figure 8 shows the results obtained on two samples during a longer experiment. Two pairs of GL-O ($K_2O = 7.9\%$) aliquots have been weighed regularly for one week together with two aliquots of a Recent glaucony from the Gulf of Guinea ($K_2O = 4.2\%$). The weight losses are similar to those observed in Figure 6. Humidity was 55–60%. Even after 4–6 days, the initial weight was not recovered in GL-O. In conclusion, we recommend never drying glauconies at more than 70°C, especially little-evolved ones. Even after such a moderate treatment, a period of reequilibration of at least 24 hr should be allowed before putting glaucony into a waterproof bag. Weighing of glaucony will be more reproducible under normal conditions of hygrometry and temperature. In the geochronological K–Ar studies we have done there are often several weeks or months between measurements of K and Ar; due to the possible difference of conditions of hygrometry we systematically add to the normal analytical uncertainty on the potassium datum 0.5–1.0% to account for this extra possible source of error.

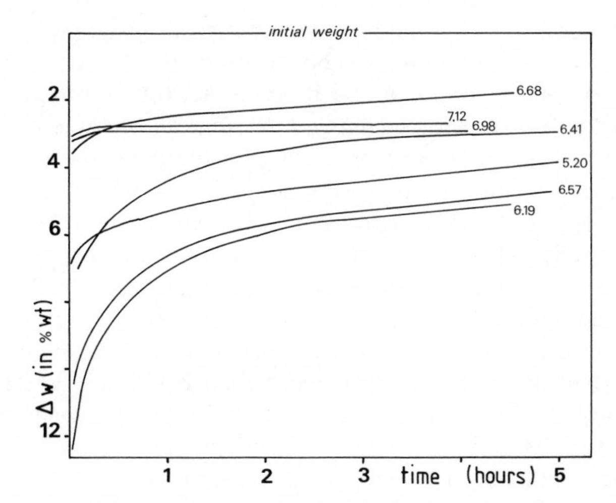

Figure 7. Loss of weight and readsorption of water after heating of glauconies of various mineralogical compositions at 100°C (according to data by E. Elewaut, Brussels). The mineralogical composition is shown on the right (potassium content in wt % K).

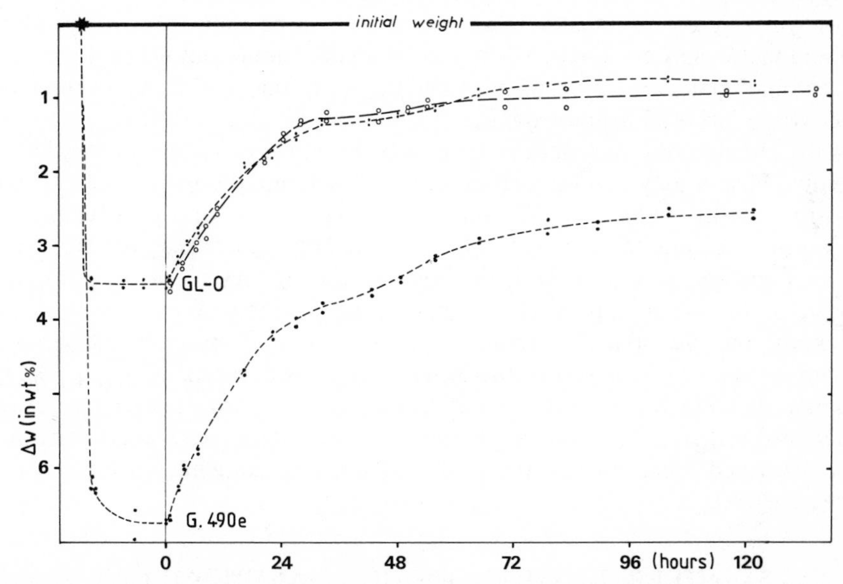

Figure 8. Loss of weight and readsorption of water after heating two glauconies at 100°C. Two aliquots were used for each experiment; two experiments were done for GL-O (K% = 6.56) and one for the Recent glaucony from the Gulf of Guinea (G.490 e, K% = 3.5, sample P. Giresse). The loss of weight is nearly complete after one hour (see more data in the chapter by Odin and collaborators). The humidity of the laboratory was 55% ± 5.

If it is intended to remove adsorbed water from glaucony after weighing and before extraction of argon, it is not recommended to heat the sample at atmospheric pressure (see Odin and Bonhomme, this volume). This heating seems to lead to a systematic *increase of the atmospheric argon content* of the pellets which diminishes the accuracy of measurement. On the other hand, heating under vacuum in the extraction line hastens the departure of adsorbed water and the attainment of a good high vacuum. According to the numerous experiments done by the present author, heating at temperatures of 100–180°C does not lead to detectable loss of argon in the glauconies; however, the highest temperature was never routinely used. The mean preheating temperature of 150°C is recommended for all glauconies. It remains prudent not to bake out at more than 200°C, as detectable loss of radiogenic argon has been observed at 250°C under vacuum even for well-closed minerals.

Now, as far as the extraction of gases is concerned, it is better to heat the crucible very slowly at the beginning of the procedure. This is to avoid green pellets popping out of the crucible. Some high-frequency generators used for induction heating cannot be used at very low power. In order to obtain low temperatures we have tried two solutions. The easier one is to heat with the high-frequency inductor *above* the crucible for 3–5 mn, and to observe the gas loss on the Pirani gauge to adjust to the best position for heating. Then the inductor may be lowered gradually to its maximum output position. For a 'normal' glaucony (more than 20 Ma old, more than 6–7% K_2O), the time necessary for a full fusion from the beginning of heating is 8–12 mn. As soon as the crucible becomes dark red, virtually no more gases are present in the grains. Fusion may be obtained quickly at low temperature (1200°C). If the position of the induction coil cannot be changed, a second possibility is to arrange a second coil connected in parallel. The greater part of the power may then be used to heat a dummy crucible (a big piece of iron), the position of which may gradually be changed. Complete removal of the dummy crucible from the second coil will return full power to the sample. With a view to calibrating the time scale, a minimum of three GL-O standards were tested before each series together with at least one other standard mineral. The results obtained on these standards should generally be published together with the analytical data on the glauconies measured (see part II).

5 INTERPRETATION OF THE ANALYTICAL DATA

5.1 Choice of analytical results in the literature

Since the beginning of the application of radiometric methods for dating rocks, much progress has been achieved in obtaining more reliable ages from the analytical point of view. As a general rule, the most recently obtained results proposed in the literature *must replace the older ones*, as will be shown below. Baadsgaard *et al.* (1964) showed that the age of a biotite from a bentonite calculated at 75 ± 3 before 1962 yielded a date of 70.0 ± 1.7 Ma a few years later. Ghosh (1972) clearly showed the difference in analytical results obtained in the same laboratory on the one hand by Evernden *et al.* (1961) and on the other hand, more recently, by Obradovich (1964). Figure 9 shows the difference in age between ages obtained on the same samples quoted in the two papers as a function of the apparent age given by the first authors. The two series of data are given with an accepted analytical uncertainty of ±5%; half of the data fails to agree within this error margin. The explanation may lie in the analytical process: in particular, Professor G. H. Curtis believes that potassium measurements as well as the preliminary preparation of the sample may be suspect in the first paper. The calibration of the system of measurement of argon is one of the possibilities which often may provoke a systematic bias as shown by Gramann *et al.* (1975). In the same manner, Figure 10 shows the results obtained after 1971 in Berne by the K–Ar method, assumed to be correct,

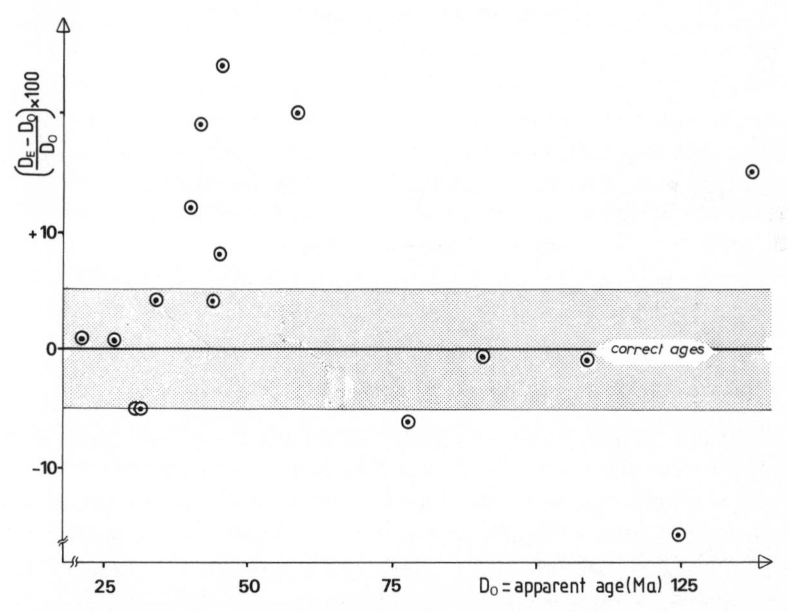

Figure 9. Apparent ages of glauconies in successive experiments. Obradovich (1964) remeasured samples used by Evernden *et al.* (1961). As a mean the dates obtained by Obradovich (D_O) are lower than the dates quoted by Evernden *et al.* (D_E). We ourselves measured (Odin *et al.*, 1978) several glauconies from the same outcrops; our results all agree with the dates by Obradovich, which are therefore assumed to be correct ages.

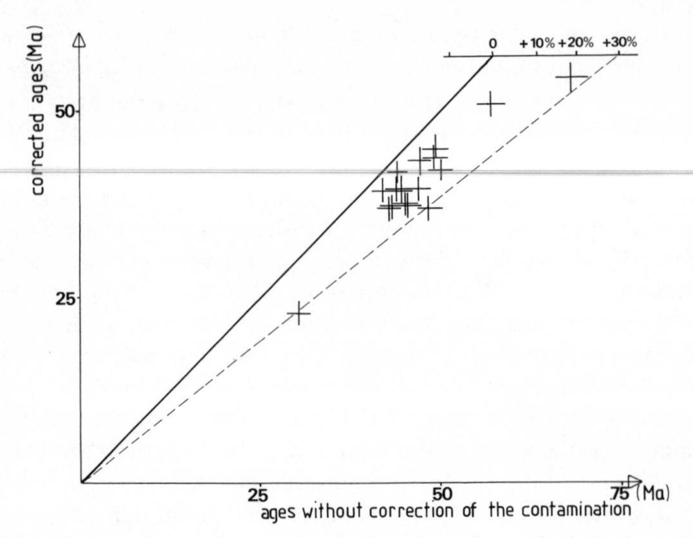

Figure 10. Change of apparent age as a function of a technical error. The x-axis shows the apparent age calculated using an apparatus where atmospheric argon contamination was not measurable (too small peaks); the y-axis shows the corrected ages obtained in Berne four years later.

compared with those proposed earlier by another laboratory on the same samples. One sees that the earlier work appears to yield older ages. This has been linked with systematic underestimates of the correction for atmospheric contamination, which was corrected later but too late for the first paper published (see comments of NDS 1 and others). This kind of error is not unusual. It is right to publish new results when more reliable data have been obtained; the worst practice is to always support obsolete data.

5.2 Interpretation of an analytical datum

The interpretation of an analytical datum preceded by a perfect study leading to a complete knowledge of the history of the chronometer presents no special problems. One just has to successively add to the analytical datum with its own analytical uncertainty the various uncertainties on the stratigraphic position, on the genetic conditions and on the subsequent history of the chronometer.

Unfortunately, perfectly studied chronometers are scarce and it is generally useful to adopt a method for improvement of the results. Three approaches are employed: the use of several 'best glauconies' available; the use of various fractions of glaucony of one sample; and the use of two different methods of age measurement.

The use of the best glaucony available from an outcrop is the solution we adopt to diminish the number of analyses; this solution gives reliable results when two or three outcrops are treated in the same manner.

—if the K_2O content is less than 6% we never use the sample; results would be *a priori* suspect.

—if the K_2O content is from 6 to 7% there is still the possibility of unreliable results, but if several outcrops, palaeogeographically independent, give equivalent ages these apparent ages may tentatively be used, in the absence of better data, for calibrating the time scale. Obviously, low K glaucony does not necessarily always give an age different from that of deposition, but in the case of any disturbance the apparent age change will be higher in such.

—if the K_2O content is from 7 to 8% there are fewer problems concerning the genesis of the glauconies and there is less likelihood of isotopic exchange, except for those clearly visible in the field. With these glauconies the zero age, the moment of closure, occurs *after the deposition* of the glauconitized substrate and corresponds better with the deposition of the horizon which buried the glauconitic layer.

—if the K_2O content is high to very high (8% or more) the glaucony may have been closed before burial by the next deposits. A geologically significant break in deposition may then be detected on sedimentological or palaeontological grounds.

The use of several parts of the same sample has been tested by various authors after Obradovich (1964), while H. Kreuzer systematically measured samples treated with various leaching reagents such as 0.01 N ammonia, ethanol, 10% formic acid, 5% barium acetate (Gramann *et al.*, 1973). According to these authors an apparent age will approach the deposition time if the various apparent ages are coincident. The principle used in dating several fractions is always identical: we have shown that the amplitude of the age disturbances is frequently related to the potassium content, so that if equivalent results are obtained on fractions of very different K content this proves that no geochemical disturbance has occurred.

Thus, when various fractions of a sample give concordant apparent ages, this is an argument for reliability. When ages are discordant, the fraction richest in K is the most likely to yield a date which approaches the time of deposition.

The use of both K–Ar and Rb–Sr methods may be of great help in estimating the reliability of the apparent ages of glaucony. Unfortunately, many geochemical aspects of the Rb–Sr isotopic equilibria on glauconitic minerals have not been investigated yet. From the available results, and particularly with the conventional ages, a greater scattering is obtained compared with K–Ar ages. Further data are needed, and we prefer K–Ar data when discrepancies between the two methods occur.

6 CONCLUSION

The recent data obtained on glaucony, thanks to improved geochemical research methods together with a better understanding of the stage of genesis of this authigenic marine facies, permit the design of a specific method of sampling.

The preliminary studies done on these samples will help to *presume greater reliability* for the selected chronometers. This method alone will establish the numerical age of a stratigraphic horizon in the absence of preconceived ideas. The proposed method will allow *a priori* assessment of the reliability data and tends to eliminate the often used procedures which deduce *after* the analysis the geochemical history of glaucony according to whether the obtained age 'appears correct' or not. We do not imply that it can be known with certainty beforehand that an apparent age will be that of deposition or not, but we know better today how to increase the probability of measuring glaucony ages which do accurately reflect depositional ages.

Acknowledgements

Improvement in the English version of this chapter by the publishers is gratefully acknowledged. This is a contribution to IGCP Project 133.

Résumé

Ce travail propose les recommandations auxquelles on aboutit à la suite des études méthodologiques récentes concernant les mesures d'âge sur glauconies en grain destinées à connaître l'âge numérique du dépôt d'une formation sédimentaire. Il apparaît, en effet, que des analyses effectuées au hasard de la récolte d'un échantillon glauconieux conduisent rarement à des résultats ayant une valeur scientifique: les diverses possibilités de modification d'un âge apparent sur glauconie permettent toujours *a posteriori* d'expliquer un âge mesuré différent de celui que l'on aurait souhaité obtenir.

Le principe fondamental commun à toutes ces recommandations est l'élimination méthodique des échantillons, ou de leurs parties, les plus susceptibles de ne pas remplir les conditions requises par le modèle de comportement d'un bon géochronomètre. Les études paléontologiques, sedimentologiques et paléogéographiques aident à choisir *avant datation* une série d'échantillon pour laquelle on peut présumer qu'elle constitue l'échantillonnage le plus digne de confiance. Avant les mesures on essaie d'obtenir une connaissance claire des probabilités de modification des âges apparents, les analyses isotopiques *sur divers échantillons* n'interviennent que pour confirmer l'ampleur ou l'absence des altérations.

Lorsqu'on dispose de nombreux échantillons, on sélectionne les glauconies les plus évoluées, les moins enfouies, les mieux correlées, celles récoltées à l'abri des altérations superficielles ou des influences tectoniques. Dans tous les cas, on sélectionne la fraction des grains la plus pure possible mais aussi la plus riche en potassium (la moins altérable).

Lorsqu'on peut effectuer de nombreuses mesures on compare les diverses fractions d'une glauconie, on les soumet à divers agents susceptibles de s'attaquer sélectivement, dans un lot de grains, à ses divers composants. C'est la comparaison des résultats analytiques obtenus qui permet d'évaluer la confiance que l'on peut avoir dans les âges obtenus et leur signification par rapport à l'âge de la formation dont ils sont extraits.

(Manuscript received 27-6-1980)

Section IV

Utilization of high-temperature rocks as chronometers

Numerical Dating in Stratigraphy
Edited by G. S. Odin

21
The genesis of bentonites

Alain Person

1 INTRODUCTION

Bentonitic deposits can be adopted as stratigraphic references for two reasons.

1.1 A means of correlation

A bentonitic deposit reflects the immediate impact of a geological event (for example an eruption) which can cover a very wide area up to 5000 km at its widest point. Such a formation can be very thin (variations in thickness will be discussed later). In all cases, the time needed for a bentonitic deposit to settle has no relation to the sedimentation rate of the environment. Therefore the determination of a characteristic bentonitic level makes it possible to establish correlations within stratigraphic series from drillings several hundred kilometres apart.

1.2 A means of radiometric dating

A bentonitic deposit contains minerals which allow radiometric dating (radiochronology, fission tracks). But these minerals can have different origins that we shall enumerate. Furthermore, bentonites are often remarkably well preserved, sometimes even better than the volcanoes from where they originated. In this regard, two types of problems will be discussed.

The process of evolution of volcanoclastic material to bentonitic minerals: the role of the environment (sedimentation basins) in which the eruptive material settles is more influential than the chemical properties of the original magma. The energy available for processes of hydrolysis and neoformation determines the type of 'bentonitization'. One can distinguish bentonite formed by hydrothermal or deuteric processes from bentonite evolved by a process of halmyrolysis.

The eruptive mechanisms (and the prevailing winds at high altitudes for volcanic ash) are important in determining the location and spread of bentonite layers. Small-scale sedimentary reworkings may sometimes occur.

1.3 Definitions

Knight (1898) suggested the name *bentonite* for clay mineral seams with soapy properties in the Fort Benton unit of Cretaceous age formations in Wyoming (USA). The original definition of this term referred mainly to the physical properties of these clay minerals. Further studies established that this clay is a volcanic ash alteration product. At present the meaning of the term bentonite depends on the field in which it is used.

(a) The engineering and commercial meaning refers to clay muds with high colloidal and adsorbent properties, without any regard to their mineralogical components or their geological origin.

(b) The petrological meaning results from the early petrological studies by Hewit (1917) and Wherry (1917) showing that these clay deposits originate from volcanoclastite alteration. Ross and Shannon (1926) gave the following definition, which is still widely accepted: 'Bentonite is a rock composed essentially of a crystalline clay-like mineral formed by the devitrification and the accompanying chemical alteration of a glassy igneous material, usually a tuff or volcanic ash'.

Millot (1964) compares the mineralogical definitions of the terms 'bentonite' and 'kaolin'; several authors, in fact, consider a formation essentially smectitic as a bentonitic formation regardless of its origin. As 'true' kaolins are produced from hydrothermal alteration of silica-aluminous rocks, so 'true' bentonites are produced from *in situ* alterations of volcanic material. Wright (1968) defined it thus: 'Any clay which is composed dominantly of a smectite clay mineral and whose physical properties are dictated by this clay mineral'. In the same way, Dunoyer de Segonzac (1970) and Grim and Güven (1978) retain a definition of the term bentonite which does not refer to any volcanic origin.

However, in the present review we find it necessary to restrict the meaning of this term, retaining the petrogenitic aspect of the definition. The formation of a typical bentonite comprises two stages. The first stage comprises the formation, aerial transport and deposition of volcanoclastites from which the geometry of the bentonitic rock results. The second stage corresponds to the hydrothermal or meteoric alteration of the endogenous components, leading to either a slow or a rapid neoformation of clay minerals, mainly smectitic or occasionally kaolinitic, depending on the intensity of drainage.

After this stage there is no transport of the clay neoformations and only slight reworkings take place, which differentiates bentonites from the whole volcano-sedimentary series.

2 PETROLOGICAL PROPERTIES OF RECENT TEPHRA AND ASH BEDS SUITABLE FOR BENTONITIZATION

The name *tephra* was suggested by Thorarinson (1954) (in Rittmann, 1963), from Aristotle, for incoherent material torn out and ejected in solid and liquid form by volcanic gas.

The mineralogical components of tephra have two origins: they derive directly from juvenile magma and from rock fragments which have been torn out from the walls of the neck during the gaseous eruption. These last can be magmatic or not. One classic example of such formations is the crater sides of Naka-Dake volcano (Japan).

Magmatism producing explosive volcanic emissions whose products are likely to turn into bentonites is generally acid. But petrographic types are very diverse. Keller *et al.* (1978) studied tephra from Mediterranean Upper Quaternary series: 'Petrographical examination based on refractive index, phenocryst content, and chemical composition of the volcanic glass distinguishes the parent magma types: tephritic, alkalic-trachytic, peralkalic-pantelleritic, and calc-alkalic andesitic to rhyodacitic and alkali-basaltic'. Rose *et al.* (1978) give the following petrographic composition for historical tephra from Fuego volcano (Guatemala): 'Olivine-bearing high-alumina basalts, made up of phenocrysts (generally >0.1 mm), microphenocrysts (10–60 μm), glass and round subspherical vesicles (1–400 μm). Phenocrysts are dominantly plagioclase (An_{95-80}) and olivine (Fo_{76-66}). Magnetite (4.3–15.0% TiO_2), augite ($En_{45}Fs_{13}Wo_{42}$) and oxyhornblende are minor phenocrysts'. It is likely that these examples represent the original petrographic nature of bentonitic levels in stratigraphic series.

In addition, contact between volcanoclastites and sea-water may modify the physical nature of the deposit: this is the case with hyaloclastites produced during shallow submarine eruptions. Honnorez (1963) reports three main types of hyaloclastites. The first originates from the crumbling of the vitreous crust of the pillow lavas when they cool rapidly in water. These glass fragments are often altered into palagonite by means of hydratation and devitrification. The second type proceeds from the crushing of the lava itself when it enters sea-water, which gives a clastic, massive and friable rock crossed by dykes whose alteration leads to a massive bentonite. (Bentonites of this type are also described in a continental lacustrine context on the site of Murol in the French Massif Central, Person, 1976). The third species includes stratified hyaloclastites not visibly related to a lava flow, exhibiting disturbances due to currents. Such formations are found in volcano-sedimentary series. Glass fragments quenched at the moment of emission into sea-water are subjected to a distinctive alteration process (Bonatti, 1967). Self and Sparks (1978) have recognized a type of pyroclastic deposit formed by the interaction of water and silicic magma during explosive eruptions. These deposits have a widespread dispersal similar to plinian tephra, but the overall grain size is much finer. The work was carried out on 'the Oruanui Formation (New Zealand) and the Askja 1875 deposit'. These authors propose the name 'phreatoplinian deposits' for such formations.

The occurrence of explosive eruptions under water may induce the formation of the volcanosedimentary type, which, in the form of ash emissions, are *instantaneous.* Mutti (1965) describes in the island of Rhodes (Greece) 'igniturbidites' to characterize pyroclastic deposits of turbidity currents genetically related to submarine ash emissions. The sedimentary deposition of igneous biotites is strictly contemporaneous with eruptive emission. Fiske and Matsuda (1964) also describe the submarine equivalent of a dacitic ash emission in Miocene formations of Tokima (Japan). Volcanic fragments are immediately sorted and carried laterally into deep water, where they are presently interbedded with fossiliferous marine clays. Five main sequences of ashes are known: the most widespread comprises two parts, each originating in one great submarine eruption.

These pyroclastic formations abruptly emitted into sea-water display actual sedimentary patterns: horizontal sorting of particles, graded bedding, sedimentary figures, ripple-marks, fossils. Their composite features (volcanic and sedimentary) cause them to be taken for stratigraphic references according to which radiometric dating is possible, with or without bentonitic alteration. Following wind deposition, volcanic ashes might be mixed with sediments after being transported by floating ice. Ruddiman and Glover (1972) have shown that sand-sized bubble-wall shards were initially deposited on ice cover on and near Iceland, from Icelandic or Jon Mayen volcanoes. These highly silicic ashes later detached from melting ice over the last 620,000 years, as far as 1800 km to the south in the North Atlantic Ocean. These elements are mixed with sediments according to Berger and Heath's model (1968). The geologically instantaneous ash-falls are spread vertically through well-defined zones in cored sediments, providing models for vertical mixing of sand-sized microfossils. The transport and deposition of tephra by wind also varies the relative proportion of minerals to glass as a function of distance. Larsson (1937) has shown that the process results in a change in the bulk chemical composition of tephra.

3 GEOLOGY OF FORMATIONS LIKELY TO BE ALTERED INTO BENTONITE

The occurrence of such formations is common in the eastern Mediterranean. As early as 1948, the Swedish oceanographic ship 'Albatross' performed some cores with the purpose of searching for volcanoclastites in marine sediments. Mellis (1954) has shown in this material the existence of volcanic ash levels which he characterizes with the occurrence of plagioclase, diopside, hornblende and biotite.

Ninkovich and Heezen (1965) have also shown the spreading of tephra deposits in the eastern Mediterranean by studying 21 deep-sea cores. Two ash-falls can be distinguished through the refractive indexes of volcanic glasses ($n = 1.521$ and $n = 1.509$) related to the stratigraphic series of the Santorini volcanic complex (Greece).

The lower tephra, interbedded with Upper Pleistocene carbonaceous sediments, covers about 1000 km west of Santorini, its thickness decreasing away from the source (Figure 1). It appears that it was transported by high-altitude winter winds. These authors give an age of 25,000 years. The upper tephra is interbedded in carbonaceous post-Pleistocene sediments and spreads over 700 km south-east of Santorini, covering an area of about 20,000 km^2. It appears to have been transported by high-altitude summer winds. It can be related to a 30 m thick pouzzolane (ash-bed) covering the islands of Santorini archipelago. This formation originated during the eruption which produced the partly immersed caldera during Minoan times about 1400 BC. This level, as shown by archaeological dating onshore, is an example of a good stratigraphic reference in the eastern Mediterranean.

Near the eruptive vents, the sequence of beds related to explosive emissions is distributed by formations resulting from progressive erosion of the eruptive structures themselves. In addition, part of this material can be affected by alteration–neoformation processes related to early diagenesis. Chamley (1971) traces a rapid evolution of the erosive products from Santorini into smectites (and therefore a rapid formation of bentonite). In the same way, exceptional concentrations of radiolaria and diatoms can occur in the nearest marine surroundings of volcanic islands. Chamley and

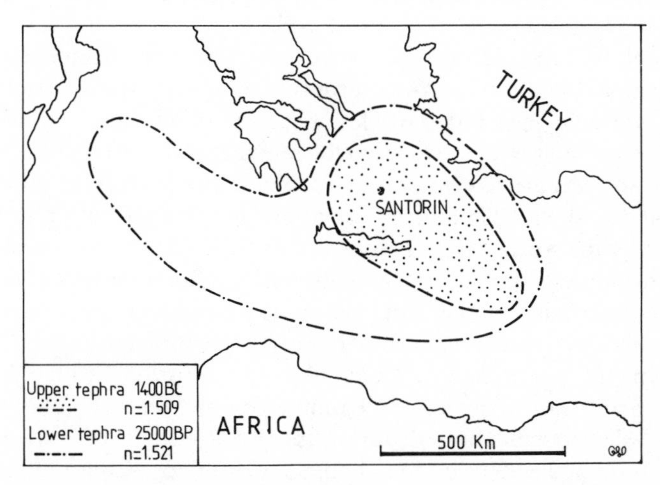

Figure 1. Area covered with Santorini tephra, according to Ninkovich and Heezen (1965).

Millot (1972) have described the formation of smectitic clay minerals from such accumulations of diatoms. A similar process is presently taking place in diatomitic deposits in the Bolivian Altiplano lakes (Badaut *et al.*, 1980). Associated detrital minerals, although having a volcanic origin, observed in both cases, cannot be used by geochronologists in order to date these bentonitic-like formations.

4 DEPOSITION OF VOLCANIC ASH AND RESULTING BENTONITES

Volcanic material yielding a bentonite by alteration has various petrographic features: the chemistry and mineralogy of the volcanoclastites have no direct influence on the authigenic clay features. But the type of magmatism is related to volcanic material settling, and decides the relation between the volcanic particles and the environment. Therefore, the magmatism does have an indirect influence on the bentonitization process. The bentonite structure should differ according to the size of the volcanic material (compact lava flows, pyroclastites, ash-falls). In this chapter, only depositions of volcanic ash have been described. They are the origin of bentonites which are of interest to geochronologists.

A theory of vulcanian explosions is suggested by Self *et al.* (1979). These phenomena have been studied from several recent eruptions. (The last one is the andesite stratovolcano Ngauruhoe in North Island, New Zealand, in February 1975). The common characteristics are: the rock mass ejected per explosion is usually 10^2–10^6 tonnes, and *often contains a high proportion* (>50%) *of non-juvenile material;* the life of the explosion varies from less than one minute (Anak, Krakatoa) to about one day (Asana, Sakurazima); pyroclastic avalanches are often produced.

A model for the deposition of volcanic ash has been suggested by Slaughter and Hamil (1970) for an eruption of a high explosive volcano. The cloud of volcanic gas and particles assumes a flat-topped mushroom or umbrella shape as it rises up to a height ranging from 6 to 45 km. This rising ash cloud is accelerated laterally by high-altitude winds, producing an expanding disc shape. Therefore the pattern of ash distribution is essentially controlled by the winds. The particle size distribution affects significantly the dimensions of the deposits by aggregation of the finest particles due to water and static electricity in the volcanic cloud. Assuming these features, this model yields elongated tongue-like deposits as shown in Figure 2.

This type of phenomenon looks like ash emissions which have been observed many times since the beginning of the century. Larsson (1937) describes ash deposits emitted during the April 10th 1932 explosive eruption of Quizapu located in the Andean Cordillera (Chile). Pacific winds carried the volcanic materials up to Argentina and Uruguay. The ash beds,

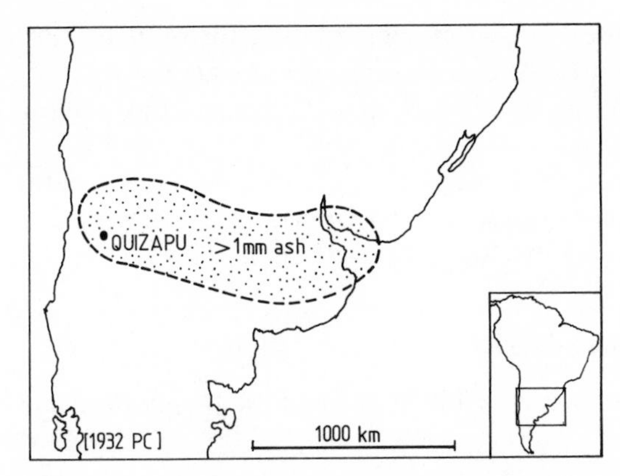

Figure 2. Area covered with Quizapu tephra, according to Larsson (1937).

up to 1 mm in thickness, covered 1200 km (Figure 2). Likewise the two Bezimianny (Kamtchatka) explosions studied by Gorshkov (1959) flung ashes respectively up to 35 and 45 km high. The ash clouds have moved over the Bering Sea, where the material was deposited. In the last 10 years two similar eruptions have occurred: Fuego (Guatemala, 1974) and Mt St Helens (USA, 1980). The first reviews (XXth Geological Congress of Paris, 1980) describe the drastic ash fall (exclusively plagioclase) as over 350 km and 2 cm thick (with 75 cm maximum). Kennedy (1980) described, at 400 km from the volcano, ash clouds which reduced by 90% the light intensity necessary to the phytosphere and which virtually stopped photosynthesis.

5 MINERALOGICAL COMPONENTS OF BENTONITES AND THEIR ORIGIN

The mineralogical components of bentonites depend on the method of deposition of the volcanoclastites and on the nature and intensity of the alteration process of this material. They are also related, to a lesser extent, to the chemistry of the original magma. These components can be classified as follows:

Authigenic minerals: neoformations, mostly clay minerals, especially smectite (but also kaolinite), with occasional association of zeolite, cristobalite, allophane and others (such as barite or Mn-oxide). These strictly bentonite minerals result from the alteration–neoformation process.

Volcanic minerals, and therefore residual at the level of bentonitization but contemporaneous with the eruptive stage.

Minerals of any origin anterior to the eruption derived from substratum fragments of the volcanic framework and torn out by the eruption (more than 50% of initial volcanoclastites, on average).

Detrital minerals resulting from sedimentary contributions or from aerial contamination.

In this respect, geochronologists use rare minerals (apatite or zircon) and fragile minerals (biotite): compared with the authigenic minerals, all the residual minerals represent much less than 1%.

5.1 Authigenic minerals

Authigenic minerals result from three main alteration–neoformation processes.

5.1.1 Hydrothermal alteration of volcanic material

Hydrothermal alteration takes place in rocks which exhibit a wide chemical range, such as basaltic to rhyolitic lavas or tuffs. In some cases, the lava or tufflava structure is completely retained in the alteration products and the final volume is unchanged. Sadran *et al.* (1955) have described a now classic example of bentonite produced by hydrothermal alteration: the Lalla Maghnia rhyolitic plug in Algeria. Smectites are the main mineral in alteration products but kaolinite is also present in amounts of about 20%. Often a subsequent weathering alteration stage is superimposed on the hydrothermal alteration. This is the case in Lalla Maghnia, where the circulation of magnesium-rich meteoric water, leaching some silica and alkalies, is believed to have followed the hydrothermal stage. This neoformation process is comparable to the direct clay–mineral deposit from the Recent hot springs (Person, 1976; Sudo and Shimoda, 1978). Sometimes, a single clay mineral has developed replacing a large mass of host rock (Grim and Güven, 1978). In this case, residual volcanic minerals are available material for dating. With clay minerals (smectite or kaolinite groups), we found zeolite and calcium carbonate (calcite or aragonite). Kaolinite is the main authigenic mineral in several examples of hydrothermal alteration described in northeastern Japan (Iwao, 1970).

5.1.2 Deuteric alteration of volcanic material

Deuteric is used to indicate that the energy available for mineralogical change is produced by the volcanic process itself. The authigenic smectite or kaolinite appears in igneous rocks immediately after its emplacement. An example of an authigenic iron-bearing smectite formation through deuteric alteration occurs in the French Massif Central (Mélières and Person, 1978); a rapid deuteric alteration of lava flow is described—a natural autoclave effect is thought to occur with the lava flows onto a substratum charged with water.

5.1.3 Alteration of volcanic ash or tuff by halmyrolysis

When the ash-fall settles in marine or lacustrine basins, the volcanic particles are slowly hydrated and hydrolysed. The volcanic mineralogical composition is changed to authigenic clay minerals. Meanwhile, relict structures of the parent ash or tuff can be seen by microscopic examination. The essential minerals produced in this way are smectites. But when the drainage is better, the clay mineral components are mainly kaolinite or halloysite. Halloysite is above all a soil mineral produced by weathering of the volcanic ash coat, associated with allophane and imogolite (hydrated aluminosilicate) (Wada, 1978). Thus, the same volcanic ash-fall can change to smectitic bentonite, kaolinitic bentonite or halloysite soil according to the environmental gradient of water content and drainage. Millot (1964) reports an example of a volcanic ash bed altered to illite in a lacustrine environment near Denver (Colorado). Bentonite formed by alteration of Upper Miocene basaltic ashes deposited in a fresh-water lake has been described by Carlson and Rodgers (1974) from New Zealand: ferriferous beidellite is the dominant component and the alteration is believed to have occurred in a mildly alkaline environment in which initial reducing conditions became oxidizing.

Halmyrolytic alteration of the volcanic material is indicated by the high concentration of clay minerals, by the absence of detrital minerals and by a transition to volcanic ash or tuff. Alteration of ash or tuff *in situ* is by far the commonest origin of bentonite (Grim and Güven, 1978).

The widest range of associated marine or non-marine sedimentary beds indicates the environments in which the alteration took place; it reveals that there are no narrowly restricted conditions necessary for the formation of bentonites from volcanic ash or tuff: the geological environment can be lacustrine, coal-forming, estuarine, shallow marine (lagoonal or glauconitic) or deep-sea. The only evidence is that a wet ash favours the formation of bentonite. And the devitrification process has generally taken place in water: this is the defining factor for halmyrolysis (Hummel, 1922). These authigenic clay formations cannot be considered as the result of superficial weathering processes, but a little further alteration is frequently the case. The bentonitization process is essentially the devitrification and hydratation of glass shards and of associated volcanic minerals, and the crystallization of the smectite or kaolinite around many nuclei.

The alteration of the ash or tuff is probably contemporaneous with the deposition of the igneous material. Grim and Güven (1978) report that in Mississippi a bentonite grades upwards into a glauconitic sand which contains nodules of bentonite, indicating a rapid alteration of ash.

5.2 Residual volcanic minerals

The non-altered mineralogical components of parent tephra are the residual minerals of bentonite: essentially feldspar, quartz, pyroxene, olivine, amphibole, mica, opaque minerals, and such minor components as sphene, apatite, allanite and zircon. Some of them are datable, the others indicate by their various features (habitus, crystalline void, . . .) that the clay formation is a bentonite.

Apatite (generally hydroxyfluorapatite [$Ca_5(PO_4)_3(F, OH)$]) is a frequent accessory mineral of andesitic volcanism; Coulon (1977) describes it as thick-set prisms overlapped with titanomagnetite phenocrysts in mineralogical components of andesitic lavas in Sardinia. This lava also contains opaque minerals and zircon. However, apatite is very common in both regional and contact metamorphosed rocks, where it is associated with sphene, zircon, pyroxene, amphibole, spinel, idocrase and mica. An example of apatite-bearing tephra is described by Ninkovich *et al.* (1978) in the Late Quaternary sequences of deep-sea cores from the eastern Mediterranean. Amongst 20 air-borne tephra layers, there are 14 apatite-bearing tephra and two apatite- and zircon-bearing tephra.

Zircon is the earliest mineral to crystallize and may therefore be included in all the other minerals in the rocks. This mineral may be a component of the volcano substratum, volcanic throat rock or juvenile magma in the mineralogical composition of the tephra.

Biotite and *sanidine* are the most resistant potassic minerals of the residual volcanic bulk. Biotite sometimes presents solution cavities inside apparently intact automorphic crystals (Figure 3). These cavities, observed by fragmentation of the crystal, have been studied by scanning electronic microscopy (SEM) in work on bentonites of the Mont Dore, France (Person, 1976). The insides of cavities in some biotites exhibit the presence of authigenic smectites similar to the bentonitic smectites.

PALAEOZOIC BENTONITIC FORMATIONS (tonsteins and volcanic ash layers from the European Carboniferous)

The Carboniferous series of northern France and Belgium, South Germany and England display a typical stratigraphic facies of bentonites: *tonsteins* and acid cinerites.

The mineralogical composition of these formations was first described by Termier (1890), who defines a potassic aluminous silicate, the *leverrierite.* A detailed study made by Kulbicki and Vetter (1955) showed that this was a vermicularous facies corresponding to kaolinite, with a possible alternation of mica layers and kaolinite: the kaolinite develops by epitaxic growth between the mica layers.

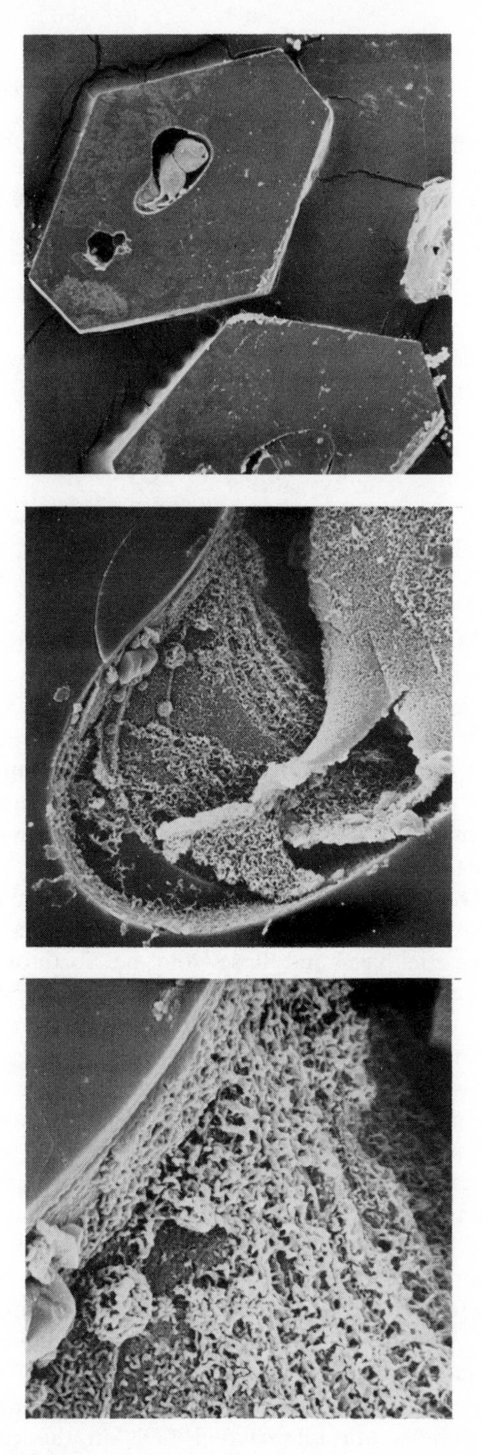

Figure 3. Authigenic smectite inside a solution cavity of a biotite crystal observed by SEM. Bentonite from the French Massif Central.

Millot (1964) considers and rejects the classic hypotheses of non-volcanic origins of tonstein: detritic kaolinite inheritance (Pruvost, 1934); chemical or biochemical sedimentary neoformation (Shuller 1951); pedological inheritance from a marshland soil (Erhart, 1962). He supports the hypothesis of a volcanic origin of tonsteins, referring to the studies of Chalard (1952) and Bouroz *et al.* (1958). These authors compare the petrography of tonstein from the North France Basin with the 'Baldur' tonstein from the Ruhr Basin. They point out the constancy of petrographic features for these tonstein, which are several hundred kilometres apart: only a volcanic ash deposit exhibits such an homogeneity. The transition from a volcanic ash tuff to a kaolinitic horizon (Francis, 1961, in Millot, 1964) confirms the volcanogenetic hypothesis.

The Carboniferous tonstein and cinerites have been used as stratigraphic markers to establish correlations between the Jura Basin and the Cévennes Basin in France (Bouroz, 1972). Datings have been made on the cinerites by Bouroz *et al.* (1972).

Loughnan (1978), on the other hand, distinguishes the cinerites from tonstein, to which he attributes an essentially allochtone and non-volcanic origin. However, Spears and Kanaris-Sotiriou (1979), taking into account mineralogical and chemical criteria (study of the ratios: Ti/Al; Cr/Al; Zr/Al and Ni/Al), show that Britannic tonstein are produced from an *in situ* alteration process of volcanic ashes: this is therefore kaolinitic bentonite according to the classic definition of Ross and Shannon (1926). Among those layers, two groups are distinguished.

The first group corresponds to an acid magmatism giving rise to high explosive eruptions and allows stratigraphic correlations to be made between the Britannic coal basin and those from the North of France and from the Ruhr. The occurrence of zircon and apatite is reported in these formations.

The second group comprises tonstein which exhibit TiO_2 concentration expressed as anatase, which is a typical feature of a more basic volcanism. The extent of the basic tonstein is restricted to the Britannic Basin.

As a recent example, Pevear *et al.* (1980) confirm the evolution of volcanic ashes in kaolinitic bentonite, as demonstrated in studies of Eocene non-marine basins in the Tulameen coal field (British Columbia). The volcanic glass was transformed either to smectite–cristobalite–clinoptilolite or to smectite–kaolinite. Some beds are nearly 100% kaolinite and can be designated as tonstein. These are associated with coal ranks. Carboniferous bentonites can also occur in non-coal Lower Carboniferous marine sediments: Thorez and Pirlet (1978) note the occurrence of several thin, but widespread K-bentonite layers in the Namur and Dinant synclines of Belgium. These formations are intercalated beds in the calcareous Visean and Tournaisian stages. K-bentonite layers exhibit various types of coarse particles and phenocrysts: pseudomorph biotite and feldspar, quartz shards, idiomorphic apatite and zircon. The clay composition comprises predominantly illite–smectite or illite–vermiculite mixed-layer minerals. They are associated with chlorites, kaolinite–smectite and illite-chlorite mixed-layers. These formations are designated *K-bentonite* in agreement with Weaver (1958).

7 CONCLUSION

The occurrence, in Recent or ancient sedimentary series, of non-altered volcanic ash horizons (tephra) or altered volcanic ash horizons (bentonite or tonstein) is related to an instantaneous event in geological time when there is evidence that a high explosive eruption occurred. Deposits formed by the alteration of high explosive vulcanism products from a single-stage settling must be differentiated from the results of progressive erosion of volcanic framework, scattered through the volcano sedimentary series. These two kinds of clay mineral bearing formations will not have the same stratigraphic and chronological significance. Mineralogical criteria can be used to establish this differentiation: homogeneity of the mineralogical composition of the residual volcanic fraction, nature and homogeneity of the clay neoformation in the case of an halmyrolitic alteration. The constancy of the refractive index or chemical composition of glass shards remains the better argument for a monogenic and isochronous layer deposited for an infinitesimal interval of geological time (days or perhaps weeks).

For time-scale purposes, smectitic bentonite, tonstein and tephra fulfill the same role. The mineral fragments for analysis must be formed in the magma chamber at the time of eruption (except for apatite when the dating method is fission tracks), and it must have not been altered or contaminated.

The extent of tephra layers depends on several factors, including:

—the nature of the volcanic eruption (such as the size of the particles in the volcanic cloud);

—the wind circulation at high altitude (subject to seasonal variations);

—the possible induction, by the eruption, of strong rains which scatter more rapidly the volcanic particles of the cloud.

The extent of bentonite layers is related to the same features as the tephra layers, but restricted to the alteration field.

Thus, Recent volcanic phenomena reveal the possibility of correlations being established on places several hundred kilometres (up to a thousand kilometres) apart. In addition, the information obtained from radiometric dating, where the bentonite is rich in datable (but rare) minerals (biotite, sanidine, zircon, apatite), can be extended to the whole original ash deposit, the homogeneity being demonstrated by elements such as glass shards, volcanic quartz, pyroxene, etc.

The geographical repartition of high explosive Recent volcanoes (continental edge) suggests that the bentonite studied as a time-stratigraphic horizon allows correlations for the marine and non-marine series to be made by radiochronological or classic stratigraphic methods.

In this manner, Obradovich and Cobban (1976) established from bentonites a time scale for the Late Cretaceous of the Western Interior of North America, and Keller *et al.* (1978) did the same from tephra of the Upper Quaternary of the Eastern Mediterranean.

Résumé du rédacteur

Les bentonites, niveaux de cendres volcaniques altérées, sont des formations ubiquistes fréquement utilisées pour l'établissement d'échelles géochronologiques, en particulier dans les séries continentales et épicontinentales du Carbonifère européen, dans les séries crétacées d'Amérique du Nord, comme dans les sédiments récents du domaine océanique profond (Méditerranée, Atlantique, Pacifique Nord). Il s'agit d'horizons peu épais, homogènes, de faciès très tranché dans les séries sédimentaires, donc d'excellents *marqueurs stratigraphiques.* Ces caractéristiques leur sont conférées par un mode de formation en deux phases de nature radicalement différente, parfois presque contemporaines, mais généralement bien dissociées dans le temps.

La première phase, essentiellement mécanique, correspond à la mise en place brutale (instantanée à l'échelle géologique) de l'ensemble de la couche, à partir d'une éruption volcanique paroxismale. Les vents dominants de haute altitude déterminent la géométrie et l'extension (plusieurs centaines, voire milliers de kilomètres) de ce dépôt de cendres volcaniques; le phénomène peut se limiter à cette première phase, il s'agit alors de 'tephra' = bentonite non évoluée.

Les particules déposées par l'éruption ont plusieurs origines possibles. Il s'agit soit de fragments de *lave juvénile* provenant du contenu de la chambre magmatique, soit de débris d'*épanchements plus anciens* obstruant le conduit volcanique, soit d'*éléments du socle* de l'appareil éruptif éjectés avec le magma et les gaz de l'explosion. Ces composants sont donc de nature pétrographique variée et leur âge radiométrique s'interprètera différemment. Les eléments non juvéniles constituent généralement plus de 50% d'une telle formation. Parmi les particules volcanogéniques: fragments de lave, esquilles de verre, se trouvent des débris de minéraux: feldspath potassique, plagioclase, biotite, apatite, zircon, plus favorables aux mesures géochronologiques.

Les éruptions catastrophiques de volcans actuels tels que le Quizapu (1932, Chili), le Fuego (1974, Guatemala) ou le St Helens (1980, USA) sont des exemples permettant de comprendre les mécanismes de mise en place des bentonites.

La seconde phase de la constitution d'une bentonite comporte un processus d'altération–néoformation donnant lieu à une modification très prononcée de la composition minéralogique originelle. Ce mécanisme se produit soit 'à chaud' (altération deutérique contemporaine de la mise en place, ou altération hydrothermale); soit 'à froid' (halmyrolyse). Il est dans tous les cas caractérisé par une évolution *in situ* du matériel volcanique, ce qui différencie les niveaux bentonitiques des formations volcano-sédimentaires. L'argilisation des cendres volcaniques conduit généralement à une roche globalement monominérale (smectite ou kaolinite suivant les conditions de drainage). Les datations sont alors à effectuer sur les minéraux résiduels (souvent moins de 1%) dont l'habitus volcanique est caractéristique. Dans certains cas cependant (apatite datée par la méthode des traces de fissions), l'origine du minéral est indifférente, dans la mesure où il a été atteint par l'élevation de température provoquée par l'explosion (remise à zéro de l'horloge 'traces de fission').

(Manuscript received 8-5-1981)

Numerical Dating in Stratigraphy
Edited by G. S. Odin

22
The dating of bentonite beds

HALFDAN BAADSGAARD and JOHN F. LERBEKMO

1 INTRODUCTION

Bentonite beds represent *altered and residual weathering products* of deposits from explosive vulcanism (as discussed by A. Person in the preceding chapter). Volcanic glass is usually converted to montmorillonitic clay and silica, while the crystalline igneous minerals in the original tuff or ashfall may or may not be altered, according to the susceptibility of the respective minerals to weathering and the nature of the local sedimentary conditions. Single bentonite beds represent a very short time of deposition and are usually negligibly contaminated with normal detrital sedimentary material. The presence (or absence) of detrital or other contaminant material in a bentonite must be established, however, because contaminated bentonite is usually of dubious geochronological value. When the bentonite comprises essentially pure pyroclastic material, those igneous minerals which contain K, Rb or U may be used to date the time of deposition of the original volcanic material. How well these dates give the time of deposition is dependent upon a number of factors such as suitability of mineral, purity of mineral, presence of contamination, degree and nature of alteration of mineral and time of alteration. Besides analytical uncertainties, there are also indeterminate variations in the purity and suitability of the mineral separates, so any dates should be further tested by comparison between as many dating methods as possible.

2 METHODS OF DATING AND SAMPLE PREPARATION

The available igneous minerals in the bentonite will determine the number and kinds of applicable dating methods. In approximate order of general abundance, the minerals are: plagioclase, sanidine, biotite, zircon, apatite, hornblende and glass. Some or all of these minerals may be absent in the bentonite, depending upon the composition of the initial magma source and the extent and nature of the secondary alteration. The mafic minerals, except for biotite, are seldom present since explosive vulcanism occurs when volatiles build up in relatively acid magmas, and because such minerals are very susceptible to weathering or secondary alteration. Glass is very rare in bentonitic material because the bentonite clay forms from it. A relict glass is usually partially altered. The simple presence of a mineral is not always sufficient for dating because the separation may not yield a pure enough sample or the quantity may be too small. This is often the case with apatite. Slight alterations or coatings of secondary alteration products on the mineral grains may be impossible to remove completely.

2.1 Applicable radioactive transformations

It is assumed that the reader is familiar with the basic principles, assumptions and techniques of radiometric dating (see Faure, 1977).

In addition to the presence of sufficient parent isotope in suitable minerals, time for the generation of sufficient radiogenic daughter isotope must have occurred. That is, the daughter isotope determination will become very difficult when the bentonite horizon is too young. In general, one may expect to be able to use the K–Ar method into the Pleistocene. The Rb–Sr method can apply to the Pliocene (given suitable isochron minerals). The U–Pb method will give good $^{238}U/^{206}Pb$ dates to the Pliocene, $^{235}U/^{207}Pb$ dates to the Miocene and reasonable $^{207}Pb/^{206}Pb$ dates only in Palaeozoic bentonites. Model Pb dates will be usable only for Early Palaeozoic bentonites with abundant apatite.

2.1.1 *K–Ar dating method (conventional and $^{40}Ar/^{39}Ar$ techniques)*

The K–Ar method is the most widely applicable dating method for bentonitic materials. Because of the sensitivity of the method, even K-poor minerals such as plagioclase can be dated in Tertiary bentonites. Biotite, sanidine, plagioclase and hornblende are all very satisfactory for K–Ar dating provided they are clean and unaltered. Unlike plutonic feldspars which can lose Ar, pyroclastic sanidine and plagioclase have been rapidly cooled and the relatively disordered lattice retains Ar well (Baadsgaard *et al.*, 1961; Fechtig *et al.*, 1961). The volcanic feldspars have another advantage over plutonic feldspars in that they have a much lower content of contaminant Ar (Evernden and James, 1964; Dalrymple and Lanphere, 1969). Sanidine, with its high K content, resistance to weathering and low initial Ar content, can be dated down to 50,000 years, and is the first choice of a bentonitic mineral for reliable K–Ar dating. The common K-rich bentonitic mineral is biotite, which has rather more contaminant Ar and is decidedly more susceptible to secondary alteration than is sanidine. However, secondary alteration in bentonitic biotites does not always have a marked effect on the K/Ar ratio of the mineral, even when more than half of the K has been lost from the original mica (see Table 3). Clean, unaltered mica with $K_2O > 7\%$ is often obtained from bentonites and is an eminently suitable material for K–Ar dating down to approximately 10 million years. Plagioclase, though it has much less K_2O than sanidine or biotite, gives good results down to 10 million years. Unaltered hornblende is an excellent mineral for K–Ar dating, but is encountered only very rarely in bentonites. In summary, K–Ar is an excellent method for bentonite dating because there are several useful minerals, the Ar-contaminant levels are very low and the range of measureable ages extends to very young (Pleistocene) materials.

2.1.2 Rb–Sr dating method

The Rb–Sr method is, in principle, applicable to the same minerals as K–Ar since Rb is a geochemical companion of K. However, unlike Ar, original Sr contamination can occur in relatively large amounts. Both sanidine and plagioclase contain large amounts (200–2000 ppm) of initial Sr, but biotite excludes most Sr from its lattice (see Table 4, p. 435). Biotite is the best single common bentonitic mineral for Rb–Sr dating and can be useful down into the Tertiary. Determination of the initial $^{87}Sr/^{86}Sr$ ratio becomes important for young biotites and may be easily found by using a sanidine and/or plagioclase plus biotite mineral isochron method. That is, the expression:

$$(^{87}\mathrm{Sr}/^{86}\mathrm{Sr})_{\mathrm{biotite}} - (^{87}\mathrm{Sr}/^{86}\mathrm{Sr})_{\substack{\text{sanidine or}\\ \text{plagioclase}}} = [(^{87}\mathrm{Rb}/^{86}\mathrm{Sr})_{\mathrm{biotite}} - (^{87}\mathrm{Rb}/^{86}\mathrm{Sr})_{\substack{\text{sanidine or}\\ \text{plagioclase}}}](e^{\lambda t} - 1)$$

can be solved for t ($\lambda t \simeq e^{\lambda t} - 1$ for younger ages). This Rb–Sr date is essentially a biotite Rb–Sr mineral date. Because the original pyroclastic material is a partially open chemical system during the clay-forming period, reliable whole-rock Rb–Sr determinations are not possible. If altered biotite is present in the bentonite, various degrees of alteration give variable Rb/Sr ratios so that a mineral isochron is sometimes obtained (see Figure 1 and Table 4). The interpretation and evaluation of this type of Rb–Sr mineral isochron is still under investigation.

2.13 U–Pb dating methods

The only applicable bentonitic mineral commonly available is zircon. Although original contaminant Pb is generally very low in zircon, the small amounts of radiogenic Pb present necessitate careful determination of the

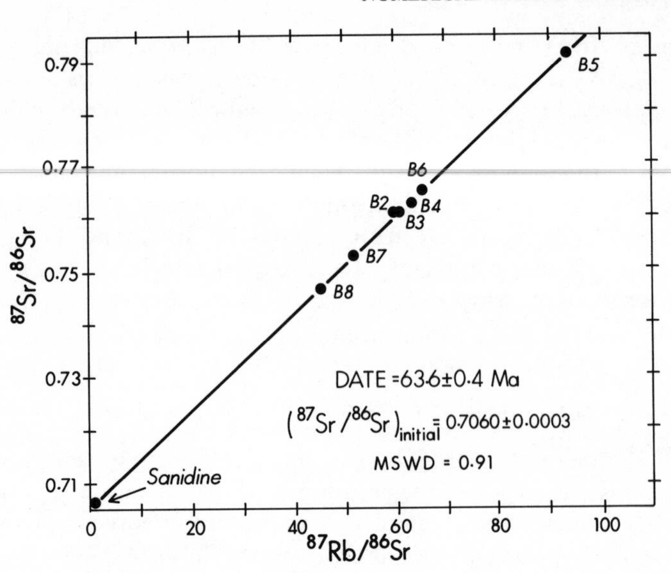

Figure 1. Isochron plot of Rb–Sr data (Table 4) for biotites and a sanidine from a bentonite in the 'Z' coal horizon in Montana, USA.

common Pb correction for samples of Mesozoic or younger age. Pb from co-crystallized sanidine or plagioclase can be isotopically measured and used for the isotopic composition of the contaminant Pb in the zircon. The concordia plot ($^{206}Pb/^{238}U$ *vs.* $^{207}Pb/^{235}U$) interpretation of discordant data cannot be easily used in Mesozoic and younger zircons. The zircon U–Pb analysis often yields two different mineral dates for younger zircons and the degree of discordance must be estimated from the difference betwen these two mineral dates. If the two zircon U–Pb dates are within analytical uncertainty, they may be utilized for geochronological purposes. In practice, U–Pb zircon dates are found to be either accurate or slightly low because of Pb-loss discordance.

2.1.4 Model Pb dating method

If enough pure apatite can be separated from the bentonite, it is possible to measure $^{207}Pb/^{204}Pb$ *vs.* $^{206}Pb/^{204}Pb$ to get a $^{207}Pb/^{206}Pb$ model age for various co-crystallized igneous minerals. Pb isotope compositions may be measured for biotite, sanidine, plagioclase and apatite; but only apatite will show appreciable radiogenic Pb because of higher uranium content. This method is often impracticable for Late Phanerozoic samples because of the small $^{207}Pb/^{206}Pb$ variation in younger geological time and the need for very pure mineral separates. In addition, the precision of the method is relatively poor; it serves better as an indicator of purity of mineral separation and lack of contamination.

2.1.5 *Fission track dating method*

The spontaneous fission of ^{238}U occurs at a very low rate and gives rise to linear damage paths (or fission tracks) in the ^{238}U host lattice (see Chapter 11 by Storzer and Wagner, this volume). Each fission event may be individually counted by making these fission tracks visible. From the number of fission events and the concentration of uranium in the host solid, the time of accumulation of the tracks may be calculated. Annealing and disappearance of original fission tracks during metamorphism may partially or completely 'reset' the fission track date and constitutes one of the major problems in obtaining accurate dates of crystallization of the host mineral. The host minerals vary in their susceptibility to annealing of fission tracks. Of the uranium-bearing minerals in bentonites applicable to fission track dating, zircon is relatively resistant while apatite is very sensitive to annealing of fission tracks. If cogenetic apatite and zircon from a bentonite give the same date, this is taken as good evidence for the lack of any loss of tracks by annealing and the dates are those of crystallization. The occurrence of a flat-lying bentonite bed usually indicates a lack of post-depositional tectonism and metamorphism. Original igneous apatite and zircon from bentonites are therefore excellent material for fission-track dating over the whole of the Phanerozoic. The reader is referred to Naeser (1978) for details of fission track dating techniques.

2.2 Mineral separation and purification

Pure mineral separates are essential for reliable dating of bentonite, but because of extensive secondary chemical activity they are not easily obtained. Also, the grain size of the minerals may be very small if the pyroclastic deposit is sampled far from the volcanic source. The initial step in mineral separation is to remove the major clay component and obtain the igneous minerals in a relatively coarse-grained fraction. To this end, the bentonite sample (collected in the field with great care to avoid contamination from adjacent horizons or materials) is first soaked in water to soften the clays and expand the montmorillonite (allow room for expansion in the container). The coarse fraction may be less than 1% by weight of the bentonite and enough sample should be taken to assure a sufficient yield of minerals. This may necessitate 50–100 kg samples on occasion, though 5–10 kg will sometimes be sufficient. The hydrated bentonite is then dispersed in three to four times its volume of water in a Waring blender or an equivalent mixer with strong shearing action of the mixer blades. The clay sample is suspended in a large (100×) volume of cold water and three or four minutes allowed for the coarse fraction to settle out. The suspended clay is then decanted from the settled coarse fraction, which is stored for yet another pass through the mixer if clay is still present. The final clay-free fraction is collected and dried by acetone wash and air-drying in preparation for magnetic and heavy-liquid mineral separation procedures.

2.2.1 *Biotite*

Biotite may be initially removed from the grit fraction by magnetic separation in a Franz Isodynamic separator or its equivalent. Such a biotite concentration is not pure, so it is barely sunk in a methylene iodide–acetone mixture (sp. gr. ~2.95). It is often found at this stage that only a small portion of the biotite sinks at the highest specific gravity to yield a biotite sink fraction. This is because alteration of the biotite tends to make it less dense, and in taking the heaviest biotite fraction the least-altered biotite in the bentonite is obtained. After removing the heaviest biotite, the methylene iodide–acetone liquid may be made successively lighter with additions of acetone and a series of increasingly altered biotite fractions separated. Sometimes the biotite is unaltered and gives a homogeneous, clean separate, but more often it is at least partially altered and still contains pyroclastic contaminants. Careful magnetic separation under closely controlled conditions effects more purifications. A one-minute wash with cold vapour-distilled 6N HCl strips contaminant coatings of carbonate or haematitic materials without significantly leaching the elements of the biotite lattice. In some cases it may be necessary to resort to hand-picking of impurities for the final stage of purification.

2.2.2 *Sanidine*

The non-magnetic fraction from the biotite concentration is further subjected to the highest degree of magnetic separation and the least magnetic fraction taken for sanidine (and plagioclase) purification. Ethylene tetrabromide diluted with acetone is used as a heavy liquid and the specific gravity of the liquid adjusted to barely sink quartz. A layer of acetone is added to the top of the heavy liquid and allowed to diffuse into the heavy liquid to form a gradient of specific gravity. The sanidine ($D \simeq 2.56$–2.62) is generally the lightest igneous mineral (some pyroclastic impurities may be lighter) and may be separated in the subtle gravity gradient. After separating the sanidine fraction (it is checked under the microscope to make sure it *is* sanidine), it is purified again in ethylene tetrabromide–acetone heavy liquid, taking a cut of sample of narrow specific gravity. The sanidine separate is next acid-washed with boiling 6N HCl, rinsed with distilled water and again run through the magnetic separator, rejecting all but the least magnetic fraction. Grain counting should give >99.5% clear, anhedral fragments of sanidine (with the other impurities mainly plagioclase) before using the separate for dating. The sanidine fraction is sometimes cloudy, probably from secondary reaction, and seems to give higher K–Ar ages than cogenetic clear sanidine (clear → 62 Ma *vs.* cloudy → 66 Ma in one unpublished test). The cloudy sanidine is just perceptibly lighter than the preferred clear sanidine and can be separated by heavy liquids using a very small gradient of specific gravity. Hand-picking may have to be done for some samples.

2.2.3 Plagioclase

The intermediate-gravity fraction from the sanidine separation is taken for plagioclase purification. The mineral in this fraction should be verified under the microscope, and if clear grains of plagioclase of a narrow range of An values are found these should be sought in further heavy liquid purification. When a reasonably clean plagioclase separate has been obtained, it is acid-washed with boiling vapour-distilled 6N HCl and run through the magnetic separator at the highest magnetic field conditions. The non-magnetic acid-washed plagioclase is again inspected under the microscope and further gravity separation carried out as necessary. The most suitable plagioclase for dating purposes comprises clear anhedral grains of homogeneous An–Ab content without inclusions or coatings and at least 99.5% pure plagioclase. Not infrequently, secondary alteration of plagioclase makes it impossible to obtain the ideal separate. For K–Ar dating of young plagioclase or sanidine, a brief HF wash to strip off the outer hydrated skin of the mineral helps to reduce contaminant argon to a low level (Evernden and Curtis, 1965).

2.2.4 Zircon and apatite

The quartz-sink fraction from the ethylene tetrabromide–acetone separation of the non-magnetic fraction contains the zircon and apatite. This fraction is placed in pure ethylene tetrabromide ($D = 2.96$), in which apatite and zircon sink. The apatite and zircon are next suspended in pure methylene iodide ($D = 3.33$), wherein zircon sinks but apatite floats. A little acetone added to the top of liquid in the separatory funnel diffuses in slowly to produce a gravity gradient which serves to separate apatite from floated impurities. The apatite concentrates may be run through the magnetic separator at highest field strength to take out impurities from the non-magnetic apatite. Apatite is usually very clean at this stage and may be hand-picked to yield a high-purity separate. The zircon concentrate is

Table 1. U-Pb dates on zircon from a bentonite in the 'Z' coal, Montana. Lead extracted from cogenetic sanidine having 204:206:207:208 equal to 1:19.148±4:15.649±4:38.785±8 was used for the common Pb correction. The decay constants used were λ-U-235 = 9.8485×10^{-10} a^{-1} and λ-U-238 = 1.55125×10^{-10} a^{-1} (Jaffey *et al.*, 1971): ICC.

Sample	Pb ratios, measured			^{238}U	^{206}Pb	Dates (Ma)	
	206/204	207/206	208/206	(ppm)	(ppm)	206/238	207/235
JFL-500 C, picked, *not* acid-treated	612±2	0.07189±4	0.2007±2	1275	10.250	59.5	60.0
JFL-500 A(1), >325 mesh, HI–HCl–H_3PO_2 treated, picked	669±2	0.07007±1	0.18996±4	1134	9.806	64.0	65.1
JFL-500 A(2), <325 mesh, HI–HCl–H_3PO_2 treated, picked	1500±500	0.05770±13	0.1663±3	1304	11.158	63.4	64.5
JFL-500 A(3), <325 mesh, HI–HCl–H_3PO_2 treated, picked	(818±3	0.06742±1 (spiked Pb run only)	5.7004±7)	1319	11.003	62.0	
JFL-500 A(4), <325 mesh, HI–HCl–H_3PO_2 treated	1555±150	0.05755±9	0.1648±2	1289	10.961	63.0	64.1
JFL-500 A(5), <325 mesh, HI–HCl–H_3PO_2 treated	(95.4±2	0.203±1 (spiked Pb run only)	2.268±2)	1296	11.111	63.5	
JFL-500 A(6), <325 mesh, HI–HCl–H_3PO_2 treated	1150±480	0.06043±16	0.1710±8	1313	11.122	62.8	63.2
Mean values (1)–(6)						63.1 (2σ)±1.4	64.2 ±1.6(2σ)

heated to ~80–85°C in 6N HNO_3 to remove pyrite and clean the zircons. In some bentonites, zircon is contaminated with barite which is impossible to remove completely from zircon by gravity and magnetic separation. A reagent mixture, HCl+HI+H_3PO_2 (Thode *et al.*, 1961), used by S-isotope geochemists to reduce solid sulphate, successfully destroys all the barite while leaving the zircon essentially unattacked. It is not always necessary to hand-pick impurities from the zircons, but this may be employed as a final purification step if the use of Clerici solution is to be avoided. Table 1 gives U–Pb data on zircons recovered from a bentonite in the 'Z' coal in Montana, USA. The barite contaminant was equal in amount to the zircon. The sample not treated with the HCl–HI–H_3PO_2 mixture to destroy barite (but picked) gives a different date than the other samples, all of which have been acid-treated. Picking the zircon fraction or using a different size fraction does not seem to cause a notable difference in such dates. The variability of the U content seems to be a sample property, because the radiogenic Pb content varies along with the changes in the amount of U.

The data in Table 1, Nos (1)–(6), may be used to show the precision of U–Pb dates at values around 60 Ma. The 2σ standard deviation of the $^{238}U/^{206}Pb$ dates is ±1.4 Ma or ±2.2%, while that for the $^{235}U/^{207}Pb$ is ±1.6 Ma or ±2.5%.

If samples are to be obtained from fission track dating, only 5–10 mineral grains of each are needed. These are best obtained by hand-picking, and much of the separation and purification procedure necessary for the U–Pb method can be avoided.

3 SOURCES OF UNCERTAINTIES RELATED TO THE USE OF BENTONITIC MINERALS IN DATING

3.1 Geochemical factors

In dating minerals from bentonites major uncertainties may arise from the effects of variable contamination by a variety of materials associated with the process of formation and deposition of the original pyroclastic sediment. Subsequent alteration of the sediment can lead to variations in dates obtained on the igneous minerals.

3.1.1 Contamination

Material not derived from the volcanic source can find its way into the eventual bentonite horizon from a number of sources. (1) Inherited crystal nuclei (contaminated magma), (2) wall and throat rock from the volcanic vent, (3) normal detrital material and (4) constituents introduced into the bentonite horizon during the hydrous alteration of the glass to clay minerals can all be present.

Inherited crystal nuclei in magma source. These have been observed occasionally in zircons. That is, pyroclastic zircons can have a faint cloudy core, which may be ascribed to a nucleus of crystallization. These zircons may show short-period linear discordance at an episodic secondary concordia intercept. This is exactly the concordia plot which would be obtained from newly crystallized zircons containing a variable amount of the same older contaminant. The presence of such contamination presupposes partial melting of country rock by the volcanic magma (or even a source by remelting?). If this is the case, then other mineral contamination has probably occurred.

Wall or throat rock from volcanic vent. A volcanic explosion will commonly blast out some preexisting rock which may accompany the molten ash and gases. If the explosion is violent enough a considerable amount of country rock can be involved. This material is very difficult to detect except when definite multimineralic rock fragments can be found (see Hills and Baadsgaard, 1967). If the glassy component of preexisting volcanic rocks alters to clay and releases mineral phenocrysts, they would be difficult or impossible to recognize as contaminants.

Normal detrital material. Conventional sediments contain a number of serious contaminants, especially if the source of the sediments is much older than the bentonite horizon. Secondary alteration products of normal weathering will not usually be severe contaminants since these materials are separated and discarded from the desired igneous minerals. Detrital igneous minerals may usually be recognized by the rounding and abrasion of the grains compared with the sharp, euhedral pyroclastic grains. Contaminant detrital biotite is not usually found, but detrital orthoclase and microcline can be very serious contaminants of sanidine and may be impossible to separate except by picking under a microscope. If the pyroclastic deposition has taken place in a marine environment far from the shore or in very quiet water, only very fine-grained detrital components are present as contaminants. The pyroclastic deposits usually overwhelm the normal sediment, and negligible detrital components are found in the bentonite mineral separates. A rule of thumb for collecting detritus-free bentonitic material is to sample in areas of flat-lying, even-thickness bentonites with great lateral persistence. This is not always easy to establish: good horizons may occur within coal seams and, in general, from within very fine-grained sediments.

Contaminants introduced during glass to clay alteration. The conversion of the original glass to clay takes place in the presence of abundant water, which will contain a variable amount of dissolved ions. In a marine location sea-water will be the solution; but in continental locations the dissolved solid content of the alteration solution will be variable and most likely dominated by calcareous waters. In any event, the solutions will bring to the eventual bentonite horizon a number of possible contaminants. Those original igneous minerals which are partially altered or coated with secondary alteration products will be contaminated. This contamination will be most severe in the case of the daughter elements Sr and Pb, since the contaminant will almost certainly have a different isotopic composition from the original igneous contaminant.

For U–Pb dating of zircons the coatings are not thick and may be stripped from the highly acid-resistant zircon with hot 8N HNO_3, then hot 6N HCl followed by thorough washing with triple-distilled H_2O. Barite contaminant will contain much Pb and can be removed from the zircon concentrate with the HI–HCl–H_3PO_2 mixture noted previously in the section on mineral separations. Sanidine and plagioclase are also resistant to brief acid washing and may be cleaned with hot 6N HCl. Biotite, especially altered biotite, is slowly decomposed by non-oxidizing acids, so to clean off surface coatings a 30-second washing with cold 6N HCl followed immediately by thorough washing with triple-distilled H_2O is recommended. Even after washing with cold HCl, altered biotite usually contains more contaminant Sr than unaltered biotite, and the amount changes with increasingly strong alteration (see Figure 2).

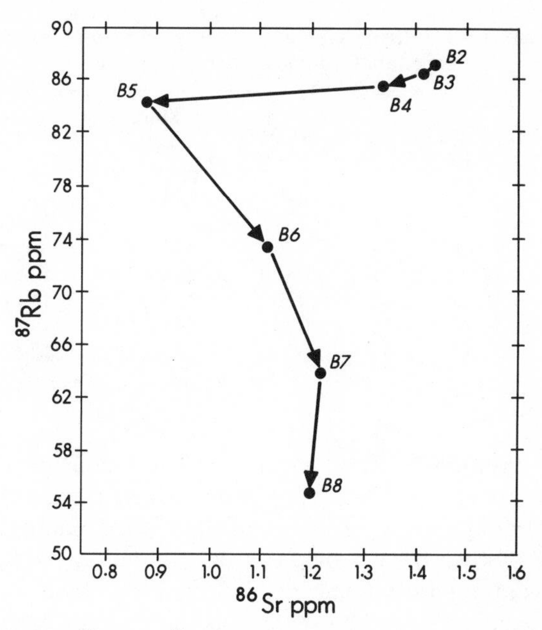

Figure 2. Variations in ^{87}Rb *vs.* ^{86}Sr content with progressive alteration (decrease in density) of biotite samples from the 'Z' coal bentonite.

3.1.2 Alteration of datable minerals

Besides determining contamination during the alteration of glass to bentonitic clay, it is necessary to know when the alteration took place, the duration of open chemical systems and the nature of the alteration of the igneous minerals.

If the alteration of glass to clay and the partial attack on datable minerals took place immediately after pyroclastic deposition ($<1\%$ of total elapsed time to present), the geochronological relations essentially begin with altered samples. If alteration occurs after an appreciable portion of the radiogenic daughters have accumulated in the datable minerals, the dates will be affected—usually decreased below the 'true' value. The same relative effect on the dates will be obtained if diagenetic alteration occurs over a sizeable portion of the age of the bentonite horizon. Recent alteration or weathering, on the other hand, will change the dates only if there is a marked effect on the ratio of the daughter element *vs.* the parent element.

We may try to decide which process dominates by looking at geochronological data obtained on variably altered samples. In Table 2, K–Ar data on biotite and sanidine show no definite effect on the K–Ar ratio in biotite as its degree of alteration (and K content) changes. Moreover, the dates match those of sanidine within analytical uncertainty. Table 3 gives

Table 2. K–Ar dates (±2σ) from cogenetic biotite and sanidine of the Kinnekulle, Sweden, Ordovician bentonite beds.

Ak-No.		Horizon	K_2O%	Ar^{40}/K^{40}	Date (Ma, ICC)	Comments
Biotite	82	A_1	5.37	0.0291	441±22	Partially altered
	77	A_2	4.43	0.0311	468±23	More altered
	59	B	3.48	0.0292	443±22	Strongly altered
Sanidine	61	B	10.91	0.0308	464±23	Impure sanidine
	90	B	11.16	0.0300	453±23	Impure sanidine
	146	B	12.08	0.0295	447±22	Pure sanidine
Illite	69	A_1	4.94	0.0216	338±17	Illitic clay

K–Ar data on bentonite strata within 380 m (about two million years difference) and shows variably altered biotite giving correlative dates except for the most altered sample. This set of data is for samples from a *single section*, yet the biotites show widely variable alteration. This would seem to imply that recent alteration has not affected the biotites, but that the alteration occurred to a variable extent at the time of original pyroclastic deposition. The Kinnekulle K–Ar results in Table 2 could also be explained by initial alteration with subsequent alteration limited by the impermeability of the bentonite horizon to water.

Figure 1 and Table 4 give Rb–Sr data on minerals from a bentonite in a coal seam in the Hell Creek area of Montana (Baadsgaard and Lerbekmo, 1980). Despite very large variations in the Rb and Sr content with changes in specific gravity (Figure 2), the biotites give the same date, with or without the sanidine point to ‘anchor’ the isochron. In view of the two types of variation in Rb and Sr as alteration increases from B2 to B8, it is likely (but not *proved*) that the alteration took place soon after deposition of the

Table 3. Effect of weathering on biotite K–Ar dates. Colt Creek section, Alberta, Canada.

Sample No.	Stratigraphic position in ft (m)	K–Ar date (Ma)	K_2O %	Appearance of biotite
4872	1250(381)	59.6, 59.8, 58.7	5.23	Fresh
4863	1200(366)	58.7	1.99	Moderately weathered
4869	1030(314)	60.0, 59.3, 61.5	4.18	Fairly fresh
4865	75(23)	55.8	1.25	Strongly weathered
4859	45(14)	61.1, 59.9	2.87	Slightly weathered
	0 = Palaeocene–Cretaceous boundary			

Table 4. Rb–Sr analytical results for biotites (B-samples) and sanidine from a bentonite in the 'Z' coal, Montana, USA

Sample No.	^{86}Sr(ppm)	^{87}Rb(ppm)	^{87}Sr/^{86}Sr*	^{87}Rb/^{86}Sr
B2	1.436	87.06	0.7609±2	59.92
B3	1.415	86.55	0.7610±2	60.48
B4	1.335	85.47	0.7627±2	63.30
B5	0.879	83.37	0.7910±1	93.77
B6	1.113	73.29	0.7646±3	65.08
B7	1.215	63.67	0.7528±5	51.88
B8	1.196	54.66	0.7464±4	45.18
Sanidine	58.81	41.30	0.7068±3	0.694

* 2σ precision given for the last significant digit.

original pyroclastic material. Difficulties in dating very altered material may well stem from the great difficulty in purifying this material, rather than from alteration itself. It should be noted in Table 2 that the clay matrix of the bentonite 'retained' approximately 70% of its radiogenic Ar *versus* the sanidine. If the clay formed by late or recent degradation of the original ash, much less radiogenic argon would have been found. The clay formed very early in the history of the ash, most likely immediately after deposition . The same is true of the mineral alteration, since recent weathering could not give the variability of biotite alteration which is often found for different bentonite beds in the same outcrop or section.

3.2 Uncertainties in analysis and sampling

In the evaluation of any radiometric data obtained on minerals from bentonites, one must consider the relative analytical uncertainty, sample purity and homogeneity, trends in any variation of data and comparisons between different dating methods. The sum of these considerations permits an evaluation of the date for the bentonite horizon. This evaluation is at best semiquantitative since alteration and sample purity will tend to strongly influence the validity of most dates.

3.2.1 Analytical uncertainties

The analytical measurements are carried out to determine the present amount of radiogenic daughter isotope and the parent isotope. The date is calculated from the relationship

$$t = 1/\lambda \ln (D/P + 1) \tag{1}$$

where t is the date, λ is the decay constant of the parent isotope, P, and D is the daughter isotope. Uncertainties in the measurement of D/P give rise to analytical uncertainty in t. Using the differential of (1):

$$\delta t = [1/\lambda\{1/(D/P+1)\}]\,\delta(D/P). \qquad (2)$$

Ignoring (for the moment) uncertainties in the decay constants, the probable uncertainty in t from variations in D/P is

$$Qt = \sqrt{[1/\lambda\{1/(D/P+1)\}]^2\{\delta(D/P)\}^2}. \qquad (3)$$

The uncertainty in parent isotope determination is relatively constant over the whole range of usual values when a sensitive low-blank method such as isotope dilution is used. The *relative* uncertainty in determination of the radiogenic daughter isotope, however, is strongly dependent upon the proportion of original or initial daughter isotope present. As the sample to be dated becomes younger and younger, the proportion of radiogenic daughter isotope becomes smaller and more difficult to measure precisely. A good modern mass spectrometer can reproducibly measure mass ratios to $\pm 0.02\%$ (2σ). If the initial daughter isotope is an amount B_1 and the radiogenic daughter isotope is B_2, then the relative uncertainty in daughter measurement may be taken to be $(B_1+B_2)B_2 \times 0.02\%$. In Table 5 an average amount of parent isotope and daughter isotope contaminant (B_1) has been estimated for different bentonitic minerals with various dating methods. The amount of radiogenic daughter generated (B_2) may be calculated for various times. The uncertainty in the measurement of the daughter is then calculated. The D/P ratio for various times may be determined and the probable uncertainty in t, Qt, is calculated by means of equation (3). To the percentage uncertainty in t (from uncertainties in measurement of D/P) are added the uncertainties from the measurements of the decay constants, and the variation in the percent uncertainty in a mineral date is plotted *vs.* the date in Figure 3. The approximate minimum analytical uncertainty and range of application of a particular bentonite mineral dating method may be obtained from Figure 3.

Of course, a mineral isochron (Nicolaysen, 1957) plot may be used for both the Rb–Sr and K–Ar methods. For Rb–Sr the $^{87}Sr/^{86}Sr$ is plotted *vs.* the $^{87}Rb/^{86}Sr$, and for K–Ar the $^{40}Ar/^{36}Ar$ is plotted *vs.* the $^{40}K/^{36}Ar$. Isochron plots have the advantage of assessing a number of related analytical determinations simultaneously, showing the temporal coherence (or not) of the samples and giving a pooled and often more accurate value for a date. The linearity of the points on an isochron plot line is used as a measure of the uncertainty in the date obtained (York, 1969; McIntyre *et al.*, 1966; Brooks *et al.*, 1972).

The above simple analytical assessment of analytical uncertainty ignores

Table 5. Relative percentage uncertainty in the measurement of the radiogenic daughter isotope at various times of accumulation.

Mineral	Average amount of parent isotope (ppm)		Average amount of initial daughter isotope (ppm)(B_1)		Average amount of radiogenic daughter (B_2) 1×10^6 a B_2(ppb)	1×10^6 a $(B_1+B_2)/B_2 \times 0.02$	10×10^6 a B_2(ppb)	10×10^6 a $(B_1+B_2)/B_2 \times 0.02$	100×10^6 a B_2(ppb)	100×10^6 a $(B_1+B_2)/B_2 \times 0.02$	600×10^6 a B_2(ppb)	600×10^6 a $(B_1+B_2)/B_2 \times 0.02$
Biotite	200	^{87}Rb	1	^{87}Sr	2.84	7.06	28.4	0.724	284	0.090	1723	0.032
Sanidine	40	^{87}Rb	50	^{87}Sr	0.568	1761	5.68	176.1	56.8	17.63	344.7	2.92
Plagioclase	2	^{87}Rb	80	^{87}Sr	0.0284	56338	0.284	5634	2.84	563	17.23	93
Zircon	1000	^{238}U	0.2	^{206}Pb	155	0.046	1552	0.023	15633	0.020	97540	0.020
Zircon	10	^{235}U	0.2	^{207}Pb	9.85	0.426	98.97	0.060	1035.0	0.024	8056	0.020
Zircon	800	^{232}Th	0.4	^{208}Pb	39.2	0.224	396	0.040	3968	0.022	24104	0.020
Biotite	8	^{40}K	0.004	^{40}Ar	0.472	56.8*	4.656	11.15*	47.8	6.50*	330.9	6.07*
Sanidine	12	^{40}K	0.0002	^{40}Ar	0.708	7.69*	6.987	6.17*	71.70	6.02*	496.3	6.00*
Plagioclase	0.5	^{40}K	0.0004	^{40}Ar	0.029	88.8*	0.291	14.2*	2.99	6.80*	20.68	6.12*

The starred entries for K–Ar at the bottom of the table have been multiplied by a factor of ~300 since the ^{36}Ar is used to measure contaminant argon.

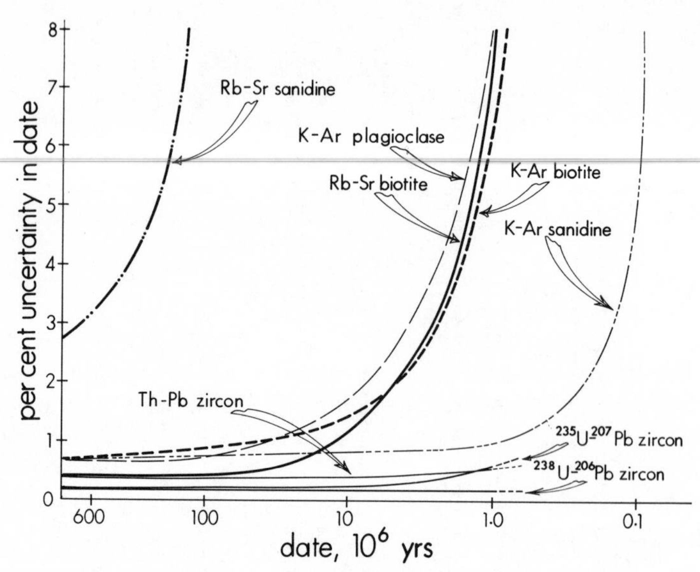

Figure 3. Estimated minimum percent uncertainty in a bentonite mineral date as a function of the magnitude of the date.

systematic indeterminate uncertainties arising from variable blanks, uncertainties in the calibration of isotope dilutants (spikes) and uncertainties in isotope ratios of contaminants. Therefore, only the măjor uncertainties of measurement in careful analysis with negligible blank levels and ultraprecise spikes are dealt with above to give *minimum* limits of uncertainty in Figure 3 at a 68% confidence level (1σ).

3.2.2 Factors of sample purity and quality

During mineral separation from the bentonite, the cleanest possible mineral sample is obtained. As noted above in the comments concerning contamination and alteration, it may not be possible to obtain a completely unaltered mineral separate. In this case, the date obtained will be suspect unless the effect of the alteration can be assessed. If a series of mineral samples with varying degrees of alteration can be obtained, trends in analytical data can be correlated with alteration. Comparison of dates with those from other geochronometers on fresh minerals (if possible) can help to evaluate and validate data from altered samples. Altered samples are also harder to free from impurities. Certain impurities, such as P€ feldspar in Mesozoic sanidine, are very serious while other impurities, such as quartz, act only as dilutants and contain essentially no parent or daughter isotopes. It is therefore very important to be able to characterize the nature of the impurities that are present so that the necessary purification procedures may be chosen. If the impurities seriously distort the parent–daughter ratio of the original igneous mineral, then they must be removed or the final date will be in doubt.

It is very difficult to quantify the uncertainties arising from sample impurity or alteration, yet these are often the most important source of uncertainties in the dating of bentonitic minerals. For this reason it is important to repeat the total dating procedure *including the mineral separation and purification.* At least two bentonite samples from adjacent places in the same horizon should be taken and run through the complete separation and analytical procedure. Dates from the same minerals by the same method can thus be compared for sampling and purification uncertainties. Mineral separates may also be separated by grain size and specific gravity to give a number of separates of the same mineral for comparison analysis. Variations in dates from these comparisons give a more realistic idea of the reliability of a mineral date than using purely analytical assessments of uncertainty (compare uncertainties for dates in Table 1 with minimum analytical assessment in Figure 3).

In like manner, the bentonite minerals should be subjected to as many dating methods as feasible. In the fortunate circumstances where three or more different methods on three or more different minerals all give dates that are all the same within the analytical uncertainties, then the date of the formation of the bentonite horizon may be considered to be well established. More often, not all the minerals are available or suitable (altered) and only the dates on the best material will have any significance.

Résumé du rédacteur

Les bentonites sont des produits résiduels d'altération de dépôts éruptifs en milieu liquide; en fonction du mode d'altération et du matériel éjecté, l'un ou l'autre ou plusieurs des minéraux suivant peuvent subsister: plagioclase, sanidine, biotite, zircon, apatite, hornblende voire verre volcanique . . . .

La méthode K–Ar est la plus utilisée sur plagioclase et sanidine, sur biotite, plus altérable, ou sur hornblende si elle est présente. La méthode Rb–Sr ne convient que pour les biotites car le strontium commun est trop abondant dans les feldspaths. La méthode U–Pb et les traces de fission sont utilisables grâce aux zircons.

L'essentiel est l'*obtention de fractions minérales pures*, toujours délicate, et qui nécessite de 10 à 100 kg de roche. La purification fait intervenir le lavage, la séparation magnétique, les liqueurs denses le lavage avec HCl (sanidine, plagioclase), HF (nettoyage des plagioclases); HNO_3 et un mélange $HCl+HI+H_3PO_2$ (nettoyage et purification des zircons).

Les incertitudes géochimiques sont liées aux diverses possibilités de contamination: lors de l'*éruption* par du matériel anciennement cristallisé, lors du *dépôt* par du matériel ancien déjà présent dans le sédiment, lors de l'*altération* du verre en argile surtout pour ce qui concerne le strontium et le plomb. Dans ce dernier cas, il est nécessaire de tenter d'estimer l'âge de l'altération par rapport au dépôt pour déterminer si l'altération a pu modifier l'accumulation normale des isotopes radiogéniques.

On ne dispose parfois que de minéraux altérés; la comparaison de l'âge apparent de fractions plus ou moins altérées permet alors de déceler une tendance éventuelle à la modification de l'âge apparent. De même, on doit comparer les âges obtenus sur différentes fractions, voire par différentes méthodes de mesure d'âge si possible.

Dans tous les cas, pour estimer l'incertitude réelle sur les âges obtenus on doit comparer les âges obtenus sur des échantillons différents prélevés dans un même niveau et traités séparément. Si différents échantillons et différentes méthodes sont appliquées, des âges concordants établissent l'âge du dépôt. Si l'on n'utilise qu'une méthode, ce sont les âges obtenus sur le chronomètre considéré comme le plus digne de confiance qui seront sélectionnés.

(Manuscript received 21-7-1980)

Numerical Dating in Stratigraphy
Edited by G. S. Odin

23
The dating of plutonic events

NOËL H. GALE

1 INTRODUCTION

It was only with the advent of modern geochronology in the 1950s that a first step could be taken in the task of calibrating the relative time scale in absolute units of time by using the K–Ar, Rb–Sr and U, Th–Pb methods to date rocks whose position in the relative geological time scale is well established. Such a numerical calibration is highly desirable, partly because it provides the foundation for estimates of rates of geological processes *per se* and partly because it allows estimates of the time scale for evolutionary processes.

Many difficulties stand in the way of establishing an accurate numerical calibration of the relative time scale through geochronological measurements. We have even to be critical of the relative time scale itself, which is based on several different stratigraphic or zonal schemes. In this situation it is necessary to accumulate sufficient geochronological age data to calibrate *each scheme independently*; the necessity for this has been forcibly demonstrated by the discovery, through isotopic dating, that correlations between the marine sequences and the vertebrate-bearing continental succession in North America were considerably in error (Berggren, 1969a; Page and McDougall, 1970). How much more hazardous correlation from continent to continent may be in some cases, is a factor which ought constantly to be borne in mind. At the least, it suggests that the numerical calibration of the time scale is a project in which collaboration between geochronologists, palaeontologists and stratigraphers is vital.

In calibrating a particular stratigraphic scheme it is desirable to measure radiometric ages on samples which are controlled as precisely as possible in that stratigraphic scheme. The ideal situation is that of an undeformed, unmetamorphosed sequence of radiometrically datable volcanic or sedimentary rocks interbedded with sediments containing a diagnostic fossil assemblage which enables the position of the volcanic rocks on the relative time scale to be fixed precisely. Tuffaceous deposits containing separable minerals suitable for isotopic dating can also be very suitable for time-scale studies if they are interbedded with suitable fossiliferous sediments, provided that contamination of the tuffs by older material is absent; an example is that of zircon-bearing bentonites treated in the preceding chapter, but fission track dating of such bentonitic zircons must be treated with great reserve for time-scale calibration purposes (see Storzer and Wagner, this volume). A good example of radiometrically datable extrusive lava which is very precisely fixed in a sedimentary stratigraphic sequence is the Stockdale rhyolite (Gale *et al.* 1979b). The next best situation for time-scale calibration purposes is based on isotope dating of extrusive volcanics which are dated by lenses of fossiliferous sediment. It is subject to uncertainties because there may be hidden time gaps at easily overlooked disconformities in such a sequence, but it is often the case that the stratigraphic uncertainties are likely to be small in comparison with the unfortunately smaller precision attainable by radiometric dating compared with palaeontological dating. Examples of this sort of stratigraphic control of extrusive volcanics important for time-scale calibration are the Pembroke and Eastport formations of North America (Bottino and Fullagar, 1966; Fullagar and Bottino, 1968c; Berdan, 1971a,b).

Unfortunately, examples of the ideal and semi-ideal cases described above are not overwhelmingly available, so that recourse must sometimes be made to *intrusive rocks.* Radiometric ages of intrusive rocks clearly provide a younger limit to the age of the rocks intruded. If the intrusive body has been eroded and overlain by sediments, the radiometric age for it gives a maximum absolute age for the overlying sediments. The value for time-scale calibration of radiometric ages on such intrusive bodies depends on the time gaps between sediment deposition, emplacement of the intrusive body, subsequent erosion and deposition of more sediment being sufficiently short. This is often not so, a case in point is the Shap granite, which has been very accurately dated by the Rb–Sr method (Wadge *et al.*, 1978) but which cannot as yet be more strictly controlled stratigraphically than to say that it lies between Early Gedinnian and mid-Devonian. However, the information available from the radiometric dating of intrusive rocks is, for time-scale calibration purposes, of considerably more use than the positively confusing attempt to use dates on metamorphic rocks or mineral veins for this purpose; it is practically never possible to tie such rocks with sufficient precision to the fossiliferous sequence, as is evident from the involved, uncertain and contorted geological arguments which are often invoked in the attempt to do so (examples will be found in Ross *et al.*, 1978*a*).

Apart from stratigraphical uncertainties arising from the problem of correlation between the apparent radiometric age and the biostratigraphical sequence, there are three major uncertainties in the determination of the radiometric age itself which may vitiate its usefulness for time-scale calibration purposes; they are:

(1) The *time of closure* of the relevant radiometric clock relative to the time of emplacement of the plutonic rock, which involves the closure temperature of the minerals involved and, for example, the rate of cooling of a pluton after emplacement.

(2) The question of possible partial or complete *resetting* of the radiometric system due to heating events subsequent to emplacement of the pluton.

(3) The *accuracy of the determination* of the apparent radiometric age, which depends on the selection of fresh and appropriately representative rock samples, on attention to detail in achieving high-purity separates for any mineral separations which may be performed, on careful analyses of controlled accuracy and on proper statistical combination of the individual errors in assessing the overall error of the quoted radiometric age.

2 THE ZERO AGE

2.1 Closure temperatures

The closure or 'blocking' temperature of a geochronological system may conveniently be defined as its temperature at the time corresponding to its apparent radiometric age. When the apparent age of a rock or mineral is calculated from the accumulated daughter products of radioactive decay, whether those products be radiogenic isotopes such as ^{207}Pb, ^{206}Pb, ^{87}Sr or ^{40}Ar or whether they be radiation damage tracks ('fission tracks') caused by fission fragments from ^{238}U, the resulting radiometric age ideally represents a point in time at which a completely mobile daughter product became completely immobile. The effective closure temperature is in part a function of cooling rate. Slow, intermediate, fast and very fast cooling rates will be taken to reflect cooling periods of 10^8, 10^7, 5×10^6 and 10^6 years respectively.

Not so long ago it was believed that the closure of a geochronological system could always be identified with either the crystallization of an igneous rock from a melt or recrystallization during metamorphism. Recently it has become increasingly clear that such a simple interpretation is inadequate for some methods of age determination, notably the dating of separated minerals by the Rb–Sr, Ar and fission track methods. Radiogenic argon and strontium are clearly mobile in some minerals at temperatures well below that of crystallization and, as is discussed in detail by Storzer and Wagner in this volume, thermal annealing of fission tracks is very important and track fading can occur within one hundred million years at temperatures as low as *c.* 50°C for apatite and *c.* 70°C for zircon. The best evidence that ^{87}Sr and ^{40}Ar become mobile at relatively low temperatures comes from the Rb–Sr and K–Ar age patterns for micas from the central Alps (Jäger, 1973; Purdy and Jäger, 1976), and it is from fission track studies in the central Alps that good evidence for thermally induced track fading also comes (Wagner and Reimer, 1972). The simplest interpretation of all these data is that closure of the fission track, Rb–Sr and K–Ar systems occurred during post-metamorphic cooling. From their work, Jäger and Purdy were able to conclude that the closure temperatures of the Rb–Sr system in muscovite and biotite are 500 ± 50°C and 300 ± 50°C respectively, whilst the closure temperatures of the K–Ar system are 350 ± 50°C for muscovite and 300 ± 50°C for biotite. Evidence accumulated by Hunziker and Jäger (quoted in Purdy and Jäger, 1976) suggests that in monazite, the blocking temperature for the U–Pb system is about 530°C. Work by Harrison *et al.* (1979) yields estimates for closure temperatures, at an intermediate cooling rate, for ^{40}Ar in plagioclase and K feldspar of ~260°C and ~160°C respectively, whilst for fission tracks in epidote, the closure temperature is ~240°C. Glassy phases are notorious both for easy loss of ^{40}Ar and easy annealing of fission tracks. Theoretical studies of closure temperatures have been made by Dodson (1973). Further discussions of this matter, both empirical and theoretical and including the effect of cooling rate, appear in Jäger and Hunziker (1979) and in Harrison *et al.* (1979). *The quite low closure temperatures now established show that for time-scale calibration purposes radiometric dates established for minerals in rocks which may have suffered thermal events subsequent to emplacement are of little value.*

2.2 Cooling rate of plutons

The realization that the closure temperatures for particular geochronological systems are, in some minerals, quite low and that the apparent radiometric age dates the time at which the pluton cooled to the appropriate closure temperature makes it important to have good data on the rates of cooling of intrusive bodies if an estimate is to be made of the date of intrusion. The theory of the cooling of intrusive bodies has been studied for several decades and the development of conventional heat flow models, neglecting the effects of geothermal flux, radioactive heat production and geological uplift, culminated in the work of Carslaw and Jaeger (1959). There followed a number of geochronological applications concerned chiefly with the thermal effects of intrusion on mineral ages in the country rock. Improvements in radiometric dating and progress in understanding closure temperatures nowadays allow, in favourable cases, experimental determination of cooling curves for the pluton itself; an example is the determination of the cooling curve for the Quottoon pluton in British Columbia (Harrison *et al.*, 1979). This achievement has stimulated the development of more realistic cooling models by Harrison and Clarke (1979). Their model is a time-dependent one of the two-dimensional cooling of a dyke having rectangular cross-section and intruded into a medium which is in equilibrium with a constant geothermal flux and evenly distributed radioactive heat sources, cooling being by thermal conduction taking into account also the effects of regional uplift.

Such a theory can, in an appropriate case, be compared with an empirical cooling curve established for a pluton by combining apparent mineral radiometric dates with independently established values for mineral closure temperatures. Complex cooling histories, involving perhaps episodic heating, can however introduce many ambiguities. In the simplest case, where a thermal contrast exists between the pluton and the host rock, the pluton will start to cool rapidly, resulting in a simple concave curve of temperature *vs.* time. For the case of no thermal contrast, such as an *in situ* partial melt, the pluton will, at first, cool slowly as uplift proceeds and then progressively more quickly as the surface is approached, producing a convex cooling curve. Such advective cooling is characteristic of regional metamorphic terrains. More complex cases involving a thermal pulse are considered by Harrison *et al.* (1979), who show that the pattern of radiometric mineral dates produced can easily be misinterpreted as due to an apparent gradual steady cooling. However, application of the theory to the 180 km long by 10 km wide dyke-shaped Quottoon pluton produces a good theoretical description of the empirical cooling curve, and shows that it cooled from about 700 to 200°C in approximately 5×10^6 years. It was emplaced at about 51 Ma, cooling rapidly with initial uplift which ended approximately 46 Ma ago, and records a biotite Rb–Sr mineral age of 48 ± 3 Ma, a hornblende K–Ar mineral age of 49.6 ± 1.1 Ma and a Rb–Sr whole-rock isochron age of 51 ± 2 Ma (Armstrong and Runkle, 1979). The Quottoon pluton study is useful in giving an order of magnitude for the rate of cooling of plutons not cooled additionally by convecting water, which is not taken into account by the theoretical treatment of Harrison and Clarke (1979); oxygen and deuterium isotope studies (Magaritz and Taylor, 1976) found the Quottoon to be isotopically 'normal', indicating the absence of meteoric–hydrothermal water interactions during cooling. M. Parmentier (unpublished lecture given at Oxford) has shown that the inclusion of convective water flow into cooling models for intrusive plutons can very considerably increase the cooling rate; in such a case, apparent radiometric mineral ages will often closely approximate the intrusive age, other things being equal and assuming no subsequent metamorphism.

3 DISTURBANCE OF RADIOMETRIC AGES BY THERMAL EVENTS SUBSEQUENT TO EMPLACEMENT

With the realization that closure temperatures for geochronological systems can be quite low, comes the conclusion that disturbance—complete or partial resetting—of radiometric ages is not at all a rare event in geochronology. In applying radiometric ages to the calibration of the time scale, we naturally try to avoid cases of disturbed ages, but the detection of lack of disturbance and proper age interpretation is only possible if it is based on a clear age pattern; generally, more than one radiometric dating method should be applied and a good understanding of the geology of the area is essential. Further, proper selection of rocks to be dated is essential. Fresh, unweathered rock samples are required and they must be carefully examined petrographically by thin-section microscopy, in order to decide which minerals can be separated for age measurement, whether the samples can be used for whole-rock determinations, and whether the age measurement on a sample is likely to relate to the time of original crystallization or a subsequent event such as metamorphism. *For instance, the K–Ar dating of volcanic glass, or whole rocks containing volcanic glass, must be treated cautiously, since glass readily loses radiogenic argon.* A simple example of a check on lack of disturbance is that, if K–Ar ages on biotite and sanidine phenocrysts in a volcanic rock agree with one another and with ages of whole-rock samples from the same volcanic sequence, we would be confident that the ages have geological significance which, in an undisturbed sequence, would give the age of crystallization and (rapid) cooling of the rocks. On the other hand, if K–Ar ages on biotite and hornblende from an intrusive rock disagree, the biotite giving a significantly younger age than the hornblende, one would suspect either a slow-cooling history or mild metamorphosis of the intrusion so that some radiogenic argon was lost from the biotite but none from the more retentive hornblende. It cannot be too much stressed that the interpretation of the geochronology must rely very much on the known geology.

A further check of K–Ar chronology is to plot data from assumed cogenetic samples on isochron-type diagrams such as the K *vs.* radiogenic Ar diagram (Funkhouser *et al.*, 1966; Harper, 1970) or the $^{40}Ar/^{36}Ar$ *vs.* $^{40}K/^{36}Ar$ diagram (McDougall *et al.*, 1969). In the ideal case the data points will lie on a single line whose slope gives their age; sometimes data plotted on the isochron diagrams will form a linear array even when individual samples give discordant K–Ar ages when calculated using the customary assumptions. Thus Hayatsu and Carmichael (1970) were able to demonstrate that a sequence of rocks in Newfoundland deposited in the Cambrian was reset in the Devonian and that the argon then incorporated in the rocks, differed considerably in isotopic composition from atmospheric argon. Excess radiogenic argon continues to be a problem in some cases, and no good method has yet been devised to correct for it (Jäger, 1979a). On the other hand, the step-heating $^{40}Ar/^{39}Ar$ method of dating has shown great promise in distinguishing disturbed from undisturbed rocks or minerals (Dallmeyer, 1979); though it has not yet been used extensively in physical time-scale calibration studies, it is potentially a very useful technique for future work (see Albarède, this volume).

Compared with the various ways of using the K–Ar method to date plutonic rocks, the Rb–Sr dating method has the great advantage that *the Rb–Sr geochemistry is related to the rock geochemistry.* In contrast, if radiogenic argon is expelled from a mica, the accidental location of nearby fissures and fracture zones determines whether the argon remains in the rock system. Compared with the U–Pb method, the Rb–Sr method has the advantage of being applied to the main rock-forming minerals such as micas and feldspars. Even though, a zircon suite may be well dated by the U–Pb discordia method (Gale and Mussett, 1973; Gebauer and Grünenfelder, 1979), there can still be doubt whether this date is that of the rock formation itself or whether the zircons are detrital or have inherited radiogenic lead, resulting in the U–Pb result giving an 'age' older than the rock formation. (This danger is also inherent in fission track ages of zircons from bentonites.) There is no doubt that the Rb–Sr system is also the most versatile geochronometer, being applicable to whole-rock samples and to isolated minerals and competent to date magmatic rock formation, metamorphism and sedimentation as well as mineral crystallization and cooling histories. However, the long half-life of ^{87}Rb and the rather high content in common strontium of ^{87}Sr (~7%) means that only rocks and minerals with rather high Rb/Sr ratios can be dated; in Tertiary rocks only biotite and phengite can usually be dated. In the Palaeozoic, however, it is usually easy to find rocks and minerals with a sufficient range in Rb/Sr ratio for accurate dating.

The elementary principles of Rb–Sr dating and isochron diagrams have been adequately described elsewhere (Jäger, 1979b). *In dating plutonic rocks, the major point of importance is that Rb–Sr systems in whole rocks are much more stable than in minerals*; so that several whole-rock samples from one granite body usually define a straight line on the isochron diagram. Rock systems are much less affected than minerals by later metamorphic events. Partial Sr exchange can occur at rather low temperatures between adjacent minerals, and empirical observation shows that even complete Sr exchange and Sr isotopic homogenization between the different minerals of a rock do not require a molten state, but on a sufficiently large whole-rock scale the Rb–Sr system is often not reset. In such a case of homogenization on a mineral scale the mineral points will define a straight line of smaller slope corresponding to the younger age of the metamorphism, whilst the total rock points lie on an isochron preserving the older age of formation.

Many examples of this behaviour are known, amongst which may be cited the case of the Rotondo granite in the central Alps (Jäger, 1979b). For this granite the whole-rock isochron defines a granite formation age of 269 ± 11 Ma, whilst the biotite Rb–Sr age of 15 Ma represents the time when the granite cooled to about 300°C after the Tertiary Lepontine phase of Alpine metamorphism. This case, however, where there is a clear demonstration of post-intrusion thermal effects, would not be an ideal one for time-scale calibration purposes even if it was stratigraphically well controlled. Much better for this purpose from the radiometric dating standpoint are cases like the Shap granite (Wadge *et al.*, 1978), where the biotites and feldspars were clearly demonstrated not to have been reset and to lie on the whole-rock Rb–Sr isochron, defining an age of 394 ± 3 Ma; moreover, the K–Ar ages on biotites from the granite averaged 397 ± 7 Ma, whilst a U–Pb discordia line on zircons from the Shap granite gives 390 ± 6 Ma.

4 COMPUTATION OF THE ANALYTICAL ERROR ON A RADIOMETRIC AGE

It cannot be stressed too much that *a radiometric age without a properly assessed analytical error is valueless,* especially for numerical calibration of the time scale. It should be emphasized that, if an age is quoted as

52 ± 2 Ma (2σ), it means that the apparent age of the sample has a 95% chance of being between 50 and 54 Ma and a 66% chance of being between 51 and 53 Ma. This apparently simple point is occasionally forgotten even by geochronologists; thus Ross *et al.* (1978a, p. 348), in comparing the fission track age of 451 ± 21 Ma for the Acton Scott beds in Shropshire with the biostratigraphically older Tyrone limestone in Kentucky (which yielded a K–Ar age of 435 ± 15 Ma), make the nonsensical comment: 'Ages of Caradocian samples, one from Shropshire and one from Kentucky, are within analytical error of one another, although the youngest biostratigraphically seems to be the oldest isotopically'. This is to quote, but ignore, the errors, which show that these ages cannot be distinguished from one another; this, however, is only one example of numerous cases encountered in the literature.

For conventional K–Ar dating the error of the calculated age can be computed by combining in the usual way (Campion *et al.*, 1973) the errors in determining the radiogenic ^{40}Ar contents and the K content. The formula relating the ^{40}Ar content to the amount of added ^{38}Ar spike and the mass-spectrometric measurements of argon isotopic composition is given, *inter alia*, by York and Farquhar (1972). From this formula, the error in ^{40}Ar determination can be computed; one of the most important factors leading to error, especially for young rocks, is the atmospheric argon correction, which has been considered in detail by Baksi *et al.* (1967). If the ^{40}Ar/^{36}Ar *vs.* ^{40}K/^{36}Ar plot is used, whose slope gives the age, then the statistical analysis appropriate to Rb–Sr and U–Pb isochrons may be used to compute errors, with the complication that the errors will be correlated because of the occurrence of ^{36}Ar in the denominator of the quantities plotted on each axis (see Gale, 1979, for a general discussion). For conventional K–Ar ages the discussion of attainable accuracy given by York and Farquhar (1972) essentially still obtains so that, if one includes systematic intralaboratory errors (as one must in numerical calibrations of the time scale which rely on ages from different laboratories), *it is difficult to reduce total errors in the ages below* $\pm 3\%$ (2σ) in the Phanerozoic. Somewhat smaller errors appear to be attainable by the use of the ^{40}Ar/^{39}Ar dating method, as demonstrated by Berger and York (1970); here step heating and the ^{40}Ar/^{36}Ar *vs.* ^{39}Ar/^{36}Ar plot are often used and the error of the age calculated from the slope of the line must again be assessed by a line-fitting algorithm using correlated errors (York, 1969).

For the Rb–Sr method one will either be dealing with mineral ages or, much better for time-scale purposes, *isochron ages for a number of whole-rock* (sometimes plus mineral separates) samples from the body to be dated. For mineral ages the overall error on the age can be computed by the normal rules for combining errors using the fundamental age equation as given, for example, in equation 3.10 of the book by York and Farquhar (1972). For

Rb–Sr isochrons a plot is made of $^{87}Sr/^{86}Sr$ against $^{87}Sr/^{86}Sr$, the age being given by the slope of the best fitting straight line through the data points. It is essential to fit the data by a least squares computer programme based on the algorithms developed by York (1966; 1969), Williamson (1968) and Cumming *et al.* (1972); here we should note that the programme listed by Faure (1977) contains some errors. If Rb and Sr elemental concentrations are measured separately say by isotope dilution, there will be no correlation between errors; if, however, the Rb/Sr ratio is measured by XRF there will be correlation of errors and the approach of York (1969) or Cumming *et al.* (1972) must be used. For the proper comparison of Rb–Sr isochron ages from different laboratories it is essential that isochron data reduction programmes be used which give the same results on a given data set; for this purpose we recommend Data Set 3 of Brooks *et al.* (1972) and that the authors should always report the results of applying their programme to this data for the cases $p = 0$ and $p = 0.999$, where p is the correlation coefficient. Tests for goodness of fit to a straight line are vital, and should be applied according to the recommendations of Brooks *et al.* (1972).

For Th–U–Pb dating, using for example the discordia method for zircons, the correlation between errors becomes even more vital in fitting a straight line to the data. Recent discussions have been presented by Gale (1979) and Ludwig (1980); for the sake of uniformity it is recommended that the formulae for correlation coefficients given by Ludwig be adopted. It is of course essential to use a line-fitting programme including correlated errors, and this applies to Pb–Pb whole-rock isochrons just as forcibly as to U–Pb zircon discordia, though it has been largely ignored except in applications to meteorite dating.

Résumé du rédacteur

En l'absence de roches sédimentaires ou volcaniques favorables, on utilise les roches intrusives dont la situation stratigraphique est souvent peu précise. Cependant, ces roches sont bien plus favorables que les roches métamorphiques qui donnent des âges numériques d'interprétation douteuse. Trois sources d'incertitudes doivent être envisagées systématiquement pour les roches plutoniques: l'*âge de fermeture* du chronomètre daté par rapport à la mise en place du pluton; les *évènements thermiques* postérieurs à la mise en place; l'*exactitude* de l'âge calculé.

Le moment de la fermeture d'un chronomètre est en partie fonction de la vitesse de refroidissement du pluton dont la durée peut aller de 10^6 à 10^8 années. On sait depuis peu que la température de *cristallisation* d'un minéral est différente de sa *température de fermeture*. Cette dernière est très inférieure: pour le système 'muscovite/Rb–Sr', on l'évalue autour de 500°C mais le système 'muscovite/K–Ar' ne se ferme qu'à 350°C et le système 'feldspath K/K–Ar' à 160°C.

Les âges 'roche totale/Rb–Sr' donnent l'*âge d'emplacement de la roche*, utile au stratigraphe; les autres systèmes permettent de reconstituer l'*histoire thermique* du pluton: son refroidissement mais aussi ses métamorphismes postérieurs. On connaît ainsi dans les Alpes un granite dont l'âge 'roche totale/Rb–Sr' est d'environ 269 Ma mais dont l'âge 'biotite/Rb–Sr' n'est que de 15 Ma: celui de métamorphisme alpin. Pour l'échelle des temps, on essaie de trouver des plutons dont tous les systèmes donnent le même âge: celui de son emplacement dans la série stratigraphique suivi d'un refroidissement rapide et d'une absence d'évènements thermiques postérieurs.

La sélection d'un échantillonnage adéquat et une étude pétrographique détaillée sont nécessaires pour comprendre la signification des âges mesurés par différentes méthodes sur différents chronomètres et proposer ainsi un âge utilisable en stratigraphie.

On dispose aujourd'hui d'un arsenal mathématique élaboré pour calculer l'erreur analytique sur un âge provenant d'une série de données (isochrones, âges plateaux). Pour ce qui est des âges conventionnels K–Ar, si l'on tient compte des incertitudes interlaboratoires, une incertitude de ±3% (2σ) paraît un minimum.

(Manuscript received 15-3-1981)

Calibrating the time scale

Section V

The Cambrian to Triassic times

Numerical Dating in Stratigraphy
Edited by G. S. Odin

24
Age data from Scotland and the Carboniferous time scale

HUGH A. F. DE SOUZA

1 AGE DATA FROM SCOTLAND FOR THE CARBONIFEROUS TIME SCALE

1.1 Introduction

The Carboniferous of Scotland is one of the most favourable areas in the world for determining a numerical time scale for this period owing to the abundance of volcanic rocks, the stratigraphic ages of which are relatively well known. Basaltic activity of mainly alkaline character occurred almost throughout the Carboniferous in Scotland, most of it being concentrated in the Midland Valley of Central Scotland.

As a result of a long period of economic interest in the Carboniferous rocks of the Midland Valley, the distribution and relative ages of the lavas and intrusions in this area are now known in great detail and the understanding of their three-dimensional distribution is quite exceptional in places (see Francis, 1967, for a review). Palynological research into the Dinantian of Scotland (Neves *et al.*, 1973) has resulted in a much improved understanding of the stratigraphic ages of the Early Dinantian lava piles, while the Late Dinantian and Silesian lava piles are often interbedded with sediments whose biostratigraphic ages are well known.

The ages of these rocks are therefore of great value to the Carboniferous time scale, particularly as they are from a single province. This avoids the uncertainties of long-distance stratigraphic correlation and allows the use of the principle of superposition in age interpretation. In addition, the generally more reliable ages obtained from intrusive rocks may also be of value where petrology or intrusive form links them to particular lava piles.

Figures 1 and 2 show the distribution and stratigraphic ages of the major Carboniferous lava piles in Scotland. The regional stages of the Dinantian of the British Isles shown in Figure 2 have been only partly recognized in Scotland (George *et al.*, 1976).

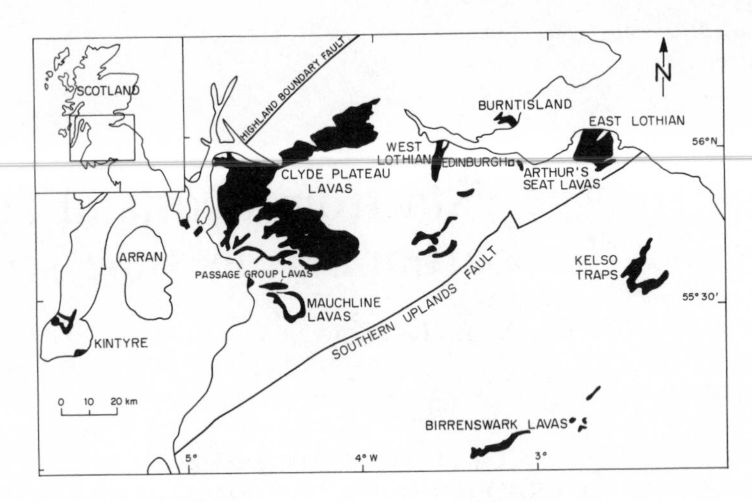

Figure 1. Sketch map showing locations of the major Carboniferous lava piles in Scotland.

The bulk of surface volcanic activity occurred in the Dinantian. The Silesian lava piles were much smaller and intrusive activity more widespread. This chapter is based on a wider geochronological study of Scottish Carboniferous volcanism (De Souza, 1979) which will be the subject of a more detailed account later. As the majority of lavas are basaltic, the K–Ar whole-rock method was used; mineral ages were obtained wherever possible. All ages quoted are ICC recalculated and the errors are 2σ reproducibility errors.

1.2 Early Dinantian

The first magmatic episode in the Scottish Carboniferous volcanic cycle occurred in the Scottish Borders with the eruption of the Birrenswark and Kelso lavas dated at 361 ± 7 Ma (NDS 165). Although their biostratigraphic age cannot be fixed precisely, they occur between the Late Devonian Upper Old Red Sandstone and the Tournaisian and are thought to have been erupted near to the Devonian–Carboniferous boundary (Francis, 1967; Lumsden *et al.*, 1967; Greig, 1971; George *et al.*, 1976). The radical palaeogeographic changes which occurred between the Late Devonian and the Tournaisian as a result of the eruption of the lavas (Leeder, 1974) offer further evidence of their Devonian-Carboniferous age. The radiometric age of these lavas suggests that the Devonian–Carboniferous must be between 365 and 360 Ma.

The early Visean volcanism in the Lothians—in the Edinburgh region and in the Garleton Hills, East Lothian—marks the start of Carboniferous volcanism in the Midland Valley. Dates of 354±5 Ma for an Arthur's Seat lava in Edinburgh (ICC, PTS 360/5, Fitch *et al.*, 1970) and 353±7 Ma for sanidines from trachytes from East Lothian (NDS 166) indicate that the episode occurred between 355 and 350 Ma.

According to George *et al.* (1976, p. 76), the East Lothian lavas are of Middle Arundian age but no evidence of this is known to the author. The East Lothian lavas overlie sediments of the Pu miospore zone, the lower part of which spans the Tournaisian–Visean boundary (Neves and Ioannides, 1974). Pu zone spores have also been found in tuffs in the Arthur's Seat lavas (Neves *et al.*, 1973). Hence these dates provide a minimum age for the Tournaisian–Visean boundary, which must be about 355 Ma or older.

The Early Visean volcanics are overlain by Asbian (TC miospore zone) sediments but it has been suggested by De Souza (1979) that there is a hiatus of considerable duration between the top of the East Lothian volcanics and the overlying TC zone sediments. Evidence for this comes from the age of the latest Dinantian volcanic event in the Scottish Borders—the Glencartholm volcanic beds—within which the base of the Asbian (which is equivalent to the base of the TC zone) has been identified (George *et al.*, 1976). Although it has not been possible to date these volcanic beds directly, they contain trachytic fragments (Greig, 1971) which must have been derived from the only Dinantian acidic activity to have occurred in the area and which has been dated at 336±7 Ma (K–Ar sanidine, Eildon Hills laccolith; De Souza, 1979). Therefore, it is proposed that the base of the Asbian (equivalent to mid-Visean) should be about 335 Ma.

1.3 Late Dinantian

Volumetrically and geographically the eruption of the Clyde plateau lavas was the most important volcanic event in the Scottish Carboniferous. Whole-rock and mineral dates of the fresher lavas and associated intrusions range between 332 and 326 Ma and it is therefore thought that the bulk of the lavas were erupted between 335 and 325 Ma—i.e. mid- to Late Visean. An Early Visean episode has also been recognized (De Souza, 1979). Dates from the base of the Clyde plateau lavas cannot be used to calibrate the numerical time scale as the lavas rest unconformably on Early Visean or older rocks. Hence the date of 367±5 Ma (ICC, PTS 360/6) of one of the lower Clyde plateau lavas described as being of possible Tournaisian age is now considered to be incorrect.

The Clyde plateau lavas are overlain by volcanic detritus followed by sediments of Brigantian age. In the Strathaven area, however, they are overlain by sediments of the Upper Oil Shale Group which may be of Late

Asbian age and it is probable that the end of volcanicity in this area occurred near to the Asbian–Brigantian boundary. Two K–Ar anorthoclase dates of 327±7 Ma (NDS 167) have been obtained from lavas from the top of the succession in this area and therefore, it is proposed that the Asbian–Brigantian should be placed at about 325 Ma.

This proposal and the dates from the Clyde plateau lavas also suggest that two PTS dates of Asbian–Brigantian rocks are incorrect. The date of 345±4 Ma (ICC, PTS 360/4, Fitch *et al.*, 1970) for one of the Lower Burntisland lavas, an episode which extended from the mid-Asbian into the Brigantian, is considered too high, as is the date of 341±17 Ma (ICC, PTS 360/3, Fitch *et al.*, 1970) recorded for the Asbian–Brigantian Little Wenlock lava flow in England.

1.4 Dinantian–Silesian boundary

The age of the Dinantian–Silesian boundary is difficult to fix exactly in the absence of good datable rocks. The top of the Clyde plateau lavas is separated from this boundary by the Brigantian stage. Thus, the ages of 327±7 Ma (NDS 167) obtained from the lavas at the top of the succession provide a maximum age for it. It is evident that earlier estimates of 330 Ma (ICC) (Francis and Woodland, 1964) for the boundary are no longer correct.

The series of Late Visean (Brigantian)–Namurian A lavas in the West Lothian region near Edinburgh span this boundary, but are unsuitable for K–Ar dating because of alteration. However, a minimum age for these lavas can be fixed by considering the ages of associated Namurian intrusions in the Lothians and West Fife. These sills do not cut strata younger than Namurian A. They are associated with a variety of tuffaceous activity and Francis (1968) has argued that the West Fife sills in particular were intruded into wet, unconsolidated sediments so that their age spans that of the Namurian A sediments into which they are intruded. In their petrography and chemistry, they resemble the West Lothian lavas. Their isotopic ages, however, indicate that intrusive activity carried on into the Westphalian. The highest ages were obtained from two teschenite sills in the Edinburgh area (at Barnton and by the River Almond), which gave K–Ar biotite dates of 318±9 and 316±7 Ma (ICC). These sills were possibly contemporaneous with or slightly younger than the lavas and their ages are regarded as fixing a minimum age for the West Lothian event, for which an age of 320±5 Ma has been assumed. This is also thought to be a good estimate for the Dinantian–Silesian boundary. The age of 329±12 Ma (ICC, PTS 191) for the Namurian Hillhouse basalt sill in West Lothian, which was previously used to estimate this boundary, is considered to be slightly too high.

The 320±8 Ma age (NDS 230) of the Dinantian Machrihanish lavas in Kintyre may also be used as a guide to the age of the Dinantian–Silesian

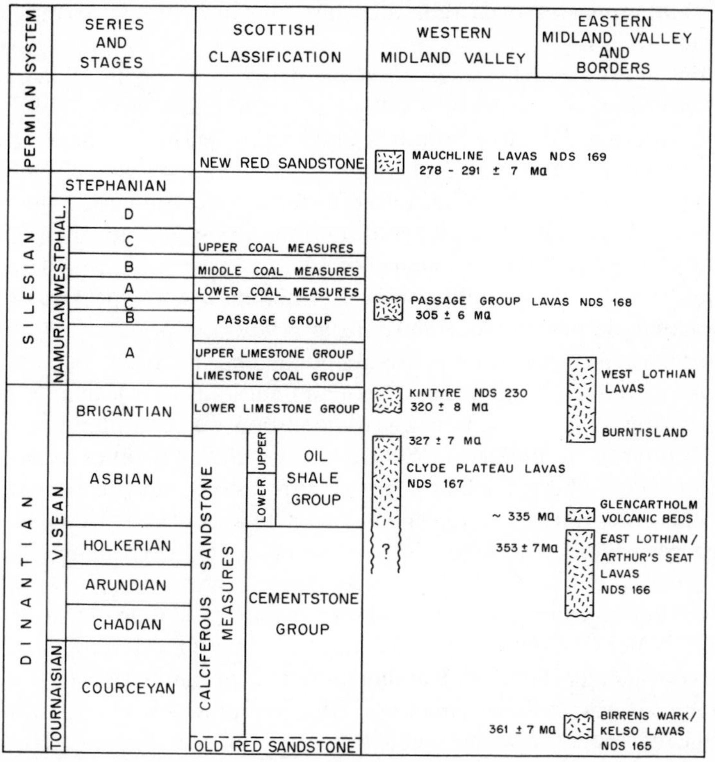

Figure 2. Stratigraphic ages of the major Carboniferous lava piles. Stratigraphic correlations from Francis (1967) and George *et al.* (1976). Note that the Dinantian and Silesian are subsystems only. The Tournaisian age of the top of the Old Red Sandstone is conjectural.

boundary as the lavas here are overlain by sediments of possibly Late Dinantian or very early Silesian age (George *et al.*, 1976). The basalts in this succession are slightly more silica-undersaturated in general than those of the Clyde plateau (MacDonald, 1975). If MacDonald's premise that the Dinantian basic lavas become increasingly silica-undersaturated with time is accepted, then the Machrihanish lavas are probably younger than those of the Clyde plateau, i.e. of Late Dinantian age.

All the evidence discussed in this section suggests, albeit circumstantially, that the Dinantian–Silesian boundary should be placed at 320 ± 5 Ma.

1.5 Silesian

Silesian volcanicity was on a much smaller scale than that of the Dinantian, with mainly tuffaceous activity and intrusion of dolerite sills. However, two volcanic episodes occurred, the ages of which are important for the Carboniferous time scale.

The Passage Group lavas of Ayrshire occur between mid-Namurian A and Westphalian A sediments. Although their exact stratigraphic age is unknown because of unconformities above and below the lavas, they are thought to be of Late Namurian age. Their apparent age of 305 ± 6 Ma (NDS 168) is considered low because of a certain amount of alteration in the groundmass of the samples dated and therefore, this data is only a minimum age for the Namurian–Westphalian boundary.

The age of these lavas could also be a minimum age for the Mississippian–Pennsylvanian boundary. As Ramsbottom (1969) has pointed out, there are many similarities between the Mississippian and Pennsylvanian of North America and the Carboniferous of Scotland and northern England, north of the Craven fault. As in America, sedimentation was continuous across the Visean–Namurian boundary in Scotland, but unconformities occur in the Late Namurian. These northern British successions, on the border of the Late Palaeozoic Northern Continent, are a link between the typical West European Carboniferous and the Mississippian–Pennsylvanian of eastern North America.

The second volcanic episode occurred in the Mauchline Basin of central Ayrshire. A floral assemblage found in sediments intercalated in the lower flows was originally dated as Stephanian but a reexamination has identified it as being of very early Permian age (Wagner in Smith *et al.*, 1974).

An age of 286 ± 7 Ma has been obtained for this episode (NDS 169). However, the amphibole date of 291 ± 7 Ma for the Carskeoch vent intrusion is considered the most reliable age. The Carboniferous–Permian boundary is therefore older than 286 Ma and an age of 290 ± 5 Ma is proposed for this boundary.

1.6 Conclusion

Age data from the Carboniferous–Early Permian volcanism of Scotland suggest the following time scale for the Carboniferous:

Devonian–Carboniferous:	360–365 ± 5 Ma
Tournaisian–Visean:	355 ± 5 Ma
Holkerian–Asbian:	335 ± 5 Ma
Asbian–Brigantian:	325 ± 5 Ma
Dinantian–Silesian:	320 ± 5 Ma
Namurian–Westphalian:	Minimum age of 305 ± 6 Ma
Permian–Carboniferous:	290 ± 5 Ma

The numerical time scale proposed here, particularly that of the upper part of the Dinantian, differs significantly from the time scale obtained by Fitch *et al.* (1970), which was largely based on ages from British Carboniferous rocks. This is a result of the systematically higher ages measured by them.

2 THE CARBONIFEROUS TIME SCALE

2.1 Introduction

Although the time scale for the Carboniferous was one of the best known of the pre-Tertiary periods in the 1964 PTS compilation, the quality of much of the data available in 1964 seems quite inadequate by present standards. The analytical and stratigraphical deficiencies of the PTS data have been discussed by Lambert (1971) and in individual contributions.

One serious criticism of current Carboniferous time scales is the over-reliance on ages of intrusive rocks for crucial points; the most recent time scale (Armstrong, 1978) is based largely on PTS items some of which are of dubious value. In addition, a few items in the data file have incorrect stratigraphic brackets (e.g. RLA 438, 2360, 3360). The Carboniferous time scale of Bouroz (1978) in the same volume is based on some unreliable ages and on dates calculated with a variety of constants.

2.2 Carboniferous boundaries

Age data for the base of the Dinantian are discussed in a following section on the Devonian–Carboniferous boundary. The 361 ± 7 Ma age for the Scottish Border volcanics (NDS 165) which occur close to the Devonian–Carboniferous boundary offers the most direct evidence on its age. Upper age limits on it are set by the K–Ar biotite dates of 369 ± 6 Ma (PTS 354) and 367 ± 2 Ma (NDS 234) for the Frasnian Cerberean volcanics in Australia. The 368 ± 6 Ma age of the Hoy lavas at the base of the Upper Old Red Sandstone provides an upper limit in Scotland. The 353 ± 7 Ma Early Visean East Lothian volcanics (NDS 165) are a reliable lower limit.

The 341 ± 6 Ma (ICC) Rb–Sr isochron age for an Early Tournaisian ignimbrite (Bouroz, 1978) is much too low in comparison with the above data and the analyses are suspect (personal communication, P. Vidal to G. S. Odin). The K–Ar dates from the mid- to Late Devonian of the USSR which range between 335 ± 10 and 355 ± 20 Ma (Afanass'yev, 1970; RLA 440–443) are also considered low. Few stratigraphical and analytical details are available for these dates. Rb–Sr isochron ages of the Early Mississippian Fisset Brook formation (PTS 347) recalculated by McDougall *et al.* (1966) at 374 (or 387) ± 32 Ma (ICC) and the Huelgoat granite at 336 ± 13 Ma (MSWD = 5) (NDS 229) are not precise enough to calibrate the time scale. Thus an age of 360 ± 5 Ma for the Devonian–Carboniferous boundary seems the best estimate at present.

Three items were heavily weighted for the Carboniferous–Permian boundary in the 1964 PTS compilation—the Brassac tuffs (PTS 63), the Castro Daire granite (PTS 122) and the Sande 'essexite' (PTS 192). However, these items were considered individually unreliable by Lambert (1971) who, nevertheless, recommended an age of 280 Ma (ICC recalculated) for this boundary on the general agreement of these dates.

The 286±7 Ma of the very early Permian Mauchline volcanic group in Scotland (NDS 169) indicates that the age of the boundary ought to be older. This is supported by the 283±5 Ma age of the Early Autunian Roubayre tuff (Détroit de Rodez) and Stephanian dates between 301±5 and 291±10 Ma from France (NDS 231). A K–Ar biotite age of 291±5 Ma (ICC) is listed by Armstrong (1978, RLA 483) for a dacite–rhyolite lava of Stephanian–Permian age from the Détroit de Rodez, but no information on this item has been found in the source cited—Fuchs *et al.* (1970). K–Ar biotite, Rb–Sr isochron and U–Pb ages of 286±15, 285±20 and 284±10 Ma (ICC) have been measured for the Carboniferous–Permian Akchatau granite in Kazakhstan (RLA 420–422, Afanass'yev, 1970). On the basis of these ages, an age of 290±5 Ma is proposed for the Carboniferous–Permian boundary.

2.3 Carboniferous subdivisions

The 353±7 Ma Early Visean East Lothian lavas (NDS 166) provide a minimum age for the Tournaisian–Visean boundary and an age of 355±5 Ma is proposed for it. For the mid-Visean, the basal Asbian Glencartholm volcanic beds have been indirectly dated at about 335 Ma, while the top of the Clyde plateau lavas at 327±7 Ma gives a Late Visean (Asbian–Brigantian) point (see section on Scotland and NDS 167).

The Rb–Sr biotite dates of 340±3 Ma (ICC) for the Late Visean (D2) Malavaux tuffs (PTS 172) used in the 1964 time scale are considered to be unreliable (too much common strontium). Roques *et al.* (1971) quote a K–Ar biotite age of 328±13 Ma (ICC) for the tuff which is in better agreement with the Scottish data. Also quoted are Rb–Sr mineral/whole-rock isochron ages of 336–341±6 Ma (ICC) for the Malavaux and other Late Visean tuffs in the region. As no analytical details have been given it is not possible to judge their reliability. The pre-Late Visean or older granitized tuff of Gien-sur-Cure (PTS 173) gave an age of 345±7 Ma (ICC).

The Hudson's Peak and Martins Creek andesites in Australia have a mean K–Ar hornblende age of 334 Ma (no error given, PTS 66). These rocks belong to the Gilmore formation which is considered to be of Late Visean age, although fossil control is poor (Jones *et al.*, 1973). Their radiometric age is in agreement with the Scottish ages for the Late Visean.

No direct evidence is available for the age of the Visean–Namurian (Dinantian–Silesian) boundary but the various Scottish data discussed previously (Section 1.4) suggest an age of 320±5 Ma for it.

Hardly any stratigraphically controlled dates are available for the Namurian. The K–Ar biotite 318±10 Ma (ICC) age of the Middle Carboniferous Keregetass volcanics in Kazakhstan (PTS 339) is acceptable as Namurian. The flora in the underlying tuffs indicate a Namurian B–C to Stephanian age but there are difficulties in correlations with Europe (see comments, PTS 338, 339).

The Namurian B–C to Westphalian A Passage Group lavas of Scotland set a minimum age of 305±6 Ma for the Namurian–Westphalian boundary (NDS 168). The most reliable Westphalian age is the 303±6 Ma K–Ar sanidine date for the Early Westphalian C Hagen 2 seam tonstein (NDS 232). This age also indicates that the 300±10 Ma (ICC) K–Ar glaucony date from the Verayan horizon (PTS 64) of basal Moskovian age (equivalent to Westphalian C) is reliable. The 314±10 Ma K–Ar age of the Barrow Hill intrusion (PTS 360/1) intruded into the Westphalian C Etruria marl is considered too high. Similarly, the 316±7 Ma Rb–Sr mineral ages from the Vosges granites (PTS 356) are too high for Westphalian C or D and the stratigraphic correlations must be incorrect (see stratigraphic comments, PTS 356).

For the Stephanian, the Rb–Sr isochron age of 311±10 Ma (ICC) for an Early Stephanian cinerite (Bouroz, 1978) is thought to be too high and the isochron is very dubious. K–Ar whole-rock ages of 301±5–291±10 Ma have been obtained for Stephanian rocks in France (NDS 231) but their stratigraphic definition is rather poor. From the USSR, the Taskuduk trachyandesite from Kazakhstan, dated at 297±15 Ma (ICC, PTS 340), is either of Early Stephanian or Late Stephanian/Autunian age. The calculated age is thought by the authors to be slightly too high.

The K–Ar whole-rock 285 Ma (ICC) date of the northern Caucasus, Dahut river andesite–basalt (RLA 436, Afanass'yev, 1970) at the base of the USSR Late Carboniferous (Westphalian–Stephanian) is judged to be low. Biotite ages of 291±10 Ma (K–Ar, ICC) and 286±11 Ma (Rb–Sr, ICC) have been measured for the granitic Kaldrymin complex (RLA 438/439, Afanass'yev, 1970) but the stratigraphical arguments for its postulated Middle–Late Carboniferous (Westphalian–Stephanian) age have not been presented.

None of the dates discussed above provide much direct evidence on the ages of the Namurian–Westphalian and Westphalian–Stephanian boundaries. However, ages of 310±5 Ma and 300±5 Ma respectively are tentatively proposed for them.

2.4 Summary

The overall length of the Carboniferous at about 70 Ma is in between the 65 Ma estimated by Francis and Woodland (1964) and the 77 Ma of Armstrong (1978) but well below the 90 Ma estimated by Lambert (1971). The length of the Dinantian is increased to 40 Ma and the Silesian reduced

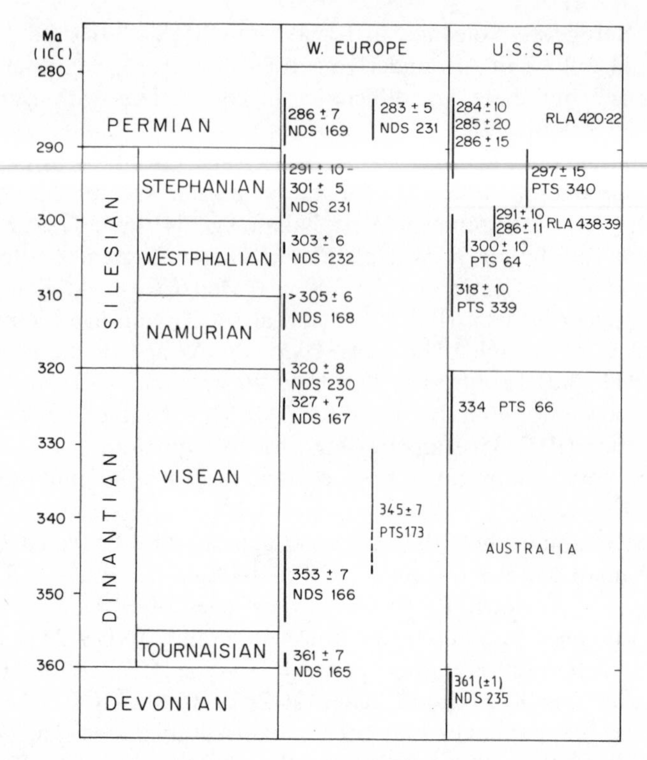

Figure 3. Proposed time scale for the Carboniferous. The data quoted are those considered as the most reliable.

to 30 Ma as a result of the lowering of the Dinantian–Silesian boundary to 320 ± 5 Ma based on the ages from Scotland. The most unexpected result is the 35 Ma length of the Visean. The Silesian stages are each about 10 Ma long; it is not possible to be more precise with the data available. Further refinement of the Silesian time scale could be achieved by careful dating of the widespread tonsteins in Western Europe.

The proposed Carboniferous time scale is shown in Figure 3 with the most reliable points. The boundaries and subdivisions of the Carboniferous are as follows:

Devonian–Carboniferous:	360 ± 5 Ma
Tournaisian–Visean:	355 ± 5 Ma
Visean–Namurian: (= Dinantian–Silesian)	320 ± 5 Ma
Namurian–Westphalian:	310 ± 5 Ma
Westphalian–Stephanian:	300 ± 5 Ma
Carboniferous–Permian:	290 ± 5 Ma

The proposed time scale is essentially a K–Ar time scale for Western Europe. Less emphasis is placed on the results from the USSR, as analytical and stratigraphical details are lacking for much of the data and stratigraphic correlations with Kazakhstan are difficult.

Résumé du rédacteur

L'Ecosse est caractérisée au Carbonifère par une grande abondance de niveaux éruptifs; ceux-ci ont été étudiés en détail sur le plan stratigraphique; ce domaine est donc favorable à l'établissement d'une échelle numérique régionale. L'étude radiométrique de H. A. F. de Souza a été réalisée par la méthode K–Ar sur des minéraux et roches totales dans les deux laboratoires de Leeds et Glasgow. Les niveaux étudiés vont des vieux grès rouges supérieurs à la base du Permien.

Les résultats obtenus complètent ou précisent certains résultats déjà disponibles et remettent en cause d'autres âges publiés d'après l'étude de niveaux équivalents de Grande Bretagne et sur lesquels l'échelle anciennement admise était fondée. Une révision s'imposait donc.

L'analyse de la littérature, à laquelle s'ajoutent les résultats présentés ci-dessus et d'autres présentés dans ce volume, montre que si la base (admise vers 360 Ma malgré divers âges plus jeunes) et le sommet du système (admis vers 290 Ma) sont bien documentés, des lacunes de connaissance demeurent (Namurien, Westphalien). La limite Dinantien–Silésien est rajeunie vers 320 Ma (encore que des âges plus anciens aient été obtenus pour le Viséen supérieur). Ceci délimiterait un Dinantien bien plus long que proposé antérieurement. L'échelle du Silésien devrait être améliorée dans l'avenir par l'étude des tonstein d'Europe de l'Ouest. De même des âges Rb–Sr seraient souhaitables pour conforter cette échelle essentiellement basée sur la méthode K–Ar.

(Manuscript received 9-3-1981)

Numerical Dating in Stratigraphy
Edited by G. S. Odin

25
Numerical dating of Caledonian times (Cambrian to Silurian)

Noël H. Gale

Subsequent to the reviews published by the Geological Society of London (1964–1971), the Cambrian, Ordovician and Silurian time scales received little serious attention until the publication in 1978 of papers given at the Geological Time Scale Symposium in Sydney in 1976; however, the conclusions drawn in the papers given at this symposium were vitiated by lack of attention to errors, failure to use uniform decay constants and by inclusion of much old data which were worthless on grounds either of poor stratigraphy or of poor geochronology. Recently there has been a revival of interest (Ross *et al.*, 1978a; Gale *et al.*, 1979; 1980; McKerrow *et al.*, 1980; Rundle, 1981) and a sufficient number of good modern dates for stratigraphically well controlled rocks to justify a reappraisal of the numerical calibration of Caledonian times, though it remains true that there is still a serious lack of data for the Silurian and the Cambrian. The chances of securing a better calibration of the time scale have recently been much improved by the international agreement (Steiger and Jäger, 1977) to accept a set of mutually consistent decay constants for the U, Th–Pb, Rb–Sr and K–Ar methods: decay constants which moreover have a good chance of being very close to the absolute ones.

Recent discussions of the Ordovician and Silurian time scales have been distorted by the inclusion of fission track ages. We exclude such ages because, as clearly shown by Storzer and Wagner in this volume, the analytical and systematic uncertainties at present associated with such ages are too large for time-scale calibration purposes. A feature of some recent reviews (Gale *et al.*, 1979; 1980; McKerrow *et al.*, 1980) is the use of an independent scale intended to be proportional to the absolute division of the stratigraphic stages, in order to make a statistical correlation between these divisions and the radiometric dates. This was at best never more than a counsel of despair to attempt to extract the maximum amount of information from very few reliable dates and has properly been criticized by Rundle

(1981), who points to the very uncertain bases used in the attempt to fix the stage lengths. In this review such an approach has been abandoned. For the most part the suggestions made here for the calibration of the stage boundaries are based on the modern data listed in part II, paying careful attention to the uncertainties of the stratigraphy and the analytical errors attached to the dates. Some older data have been used, and all isochron dates have been recalculated from the authors' original raw data using a modern line-fitting algorithm and modern methods of statistical assessment. This is vital since, as pointed out by Gale *et al.* (1980), older isochron reductions in the literature are often in error.

1 THE CAMBRIAN SYSTEM

1.1 Stratigraphic introduction

The precise biostratigraphical locations of both the Precambrian–Cambrian and Cambrian–Ordovician boundaries are still a matter of debate; the present position on these question is summarized in this book by F. Doré and M. Robardet. Although British geologists usually place the Cambrian–Ordovician boundary at the base of the Arenig series, most other geologists place it at the base of the Tremadoc; in the absence of an international convention the majority view is adopted here. The lower, Precambrian–Cambrian, boundary is arbitrarily taken as the *base of the Tommotian stage* in Siberia as defined in the article by Doré. Divisions within the Cambrian are taken provisionally as Lower, Middle and Upper, and are subject to change because of the activities of various international working groups (Cowie and Cribb, 1978). Names for these major divisions which are defined on geographical or palaeontological bases are not used here for the reasons summarized by Cowie *et al.* (1972). The boundary between the Lower and Middle Cambrian is provisionally taken between the *Protolenus* zone and the lowest *Paradoxides* zone (in Europe that of *P. oelandicus*). The base of the Upper Cambrian is taken at the base of the *Agnostus pisiformis* zone.

1.2 Earlier evaluations

Most earlier attempts to provide a numerical calibration of the Cambrian system have concluded that the base of the Cambrian is to be taken at 570–590 Ma. The estimate provided by modern data is so different that some space must be devoted to discussing the interpretation of older evidence, much of which is so intrinsically worthless that it would not otherwise merit attention.

In 1964 Cowie (1964, pp. 255–8) selected from about a hundred radiometric dates 30 for closer study, eventually accepting for the calibration of the Cambrian 15 best dates of which he wrote that 'none is entirely satisfactory'. Of these 15 dates, 11 were further selected by Cowie to provide his final calibration; of these 11 dates, 9 (PTS 42, 48, 70, 100, 116, 117, 118, 183, 185) were of potential value in fixing the base of the Cambrian. Using modern standards we must abandon PTS 48 (determined on 'illite') and PTS 70 (whose stratigraphy is uncertain). PTS 100, 116, 117, 118 and 185 are all Russian determinations on glauconies which are in varying degrees suspect, as was recognized by Cowie. Of these, PTS 185 must definitely be rejected as having too low a K content for a glaucony to yield a reliable age. If, though admitting their unreliability, we recalculate the ages (admitting only those ages on glaucony with a high enough K content) according to the ICC constants, Table 1 results.

Thus the 1964 data allow the base of the Cambrian to be set only rather loosely and uncertainty *in the interval* 522–570 *Ma*, accepting the glaucony ages; if the glaucony ages are rejected, then the base of the Cambrian rests only on the Vire-Carolles granite date, 548–578 Ma, but the radiometric data are rather poor.

The review by Lambert (1971) included two more Rb–Sr isochron dates relevant to this boundary: PTS 352 for the *Late Precambrian* Holyrood granite and PTS 353 for the *Precambrian* Hoppin Hill granite complex. The hazards of using the results of early isochron computations and merely correcting the authors' computed ages to the ICC constants are well illustrated by these two sets of data. For PTS 352 the simple conversion of the authors' age to the ICC constants yields 594 ± 11 Ma; on the other hand, if the author's original data are recomputed using a modern isochron programme tested on standard data set 3 of Brooks *et al.* (1972), then the result is 585 ± 15 Ma, MSWD = 0.2, initial $^{87}Sr/^{86}Sr = 0.7040 \pm 0.0008$. For PTS 353 only the Northbridge granite–gneiss is reliable, and recomputation of the authors' age to ICC yields 557 ± 4 Ma; recomputation of the original

Table 1

PTS item	Stratigraphy	Recalculated age (ICC)
100, Blue clay, Baltic series	Early Cambrian	522 ± 3 Ma
183, Murray shale	Early Cambrian	565 ± 30 Ma
42, Vire-Carolles granite	Late Precambrian	563 ± 15
116, Ashinsk series	Late Precambrian	558 ± ?
117, Laminarites beds	Late Precambrian	584 ± ?
118, Uksk beds	Late Precambrian	600 ± ?

data with the modern programme yields 553 ± 10 Ma, MSWD = 1.1, initial $^{87}Sr/^{86}Sr = 0.7065 \pm 0.0008$. (For both PTS 352 and 353 one aberrant data point has been omitted, as they were by the authors.) Using the authors' arguments that 15 Ma may have elapsed between the intrusion of the Holyrood granite and the beginning of the Cambrian, PTS 352 yields *560 ± 15* Ma for the base of the Cambrian, whilst PTS 353 yields a maximum of *553 ± 10* Ma for this boundary.

The earlier attempts to provide a numerical calibration of the Cambrian system, Cowie (1964) and Lambert (1971), were subsumed in the account given by Cowie and Cribb (1978), who recalculated all the available earlier data using the ICC and reported some new data, quoting nearly 30 selected radiometric dates in all for the Cambrian but unfortunately omitting proper errors in most cases. Cowie and Cribb finally proposed dates of 590 and 485 Ma (ICC) as limits of the system and suggested that the Middle Cambrian might lie in the interval 505–530 Ma, but nowhere do they present detailed evidence for these estimates (guesses?) nor do they suggest what the associated uncertainties may be. Further, for six items used by Cowie and Cribb the original papers did not quote the decay constants adopted; these data should be rejected. To illustrate the magnitude of the uncertainty for these items, note that a K–Ar age of 500 Ma assumed to be calculated with $\lambda_e = 0.557 \times 10^{-10}\ a^{-1}$ gives about 485 Ma, ICC, but if it had in fact been originally calculated with $\lambda_e = 0.585 \times 10^{-10}\ a^{-1}$ it would recalculate to about 510 Ma, ICC.

Moreover, many of the results used by Cowie and Cribb are Russian K–Ar ages of glauconies or illites subject to the dual problems of possible presence of older detrital minerals in the sediments and argon losses after deposition. In the absence of detailed sedimentological studies for each individual case these K–Ar ages, which show a wide scatter for any one stratigraphic grouping, must be regarded as of little value for time-scale calibration purposes. The other dates quoted by Cowie and Cribb as relevant to the age of Cambrian deposition are Rb–Sr whole rocks or U–Pb *maximum depositional ages,* obtained from rocks overlain by Cambrian sediments: the results range from 614 to 557 Ma.

1.3 Modern data

For the Precambrian–Cambrian boundary there are three relevant modern age determinations. That for the Vire-Carolles granite (NDS 121) is perhaps the least certain, both because the stratigraphy is somewhat uncertain and because the U–Pb data on zircon and monazite cannot be interpreted unambiguously; however, it seems that *540 ± 10* Ma corresponds to an *upper limit for the age of deposits correlated with the Tommotian stage* of the Siberian platform. In contrast, studies by Lancelot in Morocco (NDS

136) are of one of the best sections spanning latest Precambrian and Early Cambrian times. The position of successive intrusions and extrusions relative to the Upper Precambrian and Lower Cambrian sedimentary sequences was well established and U–Pb zircon studies allowed the precise dating of the successive phases. In the Bou Azzer area an age of *534 ± 10* Ma was established as a *maximum age for deposition of Early Cambrian sediments* and a minimum age for the deposition of Adoudounian sediments (Adoudounian being a rather informal equivalent of the Vendian). The third piece of evidence for the dating of the Cambrian–Precambrian boundary comes from the work of Patchett *et al.* (NDS 242) on Rb–Sr whole-rock isochron dating of the Ercall granophyre and the Rushton schist in the Wrekin area of southern Britain. The dates obtained were respectively 533 ± 13 Ma and 536 ± 8 Ma; detailed arguments given in NDS 242 show that these ages must constitute a *maximum age for the Lower Cambrian* in this area. Though both in the original paper and in NDS 242 the difficulties of extending this conclusion to other parts of the world were stressed, the close agreement between NDS 136 and NDS 242 as a result of different dating schemes applied to different chronometers in Morocco and southern Britain suggests that deposition of the Lower Cambrian must indeed post-date 534 ± 10 Ma; the date for the Vire-Carolles granite supports this conclusion. Further concordant data are discussed in NDS 249–250 and 251.

The earlier data used to estimate the top of the Cambrian (such as PTS 34 and PTS 350) are worthless by modern standards. The best modern data are those for the Křivoclàt-Rokycany volcanics in Bohemia (NDS 130), which lie on Middle Cambrian sediments and lasted until the beginning of the Tremadocian. The age of 491 ± 14 Ma is therefore a *maximum for the Cambrian–Ordovician boundary*, and is in fact probably close to the age for that boundary. The apparent age of 516 ± 15 Ma obtained for a Middle Late Cambrian micritic limestone (NDS 150) is consistent with the Bohemian evidence but there are technical difficulties which detract from the reliability of NDS 150.

There seem to be no reliable radiometric dates available which allow estimates to be made of the dates of the Lower–Middle or Middle–Upper Cambrian boundaries.

1.4 Conclusion

The rather few sets of reliable modern data suggest quite strikingly different numerical ages for the bottom of the Cambrian when compared with earlier estimates and with those of Cowie and Cribb (1978). The earlier data of PTS 352 and 353 are not in conflict with our present suggestion of *530 ± 10* Ma for the base of the Cambrian.

The date of $495 \pm ^{10}_{5}$ Ma is the best suggestion which can be made at present for the upper limit of the Cambrian system. It agrees in fact with the proposal of Cowie and Cribb and will be rediscussed with the Ordovician system. The lower limit is very different from previous estimates, though not in disagreement with PTS 352 and 353 nor with any other earlier radiometric data. We are convinced that the coincidence between ages obtained in Morocco, England and the Massif Armoricain is not fortuitous. Clearly more data are still desirable at both boundaries.

2 THE ORDOVICIAN SYSTEM

2.1 Stratigraphic introduction

The historical circumstances leading to the location of the Cambrian–Ordovician boundary at either the base of the Arenig as favoured in Britain or the base of the Tremadoc as is customary elsewhere have been discussed by Whittington and Williams (1964). Since the majority of countries include the Tremadoc, and in the absence of an international agreement, the base of the Ordovician will here be taken at the base of the Tremadoc (see Robardet, this book, for further comments).

The stratigraphic location of the Ordovician–Silurian boundary, though not seriously in doubt, can nevertheless not be confirmed by direct correlation between relevant successions in the Bala and Llandovery districts in Wales or in Kazakhstan, and some revision of the correlation between the graptolite and shelly faunas straddling the Ordovician–Silurian boundary may eventually prove necessary (Williams, 1972). Nevertheless, the terminal stage of the Ordovician is the Hirnantian, with type section in the Foel y Ddinas mudstones near Bala and a distinctive fauna easily recognized in many parts of Eurasia and North America (see M. Robardet, this book).

The Ordovician is internally divisible into six series, the Tremadoc, Arenig, Llanvirn, Llandeilo, Caradoc and Ashgill. Figure 1 gives a scheme of local divisions of the Ordovician with approximate correlations. The divisions and their correlations both within and outside the British Isles are discussed by Williams (1972), who emphasizes that no one faunal succession can serve as a ubiquitous time–stratigraphic measure, though graptolites continue to offer the most reliable means of establishing contemporaneity.

2.2 Calibration of the Ordovician system—general considerations

The 1964 and 1971 PTS reviews emphasized the difficulties of establishing a reliable Palaeozoic time scale, and the problems have proved particularly acute for the Ordovician (Fitch *et al.*, 1976). Lambert, in PTS 1971,

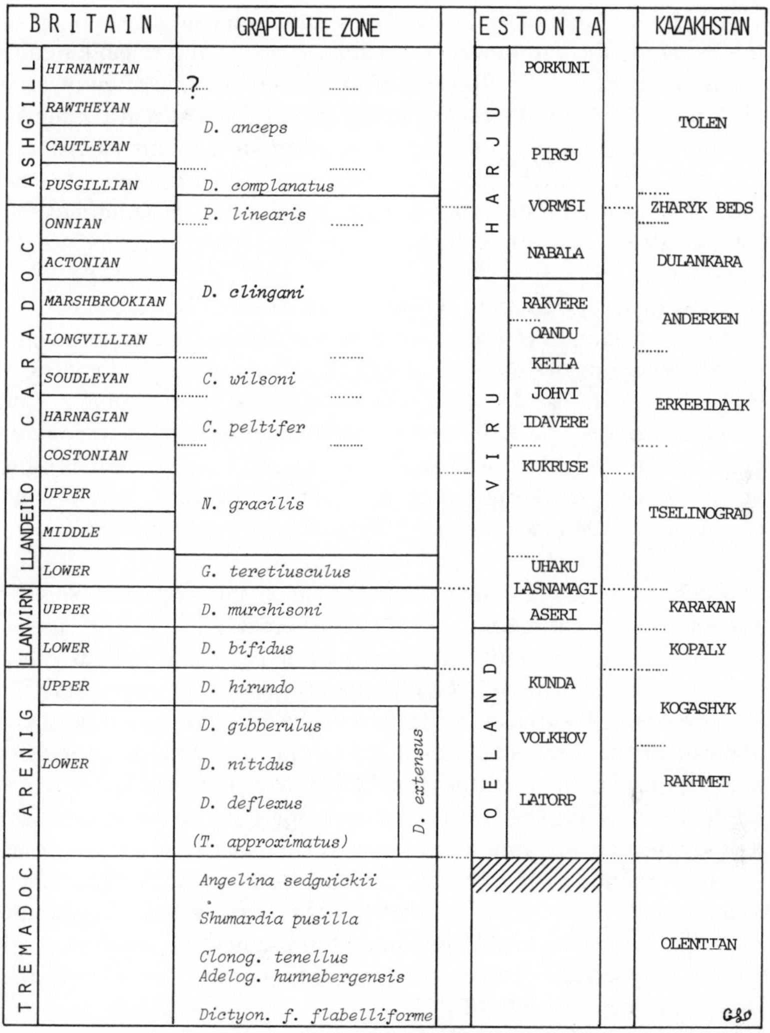

BRITAIN		GRAPTOLITE ZONE	ESTONIA		KAZAKHSTAN
ASHGILL	HIRNANTIAN	?	HARJU	PORKUNI	TOLEN
	RAWTHEYAN	D. anceps		PIRGU	
	CAUTLEYAN				
	PUSGILLIAN	D. complanatus		VORMSI	ZHARYK BEDS
CARADOC	ONNIAN	P. linearis		NABALA	DULANKARA
	ACTONIAN	D. clingani			
	MARSHBROOKIAN		VIRU	RAKVERE	ANDERKEN
	LONGVILLIAN			OANDU	
	SOUDLEYAN	C. wilsoni		KEILA	ERKEBIDAIK
	HARNAGIAN	C. peltifer		JOHVI IDAVERE	
	COSTONIAN			KUKRUSE	
LLANDEILO	UPPER	N. gracilis			TSELINOGRAD
	MIDDLE				
	LOWER	G. teretiusculus		UHAKU	
LLANVIRN	UPPER	D. murchisoni		LASNAMAGI ASERI	KARAKAN
	LOWER	D. bifidus	OELAND		KOPALY
ARENIG	UPPER	D. hirundo		KUNDA	KOGASHYK
	LOWER	D. gibberulus D. nitidus D. deflexus (T. approximatus) D. extensus		VOLKHOV LATORP	RAKHMET
TREMADOC		Angelina sedgwickii Shumardia pusilla Clonog. tenellus Adelog. hunnebergensis Dictyon. f. flabelliforme			OLENTIAN

Figure 1. Stratigraphic subdivision of the Ordovician system in Great Britain. Possible correlations with the Russian platform are given.

suggested that there were then only two reliable age determinations for the Ordovician period, PTS 156 and PTS 157. The computer data file proposed by Armstrong (1978) unfortunately includes many data reasonably rejected by Lambert (1971), and some of the newer data also included in this file are rather poorly documented.

We should like to stress that the approach of amassing large numbers of radiometric dates of dubious analytical or stratigraphic quality into a computer file is precisely the wrong way to attempt to arrive at a reliable numerical calibration of the time scale. Rather, what is needed is the attempt to *select those few analytically and geochemically acceptable* dates for material of good stratigraphic control; moreover, it is essential not only to ensure that the dates are recalculated to conform with the ICC constants but also to recompute all isochron data using consistent computer programmes and to make proper assessments of errors at the 2σ level (Gale *et al.*, 1979a; 1980; Rundle, 1981).

The review of the Ordovician by Ross *et al.* (1978a) is rather uncritical in its acceptance of data, does not recalculate the data according to a common set of decay constants or pay attention to the errors of the radiometric ages, and often attempts to use dates for rock units whose stratigraphic position is uncertain and can only be estimated by unreliable, involved and convoluted geological arguments. Lack of critical assessment of errors and too great a reliance on fission track ages has also vitiated several recent reviews (McKerrow *et al.*, 1980; Ross *et al.*, 1977; 1978b), as mentioned by Rundle (1981).

At first sight fission track dating would appear to have considerable advantages for numerical calibration of the time scale, in that it allows in principle the dating of apatite or zircon from bentonite or tuff horizons intercalated in an undisturbed, fossiliferous sequence, so that stratigraphic control is excellent. Moreover, such occurrences are often in type sections. Nevertheless, in the discussion which follows no use is made of the fission track ages reported in the literature for the following reasons:

(1) Unfortunately, though fission track dating may in the future achieve the accuracy needed for stratigraphic applications, it has not yet done so. Apart from possibilities of inherited zircons. Storzer and Wagner show in this volume that there exist relatively large uncertainties in the decay constant for spontaneous U-238 fission and in the neutron dosimetry, whilst track fading is a serious problem and no proper interlaboratory age standards have yet been introduced.

(2) The Palaeozoic fission track ages in the literature have not as yet been reported in detail, have been revised upwards by about 20 Ma between the two successive summary reports (Ross *et al.*, 1976; 1978b) and probably have absolute errors nearer 40 Ma rather than the somewhat optimistic errors quoted. Compston (1979) suggests that fission track dating is now the calibration method *par excellence.* Though it may achieve that status in the future, a salutary corrective to Compston's view is to be found in the controversy over the age of the hominid-bearing KBS tuff in East Africa, as discussed by Storzer and Wagner above.

On the other hand, we see no reason arbitrarily to reject Rb–Sr dates on stratigraphically well-defined acid volcanics. There is good evidence to support the view that it should not be assumed that resetting has occurred in every acid volcanic rock system (Gale *et al.*, 1979a; 1980; Rundle, 1981) although it is well known that any Rb–Sr whole-rock system *can* be reset, regardless of rock type. A recent review (Gale *et al.* 1979a) of modern Rb–Sr whole-rock data for acid volcanics strongly suggests that, although the Rb–Sr system is easily disturbed in acid pyroclastics by later metamorphism (when it is usually revealed by high mean square weighted residuals in the isochron computation), the Rb–Sr system in acid lavas is rather resistant to later metamorphism. This resistance to later heating events is particularly well shown by the Stockdale rhyolite (Gale *et al.*, 1979a), which is discussed further below. Rundle (1981) has shown that eight Rb–Sr whole-rock isochrons for acid lavas relevant to the time scale give ages consistent with each other and their relative stratigraphy and with Rb–Sr whole-rock dates for relevant granite intrusions in the English Lake District.

Moreover, there are examples of Rb–Sr whole-rock isochron ages for acid volcanics which are concordant with ages derived from other decay schemes. For example, the Cerberean volcanics of Victoria, Australia, yield a mean K–Ar age on biotite of 367 ± 1 Ma (NDS 234), while the basal rhyolite from the same formation gives a recalculated statistically acceptable Rb–Sr whole-rock isochron age of 362 ± 18 Ma with an MSWD of 0.24 (Gale *et al.*, 1980). In addition, Cleverly (quoted in Gale *et al.*, 1980) has dated Karoo dolerites and rhyolites from the Lebombo of Swaziland. An Rb–Sr whole-rock isochron for the rhyolites yields an age of 189 ± 7 Ma with an MSWD of 1.1; the dolerites fall on the same Rb–Sr isochron and yield a mean K–Ar age of 188 ± 5 Ma. consequently there is no *a priori* reason to consider Rb–Sr whole-rock isochrons for acid lavas as unreliable. We will therefore, in our discussion of the Ordovician and Silurian, use Rb–Sr data on acid lavas where a critical examination suggests that no later resetting is likely to have occurred.

2.3 Modern data

The numerical calibration of the Ordovician system has been unnecessarily confused by the use of unreliable data, and we shall first give briefly our reasons for rejecting some data used by others.

Lambert (1971) suggested that there were only two reliable age determinations for the Ordovician period. Of these, PTS 156 consists of five U–Pb ages on zircons, two K–Ar ages on biotites and two Rb–Sr ages on biotites, from the Carters river limestone, Bays formation and Eggleston limestone in the Caradocian succession of Tennessee. When recalculated, the dates range from 402 to 482 Ma, with a mean of 442 Ma and a 95% confidence interval of ± 18 Ma. The discordances between the $^{207}Pb/^{206}Pb$, $^{238}U/^{206}Pb$ and $^{235}U/^{207}Pb$ ages are so gross, however, that the data cannot be accepted as reliable. Lambert's second critical point (PTS 157) consists of six K–Ar ages for biotite and sanidine separates from bentonites in the Caradocian Chasmops limestone of Sweden. In the three sanidine separates, K–Ar age correlates inversely with potassium content, apparently because of illite contamination. The lowest recalculated age from these separates is 452 Ma. The biotites have low potassium contents suggesting chloritization, and they probably yield unreliable K–Ar ages. It has been shown (Obradovich and Cobban, 1976) that biotites separated from bentonites and containing less than about 5% K give apparent ages either younger or older than the true age. The single biotite from the Chasmops limestone containing more than 5% K gives an age of 442 Ma. The sanidine and biotite ages together average 449 Ma with a 95% confidence interval of ±13 Ma. However, none of the Chasmops limestone data satisfy the acceptability criteria of Obradovich and Cobban (1976). Lambert himself rejected for good reasons all of the items available in 1971 relevant to the Ordovician with the exception of PTS 156 and PTS 157; we can now see that these data also should be rejected.

The article by Afanass'yev (1970) gives few details for the dates quoted and collectively they can be given little weight for time-scale calibration. It has already been remarked that the computer file of Armstrong (1978) is heavily biased by the unreliable data of PTS 1964 and PTS 1971; it is also biased by the inclusion of data from Afanass'yev (1970) and by other items which either have unreliable stratigraphy or are analytically unacceptable. Deficiencies of some of the other data accepted by Armstrong and by Ross *et al.* (1978a) will be discussed below. The use of these data in future reviews is not recommended.

2.4 The Cambrian–Ordovician boundary

In the discussion of the Cambrian system it was shown that the best estimate for the date of the Cambrian–Ordovician boundary is about $495 \pm^{10}_{5}$ Ma, but that this estimate depends essentially on the data for the Křivoclát–Rokycany volcanics in Bohemia (NDS 130) with some support from the date of a Middle Late Cambrian micritic limestone from Texas (NDS 150). However, there is now reasonable evidence from the K–Ar dating of hornblende from the Late Tremadoc Rhobell volcanics in North Wales that the minimum age for Late Tremadoc is 475 ± 12 Ma (NDS 122). This estimate receives slight support from the rather inaccurate apparent K–Ar age (NDS 131) of 467 ± 23 Ma for a slightly evolved and possibly rejuvenated glaucony from a Late Tremadoc formation at Stora Backor, Sweden (Figure 2). Much better support comes from the K–Ar age (NDS 125) of 473 ± 17 Ma on glaucony from a basal Arenig formation in Tallinn,

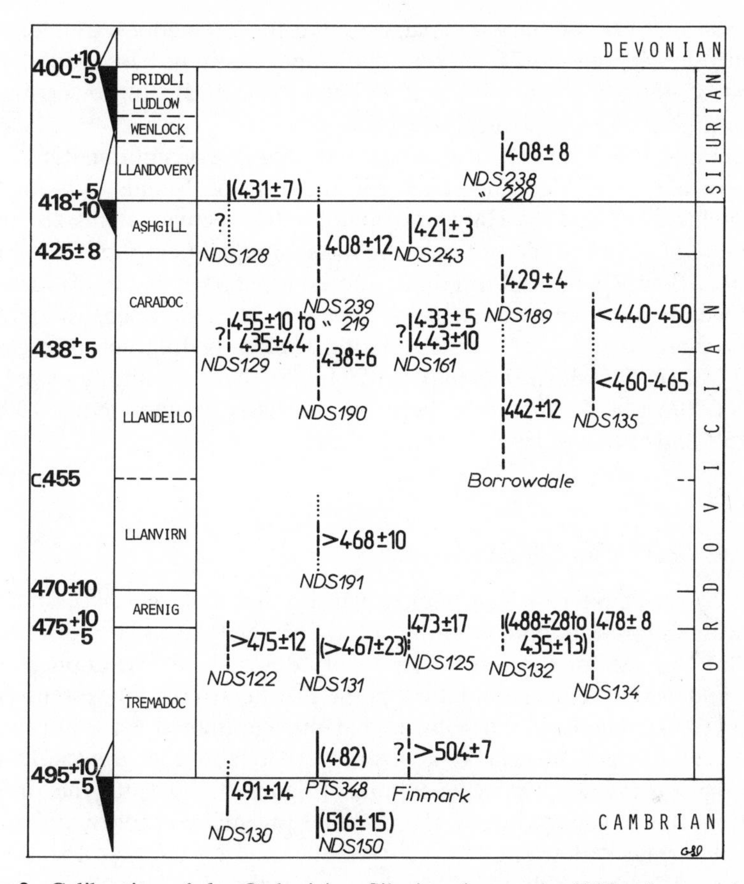

Figure 2. Calibration of the Ordovician–Silurian time scale: PTS refers to abstracts given in Harland *et al.* (1964–1971). NDS refers to the abstracts appearing in Part II of this volume. Dates given in parentheses are considered unreliable. Uncertain stratigraphy is noted with dotted lines and interrogation marks. Insufficient data are available for subdivision of the Silurian.

Estonia; this age is interpreted as a minimum for the Tremadoc–Arenig boundary but needs further analytical confirmation according to G. S. Odin. A rather poorly defined K–Ar date of 488 ± 28 Ma for glaucony from a basal Arenig formation at Stora Stolan, Sweden (NDS 132), is not in disagreement with this estimate. Rather better information comes from K–Ar, Pb–Pb and U–Pb dates for minerals from rocks from the Ballantrae ophiolite in Scotland (NDS 134) which suggest that the *Didymograptus nitidus* zone of the Lower Arenig (Figure 1) should be dated at 478 ± 8 Ma (Figure

2). Taken together, the new data suggest that the Tremadoc–Arenig boundary must be at about $475 \pm ^{10}_{5}$ Ma; the upper limit is based on the age maxima of 486 Ma (NDS 134) and 487 Ma (NDS 122), the lower limit on the age minimum of 470 Ma (NDS 134).

It remains true that more accurate data for the Cambrian–Ordovician boundary are desirable. The dates for the Bay of Islands ophiolite, the Colmonell gabbro and the Hare Bay ophiolite (respectively items 28, 30 and 32 in Gale *et al.*, 1980) do not help the discussion, all being stratigraphically ill-defined. The Rb–Sr isochron dates quoted by Sturt *et al.* (1975) for shales in a Caledonian Nappe sequence in Finmark, Norway, suggested to the authors that the base of the Tremadoc should be placed in excess of 504 ± 7 Ma (ICC), which does not conflict with the more firmly established figure of $495 \pm ^{10}_{5}$ Ma suggested here on the basis of items NDS 130 and NDS 150 and the data from Finmark.

2.5 The Ordovician–Silurian boundary

There is one modern set of data which, if correct, would necessitate the Ordovician–Silurian boundary being set at greater than 431 ± 7 Ma: this is the $^{40}Ar/^{39}Ar$ age for the Early Llandovery determined by Lanphere *et al.* (1977, and NDS 128). This datum point has been critically discussed by Rundle (1981). The essential point is that the age quoted is for hornblende from a sedimentary breccia at a position 10 m above a graptolitic shale containing a *Monograptus cyphus* fauna (Figure 3). Despite the authors' claim that 'the crystals . . . are autobrecciated juvenile volcanic material' and 'are not considered detrital material' (Lanphere *et al.*, 1977, p. 15), this age can safely be taken only as a *maximum* for the deposit. Moreover, the hornblende diorite (giving an age of 433 ± 7 Ma) is only *claimed* to intrude the 'upper part of the sedimentary sequence' and the field relations only '*suggest* that the diorite is younger than the sedimentary breccia'; clearly it is not proven that the diorite age provides a minimum for the *M. cyphus* zone. Further, these $^{40}Ar/^{39}Ar$ ages were not obtained by the step-heating method so that it is not proven that they did not contain undetected excess argon causing spuriously high ages. A further technical difficulty is that the corrections for Ca-derived isotopes are usually large for hornblende samples, so that any error in them may add a significant bias to the calculated ages. For all these reasons we believe that less importance should be attached to these data compared with the weight of the other modern data which favour a lower date for the Ordovician–Silurian boundary, as shown below.

The most precise radiometric age on Ordovician rocks, which are stratigraphically very closely defined indeed, is the Rb–Sr isochron date of 421 ± 3 Ma for the Stockdale rhyolite of Cautleyan zone 2 of the Ashgill in the English Lake District (NDS 243). This isochron age is interpreted as the age of emplacement of the lava, and not as the age of a hypothetical post-emplacement isotopic homogenization, for reasons which have been thoroughly discussed by Gale *et al.* (1979a; 1980) and by Rundle (1981); it clearly implies that the Ordovician–Silurian boundary is less than 421 ± 5 Ma, when possible systematic errors in the analyses of Rb and Sr concentrations are taken into account. Such an estimate is supported by the Rb–Sr whole-rock isochron date of 408 ± 12 Ma (NDS 239) for the volcanics of the Dunn Point formation, Nova Scotia; these rocks are Late Ordovician to Early Llandovery in age and the data of NDS 239 therefore constrain the Ordovician–Silurian boundary to lie in the interval 396–420 Ma. Further support for the correctness of this conclusion is afforded by the age of 429 ± 8 Ma for the Late Caradoc Eskdale granite (NDS 189) and the age of 438 ± 6 Ma for the Late Llandeilo–Early Caradoc Threlkeld microgranite (NDS 190) and, on the other side of the boundary, by the Rb–Sr isochron age of 408 ± 8 Ma for the Late Llandovery Quoddy formation in Maine (NDS 238, Figure 2).

The interpretation of the date for the Stockdale rhyolite as that of Cautleyan zone 2 in the Ashgill remains crucial to this discussion, and it is important to note that it has now received general support from the dates established by Rundle for the Eskdale (NDS 189) and Threlkeld (NDS 190) *granites*; these dates are all incompatible with the data from Alaska (NDS 128) for the Llandovery. McKerrow *et al.* (1980), whilst noting that the Stockdale rhyolite was not reset where it outcrops within the thermal aureole of the Shap granite, nevertheless suggested that it may have been affected by the *c.* 420 Ma event which produced the Ennerdale granophyre and other minor intrusions (Rundle, 1979). Their arguments, refuted at length by Gale *et al.* (1980), have been further criticized by Rundle (1981), who points out that the Ennerdale granophyre is a hypabyssal intrusion which crops out some 20 km away from the nearest rhyolite exposure and is an altogether unlikely resetting agent. That the rhyolite was not reset by the plutonic Eskdale intrusion has the clear implication that the rhyolite was not extruded until after this granite had cooled, and must therefore be less than 429 ± 4 Ma in age. There seems to be no reason to doubt that 421 ± 3 Ma is the date of extrusion of the Stockdale rhyolite (NDS 243); it was not reset by either the Shap granite or Eskdale plutons and thus it seems extremely unlikely that it was reset by the Ennerdale. We conclude that modern evidence suggests strongly that the Ordovician–Silurian boundary is less than *c.* 420 Ma (NDS 243) and should probably be set in the interval 396–420 Ma (NDS 239). Our final proposal for the date of this boundary is

$418 \pm ^{5}_{10}$ Ma; the upper limit is set because the boundary must be less than the maximum age of 426 Ma from NDS 243, whilst the lower limit is set less certainly by the dates of *c.* 408 Ma for both NDS 238 and NDS 239.

2.6 Subdivisions of the Ordovician system

The minimum age for the Late Tremadoc appears, from K–Ar dating of hornblende from the Rhobell volcanics (NDS 122), to be 475±12 Ma; this estimate receives slight support from the K–Ar age of 467±23 Ma on glaucony from the Late Tremadoc formation in Sweden (NDS 131). Estimates for basal Arenig come from a K–Ar age of 473±17 Ma on glaucony from Estonia (NDS 125) and a K–Ar age of 488±28 for glaucony from Sweden (NDS 132), whilst combined K–Ar, Pb–Pb and U–Pb ages for minerals from rocks from the Ballantrae ophiolite in Scotland (NDS 134) suggest that the *Didymograptus nitidus* zone of the upper Lower Arenig (Figure 1) should be dated at 478±8 Ma. Taken together, the modern data suggest that the Tremadoc–Arenig boundary should be taken at about $475 \pm ^{10}_{5}$ Ma; the upper limit is set by the age maxima of 486 Ma (NDS 134) and 487 Ma (NDS 122), whilst the lower limit is set by the age minimum of 470 Ma (NDS 134), the figure of 475 Ma being largely influenced by the data of NDS 134.

For the *Didymograptus bifidus* zone of the Early Llanvirn (Figure 1), a minimum estimate of 468±10 Ma comes from K–Ar dates on hornblende from the Great Cockup picrite and biotite from the Carrock Fell gabbro, English Lake District (NDS 191). This estimate is not in conflict with the possible maximum of 460–465 Ma for the Early Llandeilo (*Nemograptus gracilis*, Figure 1) which comes from U–Pb zircon analyses and Rb–Sr mineral isochrons for the Bennan conglomerate, Scotland (NDS 135).

Information probably for the Llandeilo, and certainly pre-*Didymograptus clingani* (Figure 1), comes from an unpublished Rb–Sr whole-rock isochron age of 442±12 Ma for the Borrowdale volcanics, English Lake District, mentioned in McKerrow *et al.* (1980). Rundle (NDS 190) argues that the Rb–Sr isochron age of 438±6 Ma for the Late Llandeilo Threlkeld microgranite, Lake District, is a maximum for the Llandeilo–Caradoc boundary, which is consistent with the maximum age of 440–450 Ma for the Early Caradoc suggested by a Rb–Sr isochron age for the Tormitchell conglomerate, Scotland (NDS 135).

Bentonites from the Tyrone limestone, Kentucky, and the Carters limestone, Alabama, are Late Wilderness in age, supposedly correlative with the Harnagian of the Early Caradoc (Figure 1). Earlier data for the Carters limestone PTS 156 when properly recalculated (Gale *et al.*, 1980) yields

435 ± 44 Ma from U–Pb on zircons and 445 ± 50 Ma for Rb–Sr on biotites, dates which are compatible with the modern data of NDS 190 and NDS 135. The recent K–Ar date of 455 ± 10 Ma for biotites and sanidine from bentonites in the Carters limestone (NDS 129) is consistent with the K–Ar date of 443 ± 10 Ma obtained on a biotite from the Tyrone limestone formation (NDS 161), which in its turn agrees within error with the $^{40}Ar/^{39}Ar$ plateau age of 433 ± 5 Ma obtained by Sutter as a mean of two incremental gas release determinations on biotite from the Tyrone limestone (quoted in Gale *et al.*, 1979), but we are left in some doubt where, in the maximum interval 428–465 Ma (433–5 to 455 + 10), we are to place the date of the Wilderness stage. However, only the lower part of this range is compatible with NDS 190, NDS 135 and the date for the Borrowdale volcanics, and the analytical quality of the data proposed by Sutter and NDS 161 seems superior to that of NDS 129. We should in any case bear in mind that biostratigraphic correlation with the European series is not yet definitive; even so, it seems that the Early Caradoc probably falls somewhere in the interval 433–443 Ma. It is noteworthy that the date of 438 ± 6 Ma (NDS 190) for the Late Llandeilo is equal to the mean date of 438 ± 5 Ma for the Upper Wilderness derived from NDS 161 and the data of Sutter; on this basis, we propose that 438 ± 5 Ma should at present be taken as the date for the base of the Caradoc.

Rundle (NDS 189) argues that the Eskdale granite of the English Lake District must have been emplaced during the Late Caradoc tectonism, which is therefore dated by concordant ages using K–Ar on biotites, a Rb–Sr whole-rock isochron and U–Pb on zircons at 429 ± 8 Ma.

In contrast with the uncertainties in dating the other subdivisions, the date of 421 ± 5 Ma for Cautleyan zone 2 of the Ashgill (NDS 243) seems well established.

2.7 Conclusions

For the reasons given above, the base of the Ordovician is to be dated at $495 \pm ^{10}_{5}$ Ma and the top of the Ordovician at $418 \pm ^{5}_{10}$ Ma, so that the duration of this period is about 70–90 Ma. We propose that at present the base of the Arenig is best taken at $475 \pm ^{10}_{5}$ Ma. The base of the Llanvirn must be older than the date of 468 ± 10 Ma for the Great Cockup picrite (NDS 191), which is a minimum for the *D. bifidus* zone of the Llanvirn, and we propose an age of 470 ± 10 Ma for the base of the Llanvirn. Presently available dates do not allow a precise calibration of the base of the Llandeilo, for which we suggest a date of *c.* 455 Ma as the mean of the date of 442 ± 12 Ma for the Borrowdale volcanics and the date of 468 ± 10 Ma

(NDS 191). For the base of the Caradoc we have advanced arguments in Section 2.6 for the acceptance of a date of 438 ± 5 Ma. We propose 425 ± 8 Ma as the base of the Ashgill, as the average of the age minimum of 418 Ma (NDS 143) for Cautleyan zone 2 of the Ashgill and the age maximum of 433 Ma (NDS 189) for the emplacement of the Eskdale granite during the Late Caradoc tectonism in the English Lake District. These suggestions are summarized in Figure 2.

The firmest date for the whole of the Ordovician seems to be 421 ± 5 Ma for Cautleyan zone 2 of the Ashgill. The absolute dating of the subdivisions of the Ordovician system, though still imprecise, does seem to support the suggestion made by McKerrow *et al.* (1980) that the Caradoc has a much longer duration than the Ashgill.

3 THE SILURIAN SYSTEM

3.1 Stratigraphic introduction

The stratigraphic location of the Ordovician–Silurian boundary, not yet defined by international agreement, is discussed by M. Robardet in this book and some comments are made in the section on the Ordovician system. For practical purposes we may usually recognize it in terms of the terminal stage of the Ordovician, the Hirnantian. The Silurian–Devonian boundary has been defined by international agreement as lying at the base of the *Monograptus uniformis* zone. Within the Silurian there are about 20 well-defined, internationally recognizable zones based chiefly upon graptolites, and relative dating within the Silurian is probably better than for any other pre-Cenozoic system (Spjeldnaes, 1978). The Silurian is divided, on the basis of the graptolite zones, into four series, the Llandovery, Wenlock, Ludlow and post-Ludlow pre-Gedinnian (Downtonian or Pridoli)—see Figure 3, which also shows the stages. Further details of Silurian stratigraphy are to be found in Cocks *et al.* (1971).

3.2 Calibration of the Silurian system—general considerations

Lambert (1971, p. 31) suggested that there was then only one reliable date for the Silurian, that for the assumed post-Lower Ludlovian Stromlo volcanics at Mount Painter, near Canberra, but these data must be ignored for the following reasons. Compston (personal communication recorded in Gale *et al.*, 1979) states that the original Mount Painter Rb–Sr whole-rock data must be rejected because the sampling may have mixed different rock units. The Rb–Sr whole-rock isochron age of 431 ± 9 Ma (ICC) calculated for the State Circle shale was thought to date either Lower Ludlow folding or perhaps deposition in the Upper Llandovery; it is not therefore a reliable

GREAT BRITAIN		GRAPTOLITE ZONES	BOHEMIA	
			LOCHKOVIAN	
PRIDOLI	(DOWNTONIAN) ?		PRIDOLI BEDS	BUDNANIAN
LUDLOW	WHITCLIFFIAN	? ?	KOPANINA BEDS	BUDNANIAN
LUDLOW	LEINTWARDINIAN	*leintwardinensis*	KOPANINA BEDS	BUDNANIAN
LUDLOW	BRINGEWOODIAN	*tumescens–incipiens*	KOPANINA BEDS	BUDNANIAN
LUDLOW	ELTONIAN	*scanicus*	KOPANINA BEDS	BUDNANIAN
LUDLOW	ELTONIAN	*nilssoni*	KOPANINA BEDS	BUDNANIAN
WENLOCK	UPPER	*ludensis*	KOPANINA BEDS	BUDNANIAN
WENLOCK	UPPER	*lundgreni*	LITEN BEDS	
WENLOCK	UPPER	*ellesae*	LITEN BEDS	
WENLOCK	MIDDLE	*linnarssoni*	LITEN BEDS	
WENLOCK	MIDDLE	*rigidus*	LITEN BEDS	
WENLOCK	LOWER	*riccartonensis*	LITEN BEDS	
WENLOCK	LOWER	*murchisoni*	LITEN BEDS	
WENLOCK	LOWER	*centrifugus*	LITEN BEDS	
LLANDOVERY	TELYCHIAN	*crenulata*	LITEN BEDS	
LLANDOVERY	TELYCHIAN	*griestoniensis*	LITEN BEDS	
LLANDOVERY	TELYCHIAN	*crispus*	LITEN BEDS	
LLANDOVERY	FRONIAN	*turriculatus*	LITEN BEDS	
LLANDOVERY	FRONIAN	*sedgwickii*	LITEN BEDS	
LLANDOVERY	IDWIAN	*convolutus*	LITEN BEDS	
LLANDOVERY	IDWIAN	*gregarius*	LITEN BEDS	
LLANDOVERY	RHUDDANIAN	*cyphus*	LITEN BEDS	
LLANDOVERY	RHUDDANIAN	*vesiculosus*	LITEN BEDS	
LLANDOVERY	RHUDDANIAN	*acuminatus*	LITEN BEDS	
LLANDOVERY	RHUDDANIAN	*persculptus*	LITEN BEDS	

Figure 3. Stratigraphic subdivision of the Silurian system. The uppermost graptolite zones have not yet been recorded in Great Britain.

calibration point for the Silurian. Biotite Rb–Sr ages for the Willow Bridge tuff of the Stromlo volcanics are 421 ± 4 Ma and 418 ± 6 Ma (Compston, personal communication), but the stratigraphy is still porly understood (Talent *et al.*, 1975).

Further confusion was caused by the Rb–Sr whole-rock isochron date of 438 ± 6 Ma (ICC) reported by Gee and Wilson (1974) for the Vilasund granite, Sweden, originally thought to be post-Upper Llandovery in age. However, it is now known that an important conjunctive tectonic discontinuity separates the unfossiliferous metasediments of the Vilasund envelope from the fossiliferous rocks which had been thought to establish the stratigraphy, and the authors have now withdrawn the Vilasund granite as a time-scale calibration point (Gee and Wilson, 1977). The review by Spjeldnaes (1978) showed that later radiometric dates had still left the dating of the Silurian in some confusion, with estimates of its duration varying from 15 to 40 Ma. He emphasized also that many radiometric dates for the Silurian were for metamorphic rocks (Dewey *et al.*, 1970; Naylor, 1971; Reynolds *et al.*, 1973) and mineral veins (Mitchell and Ineson, 1975) of imprecise stratigraphy and therefore of little value for time-scale calibration. Recently measured radiometric ages, discussed below, have but slightly improved this situation.

3.3 The Ordovician–Silurian and Silurian–Devonian boundaries

The modern data for the Ordovician–Silurian boundary were extensively discussed in Section 2.5, and it was concluded that this boundary should, on present evidence, be dated at $418 \pm^{5}_{10}$ Ma.

Support for this estimate comes from NDS 189 and NDS 190; it should be noted that the majority of the data used to establish this boundary come from the English Lake District with the exception of NDS 239, which is for the Arisaig volcanics of Nova Scotia.

The calibration of the Silurian–Devonian boundary is extensively discussed in the review of the data for the Devonian system; the conclusion drawn from all the data taken together is that this boundary tentatively be set at $400 \pm^{10}_{5}$ Ma until more radiometric dates are available for units stratigraphically located below the boundary. At present such Silurian data are limited to the Late Llandovery Quoddy formation, yielding ages of 408 ± 40 Ma (NDS 220, old measurements) and 408 ± 8 Ma (NDS 238, new measurements), perhaps the post-mid-Ludlovian, pre-Early to Middle Siegennian Gocup granite, yielding ages of 409 ± 3 Ma and 402 ± 3 Ma (NDS 210), and the very imprecise old data for the Llandovery to Wenlock Dennys volcanics (NDS 221). We exclude from consideration the data from Esquibel Island (NDS 128) for the reasons given at length in Section 2.5.

It is rather disappointing to have to report that even now there is essentially only one reliable radiometric age (NDS 238) for rocks which are undoubtedly Silurian, so that the only calibration inside the Silurian system is 408 ± 8 Ma for the Late Llandovery. The upper and lower boundaries imply that the Silurian is somewhere between −2 and +28 Ma long (though most probably about 10–18 Ma long), hardly a satisfactory conclusion to reach 18 years after the symposium of the Geological Societies of London and Glasgow (1964). As Spjeldnaes (1978) has pointed out, this situation could probably be changed dramatically if dating were attempted of bentonite horizons near the Llandovery–Wenlock boundary (especially in Scandinavia and the Baltic region) and in the Lower Ludlow (in Scandinavia, Poland and Podolia).

4 GENERAL CONCLUSIONS

It is very clear that there are still far too few reliable and accurate radiometric dates of unambiguous stratigraphy for the numerical calibration of Caledonian times, and that the situation is particularly acute for the Cambrian and the Silurian. Consequently the recommendations made above for the stage boundaries are merely the best that can be made at present, and it is hoped that this review will encourage stratigraphers and geochronologists to collaborate in producing more data so as to allow a significant improvement in the certainty of the calibration. Figure 4 presents a comparison between some of the time scales proposed since 1964 for the

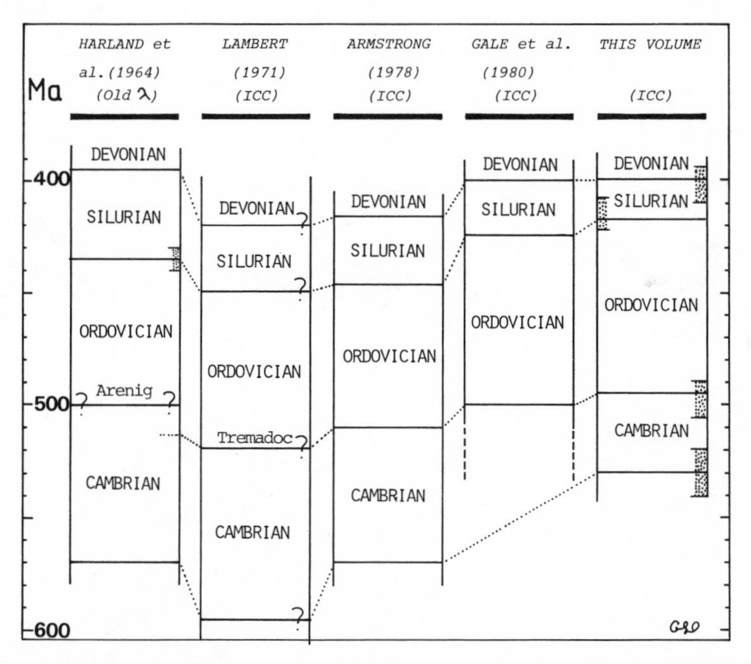

Figure 4. Comparison of time scales proposed since 1964. Except for the scale given by Harland *et al.*, for which the original values are given, all the scales were recalculated, or calculated, using the new decay constants. Note that since 1971 the accumulation of new dates has tended systematically to yield younger estimates for the boundaries. (Shaded intervals on the time axis show estimated uncertainties in the calibration of the boundaries.)

boundaries between the Cambrian, Ordovician, Silurian and Devonian periods which shows that since 1971 the accumulation of new dates has tended systematically to yield younger estimates for those boundaries.

Acknowledgements

I should like to thank G. S. Odin very much for his valuable criticism of various drafts of this review, my colleagues R. D. Beckinsale and A. J. Wadge, and those colleagues who supplied abstracts or notice of their work prior to publication.

(Manuscript received 8-4-1981)

Numerical Dating in Stratigraphy
Edited by G. S. Odin

26
Numerical dating of Hercynian times (Devonian to Permian)

GILLES S. ODIN and NOËL H. GALE

Since 1971, the Devonian to Permian time scale has not been properly discussed using all the available stratigraphical, geochemical and analytical data together. We believe that such a synthesis is needed in order to obtain a proper understanding of the question. The review by Waterhouse (1978) on the Permian time scale is probably the best available, but in the absence of analytical details one cannot properly judge the merits of the conclusions which he proposed.

The proposals below try to take into account the most recent information available, details of which are given in Section II of this book by the authors of the dates themselves.

1 THE DEVONIAN SYSTEM

1.1 Introduction

The Devonian system is situated between the Caledonian times below (Cambrian to Silurian) and the Hercynian times above (Carboniferous and Permian). This system and its related stages were defined in NW Europe as discussed by House *et al.* (1977). In this area the palaeogeography was such that two kinds of facies were developed. In the northern part the Caledonides are exposed and the *Old Red Sandstones* are deposited on the north and centre of Great Britain and the Baltic platform above a transitional stage: the Downtonian. Further to the south marine sediments are deposited especially in the Devonshire, type area of the system, in the Ardenne and the 'Massif Shisteux Rhénan', type area of the usually accepted stages (France–Belgium–Germany boundary), and in Bohemia (western Czechoslovakia).

The correlations between the Old Red Sandstone levels and the classic

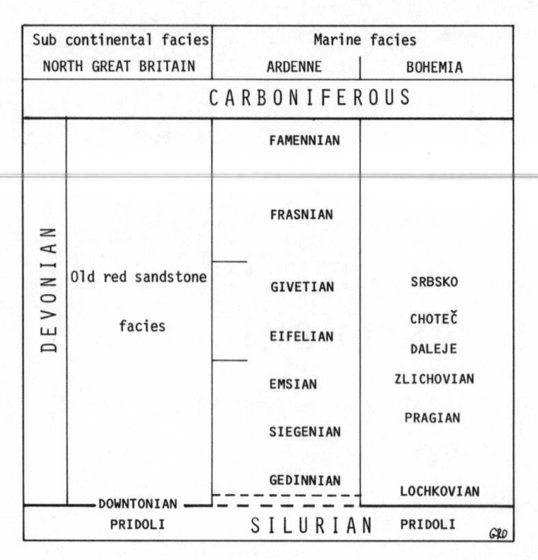

Figure 1. Stratigraphic subdivision of the Devonian. Note that the correspondence between the stages defined in the Ardenne and in Bohemia is not definite.

marine stages remain open to debate. Moreover, although the succession of stages is generally accepted (Figure 1), the exact position of each stage boundary remains open. At the moment, the situation is evolving due to the recent change in the reference fossil groups from macrofaunas—trilobites, brachiopods—to microfaunas—conodonts. The lower boundary of the system is now defined in a reference section in Bohemia at the appearance of the graptolite species *Monograptus uniformis.* This limit also defines the base of the Lochkovian local stage. The section is situated nearly 30 km SW of Prague; the Pridoli stage is also defined in the same area, about 5 km away from the same city. The correlation between the Lochkovian and the Gedinnian is rather uncertain. The top of this system which is the base of the Carboniferous, remains to be accurately defined (see Paproth, this book).

As a result, one must remember that accurate correlation between the radiometrically dated levels in the USA or Australia and the stages defined in the Ardenne may only be tentative. At the moment, not one Devonian level of the Ardenne has been radiometrically dated.

1.2 Earlier evaluations

In 1964, the proposed time scale of Friend and House was based on three dates for the top of the system and six for the base.

At the top, two dates were obtained from the stratigraphically well correlated Chattanooga shale, Tennessee, USA, PTS 2 (mid-Famennian) and PTS 94 (Late Frasnian). Unfortunately the original data were given without reference to the decay constants used (^{238}U and ^{40}K), which greatly diminishes the value of these two dates of around 250 Ma. The third calibration point was the 355 Ma old Snobs Creek rhyolite from Australia (PTS 95), which was reinvestigated later (PTS 354). Consequently the accepted age of 352 Ma (ICC recalculated) will be reevaluated below.

At the base of the Devonian system the Hecla Hoek micaschists of Late Downtonian age from Spitsbergen, possibly Gedinnian in age, always appeared to be a reliable datum (PTS 4) although the K–Ar dates of 392–397 and 433 Ma are rather scattered. The Rb–Sr ages are more concordant and average 393 Ma. The pre-Early Gedinnian Snowy River granite from Australia gives a K–Ar/biotite age of 402 Ma (PTS 97) which conforms the preceding dates. Of the remaining four data used, two appear rather badly situated in the stratigraphic column: the Calais granite (PTS 5) and the British Caledonian metamorphism (PTS 92). They will no longer be used here. The final Caledonian granites from Scotland (PTS 6 = Shap granite and PTS 93 = Creetown granite) have been reinvestigated since these early studies.

As a result, the Silurian–Devonian boundary was already fairly well documented in 1964; the age proposed by Friend and House, 403 Ma (ICC recalculated) for the Ludlow Bone beds, base of the Devonian, appears therefore to be still acceptable.

Between these ages of 403 and 352 Ma, the available data were so scarce that the authors proposed to use the interpolation system based on maximum known thicknesses of sediments to evaluate the age of the internal subdivisions. It has become clear that this is a most unreliable approach, and it is one which we reject here.

In 1971, Lambert proposed a revision of the age of the top of the system according to the new dates obtained from the Cerberean volcanics by McDougall *et al.* (1966 = PTS 354). A rounded figure of 360 Ma was proposed at that time by Lambert; it may be recalculated at nearly 363–370 Ma (Gale *et al.*, 1980).

1.3 Presently available data

1.3.1 The base of the Devonian system

The data discussed or rediscussed in this volume essentially complete the former evidence (see Figure 2). The Shap granite from Scotland (PTS 6) was reinvestigated in London and Oxford together with the Skiddaw granite (NDS 192). Their stratigraphical location is more accurately affirmed than that of the Creetown granite from the same area (PTS 93). Their mean K–Ar apparent age of 398 ± 8 Ma may be regarded as characteristic of

Gedinnian times. The Rb–Sr whole-rock isochron and Rb–Sr biotite data for the Shap granite (NDS 241) confirm this in giving a date of 394 ± 3 Ma for the intrusion of the granite.

The Hedgehog volcanics of the Late Gedinnian gave a recalculated Rb–Sr isochron age of 406 ± 10 Ma (PTS 355 → NDS 223). The reevaluation of the stratigraphic age of the volcanic rocks formerly thought to be from the Pembroke formation leads one to conclude that near Eastport, Maine, USA, all the dated rocks are inside the Eastport formation (NDS 222). The recalculated Rb–Sr isochron ages are 401 ± 12 Ma and 397 ± 24 Ma. New high-precision Rb–Sr data for the Pembroke formation (NDS 237) yield 393 ± 7 Ma as a minimum age for the base of the Gedinnian. Formations of stratigraphical age equivalent to that of the Eastport formation are intruded in SW Brunswick (eastern Canada) by the St George pluton and the Red Beach granite in Maine, USA. They give recalculated Rb–Sr and K–Ar ages around 390–405 Ma, which may be regarded as post-Early Gedinnian (PTS 5 → NDS 236). The Gocup granite from New South Wales, Australia, is post-mid Ludlovian and pre-Early to Middle Siegenian. The K–Ar mean age of 409 ± 3 Ma and the Rb–Sr isochron age of 402 ± 3 Ma obtained correspond to the Pridoli or Gedinnian stages. In Nova Scotia, Reynolds *et al.* (1973) have shown that the probably mid-Early Devonian Torbrook fm gives some ages related to a metamorphism at 398 Ma, ICC, and less, but these data cannot contribute too much to a definitive location of the boundary. The Torbrook fm is considered of Emsian age by Clarke and Halliday (1980). This formation is intruded by the South Mountain batholith, which is in turn unconformably overlain by Early Carboniferous (Tournaisian?) sediments. Whole-rock, muscovite, biotite Rb–Sr ages of 372 ± 2 (2σ) Ma, ICC (granodiorite first intrusion), and 361 ± 2 (2σ) Ma, ICC (late intrusion of aplites and porphyries), were obtained by these authors. The former age is a minimum one for Emsian age, the latter a maximum one for the probably Tournaisian overlying sediments.

According to all the dates available, the Silurian–Devonian boundary is older than 395 Ma and younger than 410 Ma. The age of $400 \pm ^{10}_{5}$ Ma will be tentatively accepted until there are more radiometric data for units stratigraphically located below the boundary (see Figure 2).

1.3.2 The top of the Devonian system

The stratigraphic location of the Cerberean volcanic rocks, Victoria, SE Australia, has been reassessed by Talent *et al.* (1975) who assigned these rocks to the Frasnian–Famennian and by Marsden (1976) who proposed a Frasnian age. The precedingly available data recalculated and *corrected* to 369 Ma, ICC (PTS 354 and J. R. Richards, personal communication, 1981), are therefore regarded as characteristic of a non-terminal Devonian time: probably Frasnian. A new, more accurate K–Ar analytical result of 367 ± 1 Ma was also obtained (NDS 234; note that the range given only refers to the short-term reproducibility: $2\sigma m$). This date, supported by others presented in NDS 234, defines an upper limit for the age of the Devonian–Carboniferous boundary.

The post-cauldron intrusions are covered by a sedimentary sequence of Early Carboniferous age. Seven intrusions were dated and gave very concordant K–Ar ages, the mean of which is near 362 Ma (NDS 235). This age may be regarded locally as a *maximum value for the Early Carboniferous* sedimentary sequence.

In Scotland, the Old Red Sandstones are covered by the Birrenswark and Kelso lavas. These lavas are covered by Early Tournaisian sediments; their maximum age is not controlled but one must remember that the ORS facies may be as young as earliest Carboniferous. The K–Ar/whole-rock basalt age of 361 ± 7 Ma is therefore a maximum age for the Early Tournaisian sediments; it is possibly itself Early Tournaisian, perhaps older (NDS 165).

In the Massif Armoricain, France, the Huelgoat granite was intruded during a plutonic event which affects older rocks including Late Famennian sediments. The overlying Strunian sediments (Early Tournaisian) are not metamorphosed. A seven-sample Rb–Sr whole-rock isochron age of 336 ± 13 Ma was obtained by R. Charlot (NDS 229). If confirmed (see also NDS 133), this age would reject the dates from the Scottish lavas (NDS 165) low in the Famennian. This age also appears younger than the Scottish lavas regarded as Early Visean, which gave K–Ar ages of nearly 353 ± 7 Ma (PTS 360–5 and NDS 166). Also, the stratigraphically younger Tournaisian–Visean Texan glauconies appear to have radiometrically older Rb–Sr ages (NDS 152–153). The much younger K–Ar ages on the same material (Morton and Long, 1980) cast some doubt on the complete closure of the dated chronometer during the time since deposition.

The often quoted dates on the Fisset Brook formation from Nova Scotia (Cormier and Kelly, 1964) have much too uncertain a stratigraphy and cannot help in clarifying the situation. Some of the acid volcanic rocks from Portugal measured by Hamet and Delcey (1971) are well bracketed between sediments with Famennian and Late Tournaisian conodonts. But according to one of the authors (J. Hamet, personal communication, 1980), the aim of the isotopic study was only to ascertain approximately the origin and verify the synchronism of the materials collected in various localities over a 600 km^2 area. The accuracy of the measurements was insufficient for establishing a time-scale calibration point; according to the author, the ICC recalculated Rb–Sr isochron age of $370 \pm n$ should be discarded.

As a result it seems to us that the date of 365 Ma is a maximum for the base of the Carboniferous system. The maximum age of the Huelgoat granite corresponds to the ages found for the Early Visean Scottish lavas; we propose this age of 350 Ma as a minimum age for the base of the Carboniferous system. The range $360 \pm {}^{5}_{10}$ Ma has the greatest likelihood of including the Devonian–Carboniferous system boundary according to the above dates, but will be rediscussed below to take into account radiometric dates obtained in Russia.

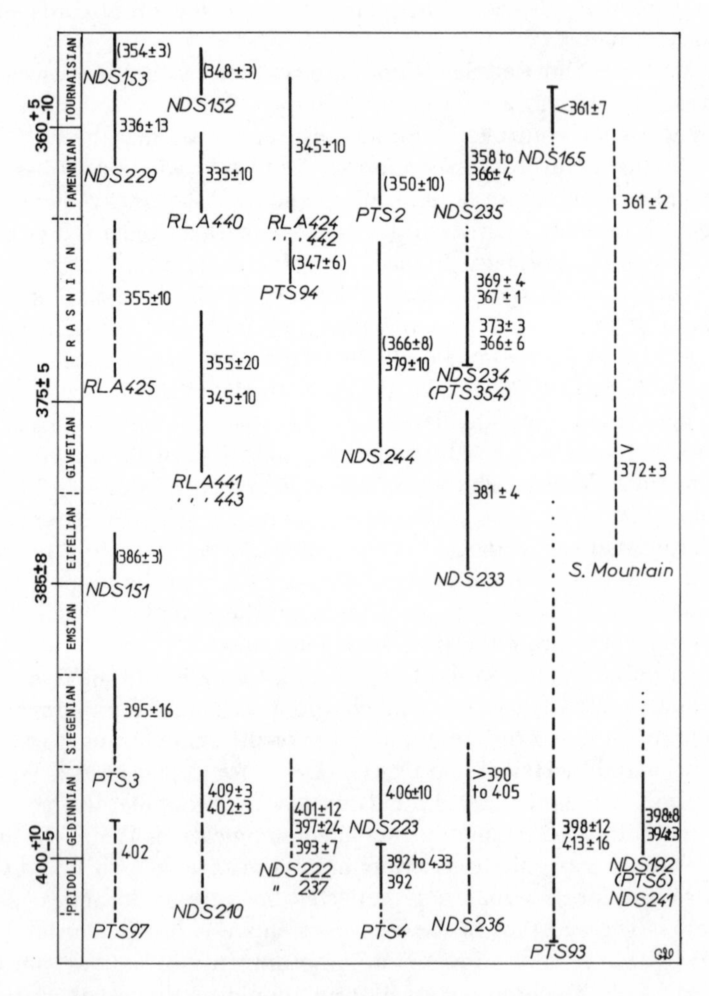

Figure 2. Presently available dates from well-identified Devonian samples (abbreviations as for Figure 4).

1.3.3 Subdivisions of the Devonian system

The presently available radiometric dates do not allow the proposal of accurate ages for each of the seven stages which subdivide the system, especially since the subsystem boundaries themselves are poorly defined.

The Early–Middle Devonian boundary may, however, be estimated younger than 390–405 Ma, the age of the post-Early Gedinnian intrusions from North America (NDS 236). The Siegenian Shiphead bentonites from Canada gave five rather scattered ages which average 395 ± 16 Ma, ICC (PTS 3; Gale *et al.*, 1980).

Above the Early–Middle Devonian boundary, the post-Emsian, pre-Frasnian Mt Buller granodiorite from Victoria, SE Australia, gave a K–Ar weighted mean age of 381 ± 4 Ma (NDS 233). In Texas, USA, glaucony from the Stribling formation, Early Couvinian in age (probably base of the Eifelian or slightly older), shows a Rb–Sr selected age of 386 ± 3 Ma (NDS 151) which may be accepted as the age of the studied boundary if there is no open-system behaviour involved.

According to Fullagar (NDS 224–225), the dates from the Kineo and Travellers volcanic rocks from Maine, USA, must be ignored for time-scale calibration purpose. In summary, an age of 385 ± 8 Ma may be proposed for the base of the Middle Devonian subsystem; the number is rounded according to NDS 151 and the range is proposed to include the minimum age of NDS 233 above and the data of NDS 236 and PTS 5 below. The post-Givetian Hoy lavas from Scotland yielded a 39/40 Ar age *corrected* to 379 ± 10 Ma (NDS 244) but the stratigraphy is rather uncertain. All the dates from Australia (NDS 234), the oldest of which are around 368–373 Ma, postdate the boundary. We may thus propose the age of 375 ± 5 Ma as the age of that boundary, halfway from the dates of NDS 233 (381 Ma) and NDS 234 (≃370 Ma).

Six ages from the USSR are also used by Armstrong (1978) according to the summary published by Afanass'yev (1970). Unfortunately very few details are available from the latter publication and we consider that, at the moment, these dates cannot be given much weight for time-scale calibration. However, these internally consistent results seem to support the datum from the Huelgoat granite leading to a younger Devonian–Carboniferous boundary, in contrast to what seems implied by the data from the Scottish lavas. To consider these Russian dates in detail, the Givetian–Frasnian K–Ar dates obtained from the Dahut river amphibole and Ugum Range pebbles, North Caucasus, USSR, are respectively nearly 355 ± 20 Ma, ICC (RLA 441), and 345 ± 10 Ma, ICC (RLA 443); the latter is in contradiction with the Frasnian dates from Australia. The Frasnian glaucony from the Kurskiy district, USSR, gives a recalculated age of nearly 355 ± 10 Ma (RLA 425 on Figure

2). Two equal dates were obtained for Central Kazakhstan granites; the second one was measured by Shanin in Moscow, in a laboratory which may be considered as analytically reliable on the basis of the data obtained on standard minerals. This granite of the Sady Adyr massif intrudes sediments with remains of Frasnian fishes and flora, and is overlain by limestone with Tournaisian brachiopods. Two dates of 345 ± 10 Ma, ICC, were obtained from these two granites; no details are available as to the minerals dated, but if biotite was used the reported potassium content of 2.18% appears low (Afanass'yev, 1970; RLA 424 and 442). Finally, the age of 335 ± 10 Ma, ICC, obtained on a Famennian subalkaline basalt from the Aksahut river , North Caucasus, USSR (Afanass'yev, 1970; RLA 440), also appears much younger than the other available dates except for the Huelgoat granite and other Russian dates. From the evidence of Late Devonian rocks it appears, however, that the Frasnian–Famennian stage boundary may be located near the top rather than near the base of this subsystem.

1.4 Conclusions

The Devonian system is nearly 40 million years long according to the numbers accepted here for the boundaries; however, the remaining uncertainties show that it may well extend for a longer duration at both sides, and accordingly we will propse a length of 45 ± 10 Ma.

The lower boundary may be accepted at $400 \pm ^{10}_{5}$ Ma; the upper limit of the accepted age has a larger uncertainty because this age is calibrated mainly with samples lying above the boundary.

The Middle Devonian boundaries may tentatively be suggested to lie at 385 ± 8 Ma at the base and 375 ± 5 Ma at the top.

The balance of evidence leads us to prefer a Devonian–Carboniferous boundary at 360 Ma using the Australian and Scottish dates. However, an age as young as 345 Ma is indicated by the dates obtained from the southern USSR and the Massif Armoricain (France). Since the Russian and French data are not yet sufficiently well documented, it appears to us that the figure $360 \pm ^{5}_{10}$ Ma embraces the various possibilities discussed above. The Frasnian–Famennian boundary probably lies nearer the top than the base of the subsystem. For the time being, its age cannot be fixed due to the scattering of available dates above, 336–361 Ma, and below, 368–345 Ma.

In conclusion, we emphasize that most results have been obtained from the continental Old Red Sandstones province (Scotland, Spitzbergen), eastern Australia, the Appalachian mountains from NW America or the south of the USSR. None of these dates are from the marine NW European province where the stages were defined; this partly explains the inaccurate figures we are able to propose in this review. For the future we recommend that datable material should be sought in the Ardennes, Bohemia or related areas the more effectively to refine the interim numerical time scale proposed here.

2 THE CARBONIFEROUS TIME SCALE

The time scale for the Carboniferous system was presented above by Hugh de Souza and we accept his arguments that the existing data favour the numbers of 360 and 290 Ma as extreme possible ages of the system. However, in view of the quite numerous younger dates obtained for the base and of the remaining stratigraphical problems concerning the definition of the Devonian–Carboniferous boundary (see Paproth, this book) we prefer to keep the question open in the following manner.

The boundary is certainly younger than 365 Ma and probably older than 345 Ma; it is possibly to be taken on the high side according to NDS 165. But the coincidence between the ages of the Russian late Devonian rocks (335 ± 10 to 355 ± 20 Ma) and the date from the Latest Devonian earliest Carboniferous Huelgoat granite (NDS 229): 336 ± 13 Ma might lead to an age younger than 350 Ma. Moreover the samples from the Chateaulin and Laval Basins (NDS 133): 342 ± 13 and 331 ± 6 Ma also indicate a younger age.

Finally, according to J. R. Richards (written communication, June 1981) 'The Victorian data would suggest rather less than 360 Ma for the Devonian–Carboniferous boundary'. The granites (post Cauldron intrusions, NDS 235) are 360 to 365 Ma old and are stratigraphically clearly younger than the volcanics quoted in the abstract NDS 234. Richards also emphasizes that 'following the granites there is quite a thickness of sediment before we come to indubitably Carboniferous fishes (the Broken River fauna). This is why Marsden (1976) proposed as young as 345 Ma for the boundary although 350–355 Ma might be correct'.

The boundary between the Dinantian–Silesian subsystems cannot properly be estimated from the available radiometric dates. From the few presently available dates it appears that the extreme values of 315 and 330 may reasonably be proposed so that the two subsystems are of equal duration. The earlier proposal of nearly 20 Ma for the Dinantian and nearly 45 Ma for the Silesian is no longer acceptable.

3 THE PERMIAN TIME SCALE

3.1 Stratigraphic presentation

The stage succession of the Permian system has been successively established in different environments of deposition. In Western Europe, and especially in France and Great Britain, the economic interest of the Carboniferous system has led to the definition of *continental stages* to which equivalent continental Permian stages succeed: the Autunian (from the Massif Central, France), Saxonian and Thuringian (from Central Europe) are regionally considered as the Early, Middle and Late Permian respectively. The lowest boundary, the Carboniferous–Permian (i.e. Stephanian–Autunian), remains inaccurately located; it appears that in some sections the top of the levels regarded as Stephanian are relevant to the Permian system elsewhere.

In northern Germany, the partly marine Permian was divided in two parts, the Rotliegende (a red sandstone facies) and the Zechstein (an evaporite facies): the Early and Late Permian respectively.

The *marine facies* were only well developed further east of Europe. In the Moscow area and the Urals, many stratotypes have been defined or re-

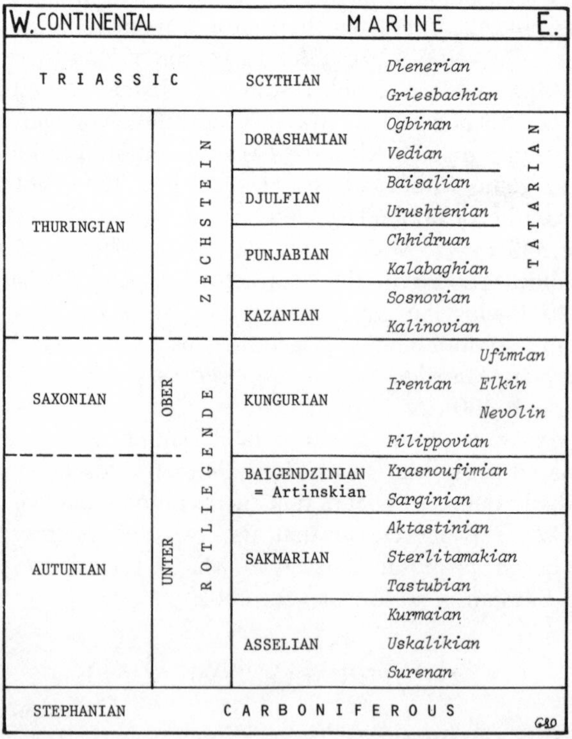

W. CONTINENTAL			MARINE		E.
TRIASSIC			SCYTHIAN	*Dienerian* *Griesbachian*	
THURINGIAN		ZECHSTEIN	DORASHAMIAN	*Ogbinan* *Vedian*	TATARIAN
			DJULFIAN	*Baisalian* *Urushtenian*	
			PUNJABIAN	*Chhidruan* *Kalabaghian*	
			KAZANIAN	*Sosnovian* *Kalinovian*	
SAXONIAN	OBER	ROTLIEGENDE	KUNGURIAN	*Ufimian* *Irenian* *Elkin* *Nevolin* *Filippovian*	
AUTUNIAN	UNTER		BAIGENDZINIAN = Artinskian	*Krasnoufimian* *Sarginian*	
			SAKMARIAN	*Aktastinian* *Sterlitamakian* *Tastubian*	
			ASSELIAN	*Kurmaian* *Uskalikian* *Surenan*	
STEPHANIAN			CARBONIFEROUS		

G80

Figure 3. Stratigraphic subdivisions of the Permian. Note that the Griesbachian and Dienerian are included in the Triassic system according to the position of Zapfe (this volume). Another proposal by Waterhouse (1978) includes both substages in the Dorashamian.

defined recently for the early and middle part of the system. The top of the system includes new stages defined in Pakistan and Armenia. We present in Figure 3 the stratigraphic succession used by Waterhouse (1978), which accepts eight stages. This author also adopts a subdivision in substages; within the latest Permian these substages are very short (less than 1 Ma long) and possibly unreliable. Note that the Griesbachian is included in the Triassic, as explained in the chapter by H. Zapfe (this book).

There is general agreement for regarding as Early Permian all the deposits from the base of the system to the top of the Baigendzinian (ex-Artinskian). The question of the definition of the Middle and Late Permian remains open. Waterhouse (1978) utilizes the expression Late Permian for the Dorashamian stage alone. This appears to us too restricted both because to separate one stage from eight does not represent a useful subdivision and because radiometric dates show that this stage is probably very short.

On the other hand, considering its possible equivalence with the Saxonian, the Kungurian may be accepted to be the Middle Permian, thus the Kazanian + Tatarian would be equivalent to a Late Permian itself equivalent to the Zechstein and the Thuringian. Because of this undefined nomenclature we will not use the expression Middle and Late Permian in the following paragraphs.

3.2 The base of the Permian system

In 1964, three internally consistent dates were selected by D. B. Smith to define the age of the base of the system: the Stephanian C Brassac tuffs from France (PTS 63: 297 ± 7 Ma, ICC), the post-Stephanian B Castro Daire granite from Portugal (PTS 122: 291 ± 4 Ma, ICC) and the Sande essexite from Norway, supposed to be Early Permian in age (PTS 192: 290 Ma, ICC). According to one of the authors of the measurements (M. G. Bonhomme, personal communication, 1981), the data actually available permit only the calculation of 'conventional' ages using an *assumed initial isotopic ratio* of $^{87}Sr/^{86}Sr = 0.712$. From this hypothesis, the recalculated ages are as follows, PTS 63: 297 ± 22 Ma, ICC; PTS 122: 291 ± 22 Ma, ICC. Other maximum ages are now available from Kazakhstan and France. In Kazakhstan, the Taskuduk acid lavas are of Late Stephanian, earliest Autunian age or possibly Early Stephanian; whole-rock K–Ar dates average 302 ± 15 Ma, ICC (PTS 340). In the Massif Central, Provence and the Pyrénées, three Stephanian samples were dated around 302 ± 5 Ma (NDS 231). The granite from the Akchatau complex (Kazakhstan) is attributed to a Late Carboniferous–Permian age by Afanass'yev (1970). U–Th, Pb and Rb–Sr dating (but not K–Ar ages as quoted by Armstrong, 1978) gave ages of 285–293 Ma, ICC, but no details are available (RLA 420–421–422, see Figure 4).

Above the boundary, a rhyolite tuff of Autunian I age from southern France gave a K–Ar age of 286±5 Ma, ICC (NDS 231). The recent results from de Souza (NDS 169) on the Early Permian Mauchline volcanics from Scotland (278–291 Ma, ICC) agree with and confirm the above results. As a result, the age of *290 Ma* is probably that of the studied boundary with a low uncertainty; a minimum age of 285 Ma and a maximum age of 300 Ma can be assumed from the above dates.

In his synthesis, Waterhouse (1978) also quoted dates from Australian rocks well correlated with the stratigraphic sequence. Figure 4 refers to W3: the pre-W2 Bulgonunna volcanics and associated intrusives, W5: the probably Early Permian Nychum volcanics, W6: the post-W5 Featherbed volcanics (Waterhouse, personal communications, 1980–1981), but in the absence of the availability of detailed analytical results these dates only show that our presently proposed age of 290 Ma is compatible with the dates from Australia.

3.3 The subdivisions of the Permian

In our opinion no one set of data may definitely help in fixing any stage boundary inside the Permian at the moment. However, some tendencies may be emphasized.

In North Caucasus, an Early Permian granodiorite gives an age of 265 Ma (RLA 435). The post-Early Permian age of 282±10 Ma, ICC (PTS 344), obtained from the USSR is roughly compatible with the two dates from Australia: (W2=Lizzie Creek volcanics and W4=Ridgelands granodiorite) for the Sakmarian. Finally, the probably post-mid Early Permian age of 269–278 Ma for the Oslo series (NDS 240) leads us to propose as the upper limit of the Early Permian an age of 265±10 Ma: younger or equal to all available data below, and older than the scarce data situated above. The Sakmarian boundaries may be proposed at $280\pm^{8}_{5}$ Ma for the base (*W2*, W3, NDS 169, *NDS 240*, PTS 341 and 344) and at 273±10 Ma for the top (NDS 240). The ages of 247±5 and 245±5 Ma obtained for the Autunian–Saxonian boundary from France (Fuchs *et al.*, 1970) appear much younger than may be accepted when compared with the other available data.

For the Kungurian stage, the age of the Salikamsk sylvinite may only be regarded as a minimum age due to the probability of recrystallization (PTS 53). If the base of the Zechstein is really equivalent to the base of the Kazanian stage, then this boundary is inaccurately approached using the data proposed by Kaemmel *et al.* (1970). According to this author the pre-Zechstein rhyolites from Querz (East Germany) yielded a minimum age of 258±19 Ma, ICC (K–Ar dates on 14 samples), and a maximum one of 269±16 Ma, ICC (on four samples). However, the high range of the dates probably indicates a low reliability. The original stratigraphical and

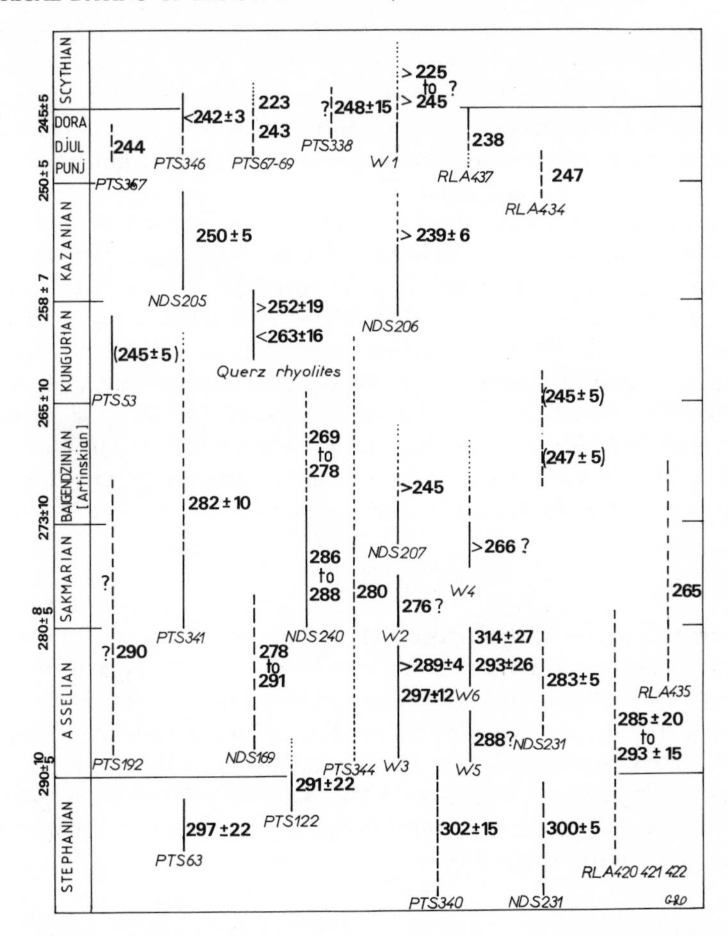

Figure 4. Presently available dates from well-identified Permian samples. The abbreviation PTS refers to the Geological Society of London Phanerozoic time scale (1964–1971); RLA refers to the dates computed by Armstrong (1978); NDS refers to the samples analysed in this volume, Section II; W refers to samples quoted by Waterhouse (1978).

radiometric ages of the Berkeley and Bombo latites (PTS 68) have been reassessed. Initially Kungurian in age, the Berkeley latite is now regarded as *above* a Kazanian fauna (J. B. Waterhouse, 1976, and personal communication, 1981). The dates recently recalculated from the original charts of the ANU (J. R. Richards, personal communication, 1981) yielded 222 ± 11 Ma for the Berkeley latite and 252 ± 4 Ma, ICC, for the Bombo latite. This last age agrees with the original publication as well as with NDS 205 and 206. The former needs confirmation using modern techniques.

From the synthesis of Afanass'yev (1970) one may quote an Aksahut river basalt said to be from the earliest Late Permian; the K–Ar age of 247 Ma, ICC, may be related to the base of the Tatarian (RLA 434). Taking into account the dates discussed to define further the base of the Triassic (see Odin and Létolle, this volume), the boundaries of the Kazanian stages may be proposed at 258 ± 7 and 250 ± 5 Ma.

The whole Tatarian superstage can definitely not be subdivided because all the data available for it are roughly between 245 and 250 Ma; we are below the limit of precision of the analytical data. To the dates discussed with the Triassic system, we have added here a pre-Triassic age quoted from Afanass'yev (1970): RLA 437. As for all the data taken from this author, insufficient analytical details are available. The same problem occurs with the Australian post-Chhidruan intrusive ages of 245–225 Ma quoted by Waterhouse: W1. Finally, the Late Tatarian or post-Tatarian granites described in PTS 67 and 69 gave ages compatible with the other dates around the Permian–Triassic boundary, which is rather well situated at 245 ± 5 Ma.

3.4 Conclusion

To summarize, we will propose that the Permian system is 45 Ma long. Although many uncertainties remain, the first five stages appear of equivalent duration, 5–10 Ma, the three younger being quite certainly very short. The boundaries of the stages, within the Baigendzinian to Kungurian period, are in fact very poorly documented. The top and the base of the system are well known at $290 \pm ^{10}_{5}$ and 245 ± 5 Ma. The use of the expression Late Permian, proposed for the Dorashamian stage alone, appears inappropriate as far as the duration factor is concerned.

Acknowledgements

We are very grateful to those colleagues who provided us with abstracts or helped us in writing these abstracts according to their published or unpublished results, and especially J. R. Richards, H. A. F. de Souza, J. B. Waterhouse and J. A. Webb. This chapter is a contribution to IGCP Project 133.

(Manuscript received 2-4-1981)

Numerical Dating in Stratigraphy
Edited by G. S. Odin

27
A calibration point in the Late Triassic: the tin granites of Bangka and Belitung, Indonesia

HARRY N. A. PRIEM and EWOUD H. BON

1 INTRODUCTION

On the island of Bangka, Indonesia (Figure 1), a sequence of low-grade metasediments has been intruded and contact metamorphosed by suites of granites (van Bemmelen, 1949; Zwart Kruis, 1962; de Roever, personal communication). The sedimentary sequence is of geosynclinal flysch facies, predominantly shales and sandstones with minor intercalations of greywackes, limestones, chert layers with radiolaria, basic volcanics and conglomerates (van Bemmelen, 1949, p. 315; de Roever, 1951). The whole sequence has been steeply folded in a number of irregularly plunging and rising synclines and anticlines (Katili, 1967). Scarce diagnostic fossils indicate that this sequence comprises the Late Carboniferous (crinoid fragments belonging to the genus *Moscovicrinus*, K. F. G. Hosking, University of Malaya, and H. L. Strimple, University of Iowa, personal communication), the Permian (fusulinids, Westerveld, 1936) and the Triassic up to and including the Norian.

The fossils indicative of a Norian age have been reported by de Neve and de Roever (1947) from a fossiliferous limestone lens exposed in the Loemoet tin mine on northern Bangka. They include crinoid fragments of the family *Encrinidae*, *Encrinus* sp. and *Entrochus* sp., the coral *Montlivaultia molukkana* J. Wanner, and *Calcispongiae* determined as *Peronidella moluccana* n. sp. All are diagnostic fossils of the Norian in fossiliferous Late Triassic strata elsewhere in Indonesia and in Malaysia. The suites of granitic intrusives are chiefly biotite granites and hornblende-bearing granites belonging to the suite of high-level plutonic complexes that are genetically related to the important cassiterite deposits of the Indonesian tin province. Among these plutonic rocks four main groups have been recognized, gabbroic, granodioritic, adamellitic and granitic (Aleva, 1960). For the genesis of the granitic magmas, processes of fractional melting of continental crust have been advocated, possibly along zones of high heat-flow from the mantle and at least in part induced by heat from ascending basaltic magmas (Priem *et al.*, 1975b).

The plutonism sets a minimum age to the Norian fossil assemblage on Bangka. The steeply folded, metamorphic Late Carboniferous/Norian sequence is discordantly overlain by a weakly folded, unmetamorphosed epicontinental or molasse-facies sequence of sandstones, shales and minor conglomerates, the Bintan formation (Bothé, 1928; Roggeveen, 1931; Adam, 1950; Jongmans, 1951). On Bangka, this formation is only present on the northern promontory of the island as deposits of sandstone and minor conglomerates with remains of limonitized tree trunks, but the sequence is well developed on neighbouring islands. So far, only a small collection of fragmentary plant remains from the formation has been studied, originating from the island of Bintan (Figure 1). Jongmans (1951) identified in this collection Cycad leaves belonging to the genus *Pterophyllum*, i.e. *Pterophyllum bintanense*, *Pterophyllum cf. contiguum* Schenk and *Pterophyllum* sp., and a few squamiform remains, *Cycadolepis* sp. This flora was correlated by Jongmans (1951) with floras of Rhaetian/(Liassic) age elsewhere in eastern Asia, but Kon'no (1972) suggests that the plant remains may be more comparable to the Neocomian flora in West Malaysia (although he did not himself study the Bintan flora). Further investigation of the plant remains in

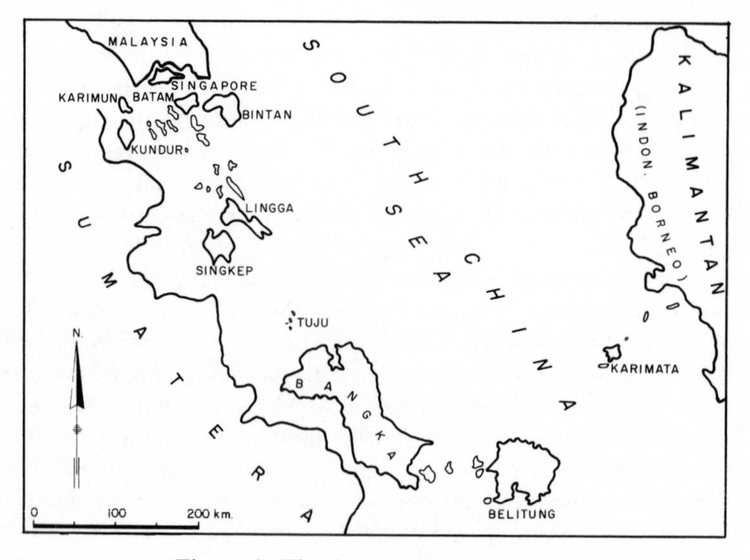

Figure 1. The Indonesian tin islands.

the Bintan formation is required in order to clarify their stratigraphical position.

No direct field evidence has been reported regarding the relationship of the Bintan formation to the plutonic complexes. A Rb–Sr whole-rock analysis of a quartz porphyry pebble from the conglomerate on northern Bangka, indicates that the pebble has been derived from much older volcanics (probably Late Carboniferous), unrelated to the plutonic complexes (ZWO Laboratory of Isotope Geology, Amsterdam, unpublished data). However, the clear tectonic break between the Bintan formation and the underlying Late Carboniferous/Norian sequence, along with the unmetamorphosed nature of the Bintan formation *versus* the regional and contact metamorphism displayed by the older sediments, strongly suggests that the deposition of the Bintan formation postdates the plutonism.

2 AGE OF THE TIN GRANITES ON BANGKA AND BELITUNG

The Indonesian tin granites form part of rather small complexes intruded at high crustal levels (Priem *et al.*, 1975b). Such complexes cool rapidly and will be exposed within a rather short time, which makes them very suitable for defining stratigraphic events. It has been already known for a long time that the intrusion of the Indonesian tin granites has taken place within a rather narrow time span (Late Triassic and/or Early Jurassic) and isotopic ages of the granites have therefore been used in early attempts to construct a geological time scale (Item 71 in the Phanerozoic time scale. Harland *et al.*, 1964). Unfortunately, a too low age of about 180 Ma was originally assigned to the granite on the basis of a single K–Ar biotite date from Belitung (Dutch: Billiton) in the late fifties (see below). Moreover, the 'Billiton granite' was rejected as a reliable item by Hutchison (1968) and Lambert (1971), since 'the stratigraphical assignment is unaccompanied by sufficient detail to support its narrow range' (a statement apparently due to ignorance of much of the Dutch literature on the subject).

Priem *et al.* (1975b) have reported the results of an isotopic age investigation of the granites in the Indonesian tin province. The study involved Rb–Sr whole-rock analyses of suites of samples from Bangka, Belitung and the neighbouring islands of Pulau Tuju, along with Rb–Sr and K–Ar analyses of separated biotites and hornblendes. A least-squares regression line fitted to all Rb–Sr data points (15 whole-rocks and four separated biotites) corresponds to an age of 213 ± 5 Ma and initial $^{87}Sr/^{86}Sr = 0.715 \pm 0.003$ (ICC recalculated; errors at 95% confidence level), which was taken as approaching the age of the magmatism. The line is not a true isochron, however, in view of the high MSWD value of 7.9, but an isochron relationship can hardly be expected in the case of a suite of samples collected from different granitic masses over an area of some 20,000 km^2.

A short cooling history after intrusion is apparent for the tin granites. The time of intrusion should therefore most nearly be approached by using from the data published by Priem *et al.* (1975b) only the Rb–Sr ages of biotite/whole-rock pairs and the K–Ar ages of the biotites and a hornblende. These data are listed in Tables 1 and 2, along with the calculated ages. Except for the lower Rb–Sr age of the biotite/whole-rock pair BIL 4, all ages lie between 214 and 217 Ma, averaging 216 ± 2 Ma. A Rb–Sr K-feldspar age reported earlier from a granite on Belitung (Edwards and McLaughlin, 1965) is about concordant, 212 ± 7 Ma. The lower Rb–Sr biotite age of BIL 4 may be related to a few other lower ages reported from Belitung, i.e. the Rb–Sr and K–Ar ages of biotite from a granite (Edwards and McLaughlin, 1965) and the K–Ar ages of whole-rocks from a diorite and a basalt (Priem *et al.*, 1975b) and two muscovites from a greisen (Jones *et al.*, 1977); such lower ages (spreading between 120 and 208 Ma) have been interpreted by Priem *et al.* (1975b) as reflecting a local partial resetting of K–Ar clocks due to a weak thermal event induced by the Late Cretaceous plutonism east of Belitung. In this connection, it is of interest to note that biotites from volcanic breccias in dykes on Singkep Laut, NW of Bangka, also yield Late Cretaceous ages (ZWO Laboratory of Isotope Geology, Amsterdam, unpublished data).

The age of 216 ± 2 Ma is within the error limits, equal to the age of 213 ± 5 Ma indicated by the regression line fitted to all Rb–Sr data points,

Table 1. Rb–Sr data and calculated ages of biotite/whole-rock pairs.[a]

Sample No.		Rb[b] (ppm)	Sr[b] (ppm)	Rb/Sr[b]	$^{87}Sr/^{86}Sr$	Initial $^{87}Sr/^{86}Sr$	Calculated age (Ma)[c]
Belitung							
BIL 3	bio	645*	12.8*		1.181	0.713	216 ± 6
BIL 3	WR	250	221	1.131	0.7230		
BIL 4	bio	667*	17.8*		1.043	0.714	206 ± 6
BIL 4	WR	202*	223	0.9065	0.7217		
Bangka							
BIL 28	bio	647*	6.95*		1.623	0.724	215 ± 5
BIL 28	WR	221	174	1.270	0.7351		
BIL 29	bio	2056*	17.1*		1.922	0.721	217 ± 5
BIL 29	WR	621	42.4	14.66	0.8532		

[a] All analytical data mean of duplicate analyses.
[b] XRF analysis, except for values marked * which were obtained by isotope dilution analysis.
[c] $\lambda(^{87}Rb) = 1.42 \times 10^{-11}\ a^{-1}$. Errors based upon estimated overall limits of relative error of 1% in XRF Rb/Sr and isotope dilution Rb and Sr, and 0.2% in $^{87}Sr/^{86}Sr$; the latter are the sum of the estimated contributions of the known sources of possible systematic error and the precision (2σ) of the total analytical procedures.

Table 2. K–Ar mineral data and calculated ages.

Sample No.		K[a] (% wt)	Radiogenic ^{40}Ar (10^{-3} ppm)[b]	Calculated age (Ma)[c]
Belitung				
BIL 3	biotite	6.59	105[a]	216±6
BIL 4	biotite	6.36	101[d]	216±6
BIL 4	hornblende	0.787	12.5[a]	216±6
Bangka				
BIL 28	biotite	6.60	105[d]	216±6
BIL 29	biotite	6.52	103[a]	214±6

[a] Mean of duplicate analyses.
[b] Atmospheric ^{40}Ar between 5 and 10% of total ^{40}Ar for all analyses.
[c] $\lambda_e(^{40}K) = 0.581 \times 10^{-10}\,a^{-1}$, $\lambda_\beta(^{40}K) = 4.962 \times 10^{-10}\,a^{-1}$ and abundance $^{40}K = 0.01167$ atom % total K. Errors based upon estimated overall limits of relative error of 1.0% in K and 2.0% in radiogenic Ar; the latter are the sum of the estimated contributions of the known sources of possible systematic error and the precision (2σ) of the total analytical procedures.
[d] Mean of triplicate analyses.

but it is preferred as a better approximation of the time of intrusion of the tin granites on Bangka and Belitung.

3 DISCUSSION

So far, few reliable isotopic dates are available in the Triassic that are well defined biostratigraphically. The Permian–Triassic boundary is poorly controlled and the age has variously been estimated at 225 Ma (Armstrong and Besançon, 1970), 240 Ma (Lambert, 1971) or even 230 Ma (Harland *et al.*, 1964; Waterhouse, 1978). Lambert (1971) has set great value on the Rb–Sr isochron age of 226±16 Ma determined for small granitic plutons cutting the Early to Middle Triassic Broowena formation in Queensland, Australia (Webb and McDougall, 1967). Ages of 240 or 225 Ma for the base of the Triassic cannot be reconciled, however, with the Rb–Sr isochron age of 238±3 Ma (K–Ar and Rb–Sr phlogopite ages of 231–237 Ma) determined on igneous complexes intrusive into Ladinian/Early Carnian sediments in the western Dolomites, Italy (Borsi and Ferrara, 1967). This age may be somewhat doubtful in view of the considerable scatter of the original data, but two other isotopic age measurements on Early Triassic sediments confirm that the base of the Triassic must be older than 224–240 Ma: the K–Ar and ^{40}Ar/^{39}Ar ages of 234±2 Ma for alkali feldspars from volcanic tuffs interbedded between fossiliferous (invertebrates) sediments deposited around the Anisian–Ladinian boundary in the southern Alps of Switzerland

(Hellmann and Lippolt, 1979), and the K–Ar age of 237±4 Ma for basalts and ignimbrites conformably overlying sediments containing fossil reptile assemblages of Scythian and Anisian age (Puesto Viejo formation) in western Argentina (Valencio *et al.*, 1975). The biostratigraphic control of both items appears to be good.

Another interesting item for the Triassic time scale, also biostratigraphically well defined, is the K–Ar age of 229±5 Ma reported for the basaltic extrusions overlying sediments containing fossil reptile assemblages of Carnian age (Ischigualasto formation) in western Argentina (Valencio *et al.*, 1975). The upper boundary of the Triassic is estimated by Armstrong and Besançon (1970) and Lambert (1971) at about 200 Ma. This estimate is consistent with the $^{40}Ar/^{39}Ar$ and K–Ar ages of around 195 Ma reported for basaltic extrusives and sills in Hettangian/Sinemurian sediments from the Newart Trend Basins in eastern North America (Sutter and Smith, 1979), but it is at odds with the Rb–Sr isochron age of 186±9 Ma of a volcanic tuff within the Falls formation in the Queen Alexandra Range, Antarctica (Faure and Hill, 1973), which tuff lies above a fossiliferous horizon containing plant remains of Anisian–Ladinian age (Retallack, 1977). However, several cases have been reported of pyroclastic volcanics that have experienced some form of Sr isotopic resetting subsequent to the time of deposition, even where no effects of metamorphism are apparent (Fairbairn *et al.*, 1966; Lanphere, 1968; Farquharson and Richards, 1975; Van Schmus and Bickford, 1976; Priem *et al.*, 1978; Gale *et al.*, 1979). The Rb–Sr isochron age of the pyroclastics of the Fall formation should thus be taken as setting only a minimum age to the volcanism.

In conclusion, the age of 216±2 Ma of the tin granites on Bangka and Belitung sets a minimum age to the fossil assemblage described from Bangka. If this fauna flourished towards the end of the Norian, a comparison of the age of the tin granites with the few reliable dates available from the Early and Middle Triassic (see above) would leave a time span for the Norian shorter than or approximately equal to that of the Rhaetian (taking the Rhaetian–Hettangian boundary at about 200 Ma). This is at odds with the fact that the Norian is known to contain eight ammonoid zones as against only one in the Rhaetian, suggesting a considerably shorter duration for the latter than for the former. However, the Bangka fossils could very well be of Early Norian age, leaving the possibility that the intrusion of the tin granites took place well within the Norian; the Late Norian and the Norian–Rhaetian boundary should then be younger. If the fossil flora in the Bintan formation does prove to be of Rhaetian age, the tin granites would also set a maximum age to the Rhaetian.

In any case, an age of 240 Ma for the Permian–Triassic boundary seems to leave an improbably short time span for the lower two stages of the Triassic, so the base of the Triassic must be older. The age of 255–260 Ma advocated by Bochkarev and Pogorelev (1967, in Armstrong and Besançon, 1970) is perhaps a more reasonable estimate.

Acknowledgements

The authors are much indebted to P. A. M. Andriessen, N. A. I. M. Boelrijk, E. H. Hebeda, E. A. Th. Verdurmen and R. H. Verschure, all of the ZWO Laboratorium voor Isotopen-Geologie, Amsterdam, for critical discussions and helpful suggestions. They thank W. P. de Roever, Amsterdam, for valuable information, and an anonymous referee for his helpful manuscript review. This work forms part of the research programme of the 'Stichting voor Isotopen-Geologisch Onderzoek', supported by the Netherlands Organization for the Advancement of Pure Research (ZWO).

Résumé du rédacteur

Des recherches par les méthodes Rb–Sr et K–Ar sur des paires roche totale–biotite et une hornblende provenant des granites de Bangka et Belitung (Indonésie) établissent un âge d'intrusion de 216 ± 2 Ma. Les granites recoupent des sédiments contenant des fossiles du Norien et donnent ainsi un âge minimum pour le Norien (peut-être inférieur). Ces formations semblent recouvertes de sédiments attribués au Rhétien ou au Lias voire au Néocomien selon les auteurs.

(Manuscript received 10-4-1980)

Numerical Dating in Stratigraphy
Edited by G. S. Odin

28
Late Triassic–Early Jurassic time-scale calibration in British Columbia, Canada

RICHARD LEE ARMSTRONG

1 INTRODUCTION

The calibration of the geological time scale requires precisely dated igneous rocks or authigenic minerals closely tied to the stratigraphic scale. The Early Mesozoic is not as well calibrated as earlier or later portions of the time scale, so that significantly different values have been suggested for the Triassic–Jurassic boundary in recent papers. Van Hinte (1976) used 192 Ma (196 Ma, ICC), in contrast to a proposal of 211 Ma by Armstrong (1978). Webb (1981) reviewed the Triassic calibration with additional data, especially from Australia, and proposed a boundary at 200 Ma.

The data for these previous calibrations are shown in Figure 1, exclusive of items from British Columbia. Also shown are the time scales of Armstrong (1978) and Webb (1981).

2 DATA FROM BRITISH COLUMBIA

In the Intermontane belt of British Columbia there are large volumes of Late Triassic and Early Jurassic volcanic rocks interstratified with marine sediments and cut by a variety of granitic plutons. Both volcanic and intrusive rocks provide material for K–Ar and Rb–Sr studies and ammonites and other fossils enable geological ages to be specified on the world-standard time scale. The work reviewed here has accumulated over the past 15 years at The University of British Columbia (UBC) and the laboratory of the Geological Survey of Canada in Ottawa (GSC). The geological background is from the Geological Survey of Canada, the British Columbia Ministry of Energy, Mines and Petroleum Resources, and student theses at The University of British Columbia.

The Late Triassic (Carnian and Norian) volcanic assemblage in British Columbia varies in name and character from south to north. In the south it

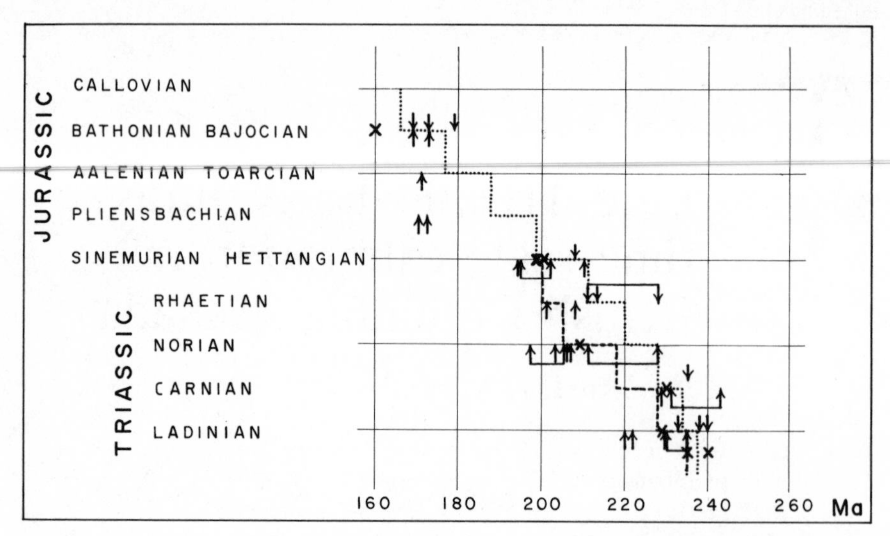

Figure 1. Late Triassic and Early Jurassic time scale. The graphic conventions are those used in Armstrong (1978). Arrows are stratigraphic brackets for pluton dates, crosses are dated stratigraphic horizons. The data are from Webb (1981) (with the Chinle formation age modified by Lupe and Silberling, 1979), Armstrong (1978), Shibata *et al.* (1978) and Westphal *et al.* (1979) (excluding all items from British Columbia). The crosses just below the Ladinian line are for Anisian–Ladinian volcanic rocks of Australia listed by Webb (1981). The two Triassic scales shown by dashes and dots are the proposed scales of Webb (1981) and Armstrong (1978), respectively. The Jurassic scale is from Armstrong (1978).

is known as the Nicola group of Carnian to Late Norian age (Frebold and Tipper, 1969; Travers, 1978; Preto, 1979). The Nicola volcanic rocks are too metamorphosed and altered to give a Rb–Sr isochron (Preto *et al.*, 1979). A whole-rock Rb–Sr date of 205 ± 8 Ma was obtained for a relatively fresh and massive rhyodacite sample by Grette (1979), but this is only a minimal estimate of the true age. Nicola is crosscut by several batholiths. Most famous for time-scale calibration purposes is the Guichon batholith, post-Norian and overlain by Hettangian conglomerate with clasts of Guichon granodiorite. The last word on Guichon geochronometry is median K–Ar date 203 Ma, the highest group of K–Ar dates 209 Ma (Northcote, 1969), and the Rb–Sr isochron for mineralization 205 ± 20 Ma (Preto *et al.*, 1979; NDS 177). Other batholiths cutting the Nicola give K–Ar dates of 200–209 Ma (Preto *et al.*, 1979) and the Tulameen ultramafic complex, which is also Late or post-Nicola (Findlay, 1969), has been dated at 208 Ma by a good $^{40}Ar/^{39}Ar$ plateau (McDougall, 1974; NDS 179). Unpublished UBC K–Ar dates for the Coldwater stock, the oldest rock so far found in the Nicola belt, are 208 Ma for biotite and 212 Ma for hornblende, with biotite Rb–Sr being 220 ± 20 Ma (NDS 178) recently revised at 207 ± 8.

In central British Columbia the Late Triassic volcanic assemblage is the Takla group, of Late Carnian to Late Norian age (with rare local continuation until Early Sinemurian time) (Tipper and Richards, 1976; Monger, 1977; Monger and Church, 1977). Metamorphosed Takla occurs in the Kutcho Creek area, where it is being studied by L. E. Thorstad (MSc thesis in progress). A composite whole-rock Rb–Sr isochron, including Kutcho rocks of L. Thorstad and Takla samples supplied to UBC by J. Monger, gives a date of 215 ± 28 Ma (NDS 173). This is almost certainly a minimum age because of the altered and metamorphosed nature of some samples. The Takla sample alone gives an isochron date of 223 ± 56 Ma (NDS 172).

The Wrede Creek zoned ultramafic complex (Wong and Godwin, 1980) occurs within Takla group volcanic rocks and may be closely related to the volcanic magmas. Wong and Godwin report a K–Ar date for hornblende from pegmatitic segregations in dunite of 225 ± 16 Ma (NDS 176).

A number of intrusive igneous rocks postdate the Takla group. Some of these are certainly plutonic equivalents of Early Jurassic volcanic rocks. The largest, the Hogem batholith, has a complex history, but the oldest K–Ar dates have a median value of 194 Ma and the best Rb–Sr isochron date for an early phase is 190 ± 50 Ma (Woodsworth, Armstrong and Eadie, in preparation; Eadie, 1976; Stock, 1974; Garnett, 1978; NDS 175). At the Kemess property (NDS 174) intrusive rocks have been dated at 204 Ma by K–Ar and 190 ± 8 Ma by Rb–Sr (Cann and Godwin, 1980). Other post-Takla intrusive bodies scattered around the northeast perimeter of the Sustut Basin give K–Ar dates ranging from 180 to 200 Ma (Gabrielse *et al.*, 1980).

In northern British Columbia the Carnian (Souther, 1971) Stuhini group is cut by the Kaketsa stock, dated by K–Ar as 222 Ma for hornblende and 218 Ma for biotite (Panteleyev, 1975; NDS 171).

In aggregate, these dates (NDS 171–179) suggest a minimum of 210–215 Ma for Norian and 220–225 Ma for Carnian time.

Early Jurassic volcanic rocks are most abundant in areas surrounding the Bowser Basin in central British Columbia. The Hazelton group is a complex assemblage of volcanic and sedimentary rocks ranging from Sinemurian to Callovian in age. The volcanic rocks are most abundant in Late Sinemurian–Early Pliensbachian and Toarcian times (Tipper and Richards, 1976). Pliensbachian and Toarcian volcanic rocks have been dated at 183 and 182 Ma by K–Ar and 189 ± 26 and 191 ± 18 Ma by Rb–Sr whole-rock isochrons (Gabrielse *et al.*, 1980; NDS 183). The Telkwa formation (Late Sinemurian–Early Pliensbachian) has been dated at 185 ± 6 Ma by Rb–Sr whole-rock isochron (NDS 180) and the Toodoggone volcanics (Pliensbachian and Toarcian) have given K–Ar dates of 189 ± 12 Ma and a Rb–Sr whole-rock isochron date of 185 ± 10 Ma (Carter, 1972; Gabrielse *et al.*, 1980; NDS 184).

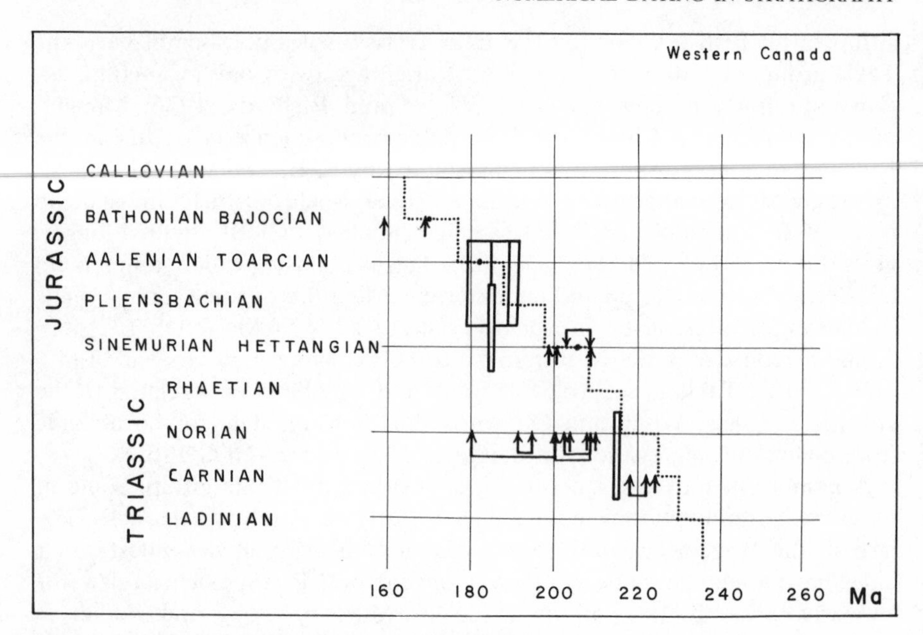

Figure 2. Late Triassic and Early Jurassic time scale. The rectangles span the stratigraphic range and time range of selected dates from different volcanic formations. All data are from British Columbia and are summarized in the text. A compromise scale is shown, taking into account data from both figures. Heavy dots on the compromise scale and vertical offsets of the dotted line are suggested stage boundaries at 235, 230, 225, 216, 208, 205, 197, 188, 182, 177, 170 and 164 Ma and system boundary at 208 Ma.

Granitic plutons that crosscut Late Sinemurian volcanic rocks have been dated at 199, 201 and 209 Ma by K–Ar (Wanless *et al.*, 1974; NDS 181). Pink hornblende–biotite quartz monzonite, thought to be associated with Bajocian volcanic rocks, has been dated by K–Ar as 159–171 Ma (Gabrielse *et al.*, 1980; NDS 182).

These data (NDS 180–184) provide a younger limit of 203 Ma for at least part of the Sinemurian and an age of at least 185–189 Ma for some part of Pliensbachian–Toarcian time.

3 DISCUSSION

The data just summarized for British Columbia are all plotted on Figure 2, along with a time-scale interpretation that takes into account all world data. The pre-Norian Triassic scale is close to that of Webb (1981) (this revised scale is 2–4 Ma younger than the Armstrong, 1978 scale; the difference is not significant). The end of the Triassic is put at 208 Ma, distinctly older than the value given by Webb (1981) and slightly younger than the Armstrong (1978) value. Jurassic time is apportioned to various stages in the manner described by Van Hinte (1976), *using palaeontological zones as an indicator of relative stage length.* The differences in relation to Armstrong (1978) are 1 or 2 Ma or less. Within limits of analytical uncertainty, all the dates discussed are compatible with the illustrated scale.

It is notable that the dates for granitic plutons coincide almost exactly with the stratigraphic range of Hazelton volcanic rocks. There is satisfying consistency in this, and there can be little doubt that the more easily dated plutons are feeders and magma chambers for the Early Jurassic volcanoes.

The illustrated compromise scale is offset by 12–13 Ma towards older dates from Van Hinte's (1976) scale. He used nothing but pre-1971 published dates, and did not even recalculate the dates to a single set of decay constants. Continued use of such an obsolete scale would only lead to erroneous contradictions that do not exist.

Future work on the volcanic rocks of British Columbia will result in further refinement of our time-scale calibration. Any interpretation such as shown on Figure 2 is only a progress report in an endless effort to refine the scale on which all geological events are organized.

Acknowledgements

This contribution to IGCP Project 133 was written while on leave from The University of British Columbia on a Killam Senior Fellowship. Geochronometry at UBC is supported by a Natural Science and Engineering Research Council grant. J. Harakal and K. Scott do most of the Ar and Sr laboratory work, respectively. H. Gabrielse has reviewed the manuscript to keep me stratigraphically honest. I thank John Webb for preprints of his paper and several useful suggestions.

(Manuscript received 20-6-1980)

Numerical Dating in Stratigraphy
Edited by G. S. Odin

29
Triassic radiometric dates from eastern Australia

JOHN A. WEBB

1 INTRODUCTION

A number of isotopic age determinations relevant to the Triassic are available from eastern Australia; the purpose of this chapter is to describe them and discuss possible ages for the Triassic stage and system boundaries in the light of this information.

Although most of the dates belong to the Triassic, a few are Late Permian, as these help to fix the Permian–Triassic boundary. All Australian dates used by previous authors to estimate the duration of the Triassic have been included in this study, although subsequent work has shown some of them to be of limited use.

Where necessary, dates have been recalculated to the constants of Steiger and Jäger (1977); they are abbreviated: ICC. Each K–Ar date is listed individually with its instrument 2σ error, which is taken to be 3% if not given by the original author.

Most of the dates described below are plotted on Figure 1 in chronological sequence, and they have been assigned individual numbers. For ease of cross-referencing, the item number for each date is given in parentheses when it is first mentioned in the text. Discrepancies between items are discussed after presentation of all the data. The stratigraphic subdivisions are shown in Figure 2.

2 DATES FROM EASTERN AUSTRALIA

In the southern Sydney Basin, New South Wales, several lava flows occur interbedded with fossiliferous Middle to ?Late Permian sediments. The Bumbo and Dapto-Saddleback latites (item 1, Figure 1) have been dated as 249 ± 8 and 251 ± 5 Ma respectively (K–Ar feldspar, whole-rock; Evernden and Richards, 1962; Facer and Carr, 1979). Interbedded clastics of the Gerringong volcanics contain Kazanian brachiopods (Waterhouse, 1976).

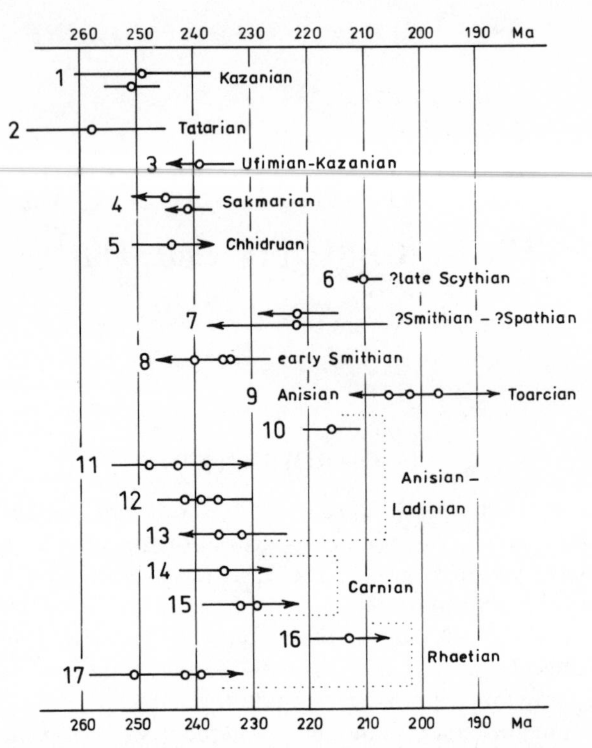

Figure 1. Radiometric data from eastern Australia, arranged in order of stages. Each date is represented as a circle, the bar associated with it indicating the 2σ uncertainty. Arrows show that the stage associated with a particular item is older or younger than the age range of that item, i.e. an arrow points to the left for post-event intrusives cross-cutting sediments and to the right for pre-event plutons unconformably overlain by sediments. Details of these items are available in the last part of this book: item 1 = NDS 205; 3 = NDS 206; 4 = NDS 207; 6 = NDS 208; 8 = NDS 195; 10 = NDS 198; 11, 12, 13 = NDS 194; 14 = NDS 197; 15 = NDS 193; 16 = NDS 199.

The Berkeley latite (item 2, Figure 1), which occurs near the base of the Tatarian (?Djulfian) Illawarra coal measures (Mayne *et al.*, 1974; Anderson, 1973), is 258 ± 13 Ma old (K–Ar feldspar; Evernden and Richards, 1962). Webb (in Ozimic, 1971) found that the Currambene dolerite (item 3, Figure 1), which intrudes the Ufimian–Kazanian Nowra sandstone (Waterhouse, 1976), has an age of 239 ± 6 Ma (K–Ar plagioclase). The Conjola sub-group, of probable Sakmarian age, is intruded by the Milton monzonite (245 and 245 ± 6 Ma) and the Termeil essexite (241 ± 4 Ma); these are all K–Ar whole-rock ages (item 4, Figure 1; Mayne *et al.*, 1974; Facer and Carr, 1979).

McDougall and Wellman (1976) found that a microgranodiorite in the Lorne Basin, New South Wales, gave an average age of 210 ± 3 Ma (item 6, Figure 1; K–Ar hornblende). It intrudes the Camden Haven group, which contains ?Late Scythian microfloras (Helby, 1973).

Dulhunty and McDougall (1966) and Dulhunty (1972) obtained ages of 197 ± 10, 202 ± 6 and 206 ± 7 Ma (K–Ar whole-rock) from the Garrawilla volcanics in central New South Wales (item 9, Figure 1), disregarding anomalously young dates. These lavas are underlain by Anisian beds and overlain by Toarcian sediments, the stratigraphic ages being based on spore–pollen assemblages (Loughnan and Evans, 1978).

Retallack *et al.* (1977) dated the Dalmally basalt member of the Nymboida coal measures, Nymboida Sub-Basin, northern New South Wales (item 10, Figure 1), as 216 ± 5 Ma (K–Ar plagioclase). This flow lies between strata containing Anisian–Ladinian floras (Retallack, 1977).

The granites in the New England and Stanthorpe areas, around the Queensland–New South Wales border, were found by Evernden and Richards (1962) and Cooper *et al.* (1963) to give ages of 250 ± 8, 252 ± 8, 248 ± 7, 259 ± 8, 242 ± 7, 226 ± 6, 230 ± 7, 244 ± 7 and 247 ± 7 Ma (K–Ar biotite, hornblende). They intrude fossiliferous Early Permian rocks and are overlain by Middle Triassic–Jurassic sediments; however, they were regarded as probably Late Permian (Cooper *et al.*, 1963), and used to fix the Permian–Triassic boundary at about 230 Ma (Smith, 1964). More recent workers have regarded at least some of these granites as Early Triassic (Webb, 1969; Olgers *et al.*, 1974).

A date of 244 ± 7 Ma (K–Ar biotite) was obtained from a tuff at the top of the Gyranda formation, in the Bowen Basin, central Queensland (Webb and McDougall, 1967; item 5, Figure 1). This formation is overlain by the Baralaba coal measures, which contain Chhidruan microfloras (Foster, 1979).

A dacite near Glasshouse Mountains township, southeastern Queensland, gave a date of 213 ± 7 Ma (K–Ar whole-rock; Green and Webb, 1974; item 16, Figure 1). This volcanic is believed to be a time equivalent of the North Arm volcanics in the Nambour Basin (Green and Webb, 1974), the uppermost sediments of which are considered Rhaetian on palynological grounds (J. McKellar, personal communication).

Ages of 239 ± 7 and 242 ± 7 Ma (K–Ar biotite) and 251 ± 8 Ma (Rb–Sr biotite) have been obtained from the Crows Nest granite (item 17, Figure 1; Webb and McDougall, 1968; Green and Webb, 1974). It is overlain by the Bundamba group in the Moreton Basin, southeastern Queensland (Cranfield *et al.*, 1976); from the lower formations of this group de Jersey (1975) recovered a Rhaetian palynoflora.

The Tarong beds occupy the Tarong Basin, southeastern Queensland, and contain palynofloras attributed to the Carnian by de Jersey (1970a). They

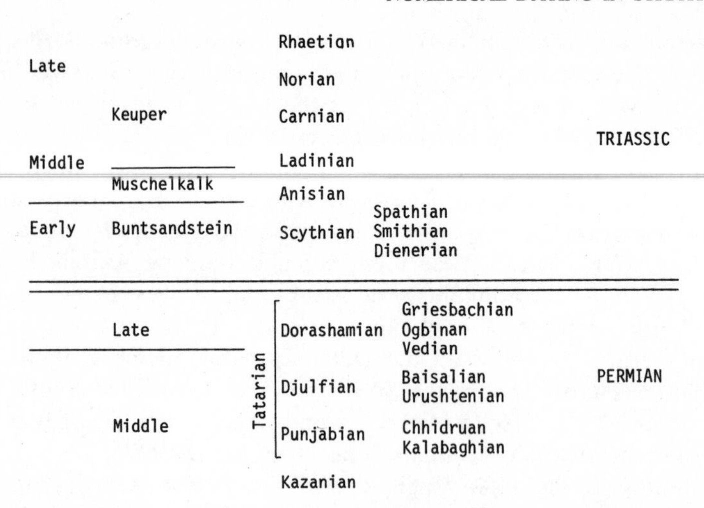

Figure 2. Subdivisions of the Triassic and Middle and Late Permian used in this chapter, from Silberling and Tozer (1968) and Waterhouse (1976). Note that the position of the Griesbachian is subject to dispute, as Silberling and Tozer (1968) originally placed it in the Triassic, and the Rhaetian may only represent the topmost portion of the Norian (Tozer, 1979a).

overlie the Djuan tonalite (235 ± 8 Ma, K–Ar hornblende; item 14, Figure 1), the Woolshed Mountain granodiorite (259 ± 8 Ma, K–Ar biotite) and the Boondooma igneous complex (Cranfield *et al.*, 1976; Murphy *et al.*, 1976). Isotopic dates on the last-mentioned intrusive are 237 ± 7, 239 ± 7, 249 ± 7, 244 ± 7, 245 ± 7, 259 ± 8 and 261 ± 11 Ma (K–Ar, Ar–Ar biotite, hornblende), with a Rb–Sr whole-rock and biotite isochron age of 284 ± 28 Ma (Webb and McDougall, 1968; Murphy *et al.*, 1976); an anomalously young date of 222 ± 7 Ma has also been recorded (Murphy *et al.*, 1976).

In the northern part of the Gympie Basin, southeastern Queensland, a bivalve fauna collected from the Brooweena formation was assigned to the Smithian–Spathian by Fleming (1966); however, faunas in New Zealand with somewhat similar species have been correlated with the Ladinian (I. Speden, personal communication). The Brooweena formation has been intruded by the Broomfield granite, the Musket Flat granodiorite (222 ± 7 Ma; K–Ar biotite; item 7, Figure 1) and the Mungore complex (212 ± 7 and 219 ± 7 Ma; K–Ar biotite; Webb and McDougall, 1967). These three granites give a Rb–Sr whole-rock isochron age of 222 ± 16 Ma (Webb and McDougall, 1967). To the south in the Gympie Basin are the Kin Kin beds, which have yielded an Early Smithian ammonoid fauna (Runnegar, 1969). Previous workers proposed a pre-Triassic age for part of this formation, but Murphy *et al.* (1976) regarded the entire unit as Triassic. The beds are intruded by the Woondum granite (223 ± 7 and 226 ± 7 Ma; K–Ar hornblende, biotite) and the Goomboorian diorite (234±7, 235±7 and 240±7 Ma; K–Ar hornblende, biotite; Webb and McDougall, 1967; Green and Webb, 1974; item 8, Figure 1).

In the Esk trough, southeastern Queensland, the Toogoolawah group consists of three formations: the Bryden formation, the Neara volcanics and the Esk formation. There is probably at least some degree of lateral equivalence between all three (Murphy *et al.*, 1976; Cranfield *et al.*, 1976). De Jersey (1972; 1973) obtained palynofloras from drillholes through the Esk and Bryden formations in the southern portion of the trough; these

Table 1. Isotopic age determinations relevant to the age of the Toogoolawah group (all are K–Ar dates except where noted otherwise); from Webb and McNaughton (1978).

Rock unit	Chronometer	Apparent age (Ma, ICC)	Reference
Units intruding the Toogoolawah group			
Dyke	Hornblende	219±5	1
Somerset Dam igneous complex	Plagioclase	213±5	2
	Hornblende	219±5	2
'Brisbane Valley porphyrites'	Hornblende	224±5	2
Station Creek adamellite	Biotite	231±7	2
	Biotite	236±7	3
Toogoolawah group			
Neara volcanics	Whole-rock	236±6	4
	Hornblende	242±5	1
	Whole-rock	239±5	1
Units overlain by the Toogoolawah group			
Kingaham Creek granodiorite	Biotite	220±6	2
	Biotite	248±8	5
	Biotite	242±10	6
Eskdale granodiorite	Hornblende	238±7	7
	Hornblende	243±8	8
	Biotite	248±7	8
Taromeo tonalite	Biotite	226±7	6
	Biotite	241±7	7
	Biotite	243±7	7
	Biotite (Ar/Ar)	248±8	6
	Hornblende	250±7	7
	Hornblende (Ar/Ar)	264±21	6

1, Irwin (1976); 2, Webb and McDougall (1967); 3, Brooks *et al.* (1974); 4, Kerr (1976); 5, McNaughton (1973); 6, Murphy *et al.* (1976); 7, Webb and McDougall (1968); 8, Cranfield *et al.* (1976).

indicated that the Esk formation was Late Anisian–Early Ladinian and the Bryden formation was the same age or slightly older. Unfortunately volcanics were not intersected in these drillholes, so the exact stratigraphic relationship between the Neara volcanics and the palynologically dated sediments is uncertain. However, it seems likely that all the formations are Anisian–Ladinian. De Jersey (1979) described a Middle Triassic spore–pollen assemblage from the Gayndah beds in the northern part of the Esk trough. This formation is probably a time equivalent of the Esk formation (Cranfield, 1979). Isotopic ages relevant to the age of the Toogoolawah group are listed in Table 1 (items 11–13, Figure 1). The younger dates obtained for the Kingaham Creek granodiorite and Taromeo tonalite appear to be anomalous, and are excluded from further discussion.

The Ipswich coal measures in the Ipswich Basin, southeastern Queensland, contain Carnian microfloras (de Jersey, 1970b; 1971). The Sugars basalt (item 15, Figure 1), which lies at the base of the sequence, has given dates of 229±7 and 232±7 Ma (K–Ar whole-rock; Webb and McNaughton, 1978). In AP–PS Matjara 1, drilled offshore from Brisbane, volcanics believed to be equivalent to those at the base of the Ipswich coal measures were encountered (Cranfield *et al.*, 1976). These were dated at 176–189 Ma (K–Ar), although the samples were weathered and the age is a minimun only (Crook and Hoyling, 1968).

The Mt Byron volcanics in southeastern Queensland have been dated as 227±7, 228±7 and 230±7 Ma (K–Ar whole-rock; Murphy *et al.*, in preparation). Cranfield *et al.* (1976) considered that the volcanics unconformably overlay the Toogoolawah Group, and Banks (1978) used this in his summary of Australian Triassic radiometric dates. However, recent mapping has shown that the boundary is faulted, and the exact age relationship between the two units is conjectural (Murphy *et al.*, in preparation).

3 DISCUSSION

Most of the dates mentioned in the previous section are plotted on Figure 1. Some have been omitted because the stratigraphic constraints were too broad or the dated rocks noticeably weathered, and when several dated intrusives cross-cut or underlie the same sedimentary formation (e.g. Table 1), only the one of most use in determining a stage boundary is shown on Figure 1.

The uncertainties associated with each age are given on Figure 1 at the 95% confidence level. For K–Ar dates this represents the instrument error, and for the single Rb–Sr isochron plotted (item 7) it indicates the error calculated using a modified linear regression analysis. No attempt has been made to average dates where several have been determined on a particular volcanic or intrusive. The arrows associated with the error bars on Figure 1 show that the stage relevant to a particular item is older or younger than the age range of the item, i.e. an arrow points to the left for post-event intrusives cross-cutting sediments and to the right for pre-event plutons unconformably overlain by sediments.

A study of the data on Figure 1 allows estimations to be made of some of the Triassic stage boundaries, although considerable interpolation is involved. Overall, the dates form a relatively consistent picture, provided the error bars and arrows are taken into account when comparing items.

An age of about 245 Ma for the Permian–Triassic boundary fits the evidence (items 1–8), if item 5 (Gyranda formation, Bowen Basin) is assumed to be anomalously young, presumably because of daughter element loss. This estimate corresponds almost exactly with the suggestions of Armstrong (1978) and Webb (1981). Smith (1964) used 230 Ma; this was derived by assuming that the New England granites were Late Permian, whereas at least some are now believed to be Early Triassic, as mentioned previously.

Webb (1981) and Armstrong (1978) proposed ages of around 240 Ma for the Scythian–Anisian boundary and 235 Ma for the Anisian–Ladinian boundary. The dates on Figure 1 are in agreement with these suggestions, with the exception of item 10 (Dalmally basalt member, Nymboida Sub-Basin). Banks (1978) had already noted that this date was aberrant, and Webb and McNaughton (1978) regarded the error associated with the age as unrealistically small.

None of the Australian dates is sufficiently precise to derive estimates for the boundaries of the Late Triassic stages.

Thus the eastern Australian data are of considerable help in calibrating the Early and Middle Triassic; however, further work, both in Australia and elsewhere, is necessary before the Triassic stage and system boundaries can be considered precisely determined.

Acknowledgements

Dr. R. L. Armstrong (University of British Columbia, Vancouver) and Dr. I. McDougall (Australian National University, Canberra) commented on an earlier version of this chapter and gave much useful advice.

(Manuscript received 8-12-1980)

Numerical Dating in Stratigraphy
Edited by G. S. Odin

30
The Triassic time scale in 1981

GILLES S. ODIN and RENÉ LÉTOLLE

1 INTRODUCTION

The time scale proposed in 1964 makes use of only two dates inaccurately located inside the Triassic system: PTS 9 and 160 from the North American 'Keuper'. An age of 230 Ma (ICC) was assumed for the base of the Triassic (Smith, 1964) on the basis of four Late Permian Australian dates: PTS 67-69-121-175. Two of them (PTS 67-175) were very inaccurately located in the stratigraphic column. The top of the Triassic system was accepted in the 195–200 Ma (ICC) range by Howarth (1964), according to the age of Hotailuh batholith in British Columbia (PTS 35). Lambert, using essentially the same results (1971), preferred boundaries located at 205 ± 5 Ma and 240 ± 5 Ma (ICC). The reader will see below that these ages, although very poorly documented at that time, were reasonably well chosen.

As a convention, we will use the abbreviation PTS for the name of an abstract from the 'Phanerozoic time scale' book of 1964; NDS numbers refer to the present volume. All ages given in this chapter are recalculated with the newly accepted International Congress of Sydney Conventional Constants (ICC).

The numerous new data will help essentially in the subdivision of the Triassic system. The Australian and western Canadian areas have been intensively studied; the syntheses proposed in the two preceding chapters by J. A. Webb and R. L. Armstrong will greatly help our comments. Scores of dates are also available from Permian–Triassic rocks from the South American area; however, in many cases the stratigraphical calibration is as yet not very well known.

The last part of this book gives details in each case on the evidence used in this chapter. Now we will essentially try to indicate what specific data were used to construct the time scale adopted here. We hope this attempt will help in confirming, modifying or obtaining more precision in future work.

2 THE STRATIGRAPHIC COLUMN

The earliest Triassic scale was proposed by von Alberti after the German series, and divides the system into three lithological units: *Buntsandstein*, *Muschelkalk* and *Keuper*; these names later became chronostratigraphic unit names. The three units correspond to three lithological facies: sandstone, limestone and gypsiferous marls. They characterize the Germanic Triassic series essentially related to lagoonal and continental environments. Establishing correspondence between these units and their homologues in other part of Europe (especially the Keuper) has obviously led to some disagreements (Ricour, 1963; and Figure 1). In England, the abbreviated

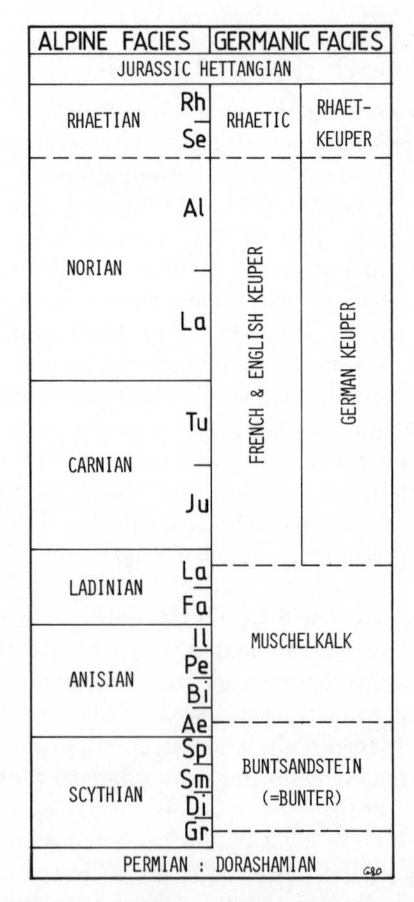

Figure 1. Correlations between Alpine and Germanic facies. The thickness of the stages is proportional to their probable duration. Stratigraphy of the Germanic facies mainly based on Kozur (1975); scheme proposed by H. Zapfe.

word *Bunter* replaces Buntsandstein. In Germany, the Rhaetian was initially included in the Keuper.

In the Alpine area corresponding to a marine environment, a parallel subdivision was developed later. The Werfernian was first defined in Austria as a basal stage corresponding to the Buntsandstein (= Early Triassic). The *Scythian* stage was later taken as synonymous with the Werfenian on the basis of some levels from the Crimea, that were richer in fossils.

Both the equivalent stages the Virglorian and the *Anisian* were defined in Austria, the second name being the most widely used in the English literature. Together with the *Ladinian*, defined in Switzerland, the Anisian is regarded as more or less synchronous with the Muschelkalk (= Middle Triassic) in France and England; in Germany, the upper part of the Ladinian (Lettenkohle) is included in the Keuper.

The equivalent of the Keuper unit (= Late Triassic) comprises three stages (the *Carnian*, *Norian* and *Rhaetian* stages respectively), defined in the Italian, Austrian and Swiss Alps. The latter stage is sometimes taken as partly Jurassic and sometimes regarded as a substage; we prefer to consider it a stage of the Triassic system, following the presently accepted custom (Jurassic Symposium, Luxembourg, 1971). It seems from both faunal and lithological considerations to correspond to a short interval of time.

This generally accepted classification (Figure 1) suggests two comments:

(1) The Early, Middle and Late subdivisions have one, two and three stages respectively corresponding to them. This disproportion will be considered further according to radiometric dates.

(2) Initial definitions of the stages from the Germanic facies were established without good references to pelagic fossil content. The Alpine sea was located in an environment very remote and different from those of Australia, Argentina and West Canada from where most of the Triassic radiometric dates are obtained. Therefore the equivalences between the dated formations and the accepted stage names must be considered with all the necessary reservations.

In Figure 2 we present two ammonite biozonations prepared by H. Zapfe for this book, according to the most recent data. There is not a complete agreement between authors to place the stage boundaries within this zonation. For our present purpose, attention must essentially be paid to the biozone position around the Anisian–Ladinian boundary (compare with discussion of NDS 196). A thorough discussion of the position of the Triassic boundaries may also be found below (Zapfe, Chapters 35).

3 THE BASE OF THE TRIASSIC SYSTEM

Several Late Permian ages are available. The Gyranda volcanics from Australia (PTS 357) are pre-Triassic but post-Kazanian, and give an age of

	SUBSTAGES	TETHYS	NORTH AMERICA
Rh	RHAETIAN s.s.	*Choristoceras marshi*	*Choristoceras marshi*
	SEVATIAN	*Rhabdoceras suessi*	*Rhabdoceras suessi*
No	ALAUNIAN	*Himavatites columbianus*	*Himavatites columbianus*
		Cyrtopleurites bicrenatus	*Drepanites rutherfordi*
	LACIAN	*Juvavites magnus*	*Juvavites magnus*
		Malayites paulckei	*Malayites dawsoni*
		Guembelites jandianus	*Mojsisovicsites kerri*
Ca	TUROLIAN	*Anatropites spinosus*	*Klamathites macrolobatus*
		Tropites subbullatus	*Tropites welleri*
		Tropites dilleri	*Tropites dilleri*
	JULIAN		*Sirenites nanseni*
		Trachyceras austriacum	*Trachyceras obesum*
		Trachyceras aonoides	*Trachyceras desatoyense*
La	LANGOBARDIAN	*Frankites sutherlandi*	*Frankites sutherlandi*
			Maclearnoceras maclearni
		Protrachyceras archelaus	*Meginoceras meginae*
			Progonoceratites poseidon
	FASSANIAN	*Protrachyceras curionii*	*Protrachyceras subasperum*
		Nevadites - Zone	*Gymnotoceras occidentale*
An	ILLYRIAN	*Aplococeras avisianum*	*Gymnotoceras meeki*
		Paraceratites trinodosus	*Gymnotoceras rotelliforme*
	PELSONIAN	*Balatonites balatonicus*	*Balatonites shoshonensis*
	BITHYNIAN	*Anagymnotoceras ismidicum*	*Anagymnotoceras varium*
		Nicomedites osmani	
	AEGAEAN	*Paracrochordiceras* Beds	*Lenotropites caurus*
Sc	SPATHIAN	*Tozericeras pakistanum*	*Neopopanoceras haugi*
		Tirolites carniolicus	
			Subcolumbites Beds
		Tirolites cassianus	*Columbites*-Beds
	SMITHIAN	*Wasatchites spiniger*	*Wasatchites tardus*
		Flemingites flemingianus	*Euflemingites romunderi*
	DIENERIAN	*Vavilovites markhami*	*Vavilovites sverdrupi*
		Meekoceras kraffti	*Proptychites candidus*
	GRIESBACHIAN	No index species	*Proptychites strigatus*
		Ophiceras tibeticum	*Ophiceras commune*
			Otoceras boreale
		Otoceras woodwardi	*Otoceras concavum*

No Ammonites in the Western Tethys

Figure 2. Substages and ammonoid zonations of the Triassic system. Preliminary draft of zonations by L. Krystyn based on Tozer (1971; 1978); scheme proposed by H. Zapfe.

244 ± 7 Ma (ICC). In the Sydney Basin, the Dapto–Saddleback latite intrudes or is interbedded in Kazanian sands; its age of 251 ± 5 Ma (NDS 205) is certainly older than the Permian–Triassic boundary. Former data from pre-Scythian rocks from the USSR (PTS 346) lead to concordant ages of 239–244 Ma (ICC). Webb (1981) also reports data from Russian laboratories leading to ages of 249 ± 7 Ma (ICC) for Late Permian rocks,

244±16 Ma (ICC) for lowermost Triassic levels, and 233–247 Ma (ICC) for Early Triassic levels. In the same area, an old result of 248±15 Ma (ICC) was obtained from the Semeitan lavas situated at or near the Permian–Triassic junction (PTS 338). One should remember that in this area, the stratigraphical correlations can generally be regarded as accurate. More recently a Scythian (Werfenian) gabbro collected from Yugoslavia was found to be 248±7 Ma old (NDS 158). From all these data, it is possible to propose an age of 245±5 Ma (that is, between 240 and 250 Ma) for the Permian–Triassic boundary. The mean age may be regarded today as the most probable, although further investigations are needed to reduce the uncertainty proposed above. According to Priem and Bon (this volume), an age as old as 255 Ma would not be surprising. The mean age of 245 Ma is slightly higher than that proposed by Lambert (1971) and obviously agrees with the proposals of Armstrong (1978) and Webb (1981 and this volume), which are mostly based on the same original data.

4 THE ANISIAN STAGE BOUNDARIES

The base of the Anisian stage can be illustrated with the data obtained from the Puesto Viejo fm of Argentina. This formation, initially regarded as Anisian, is now considered as Scythian. It includes volcanics, the K–Ar whole-rock apparent ages of which are concordant: 237±4 Ma, ICC (NDS 186). It is not clear from the original paper (Valencio *et al.*, 1975) whether the stratigraphical attribution does or does not depend on the radiometric dating results and a preliminary assumed numerical time scale. This kind of circular argument is difficult to eliminate from the results presently available from South America. The results of Anderson and Cruickshank (1978) and Bonaparte (1978) now provide confirmation for a Late Scythian age (see NDS 186).

The post-Smithian (Late Scythian) age of the plutons of the Gympie block (Australia, NDS 195) is well established and corresponds to an apparent age of 236±5 Ma. The Scythian–Anisian boundary may thus be estimated as being 239±5 Ma; this interval cannot be reduced, since an age of 234 or 244 Ma cannot be fully excluded. The data discussed in NDS 208 and 209 do not help to solve this problem. If we consider the number of biozones for the two stages Scythian and Anisian and the abstracts NDS 186 and 195, the date of 237 Ma would be more probable; but until more data are available, we prefer for the time being to put the boundary halfway between the two better calibrated boundaries of 245 and 233 Ma (see Figure 3).

The top of the Anisian stage is well documented. All the dates quoted above are older than 235 Ma. At the Anisian–Ladinian boundary, the Neara volcanics in Australia overlie dated plutons and are themselves intruded. All three units are radiometrically dated. This allows us to propose an age of between 235 and 240 Ma for the Anisian–Ladinian boundary in Australia as suggested by Webb (NDS 194). A pre-Carnian date of 235±8 Ma (NDS 197) reinforces this evaluation in the same area.

Moreover, an independent study in the Alps by Hellmann and Lippolt (1979) has established the ages of bentonites probably located very near the boundary. Although the sample location is in the vicinity of the area where stages were defined, the stratigraphy does not appear as definitive. For Hellmann (NDS 196) an apparent age of 233±5 Ma may be deduced from the numerous measurements made. A careful study of the data actually obtained shows that the scattering of the dates is greater than the assumed analytical uncertainties, even if one considers only the best chronometers used (high sanidine). If one does not reject five of the 18 values obtained from these sanidines, the apparent age becomes 231±16 Ma. If one rejects only the most deviating analytical result (253±11 Ma), the mean age becomes 230±12 Ma. Finally, four data obtained for an Early Ladinian level, just above the other samples, give a younger mean apparent age of 225±4 Ma. The Anisian–Ladinian boundary is located between these ages of 225–233 Ma in the Alps and 235–240 Ma in Australia. The time span of 4 Ma around 233 Ma appears quite certainly to include the boundary, the two extreme values being equally probable.

5 THE CARNIAN STAGE BOUNDARIES

The numerical age of the Ladinian–Carnian boundary may be approached through the plutons intruding the Neara volcanics. They gave ages of 230–235 Ma (NDS 194). The Sugars basalts, also in Australia, are regarded as Carnian to pre-Carnian in age and gave apparent ages of 229 and 232 Ma (NDS 193). The Djuan tonalite (NDS 197) also gives a maximum age of 235±8 Ma for the boundary discussed.

In Argentina, the Los Rastros and Ischigualasto formations contain an excellent plant and reptile fossil assemblage. Interbedded basalts probably related to Carnian levels gave an apparent age of 229±5 Ma (NDS 187). The level dated is not clearly related to a specific formation. The reservations made above (NDS 186) as to the rather unclear stratigraphical assignation, in the original paper, of the levels dated must be kept in mind here.

The Ladinian–Carnian boundary may be proposed as being 229±5 Ma.

The dates obtained from the Mt Monzoni rocks (NDS 201) and Predazzo granite (PTS 361), which both intrude Early Carnian levels in Italy, differ depending on authors and samples. They sometimes lead to ages older than the well-defined Anisian–Ladinian boundary: the K–Ar ages lie between 229 and 233 Ma for the Mt Monzoni and spread from 229 to 239 Ma for the Predazzo pluton. Ferrara and Innocenti (1974) proposed a 'mean age' of

220 Ma (ICC) and draw a Rb–Sr isochron of 214±8 Ma (ICC). The origin and interpretation of these internally inconsistent results remain to be clarified: we have not obtained further comments from the authors on this question.

The age of 220±7 Ma (ICC) quoted for a 'Middle Muschelkalk' basalt from Provence (NDS 200) appears rather young.

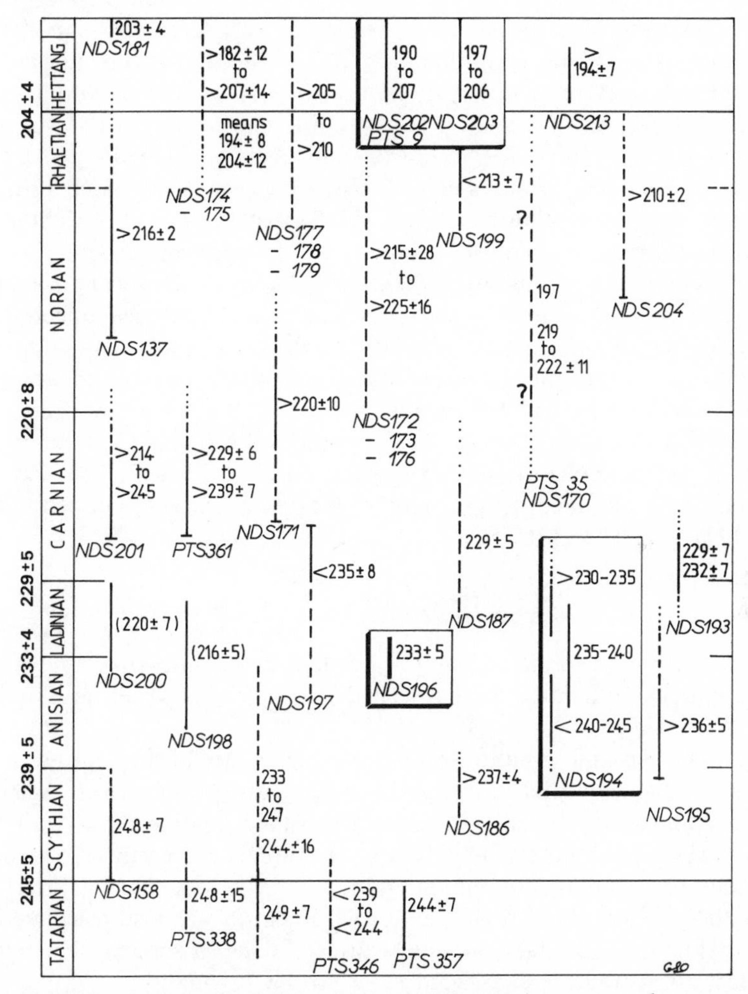

Figure 3. Radiometric ages of Triassic levels. The apparent ages shown are given with their 2σ strictly analytical uncertainty. Poorly reliable ages are given in parentheses. The best documented ages are underlined. NDS samples refer to the abstracts given in this volume, PTS samples to abstracts given in the Phanerozoic time scale, Geological Society of London, 1964 and 1971.

The Carnian–Norian boundary is documented with very inaccurate dates in British Columbia. The Kaketsa stock intrudes Carnian rocks; the two K–Ar apparent ages of 222±12 (hornblende) and 218±16 Ma (biotite) are thus minimal ages for the intruded Carnian rocks (NDS 171). Both analytical and stratigraphical uncertainties are high. The Takla volcanics of the same country were radiometrically dated by the Rb–Sr and K–Ar methods. The ages of 223±56 (NDS 172), 215±28 (NDS 173) and 225±16 (NDS 176) may be regarded as minimum ages due to low-grade metamorphism. These rocks are stratigraphically correlated with Late Carnian to Late Norian levels with local continuation up to the Early Sinemurian. The Hotailuh batholith in British Columbia was first quoted as younger than Norian (PTS 35) and revealed a K–Ar biotite age of nearly 197 Ma (ICC), but more recently new K–Ar hornblende ages of 218±11–227±11 Ma (ICC) were obtained (Wanless *et al.*, 1973; Anderson, 1980, NDS 170). This long-lived composite plutonic centre should not be heavily weighted as a time-scale calibration point according to R. L. Armstrong.

A convenient interpretation of these Canadian rock dates cannot be given because of the high stratigraphical, geochemical and analytical uncertainties, in spite of the numerous results available. Without better data from other parts of the world, the Carnian–Norian boundary cannot be accurately located. However, it will be assumed here that the time span 220±8 Ma includes this boundary because it cannot be older than 229 Ma, an age found for well-identified Ladinian–Carnian rocks (NDS 187, 193, 194), and 220 Ma is also the mean age of samples located above and below the limit: NDS 171, 172, 173, 176.

6 THE NORIAN STAGE

A dacite from Australia probably equivalent to volcanics overlain by Rhaetian sediments with microflora revealed a K–Ar whole-rock age of 213±7 Ma (NDS 199).

The Mae Sariang granite complex in Thailand locally intrudes Late Triassic (Ladinian–Norian) rocks and has been dated in Hanover. According to von Braun *et al.* (1976), the granite must be considered as post-Early Norian. The apparent age of 210±2 Ma (three concordant K–Ar biotite ages) may be taken as a minimum age for the Middle to Late Norian stage (NDS 204). A younger Rb–Sr whole-rock isochron was also calculated, but the individual Rb–Sr data do not really fulfil the geochemical conditions necessary to attribute a significant value to this calculation.

The Amsterdam group has measured the age of plutons intruding fossiliferous Norian rocks of Indonesia. K–Ar apparent ages of four biotites and one hornblende led to an accepted value of 216±2 Ma as the minimum age of Norian rocks. A Rb–Sr isochron of 213±5 Ma was also calculated

(Priem and Bon, this volume; and Priem, NDS 137). These ages are slightly older but more accurate than the old result of nearly 210 Ma obtained by Edwards and McLaughlin (1965). They supersede the very young single K–Ar age discussed in PTS 71.

In British Columbia, the batholiths intruding the Takla group volcanics (NDS 174–175) are post-Norian. Numerous radiometric measurements have given values between 182 and 207 Ma, but the 'mean' age of 190–200 Ma appears to be the best evaluation for the time of intrusion. In the same country, the Nicola group of Carnian and Norian age are intruded by batholiths, pebbles of which are found in conglomerates of Sinemurian age. Numerous concordant ages between 205 and 208 Ma were obtained by various methods and techniques (NDS 177–179). Slightly older ages were also found locally (NDS 178). All these ages are not accurately located in the stratigraphic column; they must be regarded strictly as minimum ages for some event in the Norian and can be considered tentatively as minimum ages for the top of the Norian.

In these three areas, the top of the Norian stage may be located within the 210 ± 5 Ma time span. An older age of 216 Ma is proposed by Armstrong. This is not fully refutable. We prefer here to consider the possibility of a non-terminal Norian age for the Indonesian intrusions (NDS 137) and a pre-Rhaetian age for the Australian dacite (NDS 199). But, in fact, the age we propose is crucially linked to our intuitively attributing a short duration to the Rhaetian stage, regarded as a local substage by several authors. Consequently, for the time being, the reader should keep in mind that the Norian–Rhaetian boundary is not accurately known.

7 THE TRIASSIC–JURASSIC BOUNDARY

The previously discussed post-Norian ages of British Columbia (NDS 174, 175, 177, 178, 179) may be recalled here as a possibility for dating the Triassic–Jurassic boundary.

They are only known as being post-Norian to pre-Sinemurian in age, and may therefore characterize Rhaetian as well as Hettangian levels.

In the USA, dates were obtained on the Palisade sill, first regarded as Late Triassic (PTS 9) (mean age 197 ± 3 Ma).

But the stratigraphic evidence available only led to a post-Carnian age probably contemporaneous with the Early Liassic. Numerous data from the Hartford Basin (USA) have been gathered by J. Webb (NDS 202, 203). The ages of 190–207 Ma (NDS 202) and 197–206 Ma (NDS 103) may be regarded as characteristic of the Early Jurassic rocks. Neither real Rhaetian nor well-identified Hettangian rocks have been radiometrically dated so far. But the data obtained from the Massif Central (France) probably reveal a minimum age for Hettangian levels (NDS 213, age 194 ± 7 Ma). From the above quoted evidence, it seems that the time span 204±4 Ma (200–208 Ma) probably includes the Triassic–Jurassic boundary. This span includes both estimates of Armstrong and Webb discussed in their own chapters in this book, and does not change the (ICC recalculated) proposal by R. St John Lambert (1971).

8 DISCUSSION AND CONCLUSIONS

Figure 3 gives the radiometric dates discussed here. Three series of data are especially underlined as giving the dates considered here as the best documented. As a whole, this figure shows a considerable progress in the knowledge of Triassic levels compared with the two inaccurate dates available 15 years ago.

Here we essentially used the Alpine facies stage names for their far-reaching capacity to export their faunal content. The relation between Alpine facies and Germanic facies level is known to remain tentative even in Europe. The use of these marine stage names in other continental areas such as the Gondwana Province (South America, South Africa, Australia) probably conceals great problems of correlation.

According to the results presently available, the duration of the Triassic system may be accepted as around 40 Ma between a limit somewhere within 245±5 Ma and one somewhere within 204±4 Ma. The Early Triassic subsystem, if equivalent to the Scythian stage, is rather short; its top is most probably located within 239±5 Ma.

The Middle Triassic subsystem is nearly twice as long as the Early Triassic one. The middle of the Middle Triassic is accurately located around 233 Ma, according to the coincidence between dates obtained from different areas in the world, including the area of definition of the stages in the Alps.

The Late Triassic subsystem includes three stages. According to the radiometric dates, which are not yet sufficiently precise, the duration of the Norian stage is probably equal to the duration of the two other stages, between perhaps 220 Ma and perhaps 208 Ma. The Rhaetian stage is the shortest stage of the Triassic system if one considers the small number of ammonite biozones identified in it. But the Ladinian stage also appears to be short. This, however, needs confirmation.

We have tried not to use the process of extrapolation linked with the number of biozones. If one considers the number of ammonite biozones above and below the Ladinian–Carnian boundary, here adopted at 229±5 Ma, one obtains a mean duration more than *twice as large* for the Late Triassic zones.

In summary, we have emphasized here:

—that the Carnian and the Norian are long stages;

—that the different subsystems, stages and biozones of the Triassic system are of very different durations;

—that the dates 245 ± 5, 239 ± 5, 233 ± 4, 229 ± 5, 220 ± 8, 208?, 204 ± 4 Ma include the boundaries of the five stages of the Triassic system.

Acknowledgements

We are greatly indebted to Richard L. Armstrong, Jean Claude Baubron, Klaus N. Hellmann, Harry N. A. Priem and John A. Webb for their numerous comments on earlier drafts of this chapter and the resulting time scale and for the many new data, often unpublished, with which they provided us. Leopold Krystyn and Helmuth Zapfe greatly helped in guiding us on biostratigraphic questions; their valuable contribution to this chapter is acknowledged. This chapter is a contribution to IGCP Project 133.

(Manuscript received 26-12-1980)

Section VI

The Jurassic to Palaeogene times

Numerical Dating in Stratigraphy
Edited by G. S. Odin

31 Radiometric dating of the Albian–Cenomanian boundary

GILLES S. ODIN and JOHANNES C. HUNZIKER

1 INTRODUCTION

1.1 Object of the study

In the Paris Basin, the Albian and Cenomanian type areas have recently been reinvestigated (Juignet, 1974; Rat *et al.*, 1979). The boundary is well recognized all over the basin, which is sedimentologically well known. Twenty-five glauconitic samples were selected for dating purposes by K–Ar and Rb–Sr methods.

1.2 Previous data on the boundary

The data reassembled by Casey (1964, p. 199) led this author to propose a theoretical time scale 'purely as a basis for discussion in the absence of a more positive scale of calibration'. The Albian is then situated between around 102.5 and 109 Ma (ICC recalculated) and the Cenomanian stage between 102.5 and 96 Ma. Lambert (1971, p. 15) proposed a new interpretation of the same data lowering the Albian–Cenomanian boundary to 97 Ma (ICC). The same conclusion had been proposed earlier by Obradovich (1964, p. 13). New work performed in the Paris Basin has shown that, in this area, the base of the Cenomanian must be older than 94 Ma (ICC) and situated between 94 and 96 Ma (Juignet *et al.*, 1975: Odin, 1975). In the Western Interior Basin of North America, the possible equivalent of the Albian–Cenomanian boundary proposed by Kauffman (1969) as well as by Obradovich and Cobban (1976) partly using data of Folinsbee *et al.* (1963) lies near 96 Ma (ICC). Nevertheless, due to the endemism of faunas of the Western Interior Cretaceous sea, noticeable uncertainties remain concerning the contemporaneity of the North American and NW European

Early–Late Cretaceous boundary. Recently published results obtained in northern France and southern Belgium by Elewaut and Robaszynski (1977) and Keppens *et al.* (1978b) confirm an age older than 94 Ma for the base of the Cenomanian as defined in the NW European basins.

The aims of this chapter are: (1) to present a complete view on the available analytical results obtained in close relation to the lithostratigraphic definition; (2) to show how glaucony may be used as a chronometer; and (3) to give some ideas concerning the reproducibility and uncertainties related to the use of glaucony as a chronometer.

2 STRATIGRAPHICAL UNCERTAINTIES

2.1 Nature and position of the sampled localities

The samples have been collected in four localities, from west to east: the Pays d'Auge, the Pays de Caux, the Pays de Bray and the Boulonnais (see Figure 1). The evolution of these four domains of the same epicontinental platform is not always parallel to the last detail as shoals of the often partly emerged platform may sometimes have separated one domain from another.

In the Pays d'Auge, samples have been collected from the most littoral and condensed facies (sample G.347) most likely from the Late Albian and (G.348) from the Middle to Late Albian (Juignet *et al.*, 1967; 1975). The relative imprecision in stratigraphy here is due to an extreme condensation of the Albian to one basal glaucony horizon representing the whole Albian and including the base of the Cenomanian. The fossils are sometimes reworked and give stratigraphical maxima ages. The greatest uncertainty is attained in the quarry of Gouvix. Samples G.466, G.467 and G.468 have been collected from the base to the top in a series comprising the whole condensed Albian as well as the Early Cenomanian. In addition, the fossils are visibly and extensively reworked in the uppermost layers. In this special case the main aim of isotopic analysis was to see whether all the glauconies have been reworked or if the different glaucony-rich horizons could still be distinguished. Only the lowermost glaucony is representative for a strictly Albian sedimentation. The other samples most likely reveal maximum ages for the Early Cenomanian.

In the Caux region, more to the east and north (Figure 1) the series is definitely better developed and pelagic. Juignet (1974) proposed to establish here a parastratotype, thus allowing the completion and subdivision in greater detail of the stratotype of the region of Le Mans situated in a rather littoral zone more to the south. Three samples have been collected in Cauville, three others in Octeville and one in St Jouin in the cliffs facing the English Channel. In these profiles the statigraphy has been studied in great

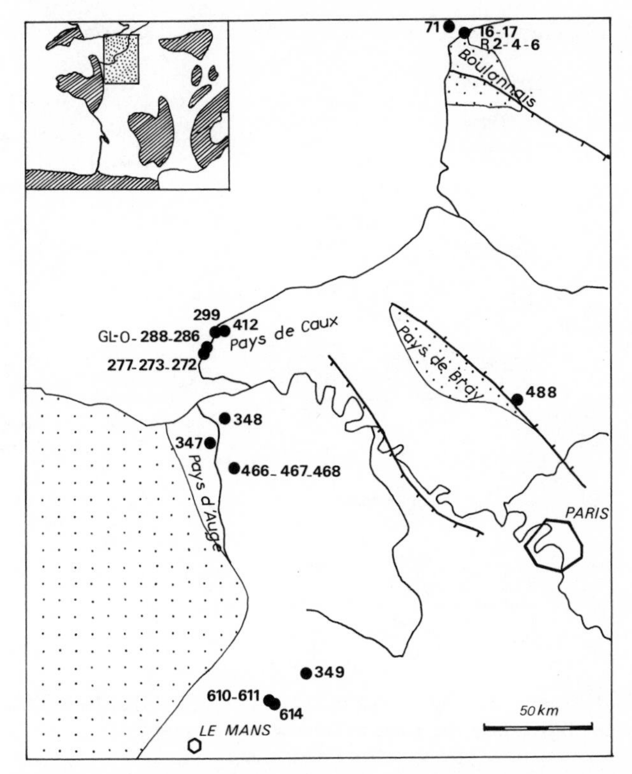

Figure 1. Area of sampling. The outcrops older than the Albian–Cenomanian boundary are shown stippled. Three main faults are shown. The folded areas are hatched on the little map.

detail (Figure 2). By means of ammonites, these samples can be located precisely and directly with respect to established biozonations.

In the Bray area as well as in the Boulonnais, the outcrops have been formed by a deformation already active at the time of formation of the sampled Cenomanian glauconies. These deformations along wide ridges oriented NW/SE led to the formation of Jurassic to Middle Cretaceous windows surrounded by more recent terrains. At St Paul, near Beauvais in the Bray area, and near Wissant, in the Boulonnais, glaucony-rich samples from the base of the Cenomanian have been collected (Destombes and Destombes, 1963). These horizons are well calibrated by rich ammonitic and molluscan faunas.

2.2 Biostratigraphical data

The succession of ammonite zones adopted here is the one used by Juignet (1974) modified in Destombes (1979), see Table 1. The Albian has

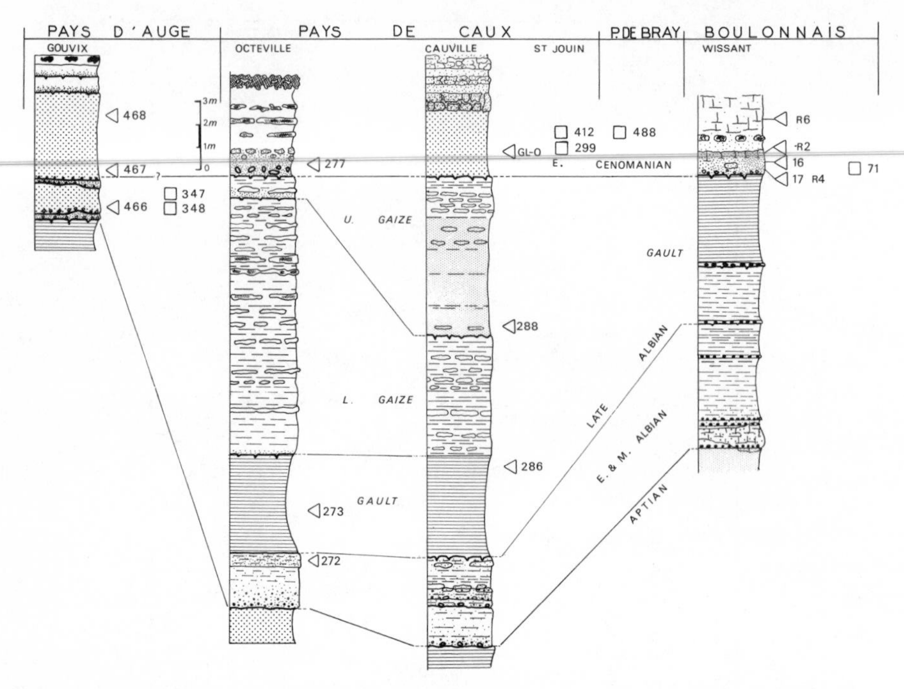

Figure 2. Correlations between the sampled outcrops. Further to the south Cenomanian transgression overlays the Jurassic limestone and may rework Albian glauconitic sand deposits.

been divided into eight zones and subdivided into 23 subzones. The mean duration of such a subzone is of the order of 0.5 Ma (accepting an Albian duration of 12 Ma). The Cenomanian has been subdivided into nine zones and terminates with a biostratigraphically unnamed horizon. The mean duration of a Cenomanian zone can be estimated to be around 0.5 Ma according to a probable total length of the Cenomanian of around 5 Ma (Odin, 1978a).

In the sedimentary sequence of the Pays de Caux, considered to be quite complete, all the present subzones of the Late Albian and the zones of the Early Cenomanian are enclosed between two depositional breaks. These depositional breaks form a hard ground. In the case of the Albo–Cenomanian hard ground, a boring fauna exists also. In addition, this boundary is marked by a sedimentational discontinuity and further amplified by rinsed furrows. However, the present data seem to suggest that, in the Pays de Caux, no faunal lacune can be found at the Albian–Cenomanian boundary. On the other hand, several faunal lacunes are obvious below.

Table 1. Ammonite zonation of the Albian and Cenomanian Paris Basin. The accepted zonation is that of Juignet's thesis (1974), recently modified by Destombes (1979). In particular, the *Douvilleiceras mammillatum* zone has been split into two zones: *S. dutempleana* and *O. raulinianus*. The Cenomanian zones and Albian subzones are of equivalent duration: 0.5 Ma.

Stage	Substage	Zones	Subzones	P. de CAUX
TURONIAN		*Mammites nodosoides*		
CENOMANIAN	Late	'Horizon A'		
		Metoicoceras gourdoni		
		Metoicoceras geslinianum		
		Calicoceras naviculare		
	Middle	*Acanthoceras jukesbrownei*		
		Turrilites acutus		
		Turrilites costatus		
	Early	*Mantelliceras dixoni*		
		Mantelliceras saxbii		
		Hypoturrilites carcitanensis		
ALBIAN	Late	*Stoliczkaia dispar*	*Mortoniceras perinflatum* *Arraphoceras substuderi*	
		Mortoniceras inflatum	*Callihoplites auritus* *Hysteroceras varicosum* *Hysteroceras orbignyi* *Dipoloceras cristatum*	
	Middle	*Euhoplites lautus*	*Anahoplites daviesi* *Euhoplites nitidus*	
		Euhoplites loricatus	*Euhoplites meandrinus* *Mojsisovicsia subdelaruei* *Dimorphoplites niobe*	
		Hoplites dentatus	*Anahoplites intermedius* *Hoplites spathi* *Lyelliceras lyelli* *Isohoplites eodentatus*	(*)
	Early	*Otohoplites raulinianus*	*Otohoplites bulliensis* *Otohoplites larcheri*	
		Sonneratia dutempleana	*Protohoplites puzosianus* *Cleoniceras floridum* *Sonneratia kitchini*	
		Leymeriella tardefurcata	*Leymeriella regularis* *Hypacanthoplites milletioides* *Farnhamia farnhamensis*	
Aptian		*Hypacanthoplites jacobi*		

(*) Time of deposition

2.3 Sedimentological environment

Four types of sediments contain the green grains.

(1) The glauconitic gravels and sands (green sands) are typical for the Early Albian levels of the eastern side: Boulonnais and Pays de Bray. In the Pays de Caux, where less deep facies lead to ferruginous sands, the Middle Albian is not glauconitic.

(2) The Gault clays begin during the Middle Albian on the eastern side in

the Boulonnais and the Pays de Bray, whereas on the western side, in the Pays de Caux, they only begin during the Late Albian.

(3) The Gaize facies is a silica–glauconitic marl of Late Albian age, present in the Pays de Caux and Pays de Bray. This facies, sometimes related to a Vraconian substage, is absent in the Boulonnais. The Gaize facies is marked by depositional breaks emphasized by phosphatization. This facies is less argillaceous than the Gault and is characterized by a fauna more rich in echinoderms than the preceding facies, indicating a more open ocean (Juignet, 1974). In the most occidental part of the basin, in the Pays d'Auge, the three above-mentioned facies are condensed to one glauconitic sand, the *glauconie de base*, a very glauconitic sand of the first type. Further to the south the Early Cretaceous deposits are absent and the Cenomanian chalk lies on Jurassic limestones, G.349, but reworking of previously altered Albian may be supposed.

(4) The fourth facies is the glauconitic chalk overlying the three preceding facies at the base of the Cenomanian. The chalk begins marly and rich in silicified horizons, but later on, towards the end of the Cenomanian, becomes more and more carbonate-rich. Thus, a detrital sedimentation with siliceous character is gradually linked to a carbonate pelagic sedimentation by an intermediate facies—the Gaize. The abundance of silica and silicic sponges throughout the Gaize leads to the conclusion of a fresh-water arrival, whereas during the Late Cretaceous, warmer water favours the development of pelagic carbonate-rich microfaunas.

On the sea-floor, at depths of over 60 m only, there is enough time for formation of glaucony during a slowdown of deposition. The ocean seems to become more and more open and deep between the Early Albian and the Cenomanian. The whole series shows many depositional gaps; these gaps are attributed to epirogenetic movements, which lift some more or less localized areas to higher energy levels, thus stopping deposition (Juignet 1974). The depositional gaps very often correspond to zone and subzone boundaries. The observed facies series leaves no possibilities of emergence during Albian and Cenomanian times: the stage boundary takes place in a marine environment.

3 GEOCHEMICAL UNCERTAINTIES

3.1 Genetic uncertainties

The closing of the glaucony chronometer can be linked to the stratigraphically dated moment by means of an examination of the sedimentological characteristics of the green pellets. The diffractometric analysis shows that the green grains are uniquely composed of highly-evolved authigenic minerals close to the glauconitic mica pole. According to our observation on Recent and old sediments (Odin, 1975), this evolution takes place over a time span of several 10^5 years. It is this time span, lacking detrital deposition, that should be localized in a lithological series. This evolution can only begin in the presence of adequate substrates for glauconitization. The end of this phenomenon, the closing of the chronometer to the surrounding marine environment, has to be fixed. The following remarks only concern K–Ar ages, as we have insufficient specific Rb–Sr data. In agreement with Lamboy (1976), we assume that in these samples the substrates were carbonate debris: molluscs, echinoderms, ... Inheritance of radiogenic isotopes therefore can be excluded, especially as the very complete evolution of the glaucony allowed it to reach equilibrium with the marine environment. Further to the west, in the continent border of the Massif Armoricain, the milieu becomes more neritic, and there, detrital quartz or micas may be used as substrates for glauconitization. For the precise timing of the end of this evolution it is assumed that *the starting of the chronometer takes place when the green pellet is isolated from the marine environment.* This isolation is a consequence of a noteworthy burial in the case of a little-evolved glaucony, while if evolution has already reached a higher stage, during a longer depositional gap, the potassium-rich glauconitic minerals do not exchange any further. In this case, the closing is progressively reached prior to burial (glauconies with more than 8% K_2O).

The highly-evolved stage of glauconitization together with the presence of hard grounds or of perforated surfaces topping the glauconitic strata leads to the conclusion that the glaucony could have been isolated from its marine environment (i.e. closed) *before* the deposition of the following bank. The apparent age of the highly-evolved glaucony chronometer, therefore, should be practically identical with the deposition of the ammonites dating the stratum (actually slightly younger by some 10^5 years), but older than the deposition of the following banks with their characteristic fauna.

The absence of a certain number of subzones (see Table 1) and the condensation of certain levels, as well as the existence of locally reworked faunas, give evidence for a possible reworking of some of the glaucony grains. This reworking could only happen in a marine environment of more than 20–30 m depth, as no trace of oxidation can be seen on the grains. Traces of such a reworking are only visible in the most occidental zones (Pays d'Auge). For the Pays de Caux, Figure 3 proposes a reconstruction of the possible chronology of glauconitization around the Cenomanian–Albian boundary. Sedimentary thickness is visualized in A and the formation base is the transitional gap Middle Albian–Late Albian.

Concerning the strata of the Late Albian two possibilities can be proposed for the glauconitization: either during the missing five subzones of the Gault clays or during one subzone beneath the Upper Gaize. The more likely time

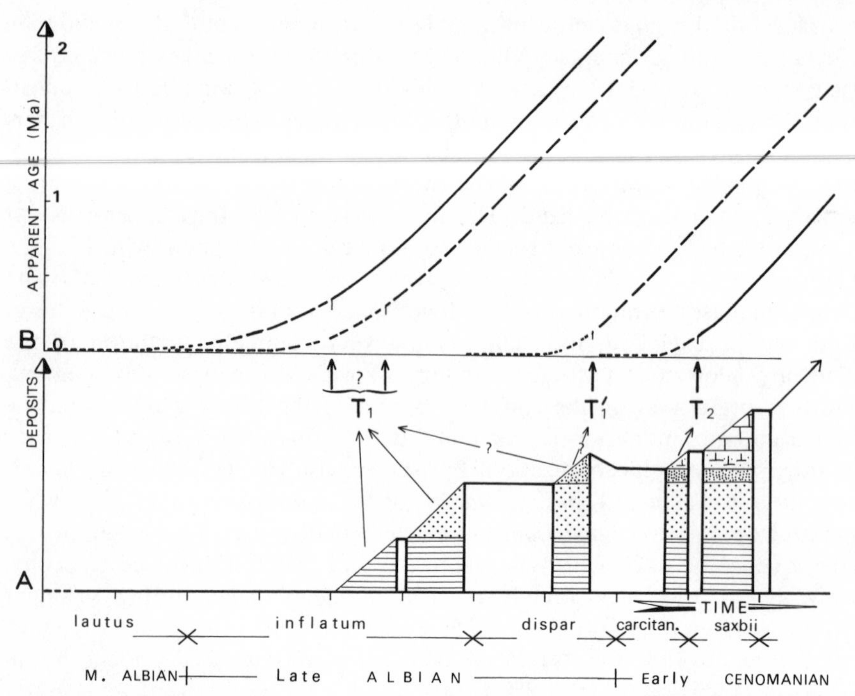

Figure 3. Correlation between accumulation of sediments (A) and the closure of the glaucony chronometer (B). The duration of the zones shown is arbitrarily proportional to the number of subzones. Breaks of sedimentation are shown open. Accumulation of sediments began during the second part of the *inflatum* zone. At this time the first glauconitization, which began a long time ago, is practically finished and the chronometer accumulates argon; the age zero is at the point T_1. A second possibility of closure is shown in T_1' but no analytical data correspond to this time. The glauconies collected from the *carcitanensis* zone began their evolution during the break of sedimentation shown between Albian and Cenomanian; the closure probably occurs in T_2 (see also Odin and Dodson, this volume).

for a glauconitization is given by the lack of five subzones at the Middle–Late Albian boundary.

This process could have started during the Middle Albian and could have terminated during deposition of the Gault clays (point T_1, Figure 3) with a certain amount of radiogenic argon already present. Glauconitization could have also started later, but certainly finished before the end of the Gault clays, stopping ionic exchanges. The missing subzone between the Lower and the Upper Gaize represents a second possibility for the glauconitization. This process would have been finished at a closing point around T_1' close to the deposition of the Upper Gaize when the traces of the characteristic fauna of the *Mortoniceras perinflatum* subzone (*S. dispar* zone) were buried (see Table 1). Reworking cannot be ruled out as no hardened surface is found at the base of the Gault.

The glauconitization of the basal Cenomanian occurred on carbonatic substrates that had been immersed during the transgression. The *verdissement* took place as soon as these carbonate substrates reached water depths of more than 60 m. The evolved grains were buried after some 10^5 years together with the characteristic fauna of the zone of *Hypoturrilites carcitanensis.* The closing of the chronometer (beginning of a noteworthy accumulation of radiogenic argon, T_2, Figure 3B) took place at the latest during the formation of the hard ground underlying the bed corresponding to the *Mantelliceras saxbii* zone. The zero age probably coincides with the top of the *H. carcitanensis* zone. The sedimentological study of the Pays de Caux leads to the conclusions: that the zero age has been reached by these evolved glauconies; that at least two glauconitization processes have taken place, the zero age of which would represent the base of the Late Albian and on the other side the end of the first Early Cenomanian zone. A third possibility, during the *S. dispar* zone, see Figure 3, cannot be ruled out. In the Boulonnais, the palaeontological studies of Destombes and Destombes (1963) show that at the top of the Late Albian one ammonite zone is missing. The Early Cenomanian, therefore, is deposited on a perforated and grooved surface of the Gault clays. Here glauconitization took place over a sufficient time span for the formation of a substrate and *verdissement* led to highly-evolved glauconies, the zero ages of which date a time during the first third of the Cenomanian. Reworking is not probable as locally the Gault is not glauconitic and only becomes glauconitic further north in the English Channel (Odin, 1967).

3.2 Historical uncertainties

Between burial and sampling, that is during the post-genetic history of the sediment, little has affected the analysed glauconies. The strata did not reach lithification except locally in the Boulonnais. Burial went as far as some tens to several hundreds of metres. Nevertheless, the Cenomanian of the Pays de Bray and of the Boulonnais have suffered modest deformation. Regarding the K–Ar system, glauconies seem to be very susceptible to deformation (Conard *et al.*, this volume).

In the recent past, the coarsest glaucony beds could have been affected by infiltrating water naps. This is definitely the case for the glaucony-rich beds overlying discontinuities at the base of the Cenomanian. With respect to surface alteration processes, all analysed samples show no traces of alteration. In relation to superficial alteration processes the Rb–Sr equilibrium is more critical than the K–Ar ages, the latter remaining unchanged under moderate climates (Odin and Hunziker, 1974).

4 ANALYTICAL UNCERTAINTIES

4.1 Potassium–argon analysis

4.1.1 Reproducibility of sampling

Potassium analyses have been performed on 300–500 mg samples of purified grains; for argon analysis, on the other hand, 100–150 mg samples were used. It has been shown by de la Roche *et al.* (1976) that for major elements even an aliquot of 50 mg is representative for GL-O. These homogeneous fractions were prepared by washing the sediment, sometimes after disaggregation with kerosene. After drying under temperatures of less than 70°C the samples were sieved into 1–0.5 mm, 0.5–0.16 mm and 0.16–0.10 mm sizes. This division into different granulometric fractions leads to a more thorough and homogeneous magnetic separation. Experience has shown that grains bigger than 0.5 mm are often mixed, whereas grains smaller than 0.16 mm are often less evolved and therefore geochemically less reliable. The magnetic separator was used to select the best grains, as described by Odin (this volume, Chapter 20). The degree of purification is controlled step by step by means of X-ray diffraction on powders (Odin *et al.*, 1975). By this procedure, it is possible to evaluate *a priori* the best possible grain class for a geochronological analysis (Odin, 1978a). The fraction 0.16–0.5 mm was used alone here. Subsequent ultrasonic cleaning in deionized water removes most of the weaker grains, late coating as well as most of the impurities held in the fissures of the pellets. After this treatment the sample is sieved again under water using a 0.16 mm sieve, rinsed in deionized water and dried during one day. For K–Ar analysis this sample is considered pure after a search for possible micas and mixed grains under the binocular microscope. A further diffractometric control allows the exclusion of all grains containing other than pure authigenic minerals.

The purified glaucony sample is then stored in a plastic bag for at least one month to attain hygoscopic equilibrium. The samples were weighed directly out of the plastic bags at normal temperature (18–25 °C), without prebaking or preliminary desiccation, in an atmosphere of around 60% humidity. As the Cenomanian and Albian samples are especially rich in K, the amount of adsorbed water is only small and the variation due to changing temperature and hygrometry can be ignored. On the other hand, desiccation or baking processes before weighing could easily lead to errors in the percent order for the weight, as reported by Flanagan (1969) for different powders (see also Odin and collaborators, Chapter 7).

4.1.2 Analytical conditions

Potassium analyses have often been performed in two different laboratories, using atomic absorption and flame photometry. As no systematic analytical deviation between the different methods could be detected, they were considered as equivalent for mean calculations. In all cases, every series of 12 analyses contained one 'glauconite GL-O' as reference. In addition, all the heavy errors can be excluded by means of a preceding X-ray diffractometric check giving preliminary K concentrations to ±5% (Odin and Hunziker, 1974; Odin and Dodson, this volume).

Concerning the analytical uncertainty on the K content, we adopt here a 2σ margin on both sides of the mean when at least three independent determinations are available. If not, we use a 2% conventional uncertainty for K including the whole analytical procedure for K and the uncertainty due to the possible difference of hygrometry between weighing for K and weighing for Ar, often performed several months apart.

Isotope dilution argon measurements have been performed at the Mineralogisches Institut, Berne, Switzerland, between 1973 and 1976. Argon was extracted in a Pyrex line; this line contains an induction-heated quartz Pyrex furnace, cleaned after every sample. A first purification with a clean U-tube, a copper–copper oxide furnace held at 550°C and a titanium sponge furnace held at 800°C is followed by a second purification with a titanium sponge furnace and an active charcoal finger. The second purification is directly linked to a Varian GD 150 mass spectrometer. The data acquisition comprises a Cary 31 amplifier and a chart recorder. Charts were evaluated manually from a set of ratios. Gas extraction was complete after 8 to 12 minutes heating, first to dim red glow, where most of the gas was liberated, and consequently turned up to around 1300°C. Ar-38 tracer (Clusius Zürich) calibration was systematically performed with each series of analyses by means of at least three standard mineral samples; GL-O standard was usually measured at least twice. The 38 spike was originally calibrated against P 207 muscovite using a value of 28.15 (nl/g). This value gave 24.69 nl/g for GL-O and 6.31 nl/g for muscovite Brione Bern 4 M, the international mean value of Brione 4 M (80 measurements in 17 laboratories) being 6.27 and of GL-O 24.8. Muscovite 4 M was initially used as internal standard to control the decrease of the ^{38}Ar spike. Later it was replaced by GL-O glaucony, adopting the recommended value of 24.8 nl/g. The error quoted on a single argon analysis (2σ at 95% confidence) is given with ±2% Ar tot./Ar rad. All samples, prior to extraction, have been baked at temperatures between 100 and 150°C for a time span of 3–24 h.

4.1.3 Isotopic data

The K–Ar analytical data are listed in Table 2. For comparison the international mean value of GL-O has been included as well as data on the basal Cenomanian of the Boulonnais published by Elewaut and Robaszynski (1977) that have also been measured in Berne. The decay constants are the conventional ones (ICC).

Table 2. K–Ar analytical results obtained on Albian and Cenomanian glauconies from the Paris Basin. Analytical uncertainties are given $\pm 2\sigma$; R2, R4 are taken from Elewaut and Robaszynski (1976). GL-O mean is the recommended value for the reference material 'glauconite GL-O (grains)' according to the data obtained in 15 laboratories. G.467 A and G.468 A are two glauconies remanié from Albian to Cenomanian which show intermediary apparent ages.

Samples	^{40}Ar rad. (nl/g)	(%)	Potassium K(%)	$^{40}Ar/^{36}Ar$ ($\pm$1%)	($^{40}K/^{36}Ar$) $\times 10^{-3}$($\pm$4%)	Conventional age ($x_0 = 295.5$)
Early Cenomanian						
R2	26.42±0.56	92.2	6.98±0.14	3813	622.2	94.8±3.1
R6	26.80±0.55	92.8	7.12±0.14	4137	683.1	94.3±3.0
R6	26.68±1.07	51.1	7.12±0.14	605.8	55.44	93.9±5.5
R4	26.64±0.62	86.5	7.40±0.15	2187	351.7	90.3±3.1
G.16 A	26.10±0.58	89.9	7.02±0.24	2915	473.5	93.2±3.8
G.17 A	26.52±0.56	94.1	7.25±0.10	5004	861.3	91.7±2.3
G.71 A	25.22±0.53	88.7	6.86±0.10	2616	422.7	92.3±2.3
G.277 A	24.72 ±0.38	91.5	6.57±0.10	3468	564.4	95.0±2.0
G.277 A	25.10	89.4	6.57±0.10	2793	438.7	
G.412 A	24.20 ±0.30	90.5	6.67±0.12	3069	511.6	91.6±2.0
G.412 A	24.49	92.0	6.67±0.12	3631	607.8	
G.488 A	23.62±0.50	94.0	6.43±0.17	4940	845.9	91.6±3.1
GL-O Berne	24.75±0.52	90.8	6.50±0.07	3217	513.9	94.8±2.0
GL-O mean	24.8±0.2	—	6.56±0.05	—	—	94.7±1.1
Clay GL-O	7.61±0.30	51.5	1.31±0.07	609.1	36.12	143±10
Late Albian						
G.288 A	25.52 ±0.33	87.3	6.41±0.18	2325	341.1	99.5±3.1
G.288 A	25.44	89.6	6.41±0.18	2830	427.6	
G.286 A	25.14±0.57	88.3	6.40±0.15	2522	379.3	98.4±3.2
G.273 A	25.17±0.58	86.9	6.42±0.21	2246	333.0	98.1±3.9
G.347 A	26.98 ±0.88	86.3	6.70±0.20	2131	305.0	99.1±4.4
G.347 A	26.09	91.3	6.70±0.20	3345	523.7	
G.348 A	25.47 ±0.30	89.2	6.44±0.19	2697	406.5	99.4±3.2
G.348 A	25.69	84.7	6.44±0.19	1910	271.1	
G.272 A	25.76 ±0.26	90.4	6.54±0.16	3082	473.5	98.6±2.6
G.272 A	25.78	92.8	6.54±0.16	4064	638.9	
G.466 A	25.42±0.76	66.9	6.38±0.26	890.7	100.1	99.6±5.0
G.467 A	25.40±0.58	87.8	6.57±0.14	—	—	96.8±3.0
G.468 A	25.26±0.57	88.7	6.50±0.11	—	—	97.4±2.7

4.1.4 *Discussion of the analytical data*

Several conclusions may be drawn from the analytical results reported in Figure 4 and Table 2. The Late Albian glauconies show only one geochemical event, precisely situated near 99 Ma (mean value of 11 analyses 99.0± 1.2 (2σ) Ma rejecting the data obtained on the reworked glauconies. The analytical reproducibility of this event in the various sections is better than

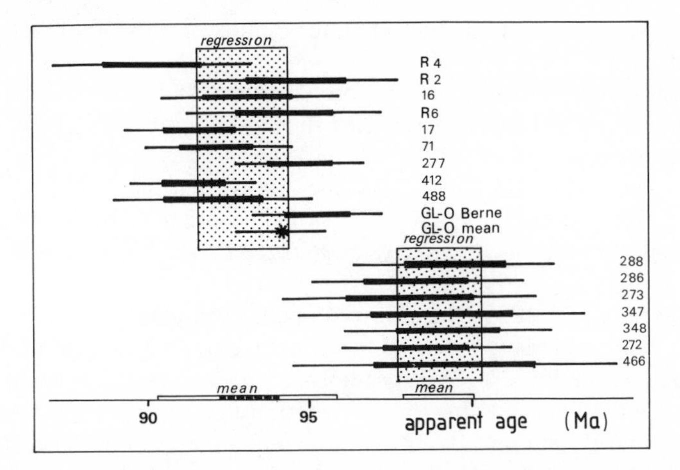

Figure 4. K–Ar apparent age of glauconies from the Paris Basin. Albian glauconies are on the right, Cenomanian on the left-hand side. 2σ analytical uncertainties are given (the heavier line is 1σ). The calculated values ($^{40}Ar/^{36}Ar$ *vs.* $^{40}K/^{36}Ar$) are also given: stippled area. These values are equivalent to the arithmetic mean values. The calculated uncertainty depends on the basic formula used. The 2σ uncertainty, 2 standard deviations, $=2\left\{\frac{\sum(x_i-\bar{x})^2}{n-1}\right\}^{1/2}$ is ±3.2 Ma for the 12 K–Ar Cenomanian glaucony ages; the mean uncertainty calculated with the formula $\left(\frac{\sum \Delta t^2}{n^2}\right)^{1/2}$ as shown in Table 5 is ±0.9 Ma (dotted bar) for the same series of samples.

the defined punctual analytical uncertainty. The apparent ages calculated for the Early Cenomanian show a higher scatter than the preceding ones, betwen 91.6 and 95.0 Ma (with the exception of R4) with a mean value of 93.2 ± 0.9 (see Figure 4). The smectite corresponding with GL-O has a higher apparent age, which demonstrates that this smectite was not in equilibrium with a marine environment during its burial, although for such clays, neoformation (crystal growth on the sea bottom) has often been proposed. The data have also been used to calculate $^{40}Ar/^{36}Ar$ plots, with the idea that the samples from one stratigraphic unit were cogenetic: they were *closed at the same time*, in equilibrium with an isotopically *homogeneous environment*, so that this part of the presumption for an isochron treatment is fulfilled. On the other hand, all argon analyses show a mixture of radiogenic and non-radiogenic argon, and we have to examine how far we are dealing with only a mixture plot and not with a true isochron. The measured argon is a mixture of *in situ* generated radiogenic argon and three other argon components:

(1) Ar initially incorporated during crystallization of the mica. This Ar

component on a continental shelf at water depths of only 60–300 m, has to be normal air Ar.

(2) Ar adsorbed during mineral preparation. This component also is normal air Ar and most of it is released during the prebaking of the sample under high vacuum.

(3) Ar blank from the line. The line blank in some cases is slightly radiogenic, but a clean line shows blanks in the order of below 1% of the measured Ar, so that this possible radiogenic component is generally negligible.

The general decrease of the non-radiogenic component of a mica with time (and here we are concerned only with micas) from about 50% in Alpine time to under 1% in Precambrian times shows that most of the non-radiogenic argon must have been incorporated during crystallization of the mineral and is not only adsorbed to the mineral after mineral separation.

From this decrease of the non-radiogenic argon with time, an initial amount of around 4 nl/g STP ^{40}Ar initially incorporated into the mica lattice under normal pressure conditions can be deduced (in high-pressure regimes this initial argon component can reach an amount 20 times higher). On the other hand, the non-radiogenic component of ^{40}Ar from line blank and adsorption to the mineral surface is more or less constant (with some exceptions due to different temperatures and duration of degassing) and not time-dependent. For a clean line it is commonly around 1% of the measured amount of gas. The position of an analytical point in the 40/36 diagram therefore is dependent primarily on the mixture between initial incorporated Ar (+argon present in the line) and *in situ* generated radiogenic argon and does not strictly fulfil the presumptions for an isochron plot treatment.

That the line was not always this clean can be seen from the repetitions (see Table 2). As all three Ar components in our special case are common air Ar, the slope of the best fit line still corresponds to the age of the samples (see NDS 62). Nevertheless, because the system is not always this clean, the use of 40/36 diagrams is critical and cannot be recommended.

4.2 The Rb–Sr method

4.2.1 Reproducibility of the sample

Normally it is recommended to grind the samples in order to reach a higher reproducibility for the Rb and Sr analyses. However, as Ar analyses have to be performed on pellets, we have not ground our glauconies in order to maintain the closest similarity between Rb–Sr and K–Ar analyses. For the same reasons the measured glauconies were not treated with acids. An ultrasonic treatment with 0.1 N acetic acid (B. Pomerol and Odin, 1974) would have greatly reduced the common Sr content of our samples and thus reduced the analytical uncertainty. Such a treatment is only recommended if afterwards the acid is also analysed as a control that no Rb or radiogenic Sr has gone into solution.

4.2.2 Analytical conditions

Eleven glauconies, two smectites and a calcite were analysed for Rb and Sr. For the chemical preparation of the samples see Jäger (1960–1962), for the laboratory blanks see Stille (1979). Rb and Sr were measured on an Ion Instruments 35 cm radius, 90° deflection solid source mass spectrometer using a triple filament. For further details see Brunner (1973). Isotopic ratios, mean values, uncertainties and age values were calculated using an on-line PDP8 computer. The conventional isotopic ratios and constants were used, as throughout the book.

4.2.3 Isotopic data

Table 3 shows the analytical data of the different mineral fractions from nine sediments from the Early Cenomanian. A conventional age for each glaucony has been calculated assuming equilibrium conditions with the marine environment, using a $^{87}Sr/^{86}Sr$ value of 0.7080. This assumption is at

Table 3. Results of isotopic dating on glauconies and related clayey and carbonate fractions. The analytical uncertainties used conform to the usual results in the laboratory. The data are nomalized with NBS 987 Sr = 0.71014. Conventional ages in Megaanna are calculated using two points isochron calculation with 0.708 as initial strontium ratio.

Sample	Rb (ppm)	Sr (ppm)	$^{87}Sr/^{86}Sr$ (±0.1%)	$^{87}Rb/^{86}Sr$ (±2%)	Conventional age ($x_0 = 0.708$)
Glaucony G.16 A	267.1	29.47	0.7450	26.24	99.3±3.4
Glaucony G.17 A	256.8	27.05	0.7465	27.50	98.5±3.3
Glaucony G.17 A	262.5	25.74	0.7474	29.53	93.9±3.1
Glaucony G.71 A	267.4	46.08	0.7305	16.80	94.4±4.7
Glaucony G.277 A	245.5	10.25	0.7958	69.31	89.1±2.1
Glaucony G.299 A	251.5	28.16	0.7389	25.87	84.1±3.3
Glaucony G.349 A	240.2	13.71	0.7782	50.70	97.4±2.5
Glaucony G.412 A	241.3	73.12	0.7207	9.547	93.9±7.7
Glaucony G.488 A	212.3	74.73	0.7179	8.224	84.7±8.8
Glaucony GL-O	234.8	18.45	0.7551	36.84	89.9±2.7
Glaucony G.500	232.7	18.80	0.7543	35.83	91.0±2.7
Clay of 71	122.9	445.1	0.7102	0.7990	194±88
Clay of GL-O	58.3	158.0	0.7103	1.060	153±67
Carbonate of GL-O	—	—	0.7088	—	—

present only an hypothesis. The $^{87}Sr/^{86}Sr$ ratio in sea-water is normally measured using a calcite phase considered as cogenetic.

The 'carbonate fraction' of the Cauville Cenomanian yielded a $^{87}Sr/^{86}Sr$ ratio of 0.7088 ± 0.0005 (compared to 0.71014 for NBS 987 Sr), which is higher than the values published by Dasch and Biscaye (1971), giving a value of 0.70730 ± 0.00014 for the Albian, whereas Peterman *et al.* (1970) propose a value of 0.70749 ± 0.00027 for the end of the Cretaceous. Veizer and Compston (1974) have measured carbonates of the Barremian–Aptian which gave a value of 0.70731 ± 0.00070. The two analysed smectites show high $^{87}Sr/^{86}Sr$ ratios. The 'carbonate' represents the leaching solution of the fine fraction including the smectite and could therefore have an increased $^{87}Sr/^{86}Sr$ ratio.

4.2.4 Discussion of the analytical data

Close inspection of the analytical data on the 11 glauconies shows that while Rb contents are rather uniform, 240 ± 30 ppm (Rb contents appear quite inaccurately linked with the K content: K/Rb ratio around 270), the Sr contents vary between 10 and 75 ppm, so that the dispersion of the $^{87}Rb/^{86}Sr$ ratio is mainly due to variations in Sr content. An acid-leached glaucony generally reveals only 5–10 ppm of strontium. The $^{87}Sr/^{86}Sr$ *vs.* 1/Sr plot, Figure 5, shows clearly that we are dealing with a mixing line between common marine Sr and a strontium-poor glaucony. Here again we have a mixture between radiogenic Sr and two, possibly three non-radiogenic marine components:

(1) Common Sr incorporated in the crystal lattice. As no inheritance could be shown, this Sr has to be marine Sr or a very low quantity.

(2) Marine Sr adsorbed on the grains during formation.

(3) Possibly carbonate fragments included in the cracks of the glaucony pellets; again this Sr has to be marine Sr.

A further component would come from the laboratory blanks, but as these are at least four orders of magnitude below the measured Sr concentrations, they seem negligible. As all Sr components are marine Sr, we can simplify the assumption to a two-component mixing between radiogenic and common marine Sr.

During the above-described cleaning procedure this Sr is only partially removed; this moves the analytical points more or less towards the initial value, depending on the amount of common Sr on the pellets. As this contamination consists of common Sr of the marine environment, it does not affect the calculated age and therefore the treatment of the data on a $^{87}Sr/^{86}Sr$ *vs.* $^{87}Rb/^{86}Sr$ plot may be envisaged. The data of the laboratories of Brussels, Leeds, Nancy and Saõ Paolo have been incorporated on the same plot (see NDS 62), but were not used for calculation here.

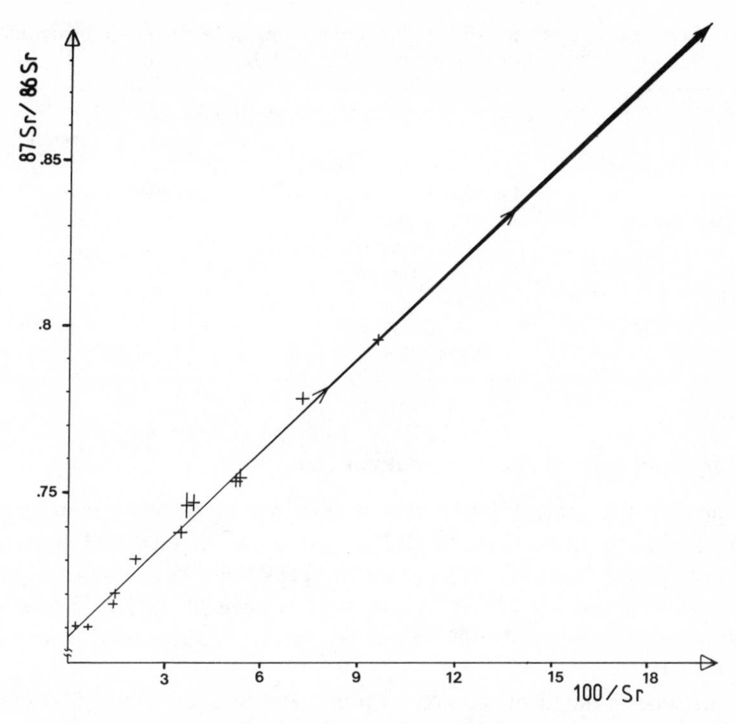

Figure 5. Isotopic ratio of the strontium of Cenomanian glauconies as a function of the total strontium. A glaucony with more strontium (adsorbed or in polluting carbonates) will be near the *y*-axis. Cleaning and leaching the grains will displace the points along the line towards right side.

5 DISCUSSION

5.1 Apparent ages of the Albian glauconies (K–Ar method)

Eleven analyses of seven different samples have now been performed. As shown in Figure 4, the apparent ages show little scatter, although taken at different levels of the Middle and Late Albian. The arithmetic mean value of these apparent ages is 99.0 ± 1.2 Ma (2σ). In Table 4 the different calculation modes of regression line analysis are given: the cubic regression of York (1969) and Brooks *et al.* (1972). According to York the K–Ar age is 97.8 ± 0.7 Ma using only the 11 glauconies. The initial $^{40}Ar/^{36}Ar$ calculated ratio is high: 319 ± 16. According to Brooks *et al.*, the 11 glauconies define a K–Ar age of 98.6 ± 2.1 Ma with a very imprecise initial ratio of 304 ± 34. All three calculation modes give ages between 98 and 99 Ma. Assuming a normal initial $^{40}Ar/^{36}Ar$ ratio leads us to accept 99.0 ± 1.2 Ma as the age of the analysed Albian glauconies.

Table 4. Apparent age of the Albian glauconies from the Paris Basin using different calculation mode (σ = standard deviation).

Method of integration	Data used	Initial ratio calculated $^{40}Ar/^{36}Ar$	Apparent age (Ma, ICC)
Arithmetic mean	11 glauconies (7 samples)	—	99.0 ± 1.2 (2σ)
Cubic regression Brooks et al. (1972)	11 glauconies	304 ± 34	98.6 ± 2.1
Cubic regression York (1969)	11 glauconies	319 ± 16	97.8 ± 0.7

5.2 Apparent ages of the Cenomanian glauconies

In Table 5, the same treatment is applied for the Cenomanian samples. The arithmetic mean value of the 12 glauconies is 93.2 ± 0.9 Ma (analytical uncertainty, see Figure 4). According to York, the 12 glauconies yield a rather high and not fixed initial value, with an age of 92.1 ± 1.3 Ma, while according to Brooks *et al.*, the initial Ar ratio is 299 ± 16 with an age of 93.0 ± 1.4 Ma.

The arithmetic mean of the conventional Rb–Sr ages of the 11 Cenomanian glauconies is 92.4 ± 1.4 Ma. In Section 4.2.4 we have seen that the $^{87}Sr/^{86}Sr$ *vs.* $^{87}Rb/^{86}Sr$ plot is not an isochron for our samples, but rather a mixing line between common marine Sr of different sources and *in situ* generated radiogenic Sr.

For our case, nevertheless, an age can be calculated from the slope of this mixing line. The 11 glauconies give 93.5 ± 1.6 Ma with an initial value of 0.7078 ± 0.0006 with the methods of both Brooks *et al.* (1972) and Wendt (1969). This line does not touch the smectites nor the calcite. By incorporating the calcite into the calculation the age drops to 92.9 ± 1.4 Ma with an initial value of 0.7082 ± 0.0005, still in agreement with Veizer and Compston's (1974) values for marine Sr of the Cretaceous. This new line touches all 14 analytical points, calcite and smectite included.

The close coincidence between the initial $^{87}Sr/^{86}Sr$ of the glauconies alone and of the glauconies including the calcite shows that the marine common Sr has to be around 0.7080. This value therefore has been used to calculate the conventional point ages.

Even if we include the calculations according to York, which give an unreasonably high scatter on the initial ratio and ages around 91 Ma, we can state that the Rb/Sr age of the Cenomanian glauconies is somewhere between 91 and 93 Ma. The coincidence between the arithmetic mean of the point ages and regression ages shows that no noticeable inherited radiogenic component is present in the system.

Table 5. K–Ar and Rb–Sr apparent ages of the Cenomanian glauconies from the Paris Basin using different calculation modes. K–Ar and Rb–Sr ages are equivalent and may be considered as the moment of closure of the various glauconies, probably at the end of the deposition of the *H. carcitanensis* zone. The ± given with the arithmetic means are the mean uncertainties usually calculated in Berne: $2s = (\sum (2\sigma \text{ on each age})^2/n^2)^{1/2}$.

Method of integration	Data used	Initial ratio calculated	Apparent age (Ma, ICC)
Rb–Sr analyses			
Arithmetic mean of conventional age	11 glauconies	(0.708)	92.4±1.4
Cubic regression of			
Brooks *et al.* (1972)	11 glauconies	0.7078±0.0006	93.5±1.6
Brooks *et al.* (1972)	11 glauconies +1 carbonate	0.7082±0.0005	92.9±1.4
York (1969)	11 glauconies	0.7086±0.0014	91.1±2.1
K–Ar analyses			
Arithmetic mean of conventional apparent ages	12 glauconies	(295.5)	93.2±0.9
Brooks *et al.* (1972)	12 glauconies	299±16	93.0±1.4
York (1969)	12 glauconies	330±44	92.1±1.3

Three other samples collected by P. Juignet from cores were selected further to the south. They were dated in Strasbourg by one of us. G.614 comes from the *Turrilites costatus* zone of basal Middle Cenomanian and gave an apparent age of 90.4±2.5 Ma; G.610 is a glaucony collected from the base of the Cenomanian and its apparent age was found to be 100.4±2.8; G.611 is a Late Albian glaucony, nearly 15 m below the boundary, and its apparent age was found to be 100.1±2.8 Ma. The calibration of the ^{38}Ar spike was done with GL-O (admitted value 24.8 nl/g). These data show that reworking is obvious at the base of the Cenomanian in this part of the basin (G.610). Concerning G.614, which is stratigraphically younger than the other Cenomanian glauconies studied above, the moment of closure also appears younger.

5.3 Interpretation of the apparent ages

The close correlation betwen K–Ar and Rb–Sr ages for the Cenomanian shows clearly that for both the Rb–Sr and the K–Ar system the closing time was almost the same. In other words, the exchanges for these elements stopped all at one time, and no futher exchanges took place after the moment of closing. As far as the Cenomanian is concerned, this time is probably situated towards the middle of the Early Cenomanian when the grains were embedded and slowly became isolated from their marine environment. Probably, as soon as a thick enough cover of sediment is deposited no further interstitial exchange is possible and thus the evolution of the grains already rich in potassium is stopped. With a calcareous ooze, probably a cover of the order of 1 m is already enough to stop further exchanges. The green Cenomanian pellets, therefore, must have formed between the Albian and the Cenomanian during a depositional gap not seen as a palaeontological gap. The grains were isotopically closed towards the end of the zone of *Hypoturrilites carcitanensis*, as proposed on Figure 3. Concerning the Albian glauconies, no apparent age difference has been found in the Albian profile, although the sampled vertical profile goes from the Middle Albian to the end of the Albian, a time span of about 5 Ma. According to sedimentological and palaeontological criteria the Albian glauconies have evolved and become closed systems during a rather long lacune situated between Middle and Late Albian. The dated time remains rather imprecise, but nevertheless is clearly pre-Cenomanian and yields an upper limit for the age of the Late Cretaceous transgression.

5.4 Conclusions

From the available data the age of the Albian–Cenomanian boundary can be estimated with good probability. In the Paris Basin this limit is certainly older than 94 Ma and younger than 97 Ma. The limit of 95 ± 1 Ma that we had proposed after preliminary work on the same beds (Juignet *et al.*, 1975) is still reasonable. This boundary seems synchronous with the Albian–Cenomanian in the Western Interior Basin of the United States assumed to be on top of the *Neogastroplites* beds.

Rb–Sr glaucony ages of this study show a greater dispersion than the K–Ar ages. A conclusion of this study is that glauconies should be treated specifically for each method. For K–Ar, glauconies are cleaned properly already with water in an ultrasonic bath, while for Rb–Sr dilute acetic acid seems indispensable. In each case the solution should be analysed as a control of possible reaction betwen glaucony and the acid.

Acknowledgements

This work was supported by the Swiss National Science Foundation. We are greatly indebted to the members of the isotope geology group in Berne for their kind help in the laboratory and discussions. Thanks are due to Professor P. Juignet who provided the samples to help us in the field, to Dr. H. Oberhänsl for critically reading this manuscript and to Dr. T. Hurford who greatly improved our English. This is a contribution to IGCP Project 133.

(Manuscript received 5-3-1980)

Numerical Dating in Stratigraphy
Edited by G. S. Odin

32 The Jurassic and Cretaceous time scale in 1981

WILLIAM J. KENNEDY and GILLES S. ODIN

1 INTRODUCTION

In 1959, Arthur Holmes proposed 180 and *c.* 135 Ma as the limits of the Jurassic period, and estimated the end of the Cretaceous period at 70 ± 2 Ma.

At the close of the joint meeting of the Geological Society of London and the Geological Society of Glasgow in 1964, Howarth estimated the lower limit of the Jurassic at 190–195 Ma and left the upper limit at *c.* 135 Ma, in the absence of any hard data (Howarth was in fact able to review 12 ages only). Casey (1964) reviewed the Cretaceous period, and proposed a time scale based essentially on glaucony dates. He took the base at 136 Ma (old constants) (without any good evidence, however) and the top at 64 Ma (old constants). This 72 Ma interval was arbitrarily divided up into 12 stages of equal duration.

More recently, Lambert (1971) reexamined the data utilized by Casey, and suggested (with remarkable foresight) that the Early–Late Cretaceous boundary lay close to 97 Ma, ICC. The criterion of equal duration of stages as a basis on which to divide up the Cretaceous time scale is clearly unsatisfactory, and is not used here; neither do we make use of magnetic reversal data for the Jurassic and Early Cretaceous because these are still poorly correlated. Much new data based on stratigraphically well-located samples have been obtained in the last decade; chronometers are better understood; and there have been substantial improvements in both instrumentation and analytical techniques which allow a reappraisal of the time scales for both periods. Previous work is reinterpreted on the basis of a better understanding of stratigraphic position and revised physical constants, and there is a body of new data with much higher confidence limits than were previously available.

Many ages published and discussed in the period since 1964 can still only be regarded as provisional, and many are, we believe, imprecise for a variety of reasons. We attempt to resolve these in terms of three areas of uncertainty: (1) stratigraphic problems, associated with the initial sampling; (2) mineralogical problems, associated with the material analysed; and (3) analytical problems, associated with the techniques used.

1.1 Stratigraphic uncertainties

Ideally, any given sample should be dated stratigraphically in terms of the type area of the stage concerned, where this category of uncertainty is (or should be) at a minimum. Unfortunately, problems arise with the definition of the limits of most of the Jurassic and Cretaceous stages in their type areas. To this, must be added correlation problems that arise because of the existence of distinct faunal provinces during these two periods. As a result, it is often difficult to correlate between northwest European Boreal faunas and those of the rest of the world. As is discussed below, this is especially difficult at the level of the Jurassic–Cretaceous boundary, whilst in the Cretaceous faunal differentiation of the US Western Interior endemic faunas means that these relatively well-dated sequences cannot always be precisely correlated with the Old World standard sequences.

In general, glauconies and bentonites are practically the only radiometrically datable materials that can be placed precisely into biostratigraphic sequences and thence referred to individual stages with confidence. Contemporaneous extrusive volcanics may also be valuable, but experience shows that data from plutons should only be used in the absence of all other chronometers.

In many cases, stratigraphic attributions are strongly influenced by the radiometric age measured. In all cases it is essential to return to the original investigation to check that stratigraphic attribution is satisfactorily established.

1.2 Geochemical uncertainties

These are peculiar to each unit dated, and are considered in detail for each formation below. In general, purified mineral fractions are much more reliable chronometers than whole rocks. Dates obtained from altered minerals are unreliable, especially those from biotites with less than 5% potassium, which are generally excluded from this discussion. Little-evolved glauconies are also unsuitable and unreliable chronometers, and samples with less than 5.5–6% potassium should not be used. Some biotites with variable potassium contents give identical ages, and these appear to be useful if considered together. Equally, when several more or less evolved glauconies give identical ages one can perhaps exclude the probability of inheritance from the initial substrate of glauconitization (Odin and Dodson, this volume). The sensitivity of these little-evolved chronometers to problems associated with original inheritance and post-depositional disturbance means that these sources of error are generally more obvious in glauconitic minerals with the lowest potassium content.

1.3 Analytical uncertainties

These now constitute only a minor element in the total uncertainty associated with a given age determination. Improvements in instrumentation mean that the poorest modern mass spectrometers are much better than the most sophisticated machines available 20 years ago; in general, more recent determinations are substituted for older ones in the present synthesis.

1.4 Terminology

We use the abbreviation PTS to designate items published in the *Phanerozoic Time Scale* (1964, 1971), and NDS for those included in the present volume. ICC indicates the *International Conventional Constants* adopted by the IUGS Sub-Commission on Geochronology in Sydney in 1976. By *glaucony* we mean the green sedimentary facies as distinct from the mica mineral usually known as 'glauconite', which is only one of the possible mineral components of the green grains, i.e. glaucony. This distinction is essential if one wishes to take into account the specific qualities of each glaucony because the reliability of this kind of chronometer is a function of its mineralogical composition.

2 STRATIGRAPHIC FRAMEWORK

The Jurassic system takes its name from the rocks of this age exposed in the Jura mountains of France and Switzerland; the name was first used by Omalius d'Halloy in 1831 as a modification of the term 'Calcaire du Jura' used by A. de Humboldt in 1795.

2.1 Lower Jurassic

The currently accepted subdivisions of the Lower Jurassic are shown in Table 1. The type localities of the stages are all within western Europe (the Hettangian at Hettange, Moselle, France; Sinemurian at Semur, Côte d'Or, France; Pliensbachian at Boll, Württemburg, Germany; Toarcian at Thouars, Deux Sèvres, France). The Pliensbachian, it should be noted, is sometimes divided into a Carixian substage (type locality Charmouth, Dorset, England), equivalent to the Lower Pliensbachian and a Domerian substage (type locality Monte Domaro [Domero], Tuscany, Italy), equivalent

Table 1. Lower Jurassic ammonite zones, based on Arkell (1956), Dean *et al.* (1961) and Mouterde *et al.* (1971).

Stage		Zone
Aalenian[1]		*Graphoceras concavum* *Ludwigia murchisonae* *Leioceras opalinum*
Toarcian	Upper Toarcian or Yeovilian	*Pleydellia aalensis* *Dumortieria pseudoradiosa* *Hammatoceras insigne* *Grammoceras thouarsense*
	Lower Toarcian or Whitbian	*Haugia variabilis* *Hildoceras bifrons* *Harpoceratoides serpentinus* *Dactylioceras tenuicostatum*
Pliensbachian	Upper Pliensbachian or Domerian	*Pleuroceras spinatum* *Amaltheus margaritatus* *Amaltheus stokesi*
	Lower Pliensbachian or Carixian	*Prodactylioceras davoei* *Tragophylloceras ibex* *Uptonia jamesoni*
Sinemurian	Upper	*Echioceras raricostatum* *Oxynoticeras oxynotum* *Asteroceras obtusum*
	Lower	*Arnioceras semicostatum* *Arietites bucklandi* *Coroniceras rotiforme*
Hettangian		*Schlotheimia angulata* *Alsatites liasicus* *Psiloceras planorbis*

1, The Aalenian is included in the Middle Jurassic by some workers.

to the Upper. The Toarcian is sometimes divided into a Lower Whitbian substage (type locality Whitby, Yorkshire, England), and an Upper Yeovilian substage (type locality Yeovil, England). The Aalenian (type locality Aalen, Württemburg, Germany) is placed in the Lower Jurassic by French workers and in the Middle by English-speaking workers. The former usage is followed here.

The basic subdivision of stages is based on ammonite faunas, but although these allow widespread correlation during the earliest Jurassic, problems arise from the Pliensbachian onwards. This is because faunas underwent progressive differentiation throughout the Jurassic into what have become known as the northern, Boreal realm and the southern Tethyan (or Mesogean) realm.

The boundary between these realms oscillated through time, but broadly followed the line of the Alpine fold belts in Europe. The zonation given in Table 1 applies to northwest Europe, and is taken from Mouterde *et al.*

(1971), which was based on Arkell's (1956) account. A comparable zonation, differing only in detail, is given by Dean *et al.* (1961). It can be applied widely up to the beginning of the Toarcian, where Donovan (1958) has proposed an alternative sequence for the Tethyan successions of:

Dumortieria meneghinii	(youngest)
Phymatoceras erbenense	
Mercaticeras mercati	
(Unnamed zone)	(oldest).

The general scheme given in Table 1 defines the base of the Jurassic as the base of the *Psiloceras planorbis* zone. Unfortunately there are many areas of the world where there is a gap in the ammonoid sequence between the first *P. planorbis* and the youngest definitely Late Triassic fossils; and the boundary has been commonly defined on lithological criteria. Some authors (e.g. Hoffman, 1962) have claimed to recognize a *Neophyllites antecedens* fauna preceding that of the *Psiloceras planorbis* zone. The matter is reviewed at length by Torrens and Getty in Cope *et al.* (1980a).

2.2 Middle Jurassic

Three stages are referred to the Middle Jurassic as defined here, the Bajocian (type locality Bayeux, Calvados, France), Bathonian (type locality Bath, Somerset, England) and Callovian (type locality the coast of Yorkshire, England). The zonation shown in Table 2 is taken from Cope *et al.* (1980a) and is largely that of Arkell (1956). The Lower Bajocian '*Sonninia sowerbyi*' zone of Mouterde *et al.* (1971) is equivalent to the *Hyperlioceras discites* and *Witchellia laeviscula* zones of Cope *et al.* The Bathonian zonation is basically that of Torrens (1974; in Cope *et al.*, 1980a) and is more refined than the fourfold division of Mouterde *et al.* (1971). The Callovian *Macrocephalites gracilis* zone of southern Europe is equivalent to the Boreal *Sigaloceras calloviense* zone.

2.3 Upper Jurassic

Waning migration of the all-important ammonites plus progressive differentiation of fauna lead to complex biostratigraphic problems that become increasingly acute through the Late Jurassic and on into the Early Cretaceous. Different zonal schemes are used in different areas, and no less than three different sets of stage names are in use for the highest Jurassic–lowest Cretaceous stages; no agreed single system is in sight, and correlation between the various stages is still unsatisfactory.

The lowest stage of the Upper Jurassic is the Oxfordian (type locality Oxford, Oxfordshire, England). Because of differentiation of ammonite

Table 2. Middle Jurassic ammonite zones, based on Arkell (1956) and Cope *et al.* (1980b). The Bathonian zonation is that of Torrens (1974; in Cope *et al.*, 1980b). Note that the *Hyperlioceras discites* + *Witchellia laeviscula* zones are together equivalent to the '*Sonninia sowerbyi*' zone of Mouterde *et al.* (1971), and that the southern European *Macrocephalites gracilis* zone is strictly equivalent to the *Sigaloceras calloviense* zone.

Stage	Zone
Callovian	*Quenstedtoceras lamberti* *Peltoceras athleta* *Erymnoceras coronatum* *Kosmoceras jason* *Sigaloceras calloviense* *Macrocephalites macrocephalus*
Bathonian	*Clydoniceras discus* *Oppelia aspidoides* *Procerites hodsoni* *Morrisiceras morrisi* *Tulites subcontractus* *Procerites progracilis* *Asphinctites tenuiplicatus* *Zigzagiceras zigzag*
Bajocian	*Parkinsonia parkinsoni* *Strenoceras garantiana* *Strenoceras subfurcatum* *Stephanoceras humphriesianum* *Emilia sauzei* *Witchellia laeviscula* *Hyperlioceras discites*

faunas three different zonations have to be used for the Middle and Upper Oxfordian of the Boreal (i.e. Greenland, northern Britain), northwest European (i.e. southern England, northern France, Germany) and sub-Mediterranean regions. These, and their probable correlation, are shown in Table 3 (modified after Sykes and Callomon, 1979).

It should be noted that the Kimmeridgian–Oxfordian boundary is probably drawn at a slightly higher level in the Tethyan realm than in areas to the north.

Above this boundary, even greater problems arise. The first of these is that the term Kimmeridgian (the type area of the stage is Kimmeridge, Dorset) is used in different ways by British and continental European

Table 3. Boreal, NW European and sub-Mediterranean ammonite zonations and correlations of the Oxfordian stage. Note that correlation of Boreal and NW European schemes is difficult above the *Cardioceras cordatum* zone and that correlation between the NW European and sub-Mediterranean schemes is difficult above the *Perisphinctes plicatilis* zone (modified after Sykes and Callomon, 1979).

Boreal	NW European	Sub-Mediterranean
		Idoceras planula
Amoeboceras rosenkrantzi	*Ringsteadia pseudocordata*	
		Epipeltoceras bimammatum
Amoeboceras regulare		
	Perisphinctes cautisnigrae	
Amoeboceras serratum		
		Perisphinctes bifurcatus
Amoeboceras glosense	*Perisphinctes pumilis*	
		Gregoryceras transversarium
Cardioceras tenuiserratum		
	Perisphinctes plicatilis	*Perisphinctes plicatilis*
Cardioceras densiplicatum		
Cardioceras cordatum	*Cardioceras cordatum*	*Cardioceras cordatum*
Quenstedtoceras mariae	*Quenstedtoceras mariae*	*Quenstedtoceras mariae*

workers such that the Kimmeridgian *sensu gallico* is equivalent to the Lower Kimmeridgian *sensu anglico* only. This dichotomy of usage arose because Alcide d'Orbigny, the author of both Kimmeridgian and Portlandian stages, equated the former with the Kimmeridge clay formation of southern England while citing as typically Portlandian ammonites (now referred to the genus *Gravesia*) that occur quite low in the same formation.

English workers (for a recent synthesis see Cope in Cope *et al.*, 1980b, p. 76) draw the boundary of the Kimmeridgian and the succeeding Portlandian stage at the boundary of the Kimmeridge clay and Portland sand.

In Tethyan marine sequences, the Kimmeridgian *sensu gallico* is succeeded by a Tithonian stage, the zonation of which is based on the succession in Bavaria for the Lower Tithonian (Zeiss, 1968) and the Ardêche region of France for the Upper (Mouterde *et al.*, 1971), the base of the Cretaceous being drawn at the base of the Berriasian stage (type locality Berrias, Ardêche, France). In Boreal regions, marine successions are divided into Kimmeridgian below, followed by the Volgian stage (type locality in the Moscow Basin), the base of the Cretaceous being drawn at the base of the Ryazanian stage (type locality Ryazan, 190 km SE of Moscow). In southern England (Dorset, the Weald) the marine Portlandian stage is succeeded by non-marine sediments referred to as a Purbeckian 'stage' or the Purbeck 'beds', whereas in East Anglia a fully marine sequence (divided into Volgian below and Ryazanian above) is recognized. The present consensus (see summaries in Rawson *et al.*, 1978 and Cope *et al.*, 1980b) is that the Purbeck 'beds' span the Jurassic–Cretaceous boundary.

These three successions can be correlated with varying degrees of success, complicated by the modification of stage boundaries to introduce supposed conformity. The Boreal Volgian stage has thus been extended downwards (Gerasimov and Mikhailov, 1966) to encompass the Upper Kimmeridgian *sensu anglico*, with the resultant presumed equivalence with the Kimmeridgian *sensu gallico*/Tithonian boundary; the Portlandian has been extended up to the top of the Jurassic by Wimbledon and Cope (1978) and Wimbledon (in Cope *et al.*, 1980b) by adding on the Volgian zones recognized in eastern England by Casey (1973). Only time can resolve these problems; Table 4 provides a summary of the zonal schemes applied to these different stage sequences.

2.4 Jurassic–Cretaceous boundary

Because of the different stratigraphic systems recognized in Boreal and Tethyan areas there are two different definitions of this boundary in marine sequences: between the Volgian and Ryazanian in the north and the Tithonian and Berriasian in the south. *Riasanites rjasanensis*, index fossil of the lowest zone of the Ryazanian, occurs with ammonites of the Upper Berriasian *Fauriella* [*Berriasella*] *boissieri* zone in the Crimea and the northern Caucasus, and it is clear that the base of the Ryazanian is younger than the base of the Berriasian: the Jurassic–Cretaceous boundary is thus drawn at a higher level in the Boreal realm than in the Tethyan realm (Casey, 1963, 1973; Hancock, 1972). It is also clear that the Volgian–Ryazanian boundary can be correlated with a quasi-marine horizon (the Cinder bed) in the middle of the English Purbeck beds (Casey, 1963; Rawson *et al.*, 1978; Cope *et al.*, 1980b).

2.5 Lower Cretaceous

Just as there is no agreement on stage nomenclature in the Upper Jurassic, so too with the Ryazanian *vs.* Berriasian. Furthermore, some workers (e.g. Saks and Shulinga, 1973) use a hybrid Volgian/Berriasian terminology. Whether the Volgian–Ryazanian boundary is moved down to a presumed equivalent level to that of the Tithonian–Berriasian boundary or whether the latter is moved up to match the former has yet to be agreed. In consequence, we give here a zonation for both stages (Table 5).

Above the Berriasian–Ryazanian, the zonations for the remaining Lower Cretaceous stages continue to present difficulties due to provincialism of ammonite faunas. Table 5 gives sequences in both areas. For the Valanginian (type area the Seyon Gorge near Valangin, Switzerland), the Tethyan

Table 4. Post-Oxfordian ammonite zonations.

Russian platform[1]			Southern England[2]			Tethyan sequences		
Lower Ryazanian		*Riasanites rjasanensis*	Lower Ryazanian		(*Hectoroceras cochi*) (*Runctonia runctoni*)	*Fauriella boissieri* *Pseudosubplanites grandis*		Berriasian
Volgian	Upper	*Craspedites nodiger*	Purbeck beds		(*Subcraspedites lamplughi*)			
					(*Subcraspedites preplicomphalus*)	*Paraulacosphinctes transitorius*	Upper	Tithonian[3]
					(*Subcraspedites primitivus*)			
		Craspedites subditus			(? *Titanites* (*Paracraspedites*) *oppressus*)			
			Portland beds		*Titanites anguiformis*			
		Kachpurites fulgens			*Galbanites* (*Kerberites*) *kerberus*			
					Galbanites okusensis	*Pseudovirgatites scruposus*		
	Middle	*Epivirgatites nikitini*			*Glaucolithites glaucolithus*			
		(gap in sequence)			*Progalbanites albani*			
		Virgatites virgatus	Kimmeridgian *sensu anglico*	Upper	*Virgatopavlovia fittoni*	*Pseudolissoceras concorsi*	Middle	
		Dorsoplanites panderi			*Pavlovia rotunda*			
					Pavlovia pallasioides	*Sublithacoceras penicillatum*		
	Lower	*Subplanites pseudoscythicus*			*Pectinatites pectinatus*	*Franconites vimineus*	Lower	
					Pectinatites hudlestoni			
		Subplanites sokolovi			*Pectinatites wheatleyensis*	*Dorsoplanitoides triplicatus*		
						Hybonoticeras hybonotum or		
		Subplanites klimovi			*Pectinatites scitulus*			
					Pectinatites elegans	*Glochiceras lithographicum*		
Kimmeridgian				Lower	*Aulacostephanus autissiodorensis*	*Hybonoticeras beckeri*	Upper	Kimmeridgian *sensu gallico*
						Aulacostephanus eudoxus		
					Aulacostephanus eudoxus	*Aspidoceras acanthicum*	Lower	
					Aulacostephanoides mutabilis	*Crussoliceras divisum*		
					Rasenia cymodoce	*Ataxioceras hypselocyclum*		
					Pictonia baylei	*Sutneria platynota*		
Oxfordian (*pars*)			Oxfordian (*pars*)		*Ringsteadia pseudocordata*	*Idoceras planula*		Oxfordian (*pars*)

1, The Volgian zonation is from Gerasimov and Mikhailov (1966); these authors extended the base of the stage down from the base of the *panderi* zone to a level that corresponds to the base of the Tithonian.
2, The lower part of this scheme is from Wimbledon and Cope (1978); the highest ammonite zone recognized in the type Portlandian and in southern England and northern France generally is the *anguiformis* zone, above which a non-marine Purbeckian facies spans the Jurassic–Cretaceous boundary. The succeeding zonation of the latest Jurassic is based on Casey's (1973) work in the Spilsby Basin of East Anglia. Extension of the Portlandian to embrace all the post-Kimmeridgian Jurassic in England and elsewhere has been proposed by Wimbledon and Cope (1978) and Wimbledon (in Cope *et al.*, 1980).
3, Zonation from Mouterde *et al.* (1971).

Table 5. Cretaceous ammonite zones.

Stage	Zonal scheme Northern Europe	Southern Europe
Albian[1]	*Stoliczkaia dispar* *Mortoniceras inflatum* *Euhoplites lautus* *Euhoplites loricatus* *Hoplites dentatus* *Douvilleiceras mammillatum* *Leymeriella tardefurcata*	
Aptian[2]	*Hypacanthoplites jacobi* *Parahoplites nutfieldiensis* *Ch. (Epicheloniceras) martinioides* *Tropaeum bowerbanki* *Deshayesites deshayesi* *Deshayesites forbesi* *Prodeshayesites fissicostatus*	*Acanthohoplites nodosocostatum/bigoureti* *Cheloniceras subnodosocostatum* *Dufrenoyia dufrenoyi* *Deshayesites deshayesi* *Procheloniceras albrechtiaustriae*
Barremian[3]	*Parancyloceras bidentatum* *Hemicrioceras rude* *'Crioceras' sparsicostata* *Paracrioceras denckmanni* *Paracrioceras elegans* *Hoplocrioceras fissicostatum* *Paracrioceras rarocinctum*	*Silesites seranonis* *Nicklesia pulchella*
Hauterivian[4]	*Simb. (Craspedodiscus) variabilis* *Simb. (Simbirskites) marginatus* *Simb. (Craspedodiscus) gottschei* *Simb. (Milanowskia) speetonensis* *Simb. (Speetoniceras) inversus* *Endemoceras regale* *Endemoceras noricum* *Endemoceras amblygonium*	*Pseudothurmannia angulicostata* *Plesiospitidiscus ligatus* *Subsaynella sayni* *Lyticoceras nodosoplicatus* *Olcostephanus jeannoti* *Crioceratites duvali loryi* *Acanthodiscus radiatus*
Valanginian[5]	*'Astierien schichten'* *Dicostella pitrei* *Neocraspedites complanatus & undulatus* *Dichotomites bidichotomus* *Dichotomites biscissoides* *Prodichotomites polytomus* *Poltyptychites middendorfi & clarkei* *Polyptychites brancoi & euomphalus* *Platylenticeras involutum* *Platylenticeras heteropleurum* *Platylenticeras robustum*	*Neocomites (Teschenites) callidiscus* *Himantoceras trinodosum* *Saynoceras verrucosum* *Thurmanniceras campylotoxum* *Thurmanniceras otopeta*
Ryazanian	*Peregrinoceras albidum* *Surites (Bojarkia) stenomphalus* *Surites (Lynnia) icenii* *Hectoroceras kochii* *Runctonia runctoni*	*Fauriella boissieri* Berriasian
Volgian (pars)		*Pseudosubplanites grandis*

1, Northern European zonation from Rawson *et al.* (1978).
2, Northern Europe after Casey (1961); southern Europe after Fabre-Taxy *et al.* (1965) and Moullade (1965a,b).
3, Northern Europe: see Rawson *et al.* (1978); southern Europe after Busnardo (1965).
4, Northern Europe from Rawson (1971a,b); southern Europe after Thieuloy (in Moullade and Thieuloy, 1967, and Thieuloy, 1973).
5, Northern Europe after Kemper (1973a); southern Europe after Busnardo *et al.* (1979).

zonation is taken from Thieuloy (1973) and Busnardo *et al.* (1979), and is based on the succession in the Vocontian trough of southeastern France. The Boreal zonation is based on the northwest German sequence reviewed by Kemper (1973). For the Hauterivian (type locality Hauterive, near Neuchâtel, Switzerland), the Tethyan zonation is again taken from the sequences in the Vocontian trough, and is based on the work of Thieuloy (in Moullade and Thieuloy, 1967; Thieuloy, 1973b). The Boreal zonation is taken from the English succession at Speeton, Yorkshire, described by Rawson (1971a,b), which differs only slightly from that proposed by Kemper (1973) for the northwest German sequence.

For the Barremian (type locality Barrême and Angles, Basses-Alpes, France), a bipartite division of the Tethyan sequence is taken from Busnardo (1965). The Boreal zonation is taken from the north German sequence (see Rawson *et al.*, 1978, p. 13).

For the Aptian (type locality Apt, Basses-Alpes), a much higher degree of correlation is possible between Boreal and Tethyan regions although two zonations are given here; the one for the type area is taken from Fabre-Taxy *et al.* (1965) and Moullade (1965a,b). The Boreal zonation is based on Casey (1961).

In contrast to the previous Lower Cretaceous stages, the type locality of the Albian is the Department of Aube, France, well within the area of predominantly Boreal faunas. The base of the stage is missing in the type area, and the highest Albian is without ammonites. The zonation used here is taken from Rawson *et al.* (1978), who discuss its historical development. It should be noted that Destombes (1973, in Rat *et al.*, 1979; see Odin and Hunziker, this volume, Table 1) divides the *mammillatum* zone into zones of *Sonneratia dutempleana* below and *Otohoplites raulinianus* above in the type area, although these species occur in the reverse order in the English sequence (Rawson *et al.*, 1978, p. 17).

Use of the term Vraconian for the *Stoliczkaia dispar* zone is to be discouraged.

The hoplitid-dominated zonal sequences of northwest Europe are extremely difficult to correlate with those elsewhere in the world (e.g. North America, Africa, Madagascar).

2.6 Upper Cretaceous

The Upper Cretaceous stages all have type areas in northwest Europe (Cenomanian around Le Mans, Sarthe, France; Turonian between Saumur and Montrichard, Touraine, France; Coniacian around Cognac, Charente, France; Santonian around Saintes, Charente-Maritime, France; Campanian at Aubeterre-sur-Dronne, Grande Champagne, Charente, France; Maastrichtian around Maastricht, Holland). The ammonite zonal scheme given here (Table 6) differs from that given for the Jurassic and Lower Cretaceous in that, for the Coniacian onwards, the zones are very broad, and indeed almost equivalent to substages. This is because of the paucity of ammonites in the post-Turonian sequences of Europe; indeed, the zonation is basically that of de Grossouvre (1895–1901), and is taken from the Aquitaine Basin.

The Cenomanian zonation is based on Kennedy (1969; 1970), Juignet and Kennedy (1976), Juignet *et al.* (1978) and Wright and Kennedy (1981). It should be noted that the Cenomanian–Turonian boundary has also been placed above, below and within the *Metoicoceras geslinianum* zone. The Turonian zonation is taken from Wright and Kennedy (1981); no index is currently available for the topmost Turonian. (See Hancock and Kennedy, 1981, for further discussion.) The Coniacian to Campanian zonations are based on de Grossouvre (1895–1901). With respect to the Turonian–Coniacian boundary, it is important to note that the supposedly Upper Turonian (*sensu germanico*) Schloenbachi Schichten yields Coniacian ammonites, and that the base of the Coniacian is drawn at a lower level in German successions than it is in France (Seibertz in Rawson *et al.*, 1978, p. 26; Ernst and Schmidt, 1979, etc.).

No acceptable ammonite zonation is available for the European Maastrichtian. Instead, the belemnite-based zonation of Schulz (1979) and others is given in Table 6.

Poor as this European Upper Cretaceous zonation is, it can be integrated with the planktonic foram/nannofossil zonations, as is shown in Table 7. Table 8 shows the much finer ammonite zonation recognized in the US Western Interior (from Kennedy and Cobban, 1976, with modifications after Hooks and Cobban, 1979). This is more closely comparable with Lower Cretaceous and Jurassic zonal sequences, and is firmly dated by pyroclastic deposits. It must be noted that the position of the Coniacian–Santonian, Santonian–Campanian and Campanian–Maastrichtian boundaries cannot be correlated with complete confidence with the European standard because of the *endemic nature of the Western Interior ammonite faunas.*

2.7 Dateable material

Having eliminated stratigraphic problems so far as possible, it is important to at least attempt to date samples from type areas and other important reference sections. We review below the potential chronometers from these sources.

In the Early Jurassic (Hettangian to Toarcian), we have seen only one level that includes a possible chronometer, a glaucony from the Early Pliensbachian (= Carixian) of the southern Paris Basin.

In the Middle Jurassic, the Bajocian is glauconitic in Normandy (Dangeard, 1940), but this takes the form of glauconitization of a hard-ground surface: dating of this type of material has not yet been attempted.

Table 6. Upper Cretaceous ammonite/belemnite zonation of the northwest European area.

Stage		Zone
Maastrichtian	Upper	*Belemnella casimirovensis*
		Belemnitella junior
	Lower	*Belemnella fastigata*
		Belemnella cimbrica
		Belemnella sumensis
		Belemnella obtusa
		Belemnella pseudobtusa
		Belemnella lanceolata
Campanian[1]		*Bostrychoceras polyplocum*
		Hoplitoplacenticeras marotti[3]
		Menabites (*Delawarella*) *delawarensis*
		Diplacmoceras bidorsatum
Santonian	Upper	*Placenticeras syrtale*
	Lower	*Texanites gallicus*[4]
Coniacian	Upper	*Paratexanites emscheris*
	Lower	*Barroisiceras haberfellneri*[5]
Turonian[2]		(Unnamed zone[6])
		Subprionocyclus neptuni
		Collignoniceras woollgari
		Mammites nodosoides
		Watinoceras coloradoense
Cenomanian	Upper	*Neocardioceras juddii*[7]
		Metoicoceras geslinianum[8]
		Calycoceras naviculare/ *Eucalycoceras pentagonum*[9]
	Middle	*Acanthoceras rhotomagense*[10]
	Lower	*Mantelliceras mantelli*[11]

1, 2, Subdivision not generally agreed.
3, *Hoplitoplacenticeras vari* zone of authors. The index species is a synonym of *H. marotti* (Coquand).
4, *Texanites texanus* zone of authors. The species does not occur in Europe.
5, Although retained as an index, *B. haberfellneri* of French authors are all *Reesideoceras petrocoriensis* (Coquand) (= *R. gallicum* Basse); the stratigraphic position of the index species in the zone that takes its name is unclear in Europe.
6, Probably equivalent to the *Reesideites minimus* zone of Japanese authors.
7, Equivalent to the upper part of the *Sciponoceras gracile* zone of recent European publications.
8, Equivalent to the lower part of the *Sciponoceras gracile* zone of recent European publications.
9, Neither of these indices is satisfactory. The type of *C. naviculare* comes from the succeeding *M. geslinianum* zone. *E. pentagonum* is rare and known from the *M. geslinianum* zone of Sarthe. Equivalent to the *robustum* and *crassum* zones of Thomel (1972).
10, Three local divisions are recognized in England, Sarthe and elsewhere (see Kennedy, 1969, etc.).
11, Three local divisions are recognized in England, Sarthe and elsewhere (see Kennedy, 1969, etc.).

Table 7. Integrated ammonite/planktonic foram/coccolith zonation for the Upper Cretaceous.

Stage	Ammonite zone	Planktonic foram zone[1]	Coccolith zone[2,3]
			Nephrolithus frequens
		Abathomphalus mayaroensis	*Arkhangelskiella cymbiformis*
Maastrichtian	No ammonite zones recognized	*Globotruncana contusa*	*Reinhardtites levis*
		Globotruncana gansseri	*Tranolithus phacelosus*
		Globotruncana tricarinata	
	Bostrychoceras polyplocum	*Globotruncana calcarata*	*Quadrum trifidum*
	Hoplitoplacenticeras marotti		*Quadrum nitidus*
Campanian			*Ceratolithoides aculeus*
	Menabites delawarensis	*Globotruncana elevata*	*Phanulithus ovalis*
	Diplacmoceras bidorsatum		*Aspidolithus parcus* s. 1.
			Phanulithus obscurus
Santonian	*Placenticeras syrtale*	*Dicarinella asymetrica*	*Lucianorhabdus cayeuxii*
	Texanites gallicus	*Dicarinella concavata*	*Reinhardtites anthophorus*
Coniacian	*Paratexanites emscheris*		*Micula decussata*
	Barroisiceras haberfellneri	*Marginotruncana schneegansi*	
	(Unnamed zone)		*Marthasterites furcatus*
	Subprionocyclus neptuni		*Lucianorhabdus maleformis*
Turonian	*Collignoniceras woollgari*		
	Mammites nodosoides		
	Watinoceras coloradoense	*Praeglobotruncana helvetica*	*Quadrum gartneri*
	Neocardioceras juddii		
	Metoicoceras geslinianum		
Cenomanian	*C. naviculare/E. pentagonum*	*Rotalipora cushmani*	*Microrhabdus decoratus*
	Acanthoceras rhotomagense		*Eiffelithus turriseifeli*
	Mantelliceras mantelli	*Rotalipora brontzeni*	

1, After Postuma (1971), Sigal (1977). Wonders (1975) and Robaszynski and Caron (1979).
2, After Thierstein (1976), Sissingh (1977) and Perch–Nielsen (1979).
3, Correlation of zonations in part based on Sissingh (1977) and Robaszynski and Caron (1979).

Table 8. US Western Interior Upper Cretaceous ammonite zonation, after Obradovich and Cobban (1976). The stage boundaries indicated follow the same authors.

Stage	Zone
Maastrichtian[1]	*Hoploscaphites* aff. *H. nicolletti*
	Baculites clinolobatus
	Baculites grandis
	Baculites baculus
	Baculites eliasi
	Baculites jenseni
	Baculites reesidei
Campanian	*Baculites cuneatus*
	Baculites compressus
	Didymoceras cheyennense
	Exiteloceras jenneyi
	Didymoceras stevensoni
	Didymoceras nebrascense
	Baculites scotti
	Baculites gregoryensis
	Baculites perplexus
	Baculites sp. (smooth)
	Baculites asperiformis
	Baculites mclearni
	Baculites obtusus
	Baculites sp. (weakly ribbed)
	Baculites sp. (smooth)
	Scaphites hippocrepis III
	Scaphites hippocrepis II
	Scaphites hippocrepis I
Santonian	*Desmoscaphites bassleri*
	Desmoscaphites erdmanni
	Clioscaphites choteauensis[2]
	Clioscaphites vermiformis
	Clioscaphites saxitonianus
	Scaphites depressus
Coniacian	*Scaphites ventricosus*
	Scaphites preventricosus
	Scaphites corvensis
Turonian	*Scaphites nigricollensis*
	Scaphites whitfieldi
	Prionocyclus wyomingensis
	Prionocyclus macombi
	Prionocyclus hyatti
	Collignoniceras woollgari
	Mammites nodosoides
	Watinoceras coloradoense
Cenomanian	*Sciponoceras gracile*
	Dunveganoceras albertense
	Dunveganoceras pondi
	Plesiacanthoceras wyomingense
	Acanthoceras amphibolum
	Acanthoceras muldoonense
	Acanthoceras granerosense
	Calycoceras (*Conlinoceras*) *gilberti*[3]

1, The base of the Maastrichtian is a much disputed boundary, drawn as low as the *Baculites scotti* zone by some authors. See Obradovich and Cobban (1975) for discussion.
2, Frerichs (1980) has recently suggested that the *Clioscaphites choteauensis* zone is Early Campanian.
3, The *Calycoceras* (*Conlinoceras*) *gilberti* zone represents a level somewhere (low?) in the Middle Cenomanian in the sense that this substage is used by European workers.

There are pyroclastic deposits in the Bathonian fuller's earth of southern England, whilst the Callovian is glauconitic in North Germany.

All three stages of the Late Jurassic are glauconitic in North Germany. The Oxfordian includes a little-evolved glaucony in the southern Paris Basin and in Switzerland, as does the Kimmeridgian (Gygi and McDowell, 1970). The Portlandian of southern England (Dodson *et al.*, 1964) and the Boulonnais includes evolved glauconies, while the Volgian of the USSR is also glauconitic and was dated by Kazakov and Polevaya in 1958.

The Early Cretaceous stages are commonly glauconitic in their type areas: the Ryazanian of Ryazan in the USSR and also Great Britain; the Valanginian and Hauterivian in the Swiss Jura, southeast France and Germany. The Aptian includes both glauconitic horizons and pyroclastic deposits (fuller's earths of Surrey, Bedfordshire, Oxfordshire (England), etc.). The Albian is glauconitic at many levels throughout northwestern Europe. The basal parts of the stage are absent in the type area of Aube (Rat *et al.*, 1979), but present and glauconitic in North Germany; the Early Albian is glauconitic in Aube and the Late Albian glauconitic (well-evolved glaucony) in many areas.

The basal parts of the Cenomanian are glauconitic throughout the Paris Basin and many other parts of Europe, with sparser glauconitic units at higher levels.

The Turonian is locally glauconitic in the type area and elsewhere in Anjou and Touraine, and also in the environs of Lille (France) and Mons (Belgium). The Mons area also includes Coniacian glauconies. The type Coniacian at Cognac (France) is locally highly glauconitic. The Turonian, Coniacian, Santonian and Campanian are all glauconitic at various North German localities, whilst Lacroix (1893, pp. 406–9) records Campanian and Maastrichtian glauconies in the Aquitaine Basin.

The Campanian and Maastrichtian around Maastricht include levels of well-evolved glaucony.

Above, the basal Tertiary (Danian) of Denmark has yielded scarce, little-evolved glaucony of no geochronological interest. Folinsbee *et al.* (1963) record a bentonite at the base of the Danish Danian, but failed to obtain sufficient material for dating purposes.

In summary, there are thus very few suitable igneous rock or mineral chronometers in the type areas of the Jurassic and Cretaceous stages—areas where stratigraphic problems are (or should be) at a minimum. There are numerous glauconies, however, and these allow dating of a wide range of units, especially in the Late Jurassic and Cretaceous. The Early Jurassic of western Europe remains much less amenable to dating by radiometric methods.

3 THE PRESENT STATUS OF THE JURASSIC—CRETACEOUS TIME SCALE

3.1 The base of the Jurassic

The recalculated ages for the four items cited by Howarth (1964) are 195–200 Ma (ICC). The stratigraphic positions of the sources of these dates (plutons in Canada, the United States and Indonesia) are imprecisely known and span the Late Triassic–Early Jurassic interval.

Revision of the data available in 1976 led Armstrong to increase the age of the boundary to 210 Ma (ICC). This author now proposes (Armstrong, this volume) an age of 208±5 Ma. Webb (1981) arrived at a similar conclusion in his most recent review: 200±5 Ma, or perhaps a little more. Examination of the data obtained from the Early Lias of the United States (NDS 202–3), the Late Triassic of Thailand (NDS 204) and the Lias of Canada (minimum age 204 Ma, NDS 181) lead to a confident estimate of 204±4 Ma as the age of the Triassic–Jurassic boundary. The conclusions of Priem and Bon in Indonesia (NDS 137) and Webb in Australia (NDS 199) are compatible with this figure, as are dates obtained by Westphal *et al.* (1979) from volcanic rocks in Oued Imini (Morocco). Only 100 km away from this locality (in the Argana corridor) what are presumed to be the same rocks overlie sediments with a Keuper fauna and are in turn overlain by beds with Sinemurian ammonites (Dutuit, 1966, and personal communication, 1980). The age of 200±7 Ma is clearly within the Hettangian or a little below this.

3.2 The subdivisions of the Early Jurassic

The Early Jurassic is divided into four or five stages (the Carixian and Domerian are here regarded as substages), depending on whether the Aalenian is placed in the Early Jurassic (as favoured by French workers) or the Middle (as favoured by English workers).

No dates presently available are strictly referable to the Hettangian (Table 9). Armstrong (NDS 181), on the basis of somewhat scant data, considered this stage as of relatively short duration because the mean age of end-Sinemurian or later batholiths is around 203±5 Ma (the youngest age given in NDS 180 is accepted to be rejuvenated). Recent data from probably Hettangian veins in the Massif Central of France give ages of 194±7 Ma (NDS 213); these lend some support to the younger end of the above-proposed 204±4 Ma age.

The very imprecise stratigraphic age established for plutonic rocks referred to the Pliensbachian and Toarcian stages in Canada, together with the associated high analytical uncertainties, means that we are still uncertain of the duration of these two stages. From the data given in NDS 183 and 184 it can only be stated that rocks dated at between 182 and 190 Ma fall within this interval; these data appear to be more precise (and are better documented) than the contradictory data from eastern Europe presented by Afanass'yev (1970)—a glaucony dated at 156±6 Ma, ICC, and a volcanic rock dated at 171 Ma, ICC, both of which were attributed to the Pliensbachian stage.

The age of 185±3 Ma, considered as pre-Bajocian and post-Toarcian by Shibata (NDS 185), can perhaps be seen as providing a date for the

Table 9. Radiometric data available within the Early Jurassic. Dates in parentheses are of low reliability; dotted lines are for a stratigraphically approximate location. Further comments on samples 'NDS' are given at the end of this volume; few details are available to us for samples 'RLA' compiled by Armstrong (1978).

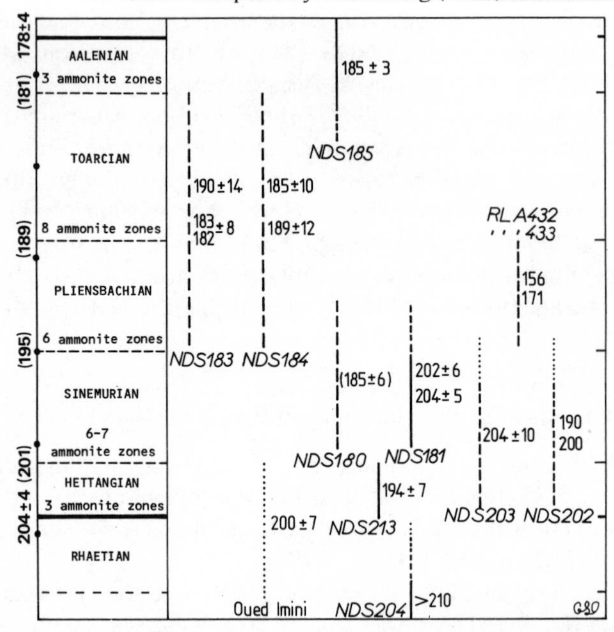

Aalenian stage, which has not yet been directly dated. Taken together with data from the Middle Jurassic, it suggests that the Aalenian–Bajocian boundary probably lies around 178 Ma, the Early Jurassic thus spanning the interval 204–178 Ma. If it is assumed that the relative duration of stages is directly proportional to the number of standard ammonite zones into which each is divided (Table 1), then the limits of the stages lie at 204, 201, 195, 189, 181 and 178 Ma respectively, with an uncertainty of ±7 Ma in each case. These can be compared with the few known dates already reviewed: the Sinemurian estimate (201–195) is compatible with the data (except NDS 180); a Toarcian spanning the interval 189–181 Ma is compatible with figures given in NDS 183 and 185 but not, however, with the less reliable data recorded by Armstrong (1978, items 432–433, Table 9).

3.3 The subdivisions of the Middle Jurassic

Three stages are included in the Middle Jurassic: Bajocian, Bathonian and Callovian.

The Bajocian was dated at 173 Ma (ICC) in 1964 on the basis of an Alaskan diorite (PTS 89). The confidence limits on this determination are low, however, because it was obtained from a single mineral separation, and this was a very altered biotite ($K = 3.6\%$); the stratigraphic attribution to the Late Bajocian is also very doubtful.

Various intrusions in Georgia (USSR: PTS 90) that are regarded as end- or post-Bajocian have provided much better chronometers that yield an age of 167 ± 2 Ma (ICC). This can perhaps be taken as a minimum age for the end-Bajocian. More recently, a probably Bajocian monzonite from British Columbia (Canada: NDS 182) has yielded a range of dates between 159 and 171 ± 7 Ma. Few details are available on the 'Bajocian–Bathonian' volcanics of the USSR that provided Armstrong (1978) with a date of 160 ± 6 Ma, ICC (Table 10: RLA 419), and this can be taken as nothing more than a date within the limits of the two stages. The basalts, syenites and dolerites of Beni Mellal in the Moroccan Atlas (Westphal *et al.*, 1979) lie discordantly on an eroded surface of limestones dated as Bajocian and perhaps Bathonian on the basis of the ammonite faunas present that were folded during the Cretaceous. Three concordant K–Ar ages have been obtained from hydrofluoric acid-washed plagioclases: 170 ± 6, 173 ± 18 and 177 ± 8 Ma, ICC. The age of 174 ± 5 Ma can be taken as the *minimum age for the base of the Bajocian.* According to the author of the determinations (Montigny, personal communication), the stratigraphic control is too imprecise for the data to be considered definitive.

The most recent pre-Bajocian dates are 180 and 185 Ma, and are pre-Aalenian. Given a relatively short Aalenian stage, as noted above, an age of 178 ± 4 Ma is proposed for the Aalenian–Bajocian boundary, although none of the data presently available is incompatible with an age of 174 or 182 Ma.

The upper limit of the Bajocian is difficult to estimate if one brings together the figures given in PTS 90 and that for the Late Bajocian of Utah (Odin and Obradovich, NDS 102). However, if one takes into account the stratigraphic and analytical uncertainties, Armstrong's proposition of 170 Ma is compatible with most results. The limit may well be a little younger or older, but only by 4 Ma at maximum.

Little documentation of the Bathonian was available in 1964. Armstrong (1978) proposed an age of 164 Ma, ICC, for the upper limit. This age is perhaps too great if one takes into account the data from the possibly Bathonian formations of Utah at 163 Ma (but with a large analytical uncertainty, NDS 102). Equally, the data obtained from Bathonian fuller's earth near Bath (England: NDS 149) give a mean apparent age of 158 Ma. The highest apparent ages are 163 ± 3 Ma. A reduction of the age of this boundary is justified because several of the apparent ages available from Oxfordian and Callovian formations are very much younger. Finally, the recently published data of Baubron *et al.* (1978; see NDS 214) suggest an age of 155±6 Ma for a pre-Late Bathonian basalt flow from the French Massif Central. We believe, therefore, that the Bathonian–Callovian boundary lies within the interval 158±5 Ma.

Evolved Callovian glauconies from Germany (PTS 78) have been redated by Obradovich, and give ages that are mostly close to 145 Ma if one uses only the most recent data (NDS 101). Glauconies from boreholes in North Germany have yielded ages that should be considered with some suspicion because of associated stratigraphic problems (the sequence is frequently disturbed by deformation) and may have been rejuvenated by deformation associated with diapiric movements. Armstrong (1978; RLA 431) cites a Callovian glaucony from Bulgaria dated at 142±5 Ma (ICC recalculated) but this must be taken with caution as stratigraphic and geochemical details are lacking. All these appear to have undergone geochemical rejuvenation, given the Early Oxfordian date of 148±3 Ma (NDS 141). We have some confidence in this last figure, and adopt an age of about 150 Ma for the Callovian–Oxfordian boundary, although this clearly remains open to revision. It indicates, however, that *the Middle Jurassic stages were of considerable duration.*

3.4 The base and subdivisions of the Late Jurassic

3.4.1 Callovian–Oxfordian boundary

The base of the Late Jurassic (base of the Oxfordian) was very little documented in 1964. As stated above, a Callovian glaucony from Germany was redated (NDS 101) and yielded an imprecise age of about 145 Ma (ICC). The base of the Oxfordian in Switzerland is dated at 148±3 Ma (NDS 141) on the basis of a high-quality chronometer. In North Germany a glaucony now considered to be Early Oxfordian yielded an age of 138.5±4.0 Ma (NDS 100). However, ages previously obtained from North German glauconies are of limited reliability because of problems of correlation (difficult in boreholes) and the salt tectonics that affect this basin. An age of 150^{+3}_{-8} Ma is adopted here for the Callovian–Oxfordian boundary with limited confidence only; it is based on the most trustworthy chronometer (NDS 141) for the value and the two other dates from NDS 100 and RLA 431 for the range values (Table 10).

3.4.2 Oxfordian–Kimmeridgian boundary

The base of the Kimmeridgian probably lies in the lower part of the evolved Swiss glaucony dated at 139.5±4.0 Ma (NDS 142). This is compatible with the date of the Horseshoe Bar gabbro (California, SA, PTS 76), which has a recalculated age of 138±4 Ma (ICC) and postdates Early Kimmeridgian ammonite faunas. The date of 140±4 Ma is proposed here.

Table 10. Radiometric data available within the Middle Jurassic. Comments as for Table 9.

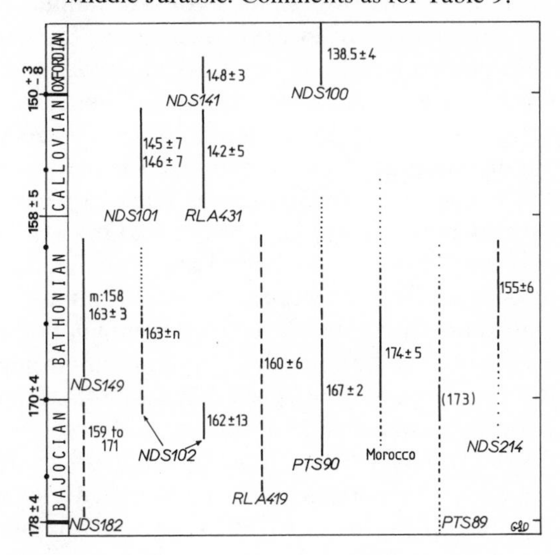

3.4.3 *The Portlandian*

The Portlandian has yielded numerous dateable samples of well-evolved glaucony. Material from the Portland sand (*glaucolithus* zone, Lower Portlandian *sensu anglico*), has given an age of 128±4 Ma (ICC) following Evernden *et al.* (1961). Obradovich has obtained new contradictory data from this sample; after slight ultrasonic treatment the potassium content of the glaucony drops and the apparent age increases to 147.0±4.5 Ma. This is unusual; an apparent age of 134±4 Ma (ICC) obtained by the same author from a comparable sample is more acceptable.

In the USSR Polevaya *et al.* (1961) have dated a little-evolved glaucony from the Late Volgian *Kachpurites fulgens* zone at 130 Ma (ICC, cf. PTS 72).

Well-evolved Portlandian glauconies from the Pays de Bray and the Boulonnais have yielded ages of 129.4±4.8 Ma (NDS 75) and 131±4 Ma (NDS 76). Anomalous dates have been obtained from evolved glaucony in a borehole sample from the Münder marls of North Germany (NDS 228). This level, attributed to the Latest Portlandian, is dated at 142 Ma (ICC) by Evernden *et al.* (1961). For Dodson *et al.* (1984) this apparent age is ‘possibly affected by older (Jurassic) glauconite’. Obradovich (1964) divided the same glaucony into two selected fine fractions which yielded much younger apparent ages of 119 and 121 Ma (ICC). This rejuvenation is not specifically attributable to severe mechanical treatment.

The Shasta Bally batholith (California, USA) has yielded apparent ages of 131 ± 4 Ma, ICC (PTS 75); its stratigraphic context is difficult to determine but it is certainly post-Bajocian and pre-Barremian. Lanphere and Jones (1978) considered this batholith to be pre-Hauterivian. New K–Ar and U–Pb measurements give somewhat disparate ages, the mean being 133.6 Ma. These authors preferred an apparent age of 136 Ma based on a U–Pb/zircon determination as the maximum age for the Hauterivian, which helps little in placing the Jurassic–Cretaceous boundary.

These authors also note an age of 136.5 ± 5.4 (2σ) Ma (ICC) obtained from biotite separated from a possibly Late Valanginian granite from Alaska. This apparent age was, however, obtained from a biotite containing only 2.2% (K) potassium. It is thus *beyond consideration in the present study.* Finally, they agree with Lambert's (1971) opinion, according to which the lower limit of the Cretaceous lay between 125 and 145 Ma, with no logical considerations supporting preference of the mean figure.

Dodson *et al.* (1964) dated several Late Jurassic glauconies: three well-evolved examples from the Early Volgian (Lower Sandringham sands and English Portland sand: PTS 178, 179, 180) gave apparent ages of 132, 131 and 135 Ma, ICC. A fourth glaucony, collected from the Late Volgian some 25 km south of Moscow (PTS 234), gave these authors an age of 122 Ma (ICC). This is of limited value because of the altered state of the outcrop and the low potassium content of the green grains.

The trustworthy data from these levels lead us to *propose a date of 130 ± 3 Ma for the end of the Portlandian* (Table 11).

There is a lack of data at the base of the stage. It can perhaps be estimated at around 135 Ma, older than the age determined from the Portland sands and younger than the Swiss Kimmeridgian date discussed above. Given the lack of precise documentation this must be regarded as tentative, however.

3.5 Jurassic–Cretaceous boundary

The age of 138 Ma (ICC recalculated) proposed for the Jurassic–Cretaceous boundary by Howarth (1964) is incompatible with some of the data available to him, including: three dates regarded as representative of the Kimmeridgian–Portlandian boundary, 128, 138 and 131 Ma (ICC), derived from the Portland sand and the Horseshoe Bar and Shasta Bally plutons respectively; two glaucony ages which Howarth placed at the 'Portlandian–Purbeckian boundary', both of these stages being accepted as Jurassic. The last two ages concern: (1) the Volgian of the USSR, for which the author accepted a date of 135 Ma, ICC, the resumé of which—PTS 52—indicates the need for a revised calculation of 130 Ma (ICC), and (2) the Münder marls—142 Ma (PTS 73). Of these five apparent ages, four are

Table 11. Relevant radiometric data from the Jurassic–Cretaceous boundary. The stratigraphical uncertainties are shown by arrowed lines; the geochemically poorly reliable samples are in parentheses; the analytical uncertainty is commonly estimated at 3–5% (2σ).

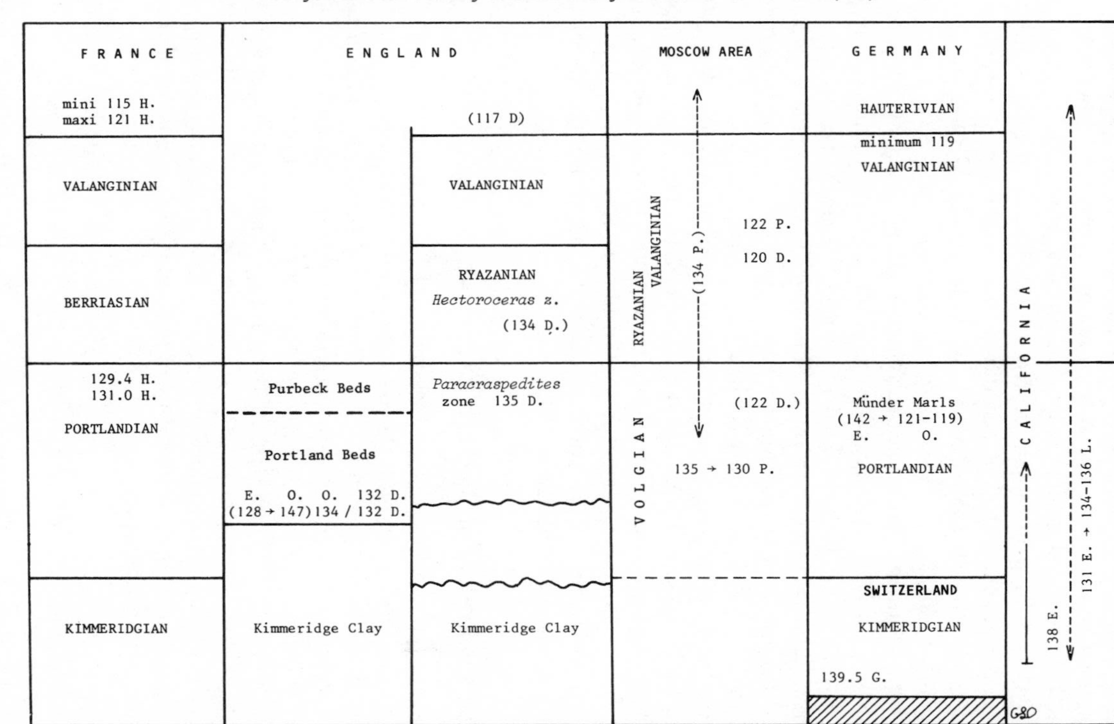

D., Dodson *et al.* (1964); E., Evernden *et al.* (1961) + Kulp (1961); G., Gygi and McDowell (1970); H., Odin (this book); K., Kreuzer (this book); L., Lanphere and Jones (1978); O., Obradovich (1964); P., Polevaya *et al.* (1961).

equal to or lower than the age proposed for the upper limit of the Jurassic, although they were obtained from samples considered as substantially older—by one or two stages. The data now available indicate that the Jurassic–Cretaceous boundary *lies close to 130 Ma.*

Dodson *et al.* (1964) obtained two dates from the Ryazanian, the lowest stage of the Cretaceous in the Boreal realm. These are: 134±4 Ma (ICC) from the base of the stage in England (associated with a *Hectoroceras* fauna) and 120 Ma (ICC) from a sample from the Moscow region, some 60 km from Ryazan (Table 11). The last dates were obtained on a very evolved glaucony; the English sample (5.5% K) is less evolved, and the date less reliable as a result.

A sample from the *Riasanites rjasanensis* zone, also near Moscow (PTS 322), included a little-evolved glaucony with a low-reliability apparent age of 122 Ma (ICC). A little-evolved glaucony of Late Volgian to Ryazanian age collected in the Lenin hills, near Moscow (PTS 215), gave a recalculated apparent age of 134 Ma (ICC) (after Polevaya *et al.*). The poor stratigraphic control means that it is of only limited value, however.

Casey (1964) discussed a range of complementary data derived from plutonic rocks collected and dated in the USSR. The arguments used to place these stratigraphically are quite inadequate. They neither confirm nor disallow the sedimentary dates presented above. Bearing in mind the data assembled in Table 11, it appears to us to be justifiable to place the boundary in the interval 130±3 Ma.

3.6 The subdivisions of the Early Cretaceous

The stage limits adoped since the summary by Casey (1964) were determined on the assumption that the six Early Cretaceous stages were of equal duration, and that they spanned the interval 135–100 Ma (old constants).

The correction associated with the revised decay constants for ^{40}K only modifies these to 138 and 103 Ma respectively (ICC). Previous discussions indicate, however, that the base of the Cretaceous is best taken at 130 Ma, while the Early/Late Cretaceous boundary is clearly younger than 103 Ma (Odin, 1975; Obradovich and Cobban, 1976). It is equally certain that the Early Cretaceous stages were not of equal duration: a revision of the 1964 time scale is thus essential.

3.6.1 Valanginian

No geochronological data are available for the Berriasian–Valanginian boundary. The Valanginian–Hauterivian boundary can be inferred thanks to a range of dated glauconies. Kreuzer (this book, NDS 148) suggests a minimum age of 119±2 Ma based on four very evolved glauconies from

Germany, although these have been affected by salt movements during the Aptian. Obradovich (1964) provided dates from the Hauterivian of south-eastern France, and these are complemented by the study by Conard *et al.* (this book; NDS 74). Following these data the boundary is situated between 115 and 121 Ma. Finally, Dodson *et al.* (1964) and Curry and Odin (this book, NDS 72) have obtained ages of between 108.3 and 117.0 Ma from very little-evolved glauconies obtained from the Hauterivian Speeton clay of Yorkshire, England, although these dates are very difficult to interpret.

Lanphere and Jones (1978) analysed a tuff collected in Central Alaska that was dated stratigraphically as Late Valanginian on the basis of the ammonites present in a unit close by. The biotite dated is strongly altered (2.2% K) and the apparent age of 137 Ma merits little confidence.

The range 119±3 Ma probably spans the interval in which the Valanginian–Hauterivian boundary should be placed on the basis of presently available data. This estimate should be compared with equivalent apparent ages of 117, 122 and 122 Ma, ICC, noted by Shibata *et al.* (1978) based on various Late Neocomian igneous rocks from Japan.

Few satisfactory data are available to allow extrapolation back to the possible date of the Berriasian–Valanginian boundary. Palaeomagnetic data combined with assumed uniform sea-floor spreading rates are not yet a suitable approach to the problem at the present time. In the absence of more convincing data, we note that the Valanginian includes more ammonite zones than the Berriasian (Table 5), and conclude that the latter was shorter than the former, and that the boundary between these two stages probably lies in the interval (126±4 Ma). The uncertainty surrounding this figure is of necessity higher than that attached to the base of the stage; the parentheses indicate that the figure is not based on direct radiometric evidence.

3.6.2 Hauterivian–Barremian boundary

The Hauterivian–Barremian boundary is again poorly documented. A moderately evolved Late Hauterivian glaucony from the Speeton clay of Yorkshire, England (NDS 72), is dated at 108.3±3.8 Ma. More recently, Kreuzer and Mutterlose (NDS 162) have obtained dates for latest Hauterivian North German glauconies. The outcrop has been heavily tectonized, and the green grains are 'mainly pieces of prediagenetically transported and broken up pellets', so that one must perhaps consider the possibility of the inheritance of some grains. In these circumstances the age of 119 Ma must be treated with caution, whereas a date of 126 Ma obtained from a glaucony very poor in potassium provides clear evidence of inheritance.

Dodson *et al.* (1964; PTS 236) dated a little-evolved (K = 5.49%) Speeton clay glaucony from above the boundary at 111±3 Ma (ICC). Apparent ages from well-evolved Late Barremian glauconies from southeastern France (NDS 73) have either been rejuvenated or indicate the stage to have been of very short duration. The base of the Barremian is situated after 112 Ma and before 116 Ma. An age of 114±2 Ma is adopted, with limited confidence, but it is certainly closer to reality than the figure of 121 (ICC) used since the survey by Casey (1964).

3.6.3 Barremian–Aptian boundary

A glauconitic horizon from the latest Barremian in southeast France yielded two well-evolved glauconies (NDS 73). One, collected from a unit that has been subjected to weathering in the soil zone, gave an apparent age of 103.4±3.3 Ma, which is interpreted as having been rejuvenated. The other gave an apparent age of 110.7±3.6 Ma, but has unfortunately suffered slight tectonism (Conard *et al.*, this volume). The latter can, however, perhaps be considered as a minimum or actual age for the top of the Barremian, the maximum age of which is 116±4 Ma.

The Aptian has yielded many figures between 106 and 114 Ma. Kreuzer (this volume, NDS 148) determined the maximum age of Aptian tectonism of a diapir in Germany at 114.0±1.6 Ma. Cowperthwaite *et al.* (1972; cf. Kreuzer, this book, NDS 147) proposed an age of 107±5 Ma based on the sanidine fractions in an Aptian fuller's earth.

The age of 108±2 Ma for two concordant (but little evolved) glauconies from the Late Aptian of Germany represents a more precise figure. European Aptian glauconies are generally little evolved, and those of the Pas de Calais, France, are no exception. Various fractions from the Aptian of this region have yielded K–Ar dates of around 106 Ma (K=5.2%) and Rb–Sr ages of around 116 Ma (Elewaut *et al.*, this book, NDS 77) but it is difficult to draw a precise conclusion. A specimen from the Late Aptian of England has been dated by Evernden *et al.* (1961, PTS 549) and repeated by Obradovich (1964, NDS 98). The equivalent ages 110.9±3.3 (ICC) and 110.8±3.3 (ICC) are usable, for the glaucony is a little more evolved than the French samples (K% = 5.9 and 5.6%). In contrast we have no confidence in a date of 118 Ma, ICC, described in PTS 50; it was derived from only slightly-evolved grains. We are also sceptical of the very variable and young ages of 91, 105 and 91 Ma (PTS 213) obtained from the USSR.

Two ages were proposed by Dodson *et al.* (1964, PTS 240) and were obtained from two glauconies derived from the Late and Early Aptian Ferruginous Sands of the Isle of Wight, England. The glaucony is moderately well evolved, but the outcrop is somewhat weathered. The apparent ages have been reduced to 97 and 104 Ma, ICC, and are the minimum ages of the formation. The glaucony from the Late Aptian Ferruginous Sands of Compton Bay, Isle of Wight, England, is moderately evolved (K=5.66%) and slightly tectonized but not weathered. The apparent age of 107.3±

3.9 Ma (Curry and Odin, NDS 71) can be used to determine the age of the top of the Aptian.

Taking the most satisfactory dates available, it would appear that the base of the Aptian is older than 108–111 Ma, but certainly no older than 115 Ma. The Barremian–Aptian boundary in all probability lies in the interval 112±2 Ma, and we suggest that the higher figure is the most likely. The Barremian is thus of very short duration.

3.6.4 Aptian–Albian boundary

Lanphere and Jones (1978) dated the Albian–Aptian boundary on the basis of two sets of results. The Alisitos formation of Baja–California, was dated as probably Aptian on the basis of fossils found in the area, but no convincing stratigraphic attribution is given to the age of 114±2 Ma obtained from a diorite. The date of around 120 Ma (ICC) for a biotite from the Middle–Late Albian Harmon shale of British Columbia (see PTS 203, Folinsbee *et al.*, 1961: AK, O) must be rejected, for, as these authors note, the biotite contains only 2.5% potassium.

We have much higher confidence in dates obtained from two bentonites from the Mowry shale of the Black Hills of Wyoming. These are dated on the basis of forams with levels between the *Neogastroplites* zones above and the *Gastroplites* zones below, corresponding to the Middle–Late Albian boundary in Europe. A mean age of 99.4±1.3 Ma is calculated on the basis of three well-preserved biotites (6.9% K), whereas the biotite from the lower bentonite gave a mean age of 104.4±1.5 Ma with a potassium content a little below 5%. The considerable difference, of 5 Ma, is perhaps too high, and we prefer the age obtained from the better preserved biotites, which are the more reliable chronometer.

A lava flow just below an Aptian limestone in Morocco yielded feldspars dated at 109.3±6.0 Ma (Westphal *et al.*, 1979; NDS 188). Although possibly rejuvenated, this is clearly from a level well below the Aptian–Albian boundary (Rolley, 1973) and is compatible with the data given below.

Seven little-evolved glauconies from Lower Saxony (3.4–4.4% K) gave closely comparable apparent ages in spite of different potassium values, suggesting an absence of inherited material (Odin and Dodson, this volume) but not absolutely disproving lack of rejuvenation. The mean age of 106.0± 0.8 Ma (Kreuzer and Kemper, this book, NDS 143) is from a basal Albian unit, and can be taken as a minimum for the base of the Albian.

The glauconitic base of the Albian in France is often stratigraphically younger than that in Germany; two evolved glauconies from the Albian of Aube have thus yielded ages of 102.3±3.1 for a higher and 108.6±3.3 for a lower glaucony level. These favourable data suggest that the base of the Albian lies between 106 and 108 Ma.

In England, a little evolved glaucony from the Late Aptian/Early Albian

Folkestone beds of Folkestone (Obradovich, 1964; NDS 98) has yielded an age of around 116 Ma, ICC, older than that obtained from the Aptian below. The same level has also produced a date of 127 Ma, ICC (Dodson *et al.*, 1964; PTS 241), which appears to reflect a systematic inheritance from the substrate of glauconitization.

The top of the Early Albian of the Boulonnais has been dated by the K–Ar and Rb–Sr methods. The glauconies are evolved, and yield K–Ar ages of around 100 Ma (Elewaut *et al.*, this book, NDS 78–79). The Rb–Sr apparent age is higher, at 104.8 Ma (NDS 78).

All the ages recently obtained from Middle and Late Albian glauconies from France (NDS 63, 65, 66, 67), Germany (NDS 97, 144, 145), Great Britain (PTS 56–242) and the USSR (PTS 212, Middle Albian, 100 Ma) are much younger than those discussed previously.

In North America, several Late Albian bentonites have been dated; they confirm the present results and are discussed further below (PTS 202–204). The Middle Albian glaucony from the Mannville formation (Sprucefield, Canada, PTS 219, AK 56: Folinsbee *et al.*, 1961) is little evolved (K = 4.9%) and the date obtained not acceptable.

In summary, the most reliable data relating to the Early Albian and Late Aptian allow us to place the boundary very precisely at no later than 106 Ma and no earlier than 108 Ma.

The dating of the subdivisions of the Albian has been approached by several authors (Elewaut *et al.*, NDS 78–79; Kreuzer and Groetzner, NDS 144; Table 12), and given that the age of the top of the stage is around 95 Ma, as discussed below, the duration of the stage—12 Ma—is exceptionally long. A number of radiometric dates suggest that the Middle–Late Albian boundary lies at around 97.5 Ma (NDS 144), although we note that in this case one must consider not only the source of the glaucony, but also the moment of closure of the minerals. In the Paris Basin, there is a concordance of ages around 99 Ma as a time of closure of glauconies, which is probably close to the Middle–Late Albian boundary (Odin and Hunziker, this volume, Figure 3). The age of 99.0 ± 1.1 Ma coincides with the figure of 99.4 ± 1.3 obtained from little altered biotites from bentonites in the Newcastle sandstones of the USA (NDS 157), which lie a little above the base of the Late Albian. One must admit, however, the difficulty of stratigraphic correlation between the Albian of France and North America.

The number of zones recognized with the Albian can also be used as a basis for determining the approximate duration of the substages; taking the most recent zonation developed by Destombes (Odin and Hunziker, this volume, Table 1), one obtains (Table 12) an Early–Middle Albian boundary date of 102.8 Ma and a Middle–Late Albian boundary of 98 Ma. Considering all the above observations together, we propose a date of 98.5 ± 1.5 Ma for the top, and 101.5 ± 1.5 for the base of the Middle Albian.

Table 12. Subdivisions of the Albian stage according to various criteria.

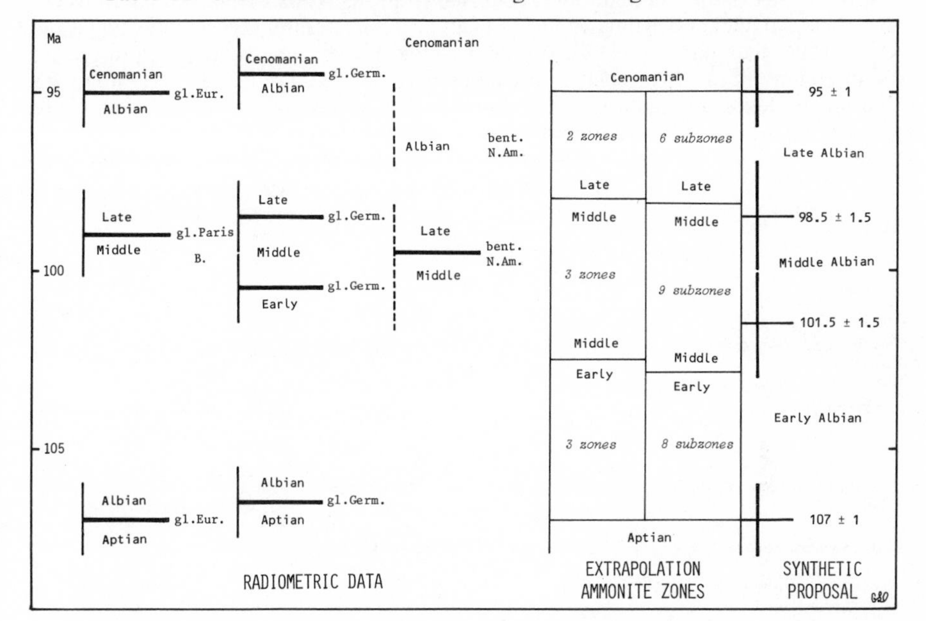

gl., glauconies; bent., bentonites; Eur., Europe; Germ., Germany; N.Am., North America; Paris B., Paris Basin.

3.7 Early Cretaceous–Late Cretaceous boundary

Folinsbee *et al.* (1963) provided an early estimate of age of the Late–Early Cretaceous boundary in North America. This was initially based on bentonites from the Crows Nest formation (Folinsbee *et al.*, 1960), first considered as Late Albian and later as Cenomanian. The apparent ages of sanidines sampled at Coleman (Canada) are 97.0 (ICC, AK 2) and 92.6 Ma, ICC (KA 22, Folinsbee *et al.*, 1961, PTS 202). Two biotite ages of 101.5 Ma, ICC (AK 97) and three sanidine ages of 104.0 Ma, ICC (AK 9; PTS 202) were determined from samples from Mill Creek. All these ages have an estimated range of error of ±5% and were obtained from units for which the palaeontological control is very poor, although the minerals concerned are fresh. The bentonites of the Mowry shale of the USA (PTS 204) are much better dated stratigraphically, thanks to the presence of the endemic ammonite genus *Neogastroplites*, currently regarded as Late Albian (Obradovich and Cobban, 1976). Folinsbee *et al.* (1963) mention ages without reporting the potassium contents of the minerals dated. But Baadsgaard (personal communication, 1980) gave us the potassium contents which indicate that the dated Cenomanian biotites are of low reliability. If one

Table 13. K–Ar ages on bentonites collected from the Albian–Cenomanian of North America. All data and estimates (±5% analytical uncertainty) according to Folinsbee *et al.* (1963) except for (1) according to Obradovich and Cobban (1976). Data between parentheses are geochemically uncertain (altered biotites). Correlations with European stages are tentative. The unpublished quoted potassium contents were kindly supplied by H. Baadsgaard.

Sample	K–Ar apparent ages (Ma, ICC) Sanidine	Biotite	Average	Biozonation (*Neogastroplites* = *N.*)
		Cenomanian		
Frontier fm (Wyoming) (1)	—	(93.2±1.8) K: 4.9 2 samples	—	
Niobrara Co (Wyoming) (1)	—	(94.2±1.8) K: 4.4	—	
Frontier fm at Casper (Wyoming)	96 K: 6.9	(88) K: 3.4	(92)	No *N.*
		Late Albian		
Clayspur bentonite (Casper)	96.5 K: 9.6 (2 samples)	92.3 K: 6.5	94.4	*N. maclearni* e. 96
Casper (Wyoming)	95 K: 9.6	92 K: 5.2	94	
Winnecook (Montana)	98.0 (2 samples)			*N. americanus* m. 98.0
Casper (Wyoming)	98 K: 10.1	(98) K: 5.0	98	*N. muelleri*
Casper (Wyoming)	99 K: 10.0	96 K: 6.7	98	e. 100.5 m. 97.6
Arrow Creek (Montana) (1)	97.6±2.0 K: 9.5	97.5±2.0 K: 6.2		*N. cornutus*

e., Estimate by Folinsbee *et al.*; m., measured.

accepts the Clayspur horizon as latest Albian (Table 13), one obtains a figure of about 93 Ma, ICC, for the Albian–Cenomanian boundary as a mean sanidine and biotite age; but according to the more reliable sanidine dates the age increases to 96 Ma.

Comparison of the estimate of 100.5±5.0 Ma (ICC) adopted by Folinsbee *et al.* (1963) for the *Neogastroplites cornutus* zone with the more recent determination of 97.6±2.0 Ma (ICC) by Obradovich and Cobban (1976; NDS 111) indicates that the numbers adopted by the former workers were on the high side. However, the large analytical uncertainty given initially actually encompasses the new figures.

The Frontier formation of Wyoming provides a reference point from just above the Albian–Cenomanian boundary; a bentonite gave a mean 'biotite–sanidine' age of 92 Ma but the biotite has a very low K content (Table 13). Obradovich and Cobban (1976; NDS 100) have published dates from altered biotites (K = 4.4–4.9%) from the Middle–Late Cenomanian. In the absence of better chronometers an age of 93–94 Ma may be tentatively taken as a minimum age for the base of the units assigned to the Cenomanian in the Western Interior Basin. This estimate is compatible with a biotite age of 98 ± 4? Ma, ICC, obtained from a stratigraphically poorly placed Californian bentonite (Evernden *et al.*, 1961; cf. PTS 226). If we take into account: (1) the sometimes large analytical uncertainties (Folinsbee and others' previous results); (2) the geochemical problems (NDS 110 and discordance between biotite and sanidine ages); (3) the stratigraphic difficulties (unsatisfactory correlations by ammonites of the Albian–Cenomanian boundary in Europe with that in North America), the Albian–Cenomanian boundary may be suggested to lie between 93 and 96 Ma in North America. The choice of the older value, an old systematic practice, appears unjustified according to Table 13.

In Europe, dating of the Albian–Cenomanian boundary can be approached from the analyses of Early Cenomanian glauconies from Devonshire, England, which gave ages of 94–95 Ma (NDS 96). Elewaut *et al.* (NDS 80) have proposed a comparable mean age of 94 Ma, based on K–Ar and Rb–Sr determinations from an Early Cenomanian glaucony.

Juignet *et al.* (1975) concluded that the boundary could be dated at between 94.5 and 96.5 Ma in the Paris Basin; these results are discussed further elsewhere in this book (Odin and Hunziker, see also NDS 62–68, 85 and 119, which are not discussed further here). From these results we can date the Albian–Cenomanian boundary in the Paris Basin at 95 ± 1 Ma with some confidence. This agrees with all the available dates, in particular the results obtained in several laboratories from the Cenomanian reference mineral 'glauconite GL-O' (Odin and colleagues, this volume).

3.8 The subdivisions of the Late Cretaceous

3.8.1 Cenomanian–Turonian boundary

Very few data were available on this boundary in 1964. The analyses from Germany (PTS 58-59-61) were repeated by Obradovich (1964), and the precise stratigraphic position of the samples is reviewed by Kreuzer *et al.* (this book, NDS 94, 95). Considering only data from suitably evolved Turonian glauconies, we have apparent ages of 87.3 ± 4.1 Ma (ICC) and 85.5 ± 3.1 Ma (ICC) for the Middle and the Late Turonian respectively. Kreuzer and Seibertz (NDS 227) have recently obtained a more precise date from a basal Late Turonian sample: 88.1 ± 1.5 Ma.

The Late Cenomanian is dated at 89.8 ± 3.6 Ma (NDS 59) in the northern Paris Basin. Elewaut *et al.* (NDS 81) obtained equivalent K–Ar and Rb–Sr apparent ages of 89.5 ± 3.3 and 91.2 ± 2.0 Ma respectively. Kreuzer *et al.*

(NDS 226) analysed five NW German glauconies from the Cenomanian–Turonian boundary and obtained ages of 91.2–89.0 Ma; a figure of 90 Ma is probably a good estimate of the true age of this limit.

Obradovich and Cobban recently obtained an apparent age of 91.0 ± 1.8 Ma for an altered biotite (3.78% K) from the Early Turonian *Inoceramus labiatus* zone of Montana (North America). In spite of the low K content and consequent low reliability, the figure is of the same order as that determined in Europe. Five less-altered biotites obtained from borehole samples from the *Inoceramus labiatus* zone of Alaska gave apparent ages of between 91.5 and 93.6 Ma (ICC; NDS 118). Ages of US Turonian bentonites and European Cenomanian glauconies are thus slightly discordant, with the former a little higher than the latter. *Inoceramus labiatus* is, however, a good indicator for the Early Turonian in both Europe and North America, and we suggest the boundary lies in the interval 91^{+1}_{-2} Ma.

3.8.2 *Turonian–Coniacian boundary*

A bentonite from the Colorado shale of Montana, USA (NDS 108), yielded little-altered biotites with coincident ages of around 88.9 ± 1.5 Ma. It was associated with *Inoceramus deformis*, which characterizes the zone of that name in Germany which is usually attributed to the Upper Turonian. Seibertz (in Rawson *et al.*, 1978) has, however, found typical Lower Coniacian ammonites in this zone, confirming it to be in fact Lower Coniacian, and that the traditional position of the Coniacian–Turonian boundary in Germany has been taken at too high a level when compared with the French type sequences.

Several glauconies from the base of the Belgian Coniacian have been dated. The least evolved (K = 5.7%) gave an age of 90.5 ± 2.1 Ma (NDS 60); three others (with 6.2–6.3% K) gave apparent K–Ar ages of 84.3–87.1 Ma and apparent Rb–Sr ages of 84.4–89.9 Ma (NDS 83). The mean K–Ar age of 85.9 Ma is near the mean Rb–Sr age of 86.6 Ma. In the same basin, an evolved glaucony gave a K–Ar apparent age of 88.1 ± 3.0 Ma and a Rb–Sr apparent age of 87.0 ± 2.3 Ma (NDS 82). Due to a slight weathering the K–Ar age is more reliable than the Rb–Sr one (Odin *et al.*, 1974; Clauer, this volume). From the same quarry a K–Ar apparent age of 88.7 ± 2.2 Ma was obtained (NDS 164).

All these ages are older than those obtained from a Late Turonian German sample (NDS 94) and are near the age of 88.7 ± 1.0 Ma obtained from a Middle Turonian sample.

In Austria a surface sample now considered as basal Coniacian yielded evolved glaucony with an apparent age of 86.8 ± 3.3 Ma; this is a minimum age for the base of the Coniacian because the formation is somewhat tectonized (Kennedy and Odin, NDS 86).

In summary, Coniacian ages of 86–87 Ma (with a maximum at 89 Ma) and Turonian ages of 88–89 Ma allow us to place the Coniacian–Turonian boundary in the interval *c.* 88 ± 1 Ma.

3.8.3 Coniacian–Santonian boundary

No certain data are available to aid in the determination of the age of this boundary. The youngest known Coniacian dates are situated around 86 Ma, and an apparent age of 84.8 Ma, ICC (PTS 229), has been obtained from a little-evolved German Santonian glaucony. This is not very trustworthy, however.

3.8.4 Santonian–Campanian boundary

Glaucony is abundant in the Early Campanian of North Germany, and Evernden *et al.* (1961) obtained an apparent age of 83.4 ± 2.4 Ma (see also PTS 62) from an insufficiently evolved sample with only 5.3% K, in spite of which the age is compatible with the stratigraphic position. These glauconies are currently under investigation by the Hanover Laboratory. In Limburg, several glauconies collected from around the Early–Late Campanian boundary have given ages of 77.6 Ma from little-evolved glauconies and 74.9 and 73.5 from evolved glauconies, the sequence of dates matching the relative ages of the samples. These dates are, however, insufficient to allow us to fix the Santonian–Campanian boundary in Europe. All that can be said is that it is younger than 86 Ma and older than 75 Ma, and perhaps older than 77.6 Ma and close to 83.4 Ma, if these ages are confirmed with better chronometers.

A lava flow from Texas attributed to the basal Campanian gave an age of 81.5 ± 3 Ma, ICC (NDS 163). Two biotites from the highest Santonian of the Western Interior were reported by Obradovich and Cobban (1976); one is somewhat altered (K = 4.5%), and gave an apparent age of 79.8 ± 1.6 Ma; the other is better preserved and gave an apparent age of 84.4 ± 1.6 Ma, ICC (NDS 107), providing these authors with a maximum age for the boundary. Four biotite ages from the lower third of the Campanian gave ages between 79.2 ± 1.6 and 80 ± 1.6 Ma, ICC (NDS 106), which can be taken as the minimum for the Santonian–Campanian boundary.

Owens and Sohl (1973; NDS 117) provided dates from stratigraphic levels between the units discussed above in the Atlantic Coast region of North America. They are based on little-evolved glauconies, and the two samples providing the highest degrees of confidence gave imprecise apparent ages of between 76.3 and 78.7 Ma, which provide minimum dates for the boundary.

The conclusion from these US data is that the Santonian–Campanian boundary is close to, but younger than 84.4 Ma, and older than 80 Ma. In our opinion it lies within the interval 83 ± 1; Obradovich and Cobban suggest *c.* 84 Ma, ICC.

3.8.5 Campanian–Maastrichtian boundary

The Campanian–Maastrichtian boundary is difficult to place with certainty in the North American sequences (cf. NDS 104–105), and correlation with the European succession is difficult. There are also problems associated with the boundary in Limburg (NDS 139), but in spite of this there is remarkable correspondence between the four independent estimates considered here. Obradovich and Cobban proposed an age older than 70.5 Ma (NDS 104) and younger than 74 Ma, ICC (cf. NDS 105) in the Western Interior, preferring a date close to 72 Ma. Evolved glauconies from the Atlantic Coast region suggested a maximum age of 71.6 ± 2.7 Ma to Owens and Sohl (1973; cf. NDS 116). Studies by Ghosh (1972; cf. NDS 54) on somewhat dubious chronometers suggested an age in excess of 71 Ma. The stratigraphic age of the evolved glaucony from the Vaalser greensand of North Germany is uncertain (NDS 93), but the glauconies from Limburg dated by Priem *et al.* (1975b; cf. NDS 139) allow us to suggest an age of 71.5 ± 2.5 Ma for the base of the Maastrichtian there.

In conclusion, we believe the Campanian–Maastrichtian boundary to be within the interval 72 ± 1 Ma.

4 DISCUSSION AND SUMMARY

4.1 Remarks on the chronometers

When possible, a comparison of results obtained using different chronometers shows reasonably good concordance of dates. There is no difference between glaucony dates and bentonite dates in some cases, as is clear from our discussion of the Albian–Cenomanian and Campanian–Maastrichtian boundaries. However, one must remember that the comparison *is* rather uncertain due to not insignificant stratigraphical uncertainties. In contrast, Cenomanian–Turonian boundary dates from bentonites in North America and from glauconies in West Europe show distinct differences although some of the dated samples appear to be well calibrated with respect to the *I. labiatus* zone, which is considered a good time marker by many palaeontologists. European dates lead to an age of the boundary of 89–90 Ma while American results would lead to 91–92 Ma, a difference of 2%. The possible explanations of this difference are several and we need more information to provide a definitive one: for example, it should be remembered that relative interlaboratory calibration differences of ±1% are a normal event. For a rock 90 million years old, this gives an error of ±1 Ma. To these analytical uncertainties must be added the geochemical ones: due to their respective process of genesis, dates from biotites and sanidines in bentonites *predate* their deposition, while those obtained from glauconitic minerals *postdate* the fossils with which they were deposited; the difference of ages observed above could well be explained by this phenomenon.

As noted above in our discussion of available data, geochronologists often use very imperfect chronometers. This problem is essentially linked to the question of time spent on each study. It is easier and less time-consuming to date the occasional sample without proper investigation and search for the best chronometer at a given level in a given basin. To this must be added a tenfold difference in the time needed for the proper picking of grains and their preparation as opposed to analysis of a bulk sample. A common fault in earlier work on the dating of the stratigraphic column was the conclusion that a low reliability K–Ar age was *necessarily younger* than the true age. This error led to the *frequent overestimate* of the age of stratigraphical boundaries. The time scale proposed in this book is a clear demonstration of this widespread error, for we know today that too high as well as too low K–Ar ages may be obtained from altered high-temperature rocks or little evolved glauconies.

4.2 Duration of the Jurassic and Cretaceous periods and their subdivisions

Table 14 gives a summary of our conclusions on the basis of the data at present available. The plus or minus given with each proposed age is an attempt to define the interval within which it seems possible to move the boundary when more accurate or reliable dates are obtained. The dates in parentheses are extrapolated (e.g. not obtained from direct radiometric measurements) and probably include a degree of error and uncertainty of plus or minus 4–5%.

The estimated duration of the stages and periods should be of some assistance to palaeontologists, as they reveal those intervals in which biostratigraphic refinement is needed, and indeed possible. As will be seen from Table 14, relative stage duration suggests that a comparable number of ammonite zones should be recognized in the Berriasian and Coniacian, but never as many as in the Albian or Campanian, whereas biostratigraphic refinement of the Campanian is clearly possible if one accepts uniform rates of faunal change during the Cretaceous. If this proves impossible, then it will indicate that evolutionary/diversification rates varied through time.

In conclusion, the Jurassic appears to have been the longest of the Phanerozoic periods, while the Cretaceous was one of the four longest periods. The Bajocian to Oxfordian stages were the four longest stages of the Jurassic, while in the Cretaceous, the Albian and Campanian were also very long. The Barremian stage was short, as were the Turonian, Coniacian

Table 14. The Jurassic and Cretaceous time scale in 1981. All the data used for this synthesis were individually reevaluated according to the present state of knowledge of the stratigraphical, geochemical and analytical uncertainties and recalculated with the conventional constants adopted in Sydney (1976).

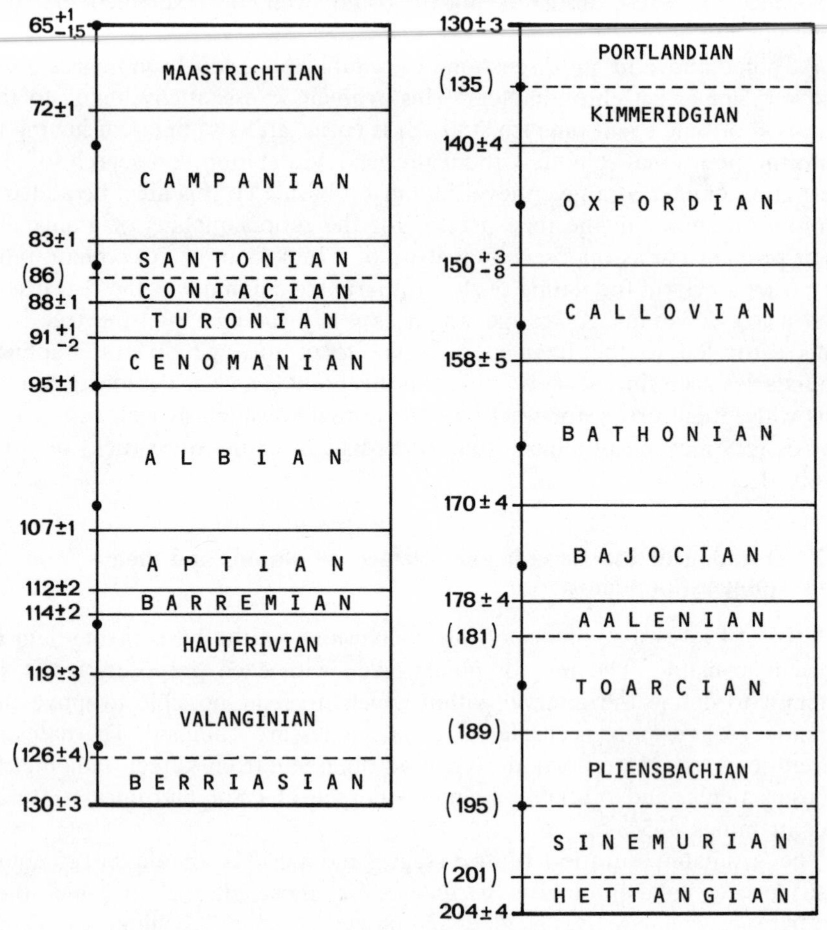

and Santonian, the combined duration of which was less than any one of the long stages quoted above.

For the future, the highest priority is to obtain better dates from the Early Jurassic (Hettangian to Aalenian) and the earliest Cretaceous (Berriasian and Valanginian); these are the intervals in which there are least reliable geochronological data at the time of writing.

(Manuscript received 16-11-1980)

Numerical Dating in Stratigraphy
Edited by G. S. Odin

33 Rubidium–strontium glaucony ages, southeastern Atlantic Coastal Plain, USA

W. BURLEIGH HARRIS

1 INTRODUCTION

This chapter summarizes previous results of Rb–Sr glaucony dating of Cretaceous and Palaeogene sediments in the southeastern Atlantic Coastal Plain, USA (e.g. Harris and Bottino, 1974b; Harris, 1976; Harris and Baum, 1977; Harris and Zullo, 1980) and presents new radiometric data on the Eocene Nanjemoy formation in Virginia.

The emerged Atlantic Coastal Plain province extends southwards from Cape Cod, Massachusetts, to central Georgia where it is continuous with the Gulf Coastal Plain (Figure 1). It consists of an oceanward thickening wedge of east to southeast dipping Mesozoic–Cenozoic sediments and sedimentary rocks that unconformably overlie an oceanward dipping pre-Cretaceous basement (Figure 1). The Coastal Plain is an area of low relief with a well-integrated network of consequent streams and rivers which flow east or southeast to the Atlantic Ocean. As a result, exposures are limited and regional stratigraphic relationships are often difficult to interpret. Coastal Plain sediments generally are intercalated alluvial–fluvial, marginal marine and continental shelf deposits which range in age from Early Cretaceous through Quaternary. These units represent a complex of depositional regimes of which glaucony forms an abundant constituent, and, therefore, offers potential as a means for regional correlation. Although glaucony-rich sediments of the northern Atlantic Coastal Plain have been dated radiometrically (Owens and Sohl, 1973), no workers have dated glauconies in the southeastern part. Therefore, Coastal Plain sediments from the Cretaceous of North Carolina, the Eocene of North Carolina and Virginia have been dated by the Rb–Sr isochron method. Coastal Plain sediments from the Palaeocene of North Carolina have been dated by the Rb–Sr conventional age method.

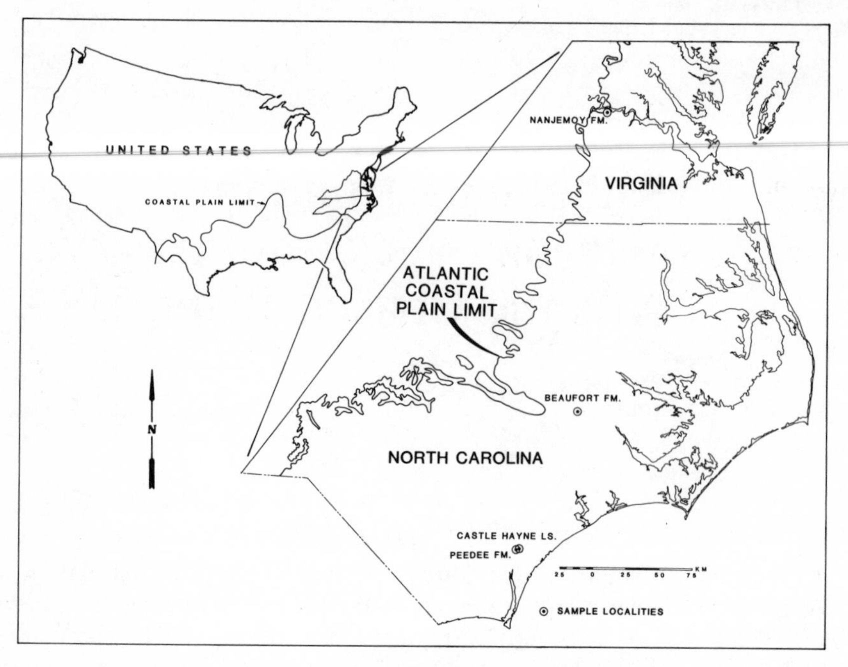

Figure 1. Generalized location map of sample localities, southeastern Atlantic Coastal Plain.

2 GEOLOGICAL SETTING

2.1 Cretaceous

Peedee formation. The youngest Mesozoic strata in the Coastal Plain of North Carolina are referred to as the Peedee formation. Although ranging in age from Santonian(?) to Maastrichtian, carbonate at the top of the formation (the Rocky Point member) and an unnamed equivalent (very fine to fine sand) contain a Maastrichtian fauna (Wheeler and Curran, 1974). In order to determine the age of the youngest Mesozoic strata in North Carolina (and hence place a maximum age on the Mesozoic–Cenozoic boundary), the Rocky Point member and stratigraphically equivalent very fine to fine sand were dated by the Rb–Sr glaucony isochron method.

The Rocky Point member consists of four lithofacies: calcareous quartz arenite; sandy, pelecypod biosparrudite; sandy biosparite; and sandy, pelecypod biosparite. The unit underlies the Palaeocene Beaufort formation or the Eocene Castle Hayne limestone in southeast North Carolina. Stratigraphically equivalent very fine to fine Peedee sands intertongue with the sandy, pelecypod biosparrudite lithology in the extreme southeast part of the state. Typical Late Cretaceous megafossils that occur in the Rocky Point member and the unnamed equivalent include: *Trigonia haynensis* Stephenson; *Exogyra costata* Say; *Ostrea subspatulata* Forbes; *Cardium penderense* Stephenson; *Cardium spillmani* Stephenson; *Hardouinia mortonis* (Michelin); and *Belemnitella* sp. (Fallaw and Wheeler, 1963). Late Cretaceous planktonic foraminifera include *Guembelitria cretacea* Cushman; *Heterohelix globulosa* (Ehrenberg); and *Globotruncana* sp.; benthonic foraminifera include *Planulina correcta* (Carsey); *Anomalinoides pseudopapillosa* (Carsey) and *Pseudouvigerina seligi* (Cushman) (Wheeler and Curran, 1974). All foraminifera occur in the Campanian or Maastrichtian stages; however, most are restricted to the Maastrichtian. A complete discussion of the stratigraphic and palaeontological evidence for placement of the Rocky Point member in the Cretaceous is presented in Wheeler and Curran (1974) and Harris (1975).

Glaucony occurs in the Rocky Point member and the unnamed sand as sand-size grains, depression and tube fillings on quartz grains, and as fillings in bryozoan zooecia and foraminifera. The largest precentage of glaucony consists of mammillated to lobate sand-size grains. Many of the other morphological types described by Triplehorn (1966) are present but they are not abundant.

Three samples of the Rocky Point member were collected between 60 and 160 cm below the Mesozoic–Cenozoic boundary at the Martin Marietta quarry located about 4.5 km northeast of Castle Hayne (New Hanover County), North Carolina (Figure 1). One sample of the unnamed member was collected along the North East Cape Fear river at Wilmington, North Carolina.

2.2 Palaeocene

Beaufort formation. Brown (1959) applied the name Beaufort formation to a fine to medium glauconitic sand interval between the depths of 32.3 and 52.1 m (subsea) in a well drilled at Chocowinity (Beaufort County), North Carolina. He assigned a Midway age to the unit on the basis of ostracods (Brown, 1958). Subsequently, the Beaufort formation was recognized in outcrop (Swift, 1964; US Geological Survey, 1972) in Lenoir County, North Carolina.

The Beaufort formation includes mudstone that intergrades with chert, mudstone–chert intercalated with sandstone, glauconitic sand and phosphatic conglomerate (Brown *et al.*, 1977). Mudstone and siliceous mudstone are the dominant rock types in the lower Beaufort formation and are referred to as the Jericho Run member. Unconsolidated, sandy, foraminiferal–glauconitic sediments alternating with thinner, slightly glauconitic foraminiferal biomicrosparite, which have not been named, characterize the upper

Beaufort formation. These sediments contain an abundant and diverse benthonic and planktonic foraminiferal assemblage. Typical benthonic foraminifera include *Bulimina quadrata* Plummer, *Pseudonodosaria pygmaea* (Reuss), *Dentalina pseudoobliquestriata* (Plummer), *Sarcenaria trigonata* (Plummer), *Lenticulina midwayensis* (Plummer), *L. insulsus* (Cushman), *L. degolyeri* (Plummer), *Cibicidoides alleni* (Plummer), *Vaginulopsis longiforma* (Plummer), *Nodosaria affinis* (Reuss), *Alabamina midwayensis* Brotzen and *Gyroidinoides subangulata* (Plummer). Planktonic foraminifera include *Globorotalia aequa* Cushman and Renz, *G. pseudomenardii* Bolli, *G. pseudobulloides* (Plummer) and *Globigerina triloculinoides* Plummer. The presence of *Globorotalia pseudomenardii* delineates Berggren's (1971a) P4 planktonic foraminiferal zone. In addition, *Globorotalia pseudobullinoides, G. aequa* and *Globigerina triloculinoides* occur in the P4 zone (Berggren, 1971a). The well-established P4 planktonic foraminiferal zone identifies the Middle Thanetian.

Glaucony is an abundant component, averaging 20% by weight, of the unconsolidated layers in the upper part of the Beaufort formation and generally is concentrated in burrows. The glaucony ranges from very fine to medium sand size and is principally mammillated to lobate. A single Beaufort formation sample was collected 1.5 m below the Palaeocene–Eocene boundary along Moseley Creek (Lenoir-Craven County-line) about 10 km east of Kinston, North Carolina (Figure 1).

2.3 Eocene

Nanjemoy formation. The Nanjemoy formation was described by Clark and Martin (1901) along Nanjemoy Creek in southern Maryland and is considered the upper part of the Pamunkey group. It is classically considered Eocene in age; however, detailed analyses of exact correlations of the unit with European stages have not been made.

The Nanjemoy formation consists of olive black to greenish black, very fine to fine, micaceous glauconitic sand. Several prominent clayey silts, which are lighter in colour and less glauconitic, occur in the unit. The unit overlies the Marlboro clay and unconformably underlies the Miocene St. Mary's formation or the Pliocene Yorktown formation.

Typical megafossils that occur in the Nanjemoy formation are *Ostrea sellaeformis* Conrad, *Venericardia potapacoensis, Meretrix ovata* Conrad, *Meretrix subimpressa* Conrad, *Corbula subengonata* Dall, *Nuculana improcera* (Conrad) and *Calyotrophorus jacksoni* Clark (Dischinger, 1979). T. R. Worsley (personal communication) has identified nannofossils *Discoaster lodoensis* and *Chiasmolithus solithus*, which, together with the marked absence of several other indicative forms, including *D. sublodoensis*, suggests that the Nanjemoy formation where studied belongs to nannoplankton zone NP-13, which is Early to Middle Eocene.

Glaucony in the Nanjemoy formation is principally mammillated to lobate fine sand size and is concentrated in burrows. Although near the base of the unit the external morphology of the glaucony suggests transportation and abrasion, upward in the unit the glaucony is authigenic with well-preserved external morphologies. Three samples of the Nanjemoy formation were collected from 2.5–11.5 m below the upper contact with the Pliocene Yorktown formation in Wilson's Gulley, approximately 3 km east of Hopewell (Prince George County), Virginia (Figure 1).

Castle Hayne limestone. The Castle Hayne limestone is one of the principal units that occurs in the Coastal Plain of North Carolina. The unit was named by Miller (1912) for exposures in the vicinity of Castle Hayne (New Hanover County), North Carolina. Because Miller (1912) did not designate a type section, Baum *et al.* (1978) designated the Martin Marietta quarry located 4.5 km northeast of Castle Hayne the lectostratotype.

The Castle Hayne limestone has traditionally been correlated with the Jacksonian stage (Clark, 1909; 1912; Canu and Bassler, 1920; Kellum, 1925; 1926; Cheetham, 1961; Copeland, 1964) of the Gulf Coast; however, recent workers consider the unit Claibornian (Brown *et al.*, 1972; Ward *et al.*, 1978).

The Castle Hayne limestone consists of lower phosphate pebble biomicrudite, middle bryozoan biosparrudite and upper bryozoan–sponge biomicrudite. The unit disconformably overlies the Cretaceous Rocky Point member of the Peedee formation, the Peedee formation, or the Palaeocene Beaufort formation. The Castle Hayne is disconformably overlain by the Eocene New Bern formation, the Oligocene Trent formation, or younger sediments. A complete palaeontological description is beyond the purpose of this chapter and is presented in Harris and Zullo (1980). However, calcareous nannofossils from the glaucony-bearing horizon are reported by Worsley and Turco (1979) as indicative of nannofossil zones NP-19 and NP-20, which is uppermost Jacksonian. Typical calcareous nannofossils found in this part of the Castle Hayne limestone include *Zygolithus dubius, Chiasmolithus grandis, Discoaster barbadiensis, D. saipanensis, Sphenolithus predistentus, Helicopontosphaera reticulata, Cyclococcolithina formosa* and *Micrantholithus procerus* (Turco, 1979).

Glaucony occurs as sand-size grains scattered throughout the Castle Hayne limestone in concentrations generally less than 1–2%. It principally consists of mammillated to lobate types, fossil and faecal replacements. A concentrated zone of 30–40% glaucony occurs approximately 1.6 m above the Mesozoic–Cenozoic boundary at the lectostratotype in the upper bryozoan–sponge biomicrudite lithofacies. A sample of the concentrated glaucony zone was collected at the lectostratotype (Figure 1).

3 METHODS AND PROCEDURES

Whole-rock samples were crushed in a jaw crusher (only when lithified), sieved and the 0.625–2 mm fraction collected. Clay- and silt-size material adhering to grain surfaces was removed by washing and decanting in demineralized water. The samples were then dried in reagent grade acetone. A Frantz Isodynamic Separator was used to separate and concentrate glaucony from quartz, calcite, clay and heavy minerals. The optimum inclination and magnetic field strength varied with each sample. Generally a field strength of 0.9 A and an inclination of 25° concentrated the greatest percentage of glaucony. Final concentration consisted of the removal of all foreign material by hand-picking under a binocular microscope. Those glaucony grains that were incompletely glauconitized, non-authigenic or with a mica precursor were also removed by hand-picking. Only mammillated to lobate or replaced fossil varieties were concentrated. Originally, a 0.075 g sample (0.025 g for mass spectroscopy and 0.05 g for X-ray diffraction analysis) was considered the minimum quantity needed for analyses; however, because foreign material (generally quartz and calcite) commonly adhered to many glaucony grains, a larger sample was hand-picked in order that after acid washing the quantity of glaucony would approach the original desired quantity. Each sample was then washed in reagent grade 0.1 N HCl (in one case 0.2 N HCl) and dried in reagent grade acetone. Any foreign material observed by binocular microscope examination was removed by repicking.

Only ordered to disordered mineral glaucony as defined by Bentor and Kastner (1965) on the basis of X-ray diffraction analysis was prepared for Rb–Sr analysis. An approximately 0.025 g (this varied with each sample) sample of purified glaucony was weighed and placed in a clean Teflon or platinum dish and Rb and Sr spikes were added by pipette to each sample and a blank. After evaporation to dryness on a steam bath, the minimum necessary volume of reagent grade HF and Vycor distilled 2 N HNO_3 was added to each sample and the blank for digestion. After the samples had been dissolved, the dishes were filled to volume with Vycor distilled 2 N HCl and evaporated to near dryness twice on a steam bath. A small quantity of Vycor distilled 2 N HCl was added to each sample to bring the volume up to about 10 mil for passage through the ion exchange column. Standard isotope dilution procedures were employed for separation of Rb and Sr (Fullagar and Odom, 1973). In order to monitor contamination encountered in handling and preparing the samples for analysis, blanks were collected for each series prepared. Glaucony samples from the Rocky Point member and the Beaufort formation have been corrected for blank as follows: (Rocky Point member) μg Sr = 0.03, $^{87}Sr/^{86}Sr = 0.710$; (Beaufort formation) μg Sr = 0.068, $^{87}Sr/^{86}Sr = 0.7083$. Blank corrections for these samples were made because of the small sample size, the low Sr concentrations and the high blank concentration. However, the effect of correction for the blank is slight and does not affect the ages because the $^{87}Sr/^{86}Sr$ blank value is close to the initial $(^{87}Sr/^{86}Sr)_0$ for the samples. Samples from the unnamed member of the Peedee formation, Castle Hayne limestone and Nanjemoy formation have not been corrected for blank because larger sample size and improvements in procedure have resulted in lower blank concentrations. On the basis of analyses of the NBS-70a K feldspar, the one standard deviation experimental errors are ±0.0005 for the $^{87}Sr/^{86}Sr$ and 1.0% for the $^{87}Rb/^{86}Sr$ ratios.

The $^{87}Sr/^{86}Sr$ values represented in Table 1 have been normalized to $^{86}Sr/^{88}Sr = 0.1194$. All ages were calculated using $\lambda\ ^{87}Rb = 1.42 \times 10^{-11}\ a^{-1}$ and are given at 1σ error. Isochron ages were calculated using the least squares cubic method of York (1966). The Rb–Sr mass spectrometry was performed with a single focusing, 30 cm, triple-filament mass spectrometer at the University of North Carolina at Chapel Hill. Data were collected and analysed with a Nuclide DA/CS-III automation and data reduction computer system.

4 RADIOMETRIC RESULTS

4.1 Cretaceous

Five glaucony concentrates from the Peedee formation, designated RP-36-3, RP-4, RP-2a, RP-2b and NCF4-2, were prepared for analysis from the four samples collected using the technique described by Harris and Bottino (1974b). Rb–Sr analytical data for the five glaucony samples are given in Table 1.

The isochron plot for the five glaucony samples (Figure 2) suggests an isochron age of 66.7 ± 1.0 Ma for the youngest Cretaceous beds of the Peedee formation in North Carolina. The interpreted initial $(^{87}Sr/^{86}Sr)_0$ ratio of 0.7070 ± 0.0004 is analytically in good agreement with the suggested $^{87}Sr/^{86}Sr$ ratio of 0.7075 for sea-water reported by Peterman *et al.* (1970) for the Late Cretaceous. This value is also in good agreement with the $^{87}Sr/^{86}Sr$ value of 0.70735 determined from an *Ostrea* sp. collected from the same beds reported by Harris and Bottino (1974b).

Obradovich and Cobban (1975) established the boundaries of the Maastrichtian on the basis of K–Ar ages of bentonites in the western interior of North America at 64–65 Ma and 70–71 Ma. In addition, Lanphere and Jones (1978) reported that the best current estimate for the age of the Cretaceous–Tertiary boundary is 65–66 Ma. Therefore, the isochron age of 66.7 Ma fits well into the best estimates of the age of the Cretaceous–Tertiary boundary.

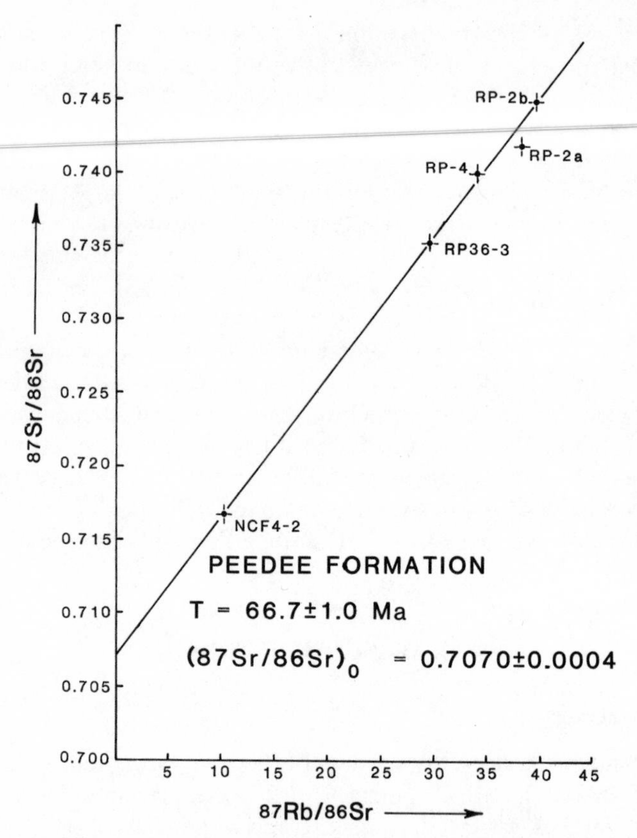

Figure 2. Plot of $^{87}Sr/^{86}Sr$ *vs.* $^{87}Rb/^{86}Sr$ of glauconies from the Peedee formation and equivalents, New Hanover County, North Carolina.

4.2 Palaeocene

Three mammillated to lobate glaucony concentrates from the single Beaufort formation sample, designated LEN 2A, LEN 2B and LEN 2C, were separated on the basis of size differences between the grains. Rb–Sr analytical data for the three glaucony samples are given in Table 1.

Model age calculations were made using an initial $^{87}Sr/^{86}Sr = 0.7078 \pm 0.0003$. This value was derived from the calculated arithmetic average of four published initial $^{87}Sr/^{86}Sr$ ratios of Palaeogene fauna (three analyses from Peterman *et al.* (1970) and one analysis from Dasch and Biscaye (1971)) and one analysis of an *Oleneothyris harlani* shell collected from the same horizon ($^{87}Sr/^{86}Sr = 0.7081 \pm 0.0005$). The three glaucony samples yield conventional ages of 63 ± 12 Ma, 54.6 ± 6.7 Ma and 56.6 ± 3.4 Ma (Table 1).

Table 1. Rb–Sr analytical data for Cretaceous and Palaeogene units, southeastern Atlantic Coastal Plain, USA.

Sample	Rb (ppm)*	Sr (ppm)*	$^{87}Rb/^{86}Sr$*	$^{87}Sr/^{86}Sr$*	Conventional age (Ma)
Castle Hayne limestone					
MM1-100HT	202.1	13.39	43.77	0.7301	35.1±1.1
MM1-100HM	195.9	26.85	21.14	0.7182	33.0±2.0
MM1-100HF	199.8	29.66	19.52	0.7188	37.9±2.2
MM1-70HT	189.8	50.25	10.94	0.7135	33.5±3.6
MM1-70HF	197.0	19.48	29.31	0.7223	33.6±1.6
Nanjemoy formation					
W2-B1-80	212.7	25.88	23.81	0.7239	47.6±2.0
W2-B1-100	205.6	17.90	33.30	0.7294	45.7±1.5
W2-B2-60	198.2	21.54	26.67	0.7264	49.1±1.8
W2-B2-100	211.3	21.28	28.78	0.7268	46.5±1.7
W2-B4-60	205.7	26.81	22.23	0.7236	50.1±2.1
W2-B4-80	212.4	24.51	25.12	0.7237	44.6±1.8
W2-B4-100	206.4	27.87	21.45	0.7209	43.0±2.1
Beaufort formation					
LEN 2A	240.3	135.57	5.13	0.7124	63 ±12
LEN 2B	250.9	76.13	9.55	0.7152	54.6±6.7
LEN 2C	238.6	33.10	20.90	0.7246	56.6±3.4
Peedee formation					
RP-36-3	198.2	19.53	29.40	0.7352	67.5±1.9
RP-4	190.3	16.19	34.13	0.7400	68.1±1.7
RP-2a	174.3	13.29	38.08	0.7418	64.4±1.5
RP-2b	182.1	13.42	39.41	0.7447	67.4±1.5
NCF4-2	199.5	56.50	10.22	0.7167	66.8±4.2

* The one standard deviation uncertainties for Rb and Sr concentrations are no greater than 1% for the $^{87}Rb/^{86}Sr$ ratios; for $^{87}Sr/^{86}Sr$ ratios they are no greater than 0.05%.
Size of grains: 100 is 0.15–0.21 mm; 80 is 0.18–0.25 mm; 70 is 0.21–0.50 mm; 60 is 0.25–0.50 mm; LEN 2A is 0.35–0.50 mm; LEN 2B is 0.25–0.35 mm; LEN 2C is 0.10–0.25 mm.

The three glaucony conventional ages support a Palaeocene age for the sample of the Beaufort formation studied; however, disregarding analytical uncertainty, there is an apparent discrepancy between the two younger ages (LEN 2B, LEN 2C) and the older age (LEN 2A). The average of the two younger model ages (55.6 Ma) is in close agreement with the suggested age range (56 and 58 Ma) of the P4 planktonic foraminiferal zone (Berggren, 1971a). Therefore, Rb–Sr glaucony conventional ages support the faunally determined age of the formation and fit reasonably well into current radiometric age limits placed on the P4 planktonic foraminiferal zone. The age of sample LEN 2A (63 Ma) is about 10% too old and does not agree with the microfaunal age; however, if the analytical uncertainty is considered, this age agrees well with the other two.

4.3 Eocene

Seven glaucony concentrates designated W2-B1-80, W2-B1-100, W2-B2-60, W2-B2-100, W2-B4-60, W2-B4-80 and W2-B4-100 were separated on the basis of size from three samples of the Nanjemoy formation. Rb–Sr analytical data for the seven glaucony samples are given in Table 1.

The isochron plot for the seven glaucony samples (Figure 3) suggests an isochron age of 46.7 ± 3.0 Ma for the Nanjemoy formation. The interpreted initial $(^{87}Sr/^{86}Sr)_0$ ratio of 0.7078 ± 0.0011 is analytically in agreement with the suggested sea-water $^{87}Sr/^{86}Sr$ ratio of Early Eocene seas of 0.7074 reported by Peterman *et al.* (1970). However, the analytical uncertainty associated with this value does not allow for an accurate evaluation of it.

The Hardenbol and Berggren (1978) Palaeogene time scale suggests that the Early Eocene covers the span of time between 49 and 53.5 Ma and that nannoplankton zone NP-13 has a numerical age of 49–50.5 Ma. Disregarding analytical uncertainty, the isochron age of 46.7 Ma for nannoplankton zone NP-13 of the Nanjemoy formation suggests that the current ages placed on the zone by Hardenbol and Berggren (1978) are too old. The age does agree with the revised Cenozoic polarity time scale proposed by Tarling and Mitchell (1976).

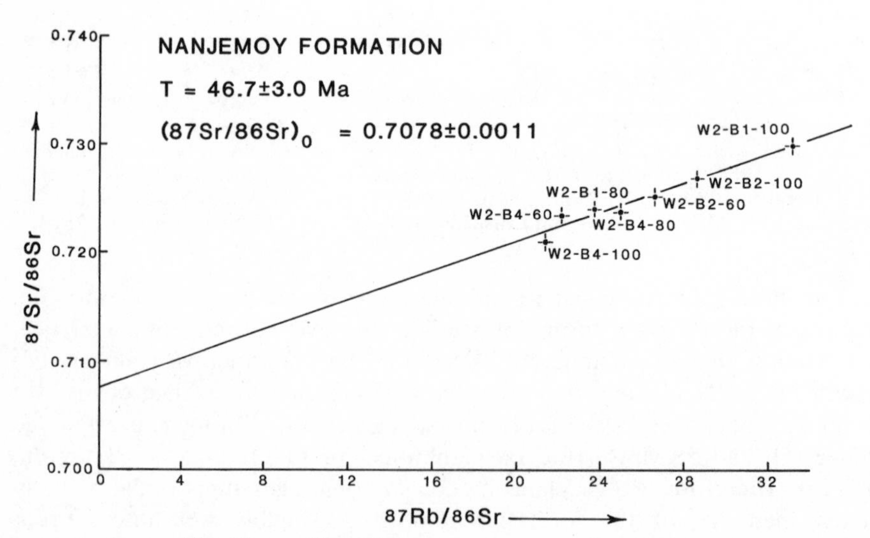

Figure 3. Plot of $^{87}Sr/^{86}Sr$ *vs.* $^{87}Rb/^{86}Sr$ of glauconies from the Nanjemoy formation, Prince George County, Virginia.

Five glaucony concentrates designated MM1-100HT, MM1-100HM, MM1-100HF, MM1-70HT and MM1-70HF were separated from the single Castle Hayne limestone sample on the basis of size and morphology. Rb–Sr analytical data for the five glaucony samples are given in Table 1 (Harris and Zullo, 1980).

The isochron plot for the five glaucony fractions (Figure 4) suggests an isochron age of 34.8 ± 1.0 Ma for the upper part of the Castle Hayne limestone at the lectostratotype. The interpreted initial $(^{87}Sr/^{86}Sr)_0$ ratio of 0.7083 ± 0.0004 is in good agreement with the suggested $^{87}Sr/^{86}Sr$ ratio of 0.7082 ± 0.0002 determined from pelagic foraminifera (Dasch and Biscaye, 1971) and 0.7078 ± 0.0004 determined from fossil shells (Peterman *et al.*, 1970) for the Middle Eocene. There are no reported Sr ratios for Late Eocene sediments.

The Rb–Sr isochron age of 34.8 Ma is younger than the upper age limit (37 Ma) for the Eocene accepted by Hardenbol and Berggren (1978). However, it is in excellent agreement with recent revisions of the Cenozoic time scale based on polarity correlations by Tarling and Mitchell (1976), glaucony ages reported by Odin *et al.* (1978) from northwest Europe, and microtektite ages reported by Glass and Zwart (1977) from the southern Atlantic Coastal Plain, Gulf of Mexico, and Caribbean.

Tarling and Mitchell (1976) used recent isotopic age determinations of

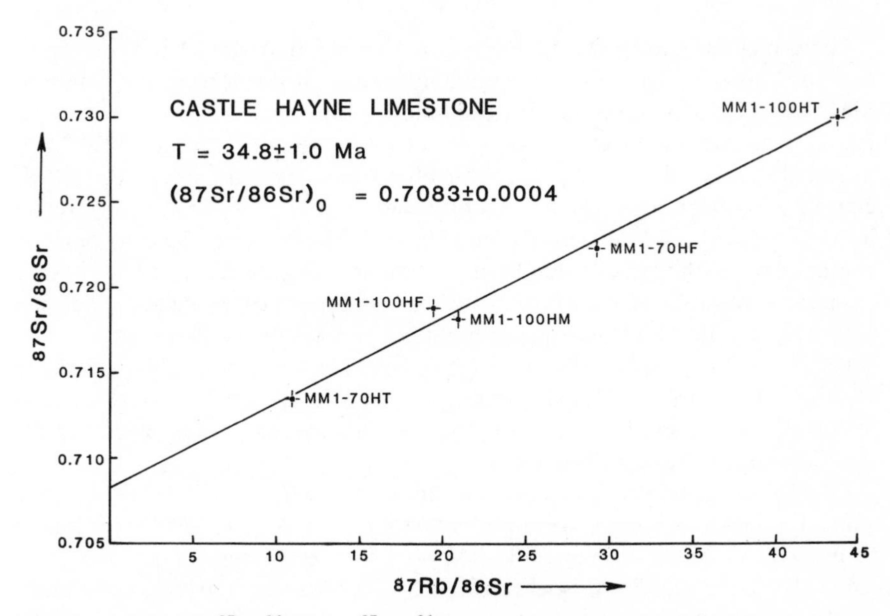

Figure 4. Plot of $^{87}Sr/^{86}Sr$ *vs.* $^{87}Rb/^{86}Sr$ of glauconies from the Castle Hayne limestone, New Hanover County, North Carolina.

sediments overlying oceanic magnetic anomalies to revise the Cenozoic time scale. They suggested that the Eocene–Oligocene boundary is close to 35 Ma.

Odin *et al.* (1978) suggested that the Eocene–Oligocene boundary based on glaucony ages from northwest Europe is about 33 Ma. This figure is based upon one radiometric age of 30.9 Ma for the Neerrepen beds of Belgium which contain nannofossil zone NP-21 and two ages of 38.9, 39.6 Ma for the Lower Barton beds in England which contain nannofossil zone NP-16.

Glass and Zwart (1977) correlated published fission track and K–Ar ages of tektites and microtektites from North America to microtektite-bearing Late Eocene sediments in cores from the Gulf of Mexico and Caribbean. North American tektites have an average K–Ar age of 34.2 Ma (Zahringer, 1963; Gentner *et al.*, 1969) and fission track age of 34.6 Ma (Fleischer and Price, 1964; Fleischer *et al.*, 1965a; Storzer and Wagner, 1971; Storzer *et al.*, 1973). As these microtektites occur with nannofossil zones NP-19 and NP-20 and planktonic foraminiferal zones P16 and P17 (the same zones as the Castle Hayne samples), the isochron age of Castle Hayne limestone of 34.8 Ma is in excellent agreement with placement of the Eocene–Oligocene at about 33 Ma.

5 SUMMARY

Hand-picked, authigenic glaucony concentrates that yield Rb–Sr isochron and conventional ages provide a reliable means of converting the standard Mesozoic–Cenozoic column to a radiometric time scale. However, because Rb–Sr conventional ages are dependent upon assumed initial $(^{87}Sr/^{86}Sr)_0$ ratios, they may be suspect and difficult to evaluate and interpret. Rb–Sr glaucony isochron ages, which determine the initial $(^{87}Sr/^{86}Sr)_0$ ratio, result in more reliable age determinations. Glaucony ages determined to date on southeastern Atlantic Coastal Plain sediments suggest the need for minor numerical revisions of the Hardenbol and Berggren (1978) Palaeogene time scale (Figure 5). The following summarizes the results of this study:

(1) A Rb–Sr glaucony isochron age of 66.7 Ma for Maastrichtian rocks in North Carolina (Rocky Point member and stratigraphically equivalent sand of the Peedee formation) supports the currently accepted age of 65 Ma for the Cretaceous–Tertiary boundary.

(2) Rb–Sr glaucony conventional ages of 54.6 and 56.6 Ma for the Thanetian P4 planktonic foraminiferal zone in North Carolina (Beaufort formation) supports the currently accepted 56–58 Ma age for the zone.

(3) A Rb–Sr glaucony isochron age of 46.7 Ma for Early Eocene sediments in Virginia (Nanjemoy formation) suggests that the Ypresian–Lutetian boundary is younger than the 49 Ma age currently placed on it.

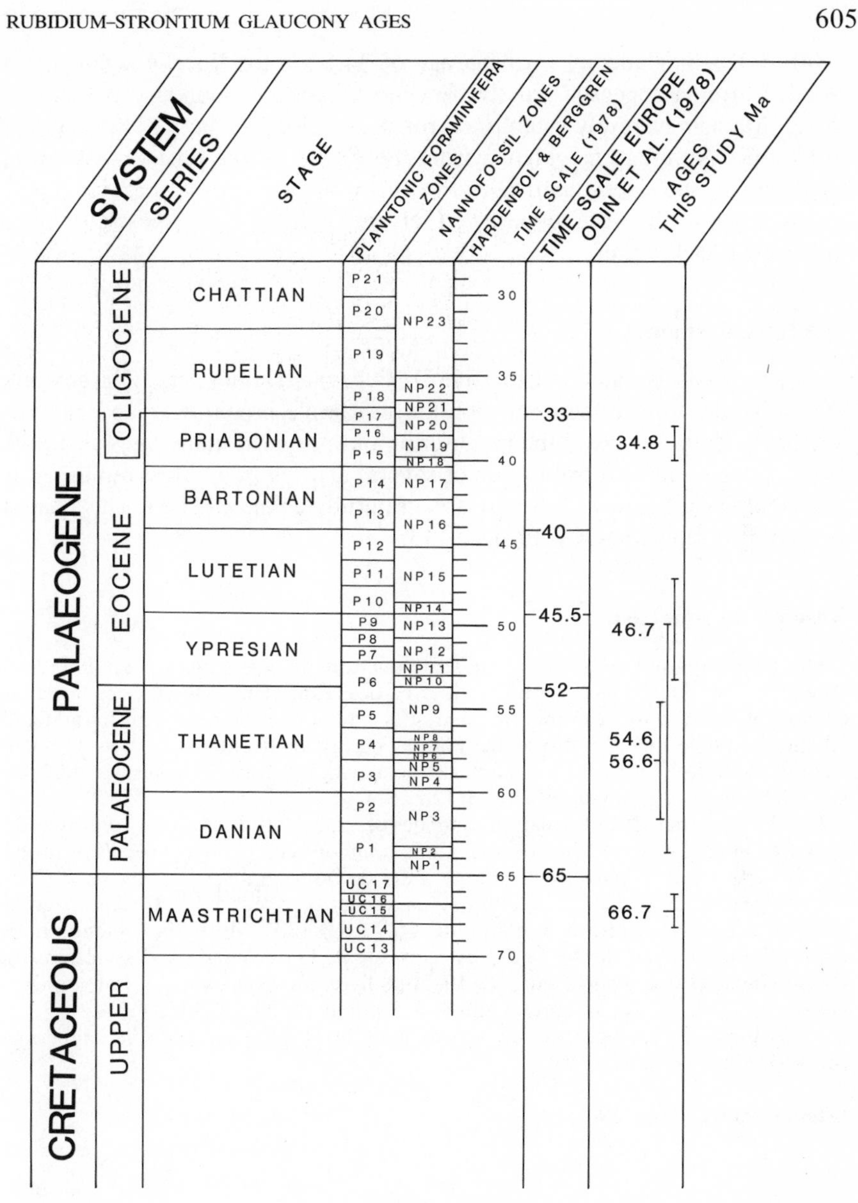

Figure 5. Summary of Cretaceous and Palaeogene glaucony ages from the south-eastern Atlantic Coastal Plain.

(4) A Rb–Sr glaucony isochron age of 34.8 Ma for Eocene sediments in North Carolina suggests that the Eocene–Oligocene boundary is less than the 37 Ma age currently established for it and close to the 33 Ma suggested in Europe. Although megafauna from the Castle Hayne limestone are both Claibornian and Jacksonian age, the isochron age and calcareous nannofossils suggest that the upper part of the unit at the lectostratotype is uppermost Jacksonian.

Acknowledgements

Appreciation is extended to Paul D. Fullagar, Delma Jean Glasgow and Debra Tobiassen for their contribution in sample preparation and analysis. Victor A. Zullo reviewed the manuscript and provided many helpful suggestions. Permission to reprint parts of several articles previously published by the Geological Society of America is gratefully acknowledged. This chapter is a contribution to IGCP Project 133.

Résumé du rédacteur

Des glauconies ont été séparées en petite quantité de divers niveaux fossilifères du Paléogène et du Crétacé terminal du SE des Etats-Unis. Toutes les fractions convenablement purifiées ont été analysées radiométriquement par la méthode Rb–Sr. La faible quantité disponible n'a pas permis d'effectuer d'analyse de potassium de contrôle. La formation maastrichtienne de Peedee a livré un âge apparent sur isochrone à 5 points de 66.7 ± 2.0 (2σ) Ma.

Les 3 glauconies de la formation thanétienne (zone P4) de Beaufort ont livré des âges très imprécis (abondance de strontium commun) compatibles avec l'âge probable de cette zone: le plus précis obtenu a été de 56.6 ± 6.8 (2σ).

La formation de Nanjemoy a livré des glauconies corrélées avec l'Eocène inférieur (NP13). L'âge isochrone à 7 points de 46.7 ± 6.0 (2σ) Ma a été obtenu. L'âge conventionnel moyen de 46.7 ± 5.0 est équivalent. Un échantillon des calcaires de Castle Hayne (Eocène supérieur; NP 19–20) a livré une glauconie qui, fractionnée, a permis d'établir un âge isochrone Rb–Sr à 5 points de 34.8 ± 2.0 (2σ) Ma.

L'ensemble de ces âges est en accord avec la révision de l'échelle numérique proposée dans ce livre.

(Manuscript received 29-1-1980)

Numerical Dating in Stratigraphy
Edited by G. S. Odin

34
Dating of the Palaeogene

DENNIS CURRY and GILLES S. ODIN

In this chapter we examine the conceptual and factual basis for the numerical scales used by different authors in their stratigraphical syntheses. Many versions of such syntheses have been published, differing only in detail. They fall into two fundamentally different groups, however: one based on biostratigraphy, the other essentially on lithostratigraphy. In the first case the approach tends to be conservative, and discordant data are treated as suspect and may be discarded: in the second, the view is taken that the most recent analytical and geochemical data must be allowed to supplant earlier results, which are regarded as inherently less reliable both analytically and geologically.

1 INTRODUCTION

Following the publication of the earliest systematic studies of isotope dating, Funnell (1964) presented the first well-documented synthesis of information relating to the Palaeogene. In this he assembled 65 dates, of which 17 were rejected because they appeared inconsistent with the stratigraphical position of the rock concerned. Amongst the remaining 48, 14 were based on glauconies and most of the remainder on volcanic rocks collected in North America and correlated with European sequences by means of the associated mammal faunas. The resulting rather imprecise correlations have subsequently been discussed and questioned, as in Evernden *et al.* (1964) and Evernden and Evernden (1970).

Previous invstigations have shown that the reliability of radiometric data is a function of the nature of the chronometer used. We attach much importance to this point in the present work. The word glaucony (plural, glauconies), as used above, is employed to designate green granules of sedimentary origin, whatever their mineralogical composition, to avoid confusion with the mineral glauconite, which is a mica of rare occurrence in nature.

The data used by Funnell to date the Palaeocene are nine in total (six based on glaucony), for the Palaeocene–Eocene boundary four (all glaucony), and for the Eocene thirteen (eight on glaucony). Above these levels his dating was based on the North American mammal stages: Duchesnean, Chadronian, Whitneyan and Arikareean, considered by him as approximately equivalent respectively to the Ludian (Late Eocene in part), Tongrian (Early Oligocene), Chattian (Late Oligocene) and Early Miocene. The scale of Funnell is included in Figure 1, column 1. As throughout this publication, the original dates have been recalculated on the basis of the conventional constants adopted at Sydney (ICC). Depending on the original data this recalculation decreases or increases the original dates by 2–5%. Fortunately, almost all of the data used by Funnell and succeeding writers on the Palaeogene were based on K–Ar analyses and were originally calculated using the same (old) series of constants. As a result all these dates must be increased slightly before they and their associated time scales can be compared directly with those based on recent data, which use the new constants. Funnell's original dates are included in parentheses in Figure 1 as some authors still use them in spite of the most recent proposals, in order, it seems, to provide greater consistency. Thus the original scale of Funnell (1964) is still reproduced in its essentials by many authors.

The various publications of Berggren (1969; 1971b; 1972), Berggren and van Couvering (1974), etc., are in this tradition and almost all include practically the same figures. There is one important difference from the 1964 scale, however: the age of the base of the Miocene has been reduced to accord with the data of Turner (1970a). Turner's data (NDS 154–155) were based on volcanic rocks which could be correlated with the foraminiferal biozonation of California and thence with the rest of the world and so provide an important reference point for the base of the Miocene at around 23 Ma (ICC). The acceptance by Berggren and his collaborators of this date involved an increase in the duration allotted to the Chattian and the establishment of a date for the end of the Rupelian (Early Oligocene) of 30 or, alternatively, 32 Ma.

Changes in detail in their analyses have been based on 'the fundamental assumption ... that, in general, the *average* length of a planktonic foraminiferal zone remains about the same (1.5–2 Ma)' (Berggren 1971b, p. 766: version 1 of that author's time scale, reproduced here as Figure 1, column 2). Minor variations are shown by succeeding versions: Berggren 1969a (version 2), 1969b (version 3), 1972 (version 4), van Couvering and Berggren 1974 (version 5). Nevertheless, the various presentations closely resemble the 1964 scale so far as the stage and series limits outside the Oligocene are concerned. In the last of the series (Hardenbol and Berggren, 1978, version 6) the authors have related their conclusions to the biozonal sequences established for the Late Eocene. This time span is classically referred to the Bartonian stage, the Priabonian, defined in Italy, being thought to be an approximate equivalent. They note (p. 221) that 'planktonic microfossils in the Bartonian type area (NP17 and P14-P13) are by

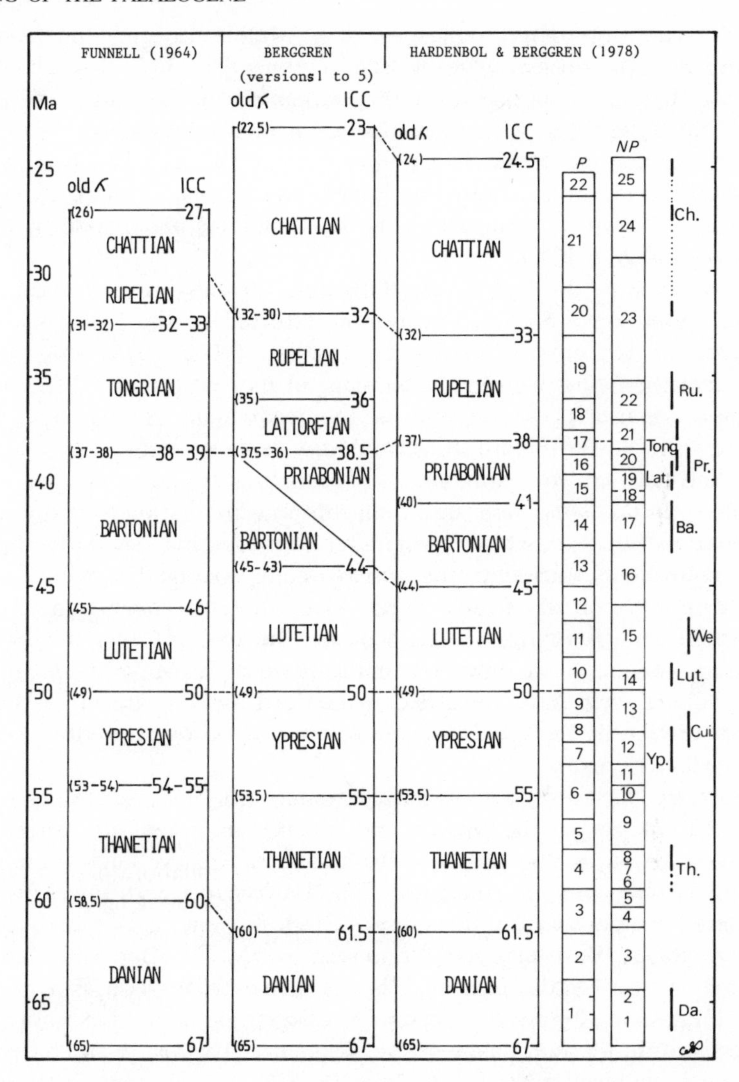

Figure 1. Time scales of the Funnell lineage. The dates given by the original authors are given in parentheses (left-hand side); they are recalculated using conventional constants (right-hand side). The biozonal correlations are given: P = planktonic foraminiferids (Blow, 1969; Berggren and Van Couvering, 1974); NP = calcareous nannofossils (Martini, 1971; Müller, 1974). The positions of stratotype sections are shown according to Hardenbol and Berggren (1978). Da.: Danian; Th.: Thanetian; Yp.: Ypresian; Cui.: Cuisian; Lut.: Lutetian; We.: Wemmelian; Ba.: Bartonian; Lat.: Lattorfian; Pr.: Priabonian; Tong.: Tongrian; Ru.: Rupelian; Ch.: Chattian.

general usage considered indicative of a Middle Eocene age' and 'the Priabonian... [is considered] to include... zones P15–P17'. Hardenbol and Berggren therefore superposed the Priabonian on the Bartonian and placed the Middle–Late Eocene boundary between them because the 'alternative solution, moving planktonic foraminiferal zones P13 and P14 from Middle to Late Eocene, would cause far greater confusion...'. Accordingly this boundary is drawn by them within the Bartonian (as understood by French authors) at 41 Ma (ICC).

The position of the base of the Oligocene has also been revised. In its original definition it included beds now referred to the Lattorfian stage. However, on the basis of studies by Cavelier (1972–1979) and others it seems that the nannoplankton assemblage of the Lattorfian (NP21, zone of *Ericsonia subdisticha*) is present in the Priabonian of Italy (previously equated with the Bartonian). In addition this biozone, defined by Martini as the period between the extinction of *Discoaster saipanensis* and the extinction of *Cyclococcolithus formosus*, is diachronous in relation to latitude when compared with the planktonic foraminiferal biozonation. The Lattorfian and all its equivalents, including the corresponding mammal sequences which have been dated in the United States, have therefore been transferred by Hardenbol and Berggren to the Eocene. In spite of this the Eocene–Oligocene boundary has been left undisturbed at 38 Ma (version 6), as in 1964. All the new dates recorded in this last version are either used to support previously accepted data or rejected as unrepresentative because analytically erroneous.

To sum up, the time scale established by Funnell (1964) has been retained in its entirety up to the present day by Berggren and his various collaborators insofar as the time span from the base of the Palaeocene to the base of the Oligocene is concerned. The Bartonian (*sensu lato*), originally considered as equivalent to the subseries Late Eocene, has been subdivided into two stages, Bartonian and Priabonian, whilst the Bartonian has been relegated to the Middle Eocene. The Lattorfian–Early Tongrian, formerly Early Oligocene (Figure 1, column 1), disappears as a time unit and is replaced within the same dates by the Rupelian. The age of the base of the Miocene is reduced from 27 to 24.5 Ma (ICC). This time scale is very convenient because the dates it includes have been kept essentially constant throughout, and no doubt for that reason it is the one which is most commonly used in publications. But, as we have shown, different formations have been included from time to time between these dates, so that the appearance of constancy is somewhat illusory.

It is important to remember that this scale is based on a *selected series* of dates which includes 26 (21 from glauconies) from the United States, 46 (all from glauconies) from Europe and two from basalts from DSDP cores (see NDS 52) which are not easily interpretable. Finally, it should be noted that the planktonic foraminiferal zones are scattered fairly regularly along this scale—in fact they vary in duration by a factor of 3—which is a reflection of the basic hypothesis of Berggren (version 1) as noted earlier.

Since 1968 one of the present authors (Odin) has carried out research on similar lines but, instead of concentrating on the interpretation of currently available analytical data in the light of advancing knowledge of the biostratigraphy of the rocks dated, he has tried to acquire additional and more reliable analytical data from the classical reference sequences of Europe. As a result 35 samples from Belgium, France, Germany, England and Italy were dated at a new geochronological laboratory established in France in 1969 and 1970. The results obtained from glauconies were used to construct a numerical scale (Odin, 1973b) which is listed in Figure 2, column 1. This scale, established for the most part on results from NW Europe, complements that of Funnell and the analytical data on which it was founded have been used by Berggren since 1972. The age of the base of the Miocene was lowered, as in the United States, as was that of the base of the Bartonian. The new scale was not published without reservations in spite of the agreement between the ages provided by the new measurements and those accepted by Funnell (1964). For instance, Odin *et al.* (1969) note: '*le pic de masse* 36 *n'est pas décelable sur nos diagrammes après un dégazage* (*des glauconies*) *de* 20 *heures à* 130°*C*; nous admettons donc que tout l'argon extrait est radiogénique' and, again, (Odin, 1973) '*du fait du manque de sensibilité du spectromètre... ces analyses restent à confirmer dans chaque cas... les analyses refaites à Hanovre* (*H. Kreuzer*) *et par l'un de nous* (*GSO*) *à Berne montrent un* excès systématique de l'évaluation de l'argon radiométrique *et par conséquent des âges apparents* trop vieux de 5 à 15% *dans les analyses citées* (*par Odin,* 1973*b*)'. Further experiment and analysis were clearly needed to resolve the doubts about the quality of the experimental method. Portions of the original material were therefore remeasured in two additional laboratories (Strasbourg and Berne) and other checks were made independently (Hanover, Gif, etc.). At the same time, some of the samples dated in America in 1961 were reassessed. The analytical problems involved both the measurements themselves and the methods used in the selection and preparation of the materials tested. Following three years of additional work a new time scale was proposed (Odin, 1975) which was based on 33 new age determinations of glauconies from the NW European Tertiary basins, together with additional data for the Eocene (Obradovich, 1964) and further new data for the Oligocene (Kreuzer *et al.*, 1973). By comparison with Funnell (1964), the new time scale proposed reductions in the age of various boundaries as follows: Early–Middle Eocene 10%, Middle–Late Eocene 13%, Oligocene–Miocene 15% and Eocene–Oligocene 8% (the last as a result of the transfer of the Lattorfian to the Eocene). The new proposal was based on samples collected from type

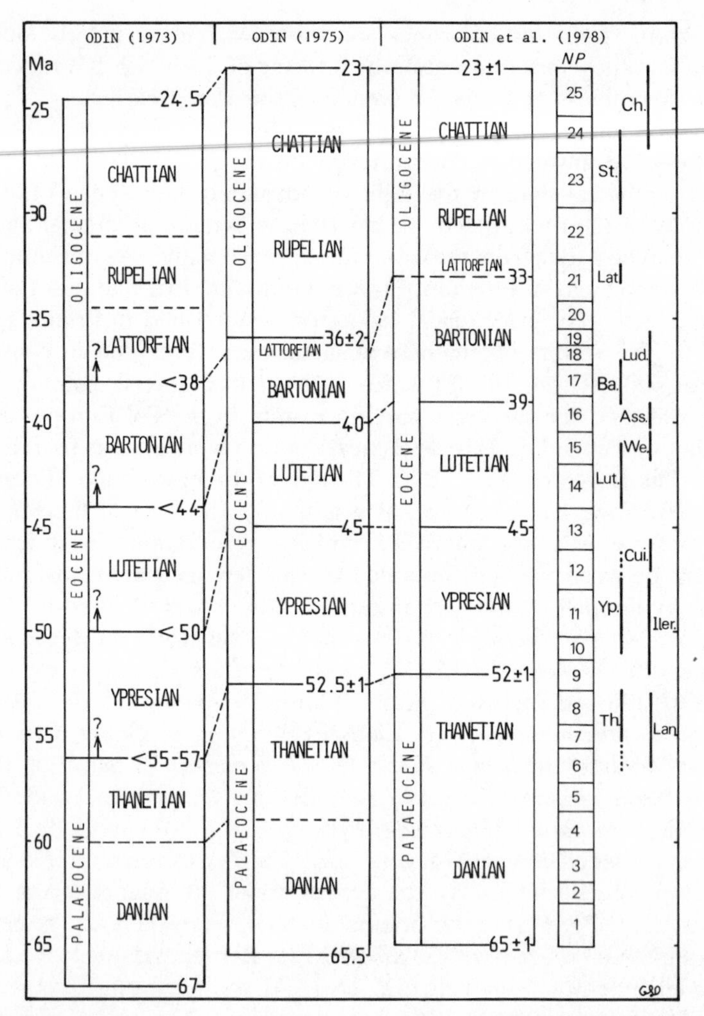

Figure 2. Time scales drawn on the basis of marine formations dated at that time in Europe. Abbreviations as for Figure 1, also Lan.: Landenian; Ilerd.: Ilerdian; Ass.: Assian; Lud.: Ludian; St.: Stampian. In the column Odin (1973) all the boundaries are shown with an arrow and an interrogation mark as a reminder that the values proposed were considered by the author as tentative.

sequences or from formations whose stratigraphical relations are well established and which are correlated with type sequences by both lithological and faunal criteria (mostly by molluscs but, at some levels, by nannoplankton or larger foraminiferids).

The data so assembled (40 determinations in all) were discussed in Odin *et al.* (1978) and were compared with the large number of determinations made independently by Obradovich (1964), Ghosh (1972) and Elewaut *et al.* (1976). The main alteration to the scale which resulted was a reduction in the age proposed for the base of the Oligocene. This was based, not on new analytical data, but on a revised sedimentological and biostratigraphical analysis. Sedimentological studies have revealed that the glaucony from the Silberberg beds (Lattorfian) is probably reworked. As no reliable data are then available close to the Eocene–Oligocene boundary, recourse was made to the system of interpolation used by Berggren. It was assumed that the mean durations of the zones based on planktonic foraminiferids and nannoplankton tend not to vary with time. The sequence of ages obtained within the span 65–40 Ma, where many concordant results have been recorded, was projected towards the Oligocene–Miocene boundary at 23 Ma. The pattern of biozones suggested an age in the region of 32–33 Ma for the base of the Lattorfian (and of the classical Oligocene). The latter figure was chosen because it seemed to be more consistent with the few results which were available around that limit. The age so proposed is clearly relatively unreliable in view of the method of extrapolation on which it is mainly based, but it shares this unreliability with many ages proposed elsewhere which rely on the same principle.

It is regrettable that there are considerable differences between the 'conservative' scale summarized in Figure 1 and that proposed at about the same time on the basis of 'new' data (Figure 2). It would have been much more satisfactory to have been able to confirm the older dates.

In the following paragraphs, the data used in the two cases will be analysed in detail to enable the reader to make a personal judgement. This analysis will explore the quality of the material used, the analytical methods employed and the correlations proposed: that is, the totality of the uncertainties associated with the data employed. This approach is somewhat different from that of some other writers, who rely on those results which fit their proposed scale although 'no attempt has been made in this paper to evaluate the overall quality of the published (radiometric) ages or to compare the measurement techniques used for the European and American (radiometric) ages. Quality of glauconites, possible reworking, potassium content, atmospheric argon, argon loss, and decay rates, have not been seriously considered.' (Hardenbol and Berggren, 1978, p. 228).

.2 THE STRATIGRAPHICAL DATA

2.1 Fundamental steps in correlation

The construction of a numerical time scale must of course be related to the stratigraphical time framework, that is, to the succession of recognized stages and their accepted boundaries. However, field evidence clearly indicates that, in the Palaeogene at least, such stages, defined in epicontinental sequences of NE Europe, have almost all been based on sedimentary cycles in shallow-water marine deposits which have a transgressive base (preceded by a diastem) and a regressive top (succeeded by a diastem). The stratigraphical column is thus represented in rock only by a series of relicts which are more or less limited in their time span. The problem then arises of deciding where precisely a limit should be set within the diastem separating two such relicts when more complete sequences become available for study.

Classical stratigraphy, based on lithostratotypes, has slowly become modified by the incorporation of biostratigraphical information based on successions of marine organisms. Appearances and extinctions amongst these, mostly identified in Cenozoic times in regions where the climate was warmer than in NW Europe, have been used to define biozones, whose boundaries have tended to supplant those on which the original (litho)stratigraphical framework and nomenclature were based. When as a result it is proposed to change the original definition of a stage (and, in consequence, of stratigraphical divisions of higher order) to one based on biostratigraphy the original stratotypes in most cases become ones which are correlated at second hand. This fact tends not to be openly recognized, however.

Our method of approach has been to consider initially the radiometric ages determined on rocks from the areas in which the classical stages have been defined. This has been done purely because of the reliability with which these rocks can be correlated with the standard chronostratigraphical scale. It frequently happens that the span of such a stage in terms of biozonations is uncertain, but it is a representative of that stage, or some near equivalent, which has been dated directly by the use of glauconies. The correlation is a *first-degree* one. If now a distant marine formation is dated it becomes necessary to use some system of biozonation. The distant formation contains fossils which are compared with those (often of different taxa, unfortunately) yielded by the type stages and the correlation then becomes of the *second degree*. Finally, when a continental volcanic sequence is dated (in order, for example, to check radiometric data obtained from a sedimentary rock) it is normally necessary to pass *via* the local continental faunas, usually mammals, to the nearby marine equivalents and thence finally to the marine faunas of the internationally recognized stages. In such a case correlation is at best of the *third degree*. Every step in the correlation process has its own uncertainties, and the resulting possibilities of error are cumulative. In our opinion therefore, it is unwise to use dates based on mammals as part of the basic data for the construction of a time scale in view of the amount of other data which, currently available. This is particularly true in view of the fact that very few new studies have been carried out since the pioneer work of Evernden *et al.* (1964). It would be very valuable to able to integrate radiometric data based on the American mammal zonations with those based on the classical marine stages of Europe. However, it seems unwise to attempt this at the moment in view of the uncertainty of the detailed correlation between the two (Figure 5). Additional analyses and more detailed comparative palaeontological studies are still needed before the integration of these 'continental' results can safely be attempted.

2.2 Stratigraphy of NW European sequences

Figure 3 shows the lithological correlations currently accepted between those West European sequences which have yielded radiometric dates. These correlations were based originally on marine macrofaunas and are supported by more recent work on microfossils (see Curry *et al.*, 1978 for references), both animal and plant (e.g. Chateauneuf, 1980). In recent years, too, studies of calcareous nannoplankton (and to some extent of dinoflagellates) have provided means of correlation with an international biostratigraphical scale.

3 RADIOMETRIC DATA

3.1 Earlier data and more recent results

We noted earlier that the data employed to construct the time scale of Funnell (1964) and succeeding authors were based essentially on the first results provided by Evernden *et al.* (1961) at Berkeley. We have shown also how these data appeared to be confirmed by our own publications (Odin, 1973). We shall now recall several discrepancies of a strictly analytical nature between these first results and ones now available. The reader is also referred to the section on methodology (Odin, this volume, chapter 20 and especially Figures 9 and 10), and to the discussions on data (NDS 1, etc.).

The glaucony of the Bashi marl (base of Hatchetigbee fm, zone P6b) was dated (Evernden *et al.*) in 1961, the recalculated age being 53.5 Ma (ICC) for a glaucony with a potassium content of 4.46% (% K). Two further samples were taken from this level and yielded recalculated ages of 49.8 and 49.3 for glauconies containing 5.62 and 5.98% of potassium respectively (Ghosh, 1972). Only the first of these ages (53.5 Ma) was used in the scale of Funnell, and that of Hardenbol and Berggren (see NDS 56).

A Palaeocene glaucony from California was dated in 1961 and its apparent age was determined as 60.7 Ma (ICC), with a potassium content of 3.95%. Obradovich (1964) treated this sample ultrasonically and dated the coarse fraction separately from that of grain size between 4 and 53 μm; the calculated ages proved to be 46.4 and 47.7 for potassium contents of 5.23 and 4.87% respectively (Table 1). A series of similar cases of a strictly

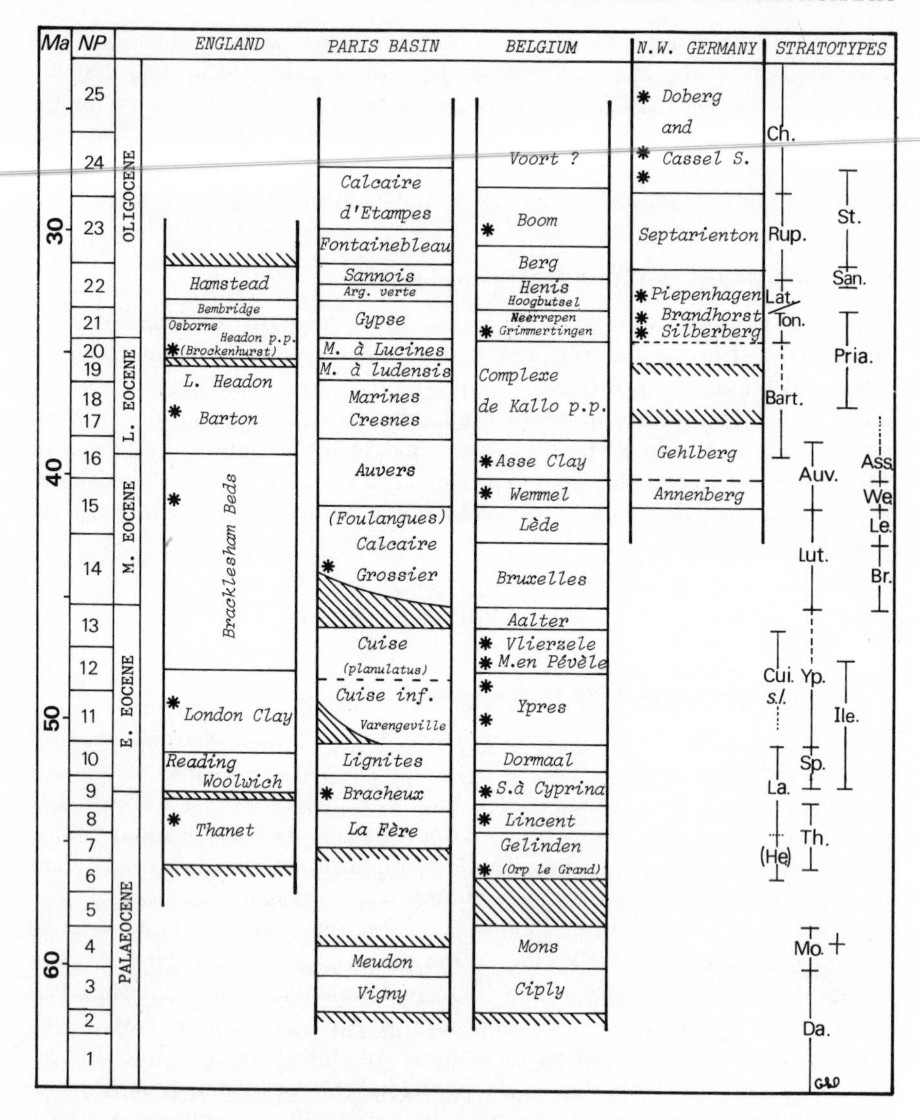

Figure 3. Palaeogene stratigraphic sequences of the NW European basins. The levels where nannoplankton was found are marked with an asterisk. The positions of the stratotypes accepted in this work are indicated. Abbreviations as for Figures 1 and 2, also Mo.: Montian; He.: Heersian; Br.: Bruxellian; Le.: Ledian; Auv.: Auversian; San.: Sannoisian. The type Montian has been correlated with planktonic foraminiferal zone P2 (see Salaj *et al.*, 1976). Correlation in this diagram with NP4 is inferential and depends on Martini (1971). Berggren would correlate with NP3 (cf. note to Figure 5).

analytical nature is reported in relation to the glauconies of the Kreyenhagen fm (see Odin, this volume chapter 15, Table 3). Redetermined ages have frequently turned out to be very different from, and usually younger than, earlier ones.

We discuss finally the typical example of the glaucony from the Lutetian of Fosses (France), studied by various authors on a total of five occasions. A sample was dated in Berkeley in 1961 as 48.1 Ma (ICC), but two redeterminations with improved equipment and techniques gave, in 1964, a calculated age of 43.8. Odin (1967) measured the same sample using the Rb–Sr method to provide a calculated age of 49.0 Ma (ICC), after which the sample was dated at a laboratory which was starting up K–Ar determinations in 1969, from which an average of 46.0 was obtained. Finally the sample was remeasured by Odin (Odin *et al*; 1978) on specially adjusted equipment, this time at Berne, to provide an age of 44.4 Ma (ICC), based on several measures both of potassium and of argon to reduce the analytical uncertainty. This last age coincides with the preferred age determined at Berkeley. Unfortunately, however, only the first-published result was used in the scales of Funnell and his successors.

These examples are quoted to illustrate the fact that early analytical data are less reliable than those of more recent date. Whilst it is perhaps unnecessary to point out that the experience of a laboratory at the moment of publication is more important than the date of the publication itself, it does seem evident to us that the most recent data are necessarily more reliable analytically than earlier ones. However, this fact is not evident to

Table 1. Comparison of data obtained from specified glauconitic horizons at different times, following the refinement of techniques.

	First measurements		Corrected measurements	
Samples	Potassium (% K)	Apparent ages (Ma, ICC)	Potassium %	Apparent ages (Ma, ICC)
Bashi marls	4.46	53.5 (Berkeley, 1961) →	5.62 5.98	49.8 49.3 (Austin, 1972)
Palaeocene California	3.95	60.7 (Berkeley, 1961) →	5.23 4.87	46.4 47.7 (Berkeley, 1964)
Kreyenhagen (6 samples)		27.7 to 42.5 (Berkeley, 1961) →		27.5 to 35.6 (Berkeley, 1964)
'*Glauconie grossière*' Fosses (NDS 90–91)	6.02	48.1 (Berkeley, 1961) →	6.75	43.8 (Berkeley, 1964)
		46.0 K–Ar old spectrometer → (Orléans, 1969)		44.4 (Berne, 1974)

some authors and Hardenbol and Berggren (1978) have suggested, in justifying their preference for our earlier determinations, that our later preparation techniques were faulty and that our samples had been heated unduly, with a consequent loss of radiogenic argon of as much as 10–30%. We have demonstrated earlier that losses of that magnitude begin to occur only at around 300–400°C (Zimmermann and Odin, this volume). In addition, detailed experiments were carried out in Berne before 1975 to check on the influence of drying temperatures under vacuum. In order to recheck what, in our opinion, was already established fact, we have carried out a systematic experimental study whose results have been reported earlier (Odin and Bonhomme, this volume). This demonstrates that losses by diffusion of the order of several percent occur only above 200°C, a temperature never attained during routine drying procedures. Finally, in the hope of closing the matter, we refer the reader to the results obtained at Gif by Cassignol and Gillot (this volume), who do not heat their samples before analysis. These show no systematic difference from those obtained on the same samples dried before analysis at 140–160°C. As was stated when our first time scale was published (Odin, 1973b) and has subsequently been confirmed, we now insist that it is necessary to give preference to the most recent data, which are analytically more reliable.

In practical terms, the above considerations bring out the fact that 39 of the 74 datings selected to prepare the six versions of the Berggren time scale are analytically unacceptable and should be replaced by dates which in part disagree with (and are younger than) those used by many authors to fix the dates of chronostratigraphic boundaries since 1964.

3.2 Problems in the interpretation of analytical data

We have demonstrated earlier (Odin and Dodson, this volume) the great importance of taking into account the nature and state of evolution of the glauconitic minerals present in green granules when interpreting analytical data based on glaucony. Obradovich and Cobban (1976) and Baadsgaard and Lerbekmo (this volume) have shown, on the other hand, that it is necessary to be very cautious in using analytical data based on potassium-poor minerals isolated from bentonites. Stated simply, a mineral which after purification is poor in potassium is geochemically unreliable and should preferably not be used to calculate an age in cases where it is proposed that the age should be regarded as the age of the formation from which it was obtained. The work of Ghosh (1972) in particular, deals with a high proportion of micas, sanidines and glauconitic minerals whose potassium content is less than 5% (see NDS 55–58). Ghosh has demonstrated the existence of calculated ages which are *too old* because of insufficiently high percentages of potassium due to weathering. The three glauconies from the

Midway stage are 'unreliable'. Of the three determinations from the Hatchetigbee fm, one only is in fact reliable, and this provides the youngest age. The age of the younger Tallahatta fm (correlated with the P12 biozone) is based on a 'favourable' glaucony also but there is only 1 Ma difference between it and that of the Bashi marl (referred to P6). The other six glauconies from the Claiborne stage are definitely unfavourable as chronometers, their potassium content lying between 1.6 and 5.1%. Of the six mineral fractions separated from the Cook Mt. fm (Claiborne stage), two biotites are favourable and yield apparent ages between 40 and 44 Ma. Unfortunately the bentonites from which they were extracted are poorly correlated stratigraphically. The five apparent ages proposed within the Jackson stage are based on four glauconies with K less than 4% and a mica with K of 4.6%. Within the Vicksburg stage, the Catahoula fm was dated by means of a whole-rock vitric tuff which cannot be regarded as an excellent chronometer. As a whole, of the 21 data selected in the work by Ghosh (1972), 16 must be considered as *very tentative* if not definitely unreliable. The apparent ages calculated may be older than the date of deposition (due to inheritance) or, alternatively, younger (open system). In conclusion, it appears that in view of the very small number of reliable data which are available from the Gulf Coast region it is hardly possible to develop a definitive analysis from them.

For all these reasons, we are led to the conclusion that the time scale of Funnell, together with all those based upon it, is supported by data which are for the most part highly suspect, either because of stratigraphical or geochemical unreliability or because of analytical error. We believe that it should be abandoned *in all respects where more reliable data have become available.* Such data are discussed in the following paragraphs.

3.3 The available data

In the summaries contained in the second part of this book we have assembled the stratigraphical, geochemical and analytical data obtained by various researchers since 1964, the date of the synthesis presented by Funnell. We have demonstrated that whilst some early data have proved to be unreliable, others are satisfactory and have in many cases been confirmed by recent work. In Figure 4, we have assembled the apparent ages obtained from the sequences in NW Europe. These alone give a good picture of the stratigraphical position of the classical stages in relation to the nannoplankton biozonation. However, the earliest part of the Palaeogene provides few reliable data and is dealt with separately. Figure 5 includes those data obtained elsewhere which have been discussed earlier or which will be utilized later. Stratigraphical correlations have in some cases been of the third degree, as for instance when they have been carried out by the use of

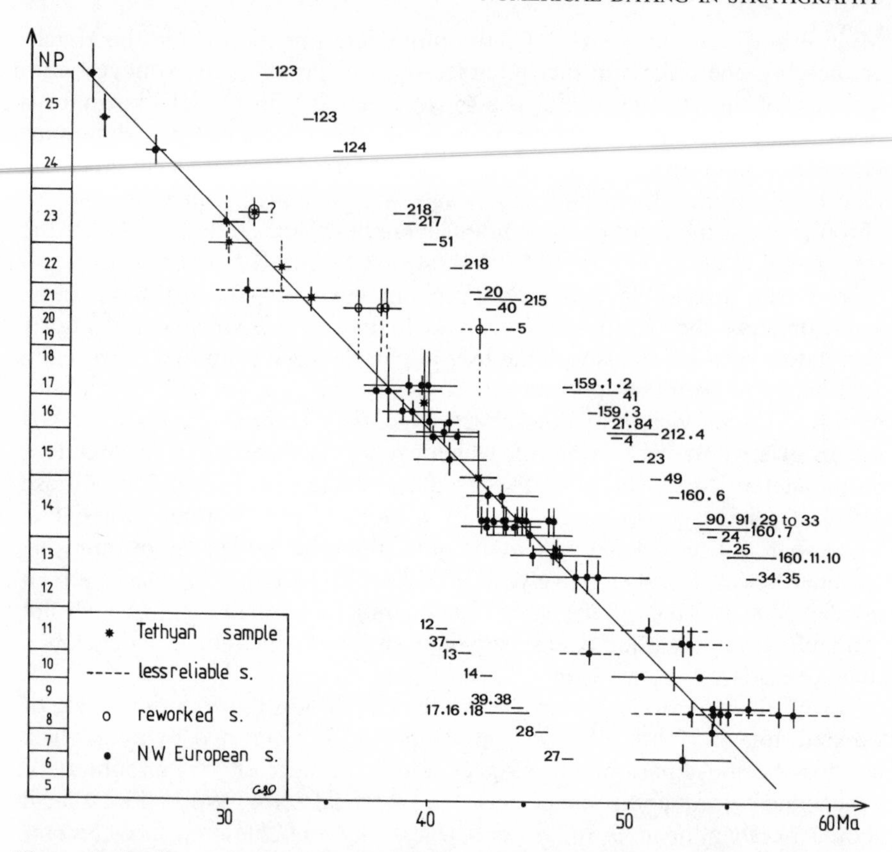

Figure 4. Radiometric ages provided by European samples in relation to the time scale and biozonal correlations here adopted. The numbers on each side of the mean curve refer to the abstracts where the data are discussed in this volume. For example, the three points stratigraphically correlated around the NP12–NP13 nannofossil biozone boundary are discussed in the abstracts NDS 34 and 35.

planktonic foraminiferid zonations correlated with the standard nannoplankton scale, itself correlated with the NW European sequences. The correlations accepted in this case are those used in the last biozonal scale of Hardenbol and Berggren (1978) in order not to introduce new sources of discrepancy between various presentations. Those apparent ages which we believe to be relatively unreliable for dating purposes, are enclosed in brackets.

Before proceeding to the construction of a time scale, we draw attention to a fundamental idea which is illustrated above in the course of the study of the genesis of those chronometers which are particularly useful for that purpose: the minerals from glauconies and from bentonites. 'Glauconitic minerals' are triggered for dating purposes at the time when the grains become isolated from the marine environment. Except for highly-evolved glauconies this is the moment of deposition of the *overlying* stratum, which isolates the grains from their source of potassium. Thus, strictly speaking, it is this overlying stratum with its associated faunas which is being dated, and not the bed which yielded the glaucony. On the other hand, and by definition, bentonite minerals crystallize in the magma chamber or near to it *before* extrusion, and thus before the deposition of the sedimentary sequence, at which time crystallization is complete and only alteration can ensue. No data are available to provide an estimate of this time difference in the case of bentonites but, in principle, it must exist.

4 REVISION OF THE NUMERICAL TIME SCALE OF THE PALAEOGENE

More than 65 determinations of acceptable quality are available to construct this scale. This would be a sufficient number if these were spaced regularly throughout the period of 40 Ma or so which it represents. Unfortunately this is not the case and it must be admitted that the age of Early Palaeocene and Early Oligocene sequences is still imprecisely known.

4.1 The base of the Palaeogene system

The only date available from NW Europe is from the province of Limburg, but the glaucony dated is appreciably older than the date of deposition, as explained in item NDS 87. In beds correlated with marine faunas the sequences of New Jersey seem to be the most reliably dated biostratigraphically. There the glauconies of the Hornerstown fm, the lowest post-Cretaceous unit, have yielded an apparent age of 62.1 Ma. However, identical ages have been obtained from underlying beds which are dated as Maastrichtian on their ammonite fauna. This throws doubt on these various ages in the absence of reliable data on the history of these chronometers. In South America various ages have been obtained from basalts overlying levels with dinosaurs and underlying Danian beds dated by foraminiferids (NDS 120, Table 2). These whole-rock determinations can only provide an approximate date, however.

The continental sequences of Canada (NDS 126, 127) and the central United States (NDS 103) have provided data on good-quality minerals. The Kneehills tuff zone of the Upper Edmonton or Battle fm is associated with the appearance of *Triceratops*, *Tyrannosaurus*, *Ankylosaurus* and *Thescalosaurus* in Maastrichtian times (Folinsbee *et al.*, 1961). According to the latest data, minerals from bentonites of this zone provide an age of 67 ± 1 Ma (Folinsbee *et al.*, 1966). Table 2 indicates that the Cretaceous–Tertiary

Table 2. Apparent ages obtained from samples near the Cretaceous–Tertiary boundary.

Dated level	Stratigraphy	Method–chronometer	Apparent age (Ma, ICC)
Hornerstown fm (NDS 92–114)	Early Danian Foraminifera	K–Ar Glauconies (evolved)	62.1 ± 3.1
Maastrichtian (NDS 115)	Ammonites	K–Ar Glauconies (evolved)	60.5–64.9
Salamanca fm (NDS 120)	Danian Dinosaurs Foraminifera	K–Ar Whole-rocks (basalts)	62.8–64.0 ± 0.8
Denver fm (NDS 103)	Puercan (Danian) Mammals	K–Ar Biotite K–Ar Plagioclase	65.8 ± 1.4 <66.4 ± 2.5
Scollard fm (NDS 126)	Base Danian Pollen Dinosaurs extinction	K–Ar Sanidines	63.1 ± 0.5
Ft Union fm (NDS 127)	Base Danian Pollen Dinosaurs extinction	Rb–Sr Biotites, Sanidines U–Pb Zircons	 63.6 ± 0.8
Kneehills tuff	Late Maastrichtian Dinosaurs	K–Ar Sanidines	67 ± 1

boundary may either be at about 63.5 Ma or at about 66 Ma (Denver fm). This discrepancy probably results from correlation problems, which can be serious in continental sequences, and in this situation we suggest an age of $65^{+1}_{-1.5}$ Ma, ICC, which is close to that proposed by Funnell but, once again, somewhat younger than that original proposal as a result of incorporating the revised data from the Edmonton laboratory. It should be noted that the exact position of the Cretaceous–Tertiary boundary in continental sequences has been queried by some palaeomagnetists. (Butler *et al.*, 1977; Channell, this volume). On the basis of the magnetostratigraphy it seems possible that dinosaurs were still in existence at levels above that of the Cretaceous–Tertiary boundary as determined from marine organisms. This suggestion might explain ages of around 63 Ma obtained for beds immediately overlying the youngest dinosaur-bearing rocks.

4.2 The Palaeocene

The marine Danian has provided few dates. In the case of Denmark, samples containing glaucony have been examined on two occasions but none of these have yielded material of a quality acceptable for dating. In Poland, a potassium-rich glaucony from near the Danian–Montian boundary was recently investigated (NDS 247). The date of 60.5 ± 0.7 Ma probably represents an age very near this boundary, but possibly below. Figure 5 presents other data available at around the lower boundary of the Palaeocene. The age of 57.5 ± 2.0 Ma of the Early Thanetian New Jersey glaucony (Owens and Sohl, 1973) appears to be more trustworthy (NDS 113) than those obtained in Texas (NDS 55). The Early–Late Palaeocene boundary (above the Montian) may tentatively be fixed at around 59^{+1}_{-2} Ma. So far as the Thanetian is concerned, levels included within the biozones NP6–NP8 have been dated in various regions (NDS 16, 17, 18 in England; NDS 27, 28 in Belgium; NDS 38, 39 in France; NDS 22 in the United States). The most reliable dates seem to be those lying between 52.6 and 56.6 Ma (K–Ar method). The glauconies from the Thanet sands are not ideal from a dating point of view, and those from the Gelinden marls (which are probably amongst the earliest Thanetian deposits of NW Europe) provide an age which is younger than their stratigraphical position would suggest.

The upper boundary of the Thanetian is here taken to be the base of the Ilerdian stage, which corresponds to the middle part of the NP9 biozone. Its age lies between those obtained from Thanetian chronometers on the one hand and on the Woolwich and Reading beds, correlated by dinoflagellates with the type Early Ilerdian, on the other.

4.3 The Eocene

Glauconies from the Varengeville fm (NDS 37) and the middle part of the London clay (NDS 12) seem more reliable than those from its lower part (NDS 13). Thus the Thanetian–Ilerdian boundary in Europe can probably be placed at about *53 ± 1* Ma, a figure which is supported by the glauconies of the Bashi marls of the Gulf Coast of the United States. One of these is evolved and yields an age of 49.3 Ma (NDS 56); the other (with K = 5.6%, and less reliable) yields 49.8 Ma. All these results agree with our earlier proposal, which assigned to the *Pseudohastigerina* (appearance) datum of Berggren (1971b) an age of about 50–51 Ma. This datum correlated with the base of zone NP10 by Berggren (1972) is placed slightly above the base of NP11 biozone by Odin *et al.* (1978). By contrast, the radiometric age of the evolved glaucony of the Manasquan fm (NDS 112) is significantly older and does not agree with that of the glauconies of the Bashi marls (NDS 56) attributed to the same biozone (P6).

The figure of 55 Ma (ICC recalculated) proposed in 1964 for the base of the Eocene was based on four determinations. The first was on the Bashi marls, which subsequently yielded a younger age; the second on a glaucony from Austria, dated rather imprecisely as Early Eocene. The other two were from units dated as Late Palaeocene (NP8): glaucony from the type Thanet sands (KA274, 530, see NDS 16), subsequently redated in the same

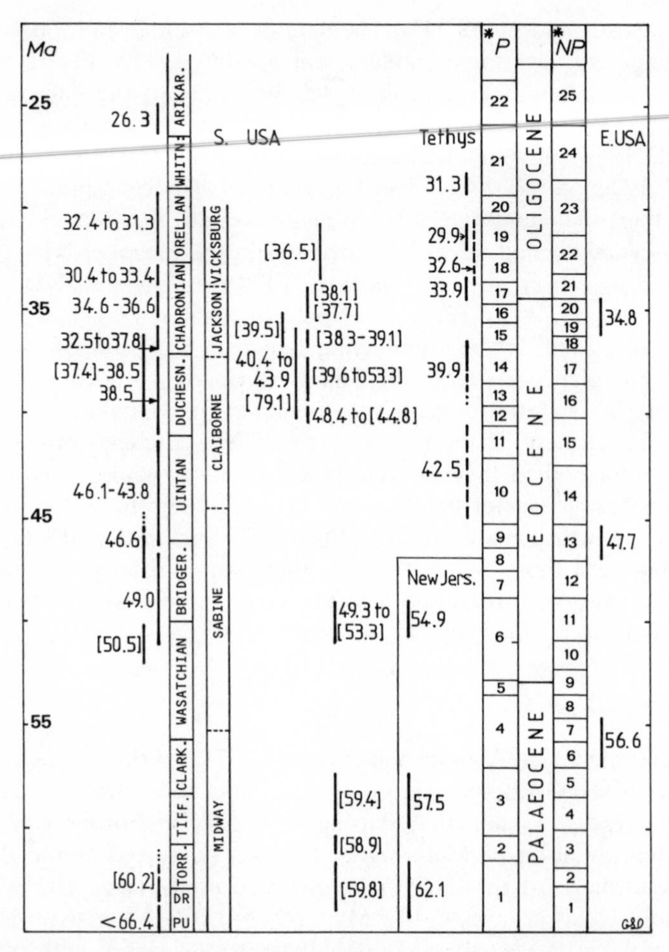

Figure 5. Radiometric data from outside NW Europe. Only the glauconies dated by Harris (NDS 8, 15, 22; NE USA) are correlated by nannoplankton. Data from the Tethys are by Savelli and Lipparini (NDS 49–51), Baubron and Cavelier (NDS 215) and Hartung *et al.* (NDS 218); data from New Jersey (NDS 92, 112–114) are by Obradovich (1964) and Owens and Sohl (1973). The ages from the thesis of Ghosh (1972) 'S USA' used elsewhere are very difficult to interpret because of the poor quality of the chronometers dated; the ages, which are of doubtful reliability, are included in brackets (NDS 55–58). The correlation between nannofossil and planktonic foraminiferal zonations is the same as that used by Hardenbol and Berggren (1978). Mammalian stages are based on Evernden *et al.* (1964) and Evernden and Curtis (1968). Whitn.: Whitneyan; Duchesn.: Duchesnean; Bridger.: Bridgerian; Clark.: Clarkforkian; Tiff.: Tiffanian; Torr.: Torrejonian; Dr.: Dragonian; Pu.: Puercan. Radiometric ages were obtained from separated minerals; these ages have not been used to define our scale because the original correlations with European stages were made *on the basis of radiometric ages determined at that time.* Ages based on glaucony are to the right; ages based on volcanic rocks are to the left of the

laboratory as 6% younger, and, finally, a poorly-evolved (5.3% K) glaucony from the Lodo fm, California. Revision of the data on which the original proposal was made, together with additional determinations, thus led to the proposition that the age of this boundary (and of the base of the Ilerdian) should be reduced to 53 Ma. An extrapolated age of 51 ± 1.5 Ma seems appropriate for the *Pseudohastigerina* datum, 1–2 zones above.

The base of the Middle Eocene (base of the Lutetian stage) is generally well defined in NW Europe; it is equated with the base of two biozones: that of nannoplankton (NP14) and planktonic foraminiferids (P10). The data assembled by Funnell are from glauconies which, for the most part, are not stratigraphically close to this boundary, or are based on mammal-bearing sequences in the United States. Radiometric dates are now available from strata whose chronostratigraphic age is much more precisely known, especially from western Europe (Belgium, France, England). These dates have been determined in four different laboratories (Berkeley, Berne, Gif and Strasbourg) and cannot, as a group, be dismissed as 'too young because of argon loss due to high bake-out temperatures' (Hardenbol and Berggren, 1978, p. 228) because some of the samples were in fact not heated at all before analysis, as already pointed out.

Ages immediately beneath this boundary are higher than 46 Ma (NDS 10–13, 25, 34, 35; Figure 4). Immediately above it, they are lower than 45 Ma (NDS 29–33, 90, 24, 6, 7, 9) with one exception (NDS 32). An age on basalt of 42.5 ± 1.5 Ma has been obtained on a Lutetian level in Italy (NDS 49), whilst in the United States there is a single Rb–Sr determination on a rock correlated with the NP13 biozone (NDS 15).

Bearing in mind the mode of genesis of the glauconies used, the numerical age of the base of the Lutetian may be placed fairly precisely at slightly more than 45 Ma; we propose a figure of $45^{+1}_{-0.5}$ Ma. This age agrees with several obtained in the United States on rocks correlated with the mammal faunas of the Bridgerian and Uintan stages (Figure 5). An age older than 46 Ma can definitely be eliminated.

The age of the boundary between the Lutetian and Bartonian stages is well documented by glauconies in England and Belgium (NDS 1–4, 21, 23); some data are also available from Germany (NDS 40, 41). As a whole they suggest that the age of the base of the Barton beds (here regarded as the

---

corresponding line showing the stratigraphical location. Complementary dates were recently obtained from SE Atlantic Coastal Plain sediments (eastern USA). A five-point Rb–Sr isochron glaucony age of 34.1 ± 3.0 Ma was calculated for a level assigned to nannofossil zone NP20. A six-point Rb–Sr isochron glaucony age of 36.7 ± 1.2 Ma was calculated for a formation correlated with the nannofossil zones NP16–17 and planktonic foraminiferal zone P11–14 (Fullagar *et al.*, 1980). *Note.* The correlations between P and NP zonations included in this figure would not be accepted by many workers (e.g. Martini, 1971).

base of the Bartonian) is a little over 39 Ma. This boundary probably lies in the upper part of the nannoplankton zone NP16. The lowest part of this zone is at around 41 ± 0.5 Ma on the basis of K–Ar data on glauconies from Belgium and England. In Italy a pre-Priabonian age (based on basalt) of 39.9 ± 1.1 Ma does not conflict with this proposal (NDS 50). The base of the Auvers sands of the Paris Basin (= base of the Bartonian *sensu gallico*) may be proposed at $41 \pm ^{+1}_{-0.5}$ Ma, which means that an age older than 42 Ma is highly improbable.

Amongst the analytical data obtained from the Claiborne stage of the Gulf Coast of America, only one item is geochemically acceptable, bearing in mind the state of evolution of the glauconies in question. The age obtained on the glaucony of the Tallahatta fm, dated on planktonic foraminiferids as P12, is 48.4 Ma, which seems to be far too high, especially in comparison with the almost equivalent apparent age of the Bashi marls, a level which appears to be six biozones earlier.

Insufficient data are available on the base of the Priabonian, which must, however, be younger than the basalt mentioned above, dated at 39.9 Ma. Extrapolation on the basis of the average duration of biozones suggests a date in the region of *37 ± 1.5* Ma.

4.4 The base of the Oligocene

The only data at this level which were available to Funnell (1964) were from high-temperature rocks and minerals correlated by reference to vertebrate faunas or (more rarely) continental molluscs. The Chadronian was regarded as the lowest stage of the Oligocene and Funnell selected the *maximum age* given by undoubted Chadronian samples as the minimum for the base of the Oligocene. This frequently used principle, which chooses the maximum age available with the hidden thought that any rock dated by the K–Ar method must yield an apparent age which is too young because of argon loss, ignores the possibility, now considered to be of frequent occurrence, of initial heritage. Ages from continental rocks are, in our opinion, of little value without a detailed study of the possibilities and problems of correlating them either by pelagic or by continental faunas with the type sequences studied in Europe. Such a study is lacking in the present case.

Information is now available on the Silberberg beds (NDS 40), whose glaucony is probably relict or inherited and which provides maximum ages for a formation whose biostratigraphical age is the range NP19–21. From this the base of the Oligocene must be *appreciably younger than* 37 *Ma*. The Rb–Sr age of the glauconies of the Castle Hayne limestone of the United States (Harris, this volume, and NDS 8) suggests an alternative maximum age for the boundary of 34.8 ± 2.0 Ma. In our opinion the basalt ages of

DSDP Leg 31 provide only a wide bracket for the boundary of 35^{+2}_{-n}, n lying between 2 and 5 (NDS 26). The glaucony of the Neerrepen sands, regarded by us as Oligocene, is of poor quality (NDS 20), but is nevertheless geochemically more reliable than the glauconies of the Jackson formation and the bentonite minerals from the Gulf Coast region of the United States. The latest Priabonian basalts from SE France yield a well-documented date of 33.9 ± 1.5 Ma (NDS 215). The pre-NP23 or pre-NP22 ash of the Island of Rhodes (NDS 218 = 32.6 ± 0.3 Ma) agrees with this date.

An approximation based on the mean duration of biozones suggests that the base of zone NP21 is at about 33 ± 2 Ma. The radiometric data above permit the associated margin of error to be reduced somewhat.

If the base of the Oligocene is taken above the Bembridge marls of England and the Latdorf beds (= *Ostrea queteleti* beds) of Germany and their equivalents in Belgium, and below the Sannoisian sequences in the Paris Basin, that is, at or near the base of nannoplankton zone NP22 (cf. Cavelier, 1979, pp. 220–1), then an age of 33^{+2}_{-1} Ma seems appropriate. An age for those levels older than 35 Ma can definitely be eliminated in view of the radiometric data assembled in this volume for the first time. This younger age was initially proposed by Odin (1975) and indirectly documented by Odin *et al.* (1978). In this chapter, however, we accept the classical view based on Beyrich (1854) as to the position of the base of the Oligocene (base of Latdorf sands in Germany and Grimmertingen sands in Belgium). These units are approximately equivalent to the Marnes à Lucines of the Paris Basin and the Brockenhurst beds in England (cf. Châteauneuf, 1980, Figure 50) and so we propose for this boundary an age of 34^{+2}_{-1} Ma. The Rupelian–Chattian boundary, as indicated in Figure 3, would appear to be at about 27–28 Ma. There is no detailed confirmation for this proposal but no German Chattian glauconies have yet yielded an age higher than that (NDS 123–124), while a single Rupelian glaucony yields an age of 30 ± 1 Ma (NDS 217).

The result from the Middle Chattian Dali ash from Greece (NDS 218–31.3 ± 0.3 Ma) is clearly discrepantly old and the age obtained is now regarded as problematic and possibly inherited. As a result the Rupelian–Chattian boundary here proposed at 27^{+2}_{-1} Ma remains inaccurately documented.

4.5 The Palaeogene–Neogene boundary

The age which was adopted for the base of the Miocene in 1964 on the basis of continental formations in the United States has been superseded by data obtained from basalts in California which can be correlated with units containing planktonic foraminiferids (Turner, 1970a). The boundary dated is

Table 3. Apparent ages obtained from samples close to the Palaeogene–Neogene boundary.

Aquitanian/Burdigalian (N Italy)	Foraminifera	(NDS 46)	Evolved glauconies	(18.5–20.0)
Aquitanian/Burdigalian (Aquitaine)	N5–N7	(NDS 48)	Highly-evolved glaucony	20.5 ± 1.9
Middle Aquitanian (Sicily)	Foraminifera	(NDS 47)	Evolved glaucony	22.6 ± 1.2
Chattian/Miocene (Germany)	NP25–NN1	(NDS 123)	Highly-evolved glauconies	Max. 23.5 ± 24
Late Eochattian (Germany)		(NDS 89)	Evolved glauconies	25.2
Early Eochattian (Germany)		(NDS 124)	Highly-evolved glauconies	26.2 ± 0.5
Latest Oligocene (California)	N4	(NDS 154)	Plagioclases, basalts	23.8 ± 0.4
Earliest Miocene (California)	N5	(NDS 155)	Plagioclases—biotites, basalts	22.5 ± 0.2
Earliest Miocene (Australia)	Foraminifera	(NDS 156)	Whole-rock, basalt	Min. 22.0 ± 0.6
Late Oligocene (Australia)	Foraminifera	(NDS 156)	Whole-rock, basalt	26.9

that between the Zemorrian and Saucesian stages, a level placed by van Couvering and Berggren (1976) 'somewhat below the planktonic foraminiferal zone boundary N4/N5'. The San Emigdio basalts (22.5 ± 0.2 Ma, ICC, cf. NDS 155) are younger than this boundary, whilst the Iversen Point basalts (23.8 ± 0.4 Ma, ICC, cf. NDS 154) are older. Turner proposed an age of about 23 Ma, ICC, for this boundary. Similarly, the boundary between the Janjukian and Longfordian in Australia is equated with the base of the Miocene and is indirectly correlated with the lower or middle part of planktonic foraminiferal zone N4 (NDS 156). The limit is postdated by a basalt at 22.0 Ma. In Germany the Chattian–Vierlandian boundary is believed to correspond to the NP25–NN1 nannofossil zone boundary and gives a maximum age of 23.5–24 Ma (NDS 123). In southwest France, where the type sections of the Aquitanian and Burdigalian stages are located, we have only obtained a minimum age on Late Aquitanian levels (NDS 45). Analysis of these results (Table 3) leads us to conclude that the Oligocene–Miocene boundary is situated within the time span $23^{+1}_{-0.5}$ Ma, ICC.

5 CONCLUSION

On the basis of an analysis of published data, together with others announced in the present work, we have endeavoured to evaluate the data

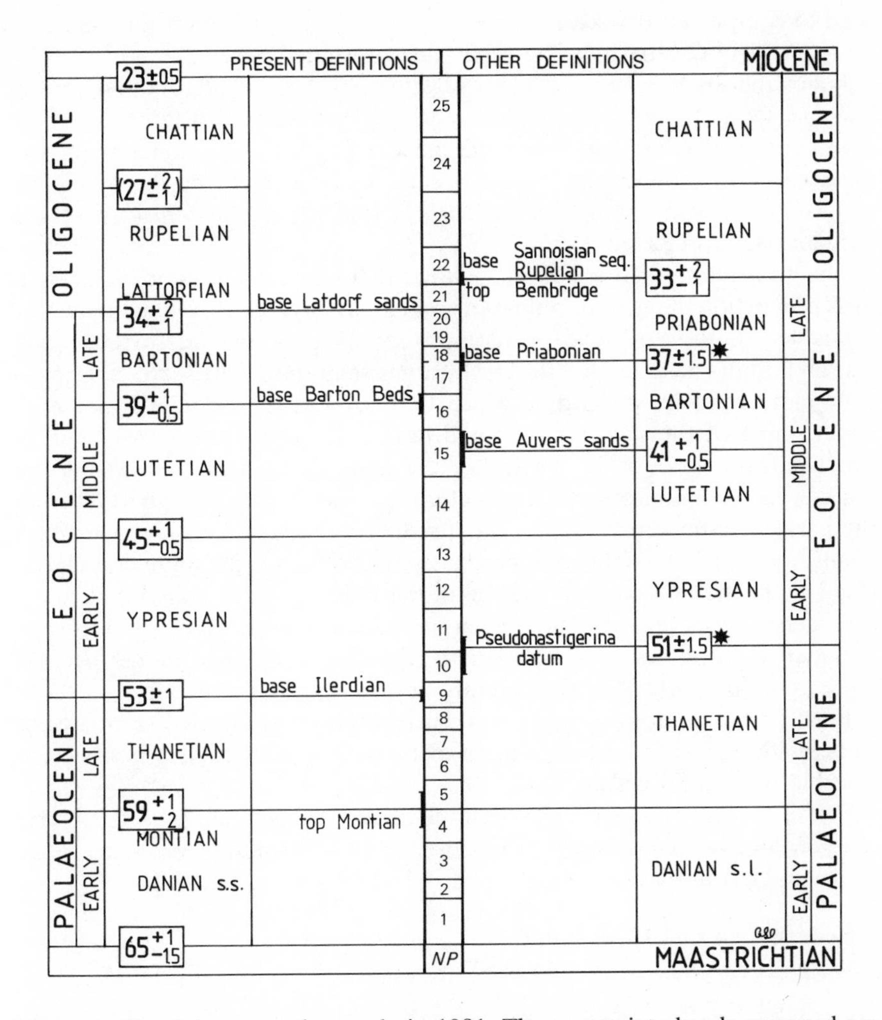

Figure 6. The Palaeogene time scale in 1981. The uncertainty bands proposed are not related to statistical errors of a Gaussian type, but indicate that in view of the problems of stratigraphical correlation and geochemistry (genesis and history of the chronometers dated) associated with a particular boundary, the age of that boundary may ultimately be modified within the limit indicated. It would then be the purpose of subsequent research to endeavour to reduce the width of the uncertainty band by improving on the data available at or near the boundary. Asterisks indicate extrapolated ages; correspondence between planktonic foraminiferal and nannofossil zonations is given on Figure 5.

used to compile earlier scales and to justify our choices both from older and newer data in constructing the time scale presented here.

Following studies in parallel of stratigraphy and geochronology it has become apparent that it is necessary to give preference to later analytical data over earlier ones; to select the most trustworthy chronometers at the expense of rejecting many classical data; and to interpret the analytical results in the light of new knowledge of the genesis and history of the chronometers preferred.

The application of these choices shows that the 1964 scale is based on a high proportion of age determinations which are less reliable than many now available. In the course of verifying and updating this early time scale certain authors have tended to select (rather than those which are geochemically more reliable) those data which best support the original scale. As a result, most of the age data adopted have been analytically erroneous or geochemically suspect. In particular, the general practice of choosing the highest values amongst groups of determinations, a practice inherited from the earliest radiometric studies, has produced overstatements of age. As a result, we have been obliged to propose reduced figures for some boundaries (Figure 6). The reduction of the age of the Oligocene–Miocene boundary is now generally accepted. The similar reduction for the Eocene–Oligocene boundary is accepted by some workers, but this acceptance should become more general considering the recently obtained radiometric data.

Finally, the adoption of conventional constants has enabled comparisons to be made between different dating methods and as a result it has become possible to confirm certain data.

In this context it is especially desirable, when a time scale is used, to refer both to the research on which it was based and to the decay constants which were adopted in its calculation.

(Manuscript received 10-10-1980)